JACARANDA

MATHS QUEST

ESSENTIAL MATHEMATICS 11 FOR QUEENSLAND

UNITS 1 & 2 | SECOND EDITION

JACARANDA

MATHS QUEST

ESSENTIAL MATHEMATICS 11 FOR QUEENSLAND

UNITS 1 & 2 | SECOND EDITION

MARK BARNES

PAULINE HOLLAND

REVIEWED BY

Arrian Hannebach | Kamlesh Kumar | Craig Otto | Emilia Sinton

jacaranda
A Wiley Brand

Second edition published 2024 by
John Wiley & Sons Australia, Ltd
Level 4, 600 Bourke Street, Melbourne, Vic 3000

First edition published 2018

Typeset in 10.5/13 pt TimesLTStd

ISBN: 978-1-394-26958-7

Front cover images: TWINS DESIGN STUDIO/Adobe Stock Photos, veekicl/Adobe Stock Photos, mrhighsky/Adobe Stock Photos, sapunkele/Adobe Stock Photos, NARANAT STUDIO/Adobe Stock Photos, Kullaya/Adobe Stock Photos, vectorplus/Adobe Stock Photos, WinWin/Adobe Stock Photos, valeriya_dor/Adobe Stock Photos, Ludmila/Adobe Stock Photos, Анастасия Трофимова/Adobe Stock Photos, katarinalas/Adobe Stock Photos, izzul fikry (ijjul)/Adobe Stock Photos, Tatsiana/Adobe Stock Photos, nadiinko/Adobe Stock Photos

Illustrated by diacriTech and Wiley Composition Services

Typeset in India by diacriTech

A catalogue record for this book is available from the National Library of Australia

Printed in Singapore
M130171_080824

The publisher of this series acknowledges and pays its respects to Aboriginal Peoples and Torres Strait Islander Peoples as the traditional custodians of the land on which this resource was produced.

This suite of resources may include references to (including names, images, footage or voices of) people of Aboriginal and/or Torres Strait Islander heritage who are deceased. These images and references have been included to help Australian students from all cultural backgrounds develop a better understanding of Aboriginal and Torres Strait Islander Peoples' history, culture and lived experience.

It is strongly recommended that teachers examine resources on topics related to Aboriginal and/or Torres Strait Islander Cultures and Peoples to assess their suitability for their own specific class and school context. It is also recommended that teachers know and follow the guidelines laid down by the relevant educational authorities and local Elders or community advisors regarding content about all First Nations Peoples.

All activities in this resource have been written with the safety of both teacher and student in mind. Some, however, involve physical activity or the use of equipment or tools. **All due care should be taken when performing such activities**. To the maximum extent permitted by law, the authors and publisher disclaim all responsibility and liability for any injury or loss that may be sustained when completing activities described in this resource.

The publisher acknowledges ongoing discussions related to gender-based population data. At the time of publishing, there was insufficient data available to allow for the meaningful analysis of trends and patterns to broaden our discussion of demographics beyond male and female gender identification.

Contents

Learning with learnON

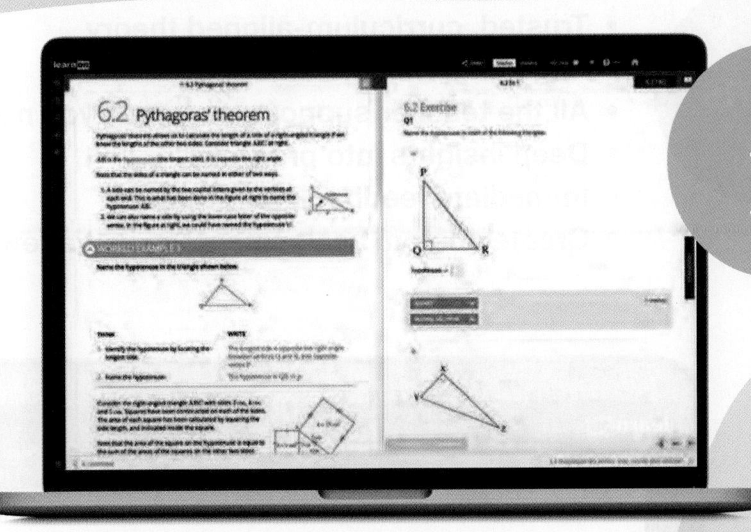

Everything you need for your students to succeed

JACARANDA MATHS QUEST
ESSENTIAL MATHEMATICS 11
UNITS 1 AND 2 FOR QUEENSLAND
SECOND EDITION

Developed by expert teachers for students

Tried, tested and trusted. The completely revised and updated second edition of *Jacaranda Maths Quest 11 Essential Mathematics Units 1 & 2 Queensland* continues to focus on helping teachers achieve learning success for every student — ensuring no student is left behind and no student held back.

Because both *what* and *how* students learn matter

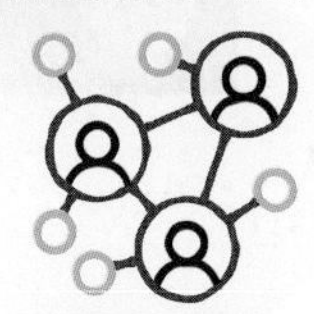

Learning is personal

Whether students need a challenge or a helping hand, you'll find what you need to create engaging lessons.

Whether in class or at home, students can get unstuck and progress! Scaffolded lessons, with detailed worked examples, are supported by teacher-led video eLessons. Automatically marked, differentiated question sets (including brand-new quick quizzes) are all supported by detailed worked solutions.

Learning is effortful

Learning happens when students push themselves. With learnON, Australia's most powerful online learning platform, students can challenge themselves, build confidence and ultimately achieve success.

Learning is rewarding

Through real-time results data, students can track and monitor their own progress and easily identify areas of strength and weakness.

And for teachers, Learning Analytics provide valuable insights to support student growth and drive informed intervention strategies.

Learn online with Australia's most

Everything you need for each of your lessons in one simple view

- **Trusted, curriculum-aligned theory**
- **Engaging, rich multimedia**
- **All the teacher support resources you need**
- **Deep insights into progress**
- **Immediate feedback for students**
- **Create custom assignments in just a few clicks.**

Practical teaching advice and ideas for each lesson provided in teachON

Each lesson linked to content points from the QCAA Essential Mathematics 2025 Applied senior syllabus

Reading content and rich media including embedded videos and interactivities

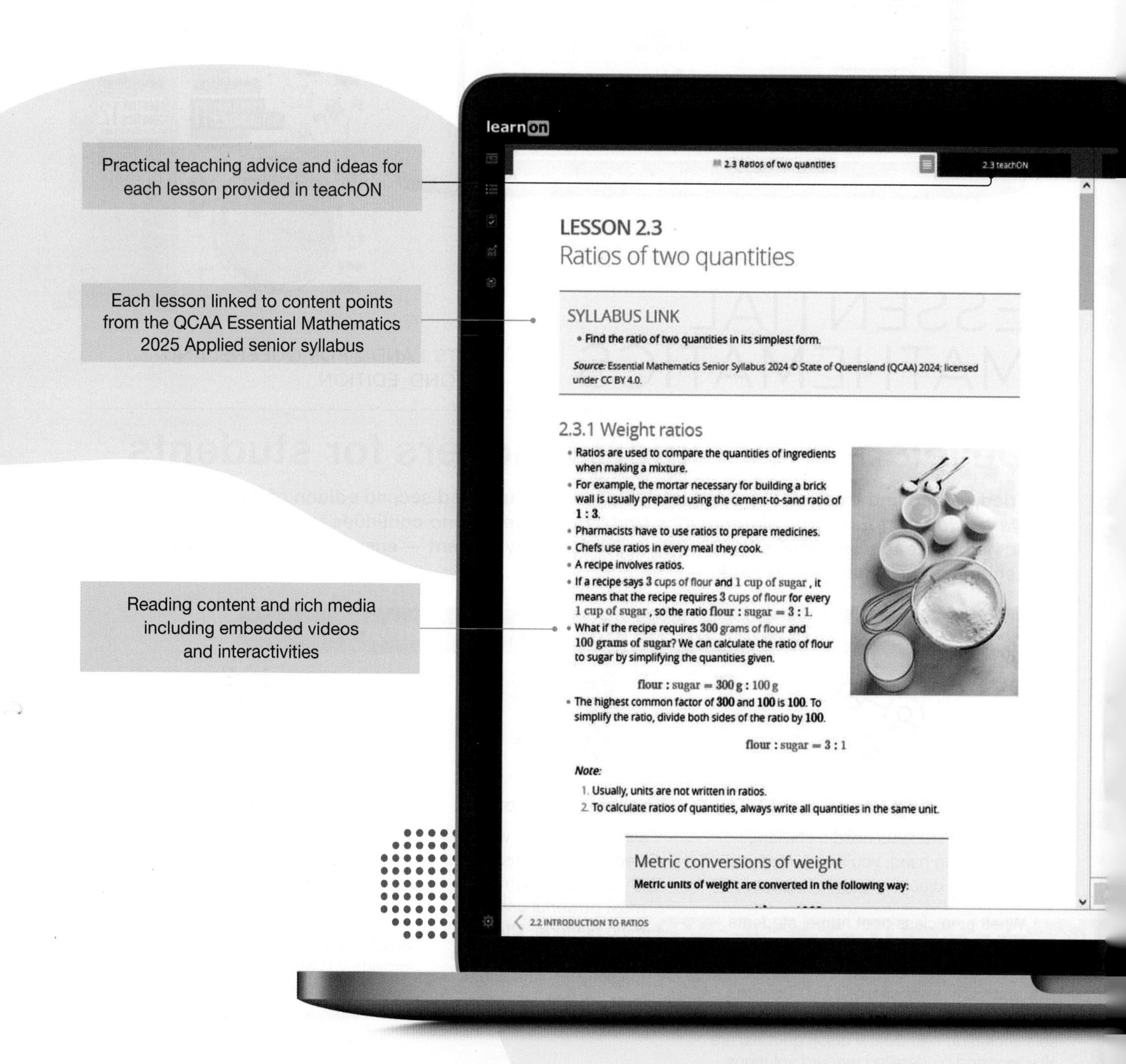

powerful learning tool, learnON

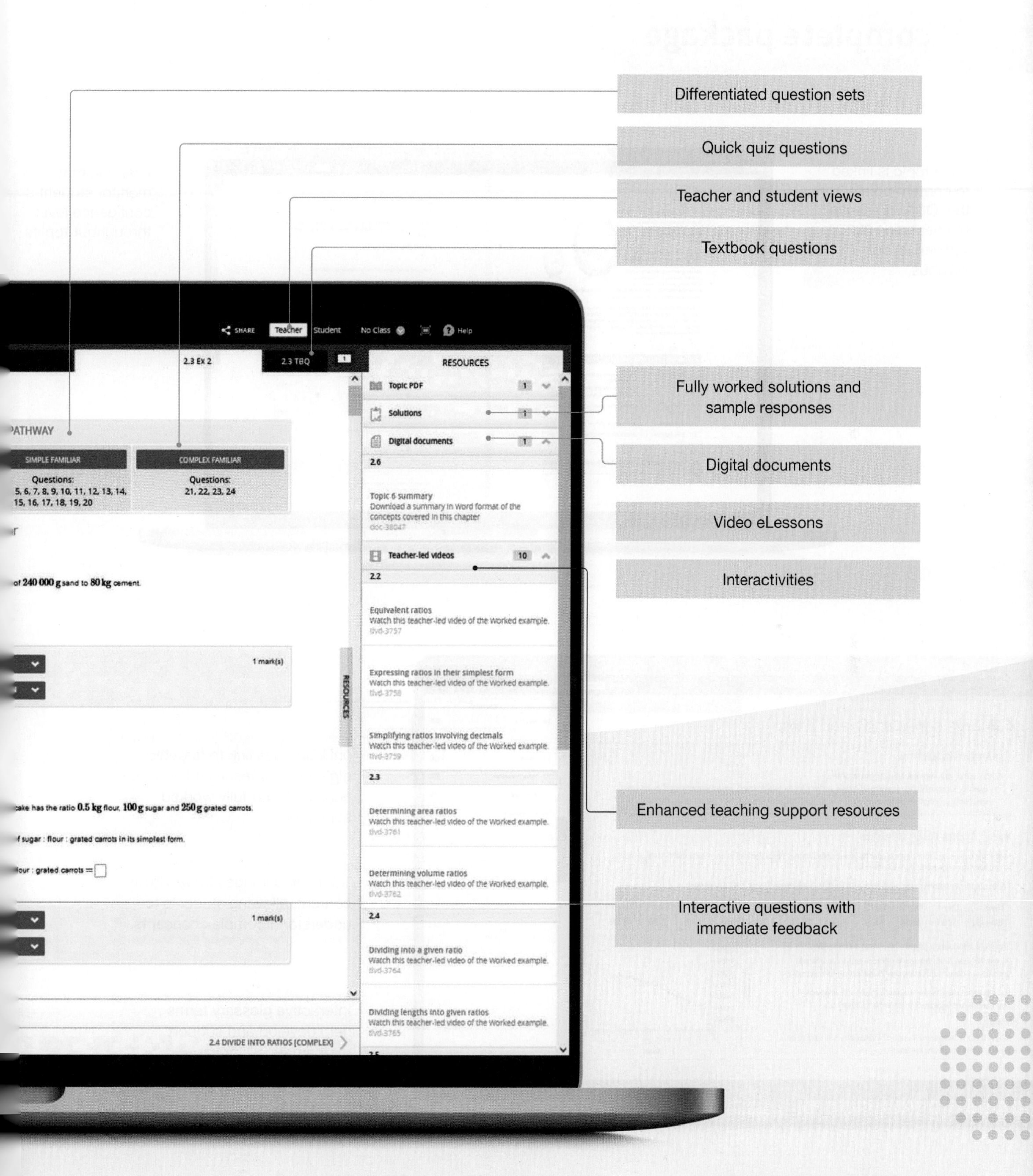

Online, these new editions are the **complete package**

Trusted Jacaranda theory, plus tools to support teaching and make learning more engaging, personalised and visible.

Each topic is linked to content points from the QCAA Essential Mathematics 2025 Applied senior syllabus.

Learning matrix to monitor student's confidence level throughout topics

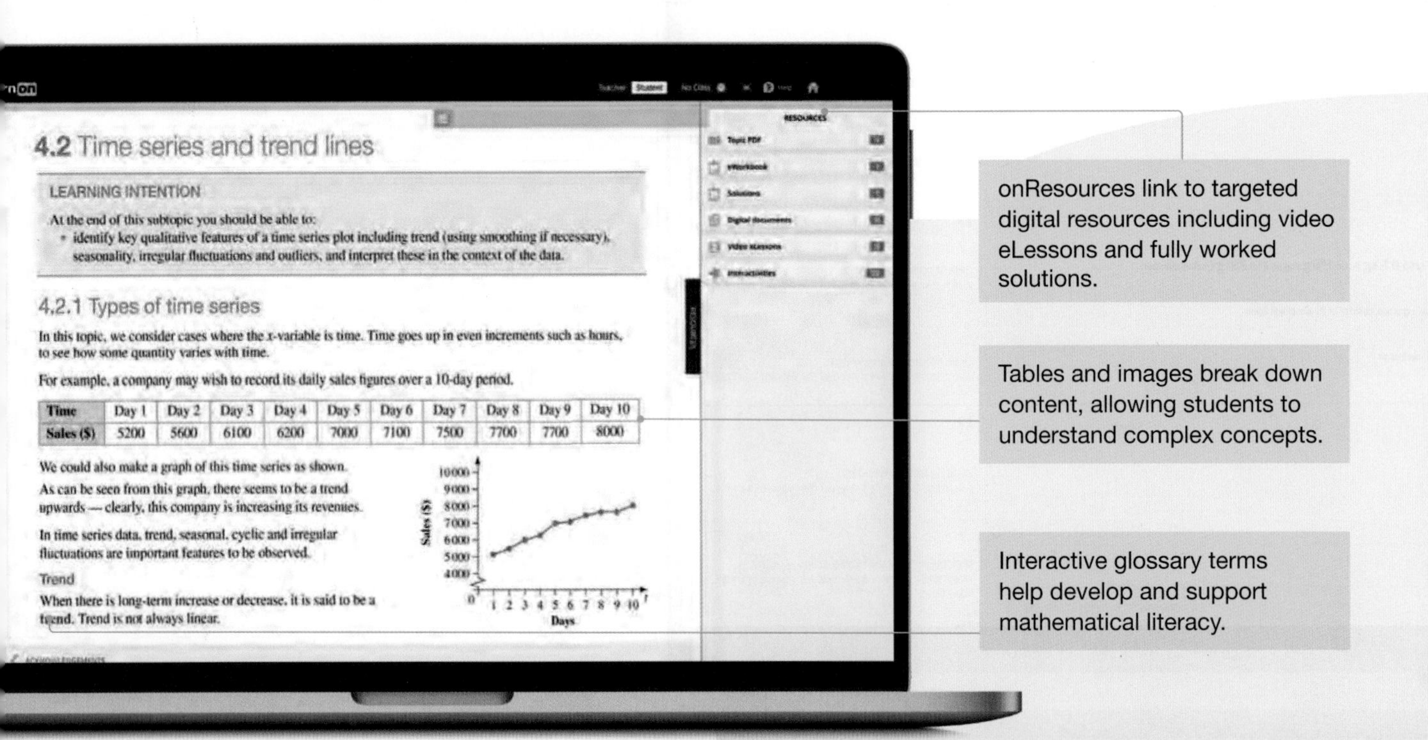

onResources link to targeted digital resources including video eLessons and fully worked solutions.

Tables and images break down content, allowing students to understand complex concepts.

Interactive glossary terms help develop and support mathematical literacy.

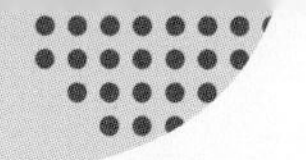

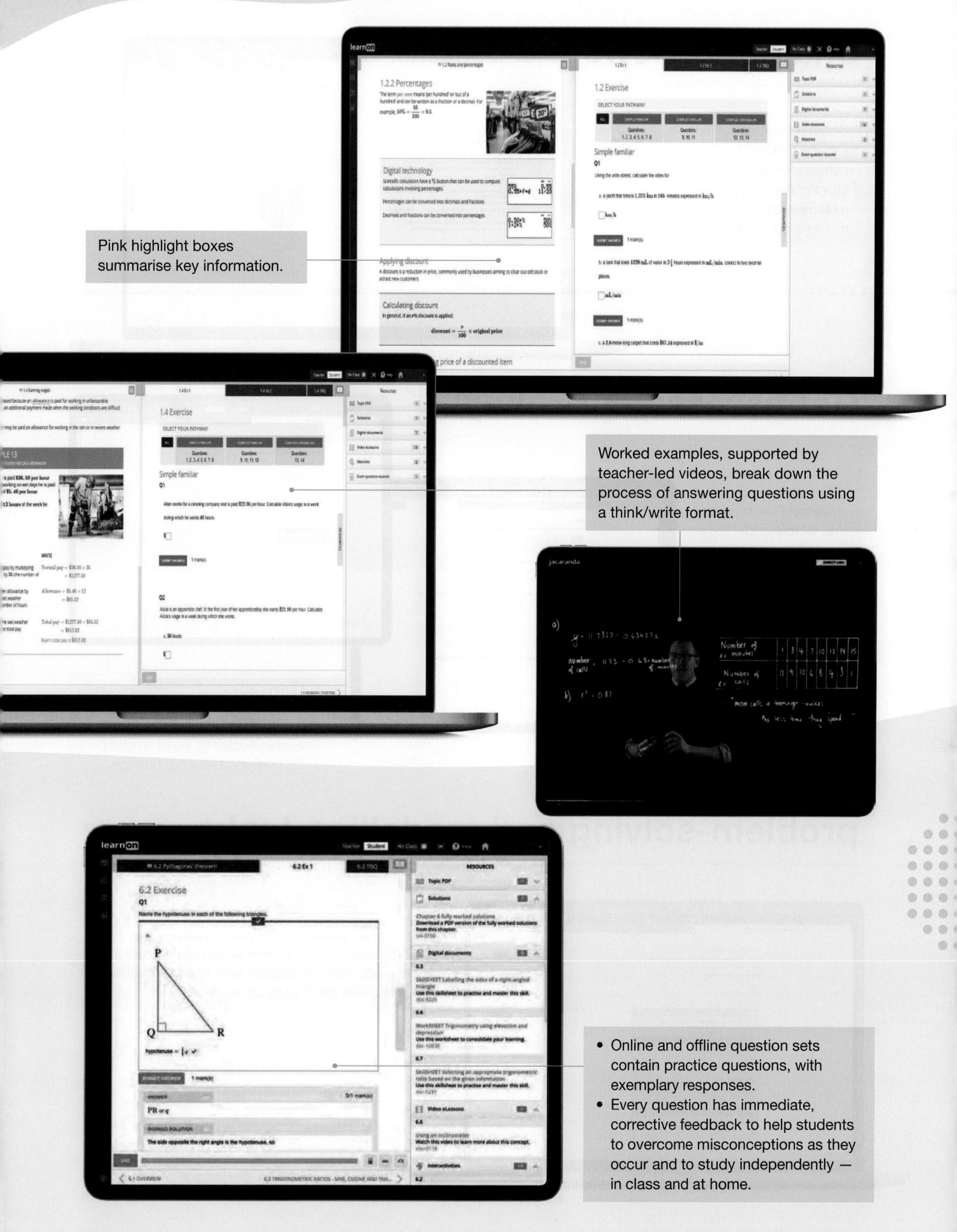

Pink highlight boxes summarise key information.
Worked examples, supported by teacher-led videos, break down the process of answering questions using a think/write format.
• Online and offline question sets contain practice questions, with exemplary responses.
• Every question has immediate, corrective feedback to help students to overcome misconceptions as they occur and to study independently — in class and at home.

Chapter reviews

Chapter reviews include online summaries and chapter level review exercises that cover multiple concepts.

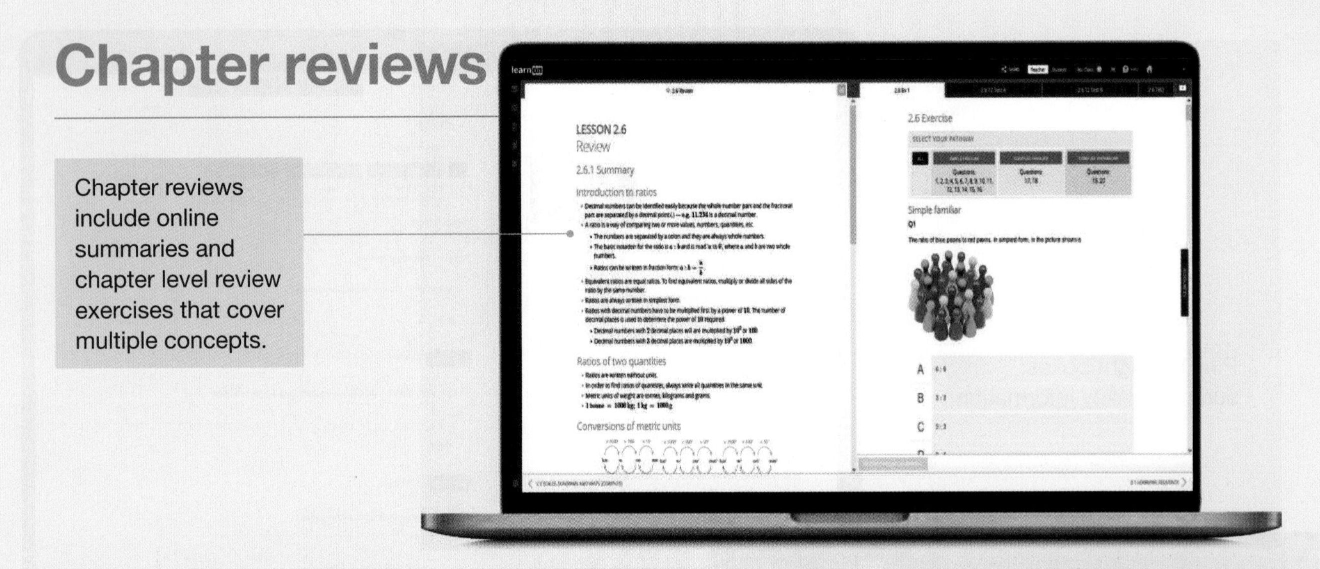

Get assessment-ready!

Unit-level review questions are broken into simple and [complex] content to prepare for assessments.

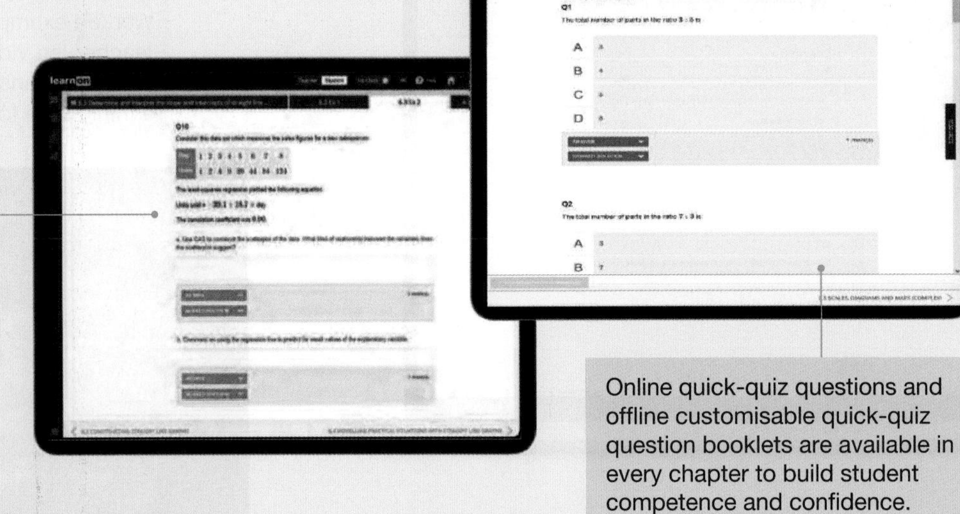

Online quick-quiz questions and offline customisable quick-quiz question booklets are available in every chapter to build student competence and confidence.

Expert advice for problem-solving and modelling tasks

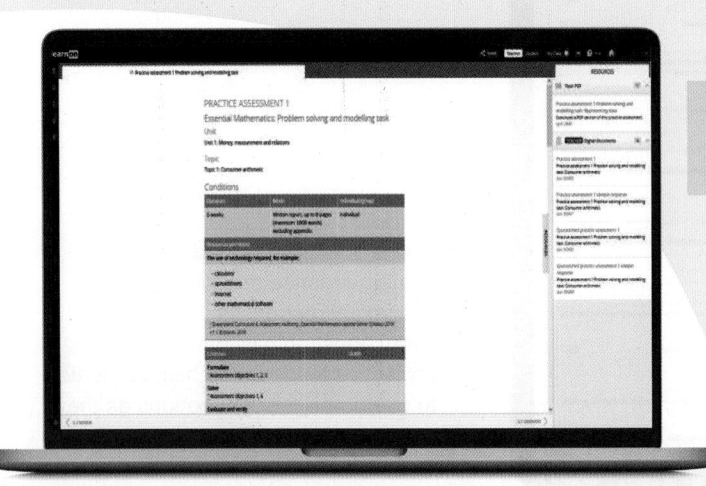

Step-by-step guide on how to complete problem-solving and modelling tasks with tips for teachers on how to create good assessments

Teaching with learnON

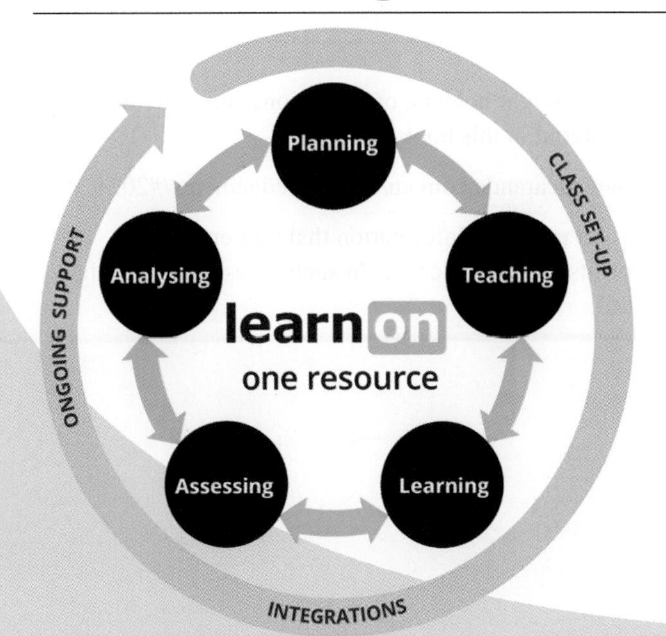

Enhanced teacher support resources, including:

- work programs and curriculum grids
- teaching advice and additional activities
- quarantined topic tests (with solutions)
- Unit reviews
- Quarantined PSMTs and examinations
- Custom exam-builder with question differentiation (SF/CF/CU) question filters

Customise and assign

A testmaker enables you to create custom tests from the complete bank of thousands of questions.

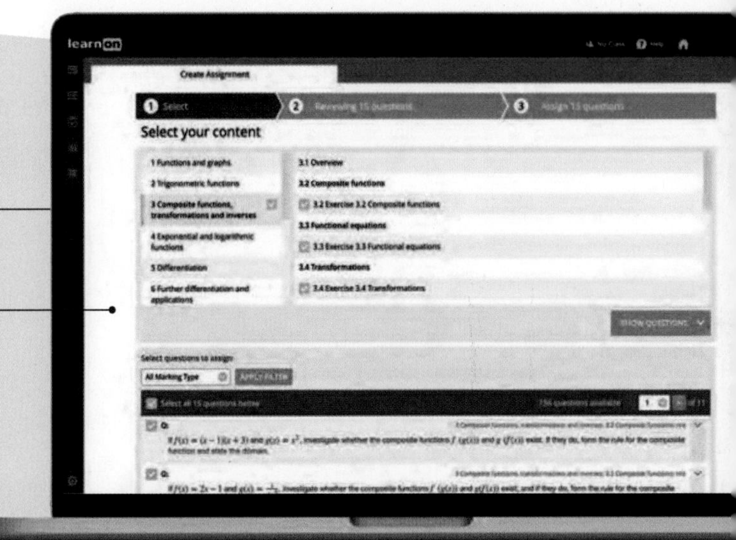

Reports and results

Data analytics and instant reports provide data-driven insights into performance across the entire course.

Show students (and their parents or carers) their own assessment data in fine detail. You can filter their results to identify areas of strength and weakness.

Acknowledgements

The authors and publisher would like to thank the following copyright holders, organisations and individuals for their assistance and for permission to reproduce copyright material in this book.

The full list of acknowledgements can be found here: www.jacaranda.com.au/acknowledgements/#2024

Every effort has been made to trace the ownership of copyright material. Information that will enable the publisher to rectify any error or omission in subsequent reprints will be welcome. In such cases, please contact the Permissions Section of John Wiley & Sons Australia, Ltd.

UNIT

1 Number, data and money

UNIT 1 Number, data and money
TOPIC Fundamental topic

1 Calculations

LESSON SEQUENCE

Fully worked solutions for this chapter are available online.

on Resources

Solutions Solutions — Chapter 1 (sol-1254)

Digital documents Learning matrix — Chapter 1 (doc-41747)
Quick quizzes — Chapter 1 (doc-41748)
Chapter summary — Chapter 1 (doc-41749)

LESSON
1.1 Overview

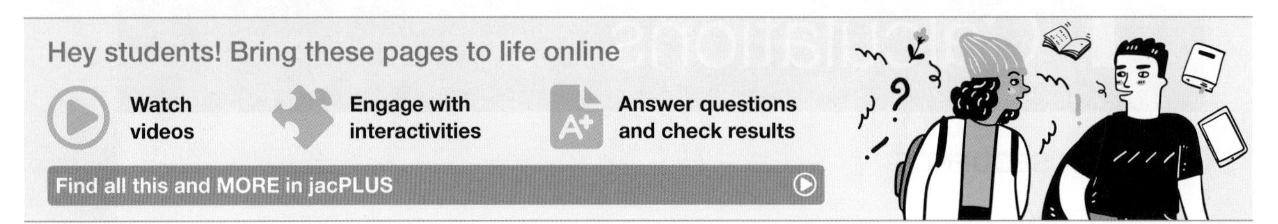

1.1.1 Introduction

Being able to add, subtract, multiply and divide integers is important in many parts of everyday life. It helps with budgeting and knowing what you can and can't afford to buy — or how much of something to buy. It also helps to keep you safe; for example, being able to read the integers on the speedometer helps you stay within the speed limit. It is also useful for understanding temperatures, weights and measures when you're cooking or you need to store food safely.

In finance, positive numbers are used to represent the amount of money in someone's bank account, while negative numbers are used to represent how much money someone owes (for example, how much they have to pay back after borrowing money or taking out a loan from a bank).

These calculations are the building blocks in mathematics and will be covered in this chapter as well as across all of Units 1 & 2.

1.1.2 Syllabus links

Lesson	Lesson title	Syllabus links
1.2	Number operations	◯ Solve practical problems requiring basic number operations.
1.3	Order of operations and reasonableness	◯ Apply arithmetic operations according to their correct order. ◯ Ascertain the reasonableness of answers to arithmetic calculations.
1.4	Estimation	◯ Use leading-digit approximation to obtain estimates of calculations. ◯ Check results of calculations for accuracy. ◯ Round up or round down numbers to the required number of decimal places.
1.5	Decimal significance	◯ Recognise the significance of place value after the decimal point. ◯ Evaluate decimal fractions to the required number of decimal places. ◯ Round up or round down numbers to the required number of decimal places.
1.6	Calculator skills	◯ Use a calculator for multi-step calculations. ◯ Check results of calculations for accuracy. ◯ Ascertain the reasonableness of answers to arithmetic calculations.
1.7	Approximation	◯ Apply approximation strategies for calculations.

This chapter only contains Simple familiar questions from the Fundamental topic Calculations in the syllabus.

LESSON
1.2 Number operations

SYLLABUS LINKS

- Solve practical problems requiring basic number operations.

1.2.1 Addition and subtraction of integers using a number line

- An **integer** is a whole number. It can be positive (2, 4, 89, 1035) or negative (−2, −4, −89, −1035).
- Zero is also an integer because it is a whole number.

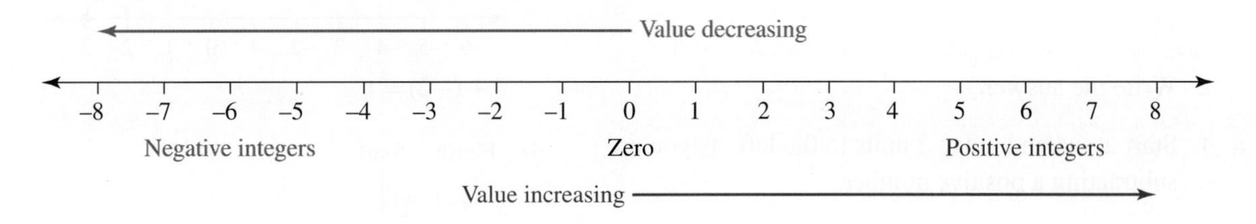

- The positive numbers do not usually have a symbol to show that they are positive (+2 is the same as 2).
- The negative numbers must include the negative sign to show that they are negative (−2, −27).
- A whole number is a number that does not include any fractions or parts of a number.
  For example, the following are *not* whole numbers or integers:
  - Fractions $\left(\frac{1}{2}, -12\frac{3}{4}, 5\frac{1}{2}\right)$ because they are or include parts of a whole number
  - Numbers that continue after a decimal point $(0.5, -12.75, 5.5)$

Addition and subtraction using a number line

Start at the first number.

To add a:
- **positive integer, move to the right**
- **negative integer, move to the left.**

To subtract a:
- **positive integer, move to the left**
- **negative integer, move to the right.**

- For example, to show $(-3) + (+2)$ on a number line, start at the first number (-3) and place a pointer on the number line at -3.
  To add the positive integer $(+2)$, move 2 places to the right.

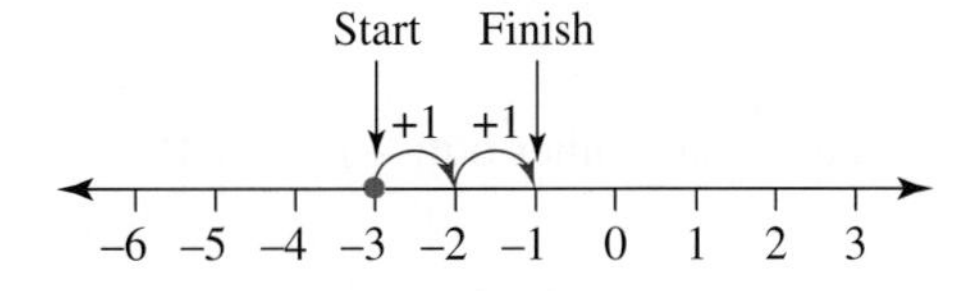

Therefore, using the number line, we can see that $(-3) + (+2) = -1$.

tlvd-3549

WORKED EXAMPLE 1 Solving addition and subtraction of integers using a number line

Use a number line showing intervals from −6 to 3 to calculate each of the following.

a. $2+(+1)$ b. $3+(-2)$ c. $-4-(+2)$ d. $-3-(-5)$

THINK	WRITE
a. 1. Start at 2 and move 1 unit to the right, as you are adding a positive number.	a. 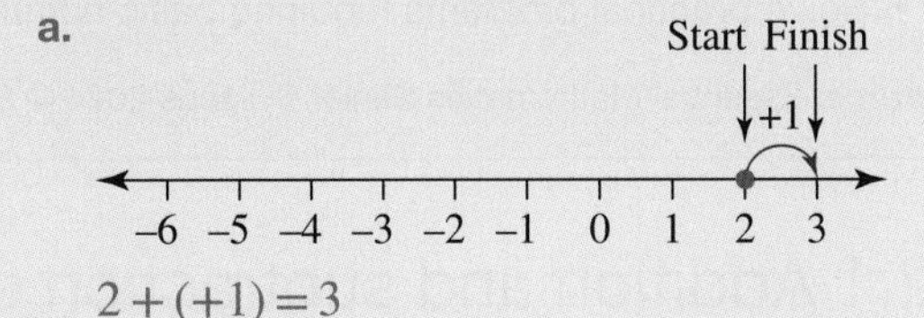
2. Write the answer.	$2+(+1)=3$
b. 1. Start at 3 and move 2 units to the left, as you are adding a negative number.	b. 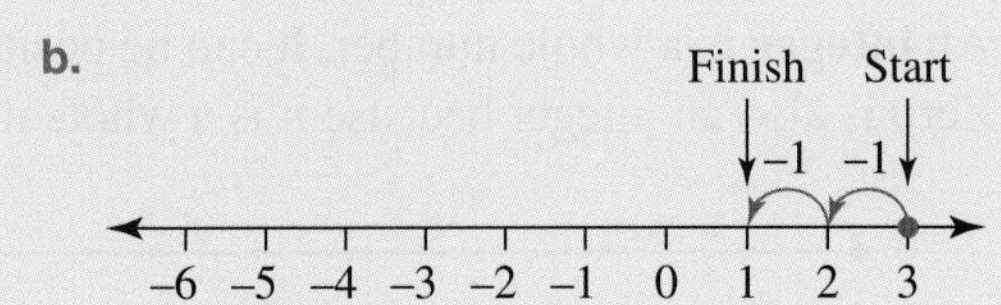
2. Write the answer.	$3+(-2)=1$
c. 1. Start at −4 and move 2 units to the left, as you are subtracting a positive number.	c. 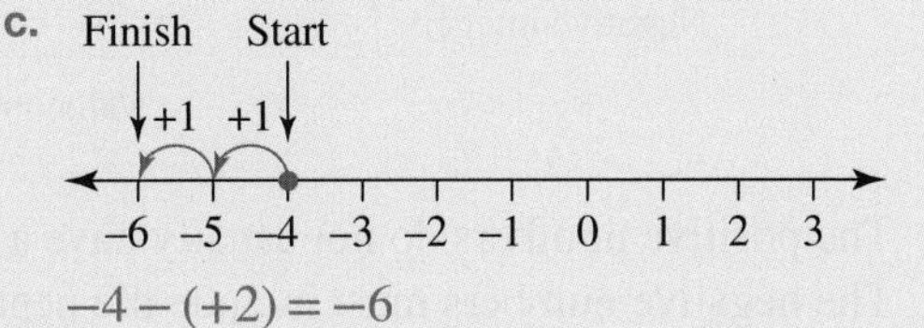
2. Write the answer.	$-4-(+2)=-6$
d. 1. Start at −3 and move 5 units to the right, as you are subtracting a negative number.	d. 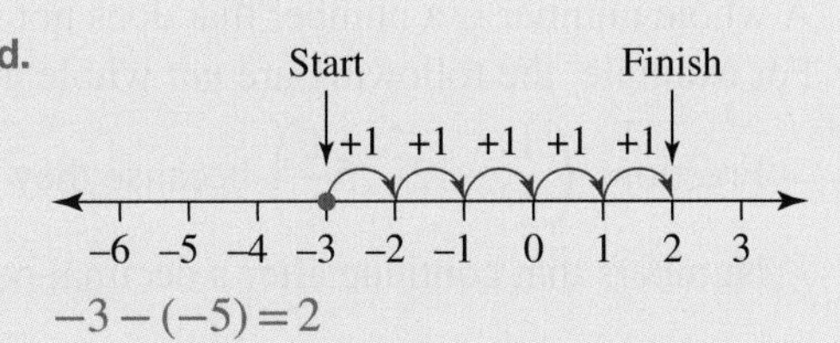
2. Write the answer.	$-3-(-5)=2$

1.2.2 Addition and subtraction of integers using symbols

- Another way to remember the rules for adding and subtracting integers is to imagine positive and negative symbols (or counters) representing each of the numbers in an equation.
- A pink positive symbol represents +1 and a blue negative symbol represents −1.
- When calculating:
  - 2 is written as:

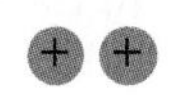

  - −2 is written as:

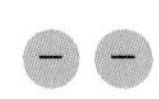

  - positive and negative symbols cancel each other out: $-1+1=0$

tlvd-3550

WORKED EXAMPLE 2 Solving addition and subtraction of integers using symbols

Calculate the value of each of the following.

a. $2+2$
b. $-2+2$
c. $-2-(3)$
d. $-4-(-2)$

THINK	WRITE
a. 1. Set the equation up using symbols.	a. $2+2=$ ⊕ ⊕ + ⊕ ⊕
2. Add all the positives.	4 positives $=+4$
3. Write the answer.	$2+2=4$
b. 1. Set the equation up using symbols.	b. $-2+2=$ ⊖ ⊖ + ⊕ ⊕
2. Cancel any pairs of positives and negatives.	⊖̸ ⊖̸ + ⊕̸ ⊕̸
3. Count how many symbols are left.	Zero symbols remain.
4. Write the answer.	$-2+2=0$
c. 1. Set the equation up using symbols.	c. $-2-(3)=$ ⊖ ⊖ − ⊕ ⊕ ⊕
2. When subtracting positive integers, change the sign of the symbols and then add them.	⊖ ⊖ + ⊖ ⊖ ⊖
3. Count all the negatives.	5 negatives $=-5$
4. Write the answer.	$-2-(3)=-5$
d. 1. Set the equation up using symbols.	d. $-4-(-2)=$ ⊖ ⊖ ⊖ ⊖ − ⊖ ⊖
2. When subtracting negative integers, change the sign of the symbols and then add them.	⊖ ⊖ ⊖ ⊖ + ⊕ ⊕
3. Cancel any pairs of positive and negative symbols.	⊖̸ ⊖̸ ⊖ ⊖ + ⊕̸ ⊕̸
4. Count how many positives or negatives remain.	Two negatives remain.
5. Write the answer.	$-4-(-2)=-2$

1.2.3 Addition and subtraction of integers by applying rules

- Follow the rules shown in this table to add and subtract positive and negative numbers.

Rules for addition and subtraction of integers

Rule		Example
Adding a positive integer is the same as adding.	$++=+$	$+2+(+5)=+7$
Adding a negative integer is the same as subtracting.	$+-=-$	$+2+(-5)=-3$
Subtracting a positive integer is the same as subtracting.	$-+=-$	$+2-(+5)=-3$
Subtracting a negative integer is the same as adding.	$--=+$	$+2-(-5)=7$

tlvd-3551

WORKED EXAMPLE 3 Solving addition and subtraction of integers by applying rules

Calculate the value of each of the following.

a. $2+2$ b. $2+(-2)$ c. $2-(+3)$ d. $-4-(-2)$

THINK	WRITE
a. 1. Adding a positive integer is the same as adding.	a. $2+2$
2. Apply the rule $++=+$ and calculate the value.	$2+2=4$
b. 1. Adding a negative integer is the same as subtracting.	b. $2+(-2)$
2. Apply the rule $+-=-$ and calculate the value.	$2+(-2)=2-2$
3. Write the answer.	$2-2=0$
c. 1. Subtracting a positive integer is the same as subtracting.	c. $2-(+3)$
2. Apply the rule $+-=-$ and calculate the value.	$2-(+3)=2-3$
3. Write the answer.	$2-3=-1$
d. 1. Subtracting a negative integer is the same as adding.	d. $-4-(-2)$
2. Apply the rule $--=+$ and calculate the value.	$-4--2=-4+2$
3. Write the answer.	$-4+2=-2$

1.2.4 Multiplying or dividing integers

- When multiplying integers, the following rules apply.

Rules for the multiplication of integers

- **When multiplying two integers with the same sign, the answer is positive.**

$$+\times+=+$$
$$-\times-=+$$

- **When multiplying two integers with different signs, the answer is negative.**

$$+\times-=-$$
$$-\times+=-$$

WORKED EXAMPLE 4 Solving integer multiplication by applying rules

Evaluate each of the following.

a. $(-4)\times+3$ b. $(-7)\times(-6)$

THINK	WRITE
a. The two numbers have different signs, so the answer is negative.	a. $(-4)\times(+3)$ $=-12$
b. The two numbers have the same sign, so the answer is positive.	b. $(-7)\times(-6)$ $=42$ (or $+42$)

- Division is the inverse (opposite) operation of multiplication.
- We can use the multiplication facts for integers to work out the division facts for integers.

Multiplication fact	Division fact	Multiplication pattern	Division pattern
$4 \times 5 = 20$	$20 \div 5 = 4$ and $20 \div 4 = 5$	positive × positive = positive	$\frac{\text{positive}}{\text{positive}} = \text{positive}$
$(-4) \times (-5) = 20$	$(-20) \div (-4) = 5$	negative × negative = positive	$\frac{\text{negative}}{\text{negative}} = \text{positive}$
$(-4) \times 5 = -20$	$(-20) \div 5 = -4$ and $20 \div (-5) = -4$	negative × positive = negative and positive × negative = negative	$\frac{\text{negative}}{\text{positive}} = \text{negative}$ and $\frac{\text{positive}}{\text{negative}} = \text{negative}$

Determining the sign of the answer when dividing integers

- **When dividing two integers with the same sign, the answer is positive.**

$$+ \div + = +$$
$$- \div - = +$$

- **When dividing two integers with different signs, the answer is negative.**

$$+ \div - = -$$
$$- \div + = -$$

- Remember that division statements can be written as fractions and then simplified.
  For example:

$$\begin{aligned}(-12) \div (-4) &= \frac{-12}{-4} \\ &= \frac{12 \times \cancel{-1}}{4 \times \cancel{-1}} \\ &= 3\end{aligned}$$

WORKED EXAMPLE 5 Solving division of integers

Evaluate each of the following.

a. $(-48) \div 6$

b. $\frac{-54}{-9}$

c. $144 \div (-6)$

d. $(-240) \div (-16)$

THINK	WRITE
a. The two numbers have different signs, so the answer is negative.	a. $(-48) \div 6$ $= -8$

b. The two numbers have the same sign, so the answer is positive.

b. $\frac{-54}{-9} = \frac{(-1)\times 54}{(-1)\times 9}$

$= \frac{54}{9}$

$= 6$

c. 1. Complete the division as if both numbers were positive numbers.

c. $6\overline{)14^24}$ = 24

2. Determine the sign of the answer. The two numbers have different signs, so the answer is negative.

$144 \div (-6) = -24$

d. 1. Complete the division as if both numbers were positive numbers.

d. $16\overline{)24^80}$ = 15

2. Determine the sign of the answer. The two numbers have the same sign, so the answer is positive.

$(-240) \div (-16) = 15$

1.2.5 Solving practical problems

- We constantly solve practical problems using basic number operations without realising that we are doing it.
- You do this when you calculate what time you need to leave home to get to school on time or how much money you need to purchase food from the school canteen.

WORKED EXAMPLE 6 Solving practical numerical problems

A USB stick can hold 512 MB of data. If 386 MB of the USB stick is already filled, determine how much space is left. Check your result for accuracy.

THINK	WRITE
1. Determine the total storage space.	Space available = 512 MB
2. Identify how much space has been used.	Space used = 386 MB
3. The space left is the difference between the total space and the space used.	Space left = 512 − 386 = 126 MB
4. Check the answer for accuracy.	Space available = space used + space left LHS = RHS
5. Substitute in each value.	512 MB = 386 MB + 126 MB 512 MB = 512 MB
6. Write the answer.	The space left on the USB stick is 126 MB.

WORKED EXAMPLE 7 Solving real-world problems using calculations

Nathan has a part-time job that pays $15.50 per hour. Nathan gets paid time and a half for hours worked on Saturdays and double time for hours worked on Sundays. Determine how much Nathan gets paid in a week in which they work 5 hours on Friday, 4 hours on Saturday and 5.5 hours on Sunday.

THINK	WRITE
1. Calculate the amount Nathan earned on Friday.	Money earned Friday $= 5 \times \$15.50$ $= \$77.50$
2. Calculate the amount Nathan earned on Saturday.	Money earned Saturday $= 4 \times \$15.50 \times 1.5$ $= \$93.00$
3. Calculate the amount Nathan earned on Sunday.	Money earned Sunday $= 5.5 \times \$15.50 \times 2$ $= \$170.50$
4. Calculate Nathan's weekly pay by adding the amounts earned on Friday, Saturday and Sunday.	Total pay $= \$77.50 + \$93.00 + \$170.50$ $= \$341.00$
5. Write Nathan's weekly pay.	Nathan earned $341.00 that week.

on Resources

Video eLesson Integers on the number line (eles-0040)

Interactivities Direct number target (int-0074)
Addition of positive integers (int-3922)
Subtraction of positive integers (int-3924)

Exercise 1.2 Number operations

learnon

1.2 Quick quiz  | 1.2 Exercise

Simple familiar	Complex familiar	Complex unfamiliar
1, 2, 3, 4, 5, 6, 7, 8, 9, 10, 11, 12, 13, 14, 15, 16, 17, 18, 19, 20, 21, 22	N/A	N/A

These questions are even better in jacPLUS!
- Receive immediate feedback
- Access sample responses
- Track results and progress

Find all this and MORE in jacPLUS

Simple familiar

1. Select the integers from the following numbers.

$$5, -2, \frac{3}{4}, 212, 12.3, -2.5, -33, -2\frac{1}{2}$$

2. WE1 Calculate each of the following using the number line method.

a. $(-5) + 2$ b. $(-6) + 1$ c. $5 + (-3)$
d. $(-8) + (-3)$ e. $21 + (+9)$ f. $18 + (-5)$

3. **WE2** Calculate each of the following using the cancelling signs method.

a. $3+(+1)$ b. $3+(-2)$ c. $4-(+5)$
d. $2+(-2)$ e. $3-(-2)$ f. $(-1)-(-3)$

4. **WE3** Calculate the value of each of the following using rules.

a. $(-7)+(-3)$ b. $8-(-7)$ c. $(-23)+(+15)$
d. $(-18)-(-17)$ e. $26-(-13)$ f. $(-72)-(-26)$

5. Calculate the value of each of the following using the method of your choice.

a. $(-37)+(-12)$ b. $42-(+7)$ c. $(-14)-(+18)$
d. $(-27)-(-15)$ e. $37-(+12)$ f. $135-(-37)$

6. Calculate the value of each of the following using the method of your choice.

a. $12+(-4)+(+6)$ b. $28+(-7)-(-10)$ c. $(-15)+(+5)-(+8)$
d. $28-(+15)-(+4)$ e. $18-(-12)+(-5)$ f. $(-42)-(-21)-(-21)$

7. **WE4** Evaluate each of the following.

a. $(-3)\times 5$ b. $3\times(-7)$ c. $(-6)\times(-5)$
d. $2\times(-10)$ e. $(-8)\times(-5)$ f. $(-7)\times 8$

8. Evaluate each of the following.

a. $(-10)\times 25$ b. $(-125)\times 10$ c. $(-6)\times 9$
d. $(+9)\times(-7)$ e. $(-11)\times(-7)$ f. $250\times(-2)$

9. Use an appropriate method to evaluate the following.

a. $(-2)\times 3\times(-5)\times(-10)$ b. $6\times(-1)\times 5\times(-2)\times 1$
c. $8\times(-3)\times(-1)\times(-2)\times 4$ d. $(-3)\times 5\times(-2)\times(-1)\times(-1)\times(-1)$
e. $(-5)\times 6\times(-2)\times(-2)$ f. $6\times(-1)\times 3\times(-2)$

10. **WE5** Evaluate each of the following.

a. $(-54)\div 9$ b. $10\div(-2)$ c. $(-8)\div(-2)$
d. $(-5)\div(-1)$ e. $99\div(-11)$ f. $0\div(-9)$

11. Evaluate each of the following.

a. $42\div(-3)$ b. $(-130)\div 5$ c. $(-56)\div(-8)$
d. $(+88)\div(-4)$ e. $(-66)\div(-6)$ f. $168\div(-8)$

12. Evaluate each of the following.

a. $184\div(-8)$ b. $(-189)\div 9$ c. $(-161)\div(-7)$
d. $(-132)\div(-2)$ e. $(-204)\div 6$ f. $1080\div(-9)$

13. Evaluate each of the following.

a. $216\div(-12)$ b. $(-345)\div 15$ c. $(-1536)\div(-24)$
d. $(-1764)\div(-49)$ e. $4096\div 64$ f. $(-2695)\div 55$

14. **WE6** The monthly data allowance included in your mobile phone plan is 15 GB. If you have already used 12 GB of data this month, determine how much data you have left to use for the rest of the month. Check your result for accuracy.

15. A UHD television was priced at $8999. It has a sale sign on it that reads: 'Take a further $1950 off the marked price.' Determine the sale price of the television. Check your result for accuracy.

16. The photographs show three tall structures.

a. Determine how much taller the Eiffel Tower is than the Chrysler Building.
b. Calculate the height difference between the Infinity Tower and the Eiffel Tower.
c. Determine the difference in height between the Chrysler Building and the Infinity Tower.
d. Explain how you obtained your answers to parts **a**, **b** and **c**.

17. This sign shows the distances to a number of locations in the Northern Territory and Queensland.

PLENTY HIGHWAY 12
Jct. Sandover Hwy 27
Harts Range 145
Tobermorey 496
Boulia 743
Central Australian Gemfields 77

a. Determine how much further Tobermorey is from Harts Range.
b. Calculate the distance between Boulia and Tobermorey.

18. **WE7** Sarah has a part-time job that pays $18.50 per hour. Sarah gets paid time and a half for hours worked on Saturdays and double time for hours worked on Sundays. Use your calculator to calculate how much Sarah gets paid in a week in which she works 4 hours on Friday, 5 hours and 30 minutes on Saturday, and 8 hours on Sunday.

19. Assuming that all the eggs in the photograph are the same size, determine the mass of each egg.

20. Bryce decides to cook meat pies for lunch. He takes the pies out of the freezer that is set at −18 °C to defrost on the bench where the room temperature is 21 °C. He sets the oven to 180 °C to cook the pies.

    a. Identify the difference between the temperature of the freezer and the room temperature.
    (*Hint:* difference = largest number − smallest number)
    b. Calculate the difference between the room temperature and the temperature of the oven.
    c. Calculate the difference of the freezer temperature and the temperature of the oven.

21. This photograph shows a mobile phone and its associated costs.

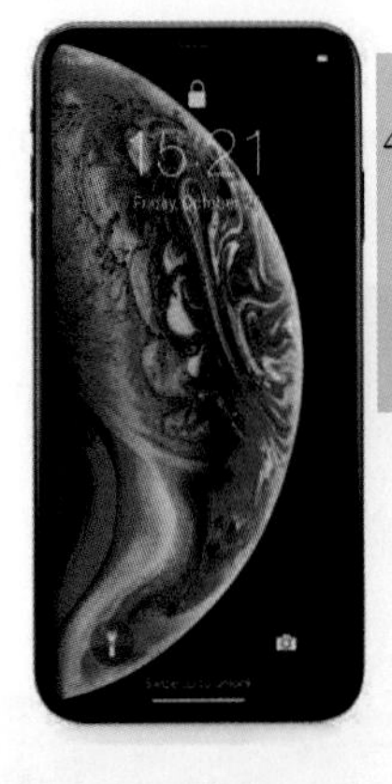

Voice calls:
47c per 30 seconds
Text messages:
15c per message
Picture messages:
50c per message

    a. Determine the cost of a 1-minute call on this phone.
    b. Determine the cost of a 35-minute call on this phone.
    c. If you use the phone to make five 3-minute calls every day, calculate how much the calls will cost you for a year. (Assume 365 days in a standard year.)
    d. Determine the cost of sending 20 text messages from this phone.
    e. Calculate the difference in cost between sending 100 text messages and sending 50 picture messages using this phone.

22. Consider the digits 2, 3, 4 and 5.

    a. Construct the largest possible number using these digits.
    b. Construct the smallest possible number using these digits.
    c. Calculate the difference between the numbers from parts **a** and **b**.

Fully worked solutions for this chapter are available online.

LESSON
1.3 Order of operations and reasonableness

SYLLABUS LINKS

- Apply arithmetic operations according to their correct order.
- Ascertain the reasonableness of answers to arithmetic calculations.

1.3.1 Order of operations

- The **order of operations** is a set of rules that determines the order in which mathematical operations are to be performed.
- The acronym BIDMAS is one example used to remember the correct order in which to complete operations within an equation.

Order of operations

The order of operations rules can be remembered as BIDMAS:

B Brackets	I Indices		D Divide		M Multiply	A Add		S Subtract
( )	$\sqrt{x}$	x^2	÷	or	×	+	or	−

Order	Operations	What does it mean?
First	Brackets	Calculate any parts of the expression that are shown in brackets. For example: $(3+1)\times 4 = 4\times 4$
Second	Indices	Multiply or divide out any indices. Indices are: • powers (3^2); a number raised to a power is multiplied by itself $3^2 = 3\times 3$ $= 9$ $3^3 = 3\times 3\times 3$ $= 27$ • roots ($\sqrt{9}$); a root is the opposite of a power. $\sqrt{9} = 3$ $\sqrt[3]{27} = 3$

(continued)

(continued)

Order	Operations	What does it mean?
Third	**D**ivision and **M**ultiplication	Calculate any parts of the expression that involve division or multiplication. If the expression contains both multiplication and division, start from the left and work across to the right. For example: $3 \times 4 + 36 \div 6 = 12 + 6$ $= 18$
Last	**A**ddition and **S**ubtraction	Calculate any parts of the expression that involve addition or subtraction. If the expression contains both addition and subtraction, start from the left and work across to the right. For example (remember division comes before addition or subtraction): $4 + 24 \div 4 - 2 = 4 + 6 - 2$ $= 10 - 2$ $= 8$

tlvd-3553

WORKED EXAMPLE 8 Applying the order of operations

Calculate the value of each of the following.

a. $\mathbf{23 - 6 \times 4}$ b. $\mathbf{(12 - 8) + 5^2 - (10 + 2^2)}$ c. $\mathbf{\dfrac{3(4 + 8) + 4}{4 + 2(3^2 - 1)}}$

THINK	WRITE
a. 1. Apply BIDMAS to determine the first step (in this case, multiplication).	a. $23 - 6 \times 4$ $= 23 - 24$
2. Complete the next step in the calculation (in this case, subtraction) and write the answer.	$= -1$
b. 1. Apply BIDMAS to determine the first step (perform the calculations in brackets first, then remove the brackets).	b. $(12 - 8) + 5^2 - (10 + 2^2)$ $= 4 + 5^2 - (10 + 4)$ $= 4 + 5^2 - 14$
2. Complete the next steps in the calculation (work out the powers, then carry out addition and subtraction from left to right).	$= 4 + 25 - 14$ $= 29 - 14$
3. Complete the final step and write the answer.	$= 15$
c. 1. Apply BIDMAS to determine the first step (brackets).	c. $\dfrac{3(4 + 8) + 4}{4 + 2(3^2 - 1)} = \dfrac{3 \times 12 + 4}{4 + 2(9 - 1)}$
2. Complete the next step in the calculation (multiplication).	$= \dfrac{3 \times 12 + 4}{4 + 2 \times 8}$ $= \dfrac{36 + 4}{4 + 16}$
3. Perform the calculations on the numerator and denominator separately.	$= \dfrac{40}{20}$
4. Complete the division and write the answer.	$= 2$

WORKED EXAMPLE 9 Applying the order of operations to real-world problems

Sean buys 4 oranges at \$0.70 each and 10 pears at \$0.55 each.

a. **Calculate the price Sean paid for the oranges and pears.**

b. **Calculate the average price Sean paid for a piece of fruit.**

THINK	WRITE
a. 1. Calculate the total amount of money Sean spent on fruit.	a. $(4 \times \$0.70) + (10 \times \$0.55) = \$2.80 + \$5.50$ $= \$8.30$
2. Write the answer.	The total amount Sean spent on fruit is \$8.30.
b. 1. Count the total number of pieces of fruit that Sean purchased.	b. 4 oranges + 10 pears = 14 pieces of fruit
2. The average price Sean paid for a piece of fruit is equal to the total cost divided by the total number of pieces of fruit.	$\text{Average price} = \dfrac{\text{total cost}}{\text{total number of pieces}}$ $= \dfrac{\$8.30}{14}$ $= \$0.59$
3. Write the answer.	The average price Sean paid for a piece of fruit is \$0.59.

1.3.2 Reasonableness

- When calculating mathematical answers, it is always important to understand the question so you have an idea of what a reasonable answer would be.
- Checking the reasonableness of answers can help you see possible mistakes in your working.

WORKED EXAMPLE 10 Evaluating the reasonableness of an answer

Cathy goes shopping to purchase an outfit for a hike. She buys a \$110 pair of waterproof pants and a \$180 jacket, and gets \$55 off the combined price by purchasing the two together. She also buys a pair of boots for \$120 and a pair of socks for \$15.
Calculate how much Cathy spent in total and check your answer for reasonableness.

THINK	WRITE
1. Read the question carefully to understand what it is about.	The question asks you to add up Cathy's total shopping bill including the discount.

2. Have an idea of what sort of answer you expect.	Cathy purchased three items, each over \$100 dollars, so the answer should be over \$300.
3. Write a mathematical expression to calculate the total amount that Cathy spent, including the discount.	Amount spent $= \$110 + \$180 - \$55 + \$120 + \$15$ $= \$370$
4. Check the answer for reasonableness.	This answer is above \$300, so it seems reasonable.

Digital technology

Using a calculator to evaluate expressions involving the order of operations

Many scientific calculators can input and evaluate expressions in one line.

Consider the expression $\left[(4+2)^2+6\right] \div 7$.

To determine its value on a calculator, follow the steps below.

- Input the expression $\left((4+2)^2+6\right) \div 7$ in one continuous line into your calculator using only round brackets.
- Press = or ENTER to evaluate.

D Math ▲

((4+2)²+6)÷7

6

Now consider the expression $\dfrac{3 \times \left(\sqrt{40-4}-3\right)^2 + (10 \div 2)}{12 \div 3}$.

To determine its value on a calculator, follow the steps below.

- Press the fraction template button. This will set up two entry boxes — one for the numerator and one for the denominator.
- Input $3 \times \left(\sqrt{40-4}-3\right)^2 + (10 \div 2)$ into the entry box for the numerator.
- Input $12 \div 3$ into the entry box for the denominator.
- Press = or ENTER to evaluate.

D Math ▲

$\dfrac{3\times(\sqrt{40-4}-3)^2+(\ldots}{12\div3}$

8

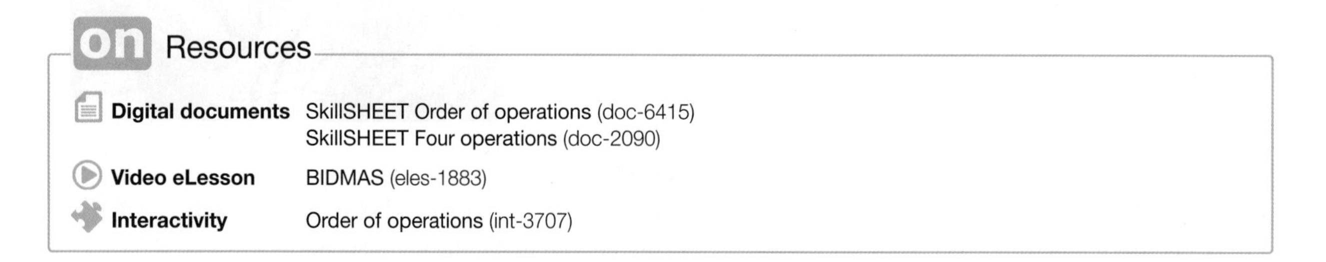

on Resources

Digital documents	SkillSHEET Order of operations (doc-6415) SkillSHEET Four operations (doc-2090)
Video eLesson	BIDMAS (eles-1883)
Interactivity	Order of operations (int-3707)

Exercise 1.3 Order of operations and reasonableness

learnon

1.3 Quick quiz on	1.3 Exercise

Simple familiar	Complex familiar	Complex unfamiliar
1, 2, 3, 4, 5, 6, 7, 8, 9, 10, 11, 12, 13, 14	N/A	N/A

These questions are even better in jacPLUS!
- Receive immediate feedback
- Access sample responses
- Track results and progress

Find all this and MORE in jacPLUS

Simple familiar

1. WE8 Calculate the values of the following.
   a. $(-4) \times 2 + 1$
   b. $8 \div (2 - 4) + 4$
   c. $9 \times (8 - 3)$
   d. $(-3) - 40 \div 8 + 2$
   e. $4 + 12 \times (-5)$
   f. $(-5) \times 12 + 2$
2. Calculate the values of the following.
   a. $12 - 6 \div 3$
   b. $45 \div (27 \div (-3))$
   c. $(17 - +7) \div (-5)$
   d. $-12 + 8 \times 7$
   e. $100 \div ((-50) \div (-2)) + 10$
   f. $9 + \dfrac{24}{-6} \times 3$
3. Calculate the values of the following.
   a. $(-7) + 4 \times -4$
   b. $((-63) \div (-7)) \times (-3) + (-2)$
   c. $(-5)^2 - 3 \times (-5)$
   d. $(-6) \times (-8) - \left(3 + (-6)^2\right)$
   e. $52 \div ((-9) - 4) - 8$
   f. $-6 - 64 \div (-16) + 8$
4. WE9 Pat buys 5 bananas at \$0.30 each and 8 apples at \$0.25 each.
   a. Calculate the price Pat paid for the bananas and apples.
   b. Calculate the average price Pat paid for a piece of fruit (rounded to the nearest cent).

5. Carrol purchases chocolate for her daughter's party. She buys 6 packets of Mars bars at \$4.85 each and 8 packets of Freddo Frogs at \$3.55 each. Calculate the average price that Carrol paid for each packet of chocolate.

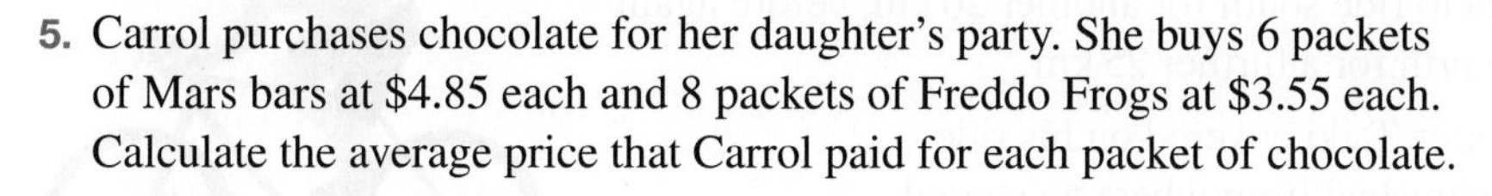

6. WE10 Fred went to a sports store to get some clothes for the gym. He purchased:
   - a tracksuit for \$120
   - shoes for \$150
   - shorts for \$30
   - a singlet top for \$25.

Given that he got a \$35 discount, calculate the amount that he spent at the sports store. Check your answer for reasonableness.

7. Erika goes shopping for some party food. She buys:
   - 6 packets of chips at \$2.50 each
   - 2 boxes of cola at \$8.50 each
   - 5 packets of biscuits at \$1 each
   - 3 dips at \$2 each.

   Calculate the amount Erika spent in total, if she got a \$2 discount. Check your answer for reasonableness.

8. Bob went to the warehouse to get some building supplies.
   He bought:
   - three 4 m-long pieces of timber at \$4.50 per metre
   - 2 packets of nails at \$5.50 each
   - 3 tubes of liquid nails at \$4 each.

   Because Bob is a regular customer, he received a \$15 discount off his total purchase.
   Calculate the amount Bob spent in total. Check your answer for reasonableness.

9. Insert one set of brackets in the appropriate place to make each of these statements true.
   a. $12 - 8 \div 4 = 1$
   b. $4 + 8 \times 5 - 4 \times 5 = 40$
   c. $3 + 4 \times 9 - 3 = 27$
   d. $3 \times 10 - 2 \div 4 + 4 = 10$
   e. $10 \div 5 + 5 \times 9 \times 9 = 81$
   f. $18 - 3 \times 3 \div 5 = 9$

10. For a birthday party, you buy two packets of paper plates at \$2 each, three bags of chips at \$4 each and three boxes of party pies at \$5 each.
    a. Explain how the order of operations helps you calculate the correct total cost of these items.
    b. Devise an equation to show the operations required to calculate the total cost.
    c. Calculate the total cost.

11. Taiki headed north on a bike ride, initially travelling 25 km. He then turned around and travelled 15 km south before stopping for a drink.
    After his drink, Taiki continued to ride south for another 20 km, before again turning around and travelling north for a further 25 km.
    a. Calculate how many kilometres Taiki covered on his ride.
    b. Determine how far north he finished from where he started.

12. Students were given the following question to evaluate.

$$4 + 8 \div (-2)^2 - 7 \times 2$$

    a. A number of different answers were obtained, including −8, −12 and −17. Determine which one of these is correct.
    b. Using only brackets, change the question in two ways so that the other answers are correct.

13. Calculate the number required to make the following equation true:

$$13 \times (15 - 1) = 180 + \square$$

14. Calculate the number required to make the following equation true:

$$\left(10^2 + 12 \div 3\right) \div (-8) = \square$$

Fully worked solutions for this chapter are available online.

LESSON
1.4 Estimation

SYLLABUS LINKS

- Use leading-digit approximation to obtain estimates of calculations.
- Check results of calculations for accuracy.
- Round up or round down numbers to the required number of decimal places.

1.4.1 Rounding

- Numbers can be rounded to different degrees of accuracy.
  For example, they can be rounded to 1, 2, 3 or more decimal places. The more decimal places a number has, the more accurate it is. However, more decimal places are not always necessary or relevant.

Rounding to the nearest 10

- To round to the nearest 10, think about which multiple of 10 the number is closest to.
  For example, 34 rounded to the nearest 10 is 30, because 34 is closer to 30 than 40.

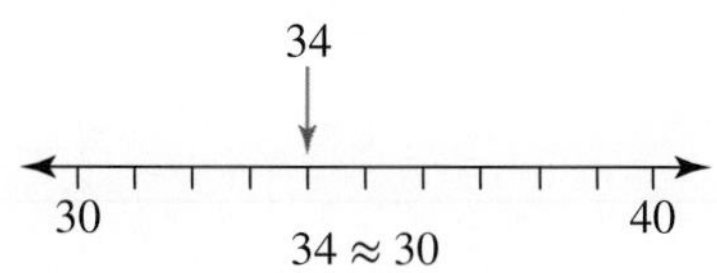

(***Note:*** The $\approx$ symbol represents 'is approximately equal to'.)

Rounding to the nearest 100

- To round to the nearest 100, think about which multiple of 100 the number is closest to.
  For example, 375 rounded to the nearest 100 is 400, because 375 is closer to 400 than 300.

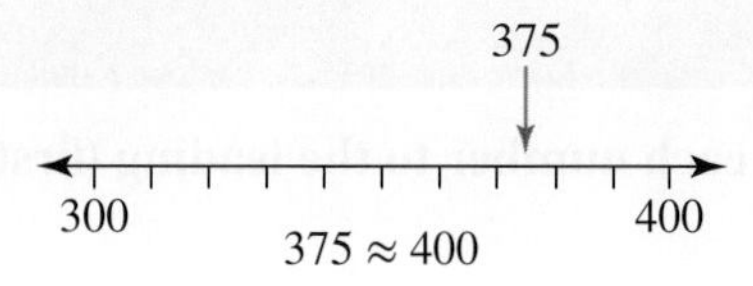

Rounding to the first digit

- When rounding to the leading (first) digit, the second digit needs to be looked at.
  - If the second digit is 0, 1, 2, 3 or 4, the first digit stays the same and all the following digits are replaced with zeros.
  - If the second digit is 5, 6, 7, 8 or 9, the first digit is increased by 1 (rounded up) and all the following digits are replaced with zeros.

  For example, if 2345 is rounded to the first digit, the result is 2000 because the second digit in the number is 3, so all digits following the first digit (in this case, 2) are replaced with 0.

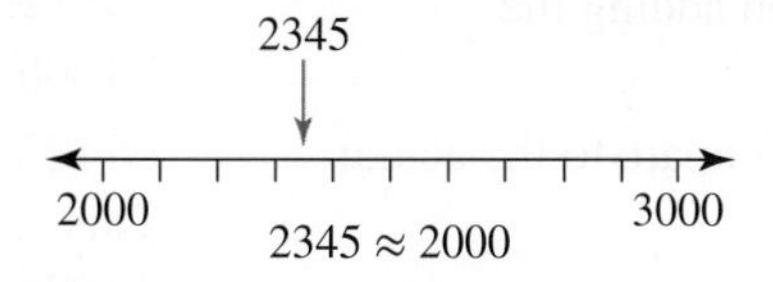

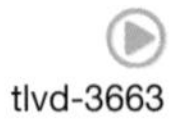
tlvd-3663

WORKED EXAMPLE 11 Applying rounding to integers

Round each of the following as directed.

a. **563 to the nearest 10**

b. **12 786 to the nearest 100**

c. **7523 to the leading digit**

THINK	WRITE
a. 1. To round to the nearest 10, look at the units place value.	a. 563
2. Is the units place value less than 5? Since 3 is less than 5, we round down. Replace 3 with 0 and write the answer.	≈ 560
b. 1. To round to the nearest 100, look at the tens place value.	b. 12 7**8**6
2. Is the tens place value greater than or equal to 5? Since 8 is greater than or equal to 5, we round up. Increase 7 by 1 to 8 and replace the following digits with zeros.	12 786 ↓↓↓ 800
3. Write the answer.	$12\,786 \approx 12\,800$
c. 1. When rounding to the leading digit, look at the second digit.	c. 7523
2. Since 5 is greater than or equal to 5, we round up. Increase the first digit (7) by 1 and replace the remaining digits with zeros.	7523 ↓↓↓↓ 8000
3. Write the answer.	$7523 \approx 8000$

1.4.2 Estimation

- An **estimate** is not the same as a guess, because it is based on information.
- Estimation is useful for checking that your answer is reasonable when performing calculations.
- This can help avoid simple mistakes, like typing an incorrect number into a calculator.

WORKED EXAMPLE 12 Solving problems of estimating by rounding numbers to the leading (first) digit

Estimate $66\,123 \times 749$ by rounding each number to the leading (first) digit and then completing the calculation.

THINK	WRITE
1. Round the first number (**6**6 123) by looking at the second digit. Since the second digit (6) is greater than or equal to 5, increase the leading digit (6) by 1 and replace the remaining digits with zeros.	**6**6 123 ≈ 70 000
2. Round the second number (**7**49) by looking at the second digit. Since the second digit (4) is less than 5, leave the leading digit and replace the remaining digits with zeros.	**7**49 ≈ 700
3. Multiply the rounded numbers by multiplying the leading digits ($7 \times 7 = 49$) and then adding the number of zeros ($4 + 2 = 6$).	$70\,000 \times 700$ $7 \times 7 = 49$ $70\,000 \times 700 = 49\,000\,000$
4. Write the estimated answer and compare to the actual answer.	$66\,123 \times 749 \approx 49\,000\,000$ This compares well with the actual answer of 49 526 127, found using a calculator.

1.4.3 Checking an answer for accuracy

- When performing calculations, you have to be very careful not to make careless or common mistakes.
- It is helpful to be able to identify mistakes to make sure you don't also make common calculation errors.

WORKED EXAMPLE 13 Testing an answer for accuracy

Check the following answers for accuracy. If the answer is not accurate, identify and explain where the mistake was made.

a. $2+5\times 2=14$ b. $12-(6\times 2)=12$ c. $\frac{12+12}{12}=12$

THINK	WRITE
a. Using BIDMAS, the multiplication needs to be done first.	a. $2+5\times 2=14$ $2+10=14$ $12\neq 14$
This answer is not accurate. Ask yourself how they got 14 as an answer.	They would get 14 if they did the calculation from left to right without following BIDMAS. $2+5\times 2=14$ $7\times 2=14$ $14=14$ This is not correct mathematics, but it identifies the error made.
b. Using BIDMAS, the brackets need to be worked out first.	b. $12-(6\times 2)=12$ $12-(12)=12$ $0\neq 12$
This answer is not accurate. Ask yourself how they got 12 as an answer.	They would get 12 if they did the calculation from left to right and ignored the brackets. $12-(6\times 2)=12$ $12-6\times 2=12$ $6\times 2=12$ $12=12$ This is not correct mathematics, but it identifies the error made.
c. You should calculate the numerator first.	c. $\frac{12+12}{12}=12$ $\frac{24}{12}=12$ $2\neq 12$
This answer is not accurate. Ask yourself how they got 12 as an answer.	One possible error could be that they cancelled 12 from the numerator with 12 from the denominator. $\frac{12+12}{12}=12$ $\frac{\cancel{12}+12}{\cancel{12}}=12$ $\frac{12}{1}=12$ $12=12$ This is not correct mathematics, but it identifies the error made.

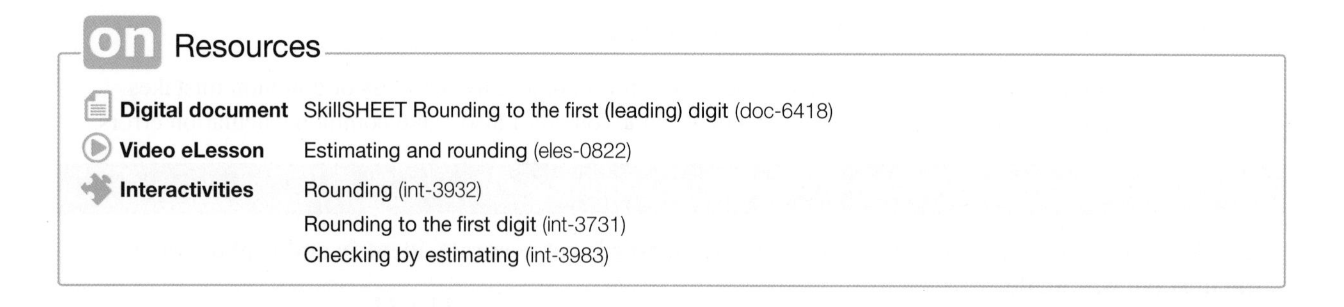

Exercise 1.4 Estimation

learnon

1.4 Quick quiz on | 1.4 Exercise

Simple familiar	Complex familiar	Complex unfamiliar
1, 2, 3, 4, 5, 6, 7, 8, 9, 10, 11, 12, 13, 14, 15, 16, 17, 18, 19, 20	N/A	N/A

These questions are even better in jacPLUS!
- Receive immediate feedback
- Access sample responses
- Track results and progress

Find all this and MORE in jacPLUS

Simple familiar

1. **WE11** Round each of the following as directed.
   a. 934 to the nearest 10 b. 12 963 to the nearest 100 c. 85 945 to the leading digit
2. Round the following.
   a. 347 to the nearest 10 b. 86 557 to the nearest 100 c. 65 321 to the leading digit
3. Round each of the following numbers to the nearest 10.
   a. 47 b. 82 c. 129 d. 162
   e. 250 f. 2463 g. 4836 h. 7
4. Round each of the following numbers to the nearest 100.
   a. 43 b. 87 c. 142 d. 177
   e. 3285 f. 56 346 g. 86 621 h. 213 951
5. Round each of the following numbers to the nearest 1000.
   a. 512 b. 3250 c. 1324 d. 6300
   e. 7500 f. 13 487 g. 435 721 h. 728 433
6. Round each of the following numbers to the leading digit.
   a. 12 b. 23 c. 45 d. 153
   e. 1388 f. 16 845 g. 23 598 h. 492 385
7. **WE12** Estimate $74\,852 \times 489$ by rounding each number to its leading digit.
8. Estimate $87\,342 \div 449$ by rounding each number to its leading digit.
9. Estimate the answer to the following expressions by rounding each number to the leading digit and then completing the calculation.
   a. $482 + 867$ b. $123 + 758$ c. $1671 - 945$ d. $2932 - 1455$
   e. 88×543 f. 57×2632 g. $69\,523 \div 1333$ h. $3600 \div 856$

10. In your own words, explain how to round using leading digit approximation.

11. Emily purchased a number of items at the supermarket that cost \$1.75, \$5.99, \$3.45, \$5.65, \$8.95, \$2.35 and \$7.45. She was worried she didn't have enough money, so she used leading digit approximation to do a quick check of how much her purchases would cost.
Determine what approximation she came up with.

12. WE13 Check the following answers for accuracy. If the answer is not accurate, identify and explain where the mistake was made.
a. $15 - 3 \times 4 = 48$
b. $13 + (2 \times 2) = 30$
c. $3^2 - 2 = 1$

13. Check the following answers for accuracy. If the answer is not accurate, identify and explain where the mistake was made.
a. $27 - 3 \times 2 + 5 = 53$
b. $6 \times 7 + (2 \times 2) = 52$
c. $2^3 - 2 = 2$

14. Shae went to the shop to purchase two steaks for dinner. One steak was \$12.50 and the other steak was \$8.50. The difference in the prices of the two steaks is \$21. Determine whether this is accurate, and if not, identify the error.

15. A bank teller was counting \$10 notes. They counted 100 \$10 notes and said they had \$10 000 worth of money. Determine whether this is accurate, and if not, identify the error.

16. Give three examples of situations where it is suitable to use an estimate or a rounded value instead of an exact value.

17. The crowds at the Melbourne Cricket Ground for each of the five days of the Boxing Day Test were: 88 214, 64 934, 55 349, 47 567 and 38 431.
Using leading digit approximation, determine the estimated total crowd over the five days of the test.

18. The crowds at the Melbourne Cricket Ground for each of the five days of the Boxing Day Test were: 88 214, 64 934, 55 349, 47 567 and 38 431.
Determine the difference between the leading digit estimate and the actual crowd over the five days.

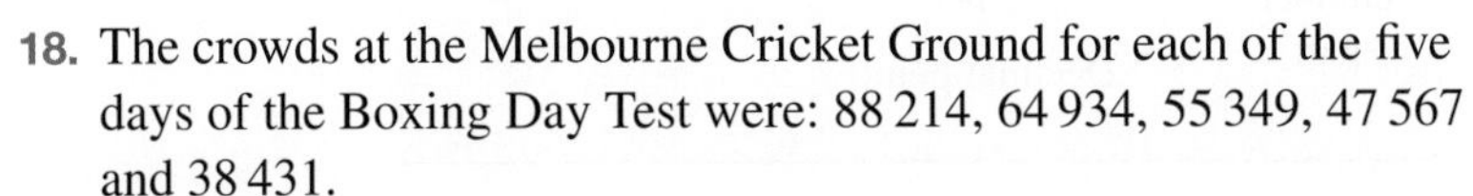

19. A car's GPS estimates that a family's trip for Christmas lunch is going to take 45 minutes.
a. Determine if the trip is likely to take exactly 45 minutes. Explain your answer.
b. Explain how this time is estimated by the GPS.

20. Rhonda went to the shop to buy a new school uniform. She decided she needed a new blazer, a new winter dress and a new shirt costing \$167, \$89 and \$55 respectively.
Determine the difference between her actual total and a leading digit estimate.

Fully worked solutions for this chapter are available online.

LESSON
1.5 Decimal significance

SYLLABUS LINKS

- Recognise the significance of place value after the decimal point.
- Evaluate decimal fractions to the required number of decimal places.
- Round up or round down numbers to the required number of decimal places.

1.5.1 Whole numbers and decimal parts

- Each position in a number has its own place value.

Hundred thousands	Ten thousands	Thousands	Hundreds	Tens	Units
100 000	10 000	1000	100	10	1

- Each place to the left of another position has a value that is 10 times larger.
- Each place to the right of another position has a value that is $\frac{1}{10}$ of the previous position.
- The position of a digit in a number gives the value of the digit.
- The following table shows the value of the digit 3 in some numbers.

Number	Value of 3 in the number
132	3 tens or 30
3217	3 thousands or 3000
4103	3 units (ones) or 3

Decimal numbers

In a decimal number, the whole number part and the fractional part are separated by a decimal point.

Whole number part — 73.064 — Fractional part

Decimal point

- A place value table can be extended to include decimal place values. It can be continued to an infinite number of **decimal places**.

Thousands	Hundreds	Tens	Units	.	Tenths	Hundredths	Thousandths	Ten thousandths
1000	100	10	1	.	$\frac{1}{10}$	$\frac{1}{100}$	$\frac{1}{1000}$	$\frac{1}{10\,000}$

- The following table shows the value of the digit 3 in some decimal numbers.

Number	Value of 3 in the number
14.32	3 tenths or $\frac{3}{10}$
106.013	3 thousandths or $\frac{3}{1000}$
0.000 03	3 hundred thousandths or $\frac{3}{100\,000}$

- The number of decimal places in a decimal is the number of digits after the decimal point.
- The number 73.064 has 3 decimal places. The zero (0) in 73.064 means that there are no tenths.
- The zero must be written to hold the place value; otherwise the number would be written as 73.64, which does not have the same value.

Expanded notation

A number can be written in expanded notation by adding the values of each digit.

The number 76.204 can be written in expanded notation as:

$$(7 \times 10) + (6 \times 1) + \left(2 \times \frac{1}{10}\right) + \left(4 \times \frac{1}{1000}\right)$$

WORKED EXAMPLE 14 Identifying the place value

Write the value of the 3 in each of the following decimal numbers.

a. 59.378 **b. 1.0003** **c. 79.737**

THINK	WRITE
a. 1. Identify at which place after the decimal point the 3 lies.	**a.** 59.**3**78
2. The 3 is the first number after the decimal point, which represents tenths.	$\frac{3}{10}$
b. 1. Identify at which place after the decimal point the 3 lies.	**b.** 1.000**3**
2. The 3 is the fourth number after the decimal point, which represents tens of thousandths.	$\frac{3}{10\,000}$
c. 1. Identify at which place after the decimal point the 3 lies.	**c.** 79.7**3**7
2. The 3 is the second number after the decimal point, which represents hundredths.	$\frac{3}{100}$

1.5.2 Rounding

- When rounding decimals, look at the first digit after the number of decimal places required.

Rules for rounding

- **If the first digit after the number of decimal places required is 0, 1, 2, 3 or 4, write the number without any change.**
- **If the first digit after the number of decimal places required is 5, 6, 7, 8 or 9, add 1 to the digit in the last required decimal place.**

- We can use the symbol ≈ to represent approximation when rounding has occurred, as shown in Worked example 15.
- *Note:* If you need to add 1 to the last decimal place and the digit in this position is 9, the result is 10. Zero is put in the last required place and 1 is added to the digit in the next place to the left. For example, 0.298 rounded to 2 decimal places is 0.30.

tlvd-3678

WORKED EXAMPLE 15 Applying rounding to decimal numbers to a given number of decimal places

Round the following to the number of decimal places shown in the brackets.

a. **7.562 432 (2)**
b. **27.875 327 (2)**
c. **35.324 971 (4)**
d. **0.129 91 (3)**

THINK	WRITE
a. 1. Write the number and underline the required decimal place.	a. 7.56̲2 432
2. Circle the next digit and round according to the rule. *Note:* Since the circled digit is less than 5, we leave the number as it is.	= 7.56(2)432 ≈ 7.56
b. 1. Write the number and underline the required decimal place.	b. 27.87̲5 327
2. Circle the next digit and round according to the rule. *Note:* Since the circled digit is greater than or equal to 5, add 1 to the last decimal place that is being kept.	= 27.87(5)327 ≈ 27.88
c. 1. Write the number and underline the required decimal place.	c. 35.324 9̲71
2. Circle the next digit and round according to the rule. *Note:* Since the circled digit is greater than 5, add 1 to the last decimal place that is being kept. As 1 is being added to 9, write 0 in the last place and add 1 to the previous digit.	= 35.324 9(7)1 ≈ 35.3250
d. 1. Write the number and underline the required decimal place.	d. 0.129̲91
2. Circle the next digit and round according to the rule. *Note:* Since the circled digit is greater than 5, add 1 to the last decimal place that is being kept. As 1 is being added to 9, write 0 in the last place and add 1 to the previous digit.	= 0.129(9)1 ≈ 0.130

1.5.3 Fractions as decimals

- A fraction can be expressed as a decimal by dividing the numerator by the denominator.
  For example, $\frac{3}{8} = 3 \div 8 = 0.375$
- If a decimal does not terminate and has decimal numbers repeating in a pattern, it is called a **recurring decimal.**
- A recurring decimal with only one repeating digit can be written by placing a dot above the digit.
- When there is more than one repeating digit, the recurring decimal can be written by placing a dot over the first and last digits of the repeating part, or by placing a bar above the entire repeating pattern.

$\frac{1}{3} = 1 \div 3 = 0.333\,333\,33\ldots = 0.\dot{3}$

$\frac{3}{7} = 3 \div 7 = 0.428\,574\,285\,7\ldots = 0.\dot{4}2857\dot{}\,$ or $0.\overline{428\,57}$

tlvd-3679

WORKED EXAMPLE 16 Applying the conversion of fractions to decimals

Change the following fractions into decimals.

a. $\frac{2}{5}$ **b.** $\frac{1}{8}$

THINK	WRITE
a. 1. Set out the question as for division of whole numbers, adding a decimal point and the required number of zeros.	**a.** $5\overline{)2.0}$
2. Divide, writing the answer with the decimal points aligned.	$\begin{array}{r} 0.4 \\ 5\overline{)2.0} \end{array}$
3. Write the answer.	$\frac{2}{5} = 0.4$
b. 1. Set out the question as for division of whole numbers, adding a decimal point and the required number of zeros. *Note:* $\frac{1}{8} = 1 \div 8$	**b.** $\begin{array}{r} 0.125 \\ 8\overline{)1.000} \end{array}$
2. Divide, writing the answer with the decimal point exactly in line with the decimal point in the question, and write the answer.	$\frac{1}{8} = 0.125$

- By knowing the decimal equivalent of any fraction, it is possible to determine the equivalent of any multiple of that fraction.

on Resources

Digital document SkillSHEET Place value (doc-6409)

Video eLesson Place value (eles-004)

Interactivities Place value (int-3921)
Decimal parts (int-3975)
Conversion of fractions to decimals (int-3979)

Exercise 1.5 Decimal significance

learnon

1.5 Quick quiz on

1.5 Exercise

Simple familiar	Complex familiar	Complex unfamiliar
1, 2, 3, 4, 5, 6, 7, 8, 9, 10, 11, 12, 13, 14, 15, 16, 17, 18	N/A	N/A

These questions are even better in jacPLUS!

- Receive immediate feedback
- Access sample responses
- Track results and progress

Find all this and MORE in jacPLUS

Simple familiar

1. **WE14** Identify the value of 7 in each of the following decimals.
   a. 45.871 b. 81.710 c. 33.007

2. Identify the value of 9 in each of the following decimals.
   a. 21.090 b. 0.009 c. 47.1059

3. Identify the value of 7 in the following decimals.
   a. 34.075 b. 1.7459 c. 0.1007 d. 1945.27
   e. 450.07 f. 2.7 g. 0.007 h. 2.170

4. Determine the number of decimal places in each of the following numbers.
   a. 2.195 b. 394.7 c. 104.25 d. 0.0003
   e. 1997.4 f. 125.333 g. 69.15 h. 2.6936

5. Determine whether the following statements are True or False.
   a. $0.125 = 0.521$ b. $0.025 = 0.0250$
   c. $45.0005 = 45.005$ d. $3.333 = 3.333\,000$
   e. $6.666 = 6.6666$ f. $20.123 = 2.123$

6. **WE15** Express each of the following rounded to 3 decimal places.
   a. 59.123 45 b. 72.216 81

7. Express each of the following rounded to 2 decimal places.
   a. 2.069 14 b. 0.7962

8. Express each of the following rounded to 3 decimal places.
   a. 39.184 23 b. 3.232 323 23 c. 99.4358
   d. 125.294 61 e. 100.0035 f. 0.0479

9. Express each of the following rounded to 2 decimal places.
   a. 58.123 b. 23.345 c. 71.097
   d. 0.8686 e. 30.9999 f. 125.8929

10. **WE16** Express the following fractions as decimals.
    a. $\frac{3}{4}$ b. $\frac{2}{3}$

11. Express the following as decimals to 3 decimal places.
    a. $\frac{7}{8}$ b. $\frac{1}{7}$

12. Express the following fractions as decimals.
    a. $\frac{3}{10}$ b. $\frac{3}{5}$ c. $\frac{3}{12}$ d. $\frac{7}{25}$ e. $\frac{33}{100}$ f. $\frac{45}{50}$

13. Express the following fractions as recurring decimals.

a. $\frac{1}{3}$ b. $\frac{5}{9}$ c. $\frac{1}{12}$ d. $\frac{4}{7}$ e. $\frac{3}{13}$ f. $\frac{1}{21}$

14. Determine which packet in the photograph contains more mince.

15. Narelle is environmentally conscious and has fitted water tanks to her house to water the garden and flush the toilets. She checks the amount of rainfall each week to see how much water the tanks have collected. In the month of March, the rainfall readings were 12.48 mm, 8.82 mm, 27.51 mm and 44.73 mm. Round the total rainfall for March to 1 decimal place.

16. The Nguyen family decided to install solar panels on the roof to reduce their electricity bill. They compared the prices per kWh of three different suppliers to see how much they would save by using their panels. The prices quoted by companies A, B and C per kWh were 26.78 cents, $25\frac{3}{4}$ cents and $25\frac{7}{9}$ cents respectively.

a. List the companies from cheapest to most expensive.

b. Express the fractions as decimal numbers rounded to 1 decimal place.

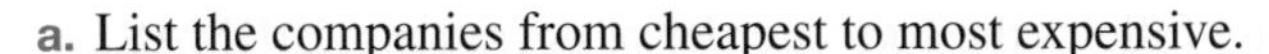

17. Using the table below, explain whether the total height of the five players is accurate to 9.91 m.

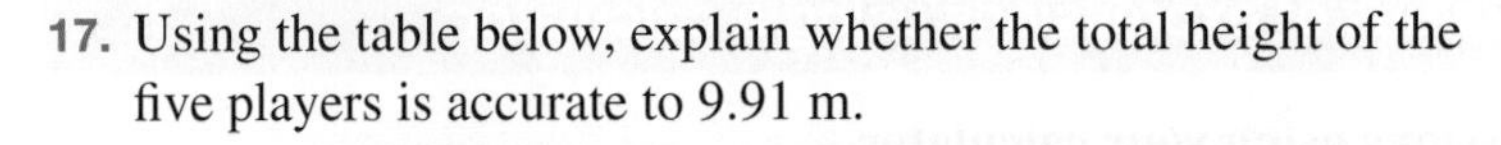

Player	Height
Kobe Bryant	1.98 m
Kevin Durant	2.06 m
Russell Westbrook	1.91 m
Carmelo Anthony	2.02 m
LeBron James	2.03 m

18. The following table shows items that were purchased at the supermarket and their price.

Item	Price
Salmon	\$4.99
Chicken fillets	\$7.35
Shampoo	\$8.95
Ham	\$4.42
Milk	\$4.29
Bread	\$3.60

Explain, using mathematical reasoning, whether the total cost of \$34 is accurate.

Fully worked solutions for this chapter are available online.

LESSON
1.6 Calculator skills

SYLLABUS LINKS

- Use a calculator for multi-step calculations.
- Check results of calculations for accuracy.
- Ascertain the reasonableness of answers to arithmetic calculations.

1.6.1 Using a calculator

- Your calculator can complete many different mathematical operations that are difficult to calculate in your head or with pen and paper. This includes dealing with decimals and fractions.
- Your calculator will always give the correct answer, but you have to input the information correctly. To make sure you get the correct answer, it's always good to estimate your answer to check for **reasonableness** and **accuracy**.
- You will notice that your calculator automatically follows the correct order of operations (BIDMAS). ***Note:*** The keys used may vary between different calculators.

WORKED EXAMPLE 17 Calculating values of expressions using a calculator

Calculate the value of the following expressions using your calculator.

a. $140 + 23 \times 8$ **b.** $7(28 + 43) - 18$ **c.** $283 - 13.78 \times 11.35$

THINK	WRITE
a. 1. Estimate your answer using leading digit approximation.	a. $140 + 23 \times 8$ $\approx 100 + 20 \times 10$ ≈ 300
2. Enter the exact equation into your calculator. Check the reasonableness of your estimate.	$140 + 23 \times 8 = 324$

b. 1. Estimate your answer using leading digit approximation.

b. $7(28+43)-18$
$\approx 7(30+40)-20$
$\approx 7(70)-20$
$\approx 490-20$
≈ 470

2. Enter the exact expression into your calculator. Check the reasonableness of your estimate.

$7(28+43)-18=479$

c. 1. Estimate your answer using leading digit approximation.

c. $283-13.78\times 11.35$
$\approx 300-10\times 10$
$\approx 300-100$
≈ 200

2. Enter the exact expression into your calculator. Check the reasonableness of your estimate.

$283-13.78\times 11.35=126.597$

- When using your calculator, first estimate the result to compare with the answer from the calculator.
- Always check that what you type into the calculator matches the question.

tlvd-3666

WORKED EXAMPLE 18 Evaluating expressions using a calculator and applying rounding

Evaluate the following expressions using your calculator, rounding your answer to 4 decimal places.

a. $(2.56+3.83)^2+45.93$ b. $\sqrt{4.39-2.51}+(3.96)^2$ c. $\frac{4}{7}+\frac{5}{9}+5$

THINK | **WRITE**

a. 1. Estimate your answer using leading digit approximation.

a. $(2.56+3.83)^2+45.93$
$\approx (3+4)^2+50$
$\approx 7^2+50$
$\approx 49+50$
≈ 99

2. Enter into the calculator. Use the x^2 key to raise the brackets to the power of 2. Compare with your estimate to check for reasonableness.

$(2.56+3.83)^2+45.93$
$=86.7621$

b. 1. Estimate your answer using leading digit approximation.

b. $\sqrt{4.39-2.51}+(3.96)^2$
$\approx \sqrt{4-3}+4^2$
$\approx \sqrt{1}+16$
$\approx 1+16$
≈ 17

2. Enter into the calculator. Use the $\sqrt{}$ and x^2 keys. Make sure the square root covers both numbers. Compare with your estimate to check for reasonableness.

$\sqrt{4.39-2.51}+(3.96)^2$
$=17.0527$

c. 1. Estimate your answer using leading digit approximation.

c. $\frac{4}{7}+\frac{5}{9}+5$

$\approx 1+1+5$

≈ 7

2. Enter into the calculator.
Use the fraction key to enter each fraction.
Compare with your estimate to check for reasonableness.

$\frac{4}{7}+\frac{5}{9}+5$

$=6.1270$

Digital technology

To evaluate the expressions in Worked example 18a using a calculator, enter $(2.56+3.83)^2+45.93$ into the calculator, by using the x^2 key to raise the brackets to the power of 2.

D Math ▲
(2.56+3.83)²+45▸
86.7621

In 18b, enter the expression $\sqrt{4.39-2.51}+(3.96)^2$ into the calculator, by using the $\sqrt{\ }$ and x^2 keys. Make sure the square root covers both numbers.

D Math ▲
√4.39−2.51 +(3.9▸
17.05273092

In 18c, enter the expression $\frac{4}{7}+\frac{5}{9}+5$ into the calculator, by using the fraction key to enter each fraction. Use the S ⟷ D button to convert from fraction to decimal form.

D Math ▲
$\frac{4}{7}+\frac{5}{9}+5$
6.126984127

Exercise 1.6 Calculator skills

learn**on**

1.6 Quick quiz on	1.6 Exercise

Simple familiar	Complex familiar	Complex unfamiliar
1, 2, 3, 4, 5, 6, 7, 8, 9, 10, 11, 12	N/A	N/A

Simple familiar

1. **WE17** Calculate the value of each of the following expressions using your calculator.
   a. $269 + 12 \times 16$ b. $9(78 + 61) - 45$ c. $506.34 - 39.23 \times 17.04$
2. Calculate the value of each of the following expressions using your calculator.
   a. $497 - 13 \times 24 \div 6$ b. $13(194 - 62) + 4 \times 8$
   c. $459.38 \div 18.5 \times 17.04 - (34.96 + 45.03)$
3. Evaluate the following expressions using your calculator.
   a. $3 + 6 \times 7$ b. $47 + 8 \times 6$ c. $285 + 21 \times 16$ d. $2859 + 178 \times 79$
4. Evaluate the following expressions using your calculator.
   a. $12 - 5 \times 2$ b. $68 - 4 \times 9$ c. $385 - 16 \times 9$ d. $1743 - 29 \times 45$
5. Evaluate the following expressions using your calculator.
   a. $4(6 + 24) - 58$ b. $5(23.5 - 18.3) + 23$
   c. $2.56(89.43 - 45.23) - 92.45$ d. $6(45.89 - 32.78) - 3(65.89 - 59.32)$
6. **WE18** Evaluate the following expressions using your calculator, rounding your answer to 4 decimal places.
   a. $(5.89 + 2.16)^2 + 67.99$ b. $\sqrt{8.77 - 3.81} + (5.23)^2$ c. $\frac{3}{5} + \frac{7}{8} + 10$
7. Evaluate the following expressions using your calculator, rounding your answer to 4 decimal places.
   a. $\left(\frac{7}{12}\right)^2 - \frac{3}{4}$ b. $\sqrt{3.56 + 8.28} - \sqrt{5.29 - 3.14}$ c. $\left(\frac{4}{5}\right)^2 + \sqrt{\frac{7}{12}}$
8. Evaluate the following expressions using your calculator, rounding your answers to 4 decimal places.
   a. $16.9 + 5.2^2 \div 4.3$ b. $9.3^2 \div 4.5$ c. $(3.7 + 5.9)^2 - 15.5$ d. $\frac{(7.2)^2}{4.2}$
9. Evaluate the following expressions using your calculator, rounding your answers to 4 decimal places.
   a. $\sqrt{5.67 + 8.34}$ b. $2.5 \times \sqrt{8.64} - 2.5$
   c. $12.8 + \sqrt{3.5 \times 5.8} \times 1.2$ d. $\frac{\sqrt{4.7 - 3.6}}{5}$
10. Evaluate the following expressions to 2 decimal places using your calculator.
    a. $\frac{1}{2} + \frac{1}{4}$ b. $\frac{3}{7} - \frac{1}{9}$ c. $\frac{18}{13} - \frac{2}{5} \times \frac{2}{7}$ d. $15 - \frac{7}{5} \times \frac{5}{3} + 7$
11. Cathy is 7 years older than Marie, and Marie is twice the age of her younger brother Fergus. Determine Cathy's age if Fergus is 3 years old.
12. After a football party night, some empty soft drink cans were left and had to be collected for recycling. Of the cans found, 7 were half full, 6 were one-quarter full and the remaining 8 cans were completely empty. Use your calculator to find out how many equivalent full cans were wasted.

Fully worked solutions for this chapter are available online.

LESSON
1.7 Approximation

SYLLABUS LINKS

- Apply approximation strategies for calculations.

1.7.1 Approximating calculations

- Approximating or estimating an answer is useful when an accurate answer is not necessary.
- You can approximate the size of a crowd at a sporting event based on an estimate of the fraction of seats filled. This approach can be applied to a variety of situations, such as approximating the amount of water in a water tank.
- When approximating size, it is important to use an appropriate unit; e.g. for the distance you drive, you would use 24 km instead of 2 400 000 cm.
- Approximations can also be made by using past history to indicate what might happen in the future.

WORKED EXAMPLE 19 Solving approximation by carrying out relevant calculations

Approximate the size of the crowd on the first day of the cricket test at the Melbourne Cricket Ground, if it is estimated to be at $\frac{8}{10}$ of its capacity of 95 000.

THINK	WRITE
1. Write down the information given in the question.	Estimate of crowd $= \frac{8}{10}$ of capacity
2. Calculate the fraction of the total capacity.	$\frac{8}{10}$ of $95\,000 = \frac{8}{10} \times 95\,000$ $= 0.8 \times 95\,000$ $= 76\,000$
3. Write the answer.	The approximate crowd is 76 000 people.

1.7.2 Approximation from graphs

- When approximating from graphs:
  - if the data is shown, read from the graph, noting the scale on the axes
  - if the data is not shown, follow the trend (behaviour of the graph) to make an estimate of future or past values.

tlvd-3668

WORKED EXAMPLE 20 Applying approximation to values from graphs

The following graph shows the farmgate price of milk in cents/litre. Approximate the price in 2025/2026.

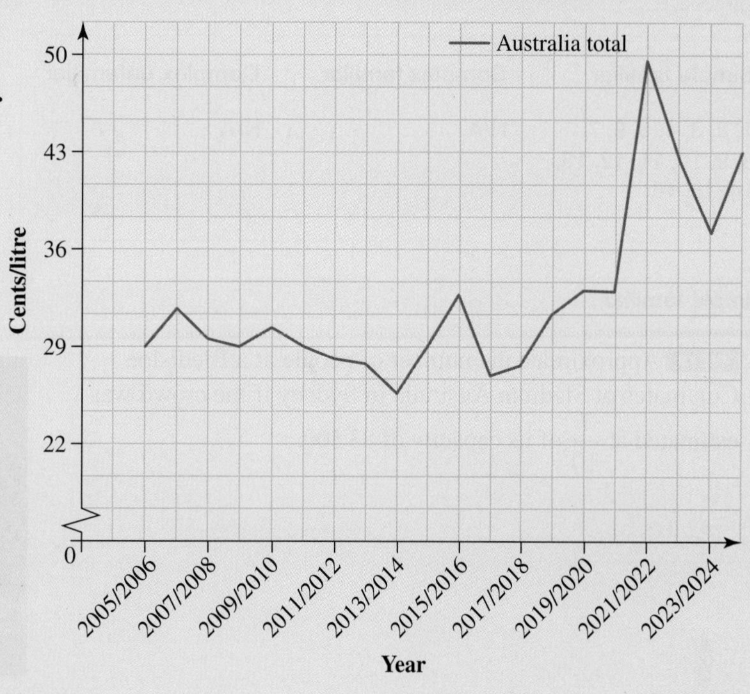

THINK	WRITE/DRAW
1. Check what each axis represents.	Vertical axis: cents/litre Horizontal axis: years
2. Look for the trend that the graph has followed in previous years. Draw a line to represent this trend and extend it to the required period of 2023/2024.	The graph has been on an upward trend for the past 8 years.

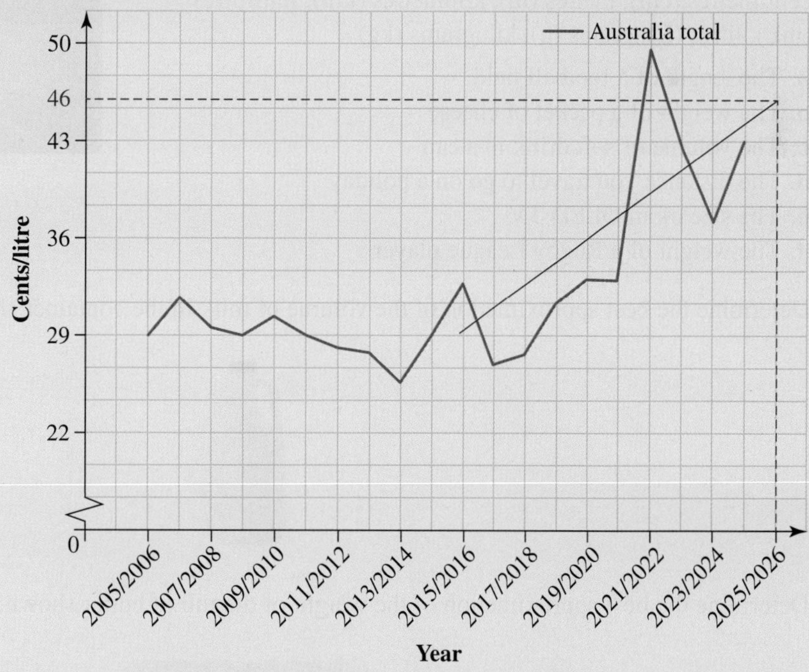

3. Read the answer from the graph.	The price in 2025/2026 is approximately 46 cents/litre.

Exercise 1.7 Approximation

learn on

1.7 Quick quiz | 1.7 Exercise

Simple familiar	Complex familiar	Complex unfamiliar
1, 2, 3, 4, 5, 6, 7, 8, 9, 10, 11, 12, 13, 14, 15, 16	N/A	N/A

Simple familiar

1. **WE19** Approximate the number of people at a Bledisloe Cup match at Stadium Australia in Sydney if the crowd was estimated at $\frac{7}{10}$ of its capacity of 83 500.

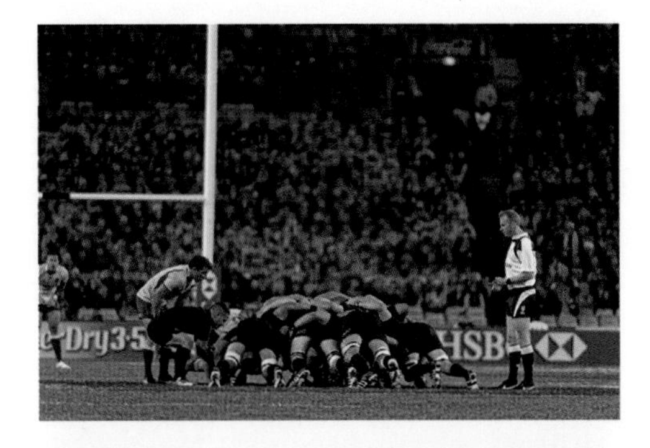

2. When full, a pool holds 51 000 litres of water. Approximate the number of litres in the pool when it is estimated to be 90% full.

3. Select the most appropriate unit from the options below for each of the following measurements.
   centimetres (cm), metres (m), kilometres (km), millilitres (mL), litres (L), grams (g), kilograms (kg)
   a. The length of a football field
   b. The weight of a packet of cheese
   c. The volume of soft drink in a can
   d. The distance you travel to go on a holiday
   e. The size of an OLED TV
   f. The weight of a Rugby League player

4. Determine the best approximation of the volume of milk in the container shown.

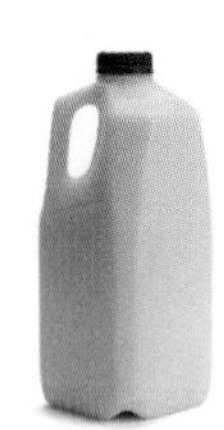

5. Determine the best approximation of the weight of the tub of butter shown.

6. WE20 From the graph of the cost of tuition fees for one school, predict the approximate tuition fees in 2025.

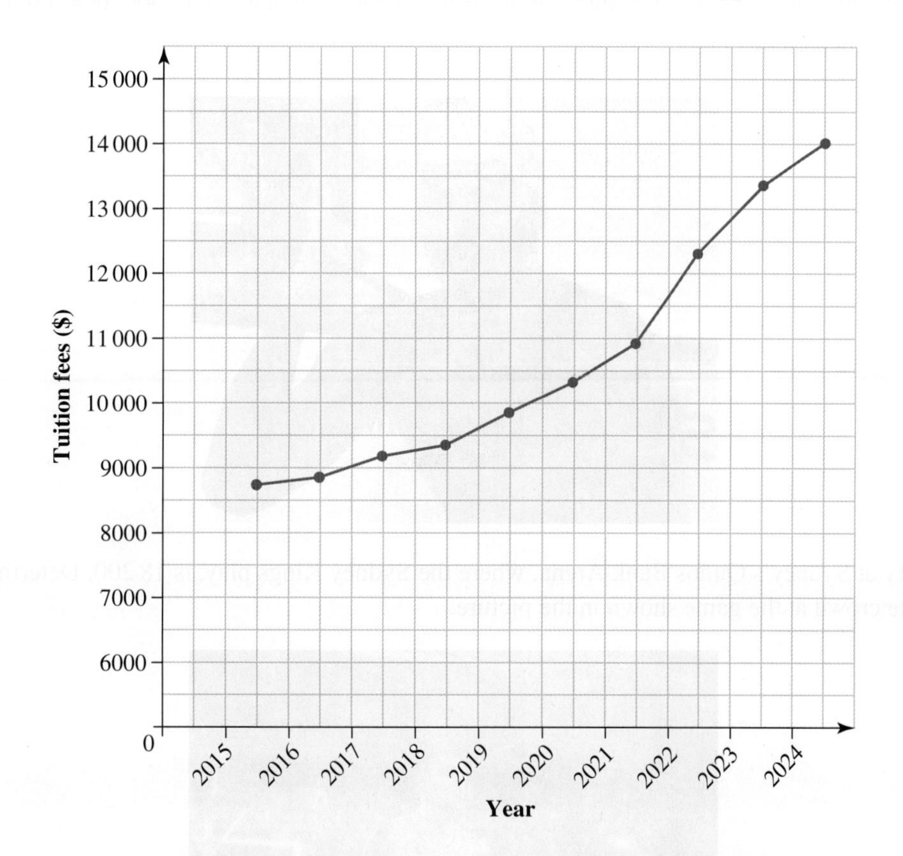

7. Use the following graph to approximate the volume of domestic air travel emissions in 2025.

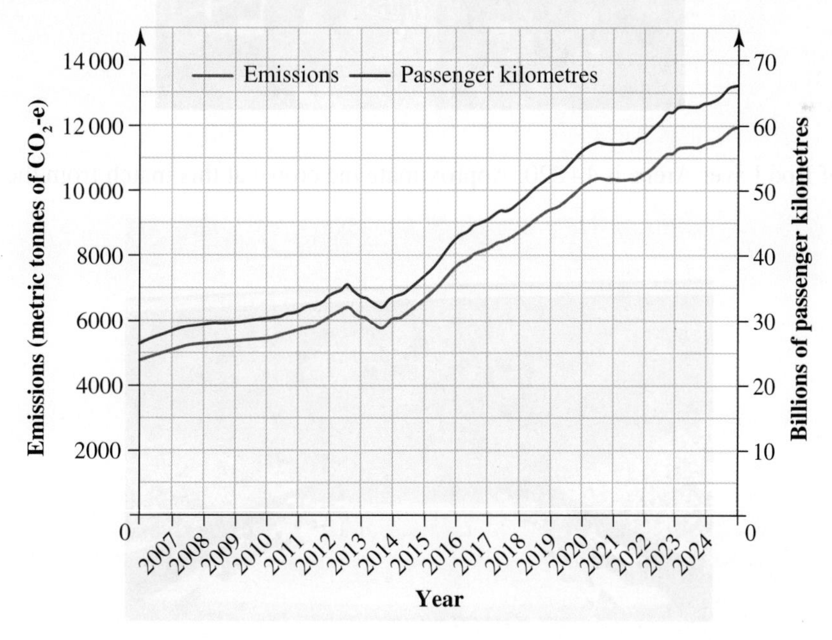

8. If the Sydney Olympics swimming venue was $\frac{9}{10}$ full and had a capacity of 17 000, determine the approximate crowd on this night.

9. The capacity at Sydney's Qudos Bank Arena, where the Sydney Kings play, is 18 200. Determine the approximate crowd at the game shown in the picture.

10. The capacity of Rod Laver Arena is 14 820. Approximate the crowd at this match from the Australian Open.

11. The picture shown is of Tiger Woods at the US Masters. Comment on why it is difficult to estimate the size of the Masters crowd on this day by looking at the picture.

12. Explain in your own words the difference between an estimate and a guess.

13. Describe a way to approximate how many students there are at your school. Investigate the actual number and compare this to your estimate. Explain how you could improve your initial estimate.

14. From the graph of the population of Australia since 1970, approximate how many people will live in Australia in 2030.

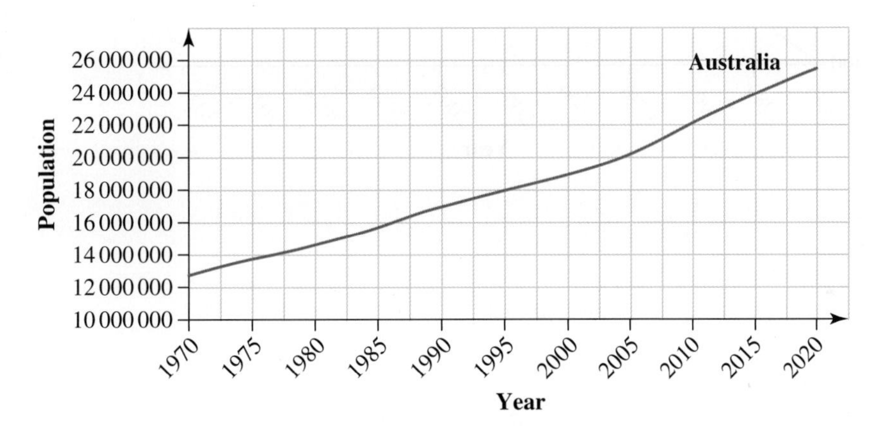

15. Use the graph of the value of the Australian dollar against the US dollar to approximate the value of the Australian dollar in January 2025.

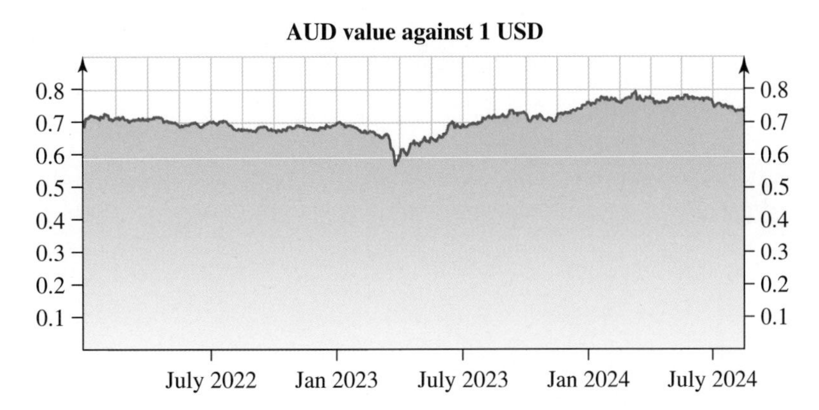

16. From the graph of Aboriginal and Torres Strait Islander population, predict the approximate population in:

a. 2025

b. 2030.

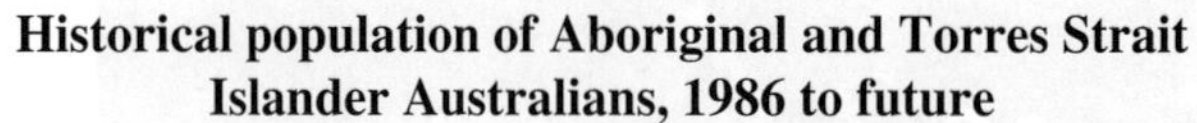

Historical population of Aboriginal and Torres Strait
Islander Australians, 1986 to future

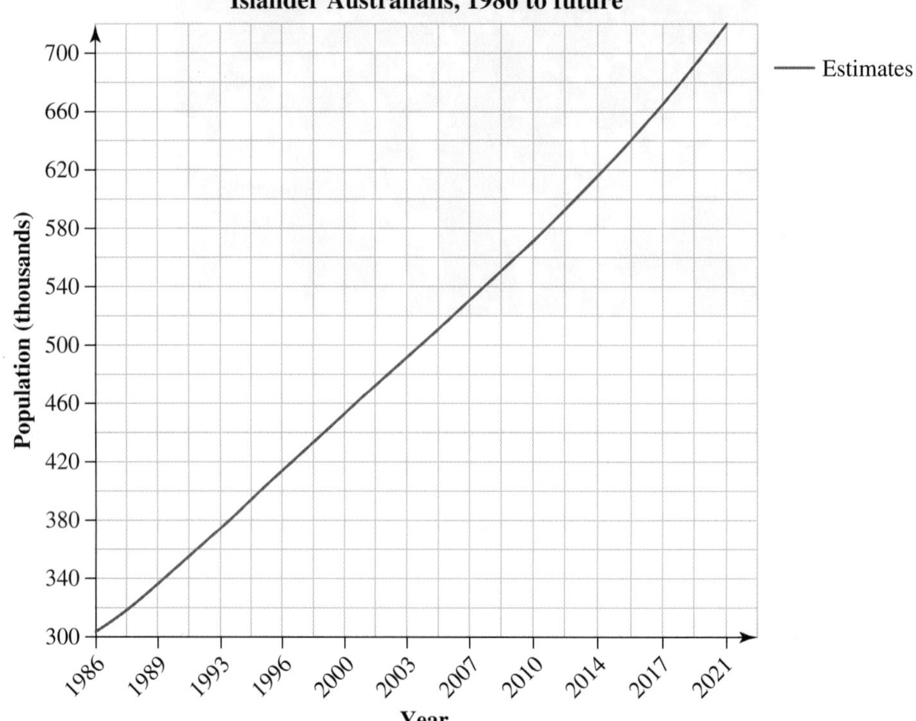

Fully worked solutions for this chapter are available online.

LESSON
1.8 Review

1.8.1 Summary

doc-41749

Hey students! Now that it's time to revise this chapter, go online to:

Access the chapter summary | Review your results | Watch teacher-led videos | Practise questions with immediate feedback

Find all this and MORE in jacPLUS

1.8 Exercise

learnon

1.8 Exercise

Simple familiar	Complex familiar	Complex unfamiliar
1, 2, 3, 4, 5, 6, 7, 8, 9, 10, 11, 12, 13, 14, 15, 16, 17, 18, 19, 20	N/A	N/A

These questions are even better in jacPLUS!
- Receive immediate feedback
- Access sample responses
- Track results and progress

Find all this and MORE in jacPLUS

Simple familiar

1. Calculate the value of the expression $17 + 3 \times 7$.
2. Calculate the value of the expression $-3(5 - -10) + 50$.
3. Estimate the number when 47 321 is rounded to the leading digit.
4. If three purchases were made with the values \$7.34, \$18.05 and \$2.69, use leading digit estimation on the three purchases to calculate the total purchase.
5. Using your calculator, calculate the value of $\sqrt{6.78 + 19.83} + 3.91$ to 2 decimal places.
6. In the following calculation, 20 is not the correct answer. Determine the most likely error in the calculation.

$$58 - 38 + (5 \times 2) = 20$$

7. Estimate the number 37.395 851 rounded to 4 decimal places.
8. Determine what fraction the number 7.9 is most accurately written as.
9. Determine the approximate volume of Coca-Cola in the can shown.

10. Calculate the following.

a. $-18 - -25$

b. -8×6

c. $36 \div 9 \times 8 - 12$

d. $-4(-4 + -6) - 33$

11. Use leading digit approximation to calculate the following.

a. 49×821

b. $1396 + 183$

c. $\frac{7563}{676}$

d. $17 \times 873 + 47$

12. Use you calculator to evaluate the following to 2 decimal places.

a. $\sqrt{3(5.94 - 1.48)}$

b. $(5.74)^2 - \frac{5}{8}$

c. $\frac{6}{9} + \frac{2}{5} \times \frac{3}{7}$

d. $\sqrt{(3.25)^2 + 1.5}$

13. Round the following fractions to 2 decimal places.

a. $\frac{1}{8}$

b. $\frac{37}{50}$

c. $\frac{6}{25}$

d. $\frac{2}{3}$

14. Round the following to 3 decimal places.

a. 0.5555

b. 1.036 39

c. 33.1047

d. 10.9999

15. Approximate each of the following.

a. A sportsground estimated to be $\frac{3}{4}$ full, with a capacity of 50 000 people.

b. A water tank estimated to be 80% full, with a capacity of 7500 litres.

16. Copy and complete the table shown.

- In the column headed 'Estimate', round each number to the leading digit.
- In the column headed 'Estimated answer', calculate the answer.
- In the column headed 'Prediction', assess whether the actual answer will be higher or lower than your estimate.
- Use a calculator to work out the actual answer and record it in the column headed 'Calculation' to determine whether it is higher or lower than your estimate.

	Question	Estimate	Estimated answer	Prediction	Calculation
Example	4200/230	$4000 \div 200$	20	Lower	1.7391 (higher)
a.	$487 + 962$				
b.	$33\,041 + 82\,629$				
c.	$184\,029 + 723\,419$				
d.	$93\,261 - 37\,381$				
e.	$321 - 194$				
f.	$468\,011 - 171\,962$				
g.	36×198				
h.	$623 \times 12\,671$				
i.	$29\,486 \times 39$				
j.	$31\,690 \div 963$				
k.	$63\,003 \div 2590$				
l.	$69\,241 \div 1297$				

17. A student is selling tickets for their school's theatre production of *South Pacific*. So far they have sold 439 tickets for Thursday night's performance, 529 for Friday night's and 587 for Saturday night's. The tickets cost $9.80 for adults and $4.90 for students.

 a. Round the figures to the first digit to estimate the number of tickets the student has sold so far.
 b. If approximately half the tickets sold were adult tickets and the other half were student tickets, estimate how much money has been received so far by rounding the cost of the tickets to the first digit.

18. Yi Rong is saving for her end-of-year holiday with her school friends. Her parents said if she did work around the house they would help contribute to the holiday expenses. They agreed to pay her $25 a week for doing the dishes and $15 a week for doing the washing, and as an extra incentive they said they would double her pay if she did each of the jobs for the next 4 weeks without missing a day.

Assuming Yi Rong didn't miss a day, calculate how much money she earned in the 4 weeks.

19. Bob the builder intends to build a new deck out the back of his house so he can entertain friends. He wants to get a rough idea of how much it is going to cost him in materials so he can see if he can afford to go ahead with the job. The hardware store gave him prices for the materials he needs, as shown.

Material	Price
Timber	$789
Ready-mix bags	$32.50
Nails	$67.25
Decking stain	$77.95
General equipment	$65.45

 a. Estimate the rough price of materials if rounding each price to the leading digit.
 b. Estimate the rough price if rounding each price to the nearest dollar.
 c. Identify which of the two rounded prices is greater. Explain why this is so.

20. A roller coaster converts its stored potential energy (PE) to kinetic energy (KE) as it picks up speed on its descent.

Given $m = 750$, $g = 9.8$, $h = 15$ and $v = 15$, calculate KE and PE, using $\text{KE} = \frac{1}{2}mv^2$ and $\text{PE} = mgh$.

Fully worked solutions for this chapter are available online.

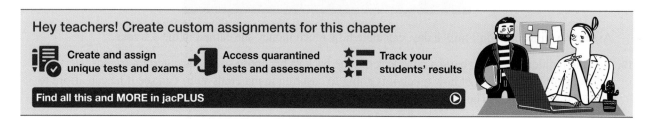

Answers

Chapter 1 Calculations

1.2 Number operations

1.2 Exercise

1. 5, −2, 212, −33
2. a. −3 b. −5 c. 2 d. −11 e. 30 f. 13
3. a. 4 b. 1 c. −1 d. 0 e. 5 f. 2
4. a. −10 b. 15 c. −8 d. −1 e. 39 f. −46
5. a. −49 b. 35 c. −32 d. −12 e. 25 f. 172
6. a. 14 b. 31 c. −18 d. 9 e. 25 f. 0
7. a. −15 b. −21 c. 30 d. −20 e. 40 f. −56
8. a. −250 b. −1250 c. −54 d. −63 e. 77 f. −500
9. a. −300 b. 60 c. −192 d. −30 e. −120 f. 36
10. a. −6 b. −5 c. 4 d. 5 e. −9 f. 0
11. a. −14 b. −26 c. 7 d. −22 e. 11 f. −21
12. a. −23 b. −21 c. 23 d. 66 e. −34 f. −120
13. a. −18 b. −23 c. 64 d. 36 e. 64 f. −49
14. 3 GB
15. $7049
16. a. 5 m
    b. 75 m
    c. 70 m
    d. To calculate the difference, subtract the height of the smaller building from the height of the taller building.
17. a. 351 km b. 247 km
18. $522.63
19. $54\frac{1}{6}$ grams
20. a. 39 °C b. 159 °C c. 198 °C
21. a. $0.94 b. $32.90 c. $5146.50 d. $3 e. $10
22. a. 5432 b. 2345 c. 3087

1.3 Order of operations and reasonableness

1.3 Exercise

1. a. −7 b. 0 c. 45 d. −6 e. −56 f. −58
2. a. 10 b. −5 c. −2 d. 44 e. 14 f. −3
3. a. −23 b. −29 c. 40 d. 9 e. −12 f. 6
4. a. $3.50 b. $0.27
5. $4.11
6. $290
7. $41
   Answer is reasonable since roughly chips ≈ $10, Coke ≈ $20, biscuits ≈ $5 and dips ≈ $5.
   This totals $40, which reduces to $38 after the discount.
   This value is close to $41, so it is reasonable.
8. $62
   Answer is reasonable since roughly timber ≈ $60, nails ≈ $10 and liquid nails ≈ $10.
   This totals $80, which reduces to $65 after the discount.
   This value is close to $62, so it is reasonable.
9. a. $(12-8)\div 4=1$
   b. $(4+8)\times 5-4\times 5=40$
   c. $3+4\times(9-3)=27$
   d. $3\times(10-2)\div 4+4=10$
   e. $10\div(5+5)\times 9\times 9=81$
   f. $(18-3)\times 3\div 5=9$
10. a. Use multiplication first to calculate the cost of each type of food, and then add these costs together to calculate the total cost of all the food.
    b. Cost = (2 × $2) + (3 × $4) + (3 × $5)
    c. $31
11. a. 85 km
    b. 15 km north
12. a. −8
    b. $4+8\div-(2)^2-7\times 2=-12$
    $(4+8)\div-(2)^2-7\times 2=-17$
13. 2
14. −13

1.4 Estimation

1.4 Exercise

1. a. 930 b. 13 000 c. 90 000
2. a. 350 b. 86 600 c. 70 000
3. a. 50 b. 80 c. 130 d. 160 e. 250 f. 2460 g. 4840 h. 10
4. a. 0 b. 100 c. 100 d. 200 e. 3300 f. 56 300 g. 86 600 h. 214 000
5. a. 1000 b. 3000 c. 1000 d. 6000 e. 8000 f. 13 000 g. 436 000 h. 728 000
6. a. 10 b. 20 c. 50 d. 200 e. 1000 f. 20 000 g. 20 000 h. 500 000
7. 35 000 000
8. 225

9. a. 1400 b. 900 c. 1100 d. 2000
e. 45 000 f. 180 000 g. 70 h. $\frac{40}{9}$

10. Since this is leading digit approximation, you look at the second digit. If it is 5 or greater, you increase the leading digit by 1 and replace the rest of the digits with zeros. If the second digit is less than 5, replace it and all of the other digits with zeros.

11. $35

12. a. Sample responses can be found in the worked solutions in the online resources.
b. Sample responses can be found in the worked solutions in the online resources.
c. Sample responses can be found in the worked solutions in the online resources.

13. a. Sample responses can be found in the worked solutions in the online resources.
b. Sample responses can be found in the worked solutions in the online resources.
c. Sample responses can be found in the worked solutions in the online resources.

14. Sample responses can be found in the worked solutions in the online resources.

15. Sample responses can be found in the worked solutions in the online resources.

16. Adding the cost of groceries when shopping.
Calculating the cost of petrol needed for a trip.
Determining the size of a crowd.

17. 300 000

18. Actual crowd = 294 495
Estimated crowd ≈ 300 000
Difference = 5505

19. a. It could be correct, but the GPS cannot predict traffic or red lights on the way, so it is only an estimate.
b. By calculating the distance of the trip and the speed limits on the roads to come up with an estimate for time

20. $39

1.5 Decimal significance

1.5 Exercise

1. a. $\frac{7}{100}$ b. $\frac{7}{10}$ c. $\frac{7}{1000}$

2. a. $\frac{9}{100}$ b. $\frac{9}{1000}$ c. $\frac{9}{10\,000}$

3. a. $\frac{7}{100}$ b. $\frac{7}{10}$ c. $\frac{7}{10\,000}$ d. $\frac{7}{100}$
e. $\frac{7}{100}$ f. $\frac{7}{10}$ g. $\frac{7}{1000}$ h. $\frac{7}{100}$

4. a. 2.195 → 3 decimal places
b. 394.7 → 1 decimal place
c. 104.25 → 2 decimal places
d. 0.0003 → 4 decimal places
e. 1997.4 → 1 decimal place
f. 125.333 → 3 decimal places
g. 69.15 → 2 decimal places
h. 2.6936 → 4 decimal places

5. a. False b. True c. False
d. True e. False f. False

6. a. 59.123 b. 72.217

7. a. 2.07 b. 0.80

8. a. 39.184 b. 3.232 c. 99.436
d. 125.295 e. 100.004 f. 0.048

9. a. 58.12 b. 23.35 c. 71.10
d. 0.87 e. 31.00 f. 125.89

10. a. 0.75 b. $0.\dot{6}$

11. a. 0.875 b. 0.143

12. a. 0.3 b. 0.6 c. 0.25
d. 0.28 e. 0.33 f. 0.9

13. a. $0.\dot{3}$ b. $0.\dot{5}$ c. $0.08\dot{3}$
d. $0.\overline{571428}$ e. $0.\overline{230769}$ f. $0.\overline{047619}$

14. The one on the right, since 0.630 g > 0.584 g

15. 93.5mm

16. a. Company B: $25\frac{3}{4}$ (25.75); Company C: $25\frac{7}{9}$ $(25.\dot{7})$; Company A: 26.78
b. $26.78 \approx 26.8$
$25\frac{7}{9} = 25.\dot{7}$
≈ 25.8
$25\frac{3}{4} = 25.75$
≈ 25.8

17. Sample responses can be found in the worked solutions in the online resources.

18. Sample responses can be found in the worked solutions in the online resources.

1.6 Calculator skills

1.6 Exercise

1. a. 461 b. 1206 c. −162.1392

2. a. 445 b. 1748 c. 343.136227

3. a. 45 b. 95 c. 621 d. 16 921

4. a. 2 b. 32 c. 241 d. 438

5. a. 62 b. 49 c. 20.702 d. 58.95

6. a. 132.7925 b. 29.5800 c. 11.4750

7. a. −0.4097 b. 1.9746 c. 1.4038

8. a. 23.1884 b. 19.2200 c. 76.6600
d. 12.3429

9. a. 3.7430 b. 4.8485 c. 18.2067
d. 0.2098

10. a. 0.75 b. 0.32 c. 1.27
d. 19.67

11. Cathy is 13 years old.

12. 5 cans

1.7 Approximation

1.7 Exercise

1. 58 450
2. 45 900 L
3. a. m b. g c. mL
   d. km e. cm f. kg
4. 2 L
5. 500 g
6. $14 500
7. 14 000 metric tonnes of CO_2-e
8. 15 300
9. Around 17 000
10. Around 11 115
11. Even though the crowd behind Tiger Woods looks large, we don't know the capacity of the golf course. Even if we did, the picture doesn't show us the crowd at the other holes. There might not be many people watching the other groups because they are not as popular as Tiger Woods.
12. A guess doesn't really have any thought or logic behind it, but an estimate does.
13. Estimate the size of each year level and add them up. This could be improved by finding how many home-room or tutor groups there are at each year level, estimating the number of students in each, and adding them all up.
14. Estimate: 28 000 000
15. Estimate: $0.70
16. a. Estimate: 750 000
    b. Estimate: 820 000

1.8 Review

1.8 Exercise

1. 38
2. 5
3. 50 000
4. $30
5. 9.07
6. Sample responses can be found in the worked solutions in the online resources.
7. 37.3959
8. $\frac{79}{10}$
9. 375 mL
10. a. 7 b. −48 c. 20 d. 7
11. a. 40 000 b. 1200 c. $\frac{80}{7}$ d. 18 050
12. a. 3.66 b. 32.32 c. 0.84 d. 4.75
13. a. 0.13 b. 0.74 c. 0.24 d. 0.67
14. a. 0.556 b. 1.036 c. 33.105
    d. 11.000
15. a. 37 500 people
    b. Between 5000 and 7000 litres (6000 L)

16.

	Estimate	Estimated answer	Actual answer
a.	500 + 1000	1500	1449
b.	33 000 + 80 000	110 000	115 670
c.	200 000 + 700 000	900 000	907 448
d.	90 000 − 40 000	50 000	55 880
e.	300 − 200	100	127
f.	500 000 − 200 000	300 000	296 049
g.	40 × 200	8000	7128
h.	600 × 10 000	6 000 000	7 894 033
i.	30 000 × 40	1 200 000	1 149 954
j.	30 000 ÷ 1000	30	32.9076
k.	60 000 ÷ 3000	20	24.325 483
l.	70 000 ÷ 1000	70	53.3855

17. a. 1500 tickets b. $11 250
18. $320
19. a. Around $1050
    b. $1032
    c. The price in part a is higher since four of the five values were rounded up using first digit rounding.
20. PE = 110 250 J
    KE = 84 375 J

2 Ratios

LESSON SEQUENCE

Fully worked solutions for this chapter are available online.

on Resources

 **Solutions** Solutions — Chapter 2 (sol-1255)

 **Digital documents** Learning matrix — Chapter 2 (doc-41561)
Quick quizzes — Chapter 2 (doc-41562)
Chapter summary — Chapter 2 (doc-41563)

LESSON
2.1 Overview

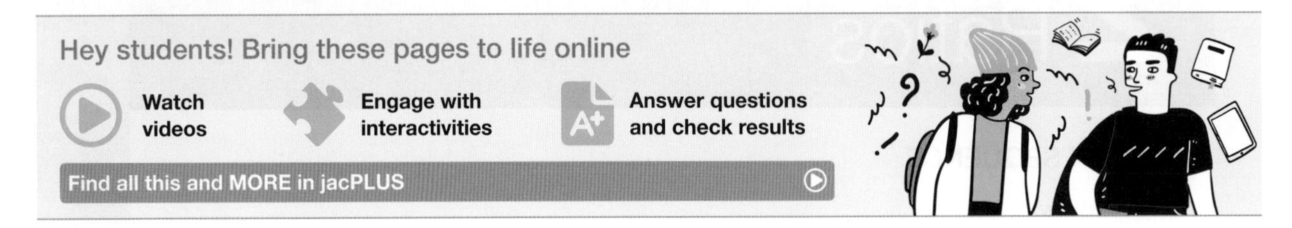

2.1.1 Introduction

Ratios and proportions are often used to compare different values or quantities. They tell us how much there is of one thing compared to another.

Ratios are used in our daily lives, both at home and at work. The use of ratios is required in areas such as building and construction, cooking, engineering, hairdressing, landscaping, carpentry and pharmaceuticals, just to name a few. Understanding ratios is important as they are used in money transactions, perspective drawings, enlarging or reducing measurements in building and construction, and dividing items equally within a group.

The golden ratio, approximately 1.618, is a proportion that is found in geometry, art and architecture. It has been made famous in the illustrations of Leonardo da Vinci (1452–1519). In the construction industry, concrete is made in different proportions of gravel, sand, cement and water depending on the particular application. Artists, architects and designers use the concept of proportion in many spheres of their work.

2.1.2 Syllabus links

Lesson	Lesson title	Syllabus links
2.2	**Introduction to ratios**	◯ Demonstrate an understanding of the fundamental ideas and notation of ratio. ◯ Understand the relationship between fractions and ratios. ◯ Express a ratio in simplest form using whole numbers.
2.3	**Ratios of two quantities**	◯ Find the ratio of two quantities in its simplest form.
2.4	**Divide into ratios [complex]**	◯ Divide a quantity in a given ratio [complex].
2.5	**Scales, diagrams and maps [complex]**	◯ Use ratio to describe simple scales [complex].

LESSON
2.2 Introduction to ratios

SYLLABUS LINKS

- Demonstrate an understanding of the fundamental ideas and notation of ratio.
- Understand the relationship between fractions and ratios.
- Express a ratio in simplest form using whole numbers.

2.2.1 What is a ratio?

- **Ratios** are used to compare two or more quantities of the same kind (e.g. two scoops of chocolate ice cream and one scoop of vanilla ice cream).
- The numbers are separated by a colon and they are always whole numbers.

Notation for ratios

The basic notation for ratios is:

$$a : b$$

where a and b are two whole numbers. This is read as 'a to b'.

- In the following diagram, the ratio of blue squares to pink squares is **2 : 4** (the quantities are separated by a colon). This is read as '**2** to **4**'.

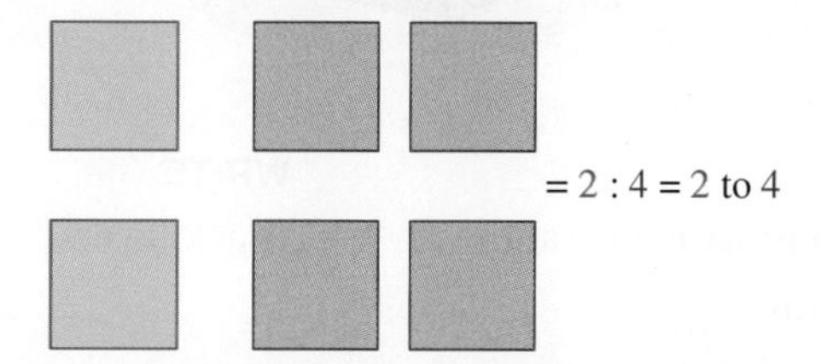

- When writing a ratio, *the order in which it is written* is very important.
  For example, the picture shows two scoops of chocolate ice cream and one scoop of vanilla ice cream. The ratio of chocolate ice cream to vanilla ice cream is 2 scoops to 1 scoop, or 2 : 1 (two to one), while the ratio of vanilla ice cream to chocolate ice cream is 1 : 2 (one to two).

- There are also ratios with more than two values.
  For example, the ratio of flour to sugar to water in a recipe could be flour : sugar : water = 3 : 1 : 4.
  In other words, for every 3 parts of flour, the recipe requires 1 part of sugar and 4 parts of water.

Expressing ratios

- **When writing a ratio, the order in which the ratio is written is very important.**
- **A ratio of 1 to 4 is written as 1: 4 (the quantities are separated by a colon) and is read as 'one to four'.**
- **Ratios can also be written in fraction form:**

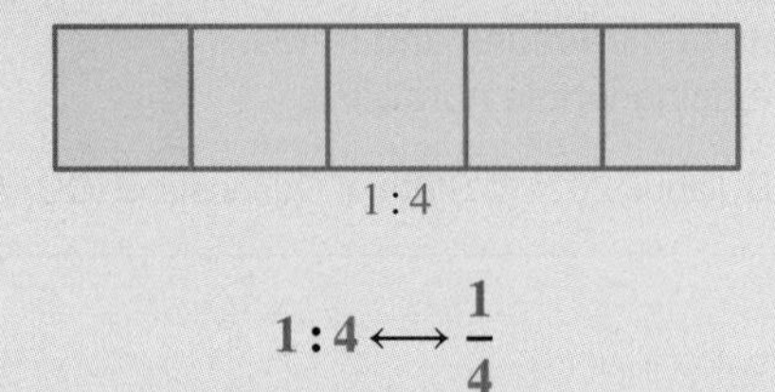

$$1:4 \longleftrightarrow \frac{1}{4}$$

- **Ratios do not have names or units of measurement.**
- **To calculate ratios of quantities, always write all quantities in the same unit.**
- **Ratios contain only whole numbers.**

WORKED EXAMPLE 1 Expressing quantities as ratios

Determine the ratio of black to coloured layers of licorice in the picture.

THINK	WRITE
1. Determine the order in which to write the ratio.	Black layers : coloured layers
2. Write the ratio in simplest form.	2 : 3

2.2.2 Ratios as fractions

- Ratios can be written in fraction form: $a : b = \frac{a}{b}$.
- Fractions should always be written in simplest form.

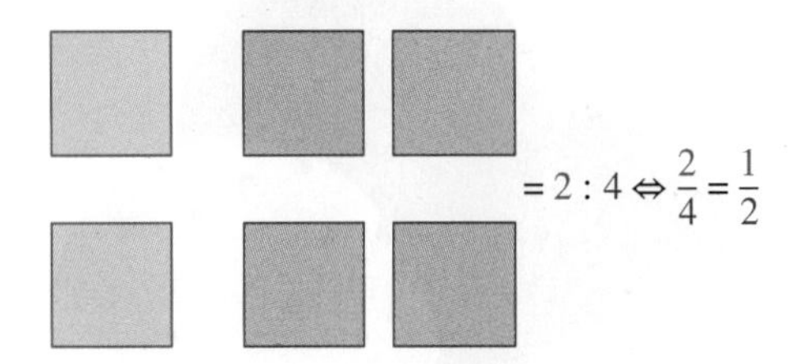

($\Leftrightarrow$ means *is equivalent to.*)

Note: This does not mean that the quantity on the left of the ratio is one half of the total.

WORKED EXAMPLE 2 Expressing ratios in fraction form

Determine the ratio of red-bellied macaw parrots to orange-bellied macaw parrots in the picture. Express the ratio in fraction form.

THINK	WRITE
1. Count the number of red-bellied macaws.	2 red-bellied macaw parrots
2. Count the number of orange-bellied macaws.	5 orange-bellied macaw parrots
3. Express the ratio in simplest form and in order.	Red-bellied macaws : orange-bellied macaws = 2 : 5 $= \frac{2}{5}$

2.2.3 Equivalent ratios

- **Equivalent ratios** are equal ratios. To determine equivalent ratios, multiply or divide both sides of the ratio by the same number. (This is a similar process to that of obtaining equivalent fractions.)
- Consider the diagram shown. A cordial mixture can be made by adding one part of cordial concentrate to four parts of water.

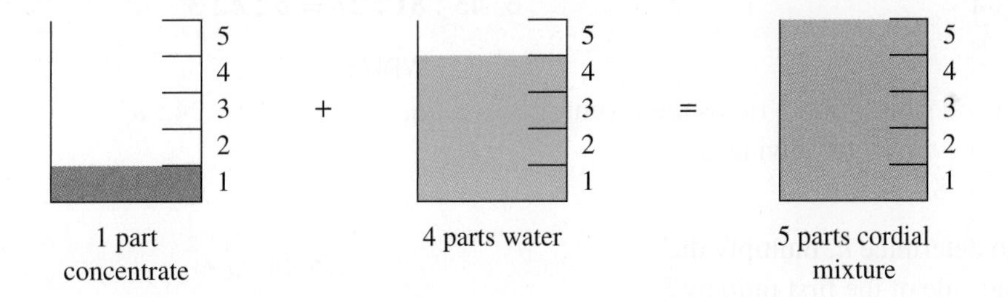

- The ratio of concentrate to water is 1 to 4 and is written as 1 : 4.
- By using the knowledge of 'equivalent ratios', the amount of cordial can be doubled. To do this, keep the ratio of concentrate to water the same by doubling the amounts of both concentrate and water.

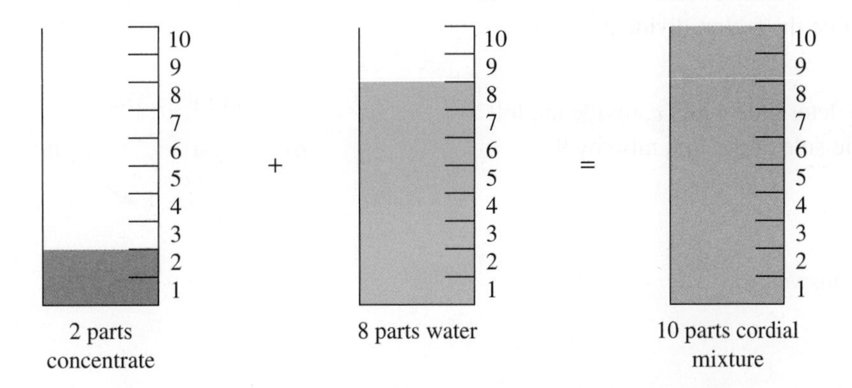

- The relationship between the amount of concentrate to water is now 2 : 8, as shown in the above diagram.

- The ratios 2 : 8 and 1 : 4 are equivalent. In both ratios, there is 1 part of concentrate for every 4 parts of water.

> **Equivalent ratios**
>
> **Ratios are equivalent if the numbers on either side of the colon have the same relationship. They work in the same way as equivalent fractions.**

- Some examples of equivalent fractions are:

Some equivalent ratios to 2 : 5	Some equivalent ratios to 100 : 40
2 : 5 × 2 → 4 : 10 ← × 2 × 3 → 6 : 15 ← × 3 × 10 → 20 : 50 ← × 10	100 : 40 ÷ 2 → 50 : 20 ← ÷ 2 ÷ 20 → 5 : 2 ← ÷ 20 ÷ 10 → 10 : 4 ← ÷ 10

- Equivalent fractions are calculated in the same way as equivalent ratios.
- Both the numerator and denominator are multiplied or divided by the same number to give an equivalent fraction.

$$\frac{2}{5} \times \frac{3}{3} = \frac{6}{15} \quad \text{and} \quad \frac{100}{40} \div \frac{5}{5} = \frac{20}{8}$$

tlvd-3757

WORKED EXAMPLE 3 Determining the values of pronumerals using equivalent ratios

Use equivalent ratios to determine the values of the pronumerals in the following.

a. $\mathbf{3 : 10 = 24 : \mathit{a}}$ **b.** $\mathbf{45 : 81 : 27 = \mathit{b} : \mathit{c} : 3}$

THINK	WRITE
a. 1. The left-hand sides of both ratios are given. To determine the factor, divide 24 by 3. The result is 8. Hence, to determine a, multiply the right-hand side of the first ratio by 8.	**a.** $3 : 10 = 24 : a$ 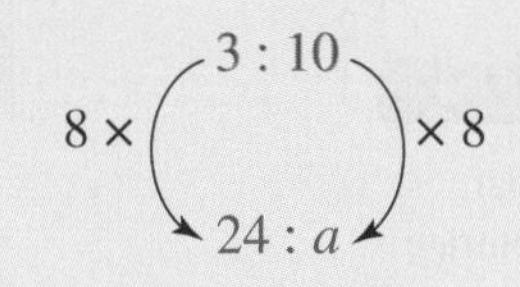 $3 \times 8 = 24$ $10 \times 8 = a$
2. Write the answer.	$a = 80$
b. 1. The right-hand sides of both ratios are given. To determine the factor, divide 27 by 3. The result is 9. Hence, to determine b and c, divide the left and middle side of the first ratio by 9.	**b.** $45 : 81 : 27 = b : c : 3$ 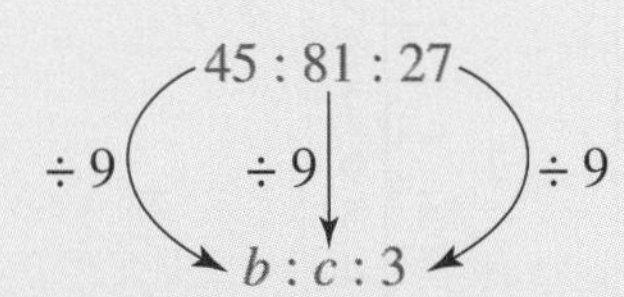 $5 : 9 : 3$
2. Write the answer.	$b = 5$ and $c = 9$

2.2.4 Simplifying ratios

- There are times when the ratio is not written in its simplest form.
- A ratio is simplified by dividing all numbers in the ratio by their highest common factor (HCF).
- In the picture shown, there are 16 chocolates: 4 white chocolates and 12 milk chocolates. The ratio between the white and the milk chocolates is 4 to 12.

white chocolates : milk chocolates = 4 : 12

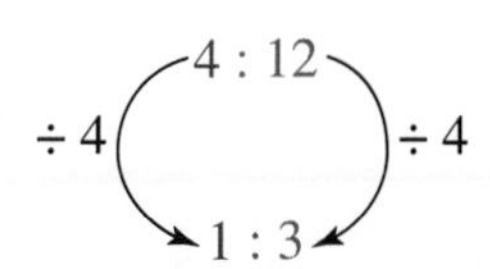

- Notice that this ratio can be simplified by dividing both values by 4, because 4 is the highest common factor of both 4 and 12. This ratio then becomes:

white chocolates : milk chocolates = 1 : 3

- We can say that 4 : 12 = 1 : 3.
- Expressed in fraction form, the ratio becomes:

$$\frac{4}{12} = \frac{1}{3}$$

- Ratios are usually written in simplest form — that is, reduced to lowest terms. This is done by dividing each number in the ratio by the highest common factor (HCF).

Simplification of ratios

Equivalent ratios are always multiplied or divided by the same number. They work in the same way as equivalent fractions.

4 : 6 ↔ 2 : 3 (× 2)

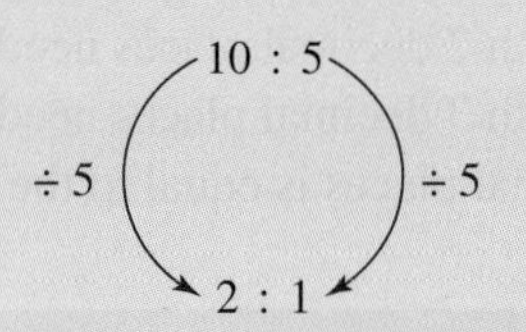

WORKED EXAMPLE 4 Expressing ratios in their simplest form

tlvd-3758

a. Examine the image and express the ratio of antique chairs to tables in simplest form.

b. Express the ratio in fraction form.

THINK	WRITE
a. 1. Count the number of the first part of the ratio (chairs).	**a.** 10 chairs
2. Count the number of the second part of the ratio (tables).	4 tables
3. Write the ratio.	Chairs : tables = 10 : 4
4. Both 10 and 4 can be divided exactly by 2. So, the highest common factor of 10 and 4 is 2. Write the ratio in its simplest form by dividing both sides by their highest common factor.	$= \frac{10}{2} : \frac{4}{2}$ $= 5 : 2$
b. Write the ratio in fraction form.	**b.** $5 : 2 = \frac{5}{2}$

2.2.5 Ratios and decimal numbers

- What do we do with ratios written in the form 2 : 3.5? We said before that ratios must be expressed as whole numbers.

  Step 1: We can convert 3.5 into a whole number by multiplying it by 10.

  $$3.5 \times 10 = 35$$

  Step 2: If we multiply the right-hand side of the ratio by 10, we have to multiply the left-hand side of the ratio by 10 also, so the ratio stays the same.

  $$2 \times 10 : 3.5 \times 10 = 20 : 35$$

  Step 3: Ratios should be expressed in simplest form, so divide both sides of the ratio by the highest common factor of 20 and 35, which is 5.

  $$20 \div 5 : 35 \div 5 = 4 : 7$$

- Decimal numbers with 2 decimal places need to be multiplied by 10^2 or 100.
- Decimal numbers with 3 decimal places need to be multiplied by 10^3 or 1000.
- The number of decimal places is equal to the number of zeros required.

tlvd-3759

WORKED EXAMPLE 5 Expressing in simplest form ratios involving decimals

Express the ratio 4.9 : 0.21 in its simplest form using whole numbers.

THINK	WRITE
1. Multiply both sides of the ratio by 100, because the largest number of decimal places is 2 on the right-hand side of the ratio.	$4.9 \times 100 : 0.21 \times 100 = 490 : 21$
2. Divide both sides of the ratio by the highest common factor of the two numbers (HCF = 7). The highest common factor of both 490 and 21 is 7.	$490 \div 7 : 21 \div 7 = 70 : 3$
3. Write the ratio in its simplest form.	$4.9 : 0.21 = 70 : 3$

Digital technology

Since simplifying ratios follows a similar process to simplifying fractions, you can use a calculator to check your answer.

Using a calculator, input 2.1 ÷ 3.5. Then, press the decimal-to-fraction button to display the answer as a fraction in simplest form.

DEG

2.1÷3.5▸f◂▸d 3/5

Exercise 2.2 Introduction to ratios

learnon

2.2 Quick quiz on | **2.2 Exercise**

Simple familiar	Complex familiar	Complex unfamiliar
1, 2, 3, 4, 5, 6, 7, 8, 9, 10, 11, 12, 13, 14, 15, 16, 17, 18, 19, 20	N/A	N/A

These questions are even better in jacPLUS!
- Receive immediate feedback
- Access sample responses
- Track results and progress

Find all this and MORE in jacPLUS

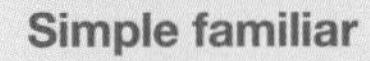

Simple familiar

1. WE1 Determine the ratio of white flowers (chamomile) to purple flowers (bluebell) in the picture shown.

2. The abacus shown has white, green, yellow, red and blue beads. Express the following as ratios:
   a. Blue beads to yellow beads
   b. Red beads to green beads
   c. White beads to red beads

3. WE2 Convert the following ratios to fractions.

   a. 3 : 5   b. 4 : 28   c. 6 : 11   d. 5 : 2

4. Using the letters in the word MATHEMATICS, express the following as ratios.
   a. Vowels to consonants
   b. Consonants to vowels
   c. Letters E to letters A
   d. Letters M to letters T

5. There are 12 shapes in this image.

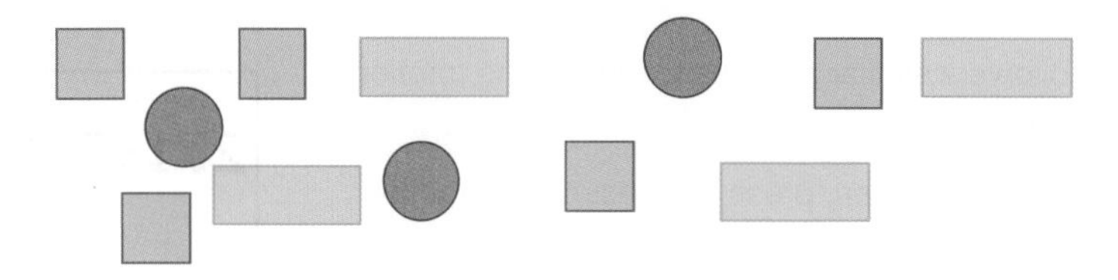

Express the following as ratios:

a. Circles to quadrilaterals
b. Squares to rectangles
c. Rectangles to circles
d. Squares to circles

6. Express the following ratios in fraction form.

a. $2:3$
b. $9:8$
c. $15:49$
d. $20:17$

7. WE3 Determine the value of the pronumerals in the following equivalent ratios.

a. $121:66 = a:6$
b. $3:2 = b:14$

8. Determine the value of the pronumerals in the following equivalent ratios.

a. $2:7:8 = 10:m:n$
b. $81:a:54 = 9:5:b$

9. Determine the value of the pronumerals in the following equivalent ratios.

a. $7:4 = x:16$
b. $a:9:b = 1:3:7$
c. $5:b = 15:30$
d. $18:12 = 3:m$

10. For each of the following ratios, determine whether it is equivalent to $12:8$.

a. $6:4$
b. $24:16$
c. $60:40$
d. $11:7$
e. $9:6$

11. Determine which of the following fractions is equivalent to $\frac{39}{65}$.

a. $\frac{78}{120}$
b. $\frac{13}{21}$
c. $\frac{6}{10}$
d. $\frac{13}{195}$
e. $\frac{6}{11}$

12. WE4 Determine the ratio of orange balls to black and white balls in the picture shown. Express your answer in its simplest fraction form.

13. Convert the ratio $56:42$ into a fraction in its simplest form.

14. Simplify the following ratios.

a. $\frac{120}{36}$
b. $1000:200$
c. $\frac{33}{2200}$
d. $58:116$
e. $36:24:60$
f. $315:180:360$

15. WE5 Express the given ratios in simplest form using whole numbers.

a. $2.5:1.5$
b. $3.6:4.2$
c. $0.11:1.1$
d. $1.8:0.32$
e. $1.6:0.24$
f. $1.15:13.8$

16. Express the following decimal ratios in simplest form using whole numbers.

a. 0.3 : 1.2 b. 7.5 : 0.25 c. 0.64 : 0.256 d. 1.2 : 3 : 0.42

17. In a class there are 28 students. If the number of girls in this class is 16, determine:

a. the number of boys in the class b. the ratio of boys to girls in simplest form.

18. a. Determine the ratio of length to width for the rectangle shown.

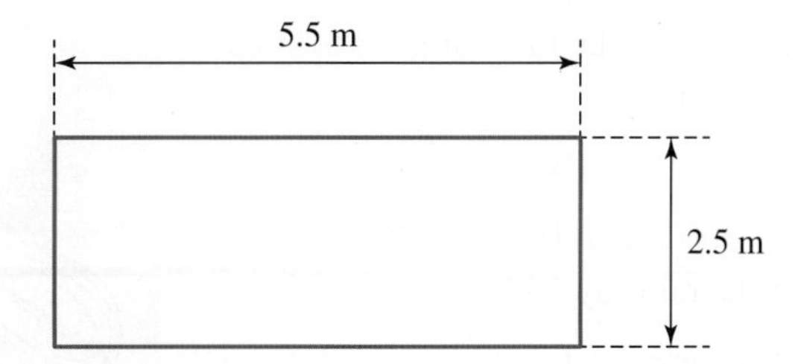

b. Determine the ratio of length to width to height for the rectangular prism shown.

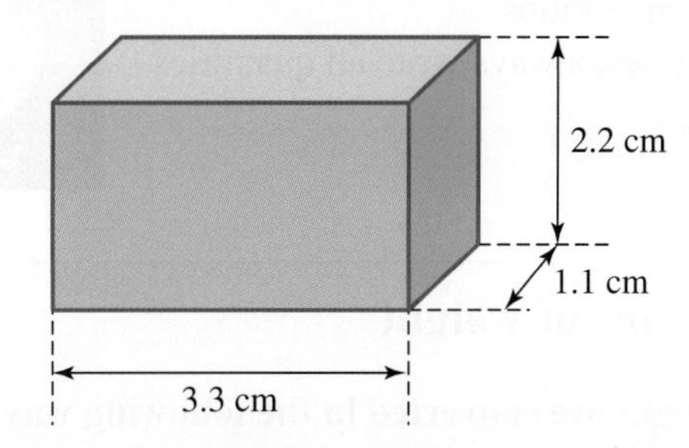

19. Using a calculator, express the following ratios as fractions in simplest form.

a. 18 : 66 b. 20 : 25 c. 30 : 21 d. 56 : 63 e. 10 : 100 f. 728 : 176

20. Using a calculator, determine ten fractions that are equivalent to the following fractions.

a. $\frac{3}{5}$ b. $\frac{1080}{840}$

Fully worked solutions for this chapter are available online.

LESSON
2.3 Ratios of two quantities

SYLLABUS LINKS

- Find the ratio of two quantities in its simplest form.

2.3.1 Weight ratios

- Ratios are used to compare the quantities of ingredients when making a mixture.
- For example, the mortar necessary for building a brick wall is usually prepared using the cement-to-sand ratio of 1 : 3.

- Pharmacists have to use ratios to prepare medicines.
- Chefs use ratios in every meal they cook.
- A recipe involves ratios.
- If a recipe says 3 cups of flour and 1 cup of sugar, it means that the recipe requires 3 cups of flour for every 1 cup of sugar, so the ratio flour : sugar = 3 : 1.
- What if the recipe requires 300 grams of flour and 100 grams of sugar? We can calculate the ratio of flour to sugar by simplifying the quantities given.

$$\text{flour} : \text{sugar} = 300\,\text{g} : 100\,\text{g}$$

- The highest common factor of 300 and 100 is 100. To simplify the ratio, divide both sides of the ratio by 100.

$$\text{flour} : \text{sugar} = 3 : 1$$

Note:

1. Usually, units are not written in ratios.
2. To calculate ratios of quantities, always write all quantities in the same unit.

Metric conversions of weight

Metric units of weight are converted in the following way:

1 kg = 1000 g
1 tonne = 1000 kg

WORKED EXAMPLE 6 Determining weight ratios

Peas and corn is a common combination of frozen vegetables. Determine the ratio of 500 g peas to 400 g corn.

THINK	WRITE
1. Ensure that the two quantities are written in the same unit (both are in grams).	500 g peas 400 g corn
2. Write the quantities as a ratio without units.	Peas : corn = 500 : 400
3. Simplify the ratio by dividing both sides by their highest common factor.	= 500 ÷ 100 : 400 ÷ 100
As 100 is the highest common factor of 500 and 400, divide both sides of the ratio by 100.	= 5 : 4
4. Write the ratio in its simplest form.	Peas : corn = 5 : 4

2.3.2 Length ratios

- Leonardo da Vinci was a famous painter, architect, sculptor, mathematician, musician, engineer, inventor, anatomist and writer.
- He painted the famous Mona Lisa, whose face has the golden ratio.
- The golden ratio distributes height and width according to the ratio 34 : 21. For Mona Lisa, this is:

height of face : width of face
34 : 21

- Ratios of lengths are calculated in the same way as the ratios for weights.

Metric conversions of length

Recall the relationship between the metric units of length. The following diagram can be used for quick reference.

km $\xrightarrow{\times 1000}$ m $\xrightarrow{\times 100}$ cm $\xrightarrow{\times 10}$ mm

mm $\xrightarrow{\div 10}$ cm $\xrightarrow{\div 100}$ m $\xrightarrow{\div 1000}$ km

WORKED EXAMPLE 7 Calculating length ratios

The window in the picture has a width of 1000 mm and a height of 1.75 m. Calculate the ratio of width to height.

THINK	WRITE
1. Make sure to convert both lengths into cm. The most appropriate unit is cm because converting 1.75 m into cm will change the decimal number into a whole number.	1000 mm = 100 cm 1.75 m = 175 cm
2. Write the lengths as a ratio without units.	Width : height = 100 : 175
3. Simplify the ratio by dividing both sides by their highest common factor (HCF = 25). As 25 is the highest common factor of 100 and 175, divide both sides of the ratio by 25.	$= \frac{100}{25} : \frac{175}{25}$ $= 4 : 7$
4. Write the ratio in its simplest form.	Width : height = 4 : 7

2.3.3 Area ratios

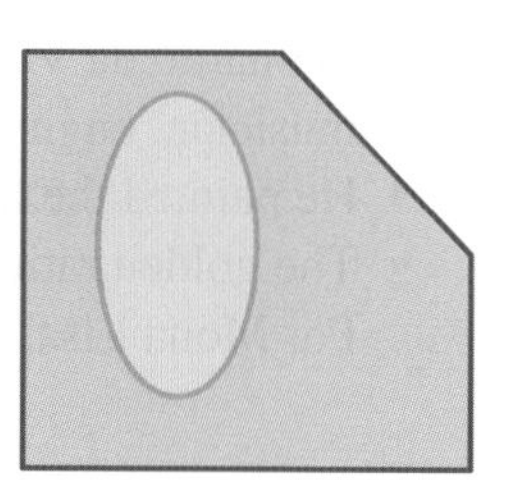

- **Areas** are enclosed surfaces.
- Landscape architects use ratios to calculate area of land to area of building, and area of pavement to area of grass.
- The diagram shown is a sketch of a paved backyard (blue) with an area of 50 m^2 and a garden (green) with an area of 16 m^2.
- The ratio of the paved area to the garden area is:

$$\begin{aligned}\text{paved area} : \text{garden area} &= 50\text{ m}^2 : 16\text{ m}^2\\ &= 50 : 16\\ &= 25 : 8\end{aligned}$$

Metric conversions of area

Recall the relationship between the metric units of area. The diagram shown can be used for quick reference.

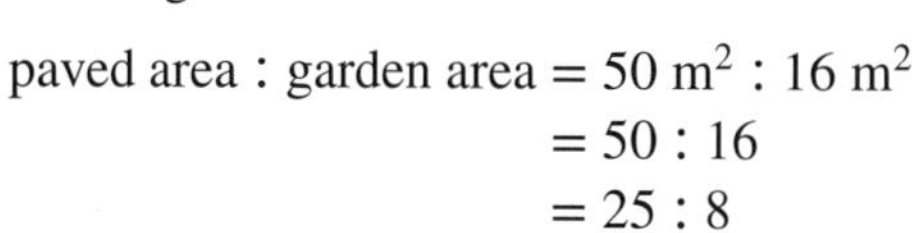

tlvd-3761

WORKED EXAMPLE 8 Determining area ratios

Kazem has a bedroom with an area of 200 000 cm². The whole house has an area of 160 m².

a. Convert the bedroom area of 200 000 cm² to m².

b. Calculate the ratio of the area of the bedroom to the area of the whole house.

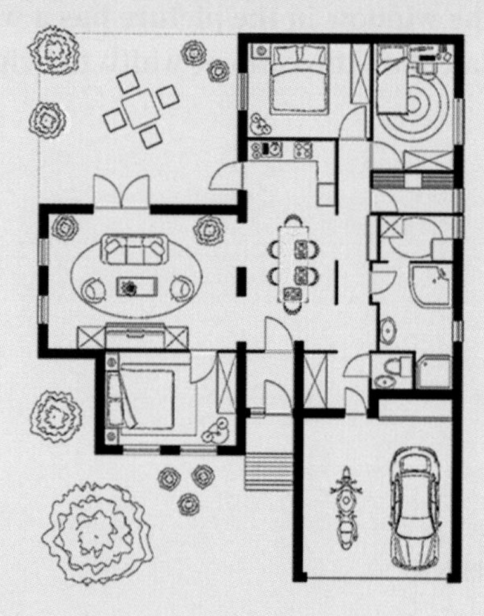

THINK	WRITE
a. To convert from cm^2 to m^2, divide the number by 100^2.	a. Area of bedroom $= 200\,000\text{ cm}^2 \div 100^2$ $= 20\text{ m}^2$
b. 1. Now that the areas are in the same unit, write the areas as a ratio without units.	b. Area of bedroom : area of house = 20 : 160
2. Simplify the ratio by dividing both sides by their highest common factor of 20 (HCF = 20).	$= \frac{20}{20} : \frac{160}{20}$
3. Write the ratio in its simplest form.	Area of bedroom : area of house = 1 : 8

2.3.4 Volume and capacity ratios

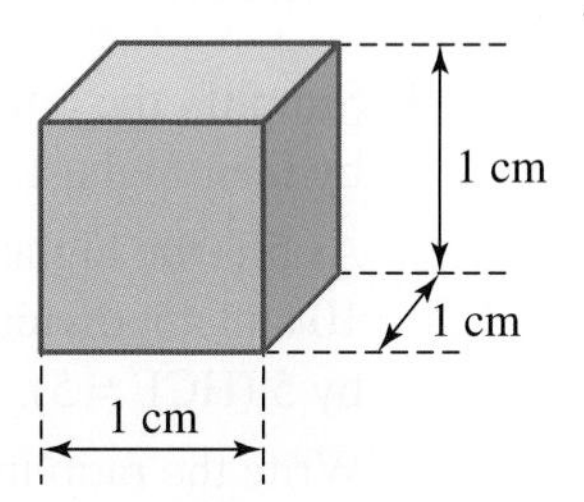

- The **volume** of an object is the space that the object takes up.
- A cube of length 1 cm, width 1 cm and height 1 cm takes up a space of $1 \times 1 \times 1 = 1\ \text{cm}^3$.
- The **capacity** of a container is the maximum volume it can hold.
- A container in the shape of a cube of length 1 cm, width 1 cm and height 1 cm can hold $1 \times 1 \times 1 = 1\ \text{cm}^3 = 1\ \text{mL}$ of liquid.

Metric conversions of volume and capacity

Recall the relationship between the metric units of volume.

The diagram shown can be used for quick reference.

$\times 1000^3$ $\times 100^3$ $\times 10^3$

km^3 m^3 cm^3 mm^3

$\div 1000^3$ $\div 100^3$ $\div 10^3$

Volume units are converted to capacity units using the following conversions.

$$1\ \text{cm}^3 = 1\ \text{mL}$$

$$1000\ \text{cm}^3 = \frac{1}{1000}\ \text{m}^3 = 1\ \text{L}$$

$$1\ \text{L} = 1000\ \text{mL}$$

tlvd-3762

WORKED EXAMPLE 9 Determining volume ratios

Salila and Devi have just installed two rainwater tanks: one for household use and one for watering the garden.
The household tank has a volume of 10 000 000 cm^3, while the tank for watering the garden has a volume of 45 m^3.

a. Calculate the ratio of the volume of the household water tank to the volume of the tank for watering the garden.

b. Calculate the ratio of the capacities of the two water tanks.

THINK	WRITE
a. 1. Convert both quantities to the same unit.	a. Volume of household tank = 10 000 000 cm^3 = 10 m^3
The most appropriate unit is m^3 because converting 10 000 000 cm^3 into m^3 will simplify this value.	Volume of garden tank = 45 m^3

2. Write the volumes as a ratio without units.

Household tank : garden tank = 10 : 45

3. Simplify the ratio by dividing both sides by their highest common factor.
As 5 is the highest common factor of 10 and 45, divide both sides of the ratio by 5 (HCF = 5).

$= \frac{10}{5} : \frac{45}{5}$

$= 2 : 9$

4. Write the ratio in its simplest form.

Household tank : garden tank = 2 : 9

b. 1. Convert both quantities to the same unit. The most appropriate unit is L because converting 10 000 000 cm^3 into L will simplify this value.

b. Capacity of household tank = 10 000 000 cm^3
= 10 000 L
Capacity of garden tank = 45 m^3
= 45 000 L

2. Write the volumes as a ratio without units.

Capacity of household tank : capacity of garden tank = 10 000 : 45 000

3. Simplify the ratio by dividing both sides by their highest common factor.
As 5000 is the highest common factor of 10 000 and 45 000, divide both sides of the ratio by 5000 (HCF = 5000).

$= \frac{10\,000}{5000} : \frac{45\,000}{5000}$

$= 2 : 9$

4. Write the ratio in its simplest form.

Capacity of household tank : capacity of garden tank = 2 : 9

2.3.5 Time ratios

- **Time** is a physical quantity we use for many of our daily activities.
- Have you ever considered the ratio of the time you spend on homework compared to the time you spend on social media?
- For example, if you spend 30 minutes completing homework and 10 minutes on social media, the ratio is:

time for homework : time on social media
30 : 10
3 : 1

Conversions of time

Recall the relationship between the metric units of time. The diagram shown can be used for quick reference.

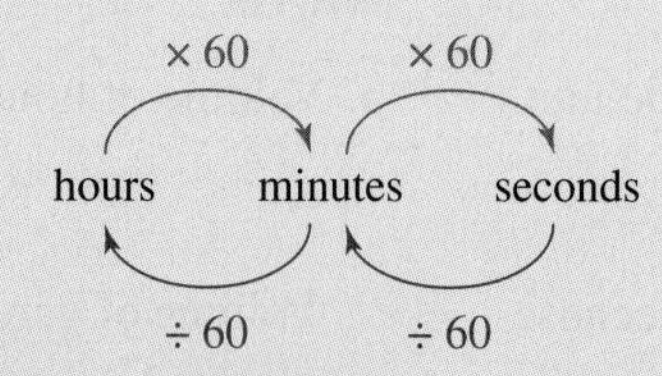

1 hour = 60 minutes
1 minute = 60 seconds
1 hour = 60 minutes (3600 seconds)

WORKED EXAMPLE 10 Determining time ratios

Emori is baking lemon cupcakes and chocolate cupcakes. It takes him $\frac{3}{4}$ of an hour to make the lemon cupcakes and 36 minutes to make the chocolate cupcakes.
Calculate the ratio of the time Emori takes to make the chocolate cupcakes to the time he takes to make the lemon cupcakes.

THINK	WRITE
1. Convert both quantities to the same unit. The most appropriate unit is minutes because converting $\frac{3}{4}$ of an hour into minutes will change the fraction into a whole number.	$\frac{3}{4}$ of an hour $= \frac{3}{4} \times 60$ minutes $= 45$ minutes
2. Write the times as a ratio without units.	Chocolate cupcake time : lemon cupcake time $= 36 : 45$
3. Simplify the ratio by dividing both sides by their highest common factor. As 9 is the highest common factor of 36 and 45, divide both sides of the ratio by 9 (HCF = 9).	$= \frac{36}{9} : \frac{45}{9}$ $= 4 : 5$
4. Write the answer.	Chocolate cupcake time : lemon cupcake time $= 4 : 5$

2.3.6 Money ratios

- A budget planner is a way of planning where money earned is going to be spent.
- Budget planners are used by people, businesses and governments.
- For example, a family budgets \$50 per week for electricity. If the family's total weekly wage is \$2100, calculate the ratio of money spent on electricity per week to the total weekly wage.

$$\begin{aligned} \text{Money on electricity} &: \text{wage} \\ 50 &: 2100 \ (\text{HCF} = 50) \\ 1 &: 42 \\ & \frac{1}{42} \end{aligned}$$

Conversions of money

Recall the relationship between dollars (\$) and cents.

$$\mathbf{\$1 = 100\ cents}$$
$$\mathbf{1\ cent = \$0.01 = \$\frac{1}{100}}$$

WORKED EXAMPLE 11 Calculating money ratios

Jackie earns $240 per month from her part-time job. She decides to spend $40 per month to buy her lunch from the school canteen. If her monthly phone bill is $20, calculate the ratio of the lunch money to her phone bill and to her income per month.

THINK	WRITE
1. Ensure that all the amounts are written in the same unit.	Monthly wage = $240 Monthly lunch money = $40 Monthly phone bill = $20
2. Write the amounts as a ratio without units.	Lunch money : phone bill : income = 40 : 20 : 240
3. Simplify the ratio by dividing all 3 parts by their highest common factor. As 20 is the highest common factor of 20, 40 and 240, divide all 3 parts of the ratio by 20 (HCF = 20).	$= \frac{40}{20} : \frac{20}{20} : \frac{240}{20}$ $= 2 : 1 : 12$
4. Write the answer.	Lunch money : phone bill : income = 2 : 1 : 12

Exercise 2.3 Ratios of two quantities

learnon

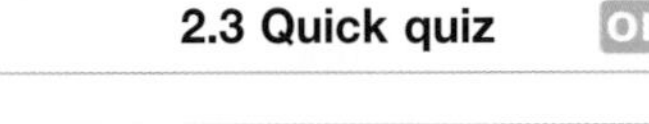

2.3 Quick quiz on | 2.3 Exercise

Simple familiar	Complex familiar	Complex unfamiliar
1, 2, 3, 4, 5, 6, 7, 8, 9, 10, 11, 12, 13, 14, 15, 16, 17, 18, 19, 20, 21, 22, 23, 24	N/A	N/A

These questions are even better in jacPLUS!
- Receive immediate feedback
- Access sample responses
- Track results and progress

Find all this and MORE in jacPLUS

Simple familiar

1. WE6 Determine the ratio of 240 000 g sand to 80 kg cement.
2. A recipe for a carrot cake has the ratio 0.5 kg flour, 100 g sugar and 250 g grated carrots. Determine the ratio of sugar : flour : grated carrots in its simplest form.
3. A puff pastry recipe requires 225 g flour, 30 g lard and 0.210 kg butter. Determine the ratio of:
   a. flour to lard
   b. butter to lard
   c. flour to butter
   d. flour to lard to butter.
4. Shane makes dry ready mix concrete using 2.750 kg of stone and gravel, 2500 g of sand, and 1 kg of cement and water.
   Calculate the ratios of:
   a. cement and water to stone and gravel
   b. sand to cement and water
   c. cement and water to sand to stone and gravel
   d. cement and water to the total quantity of materials.

5. **WE7** The wheelchair ramp shown in the diagram has a length of 21 m and a height of 150 cm. Determine the ratio of height to length.

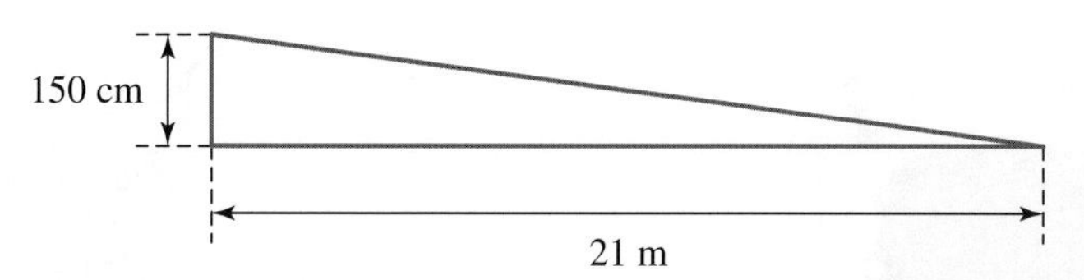

6. A rectangular swimming pool has length 50 m, width 2500 cm and depth 1000 mm.
Calculate the ratio length : width : depth in its simplest form.

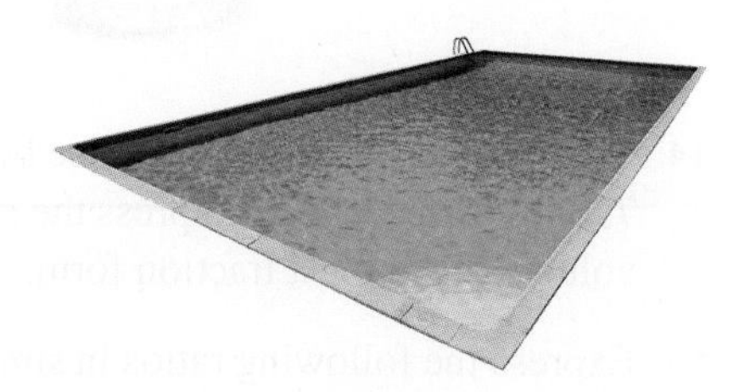

7. Determine the ratio of length to width of a football field 162 m long and 144 m wide.
8. **WE8** The area of the small triangle, ABC, is 80 000 cm^2 and the area of the large triangle, DEF, is 40 m^2.
   a. Convert the area of the small triangle, ABC, 80 000 cm^2 to m^2.
   b. Calculate the ratio of the area of the small triangle to the area of the large triangle in its simplest form.

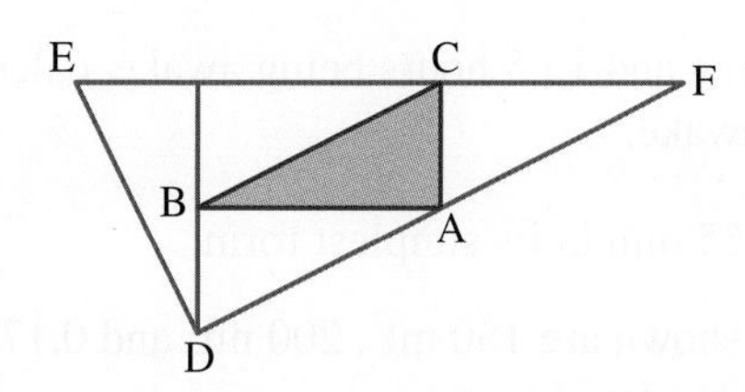

9. Earth's surface area is approximately 510 100 000 km^2. Water covers 361 100 000 km^2. Calculate the ratio of the surface area covered with water to the total surface area of Earth.
10. The landscape in the diagram represents the design of a garden with two ponds. One pond has an area of 48 m^2 and the other has an area of 640 000 cm^2. The sand area is 320 000 cm^2 and the area of the table is 4 m^2.
For the areas given, determine the ratio of:
    a. the area of the large pond to the area of the small pond
    b. the area of the table to the area of the sand
    c. the area of the sand to the area of the large pond
    d. the area of the small pond to the area of the table.

11. For each of the following conversions, determine whether it is incorrect.
    a. 3 m = 0.003 km
    b. 8.7 m^2 = 870 000 cm^2
    c. 1.5 km = 1 500 000 mm
    d. 19.6 L = 19 600 cm^3
    e. 3.25 hours = 195 minutes
12. **WE9** The capacity of the glass in the picture is 250 mL and the capacity of the carafe is 1.5 L. Determine the ratio of the capacity of the carafe to the capacity of the glass.

13. Of the two jugs shown, one holds 0.300 L of oil and the other 375 mL of water. Determine the ratio volume of water to volume of oil required for this recipe.

14. The volumes of the orange juice boxes in the picture shown are 1.250 L and 750 mL respectively. Express the ratio of the large volume to the smaller volume in simplest fraction form.

15. Express the following ratios in simplest fraction form.
   a. $70 \text{ cm}^3 : 20 \text{ cm}^3$
   b. $560\,000 \text{ cm}^3 : 0.6 \text{ m}^3$
   c. $15 \text{ cm}^3 : 45\,000 \text{ mm}^3$
   d. $308\,000\,000 \text{ cm}^3 : 385 \text{ m}^3$

16. **WE10** The time it takes Izzy to get to school by bus is 22 minutes while the time it takes her by train is 990 seconds. Determine the ratio of the time taken by bus to the time taken by train.

17. If a person spends 8.5 hours sleeping and 15.5 hours being awake, calculate the ratio of sleeping time to time awake.

18. Express the ratio of 1 h 25 min to 55 min in its simplest form.

19. The capacities of the three bottles shown are 150 mL, 200 mL and 0.175 L respectively. Determine the following ratios in simplest form:
   a. The capacity of the smallest bottle to the capacity of the largest bottle
   b. The capacity of the largest bottle to the capacity of the middle bottle
   c. The capacity of the middle bottle to the capacity of the smallest bottle
   d. The capacity of the smallest bottle to the capacity of the middle bottle to the capacity of the largest bottle

20. **WE11** Su Yi is a sales consultant and has a budget of \$154 for a mobile phone, \$198 for a computer and \$66 for lunch. Determine the ratio between the budgets for:
   a. mobile phone to computer to lunch
   b. mobile phone to computer
   c. computer to lunch
   d. mobile phone to lunch.

21. Every week Sam spends \$27.00 for petrol, \$5.40 for his favourite magazine and \$3.60 for his prepaid calls. Calculate the ratio of:
   a. petrol money to the price of the magazine
   b. the price of the magazine to the money spent on the prepaid calls
   c. the money spent on the prepaid calls to petrol money
   d. petrol money to the price of the magazine and to the money spent on the prepaid calls.

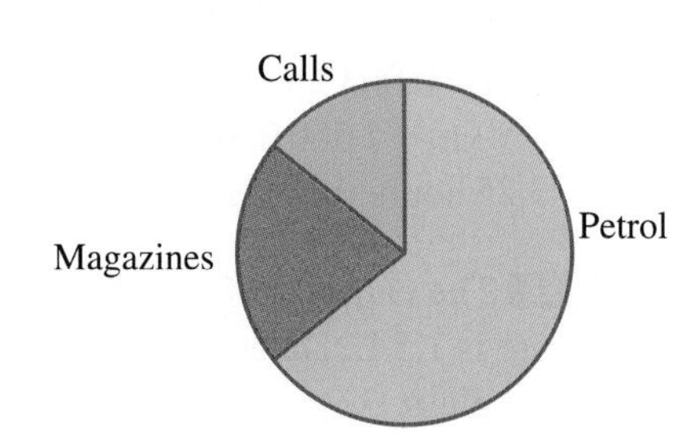

22. a. Determine the ratio of volume of tennis ball to volume of basketball.

$V = 0.0064 \text{ m}^3$

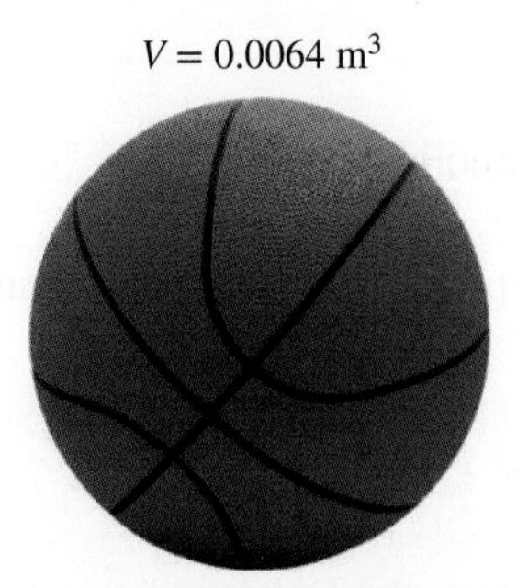

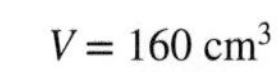

b. Determine the ratio of the capacity of the lunch box to the capacity of the drink bottle.

$V = 1000 \text{ cm}^3$ $\quad V = 400 \text{ mL}$

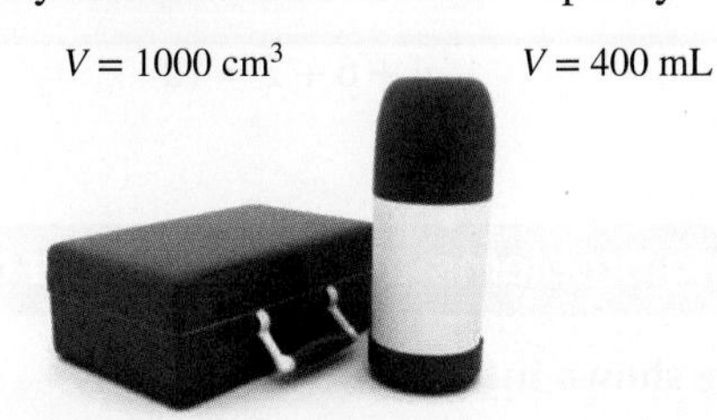

23. Using a calculator, express the following ratios as fractions in simplest form.

a. 360 : 450
b. 455 : 195
c. 2412 : 4288
d. 5858 : 32 219
e. 53 856 : 20 196

24. Using a calculator, determine 10 equivalent ratios to each of the following ratios.

a. 19 : 7
b. 720 : 480

Fully worked solutions for this chapter are available online.

LESSON
2.4 Divide into ratios [complex]

SYLLABUS LINKS

- Divide a quantity in a given ratio [complex].

2.4.1 Divide a number of items into a ratio

- There are situations when a quantity needs to be divided into a given ratio.
- The total of 12 pencils in the figure can be divided into the ratio 2 : 3 : 1. This means dividing the 12 pencils into 3 groups where the first group has twice as many pencils as the third group, and the second group has 3 times as many pencils as the third group.
- To calculate the number of pencils in each group, we need to calculate the total number of parts of the given ratio.
- Add up all parts to determine the total number of parts:

$$\text{Total pencils} = 2 + 3 + 1 = 6 \text{ parts}$$

- To calculate the number of pencils per part, divide the total number of pencils by the total number of parts required:

$$\text{Total pencils} = \frac{12}{6} = 2 \text{ pencils per part}$$

- The ratio is 2 : 3 : 1. To calculate the number of pencils in each group, multiply the parts by the number of pencils per part (in this case by 2).

$$2 \times 2 : 3 \times 2 : 1 \times 2 = 4 : 6 : 2$$

- Check your answer by adding up all the quantities in the last answer:

$$4 + 6 + 2 = 12$$

tlvd-3764

WORKED EXAMPLE 12 Solving problems of dividing into a given ratio

Divide the jelly beans in the figure shown into a ratio of 3 : 4 : 1.

THINK	WRITE
1. Calculate the total number of parts of the ratio.	$3 + 4 + 1 = 8$ The total number of jelly beans is 16.
2. As each part is equal and there are 8 parts, then one part would be worth 2. Alternatively, divide the total number of jelly beans (16) by the total number of parts (8) to calculate the number of jelly beans per part.	Total number of jelly beans = 16 Therefore, there are 2 jelly beans per part.
3. Multiply each number in the ratio by the number of jelly beans per part.	$3 : 4 : 1 = 3 \times 2 : 4 \times 2 : 1 \times 2$ $= 6 : 8 : 2$
4. Check the answer by adding up all quantities in the ratio. This number should be equal to the total number of jelly beans.	$6 + 8 + 2 = 16$
5. Write the answer.	16 jelly beans divided into the ratio 3 : 4 : 1 is 6, 8 and 2.

2.4.2 Dividing weights into ratios

- Chemists work with mixtures every day. Sometimes they need to divide a mixture into given ratios.

WORKED EXAMPLE 13 Solving division of a total weight into a given ratio

A mixture of three chemicals has a weight of 150 g.
The three chemicals must be combined in the ratio
orange : red : brown = 3 : 2 : 1.
Calculate how many grams of each chemical are required.

THINK	WRITE
1. Calculate the total number of parts of the ratio.	$3 + 2 + 1 = 6$ parts
2. As each part is equal and there are 6 parts, then one part is worth 25 g. Alternatively, divide the total weight (150 g) by the total number of parts (6) to calculate the weight per part.	The total weight of the mixture is 150 g. 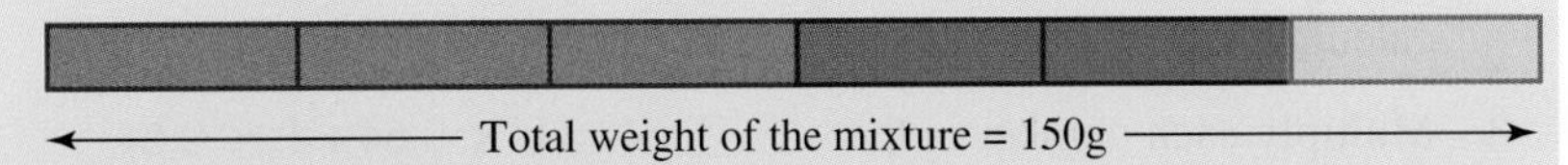 Therefore, there are 25 g per part.
3. Multiply each number in the ratio by the weight per part.	$3 : 2 : 1 = 3 \times 25 : 2 \times 25 : 1 \times 25$ $= 75 : 50 : 25$
4. Check the answer by adding up all quantities in the ratio. This number should be equal to the total weight of the mixture.	$75 + 50 + 25 = 150$
5. Write the answer.	The 150 g of chemicals should be divided into the following amounts: Orange 75 g Red 50 g Brown 25 g

2.4.3 Dividing lengths into ratios

- Carpenters use ratios to cut planks of timber, plastic pipes or metal rods to specified lengths.
- A 2-m-long piece of timber is being cut in a ratio 1 : 3.
- There are a total of four parts in this ratio, and each part is 0.5 m long.
- This makes the two pieces of timber 0.5 m and 1.5 m long respectively.

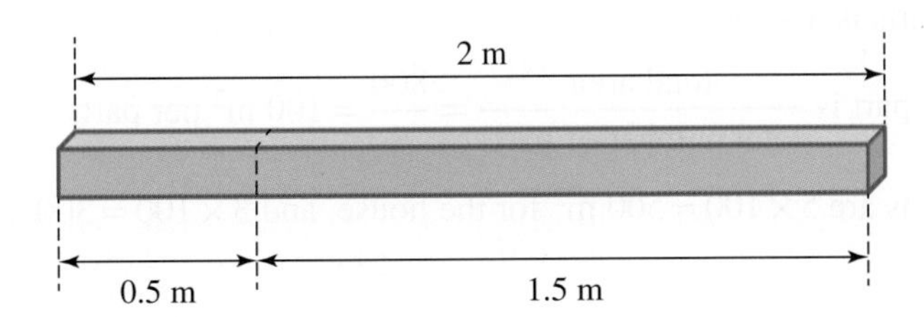

tlvd-11368

WORKED EXAMPLE 14 Solving division of lengths into given ratios

A carpenter builds a timber frame that is 3.2-m tall. The frame is built in two sections at a ratio of 1 : 4. Calculate the length of the timber frame, *x*.

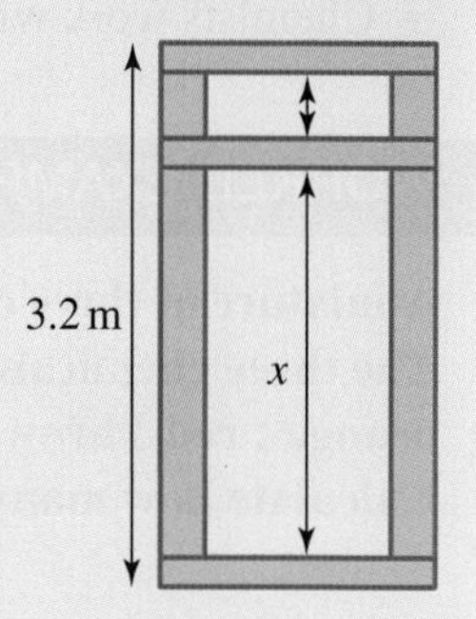

THINK	WRITE
1. Calculate the total number of parts of the ratio.	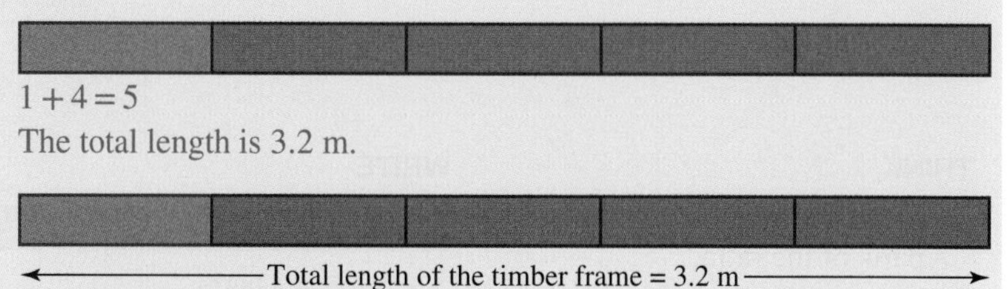 $1+4=5$
2. As each part is equal and there are 5 parts, then one part is worth 0.64. Alternatively, divide the total length of the timber frame (3.2 m) by the total number of parts (5) to calculate the length per part.	The total length is 3.2 m. Total length of the timber frame = 3.2 m. Therefore, the length per part is 0.64 m.
3. Multiply each number in the ratio by the length per part.	$1:4 = 1 \times 0.64 : 4 \times 0.64$ $= 0.64 : 2.56$
4. Check the answer by adding up all lengths in the ratio. This number should be equal to the total length.	$0.64 + 2.56 = 3.2$ m
5. Write the answer.	The length of the timber frame, *x*, is 2.56 m. 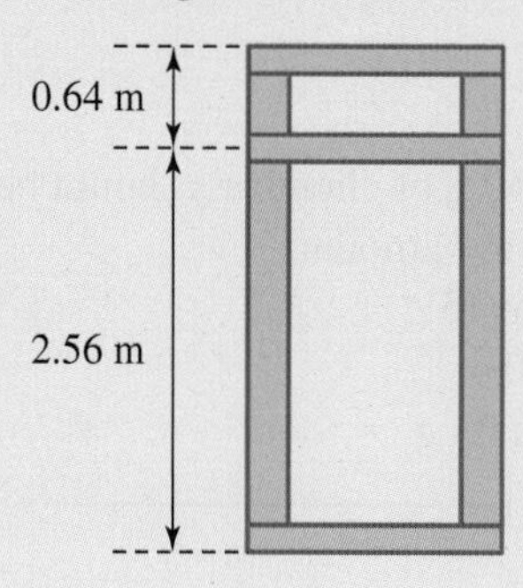

2.4.4 Dividing areas into ratios

- Surveyors divide areas into ratios. Architects use these ratios to design accurate plans.
- For example, if the area of a property is 800 m^2 and it must be divided in the ratio house : backyard = 5 : 3, then we can calculate the area of the house and the area of the backyard.
- The total number of parts is $5 + 3 = 8$.
- The number of m^2 per part is $\dfrac{\text{total area}}{\text{total number of parts}} = \dfrac{800}{8} = 100$ m^2 per part.
- The corresponding areas are $5 \times 100 = 500$ m^2 for the house, and $3 \times 100 = 300$ m^2 for the backyard.

WORKED EXAMPLE 15 Solving division of areas into given ratios

The playground area shown must be divided into a ratio of grass to sandpit of 7 : 2. If the area of the playground is 270 m^2, calculate the area required for the sandpit.

THINK	WRITE
1. Calculate the total number of parts of the ratio.	$7+2=9$
2. As each part is equal and there are 9 parts, then one part is worth 30. Alternatively, divide the total area of playground $(270\,\text{m}^2)$ by the total number of parts (9) to calculate the area per part.	The total area is 270 m^2. Total area = 270 m^2 Therefore, there are 30 m^2 per part.
3. Multiply each number in the ratio by the area per part.	Grass : sandpit = 7 : 2 = 7 × 30 : 2 × 30 = 210 : 60
4. Check the answer by adding up all areas in the ratio. This sum should be equal to the total area.	$210+60=270$ m^2
5. Write the answer.	The area of the sandpit is 60 m^2.

2.4.5 Dividing volumes and capacities into ratios

- Recipes use mass for solid ingredients and volume or capacity for liquid ingredients.
- Volume and capacity are measured in different units.
- Consider a 2-L cordial drink containing 500 mL cordial and 1.5 L water. The ratio is cordial : water = 1 : 3. The cordial drink can be made in a 2000 cm^3 jug: the volume of water fills 1500 cm^3 of the jug and the cordial fills 500 cm^3 of the jug. The ratio for the volumes filled by cordial and water in the jug is also 1 : 3.

WORKED EXAMPLE 16 Solving division of volumes into given ratios

The two containers shown have a total volume of 343 750 cm^3. The volumes of the two containers are in the ratio of 4 : 7. Calculate the volumes of the two containers.

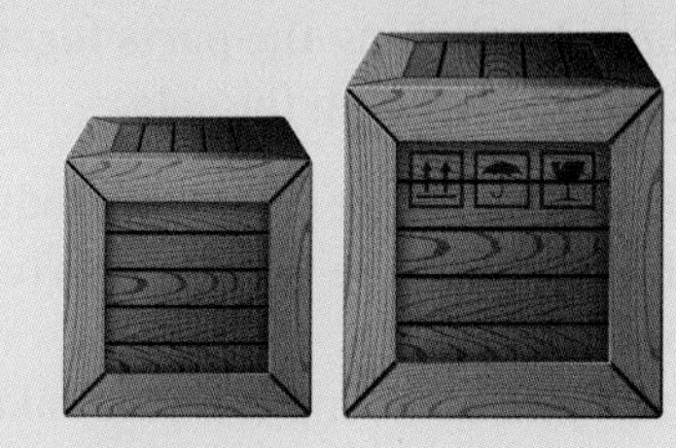

THINK	WRITE
1. Calculate the total number of parts of the ratio.	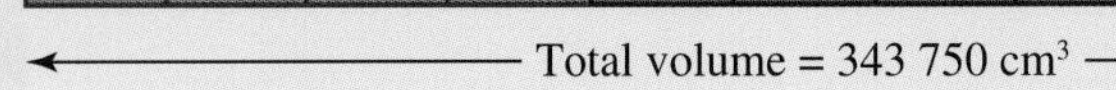 $4+7=11$
2. As each part is equal and there are 11 parts, then one part would be worth 31 250. Alternatively, divide the total volume $(343\,750\text{ cm}^3)$ by the total number of parts (11) to calculate the volume per part.	The total volume is 343 750 cm^3. Total volume = 343 750 cm^3
3. Multiply each number in the ratio by the volume per part.	Small container : large container $=4:7$ $=4\times 31\,250:7\times 31\,250$ $=125\,000:218\,750$
4. Check the answer by adding up the volumes in the ratio. This sum should be equal to the total volume.	$125\,000+218\,750=343\,750\text{ cm}^3$
5. Write the answer.	The volume of the small box is 125 000 cm^3. The volume of the large box is 218 750 cm^3.

2.4.6 Dividing time into ratios

- Suppose that you catch a bus and then a train to visit a friend.
- The whole trip takes 40 minutes.
- If the ratio of the time travelling by bus to the time travelling by train is 3 : 1, calculate the travel time on the bus.
- The total number of parts of the ratio is $3+1=4$.
- The number of minutes per part is $\frac{40}{4}=10$ minutes per part.
- Bus time : train time $=3\times 10:1\times 10$
  $=30:10$
- This means that travel time on the bus is 30 minutes and travel time on the train is 10 minutes.

WORKED EXAMPLE 17 Solving division of time into given ratios

There are three stages involved in the manufacturing of shoes:

Stage 1: cutting out all of the parts
Stage 2: stitching the parts together
Stage 3: attaching the sole

The average time taken to produce a basic pair of shoes is 1 hour.
The times for the three stages (cutting, stitching and attaching) are in the ratio 4 : 5 : 1.
Calculate how long it would take to cut out all the pieces for one pair of shoes.

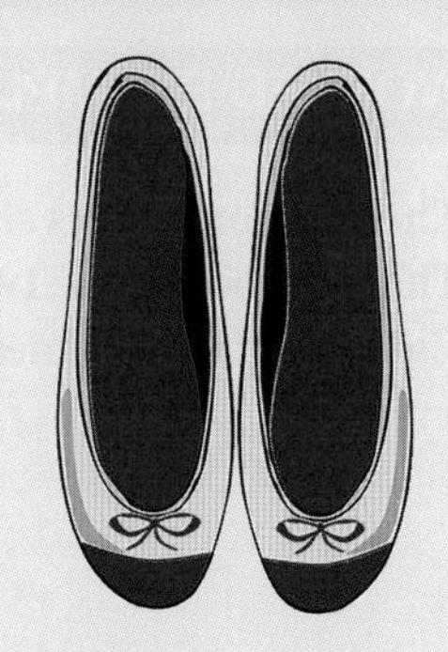

THINK	WRITE
1. Calculate the total number of parts of the ratio.	$4+5+1=10$
2. As each part is equal and there are 10 parts, then one part would be worth 6 minutes. Alternatively, divide the total time in minutes (60) by the total number of parts (10) to calculate the minutes per part. As we are going to divide 1 hour by 10 parts, it is easier to work with minutes than with hours.	The total time is 1 hour = 60 minutes 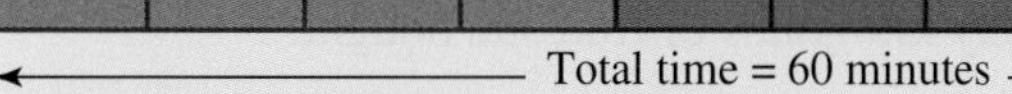 Total time = 60 minutes Therefore, there are 6 minutes per part.
3. Multiply each number in the ratio by the number of minutes per part.	$4:5:1=4\times6:5\times6:1\times6$ $=24:30:6$
4. Check the answer by adding up the times in the ratio. This sum should be equal to the total time.	$24+30+6=60$ minutes $=1$ hour
5. Write the answer.	The time taken to cut out all the pieces for a pair of shoes is 24 minutes.

Exercise 2.4 Divide into ratios [complex]

learnon

2.4 Quick quiz on | 2.4 Exercise

Simple familiar	Complex familiar	Complex unfamiliar
N/A	1, 2, 3, 4, 5, 6, 7, 8, 9, 10, 11, 12, 13, 14, 15, 16, 17, 18, 19, 20, 21, 22, 23, 24, 25	26, 27, 28

Complex familiar

1. WE12 The bouquet in the picture is made of 30 tulips. Flora wants to make two smaller bouquets in the ratio 2 : 3.
Determine how many tulips there are in each smaller bouquet.

2. The school canteen has 250 mandarins. The mandarins must be divided in the ratio 5 : 3 : 2. Determine how many mandarins there are in each group.

3. Determine the number of children in a group of 1320 people if the ratio of children to adults is:

a. 3 : 1 b. 5 : 1 c. 3 : 2 d. 7 : 4.

4. Sophia bought 5.2kg of fruits and vegetables in the fruits-to-vegetables ratio of 3 : 2.

a. Determine how many kilograms of vegetables Sophia bought.
b. Determine how many kilograms of fruit she bought.

5. Carbon steel contains amounts of manganese, silicon and copper in the ratio 8 : 3 : 2.

a. Calculate the quantity of manganese required to produce carbon steel if the total amount of manganese, silicon and copper is 936 kg.
b. Calculate the number of kilograms of silicon in this quantity.

6. WE13 A chef has a recipe for a homemade soup that requires 500 g of tomatoes and capsicum in the ratio of 3 : 2. Calculate the quantity of tomatoes and capsicum that the chef should buy to make this soup.

7. The ratio of sand to cement to make mortar is 3 : 1. A bricklayer wants to make 17 kg of mortar. Calculate the quantities of sand and cement required.

8. WE14 Monika has 4.5 m of fabric to make a top and a skirt. She wants to divide the fabric in the ratio 2 : 3. Determine the lengths of the two pieces of fabric.

9. A 1.4-m steel beam is cut into three lengths in the ratio 1 : 2 : 4. Determine the length of each of the cut beams. Give the answers in mm.

10. A picture frame has a perimeter of 123 cm. Calculate the size of the width of the picture frame if the ratio of width to length is 1 : 3.

11. The ratio of length to width to height of the container shown is 4 : 2 : 3. If the sum of the three dimensions is 6.75 m, calculate the height of the container.

12. WE15 The area of a window is 1.5 m^2. The ratio of the areas of the glass used is 1 : 2 : 3. Calculate the area of each part of glass.

13. The total area of a tennis court is approximately 260 m^2 for doubles matches. The ratio of the court area for singles matches to the extension area added on for doubles is approximately 13 : 3. Calculate the area of the court used for singles matches.

14. Three areas add up to 26 000 000 m^2. If they are in the ratio of 3 : 2 : 8, calculate the smallest area.

15. The total surface area of Victoria and Western Australia is approximately 2 760 000 km^2. If the ratio of Victoria's surface area to Western Australia's surface area is 1 : 11, calculate the approximate surface area of Victoria.

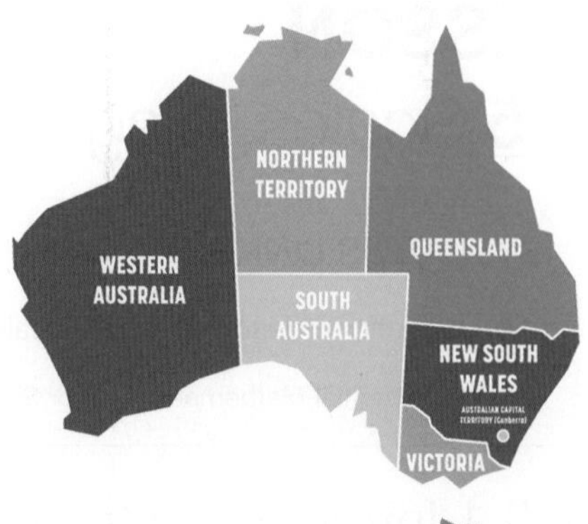

16. **WE16** The ratio of the volume of a green jar to the volume of a red jar is 5 : 6. If the total volume of the two jars is 506 cm^3, determine the individual volumes of the two jars.

17. Hot air balloons are each fitted with three propane fuel tanks, with a total volume of 171 L. Their individual volumes are in the ratio 2 : 3 : 4. Calculate the individual volumes of the fuel tanks.

18. The total capacity of three rainwater tanks is 42 500 L. Determine the capacity of each tank if their capacities are in the ratio of 2 : 5 : 10.

19. **WE17** Claire has 25 minutes to tidy up her bedroom and have breakfast before she goes to school. If the ratio of the tidy-up time and breakfast time is 3 : 2, calculate how many minutes she will have to tidy up her room.

20. Rich has 1 hour and 45 minutes to complete his Science, English and Maths homework. If the time he needs to spend on the three subjects is in the ratio 1 : 2 : 4 respectively, determine how much time he has to spend on his Maths homework.

21. Bayside College has a ratio of class time to lunch to recess of 10 : 2 : 1. If the total time for a school day at Bayside College is 6 hours and 30 minutes, calculate:
    a. the class time per day
    b. the time for lunch
    c. recess time.

22. Determine how much a family will spend on their mortgage, car loan and health insurance if they allow a budget of $1225 per month and the ratio of mortgage to car loan to health insurance is 10 : 3 : 1.

23. Three friends win a prize of $1520. Calculate how much money each of them wins if they contributed to the price of the winning ticket in the ratio 2 : 5 : 9.

24. Using a calculator, divide the quantities below in the ratios given.
    a. $420 in the ratio 7 : 3
    b. 6 hours in the ratio 1 : 5 : 6
    c. 18 293 m in the ratio 5 : 2 : 4
    d. 15 000 m^2 in the ratio 10 : 3 : 2
    e. 960 L in the ratio 7 : 13
    f. 7752 in the ratio 6 : 2 : 9

25. Using a calculator, determine the quantities obtained when $5040 is divided in the following ratios:
    a. $\frac{1}{2}$
    b. $\frac{2}{3}$
    c. $\frac{1}{5}$
    d. $\frac{2}{7}$
    e. $\frac{3}{4}$
    f. 2 : 3 : 4

Complex unfamiliar

26. Company A spends $3 875 000 on wages and cars in a ratio of 30 : 1. Company B spends $3 250 000 on wages and cars in a ratio of 24 : 1. Determine which company spends more money on wages and by how much.

27. Three people, A, B and C, purchased plants in a ratio of 1 : x : 3 respectively. If 15 of the 30 plants were purchased in total by person C, determine the value of x.

28. Two students commit to a ratio of savings to spending. Jane follows a ratio of 3 : 5 and John follows a ratio of 4 : 7, and they earn the same amount of money. Determine and justify who saves more money.

Fully worked solutions for this chapter are available online.

LESSON
2.5 Scales, diagrams and maps [complex]

SYLLABUS LINKS

- Use ratio to describe simple scales [complex].

2.5.1 Introduction to scales

- Scales are used to draw maps and represent objects such as cars and boats.
- Scales are usually used when it is not possible to construct full-size models.
- A **scale** is a ratio of the length on a drawing or map to the actual distance or length of an object. It is the length on drawing : actual length.
- Scales are usually written with no units.
- If a scale is given in two different units, the larger unit is usually converted into the smaller unit.
- A **scale factor** is the ratio of two corresponding lengths in two similar shapes.
- A scale factor of $\frac{1}{2}$ or a scale (ratio) of 1 : 2 means that 1 unit on the drawing represents 2 units in actual size. The unit can be mm, cm, m or km.

tlvd-3767

WORKED EXAMPLE 18 Calculating the scale of a drawing

Calculate the scale of a drawing where 2 cm on the diagram represents 1 km in reality.

THINK	WRITE
1. Convert the larger unit (km) to the smaller unit (cm).	1 km = 100 000 cm
2. Now that both values are in the same unit, write the scale of the drawing with no units.	2 : 100 000
3. Simplify the scale by dividing both sides of the scale by the highest common factor.	$\frac{2}{2} : \frac{100\,000}{2}$
Divide both sides of the scale by 2, as 2 is the highest common factor of 2 and 100 000.	= 1 : 50 000
4. Write the answer.	1 : 50 000

2.5.2 Calculating dimensions

- The scale factor must always be included when a diagram is drawn.
- It is used to calculate the dimensions needed.
- To calculate the actual dimensions, measure the dimensions on the diagram and then divide them by the scale factor.

WORKED EXAMPLE 19 Calculating actual dimensions using scales

Calculate the length and the width of the bed shown if the scale of the drawing is 1 : 100.

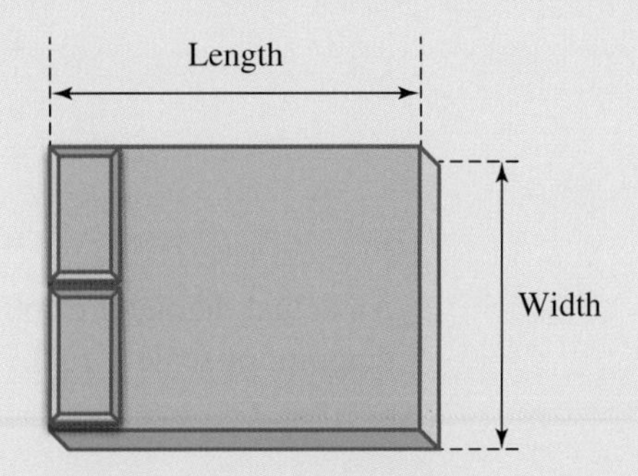

Use a ruler to measure the length and the width of the bed.

THINK	WRITE
1. Use a ruler to measure the length and the width of the diagram.	Length = 2 cm Width = 1.5 cm
2. Write the scale of the drawing.	1 : 100 This means that every 1 cm on the drawing represents 100 cm of the real bed.
3. Divide both dimensions by the scale factor.	Length of the bed $= 2 \div \frac{1}{100} = 2 \times 100 = 200$ cm Width of the bed $= 1.5 \div \frac{1}{100} = 1.5 \times 100 = 150$ cm
4. Write the answer.	The length is 200 cm and the width is 150 cm.

2.5.3 Scale drawing

- To be able to draw to scale, we need to know the dimensions on the diagram.
- To calculate these dimensions, we multiply the actual dimension by the scale factor.

tlvd-3768

WORKED EXAMPLE 20 Calculating dimensions using scale drawings

A drawing of a car uses a scale of 1 : 50. Calculate the diameter of a car wheel on the diagram if its real diameter is 60 cm.

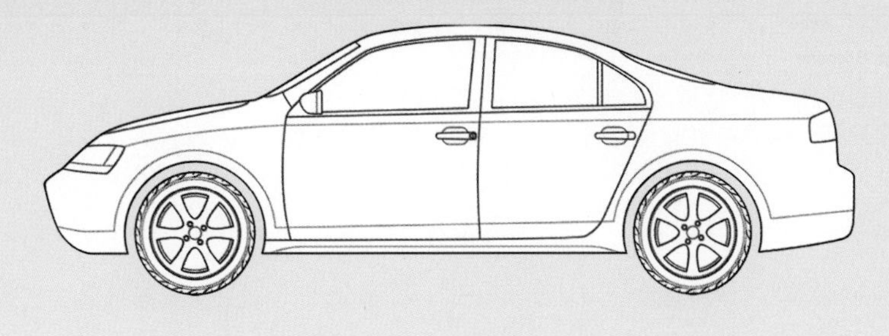

THINK	WRITE
1. Calculate the length on the diagram.	Length on the diagram = real measurement × scale factor $= 60 \times \frac{1}{50}$ $= \frac{60}{50}$ $= 1.2$ cm
2. Write the answer.	An actual diameter of 60 cm will be 1.2 cm on a diagram of scale 1 : 50.

2.5.4 Maps and scales

- Maps are always drawn at a much smaller scale.
- All maps have the scale written or drawn on the map.

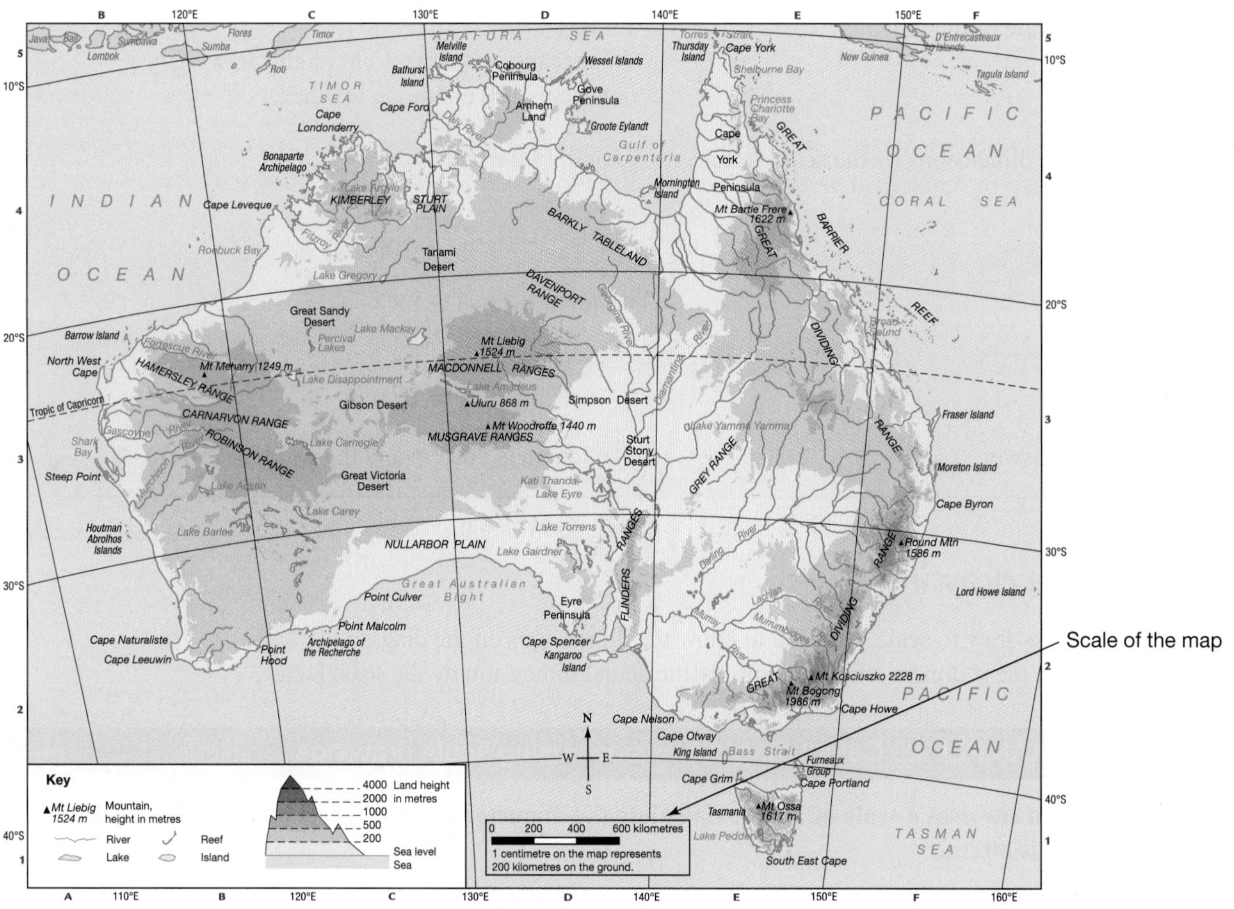

Source: © MAPgraphics Pty Ltd, Brisbane

WORKED EXAMPLE 21 Determining the scale of a map as a ratio

Determine the scale of the map as a ratio using the information in the diagram.

0 200 400 km

THINK	WRITE
1. Write the scale as a ratio.	1 cm : 200 km
2. Convert the larger unit into the smaller unit.	The larger unit is km. This has to be converted into cm. 200 km = 20 000 000 cm 1 cm : 200 km 1 cm : 20 000 000 cm
3. Write the answer.	The scale is 1 : 20 000 000.

2.5.5 Maps and distances

- Both actual distances and distances on the map can be calculated if the scale of the map is known.
- Actual distances can be calculated by measuring the lengths on the map and then *dividing* these by the scale factor.
- The dimensions on the map can be calculated by *multiplying* the actual dimension by the scale factor.

tlvd-3770

WORKED EXAMPLE 22 Calculating actual distance and distance on the map

The scale of an Australian map is 1 : 40 000 000.

a. Calculate the actual distance if the distance on the map is 3 cm.

b. Calculate the distance on the map if the actual distance is 2500 km.

THINK	WRITE
a. 1. Write the scale.	a. Distance on the map : actual distance = 1 : 40 000 000
2. Set up the ratios for map ratio and actual ratio.	Map ratio is 1 : 40 000 000. Actual ratio is 3 : x.
3. Construct equivalent fractions.	$\frac{x}{3} = \frac{40\,000\,000}{1}$ $x = 3 \times 40\,000\,000$ $= 120\,000\,000$ The distance on the map is 120 000 000 cm.
4. Convert the answer into km.	120 000 000 cm = 1200 km
5. Write the answer.	The actual distance represented by 3 cm on the map is 1200 km.
b. 1. Write the scale.	b. Distance on the map : actual distance = 1 : 40 000 000
2. The actual distance is 2500 km. Convert this to cm.	2500 km = 250 000 000 cm

3. Construct equivalent fractions.	$1 : 40\,000\,000$ $x : 250\,000\,000$ $\frac{x}{250\,000\,000} = \frac{1}{40\,000\,000}$ $x = \frac{250\,000\,000}{40\,000\,000}$ $= 6.25$
4. Write the answer.	The distance on the map is 6.25 cm if the actual distance is 2500 km.

Exercise 2.5 Scales, diagrams and maps [complex]

learnon

2.5 Quick quiz on | 2.5 Exercise

Simple familiar	Complex familiar	Complex unfamiliar
N/A	1, 2, 3, 4, 5, 6, 7, 8, 9, 10, 11, 12, 13, 14, 15, 16, 17, 18, 19	20, 21, 22

These questions are even better in jacPLUS!
- Receive immediate feedback
- Access sample responses
- Track results and progress

Find all this and MORE in jacPLUS

Complex familiar

1. WE18 Calculate the scale of a drawing where 100 mm on the diagram represents 2 m in reality.
2. If the scale of a diagram is 5 cm : 100 km, write the scale using the same unit.
3. Calculate the scale of a drawing where:
   a. 4 cm on the diagram represents 5 km in reality
   b. 20 mm on the diagram represents 100 m in reality
   c. 5 cm on the diagram represents 10 km in reality
   d. 3 cm on the diagram represents 600 m in reality.
4. WE19 Estimate the actual diameter of the tabletop shown if the scale of the diagram is 1 : 50 and the diameter of the table in the diagram is 4 cm.

5. If the scale of the diagram shown is 1 : 50 and the dimensions in the diagram are length = 3.6 cm and width = 1.8 cm, calculate the actual dimensions, in metres, of the Aboriginal flag design shown.

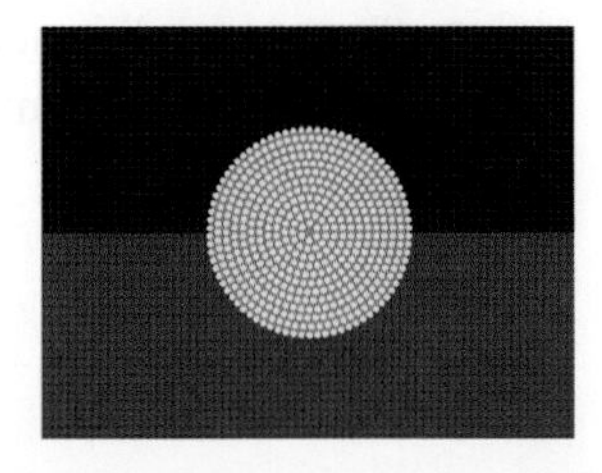

6. Calculate the actual dimensions of the nut and bolt shown if the scale of the drawing is 1 : 2 and the dimensions on the diagram are as follows:
   - Diameter of the nut is 5 mm.
   - Diameter of the bolt head is 7 mm.
   - Nut height is 10 mm.
   - Bolt height is 14 mm.

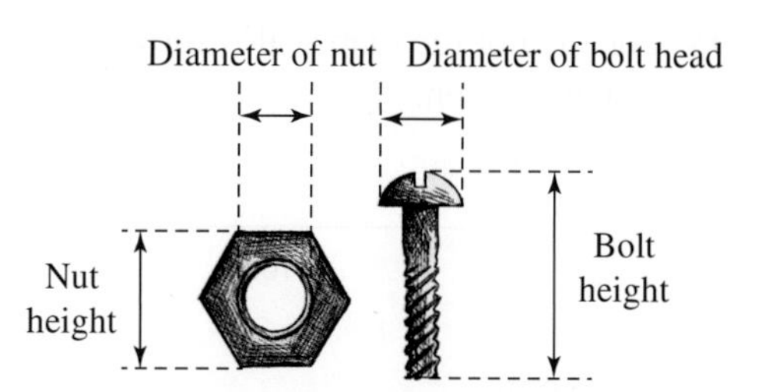

7. **WE20** The floor plan of a kitchen is drawn to a scale of 1 : 100. Determine the width of the kitchen on the plan if its real width is 5 m.

8. The real height of a building is 147 m. Calculate the height of the building on a diagram drawn at a scale of 1 : 3000.

9. The floor plan shown is planned to be drawn at a scale of 1 : 200. The actual dimensions of the house are shown on the diagram.

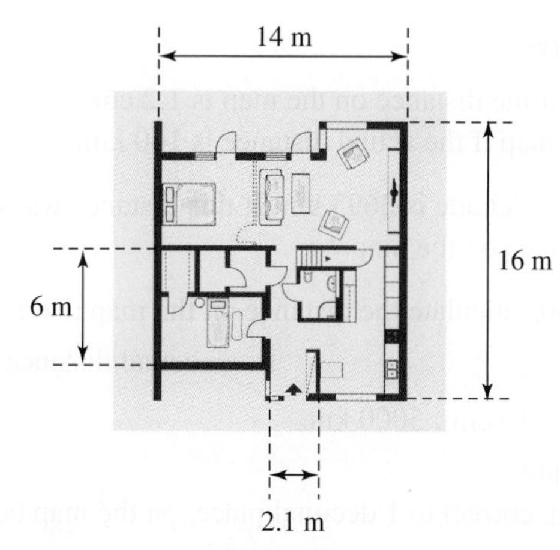

The floor plan is not yet drawn to scale. Calculate the lengths of the dimensions shown if the floor plan was drawn to scale.

10. Calculate the actual lengths for the following lengths measured on a diagram with the given scale.
    a. 12 mm; scale 1 : 1000
    b. 3.8 cm; scale 1 : 150
    c. 27 mm; scale 1 : 200
    d. 11.6 cm; scale 1 : 7000

11. **WE21** Determine the scale of the map as a ratio using the information in the diagram.

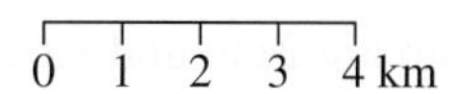

12. Determine the scale of the map as a ratio using the information in the diagram.

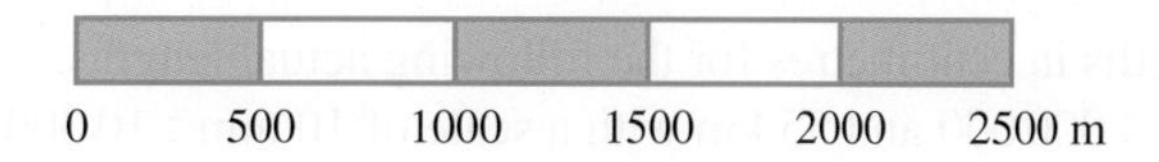

13. Determine the scale of the map shown.

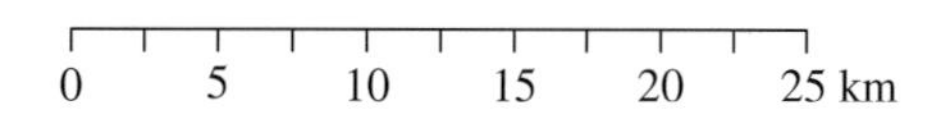

14. For each of the map scales shown, identify the scale and determine the actual distance for the map distance given.

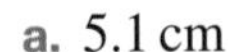

a. 5.1 cm

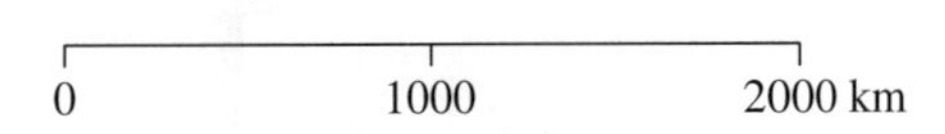

b. 27 mm

c. 38 mm

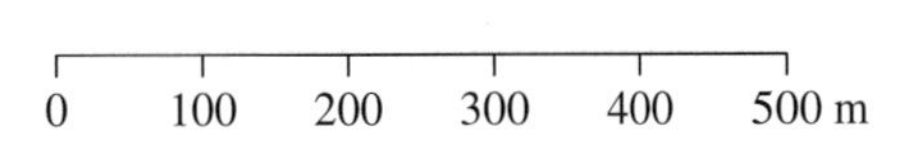

d. 9.6 cm

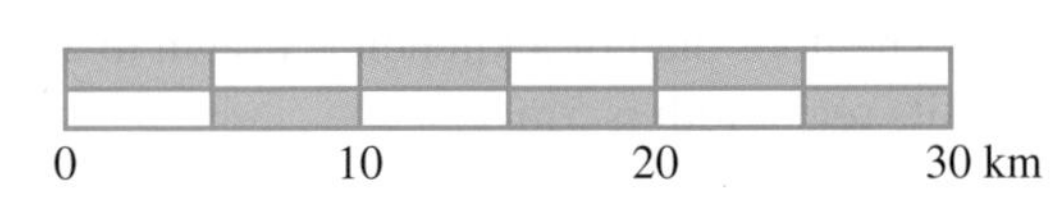

15. WE22 A street map has a scale of 1 : 500 000.

a. Calculate the actual distance if the distance on the map is 2 cm.
b. Calculate the distance on the map if the actual distance is 10 km.

16. A map has a scale of 1 : 1 000 000.

a. Calculate the actual distance if the distance on the map is 1.2 cm.
b. Calculate the distance on the map if the actual distance is 160 km.

17. The distance between Perth and Adelaide is 2693 km. If this distance was drawn on a scale of 1 : 25 000 000, calculate the distance on the map.

18. If the scale of a map is 1 : 20 000, calculate the distance on the map that represents:

a. an actual distance of 1 km
b. an actual distance of 15 km.

19. A map of Australia has a scale of 10 cm : 5000 km.

a. Write this scale in the same unit.
b. Calculate the distances, in cm, correct to 1 decimal place, on the map between:

i. Canberra and Sydney with an actual distance of 290 km
ii. Sydney and Brisbane with an actual distance of 925 km
iii. Brisbane and Darwin with an actual distance of 3423 km
iv. Darwin and Perth with an actual distance of 4042 km
v. Perth and Adelaide with an actual distance of 2693 km
vi. Adelaide and Melbourne with an actual distance of 727 km
vii. Melbourne and Canberra with an actual distance of 663 km.

Complex unfamiliar

20. A street map is drawn at a scale of 1 : 150 000. Calculate the difference in kilometres from the actual distances if the following lengths on the map are 2 cm and 1.6 cm.

21. Compare the actual dimensions in metres of the following dimensions measured on a diagram. 9.7 cm with a scale of 1 : 30 000 and 58 mm with a scale of 1 : 50 000

22. Compare the diagram lengths in centimetres for the following actual lengths. 260 km with a scale of 1 : 4 000 000 and 55 km with a scale of 10 mm : 10 000 m

Fully worked solutions for this chapter are available online.

LESSON
2.6 Review

2.6.1 Summary

doc-41563

Hey students! Now that it's time to revise this chapter, go online to:

Access the chapter summary | Review your results | Watch teacher-led videos | Practise questions with immediate feedback

Find all this and MORE in jacPLUS

2.6 Exercise

learnon

2.6 Exercise

Simple familiar	Complex familiar	Complex unfamiliar
1, 2, 3, 4, 5, 6, 7, 8, 9, 10, 11, 12, 13, 14, 15, 16	17, 18	19, 20

These questions are even better in jacPLUS!
- Receive immediate feedback
- Access sample responses
- Track results and progress

Find all this and MORE in jacPLUS

Simple familiar

1. Identify the ratio of blue pawns to red pawns, in simplest form, in the picture shown.

2. Express the ratio 385 : 154 in simplest fraction form.
3. Express the ratio of 364 m^2 to 455 m^2 to 819 m^2 in simplest form.
4. Determine the ratio between the volume of the sphere and the volume of the cube in the diagram shown, in simplest fractional form.

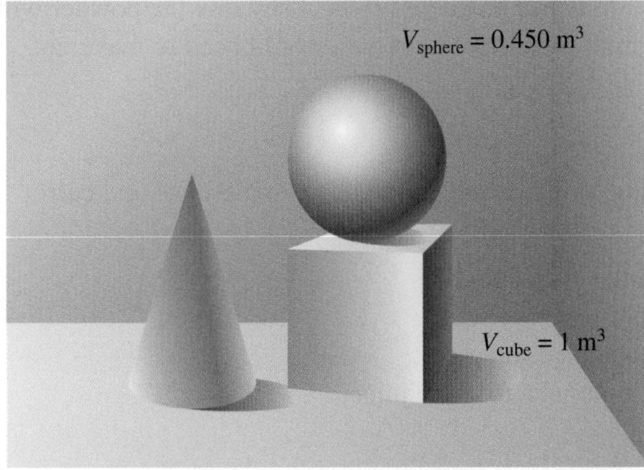

5. If an area of 2730 cm^2 is divided into the ratio 5 : 3 : 7, calculate the corresponding areas.
6. If a road map is drawn to a scale of 1 : 900 000, determine the road length represented by 3.5 cm on the map.

7. A netball court is 30.5 m long and 15.25 m wide. If the dimensions of the court in the diagram shown are 6.1 cm and 3.05 cm, determine the scale factor of the diagram.

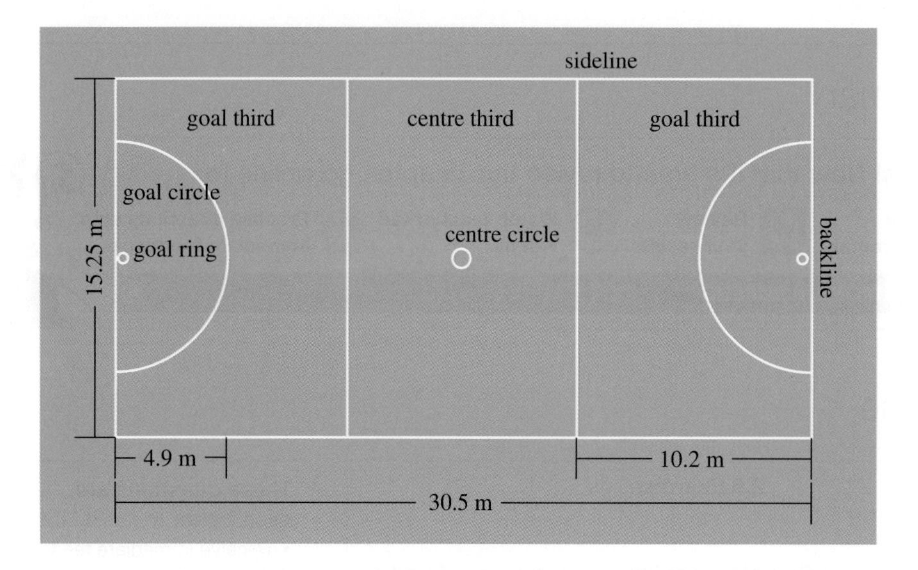

8. Express the ratio $4\frac{1}{3} : 1\frac{2}{3}$ in simplest form.

9. Three areas add up to 56 000 m^2. If they are in a ratio of 5 : 3 : 8, calculate the smallest area.

10. The ratio of length to width to height of a storage container is 3 : 1 : 5.
If the sum of the three dimensions is 6.75 m, calculate the length of the container.

11. Two bus stops, A and B, are 45 km apart. Two new bus stops, C and D, are going to be placed in a ratio 2 : 3 : 4 between A and B. Determine how far apart bus stops C and D are.

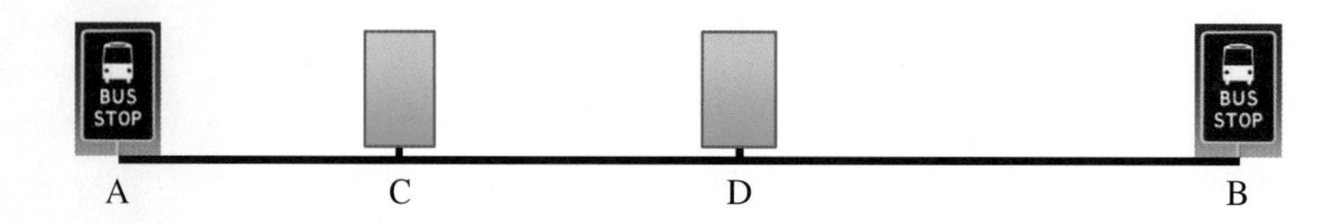

12. While camping, the Blake family use powdered milk. They mix the powder with water in the ratio of 1 : 24. Determine how much of each ingredient they would need to make up:

a. 600 mL
b. 1.2 L.

13. Yoke made 3 L of a refreshing drink using orange juice, apple juice and carrot juice in the ratio 5 : 8 : 2.

a. Calculate how much orange juice she used.
b. Calculate how much apple juice she used.
c. Calculate how much carrot juice she used.
d. Determine the ratio of carrot juice to apple juice.

14. A cake mixture has 250 g of flour to 150 g of water. Calculate the simplified ratio of flour to water.

15. A painter mixes paints to make different shades of colours. They mix 36 mL of blue paint with 54 mL of white paint. Calculate the ratio of blue to white paint in simplest fraction form.

16. Express the following ratios in simplest fraction form.
   a. $945 \text{ m}^2 : 1365 \text{ m}^2$
   b. $1386 \text{m} : 2.079 \text{ km}$
   c. 10 h 12 min : 61 h 12 min
   d. $2.592 \text{ m}^3 : 3.240 \text{ m}^3$
   e. 13.68 L : 12.24 L

Complex familiar

17. Three swimming pools have the dimensions shown in the table.

Pool	Length (m)	Width (m)	Depth (m)
Large	12.6	3.50	1.20
Medium	8.4	3.50	1.20
Small	4.2	3.50	1.20

   a. Determine the ratio between the lengths of the three pools.
   b. Calculate the volumes of the three swimming pools.
   c. Calculate the ratio between the volumes of the three swimming pools.
   d. Convert the volume measurements (m^3) into capacity measurements (L).
   e. If water is poured into the swimming pools at a rate of 5L/min, calculate how long it would take, to the nearest hour, to fill in the three swimming pools.

18. For a main course at a local restaurant, guests can select from a chicken, fish or vegetarian dish. On Friday night the kitchen served 72 chicken plates, 56 fish plates and 48 vegetarian plates.
   a. Express the number of dishes served as a ratio in the simplest form.
   b. On a Saturday night, the restaurant can cater for 250 people. If the restaurant was full, determine how many people would be expected to order a non-vegetarian dish.
   c. The Elmir family of five and the Cann family of three dine together. The total bill for the table was $268.
      i. Calculate the cost of dinner per head.
      ii. If the bill is split according to family size, calculate the proportion of the bill that the Elmir family will pay.

Complex unfamiliar

19. A recipe for shortbread cookies requires 500 g flour, 250 g unsalted butter, 125 g of sugar, 150 g walnuts and 200 g chocolate chips.
Calculate the total cost for the original recipe, correct to 2 decimal places, if 1 kg flour cost $1.50, 250 g unsalted butter $2.10, 1 kg sugar $1.25, 1 kg walnuts $9.50 and 200 g chocolate chips $3.25.

20. The floor plan shown is drawn to a scale of 1 : 200. The dimensions written represent the actual dimensions in millimetres (i.e. 4000 = 4 m).

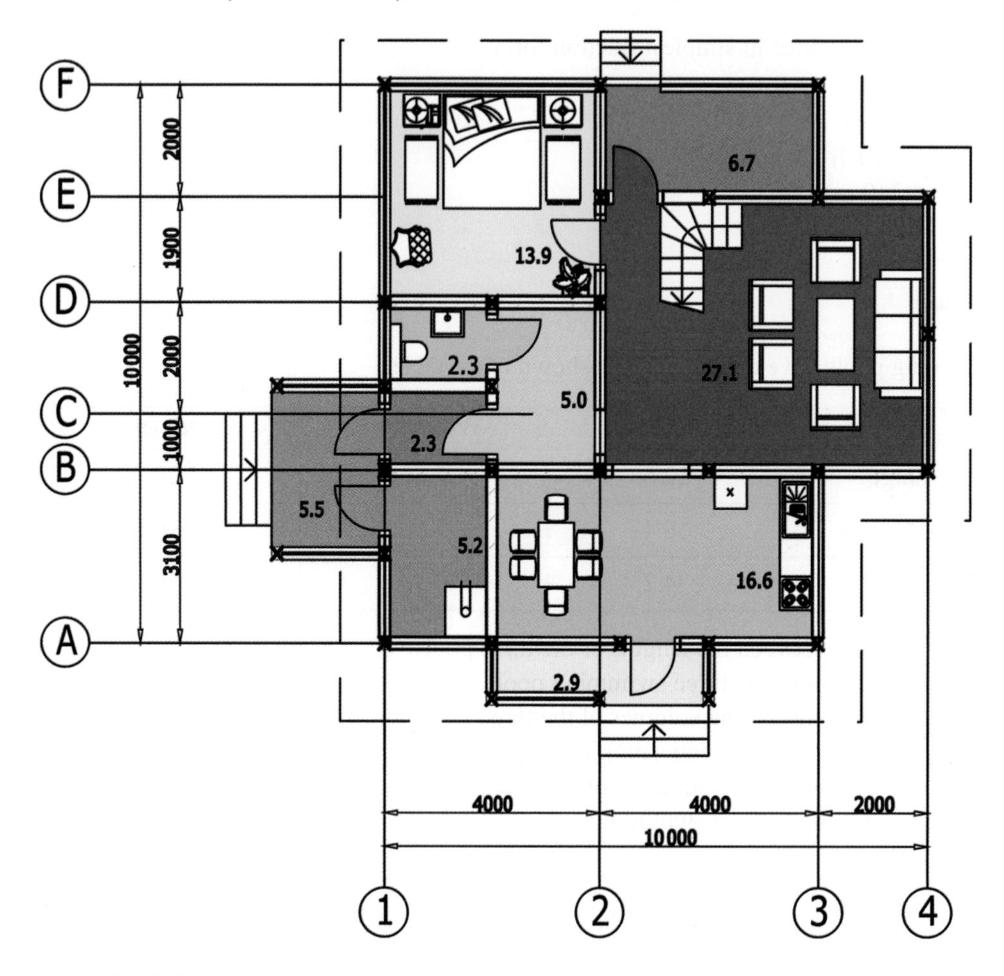

Calculate the ratio of the area of the bedroom to the area of the lounge room.

Fully worked solutions for this chapter are available online.

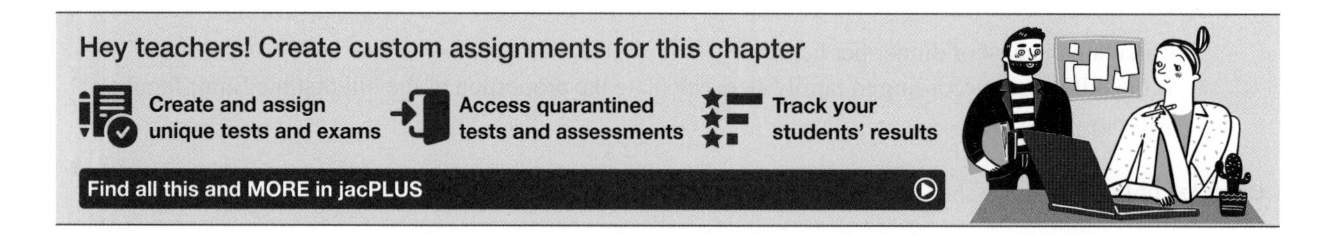

Answers

Chapter 2 Ratios

2.2 Introduction to ratios

2.2 Exercise

1. 5 : 4
2. a. 8 : 5 b. 5 : 9 c. 2 : 1
3. a. $\frac{3}{5}$ b. $\frac{1}{7}$ c. $\frac{6}{11}$ d. $\frac{5}{2}$
4. a. 4 : 7 b. 7 : 4 c. 1 : 2 d. 1 : 1
5. a. 1 : 3 b. 5 : 4 c. 4 : 3 d. 5 : 3
6. a. $\frac{2}{3}$ b. $\frac{9}{8}$ c. $\frac{15}{49}$ d. $\frac{20}{17}$
7. a. $a = 11$ b. $b = 21$
8. a. $m = 35, n = 40$ b. $a = 45, b = 6$
9. a. $x = 28$ b. $a = 3, b = 21$
   c. $b = 10$ d. $m = 2$
10. a. Equivalent b. Equivalent
    c. Equivalent d. Not equivalent
    e. Equivalent
11. a. Not equivalent b. Not equivalent
    c. Equivalent d. Not equivalent
    e. Not equivalent
12. Orange balls : black and white balls $= \frac{1}{2}$
13. $\frac{4}{3}$
14. a. $\frac{10}{3}$ b. 5 : 1 c. $\frac{3}{200}$
    d. 1 : 2 e. 3 : 2 : 5 f. 7 : 4 : 8
15. a. 5 : 3 b. 6 : 7 c. 1 : 10
    d. 45 : 8 e. 20 : 3 f. 1 : 12
16. a. 1 : 4 b. 30 : 1 c. 5 : 2
    d. 20 : 50 : 7
17. a. 12 boys b. 3 : 4
18. a. 11 : 5 b. 3 : 1 : 2
19. a. $\frac{3}{11}$ b. $\frac{4}{5}$ c. $\frac{10}{7}$
    d. $\frac{8}{9}$ e. $\frac{1}{10}$ f. $\frac{91}{22}$
20. a. Some equivalent fractions could be:
    $\frac{6}{10}, \frac{9}{15}, \frac{12}{20}, \frac{15}{25}, \frac{18}{30}, \frac{27}{45}, \frac{30}{50}, \frac{33}{55}, \frac{300}{500}, \frac{3000}{5000}$
    b. Some equivalent fractions could be:
    $\frac{540}{420}, \frac{270}{210}, \frac{180}{140}, \frac{135}{105}, \frac{108}{84}, \frac{90}{70}, \frac{54}{42}, \frac{45}{35}, \frac{18}{14}, \frac{9}{7}$

2.3 Ratios of two quantities

2.3 Exercise

1. 3 : 1
2. 2 : 10 : 5
3. a. Flour : lard = 15 : 2
   b. Butter : lard = 7 : 1
   c. Flour : butter = 15 : 14
   d. Flour : lard : butter = 15 : 2 : 14
4. a. 4 : 11 b. 5 : 2
   c. 4 : 10 : 11 d. 4 : 25
5. 1 : 14
6. 50 : 25 : 1
7. 9 : 8
8. a. 8 m^2
   b. $\text{Area}_{\Delta ABC} : \text{Area}_{\Delta DEF} = 1 : 5$
9. $\text{Area}_{\text{water}} : \text{Area}_{\text{total}} = 3611 : 5101$
10. a. 4 : 3 b. 1 : 8 c. 1 : 2 d. 12 : 1
11. a. Correct b. Incorrect
    c. Correct d. Correct
    e. Correct
12. $\text{Capacity}_{\text{carafe}} : \text{Capacity}_{\text{glass}} = 6 : 1$
13. $V_{\text{water}} : V_{\text{oil}} = 5 : 4$
14. $\frac{5}{3}$
15. a. $\frac{7}{2}$ b. $\frac{14}{15}$ c. $\frac{1}{3}$ d. $\frac{4}{5}$
16. Time by bus : time by train = 4 : 3
17. Sleeping time : awake time = 17 : 31
18. $\frac{17}{11}$
19. a. 3 : 4 b. 8 : 7
    c. 7 : 6 d. 6 : 7 : 8
20. a. 7 : 9 : 3 b. 7 : 9
    c. 3 : 1 d. 7 : 3
21. a. 5 : 1 b. 3 : 2
    c. 2 : 15 d. 15 : 3 : 2
22. a. 1 : 40 b. 5 : 2
23. a. $\frac{4}{5}$ b. $\frac{7}{3}$ c. $\frac{9}{16}$
    d. $\frac{2}{11}$ e. $\frac{8}{3}$
24. a. 38 : 14, 57 : 21, 76 : 28, 95 : 35, 114 : 42, 133 : 49, 152 : 56, 171 : 63, 190 : 70
    Other answers are possible.
    b. 3 : 2, 6 : 4, 12 : 8, 72 : 48, 720 : 480, 1872 : 1248, 5040 : 3360, 18 018 : 12 012, 32 760 : 21 840, 55 440 : 36 960
    Other answers are possible.

2.4 Divide into ratios [complex]

2.4 Exercise

1. 12 tulips and 18 tulips
2. 125 mandarins, 75 mandarins and 50 mandarins
3. a. 990 children and 330 adults
   b. 1100 children and 220 adults
   c. 792 children and 528 adults
   d. 840 children and 480 adults
4. a. 2.08 kg b. 3.12 kg

5. a. 576 kg b. 216 kg
6. 300 g tomatoes, 200 g capsicum
7. 12.75 kg sand, 4.25 kg cement
8. 1.8 m and 2.7 m
9. 200 mm, 400 mm, 800 mm
10. 15.375 cm
11. 2.25 m
12. $0.25\text{ m}^2, 0.5\text{ m}^2, 0.75\text{ m}^2$
13. 211.25 m^2
14. $4\,000\,000\text{ m}^2$
15. $230\,000\text{ km}^2$
16. Green jar volume $= 230\text{ cm}^3$; red jar volume $= 276\text{ cm}^3$
17. 38 L, 57 L, 76 L
18. 5000 L, 12 500 L, 25 000 L
19. 15 minutes
20. 60 minutes
21. a. 5 hours b. 1 hour c. 30 minutes
22. \$875, \$262.50, \$87.50
23. \$190, \$475, \$855
24. a. \$294, \$126
b. 30 minutes, 2.5 hours, 3 hours
c. 8315 m, 3326 m, 6652 m
d. $10\,000\text{ m}^2, 3000\text{ m}^2, 2000\text{ m}^2$
e. 336 L, 624 L
f. 2736, 912, 4104
25. a. \$1680, \$3360 b. \$2016, \$3024
c. \$840, \$4200 d. \$1120, \$3920
e. \$2160, \$2880 f. \$1120, \$1680, \$2240
26. Company A spends \$630 000 more on wages than Company B.
27. $x = 2$
28. Jane saves a larger fraction of her money compared to John.

2.5 Scales, diagrams and maps [complex]

2.5 Exercise

1. 1 : 20
2. 1 : 2 000 000
3. a. 1 : 125 000 b. 1 : 5000 c. 1 : 200 000
d. 1 : 20 000
4. 200 cm
5. Length = 1.8 m, width = 0.9 m
6. Nut height = 2 cm, diameter of the nut = 1 cm, bolt height = 2.8 cm, diameter of bolt head = 1.4 cm
7. 5 cm
8. 4.9 cm
9. 7 cm, 8 cm, 3 cm, 1.05 cm
10. a. 12 m b. 5.7 m c. 5.4 m d. 812 m
11. 1 : 200 000
12. 1 : 50 000
13. 1 : 500 000
14. a. Scale 5 : 200 000 000, 2040 km
b. Scale 1 : 500 000, 13 500 m
c. Scale 1 : 10 000, 380 m
d. Scale 1 : 500 000, 48 km
15. a. 10 km b. 2 cm
16. a. 12 km b. 16 cm
17. 10.772 cm
18. a. 5 cm b. 75 cm
19. a. 1 : 50 000 000
b. i. 0.6 cm ii. 1.9 cm iii. 6.8 cm iv. 8.1 cm
v. 5.4 cm vi. 1.5 cm vii. 1.3 cm
20. 0.6 km
21. 2910 m compared to 2900 m
22. 6.5 cm compared to 5.5 cm

2.6 Review

2.6 Exercise

1. 2 : 3
2. $\frac{5}{2}$
3. 4 : 5 : 9
4. $\frac{9}{20}$
5. $910\text{ cm}^2 : 546\text{ cm}^2 : 1274\text{ cm}^2$
6. 31.5 km
7. $\frac{1}{500}$
8. 13 : 5
9. $10\,500\text{ m}^2$
10. 2.25 m
11. 15 km
12. a. 24 mL of powder and 576 mL of water
b. 48 mL of powder and 1152 mL of water
13. a. 1 L b. 1.6 L c. 400 mL d. 1 : 4
14. 5 : 3
15. $\frac{2}{3}$
16. a. $\frac{9}{13}$ b. $\frac{2}{3}$ c. $\frac{1}{6}$
d. $\frac{4}{5}$ e. $\frac{19}{17}$
17. a. 1 : 2 : 3
b. $52.92\text{ m}^3, 35.28\text{ m}^3$ and 17.64 m^3
c. 1 : 2 : 3
d. 52 920 L, 35 280 L and 17 640 L
e. 7 days 8 hours, 4 days 22 hours and 2 days 11 hours
18. a. 9 : 7 : 6
b. 182
c. i. \$33.50
ii. 62.5%
19. \$7.68
20. 26 : 49

3 Rates

LESSON SEQUENCE

Fully worked solutions for this chapter are available online.

on Resources

 **Solutions** Solutions — Chapter 3 (sol-1256)

 **Digital documents** Learning matrix — Chapter 3 (doc-41698)
Quick quizzes — Chapter 3 (doc-41699)
Chapter summary — Chapter 3 (doc-41700)

LESSON
3.1 Overview

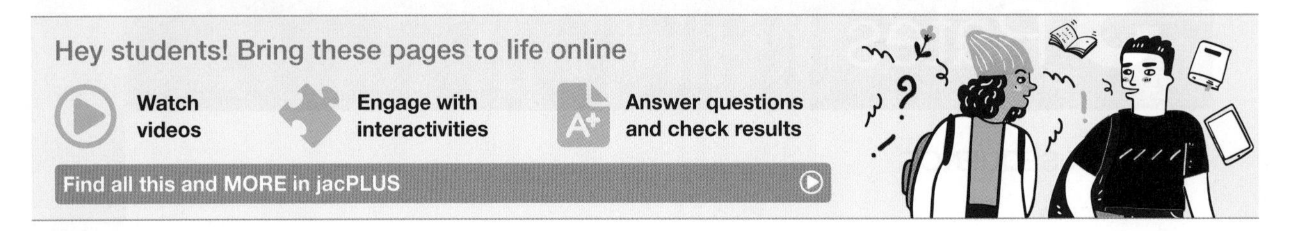

3.1.1 Introduction

Rates allow us to compare quantities expressed in different units. This is important to know when it comes to supermarket shopping, so you can ensure you are getting the best value for money. For example, is it cheaper to buy a large box of Corn Flakes (725 g for $4.50) or multiple packets (220 g for $1.95)? Using rates helps you determine which option is the best value. Other common rates are those where some quantity changes with respect to time.

For example, how far will a racing car travel on 60 litres of fuel at an average speed of 275 km/h? Or how long will it take to fill a 5000-litre water tank if water is pumped in at a rate of 15 litres per second? From these examples you can see that rates are used regularly in daily activities.

Additionally, rates are crucial in various fields such as finance, where interest rates determine the cost of borrowing money, and in medicine, where dosage rates are vital for patient safety. In engineering, rates of change are used to design systems that can handle varying loads. Understanding rates can also help in personal fitness, such as calculating the rate of calories burned during exercise. Overall, mastering the concept of rates can significantly enhance decision-making in everyday life and professional settings.

3.1.2 Syllabus links

Lesson	Lesson title	Syllabus links
3.2	**Identifying rates**	○ Identify examples of rates in the real world, e.g. rate of pay, cost per kilogram, currency exchange rates, flow rate from a tap.
3.3	**Conversion of rates**	○ Convert between units for rates.
3.4	**Calculation with rates involving direct proportion [complex]**	○ Complete calculations with rates, including solving problems involving direct proportion in terms of rate [complex].
3.5	**Comparison of rates**	○ Use rates to make comparisons.
3.6	**Rates and costs**	○ Use rates to determine costs.

Source: Essential Mathematics Senior Syllabus 2024 © State of Queensland (QCAA) 2024; licensed under CC BY 4.0.

LESSON
3.2 Identifying rates

SYLLABUS LINKS

- Identify examples of rates in the real world, e.g. rate of pay, cost per kilogram, currency exchange rates, flow rate from a tap.

3.2.1 Rates

- A **rate** is a fraction that compares the size of one quantity to that of another quantity.
- Rates have units, unlike ratios, which compare the same quantities measured in the same unit.
- A rate compares two different quantities measured in different units.
  Examples of rates are rate of pay, cost per kilogram, currency exchange rates or flow rate from a tap.
- The unit for rates contains the word *per*, which is written as /.
- For example, James works at a local cafe and his rate of pay per hour is \$20. This means that James is paid \$20 for every one hour he works.

$$\$20/\text{h} = \frac{\$20}{1 \text{ hour}}$$

Calculating rates

To calculate a rate, divide the first quantity by the second quantity.

$$\textbf{Rate} = \frac{\textbf{one quantity}}{\textbf{another quantity}}$$

$$\textbf{Exchange rate} = \frac{\textbf{money in foreign currency}}{\textbf{money in domestic currency}}$$

Common rates involving time

$$\textbf{Speed} = \frac{\textbf{distance}}{\textbf{time}}$$

$$\textbf{Time} = \frac{\textbf{distance}}{\textbf{speed}}$$

$$\textbf{Flow rate} = \frac{\textbf{litres}}{\textbf{minute}}$$

$$\textbf{Heart rate} = \frac{\textbf{number of heartbeats}}{\textbf{minute}}$$

tlvd-3773

WORKED EXAMPLE 1 Calculating the rate

A car travels 80 kilometres in 2 hours. Calculate the rate at which the car is travelling using units of:
a. kilometres per hour **b. metres per second (to 1 decimal place).**

THINK	WRITE
a. 1. Identify the units in the rate. The rate 'kilometres per hour' indicates that the number of kilometres should be divided by the number of hours taken.	a. $\text{Rate} = \dfrac{\text{distance (km)}}{\text{time (h)}}$
2. Write the rate using the correct units.	$= \dfrac{80\,\text{km}}{2\,\text{h}}$
3. Simplify the rate by dividing the numbers.	$= \dfrac{40\,\text{km}}{1\,\text{h}}$
4. Write the answer.	80 kilometres in 2 hours is equivalent to a rate of 40 km/h.
b. 1. Identify the units in the rate. The rate 'metres per second' indicates that the number of metres should be divided by the number of seconds taken.	b. $\text{Rate} = \dfrac{\text{distance (m)}}{\text{time (s)}}$
2. Convert the distance to metres by multiplying by 1000. Convert time to seconds by multiplying by 60 to change to minutes and 60 again to change to seconds.	$80\,\text{km} = 80 \times 1000$ $= 80\,000\,\text{m}$ $2\,\text{h} = 2 \times 60 \times 60$ $= 7200\,\text{s}$
3. Write the rate using the correct units.	$\text{Rate} = \dfrac{80\,000\,\text{m}}{7200\,\text{s}}$
4. Simplify the rate.	$= 11.1\,\text{m/s}$
5. Write the answer.	80 kilometres in 2 hours is equivalent to a rate of 11.1 m/s.

- When comparing the difference between two quantities, you need to subtract the smaller quantity from the larger quantity.

tlvd-3774

WORKED EXAMPLE 2 Calculating rates between two quantities

Using the growth chart shown, determine the rate at which the child has grown between the ages of 10 and 13. Give your answer in centimetres per year.

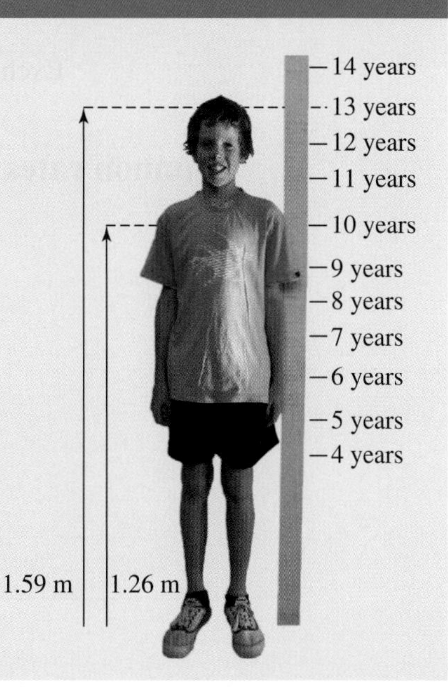

THINK	WRITE
1. Amount of growth = final height − initial height. To determine how much the boy has grown, calculate the difference in heights.	Amount of growth = 1.59 m − 1.26 m = 0.33 m
2. The final rate is in centimetres per year, so convert from metres to centimetres by multiplying by 100.	= 33 cm
3. He has grown 33 cm between the ages of 10 and 13 — i.e. in 3 years.	$\text{Rate of growth} = \frac{\text{amount of growth (cm)}}{\text{time (years)}}$ $= \frac{33}{3}$ = 11 cm/year
4. Write the answer.	The child has grown at a rate of 11 cm per year or 11 cm/year.

3.2.2 Common rates

- Some common rates that don't involve time include density, gradient and concentration.
- **Density** is a measure of how much mass the material has for a certain volume. This can be calculated by:

Formula for density

$$\textbf{Density} = \frac{\textbf{mass}}{\textbf{volume}}$$

WORKED EXAMPLE 3 Calculating common rates

A material has a mass of 4.5 kilograms and a volume of 0.75 cubic metres. Calculate the density of the material in grams per cubic metre.

THINK	WRITE
1. Density is the mass (g) divided by the volume (m^3).	$\text{Density} = \frac{\text{mass (g)}}{\text{volume } (\text{m}^3)}$
2. Convert the mass to grams by multiplying by 1000.	$4.5\text{ kg} = 4.5 \times 1000\text{ g}$ $= 4500\text{ g}$
3. Write the rate using the correct units.	$\text{Density} = \frac{4500\text{ g}}{0.75\text{ m}^3}$
4. Write the answer.	$= 6000\text{ g/m}^3$

- **Gradient** is a measure of steepness. This can be calculated by:

Formula for gradient

$$\textbf{Gradient} = \frac{\textbf{rise}}{\textbf{run}}$$

- **Concentration** is a measure of how much material is dissolved in a liquid. This can be calculated by:

Formula for concentration

$$\text{Concentration} = \frac{\text{mass of the solute dissolved}}{\text{total volume of the solution}}$$

Resources

Interactivities Gradient (int-3740)
Rates (int-3738)

Exercise 3.2 Identifying rates

learnon

3.2 Quick quiz | 3.2 Exercise

Simple familiar	Complex familiar	Complex unfamiliar
1, 2, 3, 4, 5, 6, 7, 8, 9, 10, 11, 12, 13, 14, 15, 16, 17, 18	N/A	N/A

Simple familiar

1. **WE1** A cyclist travels 120 km in 3 hours. Calculate the rate at which the cyclist is travelling using units of:
   a. kilometres per hour b. metres per second.
2. An athlete's heart beats 250 beats in 5 minutes. Calculate the athlete's heart rate in beats per minute.
3. Match the following rates with the fractions shown:
   concentration density gradient speed
   a. $\dfrac{\text{mass}}{\text{volume}}$ b. $\dfrac{\text{distance}}{\text{time}}$
   c. $\dfrac{\text{rise}}{\text{run}}$ d. $\dfrac{\text{mass of the solute dissolved}}{\text{total volume of the solution}}$

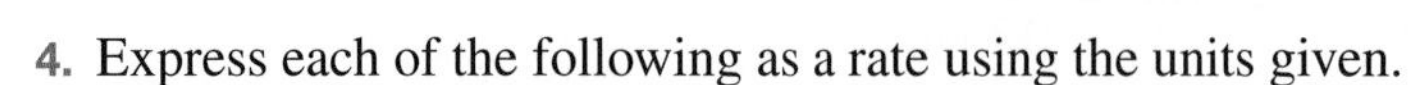

4. Express each of the following as a rate using the units given.
   a. A truck travelling 560 km in 7 hours (km/h)
   b. A 5.4-m length of timber costing \$21.06 (\$/m)
   c. A 9-litre bucket taking 45 seconds to fill (L/s)
   d. A hiker walking 9.6 km in 70 minutes (km/min)
5. A student walks 1.6 km to school in 22 minutes. Calculate the speed at which she walks to school in m/s.

6. At the 2000 Sydney Olympics, Cathy Freeman won gold in the 400-m race. Her time was 49.11 seconds.
Calculate her average speed in km/h.
7. **WE2** Noah has grown from 1.12 m at 5 years old to 1.41 m at 7 years old. Calculate Noah's growth rate in centimetres per year.
8. Calculate the growth rate in centimetres per year of a child who has grown 0.32 m in the past 4 years.
9. A child's height at 5 years of age is 85 cm. On reaching 12 years of age, the child's height is 128 cm. Calculate the average rate of growth (in cm/year) over the 7 years.
10. A school had 300 students in 2020 and 450 students in 2022. Calculate the average rate of growth in students per year.
11. **WE3** A material has a mass of 12.5 kilograms and a volume of 0.25 cubic metres. Calculate the density of the material in grams per cubic metres.

12. A 2-litre drink container was filled with a mix of 200 millilitres of cordial and 1.8 litres of water. Calculate the concentration of the cordial mix in millilitres per litre.
13. Three sugar cubes, each with a mass of 1 gram, are placed into a 300-millilitre mug of coffee. Calculate the concentration of sugar in grams per millilitre (g/mL).

14. Uluru is in the middle of Australia and is 348 m high and 3.6 km long. Calculate the average gradient (to 1 decimal place) from the bottom to the top of Uluru, assuming the top is in the middle.
15. A kettle is filled with tap water at 16 °C and then turned on. It boils (at 100 °C) after 2.75 minutes. Calculate the average rate of increase in temperature in °C/s.

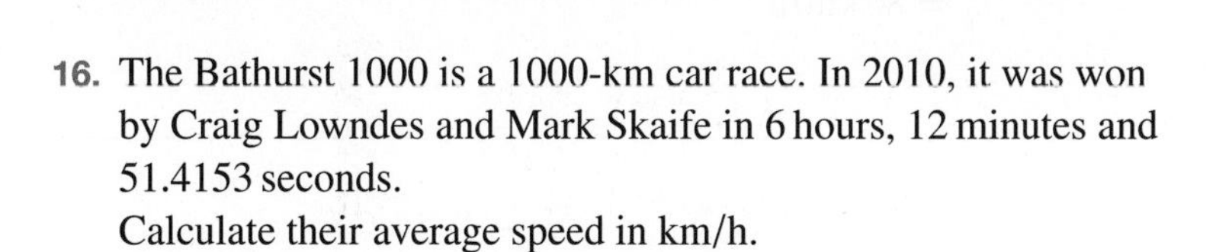

16. The Bathurst 1000 is a 1000-km car race. In 2010, it was won by Craig Lowndes and Mark Skaife in 6 hours, 12 minutes and 51.4153 seconds.
Calculate their average speed in km/h.
17. Simon weighs 88 kg and has a volume of 0.11 m^3, whereas Ollie weighs 82 kg and has a volume of 0.08 m^3.
Determine who has greater muscle density and by how much.
18. Beaches are sometimes unfit for swimming if heavy rain washes pollution into the water. A beach is declared unsafe for swimming if the concentration of bacteria is more than 5000 organisms per litre. A sample of 20 millilitres was tested and found to contain 55 organisms.
Calculate the concentration in the sample (in organisms/litre) and decide whether or not the beach should be closed.

Fully worked solutions for this chapter are available online.

LESSON
3.3 Conversion of rates

SYLLABUS LINKS

- Convert between units for rates.

3.3.1 Expressing a rate in simplest form

- A rate is a measure of the change in one quantity with respect to another.

Rate of change

$$\textbf{Rate of change of Y with respect to X} = \frac{\textbf{change in Y}}{\textbf{change in X}}$$

WORKED EXAMPLE 4 Expressing a rate in simplest from

A car travels 320 km in 4 hours. Express this rate in km/h.

THINK	WRITE
1. Write out the rate formula in terms of the two units.	$\text{Rate} = \dfrac{\text{change in Y}}{\text{change in X}}$ $= \dfrac{\text{change in km}}{\text{change in hours}}$
2. Substitute the known quantities into the formula.	$\text{Rate} = \dfrac{320\,\text{km}}{4\,\text{h}}$
3. Simplify the rate.	$= 80\,\text{km/h}$

3.3.2 Equivalent rates

- Rates can be converted from one unit to another; for example, speed in km/h can be expressed in m/s.
- To change the unit of a rate, follow these steps:
  1. Convert the numerator of the rate to the new unit.
  2. Convert the denominator of the rate to the new unit.
  3. Divide the new numerator by the new denominator.

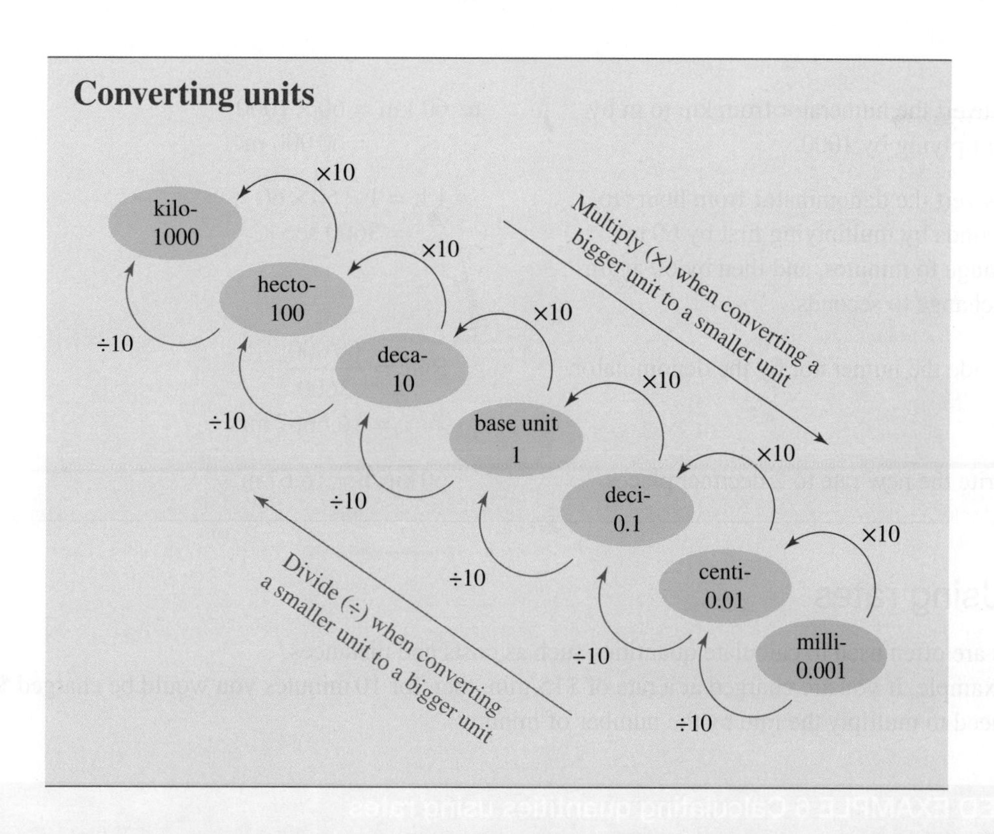

tlvd-3776

WORKED EXAMPLE 5 Solving conversion of rates

Convert the following rates, as shown, to 2 decimal places.

a. 3 km/h to m/h **b. 25 L/h to mL/min** **c. 60 km/h to m/s**

THINK	WRITE
a. 1. Convert km to m by multiplying by 1000.	**a.** $3\text{ km} \times 1000 = 3000\text{ m}$
2. Divide the numerator by the denominator.	$\text{Rate} = \dfrac{3000}{1}$
3. Write the new rate.	$3\text{ km/h} = 3000\text{ m/h}$
b. 1. Convert the numerator from L to mL by multiplying by 1000.	**b.** $25\text{ L} = 25 \times 1000$ $= 25\,000\text{ mL}$
2. Convert the denominator from hours to minutes by multiplying by 60.	$1\text{ h} = 1 \times 60$ $= 60\text{ min}$
3. Divide the numerator by the denominator.	$\text{Rate} = \dfrac{25\,000}{60}$ $= 416.6667\text{ mL/min}$
4. Write the new rate to 2 decimal places.	$25\text{ L/h} = 416.67\text{ mL/min}$

c. 1. Convert the numerator from km to m by multiplying by 1000.

c. $60 \text{ km} = 60 \times 1000$
$= 60\,000 \text{ m}$

2. Convert the denominator from hours to seconds by multiplying first by 60 to change to minutes, and then by 60 again to change to seconds.

$1 \text{ h} = 1 \times 60 \times 60$
$= 3600 \text{ sec}$

3. Divide the numerator by the denominator.

$$\text{Rate} = \frac{60\,000}{3600}$$
$$= 16.6667 \text{ m/s}$$

4. Write the new rate to 2 decimal places.

$60 \text{ km/h} = 16.67 \text{ m/s}$

3.3.3 Using rates

- Rates are often used to calculate quantities such as costs and distances.
  For example, if you are charged at a rate of \$15/min, then for 10 minutes you would be charged \$150, since you need to multiply the rate by the number of minutes.

tlvd-3777

WORKED EXAMPLE 6 Calculating quantities using rates

An athlete has an average heartbeat of 54 beats/minute. Calculate how many times on average their heart will beat in a week.

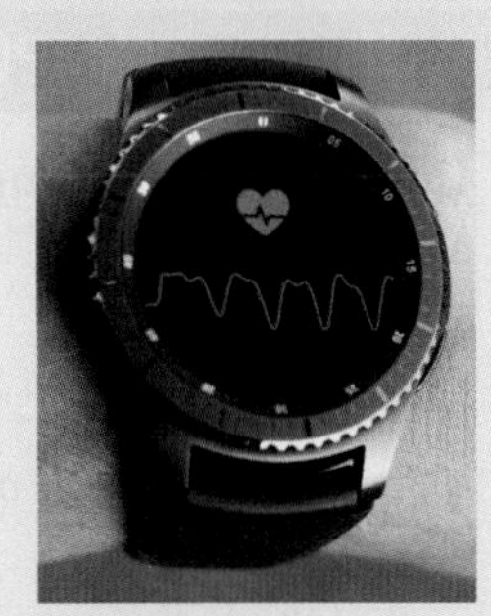

THINK

WRITE

1. Write the rate in terms of the quantities: heartbeats and time.

$$\text{Rate} = \frac{54 \text{ beats}}{1 \text{ minute}}$$

2. Calculate the number of minutes in a week.

Number of minutes in a week $= 60 \times 24 \times 7$
$= 10\,080$

3. Multiply 54 by the number of minutes in a week.

Number of beats in a week $= 54 \times 10\,080$
$= 544\,320$

4. Write your answer.

On average, the athlete's heart beats 544 320 beats/week.

Exercise 3.3 Conversion of rates

learnon

3.3 Quick quiz | 3.3 Exercise

Simple familiar	Complex familiar	Complex unfamiliar
1, 2, 3, 4, 5, 6, 7, 8, 9, 10, 11, 12, 13, 14, 15	N/A	N/A

Simple familiar

1. WE4 A cyclist travels 102 km in 3 hours. Express this rate in km/h.
2. A person purchases a 2.5-kg bag of chocolates for $37.50. Express this rate in $/kg.

3. Given the following information, express the rate in the units shown in the brackets.
   a. A car travels 240 km in 4 hours (km/h).
   b. A hiker covers 1400 m in 7 minutes (m/min).
   c. 9 litres of water run out of a tap in 30 seconds (L/s).
   d. Grass has grown 4 mm in 2 days (mm/day).
4. Given the following information, express the rate in the units shown in the brackets.
   a. A runner travels 800 m in 4 minutes (km/min).
   b. Water runs into a water tank at 800 mL in 2 minutes (L/min).
   c. A phone call costs 64 cents for 2 minutes ($/min).
   d. Apples cost $13.50 for 3 kg (cents/gram).
5. WE5 Convert the following rates as shown to 2 decimal places.
   a. 100 km/h to m/s
   b. 40 L/h to mL/min
6. Convert the following rates as shown to 2 decimal places.
   a. $5.50/m to cents/cm
   b. 10 m/s to km/h
7. Convert the following rates as shown.
   a. 70 km/h to m/h
   b. 40 kg/min to g/min
   c. $120/min to $/h
   d. 27 g/mL to g/L
8. Convert the following rates as shown, giving your answer to 2 decimal places.
   a. 80 km/h to m/s
   b. 25 m/s to km/h
   c. 50 g/ml to kg/L
   d. 360 mL/min to L/h
9. WE6 If a person has an average heartbeat of 72 beats/minute, calculate how many times their heart would beat on average in a week.
10. If a phone call to the USA costs 10c/min, calculate how much a 1.5-hour call to the USA would cost.
11. If an electricity company charges 26.5 cents/kWh, calculate how much they would charge you if you used 750 kWh.

12. An athlete has an average heart rate of 63 beats/minute. Calculate how many times their heart would beat on average in one year (assume 365 days in a year).
13. Christi wants to hire a jukebox for her party and has received a quote of \$45 per hour. If Christi has a party that goes from 8.30 pm to 1 am, calculate how much she will have to pay for the jukebox.

14. On one trip, a driver took an average of 45 minutes to travel 60 km. If the driver travelled from 11.30 am to 3.45 pm continuously, calculate how far they have travelled.
15. A gum tree grows at a rate of 80 cm/year. Calculate how long it will take to grow from the ground to a height of 10 m.

Fully worked solutions for this chapter are available online.

LESSON
3.4 Calculation with rates involving direct proportion [complex]

SYLLABUS LINKS

- Complete calculations with rates, including solving problems involving direct proportion in terms of rate [complex].

3.4.1 Direct proportion and rates

- In direct proportion, two variables change at the same rate.

Expressing direct proportion

If the y-values increase (or decrease) directly with the x-values, then this relationship can be expressed as:

$$y \propto x$$

where α is the symbol for proportionality — this means that y is directly proportional to x.

WORKED EXAMPLE 7 Testing for direct proportion

For each of the following pairs of variables, identify whether direct proportion exists.

a. The height of a stack of photocopy paper (h) and the number of sheets (n) in the stack.

b. Your Maths mark (m) and the number of hours of Maths homework you have completed (n).

THINK	WRITE
a. When n increases, so does h. When $n = 0, h = 0$. If graphed, the relationship would be linear.	a. $h \propto n$
b. As n increases, so does m. When $n = 0$, I may get a low mark but it is not necessarily zero, so $m \neq 0$.	b. m is not directly proportional to n.

- When the rate is constant, these rates represent a **directly proportional relationship**.
- This means that when one variable increases, the other increases **by the same factor**.
  - If potatoes are \$3 per kilogram, then one kilogram costs \$3, and 2 kilograms cost twice as much (\$3 × 2 = \$6. The cost is directly proportional to the weight of the potatoes.
  - When a car moves at a constant speed, the distance it travels is directly proportional to the time.
- A directly proportional relationship can be written as a fraction:

Formula for directly proportional relationships

$$\textbf{Rate} = \frac{a}{b}\textbf{, where the variables are } a \textbf{ and } b$$

Speed in km/h: rate $= \dfrac{\text{number of kilometres}}{\text{number of hours}}$ for any given time and distance

- So, if a car is travelling at 60 km/h, it will travel $\dfrac{60\text{ km}}{1\text{ h}}$ or $\dfrac{120\text{ km}}{2\text{ h}}$ or $\dfrac{180\text{ km}}{3\text{ h}}$, and these fractions are all equivalent.

$$\frac{60}{1} = \frac{120}{2} = \frac{180}{3} = 60\text{ km/h}$$

- Directly proportional relationships are essentially equivalent fractions:

$$\textbf{Rate} = \frac{a}{b} = \frac{a \times n}{b \times n}$$

- Direct proportion problems can be solved using cross-multiplication:
  - Recall that the fraction line means *divide*: $\dfrac{6}{2} = 6 \div 2 = 3$
  - You can see then that $\dfrac{6}{2} = 3$, and also $\dfrac{12}{4} = 3$, so: $\dfrac{6}{2} \times \dfrac{12}{4}$
  - Cross-multiplication means that: $6 \times 4 = 2 \times 12$

$$24 = 24$$

$$\textbf{If } \frac{a}{b} = \frac{c}{d} \textbf{ (multiplying both sides by } b \textbf{ and } d\textbf{)}$$

$$\textbf{then } ad = bc$$

WORKED EXAMPLE 8 Applying direct proportion to calculate the amount of an unknown quantity

A recipe says to use 2.1 cups of water to cook 1 cup of rice. Determine the amount of water you would need when cooking 3.5 cups of rice.

THINK	WRITE
1. Identify the equivalent fractions, putting 'water' on the top of the fractions since that is the part we are working out.	$\text{Rate} = \dfrac{\text{water}}{\text{rice}}$ $= \dfrac{w}{3.5}$ $= \dfrac{2.1}{1}$
2. Solve for w, the missing water value.	If $\dfrac{a}{b} = \dfrac{c}{d}$, then $ad = bc$. $\dfrac{w}{3.5} = \dfrac{2.1}{1}$ Therefore, $w \times 1 = 3.5 \times 2.1$ $w = 7.35$
3. Write the amount of water required.	7.35 cups of water will cook 3.5 cups of rice.

- From the formula rate $= \dfrac{a}{b}$, you can see that direct proportion can also be written as $a = rate \times b$, which is often written as:

Formula using the constant of proportionality (k)

$$y = kx$$

where the variables are x and y

and k is the 'constant of proportionality' (rate)

tlvd-11425

WORKED EXAMPLE 9 Applying the direct proportion in terms of rate

When there are 15 dogs staying at the kennels, they use 12 kg of pet mince each day. Determine the amount of dog food they will need per day if they have 20 dogs.

THINK	WRITE
1. We want to determine the amount of mince. Let mince be y and the number of dogs be x.	$y = kx$
2. $y = 12$ when $x = 15$	$12 = k \times 15$
3. Substitute into the formula and solve for k. 'k' stays the same for any number of dogs.	$k = 12 \div 15$ $= 0.8$
4. Determine the formula that compares mince with dogs.	$y = 0.8x$
5. When 20 dogs stay at the kennel, $x = 20$.	$y = 0.8 \times 20$ $= 16$ kg per day
6. Write the answer.	They will need 16 kg of pet mince per day.

Exercise 3.4 Calculation with rates involving direct proportion [complex]

3.4 Quick quiz	3.4 Exercise	
Simple familiar	Complex familiar	Complex unfamiliar
N/A	1, 2, 3, 4, 5, 6, 7, 8, 9, 10, 11, 12, 13, 14	15

Complex familiar

1. **WE7** For each of the following pairs of variables, determine whether direct proportion exists. If it does not exist, give a reason why.
   a. The distance (d) travelled in a car travelling at 60 km/h and the time taken (t)
   b. The speed of a swimmer (s) and the time the swimmer takes to complete one lap of the pool (t)
   c. The cost of a bus ticket (c) and the distance travelled (d)

2. For each of the following pairs of variables, determine whether direct proportion exists. If it does not exist, give a reason why.
   a. The perimeter (p) of a square and the side length (l)
   b. The area of a square (A) and the side length (l)
   c. The total cost (C) and the number of boxes of pencils purchased (n)
   d. The length of a line in centimetres (c) and the length in millimetres (m)
3. For each of the following pairs of variables, determine whether direct proportion exists. If it does not exist, give a reason why.
   a. The weight of an object in kilograms (k) and in pounds (p)
   b. The distance (d) travelled in a taxi and the cost (c)
   c. A person's height (h) and their age (a)
   d. The cost of an item in Australian dollars (A) and the cost of an item in Euros (E)
4. Determine whether the following statements are True or False.
   There is a direct relationship between:
   a. the total cost, $\$C$, of purchasing netballs and the number, n, purchased
   b. the circumference of a circle and its diameter
   c. the area of a semicircle and its radius.
5. **WE8** Petrol costs \$2.60 per litre. Jason needs 30 litres to fill his tank. Determine the cost of the fuel.
6. **WE9** Doughnuts cost \$13 for a packet of 6. Determine the cost of buying 30 doughnuts.
7. A plane that is travelling at a constant speed covers 597 km in 2 hours. Determine the time it will take to travel at that speed across the Pacific Ocean from Brisbane to Los Angeles, which is about 7000 km away.

8. On one trip, a driver took 45 minutes to travel 60 km. If he continues at the same speed, calculate the distance he travels from 11:30 am to 3:45 pm.

9. For many years, First Nations Australians managed the land using techniques including traditional burning, which created grassland that encouraged kangaroos to come, and also reduced the danger of intense bushfires. A fire started at 1 pm on a Friday, and burned 2750 m^2 in 5 minutes. If the fire continues burning at the same rate, and an area of 2 km^2 needs to be burnt, calculate the time and day that the fire needs to be put out.

10. A recipe that makes 12 cupcakes uses 2 cups of flour, 150 g of sugar, 120 mL of milk and 3 eggs. Nathan wants to make 20 cupcakes. Determine the appropriate measurements of flour, sugar, milk and eggs that he should use.

11. List five pairs of real-life variables that exhibit direct proportion.
12. If $y \propto x$, explain what happens to:
    a. y if x is doubled
    b. y if x is halved
    c. x if y is tripled.
13. Mobile phone calls are charged at 17 cents per 30 seconds.
    a. Determine whether direct proportion exists between the cost of a phone bill and the number of 30-second time periods. Justify your answer.
    b. If the duration of a call was 7.5 minutes, determine the cost of the call.

14. An electrician charges $55 for every 30 minutes or part of 30 minutes for his labour on a building site.
    a. Determine whether direct proportion exists between the cost of hiring the electrician and the time he spends on the building site. Justify your answer.
    b. If the electrician worked for 8.5 hours, determine how much this would cost.

Complex unfamiliar

15. Bruce is building a pergola and needs to buy treated pine timber. He wants 4.2-metre and 5.4-metre lengths of timber.
    If a 4.2-metre length costs $23.10 and a 5.4-metre length costs $29.16, determine if direct proportion exists between the cost of the timber and the length of the timber per metre.

LESSON
3.5 Comparison of rates

SYLLABUS LINKS

- Use rates to make comparisons.

3.5.1 Comparing rates

- Rates are commonly used to make comparisons.
  For example, useful comparisons for consumers could be comparing the price per 100 grams of different sizes of packaged cheese in the supermarket, or the fuel economy of different models of cars in litres per 100 kilometres.
- When you are comparing two rates, it is important to have both rates in the same unit.
- In Australia, larger supermarkets have to display rates for consumers, but consumers are required to make their own comparisons in smaller stores and market stalls.

Price per unit

$$\textbf{Price per unit} = \frac{\textbf{price}}{\textbf{number of units}}$$

Note: **Remember to represent the same units when you are making comparisons.**

tlvd-3778

WORKED EXAMPLE 10 Comparing rates for better value

Select which of the following products is the better value.
a. Brand A: \$2.15 per 100 g or Brand B: \$2.51 per 100 g
b. Brand X: \$2.64 per 250 g or Brand Y: \$5.25 per 500 g

THINK	WRITE
a. 1. Looking at the rate, you can see that they are both \$/100 g. Thus, a direct comparison can be made.	$\$2.15 < \$2.51$ Thus \$2.15 per 100 g < \$2.51 per 100 g.
2. Write the answer.	Brand A is the better value.
b. 1. To make a comparison, the same rate is needed. You can convert them to the same rate by making both weights 500 g.	$250\text{ g} \times 2 = 500\text{ g}$ $\$2.64 \times 2 = \$5.28$
2. Compare the same rates.	Brand X: \$5.28 per 500 g Brand Y: \$5.25 per 500 g $\$5.25 < \$5.28$
3. Write the answer.	Brand Y is the better value.

tlvd-11421

WORKED EXAMPLE 11 Determining the rate and comparing rates

Johan wants to purchase a new car and fuel economy is important when considering which car to buy. The first car he looked at used 34.56 litres for 320 kilometres and the second car used 29.97 litres for 270 kilometres.
Calculate the fuel consumption rate for each car in L/100 km and determine which car has the better fuel economy.

THINK	WRITE
1. The first car used 34.56 L/320 km.	First car rate = 34.56 L/320 km
2. Convert the rate to L/100 km by dividing the numerator and denominator by 3.2.	First car rate $= \dfrac{34.56 \div 3.2}{320 \div 3.2}$ $= \dfrac{10.8}{100}$ $= 10.8$ L/100 km
3. The second car used 29.97 L/270 km.	Second car rate = 29.97 L/270 km
4. Convert the rate to L/100 km by dividing the numerator and denominator by 2.7.	Second car rate $= \dfrac{29.97 \div 2.7}{270 \div 2.7}$ $= \dfrac{11.1}{100}$ $= 11.1$ L/100 km
5. Now that the rates have the same unit, a direct comparison can be made.	The first car has better fuel economy than the second car.

Exercise 3.5 Comparison of rates

learnon

3.5 Quick quiz on | 3.5 Exercise

Simple familiar	Complex familiar	Complex unfamiliar
1, 2, 3, 4, 5, 6, 7, 8, 9, 10, 11, 12, 13, 14	N/A	N/A

These questions are even better in jacPLUS!
- Receive immediate feedback
- Access sample responses
- Track results and progress

Find all this and MORE in jacPLUS

Simple familiar

1. WE10 Select which of the following products has the better value.
   a. A: \$7.25 per 50 g or B: \$7.52 per 50 g
   b. X: \$3.50 per 200 g or Y: \$7.10 per 400 g
2. Select which of the following products has the better value.
   a. Corn Flakes: \$4.50 per 250 g or \$4.05 per 250 g
   b. Corn Flakes: \$4.50 per 250 g or \$8.90 per 500 g
3. Select which of the following products is the best value.
   a. Chocolate Bullets: \$2.25 per 150 g or \$2.52 per 150 g
   b. Chocolate Bullets: \$2.50 per 150 g or \$7.35 per 450 g

4. **WE11** Two cars used fuel as shown.
   - Car A used 52.89 litres to travel 430 kilometres.
   - Car B used 40.2 litres to travel 335 kilometres.

   Determine which car has the better fuel economy by comparing the rates of L/100 km.

5. Car A uses 43 L of petrol travelling 525 km. Car B uses 37 L of petrol travelling 420 km. Show mathematically which car is more economical.

6. Josh wants to be an Ironman, so he eats Ironman food for breakfast. Nutri-Grain comes in a 290-g pack for $4.87 or an 805-g pack for $8.67.
   Compare the price per 100 g of cereal for each pack, and hence calculate how much he will save per 100 g if he purchases the 805-g pack.

7. Tea bags in a supermarket can be bought for $1.45 per pack of 10 or for $3.85 per pack of 25. Determine which of the two is the cheaper way of buying tea bags.

8. Coffee can be bought in 250-g jars for $9.50 or in 100-g jars for $4.10. Determine which is the cheaper way of buying the coffee and how much the saving would be.

9. A solution has a concentration of 45 g/L. The amount of solvent is doubled to dilute the solution. Calculate the concentration of the diluted solution.
   Explain your reasoning.

10. In cricket, a batter's strike rate indicates how fast the batter is scoring their runs by comparing the number of runs scored per 100 balls.
    Ellyse Perry scored 72 runs in 56 balls and Meg Lanning scored 42 runs in 31 balls.
    Correct to 2 decimal places:
    a. calculate the strike rate of Ellyse Perry
    b. calculate the strike rate of Meg Lanning
    c. determine who has the better strike rate and by how much.

11. Supermarkets use rates to make it easier for the consumer to compare prices of different brands and different-sized packages. To make the comparison, they give you the price of each product per 100 grams, known as unit pricing. Two different brands have the following pricing:
    - Brand A comes in a 2.5-kg pack for $3.10.
    - Brand B comes in a 1-kg pack for $1.05.

    Calculate the unit price (the price per 100 g) of each of the brands and determine which brand is better value.

12. The Stawell Gift is a 120-m handicap footrace. Runners who start from scratch run the full 120 m; for other runners, their handicap is how far in front of scratch they start.
    Joshua Ross has won the race twice. In 2003, with a handicap of 7 m, his time was 11.92 seconds. In 2005, from scratch, he won in 12.36 seconds.
    Determine in which race he ran faster.

13. From the two labels shown, determine which product has a lower amount of total fats.

Product A

Nutrition Facts	
Serving Size 1 Cake (43g)	
Servings Per Container 5	
Amount Per Serving	
Calories 200 Calories from Fat 90	
	% Daily Value*
Total Fat 10g	**15%**
Saturated Fat 5g	**25%**
Trans Fat 0g	
Cholesterol 0mg	**0%**
Sodium 100mg	**4%**
Total Carbohydrate 26g	**9%**
Dietary Fiber 0g	**0%**
Sugars 19g	
Protein 1g	
Vitamin A 0% • Vitamin C 0%	
Calcium 0% • Iron 2%	

* Percent Daily Values are based on a 2,000 calorie diet. Your daily values may be higher or lower depending on your calorie needs:

		Calories: 2,000	2,500
Total Fat	Less than	65g	80g
Sat. Fat	Less than	20g	25g
Cholesterol	Less than	300mg	300mg
Sodium	Less than	2,400mg	2,400mg
Total Carbohydrate		300g	375g
Dietary Fiber		25g	30g

Product B

Nutrition Facts	
Serving Size 1/4 Cup (30g)	
Servings Per Container About 38	
Amount Per Serving	
Calories 200 Calories from Fat 150	
	% Daily Value*
Total Fat 17g	**26%**
Saturated Fat 2.5g	**13%**
Trans Fat 0g	
Cholesterol 0mg	**0%**
Sodium 120mg	**5%**
Total Carbohydrate 7g	**2%**
Dietary Fiber 2g	**8%**
Sugars 1g	
Protein 5g	
Vitamin A 0% • Vitamin C 0%	
Calcium 4% • Iron 8%	

*Percent Daily Values are based on a 2,000 calorie diet.

14. Two car hire companies offer two different rates, as shown:
    - Company A charges an upfront fee of $45 and then 32 cents/km.
    - Company B charges an upfront fee of $85 and then 26 cents/km.

    Determine which company would be the better option if you were travelling 700 km.

Fully worked solutions for this chapter are available online.

LESSON
3.6 Rates and costs

SYLLABUS LINKS

- Use rates to determine costs.

3.6.1 Costs for trades

- Rates help us compare costs because we are comparing the same quantities.
- If we need work done around the home, we often ask for two or three quotes from different tradespeople to compare costs.
- Most tradespeople charge a call-out fee and then a rate per hour. These costs may vary from one tradesperson to another.
- Some tradespeople have fixed rates for certain jobs.

For example, an electrician may charge a \$60 call-out fee (fixed cost) and then \$75 per hour. If the job requires 3 hours to be completed, the final cost will be:

$$\begin{aligned}\text{Cost} &= \$60 + \$75 \times 3\\ &= \$60 + \$225\\ &= \$285\end{aligned}$$

Calculating cost for trades

Cost = fixed fee + (charge per hour × number of hours)

tlvd-3779

WORKED EXAMPLE 12 Calculating trade rates

Adam and Jess are both electricians. Adam charges a \$70 call-out fee and \$80 per hour, while Jess charges a \$50 call-out fee and \$85 per hour.
Determine which electrician will have the cheaper quote for a:
a. 2-hour job
b. 5-hour job.

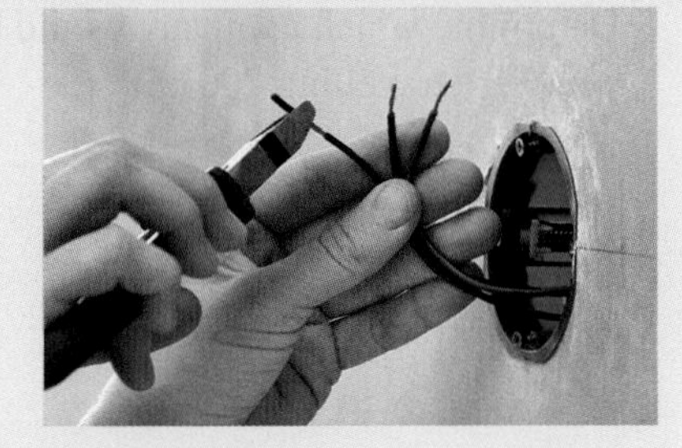

THINK	WRITE
a. 1. Write the formula.	a. Cost = fixed cost + charge per hour × hours
2. Substitute the known values for each electrician and simplify.	$\begin{aligned}\text{Adam's cost} &= 70 + 80 \times 2\\ &= \$230\end{aligned}$ $\begin{aligned}\text{Jess's cost} &= 50 + 85 \times 2\\ &= \$220\end{aligned}$

3. Compare the two costs.	Adam charges \$10 more for this two-hour job than Jess does, so Jess is cheaper.
b. 1. Write the formula.	b. Cost = fixed cost + charge per hour × hours
2. Substitute the known values for each electrician and simplify.	Adam's cost $= 70 + 80 \times 5$ $= \$470$ Jess's cost $= 50 + 85 \times 5$ $= \$475$
3. Compare the two costs and write the answer.	Jess charges \$5 more for this five-hour job than Adam does, so Adam is cheaper.

3.6.2 Living costs: clothing

- Some people like to buy clothes on a regular basis to keep up with trends.
- Other people are not concerned about the trends and buy only what they need.
- Therefore, it's important to create a budget for clothing to know how much money is spent on clothing.

tlvd-11422

WORKED EXAMPLE 13 Calculating the amount spent on clothing

Vince is creating a budget for clothing. He made four categories: small items, trousers, shirts and jumpers. He added up the money he spent on these items and recorded the amounts in the two-way table shown. Complete the table.

		Time period		
		Day (\$)	**Week (\$)**	**Month (\$)**
Category	**Small items**		12.60	
	Trousers	5.00		
	Shirts		35.70	
	Jumpers			93
	Total			

THINK	WRITE/DRAW
1. Calculate the missing entries.	Cost of small items per day $= \$12.60 \div 7$ days $= \$1.80$ Cost of small items per month $= \$1.80 \times 30$ days $= \$54$ Cost of trousers per week $= \$5 \times 7$ days $= \$35$ Cost of trousers per month $= \$5 \times 30$ days $= \$150$ Cost of shirts per day $= \$35.70 \div 7$ days $= \$5.10$ Cost of shirts per month $= \$5.10 \times 30$ days $= \$153$ Cost of jumpers per day $= \$93 \div 30$ days $= \$3.10$ Cost of jumpers per week $= \$3.10 \times 7$ days $= \$21.70$
2. Complete the table.	(table below)

		Time period		
		Day ($)	Week ($)	Month ($)
Category	Small items	1.80	12.60	54
	Trousers	5	35	150
	Shirts	5.10	35.70	153
	Jumpers	3.10	21.70	93
	Total	15	105	450

3.6.3 Living costs: food

- To calculate how much we spend per week on food, we simply add up all the money we spend on food products over a week.
- The money spent per week on food products like sugar and cooking oil could be estimated, as these are not going to be consumed in that week.
- For example, if we use 1 litre of cooking oil that costs \$6 over a period of 4 weeks, then the price estimate is \$1.50 per week. This is regardless of whether we use more cooking oil in one week than another.
  Note: Throughout this section we will use an average of 30 days per month.

WORKED EXAMPLE 14 Calculating cost-of-living rates

Georgia drinks 1 L of milk per day, eats one loaf of bread per week and uses one 500-g container of butter per month. Calculate Georgia's daily, weekly and monthly spending for the three types of food if 1 L of milk costs $1, one loaf of bread costs $4.20 and a 500-g container of butter costs $3.15.

THINK / **WRITE/DRAW**

1. Draw a table with foods listed in the first vertical column and time periods listed on the first horizontal row.

Type of food	Time period		
	Day ($)	Week ($)	Month ($)
Milk			
Bread			
Butter			
Total			

2. Fill in the table with the information given.

We know that Georgia drinks 1 L of milk per day at a cost of $1 per litre. This means that the milk costs her $1 per day.
We also know that she eats a loaf of bread per week. This means that she spends $4.20 per week on bread. She also spends $3.15 on butter per month.

Type of food	Time period		
	Day ($)	Week ($)	Month ($)
Milk	1		
Bread		4.20	
Butter			3.15
Total			

3. Calculate the missing entries.

$$\text{Price of milk per week} = \$1 \times 7 \text{ days}$$
$$= \$7$$
$$\text{Price of milk per month} = \$1 \times 30 \text{ days}$$
$$= \$30$$
$$\text{Price of bread per day} = \$4.20 \div 7 \text{ days}$$
$$= \$0.60$$
$$\text{Price of bread per month} = \$0.60 \times 30 \text{ days}$$
$$= \$18$$
$$\text{Price of butter per day} = \$3.15 \div 30 \text{ days}$$
$$= \$0.105$$
$$\text{Price of butter per week} = \$0.105\ 7 \text{ days}$$
$$= \$0.735$$

4. Fill in the rest of the table.

Type of food	Time period		
	Day ($)	Week ($)	Month ($)
Milk	1	7	30
Bread	0.60	4.20	18
Butter	0.105	0.735	3.15
Total	1.705	11.935	51.15

5. Write the answer.

Georgia spends $1.71 on food per day, $11.94 on food per week and $51.15 on food per month.

3.6.4 Living costs: transport

- Means of transport to and from school are personal cars, bicycles or public transport.
- People might use one or more forms of transport to go to work or school.
- The costs for transport when using a personal car include petrol, maintenance and insurance.

Note: Throughout this section we will use an average of 30 days per month.

tlvd-3926

WORKED EXAMPLE 15 Calculating transport costs

Tim uses his personal car as a means of transport. He spends $720 per year on car insurance, $64.40 per week on petrol and $42 per month on maintenance. Tabulate this data and calculate Tim's daily, weekly, monthly and yearly spending for running the car.

THINK | **WRITE/DRAW**

1. Construct a table with the types of service listed in the first vertical column and time periods listed on the first horizontal row.

		Time period			
		Day ($)	Week ($)	Month ($)	Year ($)
Type of service	Petrol				
	Insurance				
	Maintenance				
	Total				

2. Fill in the table with the information given.

We know that Tim spends $720 per year on insurance, $64.40 on petrol per week and $42 on maintenance per month.

		Time period			
		Day ($)	Week ($)	Month ($)	Year ($)
Type of service	Petrol		64.40		
	Insurance				720
	Maintenance			42	
	Total				

3. Calculate the missing entries.

Price of petrol per day $= \$64.40 \div 7$ days
$= \$9.20$

Price of petrol per month $= \$9.20 \times 30$ days
$= \$276$

Price of petrol per year $= \$276 \times 12$ months
$= \$3312$

Price of insurance per month $= \$720 \div 12$ months
$= \$60$

Price of insurance per day $= \$60 \div 30$ days
$= \$2$

Price of insurance per week $= \$2 \times 7$ days
$= \$14$

Price of maintenance per day $= \$42 \div 30$ days
$= \$1.40$

Price of maintenance per week $= \$1.4 \times 7$ days
$= \$9.80$

Price of maintenance per year $= \$42 \times 12$ months
$= \$504$

4. Fill in the rest of the table.

		Time period			
		Day ($)	Week ($)	Month ($)	Year ($)
Type of service	Petrol	9.20	64.40	276	3312
	Insurance	2	14	60	720
	Maintenance	1.40	9.80	42	504
	Total	12.60	88.20	378	4536

5. Write the answer as a sentence.

Tim's car costs him $12.60 per day, $88.20 per week, $378 per month and $4536 per year.

3.6.5 Living costs on spreadsheets

- Excel spreadsheets are a practical and useful tool when calculating budgets.
- For the purpose of this exercise we are going to consider the same situation as in Worked example 15.

WORKED EXAMPLE 16 Calculating living costs using spreadsheets

Tim uses his personal car as a means of transport. He spends $720 per year on car insurance, $64.40 per week on petrol and $42 per month on maintenance. Use a spreadsheet to tabulate this data and calculate Tim's daily, weekly, monthly and yearly spending for running the car.

THINK

WRITE/DRAW

1. Construct a table in a spreadsheet with the type of service in the first vertical column, column A, and time periods on the horizontal row, row 2.

	A	B	C	D	E
1		Time period			
2	Type of service	Day	Week	Month	Year
3	Petrol	4.30	30.10	129.00	
4	Maintenance	2.60	18.20	78.00	
5	Insurance	0.30	2.10	9.00	
6		7.20	50.40	216.00	

2. Fill in the table with the information given.

	A	B	C	D	E
1		Time period			
2	Type of service	Day	Week	Month	Year
3	Petrol		64.4		
4	Maintenance			42	
5	Insurance				720
6					

3. Set up the calculations required. Remember to start each formula with the = sign. Click in cell B3, type =C3/7 and press ENTER.

	A	B	C	D	E
1			Time period		
2	Type of service	Day	Week	Month	Year
3	Petrol	=C3/7	64.4		
4	Maintenance			42	
5	Insurance				720
6					

4. Click in cell D3, type =B3*30 and press ENTER.

	A	B	C	D	E
1		Time period			
2	Type of service	Day	Week	Month	Year
3	Petrol	9.2	64.4	=B3*30	
4	Maintenance			42	
5	Insurance				720
6					

5. Click in cell E3, type =D3*12 and press ENTER.

	A	B	C	D	E
1		Time period			
2	Type of service	Day	Week	Month	Year
3	Petrol	9.2	64.4	276	=D3*12
4	Maintenance			42	
5	Insurance				720
6					

6. In the bottom right corner of cell E3 you will notice a small circle. Drag it down with your mouse to copy the formula to the cell below.

	A	B	C	D	E
1		Time period			
2	Type of service	Day	Week	Month	Year
3	Petrol	9.2	64.4	276	3312
4	Maintenance			42	504
5	Insurance				720
6					

7. Click in cell B4, type =D4/30 and press ENTER.

	A	B	C	D	E
1		Time period			
2	Type of service	Day	Week	Month	Year
3	Petrol	9.2	64.4	276	3312
4	Maintenance	=D4/30		42	504
5	Insurance				720
6					

8. Drag down the small circle in the bottom right corner of cell B4 to copy the formula to the cell below.

	A	B	C	D	E
1		Time period			
2	Type of service	Day	Week	Month	Year
3	Petrol	9.2	64.4	276	3312
4	Maintenance	1.4		42	504
5	Insurance	0			720
6					

9. You will notice a '0' in cell B5. This is because at the moment there is no value in cell D5. Click in cell D5, type =E5/12 and press ENTER.

	A	B	C	D	E
1		Time period			
2	Type of service	Day	Week	Month	Year
3	Petrol	9.2	64.4	276	3312
4	Maintenance	1.4		42	504
5	Insurance	0		=E5/12	720

10. Click in cell C4, type =B4*7 and press ENTER.

	A	B	C	D	E
1		Time period			
2	Type of service	Day	Week	Month	Year
3	Petrol	9.2	64.4	276	3312
4	Maintenance	1.4	=B4*7	42	504
5	Insurance	2		60	720
6					

11. Drag down the small circle in the bottom right corner of cell C4 to copy the formula to the cell below.

	A	B	C	D	E
1		Time period			
2	Type of service	Day	Week	Month	Year
3	Petrol	9.2	64.4	276	3312
4	Maintenance	1.4	9.8	42	504
5	Insurance	2	14	60	720
6					

12. Now, we are going to add a Total row by typing 'Total' in cell A6. Click in cell B6, type =SUM (B3: B5) and press ENTER. This command will add up the values in cells B3, B4 and B5.

	A	B	C	D	E
1		Time period			
2	Type of service	Day	Week	Month	Year
3	Petrol	9.2	64.4	276	3312
4	Maintenance	1.4	9.8	42	504
5	Insurance	2	14	60	720
6	Total	12.6			

13. This time drag the small circle in the bottom right corner of cell B6 to the right to copy the formula in cells C6, D6 and E6.

	A	B	C	D	E
1		Time period			
2	Type of service	Day	Week	Month	Year
3	Petrol	9.2	64.4	276	3312
4	Maintenance	1.4	9.8	42	504
5	Insurance	2	14	60	720
6	Total	12.6	88.2	378	4536

Exercise 3.6 Rates and costs

learnon

3.6 Quick quiz on	3.6 Exercise

Simple familiar	Complex familiar	Complex unfamiliar
1, 2, 3, 4, 5, 6, 7, 8, 9, 10, 11, 12, 13, 14, 15, 16, 17, 18	N/A	N/A

Simple familiar

1. Calculate the following costs for a tradesperson.
   a. A fixed fee of $35 and $50 per hour for 3 hours
   b. A fixed fee of $45 and $65 per hour for 2 hours
   c. A fixed fee of $20 and $30 per half an hour for 1.5 hours
   d. A fixed fee of $80 and $29.50 per 15 minutes for 30 minutes
   e. A fixed fee of $70 and $37.50 per half an hour for 2.5 hours
   f. A fixed fee of $100 and $50.25 per 15 minutes for 45 minutes

2. WE12 Lynne needs a plumber to fix the drainage of her house. Lynne obtained three different quotes:
   Quote 1: Fixed fee of $80 and $100 per hour
   Quote 2: Fixed fee of $25 and $120 per hour
   Quote 3: Fixed fee of $70 and $110 per hour
   a. Determine which quote is the best if the job requires 2 hours.
   b. Determine which quote is the best if the job requires 7 hours.

3. a. Calculate the fuel consumption (in L/100 km) for:
      i. a sedan using 35 L of petrol for a trip of 252 km
      ii. a wagon using 62 L of petrol for a trip of 428 km
      iii. a 4WD using 41 L of petrol for a trip of 259 km.
   b. Determine which of the three cars is the most economical.

4. A lawyer charges a $250 application fee and $150 per hour.
   Calculate how much money you would spend for a case that requires 10 hours of the lawyer's time.

5. Calculate and compare the fuel consumption rates for a sedan travelling in city conditions for 37 km on 5 L and a 4WD travelling on the highway for 216 km on 29 L.

6. WE13 Carol has noticed that she spends too much money buying tops and wants to cut down on her spending. Carol separated the tops into three categories: singlets, casual and formal. She then added up the money spent on these items and recorded them in the table shown. Complete the table.

		Time period		
		Day ($)	Week ($)	Month ($)
Category	Singlets	3.42		
	Casual			198.30
	Formal		52.50	
	Total			

7. Elaine spends \$15 a day on food, Tatiana spends \$126 a week on food and Anya spends \$420 per month on food.
   a. Calculate how much money Anya spends on food per day.
   b. Calculate how much money Tatiana spends on food per day.
   c. Determine who, out of the three friends, spends the most on food per month. Explain your answer.
8. Compare the two quotes given by two computer technicians, Joanna and Dimitri, for a four-hour job. Joanna charges a \$65 fixed fee and \$80 per hour. Dimitri charges a \$20 fixed fee and \$100 per hour.
9. Hamish and Hannah are both tilers. Hamish charges \$200 for surface preparation and \$100 per square metre, while Hannah charges \$170 for surface preparation and \$120 per square metre.
   Compare their charges for a job that covers an area of 10 m^2.
10. WE14 A family spend \$2.80 a week on potatoes, \$45 a month on tomatoes and \$1.60 per day on cheese. Calculate their daily, weekly and monthly spending for the three types of food and construct a table to display these costs.

11. WE15 Halina spends \$4.10 per day on petrol, \$18.20 a week on insurance and \$39 per month on maintenance for her car. Calculate Halina's daily, weekly, monthly and yearly spending for the three types of services and construct a table to display these costs.
12. Andy spends \$17.50 per week on petrol, \$72 a month on insurance and \$612 per year on maintenance for his car. Calculate Andy's daily, weekly, monthly and yearly spending for the three types of services and construct a table to display these costs.
13. Ant spends \$129 a month on fruit, \$18.20 a week on vegetables and \$0.30 per day on eggs.
    a. Calculate Ant's daily, weekly and monthly spending for the three types of food and construct a table to display these costs.
    b. WE16 Construct an Excel spreadsheet for part a.
14. Nadhea made a budget for clothing. She made four categories: small items, bottoms (skirts and trousers), dresses and tops. She then added up the money spent on these items and recorded the amounts in a table. Nadhea spends \$2.65 per day on small items, \$325.50 per month on bottoms, \$22.40 per week on dresses and \$6.20 per day on tops.
    a. Tabulate this data and calculate Nadhea's daily, weekly and monthly spending for her clothing.
    b. Construct an Excel spreadsheet for part a.

15. Paris uses both her personal car and public transport. Paris spends \$630 per year on car insurance, \$9.10 per week on petrol and \$21 per month on maintenance. She also spends \$5.80 per day on public transport.
Tabulate this data and calculate Paris' daily, weekly, monthly and yearly spending for running the car and catching public transport. (Assume Paris catches public transport every day of the year.)
16. The total cost for running a car for 12 days is \$27. Determine this cost per day, per week, per month and per year.
17. Tara is budgeting for only 20% of her weekly salary to be spent on food, clothes and transport. If in a day she spends \$10 on food, \$6.70 on clothes and \$5.20 on transport, calculate what her weekly salary would need to be.
18. A cafe usually purchases coffee beans at \$3/kg. It has been offered a cheaper option at \$2/kg. If it spends \$1500 a month on coffee beans, calculate how many additional kilograms of coffee beans it can get each month.

Fully worked solutions for this chapter are available online.

LESSON
3.7 Review

3.7.1 Summary

doc-41700

Hey students! Now that it's time to revise this chapter, go online to:

Access the chapter summary | Review your results | Watch teacher-led videos | Practise questions with immediate feedback

Find all this and MORE in jacPLUS

3.7 Exercise

learnon

3.7 Exercise

Simple familiar	Complex familiar	Complex unfamiliar
1, 2, 3, 4, 5, 6, 7, 8, 9, 10, 11, 12, 13, 14, 15, 16, 17	18, 19	20

These questions are even better in jacPLUS!
- Receive immediate feedback
- Access sample responses
- Track results and progress

Find all this and MORE in jacPLUS

Simple familiar

1. If a runner travels 3 km in 10 minutes and 20 seconds, calculate the runner's average speed in m/s.
2. If an escalator rises 3.75 m for a horizontal run of 5.25 m, determine the gradient of the escalator.
3. The speed limit around schools is 40 km/h. Convert this rate to m/s.
4. On average, Noah kicks 2.25 goals per game of soccer. Determine the number of goals Noah kicks in a 16-game season.
5. If car A travels at 60 km/h and car B travels at 16 m/s, for each of the following statements, identify whether it is correct or incorrect.
   a. Both cars travel at the same speed.
   b. Car A travels 0.67 m/s slower than car B.
   c. Car A travels 44 km/h faster than car B.
   d. Car B travels 0.67 m/s slower than car A.
   e. Car B travels 0.44 m/s slower than car A.
6. At the gym, Lisa and Simon were doing a skipping exercise. Lisa completed 130 skips in one minute, whereas Simon skipped at a rate of 40 skips in 20 seconds. Given that they both skipped for 3 minutes, identify whether each of the following statements is correct or incorrect.
   a. Simon completed 20 more skips.
   b. Lisa completed 20 more skips.
   c. Lisa completed 30 more skips.
   d. Simon completed 30 more skips.
   e. They completed the same number of skips.

7. If the cost of food per week for a family is \$231 and the cost of petrol per month is \$159.60, calculate the corresponding costs for food and petrol per day, respectively.

8. For each of the following rates, identify whether it is equivalent to a rate of 86.4 L/h.

   a. 24 mL/s b. 24 L/s c. 1.44 mL/min d. 24 mL/min e. \$2.44 mL/min

9. The Shinkansen train in Japan travels at a speed of approximately 500 km/h. The distance from Broome to Melbourne is approximately 4950 km.
   Calculate how long, to the nearest minute, it would take the Shinkansen train to cover this distance.

10. A sedan, a wagon and a 4WD are travelling the same distance of 159 km. The sedan uses 20 L, the wagon uses 22 L and the 4WD uses 25 L of fuel for this trip.
    Determine the fuel consumptions for the 4WD, wagon and sedan.

11. Convert the following rates to the rates shown in the brackets.

    a. 17 m/s (km/h)
    b. 24.2 L/min (L/h)
    c. 0.0125c/s (\$/min)
    d. 2.5 kg/mL (g/L)

12. A dragster covers a 400-m track in 8.86 s.

    a. Calculate its average speed in m/s.
    b. If it could maintain this average speed for 1 km, determine how long it would take to cover the 1 km.

13. Oliver's hair grows at a rate of 125 mm/month, whereas Emilio's grows at 27.25 mm/week. If their hair is initially 5 cm long, determine how long their hair will be after 1 year (assume 52 weeks in a year).

14. Isabella worked 4 weeks at the rate of \$693.75 per week before her wage was increased to \$724.80 per week.

    a. Calculate how much Isabella was paid for the 4 weeks.
    b. Calculate how much Isabella will earn at the new rate for the next 4 weeks.
    c. Calculate how much more money Isabella is receiving per week.
    d. Determine the difference between Isabella's earnings for the first four weeks and the following four weeks.

15. Following the Australian tax rates shown, calculate the amount of tax to be paid on the following taxable incomes.

Taxable income	Tax on this income
0–\$18 200	Nil
\$18 201–\$45 000	19c for each \$1 over \$18 200
\$45 001–\$120 000	\$5092 plus 32.5c for each \$1 over \$45 000
\$120 001–\$180 000	\$29 467 plus 37c for each \$1 over \$120 000
\$180 001 and over	\$51 667 plus 45c for each \$1 over \$180 000

    a. \$21 000 b. \$15 600 c. \$50 000 d. \$100 000 e. \$210 000

16. To renovate your house, you may hire a number of different tradespeople. Given how much each tradesperson charges and how many hours they need to complete their part of the work, calculate the total cost of the renovation.
    - Carpenter at $60 per hour, for 40 hours
    - Electrician at $80 per hour for 14.5 hours
    - Tiler at $55 per hour for 9 hours and 15 minutes
    - Plasterer at $23 per half-hour for 8 hours and 20 minutes
    - Plumber with a call-out charge of $85 and $90 per hour for 6.5 hours
17. Petrol costs $2.19 per litre, and Callum needs 50 litres to fill the tank. Determine the cost of the fuel.

Complex familiar

18. A one-litre can of paint covers five square metres of wall.
    a. Determine whether a direct proportion exists between the number of litres purchased and the area of the walls to be painted. Justify your answer.
    b. Evaluate the number of cans needed to paint a wall 5 metres long and 2.9 metres high with two coats of paint.
19. Consider the nutrition information shown in the label.

NUTRITION INFORMATION		
SERVINGS PER PACKAGE: 3	SERVING SIZE: 150 g	
	QUANTITY PER SERVING	QUANTITY PER 100 g
Energy	608 kJ	405 kJ
Protein	4.2 g	2.8 g
Fat, total	7.4 g	4. 9g
– Saturated	4.5 g	3.0 g
Carbohydrate	18.6 g	12.4 g
– Sugars	18.6 g	12.4 g
Sodium	90 mg	60 mg

    a. If you ate 350 g, determine what percentage of carbohydrates you have consumed.
    b. If you ate 250 g, determine what percentage of total fat you have consumed.

Complex unfamiliar

20. It takes me 2 hours to mow my lawn. My daughter takes 2.5 hours to mow the same lawn. If we work together using two lawnmowers, calculate how long it will take to mow the lawn.
    Give your answer in hours, minutes and seconds.

Fully worked solutions for this chapter are available online.

Answers

Chapter 3 Rates

3.2 Identifying rates

3.2 Exercise

1. a. 40 km/h b. 11.11 m/s
2. 50 beats/minute
3. a. $\frac{\text{mass}}{\text{volume}} = \text{density}$
   b. $\frac{\text{distance}}{\text{time}} = \text{speed}$
   c. $\frac{\text{rise}}{\text{run}} = \text{gradient}$
   d. $\frac{\text{amount of solution}}{\text{amount of solvent}} = \text{concentration}$
4. a. 80 km/h b. \$3.90/m
   c. 0.2 L/s d. 0.14 km/min
5. 1.21 m/s
6. 29.32 km/h
7. 14.5 cm/year
8. 8 cm/year
9. 6.14 cm/year
10. 75 students/year
11. 50 000 g/m^3
12. 111.11 mL/L
13. 0.01 g/mL
14. 0.2
15. 0.51 °C/s
16. 160.92 km/h
17. Ollie is 225 kg/m^3 denser than Simon.
18. 2750 organisms/L; the beach is safe for swimming.

3.3 Conversion of rates

3.3 Exercise

1. 34 km/h
2. \$15/kg
3. a. 60 km/h b. 200 m/min
   c. 0.3 L/s d. 2 mm/day
4. a. 0.2 km/min b. 0.4 L/min
   c. \$0.32/min d. 0.45 cents/g
5. a. 27.78 m/s b. 666.67 mL/min
6. a. 5.50 cents/cm b. 36.00 km/h
7. a. 70 000 m/h b. 40 000 g/min
   c. \$7200/h d. 27 000 g/L
8. a. 22.22 m/s b. 90.00 km/h
   c. 50.00 kg/L d. 21.60 L/h
9. 725 760 beats
10. \$9.00
11. \$198.75
12. 33 112 800 beats
13. \$202.50
14. 340 km
15. 12.5 years

3.4 Calculation with rates involving direct proportion [complex]

3.4 Exercise

1. a. Yes
   b. No; as speed increases, time decreases.
   c. No; doubling distance doesn't double the cost.
2. a. Yes
   b. No; doubling the side length doesn't double the area.
   c. Yes
   d. Yes
3. a. Yes
   b. No; doubling distance doesn't double the cost (due to the initial fee).
   c. No; doubling age doesn't double height.
   d. Yes
4. a. True: as the number, n, increases, so does the cost, \$$C$.
   b. True: $C = \pi D$, as C increases with D. π is a constant number.
   c. False: $A = \frac{1}{2}\pi r^2$, so it is not linear.
5. \$78
6. \$65
7. 23.45 h or 23 h and 27 min
8. 340 km
9. 1:37 am Monday
10. $3.\dot{3}$ cups flour; 250 g sugar; 200 mL milk; 5 eggs
11. Sample responses can be found in the worked solutions in the online resources.
12. a. y is doubled.
    b. y is halved.
    c. x is tripled.
13. a. Yes, direct proportion does exist.
    b. \$2.55
14. a. Yes b. \$935
15. Direct proportion does not exist. The price per metre for the 4.2-metre length is \$5.50, and the price per metre for the 5.4-metre length is \$5.40.

3.5 Comparison of rates

3.5 Exercise

1. a. A b. X
2. a. \$4.05 per 250 g b. \$8.90 per 500 g
3. a. \$2.25 per 150 g b. \$7.35 per 450 g

4. Car A rate = 12.3 L/100 km
Car B rate = 12 L/100 km
Therefore car B has better fuel economy.
5. Car A = 8.19 L/100 km
Car B = 8.81 L/100 km
Therefore car A is more economical.
6. 290 g: rate = $1.68/100 g
805 g: rate = $1.08/100 g
805 g is cheaper by $1.68 − $1.08 = $0.60 per 100 g.
7. Pack of 10: $0.145 each
Pack of 25: $0.154 each
Technically speaking, the pack of 10 is cheaper, but when rounded off to the nearest cent, each tea bag has the same price.
8. 250 g: $3.80/100 g
100 g: $4.10/100 g
It is cheaper to buy the 250-gram jar by 30 cents per 100 grams.
9. 22.5 g/L
Since the solvent is what the solution is dissolved in, doubling the solvent halves the concentration.
10. a. 128.57 runs per 100 balls
b. 135.48 runs per 100 balls
c. Meg Lanning has the greater strike rate by 6.91 runs per 100 balls.
11. Brand B is better because it is 10.5 cents per 100 grams compared to 12.4 cents per 100 grams for brand A.
12. Speed 2003: 9.48 m/s
Speed 2005: 9.71 m/s
Joshua Ross ran faster in winning the Stawell Gift in 2005.
13. Product A has 15% fat and Product B has 26% fat, so Product A has the lower amount of total fat.
14. Company B is the better (cheaper) option when traveling 700 km.

3.6 Rates and costs

3.6 Exercise

1. a. $185 b. $175 c. $110
d. $139 e. $257.50 f. $250.75
2. a. Quote 2 b. Quote 1
3. a. i. 13.89 L/100 km ii. 14.49 L/100 km
iii. 15.83 L/100 km
b. The sedan
4. $1750
5. Sedan − 13.51 L/100 km
4WD − 13.43 L/100 km
The fuel consumption rate is better when travelling on a highway compared to city driving.
6.

Category		Time period		
		Day ($)	Week ($)	Month ($)
	Singlets	3.42	23.94	102.60
	Casual	6.61	46.27	198.30
	Formal	7.50	52.50	225.00
	Total	17.53	122.71	525.90

7. a. $14 b. $18 c. Tatiana
8. Joanna = $385; Dimitri = $420
9. Hamish = $1200; Hannah = $1370
10.

Type of food		Time period		
		Day ($)	Week ($)	Month ($)
	Potatoes	0.40	2.80	12.00
	Tomatoes	1.50	10.50	45.00
	Cheese	1.60	11.20	48.00
	Total	3.50	24.50	105.00

11.

Type of service		Time period			
		Day ($)	Week ($)	Month ($)	Year ($)
	Petrol	4.10	28.70	123.00	1476.00
	Insurance	2.60	18.20	78.00	936.00
	Maintenance	1.30	9.10	39.00	468.00
	Total	8.00	56.00	240.00	2880.00

12.

Type of service		Time period			
		Day ($)	Week ($)	Month ($)	Year ($)
	Petrol	2.50	17.50	75.00	900.00
	Insurance	2.40	16.80	72.00	864.00
	Maintenance	1.70	11.90	51.00	612.00
	Total	6.60	46.20	198.00	2376.00

13. a.

Type of food		Time period		
		Day ($)	Week ($)	Month ($)
	Fruit	4.30	30.10	129.00
	Vegetables	2.60	18.20	78.00
	Eggs	0.30	2.10	9.00
	Total	7.20	50.40	216.00

b.

	A	B	C	D
1		Time period		
2	Type of food	Day	Week	Month
3	Fruit	4.30	30.10	129.00
4	Vegetables	2.60	18.20	78.00
5	Eggs	0.30	2.10	9.00
6	Total	7.20	50.40	216.00

14. a.

		Time period		
		Day ($)	Week ($)	Month ($)
Category	Small items	2.65	18.55	79.50
	Bottoms	10.85	75.95	325.50
	Dresses	3.20	22.40	96.00
	Tops	6.20	43.40	186.00
	Total	22.90	160.30	687.00

b.

	A	B	C	D
1		Time period		
2	Category	Day	Week	Month
3	Small items	2.65	18.55	79.50
4	Bottoms	10.85	75.95	325.50
5	Dresses	3.20	22.40	96.00
6	Tops	6.20	43.40	186.00
7	Total	22.90	160.30	687.00

15.

		Time period			
		Day ($)	Week ($)	Month ($)	Year ($)
Type of service	Insurance	1.75	12.25	52.50	630
	Fuel	1.30	9.10	39	468
	Maintenance	0.70	4.90	21	252
	Public transport	5.80	40.60	174	2088
	Total	$9.55	$66.85	$286.50	$3438.00

16. Per day: $2.25
    Per week: $15.75
    Per month: $67.50
    Per year: $810
17. $30.66
18. 250 kg

3.7 Review

3.7 Exercise

1. 4.84 m/s
2. 0.71
3. 11.1 m/s
4. 36
5. Car B travels 0.67 m/s slower than Car A.
6. Lisa completed 30 more skips.
7. $33 and $5.32
8. 24 mL/s
9. 9 hours 54 minutes
10. 15.7 L/100 km, 13.8 L/km and 12.6 L/100 km
11. a. 61.2 km/h b. 1452 L/h
    c. $0.0075/min d. 2 500 000 g/L
12. a. 45.15 m/s b. 22.15 s
13. Olivia: 155 cm
    Emily: 146.7 cm
14. a. $2775 b. $2899.20
    c. $31.05 d. $124.20
15. a. $532 b. $0 c. $6717
    d. $22 967 e. $65 167
16.

		Calculation	Total
Trade	Carpenter	$\$60 \times 40 = \$2400$	$2400
	Electrician	$\$80 \times 14.5 = \$1160$	$1160
	Tiler	$\$55 \times 9\frac{15}{60} = \$508.75$	$508.75
	Plasterer	$(\$23 \times 2) \times 8\frac{20}{60} = \$383.33$	$383.33
	Plumber	$\$85 + \$90 \times 6.5 = \$670$	$670
	Total		$5122.08

17. $109.50
18. a. Yes. Since each litre of paint corresponds to exactly 5 square metres of wall, there is a direct proportion between the number of litres of paint purchased and the area of wall that can be painted.
    b. 6 cans
19. a. 12.4% b. 4.9%
20. 1 hour 6 minutes 40 seconds

4 Percentages

LESSON SEQUENCE

Fully worked solutions for this chapter are available online.

on Resources

 **Solutions** Solutions — Chapter 4 (sol-1257)

 **Digital documents** Learning matrix — Chapter 4 (doc-41750)
Quick quizzes — Chapter 4 (doc-41751)
Chapter summary — Chapter 4 (doc-41752)

LESSON
4.1 Overview

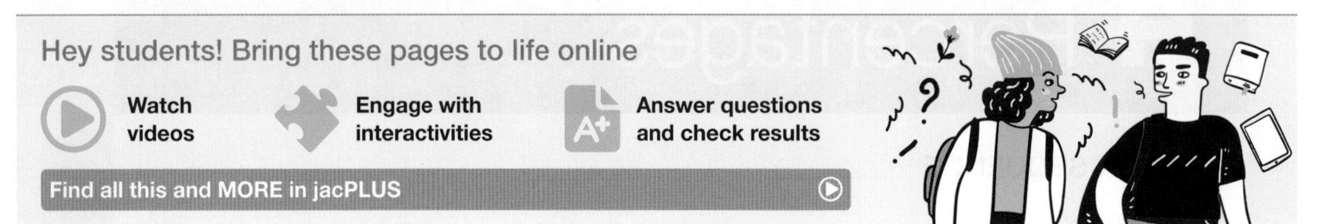

4.1.1 Introduction

Percentages are used to describe many different kinds of information and are regularly used in day-to-day activities. They are so common that they have their own symbol, %. A *per cent* is a hundredth, so using percentages is an alternative to using decimals and fractions. Percentages are a convenient way to describe how much of something you have and how meaningful information is. You are most likely familiar with percentages as a way of recording a test result, but they are used in many different areas and are therefore very important to understand.

Some examples include items advertised at a Black Friday sale at 25% discount, the percentage charged on a loan or a credit card, the percentage tax you need to pay or even the percentage chance of rain on a particular day.

4.1.2 Syllabus links

Lesson	Lesson title	Syllabus links
4.2	**Calculating percentages**	◯ Calculate a percentage of a given amount.
4.3	**Express as a percentage**	◯ Determine one amount expressed as a percentage of another for same units. ◯ Determine one amount expressed as a percentage of another for different units [complex].
4.4	**Percentage increase and decrease [complex]**	◯ Apply percentage increases and decreases in real-world contexts, including mark-ups, discounts and GST [complex]. ◯ Determine the overall change in a quantity following repeated percentage changes [complex].
4.5	**Real-world application of percentages — Simple interest [complex]**	◯ Apply percentage increases and decreases in real-world contexts, including mark-ups, discounts and GST [complex].

LESSON
4.2 Calculating percentages

SYLLABUS LINKS

- Calculate a percentage of a given amount.

4.2.1 Converting percentages to fractions

- The term **per cent** means 'per hundred' or 'out of 100'.
- The symbol for per cent is %. For example, 13% means 13 out of 100.
- Since all percentages are out of 100, they can be converted to a fraction.
- To convert a percentage to a fraction, divide by 100 or multiply by $\frac{1}{100}$ and simplify. For example:

$$35\% = \frac{{}^{7}\cancel{35}}{{}^{20}\cancel{100}} = \frac{7}{20}$$

$$12\frac{4}{5}\% = \frac{64}{5}\% = \frac{64}{5 \times 100} = \frac{{}^{16}\cancel{64}}{{}^{125}\cancel{500}} = \frac{16}{125}$$

$$42.5\% = \frac{42.5}{100} = \frac{42.5 \times 10}{100 \times 10} = \frac{{}^{17}\cancel{425}}{{}^{40}\cancel{1000}} = \frac{17}{40}$$

Expressing percentages as fractions

To convert a percentage to a fraction, divide by 100.

tlvd-3800

WORKED EXAMPLE 1 Expressing percentages as fractions

Write the following percentages as fractions in their simplest form.

a. 45% **b. 12.5%** **c. 0.75%**

THINK	WRITE
a. 1. Divide the integer value by 100.	**a.** $45\% = \frac{45}{100}$
2. Simplify by dividing the numerator and denominator by the highest common factor (in this case 5).	$= \frac{{}^{9}\cancel{45}}{{}^{20}\cancel{100}}$ $= \frac{9}{20}$
b. 1. Divide the decimal by 100.	**b.** $12.5\% = \frac{12.5}{100}\%$

2. Convert the decimal value to an integer by multiplying by an appropriate power of 10. Also multiply the denominator by this power of 10. (Here 12.5 has 1 decimal place, so multiply by 10.)

$$= \frac{12.5 \times 10}{100 \times 10}$$
$$= \frac{125}{1000}$$

3. Simplify by dividing the numerator and denominator by 125.

$$\frac{{}^{1}\cancel{125}}{{}^{8}\cancel{1000}} = \frac{1}{8}$$

c. 1. Divide the decimal value by 100.

c. $0.75\% = \frac{0.75}{100}$

2. Convert the decimal value to an integer by multiplying by an appropriate power of 10. Also multiply the denominator by this power of 10. (Here 0.75 has 2 decimal places, so to make it an integer, multiply by 10^2 or 100.)

$$\frac{0.75}{100} = \frac{0.75 \times 100}{100 \times 100}$$
$$= \frac{75}{10\,000}$$

3. Simplify by dividing the numerator and denominator by the highest common factor (in this case 25).

$$\frac{{}^{3}\cancel{75}}{{}^{400}\cancel{10\,000}} = \frac{3}{400}$$

4.2.2 Converting percentages to decimals

- In order to convert percentages to decimals, divide by 100.
- When dividing by 100, the decimal point moves 2 places to the left.

Expressing percentages as decimals

To convert a percentage to a decimal, divide by 100.

$$60\% = \frac{60}{100}$$
$$= \frac{60.0}{100}$$
$$= 0.60$$

tlvd-3801

WORKED EXAMPLE 2 Expressing percentages as decimals

Write the following percentages as decimals.

a. **67%**

b. **34.8%**

THINK

WRITE

a. 1. Convert the percentage to a fraction by dividing by 100.

a. $67\% = \frac{67}{100}$

2. Divide the numerator by 100 by moving the decimal point 2 places to the left. Remember to include a zero in front of the decimal point.

$$\frac{67}{100} = 67 \div 100$$
$$= 0.67$$

b. 1. Convert the percentage to a fraction by dividing by 100.

b. $34.8\% = \frac{34.8}{100}$

2. Divide the numerator by 100 by moving the decimal point 2 places to the left. Remember to include a zero in front of the decimal point.

$$\frac{34.8}{100} = 34.8 \div 100$$
$$= 0.348$$

4.2.3 Converting a fraction or a decimal to a percentage

- To convert a decimal or fraction to a percentage, multiply by 100%.
- When multiplying by 100, move the decimal point 2 places to the right.

Expressing fractions and decimals as percentages

To convert a fraction or decimal to a percentage, multiply by 100%.

$$0.60 = 0.60 \times 100\%$$
$$= 60\%$$

WORKED EXAMPLE 3 Expressing decimals and fractions as percentages

Convert the following to percentages.

a. **0.51**

b. $\frac{2}{5}$

THINK

a. 1. To convert to a percentage, multiply by 100.

2. Move the decimal point 2 places to the right and write the answer.

b. 1. To convert to a percentage, multiply by 100.

2. Simplify the fraction by dividing the numerator and denominator by the highest common factor (in this case 5).

WRITE

a. $0.51 = 0.51 \times 100\%$
$= 51\%$

b. $\frac{2}{5} \times 100\% = \frac{{}^{40}\cancel{200}}{{}^{1}\cancel{5}}\%$
$= \frac{40}{1}\%$
$= 40\%$

Digital technology

Scientific calculators have a % button that can be used to do calculations involving percentages.

Percentages can be converted into decimals and fractions by pressing the S<->D button.

55%

0.55

4.2.4 Determining the percentage of a given amount

- Quantities are often expressed as a percentage of a given amount.
- To calculate the percentage of an amount using fractions, follow these steps:
  Step 1: Convert the percentage into a fraction.
  Step 2: Multiply by the amount.
  For example, the percentage of left-handed tennis players is around 10%. So, for every 100 tennis players, approximately 10 will be left-handed.

$$\begin{aligned}10\% \text{ of } 100 \text{ tennis players} &= \frac{10}{100} \times 100 \\ &= 0.10 \times 100 \\ &= 10 \text{ tennis players}\end{aligned}$$

tlvd-3802

WORKED EXAMPLE 4 Determining the percentage of a given amount

Calculate the following.

a. 35% of 120

b. 76% of 478 kg

THINK	WRITE
a. 1. Write the problem.	a. 35% of 120
2. Express the percentage as a fraction and multiply by the amount.	$= \frac{35}{100} \times 120$
3. Write the answer.	$= 42$
b. 1. Write the problem as a decimal.	b. 76% of 478 kg
2. Express the percentage as a fraction and multiply by the amount.	$= \frac{76}{100} \times 478$
3. Write the answer. Remember the result has the same unit.	$= 363.28$ kg

tlvd-3803

WORKED EXAMPLE 5 Determining the percentage of a given amount

60% of Year 11 students said Mathematics is their favourite subject. If there are 165 students in Year 11, calculate to the nearest person how many students claimed that Mathematics was their favourite subject.

THINK	WRITE
1. Write the problem as a mathematical expression.	60% of 165
2. Write the percentage as a fraction and multiply by the amount.	$=\frac{60}{100}\times 165$
3. Simplify.	$=0.60\times 165$ $=99$
4. Write the answer.	99 Year 11 students said their favourite subject was Mathematics.

4.2.5 Determining a quantity shortcut

- To calculate 10% of an amount, divide by 10 or move the decimal point 1 place to the left. For example:

$$10\% \text{ of } \$23.00 = \frac{10}{100}\times 23.00 = \frac{23.00}{10} = \$2.30$$

$$10\% \text{ of } 45.6 \text{ kg} = \frac{10}{100}\times 45.6 = \frac{45.6}{10} = 4.56 \text{ kg}$$

Shortcut for determining 5%, 15%, 20% and 25%

The following shortcut can be adapted to other percentages, as listed below:

- **5% — first calculate 10% of the amount, then halve this amount.**
- **20% — first calculate 10% of the amount, then double this amount.**
- **15% — first calculate 10%, then 5% of the amount, then add the totals together.**
- **25% — first calculate 10% of the amount and double it, then calculate 5%, then add the totals together (or calculate 50% and then halve it).**

WORKED EXAMPLE 6 Determining a percentage of a quantity

Calculate the following:

a. **10% of $84**   b. **5% of 160 kg**   c. **15% of $162**

THINK	WRITE
a. To calculate 10% of 84, move the decimal point 1 place to the left.	a. 10% of $84 = $8.40
b. 1. First calculate 10% of 160. To do this, move the decimal point 1 place to the left.	b. 10% of 160 kg = 16 kg

2. To calculate 5%, divide the 10% value by 2.	$\frac{16}{2} = 8$
3. Write the answer.	5% of 160 kg = 8 kg
c. 1. First calculate 10% of 162. To do this, move the decimal point 1 place to the left.	**c.** 10% of \$162 = \$16.20
2. To calculate 5%, divide the 10% value by 2.	$5\% \text{ of } 162 = \frac{16.20}{2}$ $= 8.10$
3. Determine 15% by adding the 10% and 5% values together.	$16.20 + 8.10 = 24.30$
4. Write the answer.	15% of \$162 = \$24.30

Digital technology

The percentage button and the multiplication symbol can be used to help determine percentages of an amount.

```
D                    Math ▲
15%×162
                       24.3
```

Resources

Interactivities Fractions as percentages (int-3994)
Percentages as fractions (int-3992)
Percentages as decimals (int-3993)
Decimals as percentages (int-3995)
Percentage of an amount using decimals (int-3997)
Common percentages and short cuts (int-3999)

Exercise 4.2 Calculating percentages

learnon

4.2 Quick quiz  **4.2 Exercise**

Simple familiar	Complex familiar	Complex unfamiliar
1, 2, 3, 4, 5, 6, 7, 8, 9, 10, 11, 12, 13, 14, 15, 16, 17, 18, 19, 20, 21, 22, 23, 24, 25, 26, 27	N/A	N/A

These questions are even better in jacPLUS!
- Receive immediate feedback
- Access sample responses
- Track results and progress

Find all this and MORE in jacPLUS

Simple familiar

1. **WE1** Express the following percentages as fractions in their simplest form.

a. 48% **b.** 26.8% **c.** $12\frac{2}{5}\%$ **d.** $\frac{4}{5}\%$

2. Express the following percentages as fractions in their simplest form.

   a. 92% b. 74.125% c. $66\frac{2}{3}\%$

3. Express the following percentages as fractions.

   a. 88% b. 25% c. 0.92%

   d. 35.5% e. $30\frac{2}{5}\%$ f. $72\frac{3}{4}\%$

4. **WE2** Express the following percentages as decimals.

   a. 73% b. 94.3%

5. Express the following percentages as decimals.

   a. 2.496% b. 0.62%

6. Express the following percentages as decimals.

   a. 43% b. 39% c. 80% d. 47.25% e. 24.05% f. 0.83%

7. **WE3** Express each of the following as a percentage.

   a. 0.21 b. $\frac{4}{5}$

8. Express each of the following as a percentage.

   a. 0.652 b. $\frac{8}{25}$ c. 0.55 d. 0.83

9. Express each of the following as a percentage.

   a. $\frac{1}{2}$ b. $\frac{7}{8}$ c. $2\frac{1}{5}$ d. 1.65

10. **WE4** Calculate each of the following.

    a. 45% of 160 b. 28% of 288 kg c. 63% of 250 d. 21.5% of $134

11. Calculate the following to the nearest whole number.

    a. 34% of 260 b. 55% of 594 kg c. 12.5% of 250 m
    d. 45% of 482 e. 60.25% of 1250 g f. 37% of 2585

12. **WE5** If a student spent 70% of their part-time weekly wage of $85 on their mobile phone bill, calculate how much they were charged.

13. If a dress was marked down by 20% and it originally cost $120, calculate how much the dress has been marked down by.

14. **WE6** Calculate the following.

    a. 10% of $56 b. 5% of 250 kg c. 15% of 370 d. 20% of 685

15. If 35% of the 160 Year 11 students surveyed prefer to watch a movie in their free time, use mental arithmetic to calculate how many Year 11 students prefer watching movies. Check your answer using a calculator.

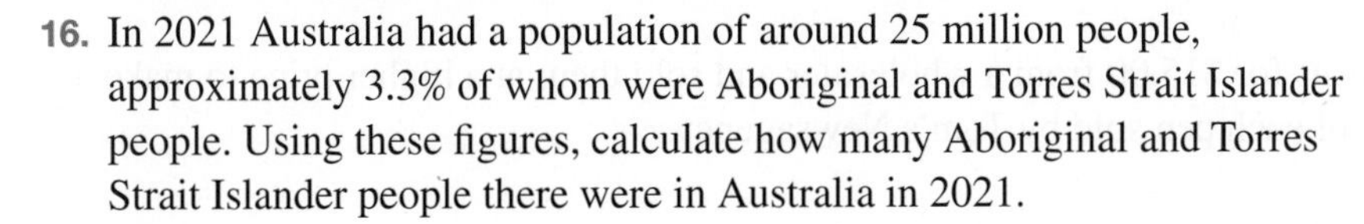

16. In 2021 Australia had a population of around 25 million people, approximately 3.3% of whom were Aboriginal and Torres Strait Islander people. Using these figures, calculate how many Aboriginal and Torres Strait Islander people there were in Australia in 2021.

17. Bella's weekly wage is $1100, and 45% of this is spent on rent. Calculate the amount Bella spends on rent each week.

18. After a good 54 mm rainfall overnight, Jacko's 7500-litre water tank was 87% full. Calculate how much water (in litres) is in Jacko's water tank following the rainfall.

19. Last year a loaf of bread cost \$3.10. Over the last 12 months, the price increased by 4.5%.
    a. Calculate how much, to the nearest cent, the price of bread has increased over the past 12 months.
    b. Calculate the current price of a loaf of bread.
    c. If the price of bread increases by another 4.5% over the next year, calculate the increase in price over the next 12 months, to the nearest cent.
    d. Calculate what a loaf of bread will cost 12 months from now.

20. Ryan has had his eyes on his dream pair of shoes. To his surprise, they were advertised with a 25% discount. If the full price of the shoes is \$175, calculate the price Ryan paid for them.

21. If a surf shop had a 20% discount sale on for the weekend, calculate what price you would pay for each of the following items.
    a. i. Board shorts with the original price of \$65.00
       ii. T-shirt with the original price of \$24.50
       iii. Hoodie with the original price of \$89.90
    b. Calculate how much you would have saved by purchasing these three items on sale compared to paying the full price.

22. Catherine wants to find the best price for a new iPad mini. She found an ad saying that if you purchase the iPad online you can get 15% off the price. The original price was advertised at \$585. Calculate how much Catherine would save if she made the purchase online.

23. If John spends 35% of his \$1200 weekly wage on rent and Jane spends 32% of her \$1050 weekly wage on rent:
    a. calculate who spends more money per week on rent
    b. calculate the difference between John's and Jane's weekly rents.

24. There are 150 students learning a subject at a school. In the last exam, 28% got an A, 54% got a B, and the remaining students got a C. Calculate how many students got a C.

25. A metal alloy used in outdoor furniture contains 25% copper. Determine how much an outdoor bench weighs (in kg) if it is known that the bench is made using 7500 g of copper.

26. Apples are transported from the farm in boxes. On the way to the supermarket, about 12% of apples go rotten or become otherwise unsuitable for sale. After a delivery, 66 apples were in good condition and suitable for sale. Determine how many apples were originally ordered from the farm.

27. Tom's Newsagency bought 50 pens for \$15.00 from a wholesaler and sold them at a higher price to make a 500% profit. Calculate the price of each pen sold by Tom's Newsagency.

Fully worked solutions for this chapter are available online.

LESSON
4.3 Express as a percentage

SYLLABUS LINKS

- Determine one amount expressed as a percentage of another for same units.
- Determine one amount expressed as a percentage of another for different units [complex].

4.3.1 Expressing one amount as a percentage of another

- To express an amount as a percentage of another, write the two numbers as a fraction, with the first number on the numerator and the second number on the denominator.
- Then convert this fraction to a percentage by multiplying by 100%.

Expressing one amount as a percentage of another

$$\frac{\textbf{amount}}{\textbf{total}} \times \frac{\textbf{100}}{\textbf{1}}\%$$

- When expressing one amount as a percentage of another, make sure that both amounts are in the same unit.

tlvd-3804

WORKED EXAMPLE 7 Expressing one amount as a percentage of another

Express the following to the nearest percentage.

a. 37 as a percentage of 50

b. 18 as a percentage of 20

c. 23 g as a percentage of 30 g

d. 2.15 g as a percentage of 17.2 g

THINK	WRITE
a. 1. Write the amount as a fraction of the total.	a. $\frac{\text{amount}}{\text{total}} = \frac{37}{50}$
2. Multiply the fraction by 100% and express as a percentage.	$\frac{37}{50} \times 100\% = 74\%$
b. 1. Write the amount as a fraction of the total.	b. $\frac{\text{amount}}{\text{total}} = \frac{18}{20}$
2. Multiply the fraction by 100% and express as a percentage.	$\frac{18}{20} \times 100\% = 90\%$
c. 1. Write the amount as a fraction of the total.	c. $\frac{\text{amount}}{\text{total}} = \frac{23\text{ g}}{30\text{ g}}$
2. Multiply the fraction by 100% and express as a percentage.	$\frac{23}{30} \times 100\% = 76.666\%$
3. Round to the nearest percentage.	77%

d. 1. Write the amount as a fraction of the total.

d. $\frac{\text{amount}}{\text{total}} = \frac{2.15}{17.2}$

2. Multiply the fraction by 100% and express as a percentage.

$\frac{2.15}{17.2} \times 100\% = 12.5\%$

3. Write the answer.

12.5%

4.3.2 Expressing one amount as a percentage of another for different units [complex]

- When expressing one amount as a percentage of another, make sure that you express both amounts in the same unit.

WORKED EXAMPLE 8 Expressing one amount as a percentage of another for different units [complex]

Write 92 cents as a percentage of $5, correct to 1 decimal place.

THINK	WRITE
1. Since the values are in different units, convert the larger unit into the smaller one.	$5 = 500 cents
2. Write the first amount as a fraction of the second.	$\frac{92}{500}$
3. Multiply the fraction by 100%.	$\frac{92}{500} \times 100\%$
4. Calculate the percentage and round to 1 decimal place as required.	$= 18.4\%$

WORKED EXAMPLE 9 Expressing one amount as a percentage of another for different units in a real-world context [complex]

If Rebecca Allen scored 12 baskets out of her 15 shots from the free throw line, calculate the percentage of free throws that she got in.

THINK / **WRITE**

1. Write the amount as a fraction of the total.

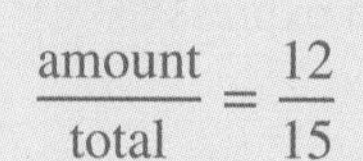

$\frac{\text{amount}}{\text{total}} = \frac{12}{15}$

2. Multiply the fraction by 100% and express as a percentage.

$\frac{12}{15} \times 100\% = 80\%$

4.3.3 Percentage discounts

- To determine a percentage discount, determine the fraction $\frac{\text{discounted amount}}{\text{original price}}$ and multiply by 100.

Determining a percentage discount

$$\textbf{Percentage discount} = \frac{\textbf{discounted amount}}{\textbf{original price}} \times \textbf{100}$$

tlvd-3806

WORKED EXAMPLE 10 Determining a percentage discount

A T-shirt was on sale, reduced from \$89.90 to \$75.45. Calculate the percentage that was taken off the original price. Give your answer to 1 decimal place.

THINK	WRITE
1. Calculate the amount saved.	Discounted amount $= \$89.90 - \$75.45$ $= \$14.45$
2. Divide the discounted amount by the original price.	$\frac{\text{discounted amount}}{\text{original amount}} = \frac{14.45}{89.90}$
3. Multiply the fraction by 100% to get the percentage.	$\frac{14.45}{89.90} \times 100\% = 16.0734\%$
4. Round to 1 decimal place.	16.1% discount on the original price

Resources

Interactivities One amount as a percentage of another (int-3998)
Discount (int-3744)

Exercise 4.3 Express as a percentage

learnon

4.3 Quick quiz 

4.3 Exercise

Simple familiar	Complex familiar	Complex unfamiliar
1, 2, 3, 4, 5, 6, 7	8, 9, 10, 11, 12, 13, 14	15, 16

Simple familiar

1. WE7 Express each of the following to the nearest percentage.
   a. 64 as a percentage of 100
   b. 13 as a percentage of 20
   c. 9 g as a percentage of 30 g
   d. 8.20 kg as a percentage of 18.6 kg
2. Express each of the following to the nearest percentage.
   a. 37 as a percentage of 50
   b. 21 as a percentage of 40
   c. 15 mL as a percentage of 23 mL
   d. 5.50 g as a percentage of 22.6 g

3. Express each of the following to the nearest percentage.
   a. 36 as a percentage of 82
   b. 45 as a percentage of 120
   c. 12 as a percentage of 47
   d. 9 as a percentage of 15
   e. 15 kg as a percentage of 44 kg
   f. 67 seconds as a percentage of 175 seconds
4. Express the following as a percentage, giving your answers to 1 decimal place.
   a. 23.5 of 69
   b. 59.3 of 80
   c. 45.75 of 65
   d. 23.82 of 33
   e. 0.85 of 5
   f. 1.59 of 2.2
5. **WE9** If Buddy Franklin kicked 6 goals from his 13 shots at goal in a game, calculate the percentage of goals he kicked. Give your answer correct to 2 decimal places.
6. If Kate Moloney has 8 of her team's 32 intercepts in a game, calculate the percentage of the team's intercepts that Kate had.

7. If a student received 48 out of 55 for a Mathematics test, calculate the percentage, correct to 2 decimal places, that the student got on the test.

Complex familiar

8. **WE8** Express 53c as a percentage of $3, to 1 decimal place.
9. Express 1500 m as a percentage of 5 km.
10. Express the following as a percentage, giving your answers to 1 decimal place.
    a. 68c of $2
    b. 31c of $5
    c. 67 g of 2 kg
    d. 0.54 g of 1 kg
    e. 546 m of 2 km
    f. 477 m of 3 km
    g. 230 mm of 400 cm
    h. 36 min of 3 hours
11. Corn Flakes have 7.8 g of protein per 0.1 kg. Calculate the percentage of protein in Corn Flakes.

12. **WE10** If a pair of boots were reduced from $155 to $110, calculate the percentage discount offered on the boots. Give your answer to 1 decimal place.

13. Olivia saved $35 on the books she purchased, which were originally priced at $187.
    a. Calculate how much Olivia paid for the books.
    b. Calculate what percentage, correct to 1 decimal place, Olivia saved from the original price.

14. Concession movie tickets sell for $12.00 each, but if you buy 4 or more you get $1.00 off each ticket.
    Calculate the percentage discount, correct to 2 decimal places.
    Hint: Find $1 as a percentage of $12.

Complex unfamiliar

15. A store claims to be taking 40% off the prices of all items, but your friend is not so sure. Calculate the percentage discount on each of the items shown and determine if the store has been completely truthful.

16. Two markets have a deal available on soaps.
    - At the local market there is a 'buy two, get one free' offer on handmade soaps.
    - At a rival market there is a 'buy one, get another half price' offer on soaps.

    Determine if one deal is better than the other, or if they are the same.
    Explain your reasoning with calculations.

Fully worked solutions for this chapter are available online.

LESSON
4.4 Percentage increase and decrease [complex]

SYLLABUS LINKS

- Apply percentage increases and decreases in real-world contexts, including mark-ups, discounts and GST [complex].
- Determine the overall change in a quantity following repeated percentage changes [complex].

4.4.1 Percentage increase (mark-up)

- Just as some items are on sale or reduced, some items or services increase in price, known as a **mark-up**. Examples of increases in price can be the cost of electricity or a new model of car.
- We calculate the percentage increase in price in a similar way to the percentage decrease; however, instead of subtracting the discounted value, we add the extra percentage value.

WORKED EXAMPLE 11 Calculating percentage increase (mark-up)

Ashton paid $450 for his last electricity bill, and since then the charges have increased by 5%.

a. **Assuming he used the same amount of electricity, determine how much extra he would expect to pay on his next bill.**

b. **Calculate his total bill.**

THINK	WRITE
a. **Method 1:**	
1. To calculate the increase amount, multiply the percentage increase by the previous cost.	a. 5% of $\$450 = 0.05 \times \$450$ $= \$22.50$
2. Write the amount of the extra charge.	The extra charge is $22.50.
Method 2:	
3. Calculate the increased price.	Increased price $= \left(1 + \frac{5}{100}\right) \times \$450$ $= 1.05 \times \$450$ $= \$472.50$
4. Calculate the difference in prices.	Difference $= \$472.50 - \$450$ $= \$22.50$
b. 1. Calculate the total cost of the bill by adding the extra charge to the original cost.	b. Total bill $= \$450 + \$22.50$
2. Write the total cost of the bill.	Total bill $= \$472.50$

4.4.2 Percentage decrease (discount)

- A percentage decrease is often used to discount goods that are on sale.
- A **discount** is a reduction in price from the original marked price.
- When the discount is expressed as a percentage, to determine the amount of the discount, we need to multiply the original price by the discount percentage expressed as a decimal or use the formula $\frac{\text{percentage change}}{100} \times \text{original price}$.

Percentage decrease

$$\textbf{Discount} = \frac{\textbf{percentage change}}{\textbf{100}} \times \textbf{original price}$$

For example, a 10% discount on an item marked at $150 gives a discount of:

$$\begin{aligned}\frac{10}{100} \times \$150 &= 0.1 \times \$150 \\ &= \$15\end{aligned}$$

Calculating the discounted price

The discounted price can be calculated in two ways.

Method 1

- Calculate the discount by multiplying the original price by the percentage expressed as a decimal.
- Subtract the discount from the original price.

Method 2

- $\text{Discounted price} = \left(1 - \frac{\text{percentage discount}}{100}\right) \times \text{original amount}$

tlvd-3807

WORKED EXAMPLE 12 Calculating the discounted price

A store has a 15%-off-everything sale. Molly purchases an item that was originally priced at $160.
a. Calculate the price Molly paid for the item.
b. Calculate how much Molly saved from the original price.

THINK	WRITE
a. Method 1: 1. Discount is 15% of $160, so multiply 0.15 (15% as a decimal) by $160.	a. $\begin{aligned}\text{Discount} &= 0.15 \times \$160 \\ &= \$24\end{aligned}$
2. Determine the discounted price by subtracting the discount from the original price.	$\begin{aligned}\text{Discounted price} &= \$160 - \$24 \\ &= \$136\end{aligned}$
3. Write the discounted price.	Molly paid $136.

Method 2:

4. Calculate the discounted price.

$$\text{Discounted price} = \left(1 - \frac{15}{100}\right) \times 160$$
$$= (1 - 0.15) \times 160$$
$$= 0.85 \times 160$$
$$= \$136$$

b. Method 1:

1. Calculate the saving as a percentage of the original price.

b. Saving $= 0.15 \times \$160$
$= \$24$

2. Write the amount saved.

Molly saved $24.

Method 2:

3. Calculate the difference between the original price and the discounted price (found in part **a**).

Saving $= \$160 - \$136$
$= \$24$

4. Write the amount saved.

Molly saved $24.

4.4.3 Investigating GST

- **GST** is a tax imposed by the Australian federal government on goods and services. (As with all taxes, there are exemptions, but these will not be considered here.)
  - Goods: A tax of 10% is added to new items that are purchased, such as petrol, clothes and some foods.
  - Services: A tax of 10% is added to services that are paid for, such as work performed by plumbers, painters and accountants.

Calculating the amount of GST

The amount of GST on an item can be determined by dividing by 10 if the price is pre-GST, or by dividing by 11 if the price is inclusive of GST.

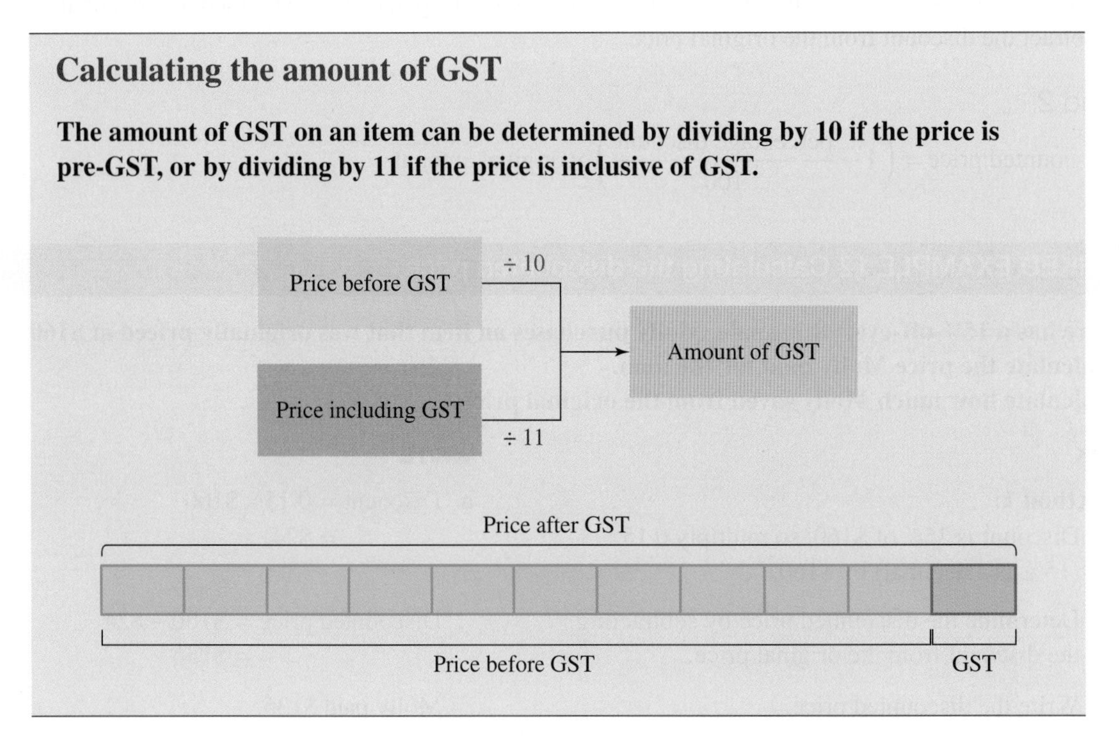

WORKED EXAMPLE 13 Calculating the amount of GST

A packet of potato chips costs \$1.84 before GST.
Calculate:
a. **the GST charged on the packet of chips**
b. **the total price the customer has to pay, if paying with cash.**

THINK	WRITE
a. 1. GST is 10%. Calculate 10% of 1.84.	a. 10% of \$1.84 = $\frac{\$1.84}{10}$ or \$0.184
2. Write the answer in a sentence.	The GST charged on the packet of chips is \$0.18 (rounded).
b. 1. Total equals pre-GST price plus GST.	b. \$1.84 + \$0.18 = \$2.02
2. Write the answer in a sentence.	The total price the customer has to pay is \$2.00 (rounded down by the seller).

4.4.4 Repeated percentage change

- In many situations, the percentage change differs over the same time period.
  For example, the population of a small town increases by 10% in one year and only by 4% the next year.
- If the initial population of the town is known to be 15 500, then percentage multipliers can be used to calculate the population of the town after the 2 years.

> **Percentage change over a time period**
>
> **The multiplying factor is $1 + \frac{i}{100}$, where i is the percentage increase,**
> **and $1 - \frac{i}{100}$, where i is the percentage decrease.**

- An increase of 10% means the population has increased by a factor of $1 + \frac{10}{100} = 1.10$ and an increase of 4% means the population has increased by a factor of $1 + \frac{4}{100} = 1.04$.
  Therefore, the population after 2 years is:

$$15\,500 \times 1.10 \times 1.04 = 17\,732$$

Overall percentage change

- To calculate the overall percentage change in the situation described previously, multiply the two percentage factors: $1.10 \times 1.04 = 1.144$. This is 0.144 greater than 1.
  Convert 0.144 to a percentage by multiplying by 100.
- That gives a percentage change of $0.144 \times 100 = 14.4\%$.

tlvd-3808

WORKED EXAMPLE 14 Calculating repeated percentage change

Calculate the overall change in the following quantities.

a. A small tree 1.2 metres tall grew by 22% in the first year and 16% in the second year. Determine its height at the end of the second year correct to 2 decimal places.

b. An investment of $12 000 grew by 6% in the first year and decreased by 1.5% in the second year. Calculate the value of the investment at the end of the second year.

THINK	WRITE
a. 1. The multiplying factors are found using the two percentages, 22% and 16%, and $1+\frac{i}{100}$.	**a.** First percentage change $= 1+\frac{22}{100} = 1.22$ Second percentage change $= 1+\frac{16}{100} = 1.16$
2. Multiply the initial height of the tree by 1.22 and 1.16.	$1.2 \times 1.22 \times 1.16 = 1.698$
3. Write the answer.	The height of the tree after two years is 1.70 m correct to 2 decimal places.
b. 1. The multiplying factors are found using 6% and $1+\frac{i}{100}$ for the increase and 1.5% and $1-\frac{i}{100}$ for the decrease.	**b.** First percentage change $= 1+\frac{6}{100} = 1.06$ Second percentage change $= 1-\frac{1.5}{100} = 0.985$
2. Multiply the initial investment by 1.06 and 0.985.	$12\,000 \times 1.06 \times 0.985 = 12\,529.20$
3. Write the answer.	The value of the investment after two years is $12 529.20.

WORKED EXAMPLE 15 Applying a percentage increase or decrease in a real-world context

You are in a surf shop and you hear, 'For today only: take 50 per cent off the original price and then a further 40 per cent off that.' You hear a customer say, 'This is fantastic! You get 90 per cent off the original price!'
Determine if this statement is correct and explain your answer.

THINK	WRITE
1. Assume an original price.	Say that the original price is $1000.
2. Calculate the 50 per cent off the original price.	Discount 1 = 50% Sale price after discount 1 = 50% × $1000 = $500
3. Calculate the 40 per cent off the sale price after discount 1.	Discount 2 = 40% Sale price after discount 2 = 60% × $500 = $300 This represents a discount of 70% in total.

4. Calculate a discount of 90 per cent off the \$1000 price.	Sale price after discounting by 90% = 10% × \$1000 = \$100
5. Write the conclusion.	No, the statement is not correct. The outcomes are quite different. You need to apply subsequent discounts in a sequence. You cannot add them and apply them as one discount.

Resources

Interactivities Percentage increase and decrease (int-3742)
Goods and Services Tax (int-3748)

Exercise 4.4 Percentage increase and decrease [complex]

learnon

4.4 Quick quiz on | **4.4 Exercise**

Simple familiar	Complex familiar	Complex unfamiliar
N/A	1, 2, 3, 4, 5, 6, 7, 8, 9, 10, 11, 12, 13, 14, 15, 16, 17, 18, 19	20, 21, 22

These questions are even better in jacPLUS!
- Receive immediate feedback
- Access sample responses
- Track results and progress

Find all this and MORE in jacPLUS

Complex familiar

1. WE11 This year it costs \$70 for a ticket to watch the Australian Open at the Rod Laver Arena. Tickets to the Australian Open are due to increase by 10% next year.
   a. Determine the amount the tickets will increase by.
   b. Determine how much you would expect to pay for a ticket to the Australian Open next year.

2. A television set costs \$2500 this week, but next week there will be a price increase of 7.5%. Calculate how much you would expect to pay for the television set next week.

3. When you purchase a new car, you also need to pay government duty on the car. You must pay a duty of 3% of the value of the car for cars priced up to \$59 133 and 5% of the value for cars priced over \$59 133.
   a. Calculate the duty on each of the following car values.
      i. \$15 500 ii. \$8000
      iii. \$21 750 iv. \$45 950
      v. \$65 000 vi. \$78 750
   b. Calculate the cost of each of the cars including the government duty.

4. A sports shop is having an end-of-year sale on all stock. A discount of 20% is applied to clothing and a discount of 30% is applied to all sports equipment. Calculate the discount you will receive on the following items.
   a. A sports jumper initially marked at $175
   b. A basketball hoop marked at $299
   c. A tennis outfit priced at $135
   d. A table tennis table initially priced at $495
5. **WE12** A store has a 25%-off-everything sale. Bradley purchases an item that was originally priced at $595.
   a. Calculate how much Bradley paid for the item.
   b. Calculate how much he saved from the original price.
6. Nicole purchases a new pair of shoes that were marked down by 20%. Their original price was $139.95.
   a. Calculate how much Nicole paid for the new shoes.
   b. Calculate how much Nicole saved.

7. A new suit is priced at $550. Calculate how much you would pay if you received the following percentage discount:
   a. 5% b. 10% c. 15%
   d. 20% e. 30% f. 50%
8. A pool holds 55 000 L of water. During a hot summer week, the pool lost 5.5% of its water due to evaporation.
   a. Calculate how much water was lost during the week.
   b. Calculate how much water the pool held at the end of the week.
9. **WE13** The pre-GST price of a packet of laundry powder is $4.50.
   a. Calculate the GST on the laundry powder.
   b. Calculate the total price, including GST, of the laundry powder.
10. The Australian government introduced the Goods and Services Tax (GST) in 2000. It added 10% to the price of a variety of goods and services. Calculate the price after GST has been added to the following pre-tax prices.
   a. $189 car service
   b. $1650 dining table
   c. $152.50 pair of runners
   d. $167.85 pair of sunglasses
11. Depreciation is a financial process that calculates the decreasing dollar value of new cars. A new car was purchased for $35 000. It depreciates from its current value each year by the percentage shown in the table.
   a. Complete the table.

End year		% depreciation	Depreciation ($)	Value ($)
End year	**First**	10%	$3500	$31 500
	Second	7.5%	$2362.50	
	Third	6%		
	Fourth	5%		
	Fifth	4%		

   b. If the car continued to lose 2% each year in the sixth, seventh, eighth and ninth year, calculate how much the car would be worth at the end of the ninth year.

12. **WE14** Calculate the overall change for each of the following situations.
   a. A new IT company of 2400 employees grew its workforce by 3% in the first year and 4.5% in the second year. Calculate how many employees the company had at the end of the second year.
   b. Roya had invested \$250 000 in her superannuation fund. The fund grew at 6% in the first year but had negative growth of 3.5% in the second year.
   Determine how much was in the superannuation fund at the end of the second year.
13. Chris purchased a new car for \$42 000. After 1 year, the value of the car had decreased by 8%. In the following year, the value of the car decreased by a further 5.2%.
   a. Calculate the value of the car at the end of the two years.
   b. Calculate the overall percentage change of the value of the car at the end of the two years. Give your answer correct to the nearest per cent.
14. An amount of \$55 000 is invested for 7 years in an account that grows by 6.5% every year. Calculate how much the investment is worth at the end of the 7 years.
15. Priscilla takes her savings of \$17 000 and invests it in a fixed term deposit that pays 5% each year on the money that is in the account. At the end of the fixed term, Priscilla's savings are worth \$20 663.60. Using trial and error, determine the length in years of the fixed term.
16. **WE15** A clothing store has a rack with a 50% off sign on it. As a weekend special, for an hour the shop owner takes a further 50% off all items in the store. Explain whether that means that you can get clothes for free and give reasons for your answer.
17. Heidi decided to purchase a bottle of perfume for her mother on Mother's Day. Heidi was able to get her mother's favourite perfume at 20% off the original price of \$115.50.
   Calculate the price Heidi paid for the perfume.
18. Alex has been saving his money for a new tennis racquet. Alex has been monitoring the website that sells top brands and notices a Yonex Ezone racquet for \$329.95.
   The next week he notices there is a 20%-off sale on all racquets on this website. After talking to some friends, Alex decides to wait until the end of financial year sale in case it is marked down further. To Alex's delight, another 15% is taken off the sale price, and he then purchases the racquet.
   Calculate how much Alex paid for the racquet.

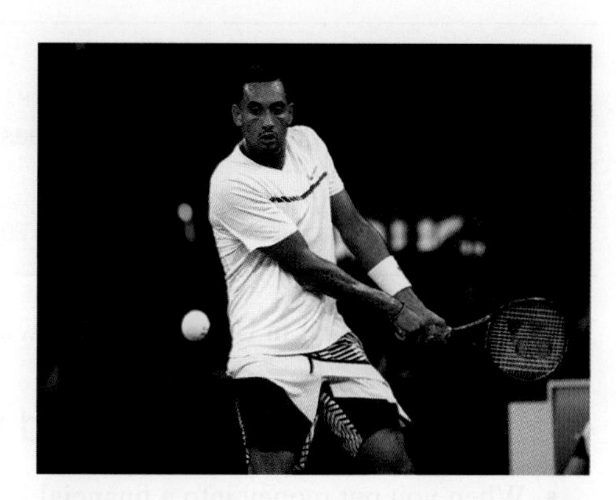

19. Store A had to increase their prices by 10% to cover expenses, whereas store B was having a 15% off sale over the entire store. Sarah wanted to purchase a soundbar and assumed it would be better to get it from store B since it was having a sale. The original price of the soundbar at store A was \$185, whereas at store B it was originally priced at \$240.
   a. Determine if Sarah's assumption was correct. Prove this mathematically.
   b. Calculate the difference in the prices of the soundbar.

Complex unfamiliar

20. A classmate was completing a discount problem where they needed to calculate a 25% discount on $79. They misread the question and calculated a 20% discount to get $63.20. They then realised their mistake and took a further 5% from $63.20. Explain if this is the same as taking 25% off $79.
Use calculations to support your answer.

21. Henry buys a computer priced at $1060, but with a 10% discount. Sancha finds the same computer selling at $840 plus a tax of 18%. Determine who has the better buy. Explain your reasoning.

22. Jane wishes to purchase a speaker. They compared two shops' offers.
    - Shop A: $250, reduced by 15%, and today only reduced a further 10%
    - Shop B: On sale for $240 not including GST

    Determine whether Jane should purchase the speaker from Shop A or Shop B.

Fully worked solutions for this chapter are available online.

LESSON
4.5 Real-world application of percentages — Simple interest [complex] (optional)

SYLLABUS LINKS

- Apply percentage increases and decreases in real-world contexts, including mark-ups, discounts and GST [complex].

This lesson is optional in Units 1 & 2 and will be useful for Units 3 & 4.

4.5.1 Principal, interest and value

- When you put money into a financial institution such as a bank or credit union, the amount of money you start with is called the **principal**.
- People who place money in a bank or financial institution (**investors**) receive a payment called **interest** from the financial institution in return for investing their money in the financial institution.
- The amount of interest is determined by the **interest rate**.
- An interest rate is a percentage for a given time period, usually a year.
  For example, a bank might offer 5.8% per year interest on its savings accounts. This is also written as 5.8% per annum, or 5.8% p.a.

Simple interest calculations

- Simple interest is the interest paid on the principal amount of an investment.
- The amount of interest paid each time period is based on the principal, so the amount of interest is constant.
  For example, $500 placed in an account that earns 10% simple interest per year earns $50 each year, as shown in the following table.

Time period (years)	Amount of money at the start of the year	Amount of interest after one year
1	\$500	\$50
2	\$550	\$50
3	\$600	\$50
4	\$650	\$50

The total interest earned is $4 \times \$50 = \$200$, so the value of the investment after 4 years is $\$500 + \$200 = \$700$.

Simple interest

$$I = Pin$$

Where:

- ***P* is the principal money invested or borrowed**
- ***i* is the interest rate per period as a decimal**
- ***n* is the total number of periods.**

- If i is given as a percentage per year, then the time n must be given in years.
- If i is given as a percentage per month, then n must be given in months.
- The total amount of money is known as the value of the investment.

Total value of the investment

Value = principal (*P*) + interest (*I*)

tlvd-3809

WORKED EXAMPLE 16 Calculating simple interest

A real estate developer offers investors a chance to invest in the company's latest development. If \$10 000 is invested, the developer will pay 11.5% simple interest per year for 5 years. Calculate the value of the investment.

THINK	WRITE
1. Identify the known quantities.	$P = \$10\,000$ $i = 0.115$ (11.5% as a decimal) $n = 5$
2. Calculate the amount of interest (I) using the formula $I = Pin$.	$\text{Interest} = \$10\,000 \times 0.115 \times 5$ $= \$5750$
3. The value of the investment after 5 years is the sum of the interest (\$5750) and the principal (\$10 000).	$\text{Value} = \$10\,000 + \$5750$ $= \$15\,750$
4. Write the answer as a sentence.	The value of the investment after 5 years is \$15 750.

- In some cases, the time period for which the money is invested is less than the time that the interest rate is quoted for — for example, if the interest rate is given as a per-year rate and the money is invested for 3 months.
- In these cases, you need to calculate the equivalent interest rate for the time for which the money is invested.

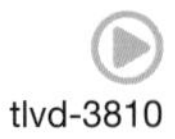
tlvd-3810

WORKED EXAMPLE 17 Calculating simple interest if the time period is not a multiple of time

A bank offers interest on its savings account at 3% per year. If a Year 11 student opens an account with $600 and leaves the money there for 4 months, calculate how much interest they earned.

THINK	WRITE
1. The interest rate period is per year and the amount of time the money is invested for is in months. Calculate an equivalent monthly interest rate. Since there are 12 months in a year, divide the annual interest rate by 12 to get a monthly interest rate.	$\frac{3}{12} = \frac{1}{4}$ 3% per year is 0.25% per month.
2. Identify the known quantities. Note that we are now dealing in months, rather than years.	$P = 600$ $i = 0.0025$ (0.25% per month) $n = 4$ months
3. Calculate the amount of interest (I) using the formula $I = Pin$.	$\text{Interest} = 600 \times 0.0025 \times 4$ $= 6$
4. Write the answer as a sentence.	The total interest earned in 4 months is $6.

4.5.2 Calculating the interest rate

- If the interest (I), principal (P) and time period (n) are known, it is possible to calculate the interest rate (i) using the simple interest formula.

tlvd-3811

WORKED EXAMPLE 18 Calculating the interest rate

A Year 11 student is paid $65.40 in interest for an original investment of $800 for 2 years. Calculate the annual interest rate.

THINK	WRITE
1. Identify the known quantities: • interest (I) • principal (P) • time period (n)	$I = \$65.40$ $P = \$800$ $n = 2$ years
2. Substitute the values into the formula $I = Pin$.	$65.40 = 800 \times r \times 2$
3. Solve the equation for i to calculate the annual interest rate by dividing both sides by 1600.	$65.40 = 1600 \times i$ $\frac{65.40}{1600} = i$ $i = 0.040\,875$
4. Write the answer.	The annual interest rate is 4.09%.

Determining P, i or n

To determine the principal: $P = \dfrac{I}{in}$

To determine the interest rate: $i = \dfrac{I}{Pn}$

To determine the number of periods: $n = \dfrac{I}{Pi}$

 Resources

 **Interactivity** Simple interest (int-6074)

Exercise 4.5 Real-world application of percentages — Simple interest [complex] (optional)

learn on

4.5 Quick quiz on | **4.5 Exercise**

Simple familiar	Complex familiar	Complex unfamiliar
N/A	1, 2, 3, 4, 5, 6, 7, 8, 9, 10, 11, 12	13, 14

These questions are even better in jacPLUS!
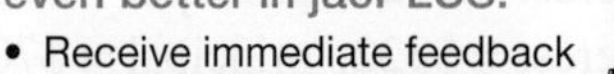
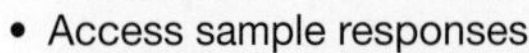
- Receive immediate feedback
- Access sample responses
- Track results and progress

Find all this and MORE in jacPLUS

Complex familiar

1. WE16 A film producer offers investors the chance to invest in their latest movie. If \$20 000 is invested, the producer will pay 22.3% simple interest per year for 2 years. Determine the value of the investment at the end of the 2 years.
2. If an investment of \$400 pays 8% simple interest per year, calculate the value of the investment at the end of 3 years.
3. Calculate the simple interest paid on the following investments.
   a. \$500 at 6.7% per year for 2 years
   b. \$500 at 6.7% per year for 4 years
   c. \$1000 at 6.7% per year for 4 years
4. Calculate the amount of interest paid on a \$1000 investment at 5% for 5 years.
5. WE17 A bank offers interest on its savings account of 6% p.a.
   If a Year 11 student opens an account with \$750 and leaves the money there for 5 months, calculate how much interest they earned.

6. If the annual interest rate is 8%, calculate the monthly interest rate.

7. Calculate the interest paid on the following investments.
   a. \$500 invested at 8% per annum for 1 month
   b. \$500 invested at 8% per annum for 3 months
   c. \$500 invested at 8% per annum for 6 months
8. Calculate the value of the following investments.
   a. \$1000 invested at 10% p.a. for 10 years
   b. \$1000 invested at 12% p.a. for 10 months
   c. \$1000 invested at 6% p.a. for 3 years
9. A bank offers investors an annual interest rate of 9% if they buy a term deposit. If a customer has \$5600 and leaves the money in the term deposit for 2.5 years, calculate the value of the investment at the end of the 2.5 years.
10. WE18 A Year 11 student is paid \$79.50 in interest for an original investment of \$500 for 3 years. Calculate the annual interest rate.
11. Bank A offers an interest rate of 7.8% on investments, while Bank B offers an interest rate of 7.4% in the first year and 7.9% in subsequent years. If a customer has \$20 000 to invest for 3 years, determine which is the better investment.
12. Determine the annual interest rate on the following investments.
   a. Interest = \$750, principal = \$6000, time period = 4 years
   b. Interest = \$924, principal = \$5500, time period = 3 years
   c. Interest = \$322, principal = \$7000, time period = 3 months

Complex unfamiliar

The following statement relates to questions 13 and 14.

A loan is an investment in reverse; you *borrow* money from a bank and are *charged* interest. The value of a loan becomes its total cost.

13. Your parents decide to borrow money to improve their boat but cannot agree which loan has better value. They would like to borrow \$2550. Your mother goes to Bank A and finds that they will lend the money at $11\frac{1}{3}\%$ simple interest per year for 3 years. Your father finds that Bank B will lend the \$2550 at 1% per month simple interest.

    Determine which bank offers the best rate over the three years and justify your answer.
14. A worker wishes to borrow \$10 000 from a bank, which charges 11.5% interest per year. If the loan is over 2 years, calculate the monthly payment for the loan.

    Hint: Most loans require a monthly payment. The monthly payment of a simple interest loan is calculated by dividing the total cost of the loan by the number of payments made during the term of the loan.

Fully worked solutions for this chapter are available online.

LESSON
4.6 Review

4.6.1 Summary

doc-41752

Hey students! Now that it's time to revise this chapter, go online to:

- Access the chapter summary
- Review your results
- Watch teacher-led videos
- Practise questions with immediate feedback

Find all this and MORE in jacPLUS

4.6 Exercise

learnon

4.6 Exercise

Simple familiar	Complex familiar	Complex unfamiliar
1, 2, 3, 4, 5, 6, 7, 8, 9, 10, 11, 12, 13	14, 15, 16, 17, 18	19, 20

These questions are even better in jacPLUS!
- Receive immediate feedback
- Access sample responses
- Track results and progress

Find all this and MORE in jacPLUS

Simple familiar

1. Express 45% as a simplified fraction.
2. Calculate 64% of 280.
3. Calculate 0.27 km of 1.5 km as a percentage.
4. The price of a new $16 000 car increased by 7.5%. Calculate the amount by which the car increased in price.
5. The annual salary of a sales assistant increased from $45 000 to $48 600. Calculate the percentage salary increase.
6. Air contains 21% oxygen, 0.9% argon, 0.1% trace gases, and the remainder is nitrogen. Calculate the percentage of nitrogen in the air.
7. A tree was planted when it was 2.3 m tall. It grew by 30% in the first year and 12% in the second year after planting. Calculate the height of the tree at the end of the second year.
8. If a house that cost $200 000 to build is sold for $10 000 less, calculate the percentage loss.
9. Determine the sale price of the car shown.

10. If an investment of \$400 pays 8% simple interest per year, calculate the value of the investment at the end of 3 years.

11. A person earns total interest of \$600 on an investment of \$6000 after 4 years. Calculate the annual interest rate.

12. The winner of a football tipping competition wins 65% of the prize pool, second place wins 20% and third place receives what is left over. The prize pool is \$860. Calculate the amount of the third prize.

13. On a weekend sale, the price of all laptops was reduced by 30%.

Determine the sale price of a laptop valued at \$1500.

Complex familiar

14. Express the following as percentages.
    a. 756 m of 2.7 km
    b. 45 g of 0.85 kg
    c. 15c of \$2.90
    d. 45 s of 1.5 min

15. A person invests \$100 each month, earning simple interest of 1% per month for 4 months.
    a. Calculate the interest on the first investment, which earns interest for 4 months.
    b. Calculate the interest on the second, third and fourth investments, which earn interest for 3 months, 2 months and 1 month respectively.
    c. Calculate the total value of the investment at the end of the 4 months.

16. The GST rate is 10%. This means that when a business sells something or provides a service, it must charge an extra $\frac{1}{10}$ of the price/cost. That extra money then must be sent to the tax office.
    For example, an item that would otherwise be worth \$100 now has GST of \$10 added, so the price tag will show \$110. The business will send to the tax office that \$10 with all the other GST it has collected on behalf of the government.
    Suppose a shopkeeper made sales totalling \$15 400. Calculate the amount of GST he should send to the tax office.

17. In the country Snowdonia, GST is 12.5%. Kira has purchased a new hairdryer that cost 111 kopeks including GST. There are 100 plens in 1 kopek.
    Calculate the amount of GST Kira paid.

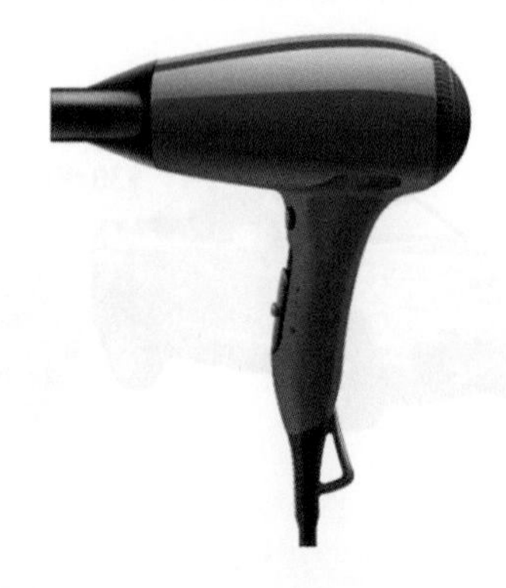

18. An expensive lamp has a regular price of $200. At that price, the shop expects to sell 4 lamps per month.
    - If offered at a discount of 10%, the shop will sell 8 lamps per month.
    - If offered at a discount of 20%, the shop will sell 16 lamps per month.
    - If offered at a discount of 30%, the shop will sell 20 lamps per month.
    - These lamps cost the shop $100 each.

    Determine which discount results in the highest total sales.

Complex unfamiliar

19. Anastasia is holding a birthday party to which she has invited 25 friends and 15 family members. On the day, not everybody turns up. Anastasia couldn't remember exactly how many turned up, but she knows that they had to reduce the amount of food by 20%.
    Determine how many people turned up.

20. Answer the following questions.
    a. If you increase $100 by 30% and then decrease the new amount by 30%, determine whether you will end up with more than $100, less than $100 or exactly $100.
    Explain your findings, using mathematics to support your answer.
    b. If you decrease $100 by 30% and then increase the new amount by 30%, determine whether you will end up with more than $100, less than $100 or exactly $100.
    Explain your findings, using mathematics to support your answer.
    c. Determine the percentage change needed in parts **a** and **b** to get back to the original $100 from the new amount.
    d. Determine if the percentage change in part **a** is greater than or less than the percentage change in part **b** and explain your answer.

Fully worked solutions for this chapter are available online.

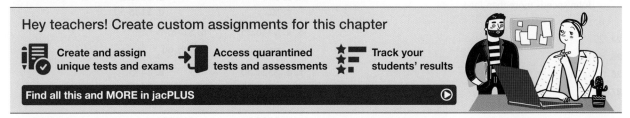

Answers

Chapter 4 Percentages

4.2 Calculating percentages

4.2 Exercise

1. a. $\frac{12}{25}$ b. $\frac{67}{250}$ c. $\frac{31}{250}$ d. $\frac{1}{125}$

2. a. $\frac{23}{25}$ b. $\frac{593}{800}$ c. $\frac{2}{3}$

3. a. $\frac{22}{25}$ b. $\frac{1}{4}$ c. $\frac{23}{2500}$
   d. $\frac{71}{200}$ e. $\frac{38}{125}$ f. $\frac{291}{400}$

4. a. 0.73 b. 0.943

5. a. 0.024 96 b. 0.0062

6. a. 0.43 b. 0.39 c. 0.80
   d. 0.4725 e. 0.2405 f. 0.0083

7. a. 21% b. 80%

8. a. 65.2% b. 32%
   c. 55% d. 83%

9. a. 50% b. 87.5%
   c. 220% d. 165%

10. a. 72 b. 80.64 kg
    c. 157.5 d. $28.81

11. a. 88 b. 327 kg c. 31 m
    d. 217 e. 753 g f. 956

12. $59.50

13. $24

14. a. $5.60 b. 12.5 kg
    c. 55.5 d. 137

15. 56

16. 825 000

17. $495

18. 6525 litres

19. a. 14 cents b. $3.24
    c. 15 cents d. $3.39

20. $131.25

21. a. i. $52.00 ii. $19.60 iii. $71.92
    b. $35.88

22. $87.75

23. a. John spends $420.
    Jane spends $336.
    John spends more than Jane.
    b. $84

24. 27

25. 30 kg

26. 75 apples

27. $2

4.3 Express as a percentage

4.3 Exercise

1. a. 64% b. 65%
   c. 30% d. 44%

2. a. 74% b. 53%
   c. 65% d. 24%

3. a. 44% b. 38%
   c. 26% d. 60%
   e. 34% f. 38%

4. a. 34.1% b. 74.1%
   c. 70.4% d. 72.2%
   e. 17.0% f. 72.3%

5. 46.15%

6. 25%

7. 87.27%

8. 17.7%

9. 30%

10. a. 34.0% b. 6.2%
    c. 3.4% d. 0.1%
    e. 27.3% f. 15.9%
    g. 5.8% h. 20.0%

11. 7.8%

12. 29.0%

13. a. $152 b. 18.7%

14. 8.33%

15. Top 1: 40% discount
    Shoes 1: 40% discount
    Top 2: 40% discount
    Shoes 2: 35.5% discount
    The claim is incorrect for one of the items.

16. Sample responses can be found in the worked solutions in the online resources.

4.4 Percentage increase and decrease [complex]

4.4 Exercise

1. a. $7 b. $77

2. $2687.50

3. a. i. $465 ii. $240
   iii. $652.50 iv. $1378.50
   v. $3250 vi. $3937.50
   b. i. $15 965 ii. $8240
   iii. $22 402.50 iv. $47 328.50
   v. $68 250 vi. $82 687.50

4. a. $35 b. $89.70
   c. $27 d. $148.50

5. a. $446.25 b. $148.75

6. a. $111.96 b. $27.99

7. a. $522.50 b. $495
   c. $467.50 d. $440
   e. $385 f. $275

8. a. 3025 L b. 51 975 L

9. a. $0.45 b. $4.95

10. a. $207.90 b. $1815
c. $167.75 d. $184.64

11. a. See the table at the bottom of this page.*
b. $23 039.83

12. a. 2583 b. $255 725

13. a. $36 630.72 b. 13%

14. $85 469.26

15. 4 years

16. No, you cannot get clothes for free.
You get 50% off the already reduced price. This means the price is first halved, then another 50% off means it is halved again, so you are paying one-quarter of the original price.
For example, if something originally costs $160:
$0.5 \times \$160 = \$80$
$0.5 \times \$80 = \$40$
$40 is one-quarter of $160.

17. $92.40

18. $224.37

19. a. Store A: $203.50
Store B: $204
No, Sarah was not correct; store A was cheaper.
b. 50 cents

20. Method 1: $60.04
Method 2: $59.25
No, it is not the same, since after the 20% discount, the 5% discount is on the reduced value of $63.20, so this method gives a smaller discount.

21. Henry pays $954; Sancha pays $991.20. Henry has the better buy.

22. Shop B is $3.75 cheaper.

4.5 Real-world application of percentages — Simple interest [complex] (optional)

4.5 Exercise

1. $28 920

2. $496

3. a. $67 b. $134 c. $268

4. $250

5. $18.75

6. 0.67%

7. a. $3.33 b. $10 c. $20

8. a. $2000 b. $1100 c. $1180

9. $6860

10. 5.3%

11. Bank A

12. a. 3.125% b. 5.6% c. 18.4%

13. Sample responses can be found in the worked solutions in the online resources.

14. $512.50

4.6 Review

4.6 Exercise

1. $\frac{9}{20}$

2. 179

3. 18%

4. $1200

5. 8%

6. 78%

7. 3.35 m

8. 5%

9. $28 500

10. $496

11. 2.5%

12. $129

13. $1050

14. a. 28% b. 5.29% c. 5.17% d. 50%

15. a. $4 b. $3, $2, $1 c. $410

16. $1400

17. 12 kopeks and 33 plens

18. 30%

19. 32

20. a. 30% of $100 is $30.
$\$100 + \$30 = \$130$
30% of $130 is $39.
$\$130 - \$39 = \$91$
This shows that the final result is less than $100.
b. 30% of $100 is $30.
$\$100 - \$30 = \$70$
30% of $70 is $21.
$\$70 + \$21 = \$91$
This shows that the final result is less than $100.
c. 23%, 43%
d. The percentage change in part a is less than the percentage change in part b. This is because percentage change values are larger when a value is increased to a set point (part b) than when it is decreased to a set point (part a).

*11a.

		% depreciation	Depreciation ($)	Value ($)
End year	First	10%	$3500	$31 500
	Second	7.5%	$2362.5	$31 500 − 2362.5 = $29 137.50
	Third	6%	0.06 × 29 137.5 = $1748.25	$27 389.25
	Fourth	5%	$1369.46	$26 019.79
	Fifth	4%	$1040.79	$24 979

5 Units of energy

LESSON SEQUENCE

Fully worked solutions for this chapter are available online.

 Resources

 **Solutions** Solutions (sol-1258)

 **Digital documents** Learning matrix — Chapter 5 (doc-41753)
Quick quizzes — Chapter 5 (doc-41754)
Chapter summary — Chapter 5 (doc-41755)

LESSON
5.1 Overview

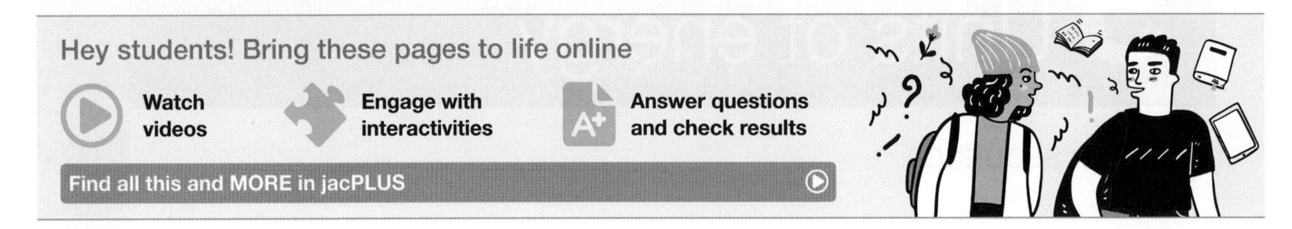

5.1.1 Introduction

The energy we intake from food and drink allows us to perform physical and mental tasks. Food serves as the primary source of energy for the body, providing the essential nutrients needed for growth, repair, and maintenance. By understanding energy intake, we can ensure that we're properly fuelled for the day ahead, whilst not having an excess of energy that can turn to fat.

Electrical appliances also require energy and have changed the way we use energy in our homes. From refrigerators and washing machines to air conditioners and light bulbs, energy-efficient models use innovative features such as improved insulation, high-efficiency motors and smart sensors to optimise energy usage. By consuming less electricity or gas, they not only help lower household expenses but also contribute to environmental conservation by reducing greenhouse gas emissions.

Additionally, governments and organisations worldwide offer incentives and rebates to encourage the adoption of energy-efficient appliances, further incentivising their use.

5.1.2 Syllabus links

Lesson	Lesson title	Syllabus links
5.2	**Units of energy**	◯ Convert from one unit of energy to another including between calories and kilojoules.
5.3	**Energy in food**	◯ Use units of energy to describe the amount of energy for foods, including calories and kilojoules.
5.4	**Energy in activities**	◯ Use units of energy to describe the amount of energy in activities.
5.5	**Energy consumption**	◯ Use units of energy to describe consumption of electricity, e.g. kilowatt hours.

LESSON
5.2 Units of energy

SYLLABUS LINKS

- Convert from one unit of energy to another including between calories and kilojoules.

5.2.1 Common prefixes of the metric system

- The International System of Units (SI) prefixes are symbols used to denote powers of ten of very large or very small numeric values.
- The prefix attaches directly to the name of a unit, and a prefix symbol attaches directly to the symbol for a unit.

Common SI prefixes

Some common SI prefixes are listed in the table below.

Factor	Name	Symbol
10^{-9}	nano	n
10^{-6}	micro	μ
10^{-3}	milli	m
10^{-2}	centi	c
10^{-1}	deci	d
10^{1}	deca	da
10^{3}	kilo	k
10^{6}	mega	M
10^{9}	giga	G

Metric units of mass

- **Mass** and **weight** are different.
- Mass describes how much matter makes up an object, and is measured in kilograms (kg).
- Weight describes the gravitational force acting on an object, and is measured in newtons (N).
- When people talk about the weight of an object, what they are really referring to is its mass.
- The mass of an object can be measured in milligrams (mg), grams (g), kilograms (kg) and tonnes (t).

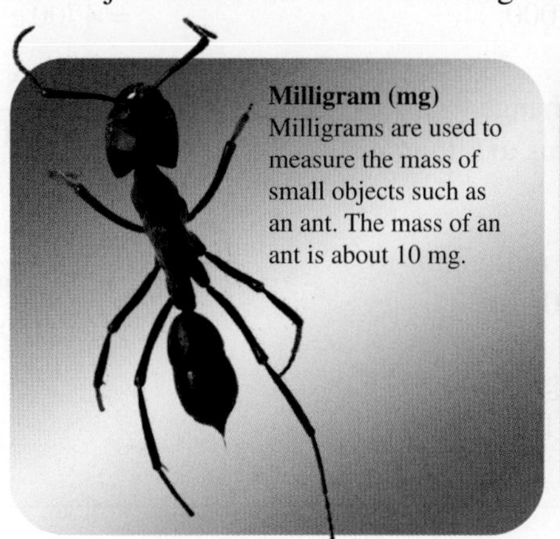

Milligram (mg)
Milligrams are used to measure the mass of small objects such as an ant. The mass of an ant is about 10 mg.

Gram (g)
There are 1000 milligrams in 1 gram. Grams are most commonly used when measuring non-liquid cooking ingredients. The mass of 1 cup of flour is about 250 g.

Kilogram (kg)
There are 1000 grams in 1 kilogram. Kilograms are used to measure larger masses such as the mass of a person.

Tonne (t)
There are 1000 kilograms in 1 tonne. Tonnes are used to measure the mass of very large objects such as a truck.

Converting between units of mass

- The relationships between the units of mass can be used to change a measurement from one unit to another.

$$1000\,\text{mg} = 1\,\text{g}$$
$$1000\,\text{g} = 1\,\text{kg}$$
$$1000\,\text{kg} = 1\,\text{t}$$

Converting between units of mass

The diagram shown can be used to assist in the conversion between different units of mass.

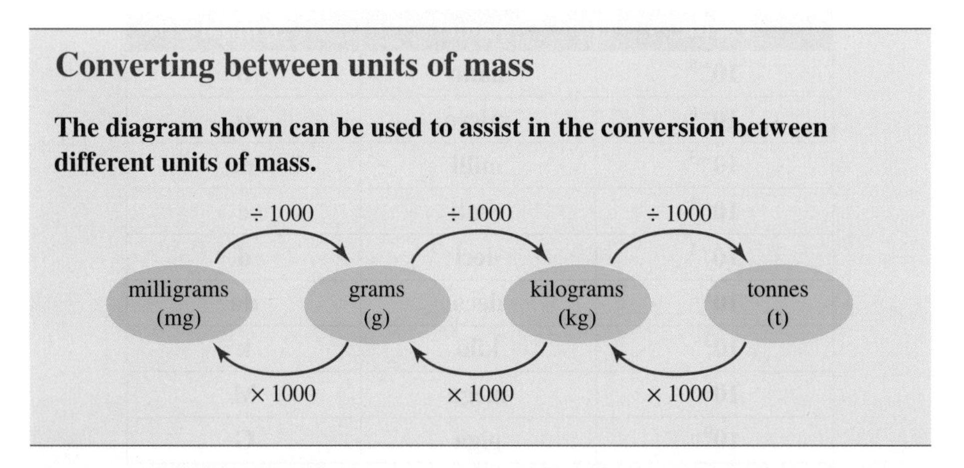

WORKED EXAMPLE 1 Applying conversion of mass measurements to the specified units

Change the following mass measurements to the units specified.

a. 4.7 kg to grams **b. 15 000 mg to kilograms**

THINK	WRITE
a. There are 1000 g in every 1 kg. 4.7 kg is 4.7 lots of 1000 g. To change from kilograms to grams, multiply by 1000.	a. $4.7\,\text{kg} = (4.7 \times 1000)\,\text{g}$ $= 4700\,\text{g}$
b. 1. To change from milligrams to kilograms, first change from milligrams to grams, and then change from grams to kilograms.	b. $15\,000\,\text{mg} = \frac{15\,000}{1000}\,\text{g}$ $= 15\,\text{g}$
2. There are 1000 mg in every 1 g; to change from milligrams to grams, group the milligrams into groups of 1000.	$= \frac{15}{1000}\,\text{kg}$ $= 0.015\,\text{kg}$
3. There are 1000 g in every 1 kg; to change from grams to kilograms, group the milligrams into groups of 1000. Write the answer as a decimal.	

5.2.2 Metric units of energy

- Energy is the ability to do work.
- There are many different forms of energy: food energy, kinetic energy, heat energy, light energy, sound energy and electrical energy.
- **Joules** (J) and **kilojoules** (kJ) are the SI units of energy. Most Australian websites and books use kilojoules.
- Food energy is commonly measured in kilojoules (kJ).
- Larger amounts of energy, such as electrical energy, are measured in **megajoules** (MJ).
- **Calories** are also units of energy.
- Calories are the older (imperial) style, non-SI unit of measurement of energy, whereas kilojoules are the modern style of measurement used more commonly today.
- The calorie (cal) is the approximate amount of energy needed to raise the temperature of 1 kilogram of water from 0 to 1 degree Celsius.
- The **kilocalorie** (kcal) is equal to 1000 calories.

Units of energy

1 kJ = 1000 J
1 MJ = 1 000 000 J
1 kcal = 1000 cal

Converting between units of energy

- The relationships between the units of energy can be used to change a measurement from one system of units to another.

Converting from one system of energy units to another

4.184 kilojoules = 4184 joules = 1 kilocalorie = 1000 calories

Or

4.184 kilojoules = 1000 calories

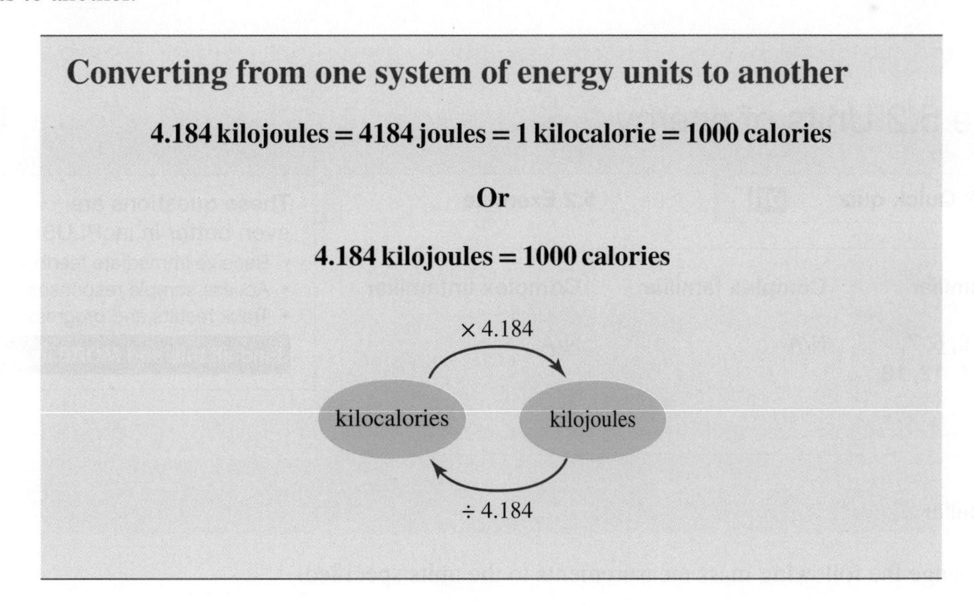

tlvd-11426

WORKED EXAMPLE 2 Applying conversion from one system of energy units to another

Change the following energy measurements to the units specified. Give your answers correct to 2 decimal places where necessary.

a. **500 kilocalories to kilojoules**
b. **10 850 joules to kilocalories**

THINK	WRITE
a. To convert from kilocalories to kilojoules, multiply the number of kilocalories by 4.184.	a. $500\text{ kcal} = (500 \times 4.184)\text{ kJ}$ $= 2092\text{ kJ}$
b. 1. To convert from joules to kilojoules, divide by 1000, since 1000 joules = 1 kilojoule.	b. $10\,850\text{ J} = \dfrac{10\,850}{1000}\text{ kJ}$ $= 10.85\text{ kJ}$
2. Then to convert from kilojoules to kilocalories, divide the number of kilojoules by 4.184.	$10.85\text{ kJ} = \dfrac{10.85}{4.184}\text{ kcal}$ $\approx 2.59\text{ kcal}$
3. Alternatively, to convert from joules to kilocalories, divide the number of joules by 4184.	$10\,850\text{ J} = \dfrac{10\,850}{4184}\text{ kcal}$ $\approx 2.59\text{ kcal}$

Resources

Interactivities Converting between units of mass (int-6907)
Converting between units of energy (int-6908)

Exercise 5.2 Units of energy

learnon

5.2 Quick quiz  | 5.2 Exercise

Simple familiar	Complex familiar	Complex unfamiliar
1, 2, 3, 4, 5, 6, 7, 8, 9, 10, 11, 12, 13, 14, 15, 16	N/A	N/A

These questions are even better in jacPLUS!
- Receive immediate feedback
- Access sample responses
- Track results and progress

Simple familiar

1. WE1 Change the following mass measurements to the units specified.
 a. 2 g to milligrams
 b. 15.1 t to kilograms
 c. 1200 mg to grams
 d. 22 g to kilograms
 e. 0.04 kg to grams
 f. 13 000 mg to kilograms

2. Change the following mass measurements to the units specified.

a. $\frac{1}{2}$ kg to milligrams

b. 20 kg to tonnes

c. 109.02 kg to grams

d. 3.8 t to kilograms

e. $\frac{4}{5}$ g to milligrams

f. 5 000 000 mg to kilograms

3. The luggage limit on many domestic plane flights is 20 kg. If the luggage that you check in for a flight to Sydney has a mass of 18.2 kg, calculate the extra mass that you can take in your luggage.

4. Determine whether each of the following is True or False.

a. 345 grams = 0.345 mg

b. 42 g = 0.042 kg

5. Calculate how many joules of energy are in:

a. 2 kJ

b. 3.5 MJ.

6. Calculate how many megajoules of energy are in:

a. 52 000 000 J

b. 370 000 000 000 J.

7. **WE2** Convert the following to kJ, giving your answer correct to 1 decimal place.

a. 100 kilocalories

b. 2200 kilocalories

8. A person has a craving for sweet food. If they require 10 150 kJ each day and eat 5 chocolate bars, each worth 1332 kJ, calculate how many more kilocalories they require that day.
Give your answer to the nearest whole number.

9. If x kilocalories are equivalent to k kilojoules and there are 4.184 kilojoules in a kilocalorie, write a formula to represent the relationship between x and k.

10. a. The treadmill at the gym shows that Alecia has used 440 kilocalories. Calculate how many kilojoules this is. Give your answer to the nearest whole number.

b. Alecia wants to use 2800 kJ on the machine. Calculate how many more kilocalories she has to burn. Give your answer to the nearest whole number.

11. Chips are sold in packets weighing 50 g, 100 g, 200 g and 250 g. These are packed in boxes of 200 packets. Each empty box weighs 4.5 kg. A supermarket orders three boxes of 50 g, four boxes of 100 g and 200 g and ten boxes of 250 g. Calculate the total mass of the supermarket order.

12. A shipping-container company has four different shipping containers to choose from. Each container has a maximum weight.
Container 1 has a maximum cargo weight of 21 460 kg, container 2 has 26 500 kg, container 3 has 26 330 kg and container 4 has 20 770 kg.
The company uses eight containers to export goods overseas. They fill to maximum cargo weight two lots of container 4, one lot each of containers 3 and 2, three lots of container 1, and they half-fill container 3. Calculate the total mass of all the containers in tonnes.

13. Consider the following nutritional information.

Product	Energy per serve (kJ)
Margarine (10 g)	240
White bread (1 slice)	280
Strawberry jam (15 g)	144
Muesli (50 g)	950
Full-cream milk (1 cup)	556
Skim milk (1 cup)	438

a. Calculate how many kilojoules there are in 2 slices of white bread spread with 5 g of margarine and 15 g of strawberry jam.

b. Calculate how many kilojoules there are in 100 g of muesli and $1\frac{1}{2}$ cups of full-cream milk.

14. a. Given that 1 kilocalorie is equal to 4.184 kilojoules, complete the following table listing some standard drink types.

Food item	Kilocalories	Kilojoules
Can of Coke	154	
Can of Fanta	195	
Apple juice	102	
Glass of milk	164	
Cappuccino	53	

b. Sami has a diet which limits her daily intake of energy to 8200 kJ. If she drinks three cans of Coke, determine what her remaining food intake (to the nearest kJ) will be limited to.

c. Calculate what percentage of her daily intake is drinking three cans of Coke. Give your answer correct to 2 decimal places.

15. As a consumer, identify three examples of situations in which it is useful to have an understanding of the units of mass. For each example, explain why it is useful.

16. Harry lives in the city. He is driven to school every day. In the evenings, he watches television or plays on his computer after doing his homework. He doesn't like sport and he prefers to play games on his computer on the weekends.

Ben lives in the country. He walks to school, which is about 2.5 kilometres away. After school, he often helps on his parents' farm and in his spare time, Ben plays footy or basketball.

Harry's standard daily diet		Ben's standard daily diet	
2 cheeseburgers	2 × 1200 kJ	3 portions of sausages and mashed potatoes	3 × 1300 kJ
2 portions of chips	2 × 1500 kJ	3 portions of vegetables	3 × 450 kJ
4 biscuits	4 × 400 kJ	3 portions of nuts and dry fruits	3 × 555
6 cans of soft drink	6 × 675 kJ	3 glasses of milk	3 × 370 kJ
1 apple pie	1 × 1300 kJ	8 glasses of water	8 × 0 kJ
Total	12 350 kJ	Total	8025 kJ

Both Harry and Ben are the same age and in Year 9. Harry's total daily energy consumption is usually 12 350 KJ while Ben's is 8025 KJ. Taking into consideration Harry's and Ben's diets and energy needs, give a recommendation for changes that they could make to their diets.

Fully worked solutions for this chapter are available online.

LESSON
5.3 Energy in food

SYLLABUS LINKS

- Use units of energy to describe the amount of energy for foods, including calories and kilojoules.

5.3.1 Energy in food

- Energy is supplied to our bodies from the food we eat and what we drink. Food is the fuel our bodies convert into energy. We need this energy to be active and to work, to build and repair cells, and to grow and fight infections.
- Energy content of food in Australia is measured in kilojoules (kJ). Kilojoules (kJ) can be found on food labels.
- In food science and nutrition, the term *Calorie* is commonly used to refer to 1000 calories, which is a kilocalorie, so 1 kilocalorie = 1 Calorie = 1000 calories
- Some examples of the amount of energy stored in some normal-sized serves of food are:

Food item	Serving size	Energy content (kJ)
Apple	1 medium (182 g)	397
Banana	1 medium (118 g)	439
Chicken breast	1 medium (120 g)	690
Egg (large)	1 egg (50 g)	293
Brown rice	1 cup cooked (195 g)	912
Whole-wheat bread	2 slices (66 g)	578
Almonds	1 ounce (28 g)	686
Salmon (cooked)	3 ounces (85 g)	733
Pasta (cooked)	1 cup (140 g)	924
Yogurt (plain, low-fat)	1 cup (245 g)	644

Note: These values are rounded approximations and may vary slightly depending on factors such as cooking methods and specific food brands.

WORKED EXAMPLE 3 Calculating calories (kilocalories) in food items

Consider the following nutritional information.
Calculate how many calories there are in 1 cheese-and-bacon roll, 1 chocolate-filled croissant and 2 toasted crumpets.

Food item	Energy per serve
1 cheese-and-bacon roll (75 g)	221 kcal
1 mixed-grain roll (70 g)	721 kJ
1 sourdough roll (55 g)	148 kcal
1 chocolate-filled croissant (125 g)	2263 kJ
1 white bagel (30 g)	78 kcal
1 toasted crumpet (50 g)	418.4 kJ

THINK

1. Look up the energy content for each food item from the table.

WRITE

1 cheese-and-bacon roll = 221 kcal
1 chocolate-filled croissant = 2263 kJ
1 toasted crumpet = 418.4 kJ

2. Convert all food items to kilocalories. To convert to kilocalories from kilojoules, divide by 4.184.

1 cheese-and-bacon roll = 221 kcal

$$1 \text{ chocolate-filled croissant} = \frac{2263}{4.184} = 540.8... \approx 541 \text{ kcal}$$

$$1 \text{ toasted crumpet} = \frac{418.4}{4.184} = 100 \text{ kcal}$$

3. Calculate the total amount of calories in the food items.

Total energy = 1 cheese-and-bacon roll + 1 chocolate-filled croissant + 2 toasted crumpets

$$\text{Total energy} = 221 + 541 + 2 \times 100 = 962 \text{ kcal}$$

In total, there are 962 kilocalories in 1 cheese-and-bacon roll, 1 chocolate-filled croissant and 2 toasted crumpets.

5.3.2 Nutritional labels

- Nutritional labelling appears on prepared and packaged foods to assist people in making informed decisions about what they eat.
- The nutritional information label on a packaged food item provides the number of kJ per serve or per 100 g.
- From the label, we can determine the number of kJ we consume by either:
  - determining how many of the manufacturer's serves you have consumed and multiplying this by the number of kJ per serve, or
  - dividing the weight of the amount you have consumed (in g) by 100 and multiplying this by the number of kJ in 100 g.

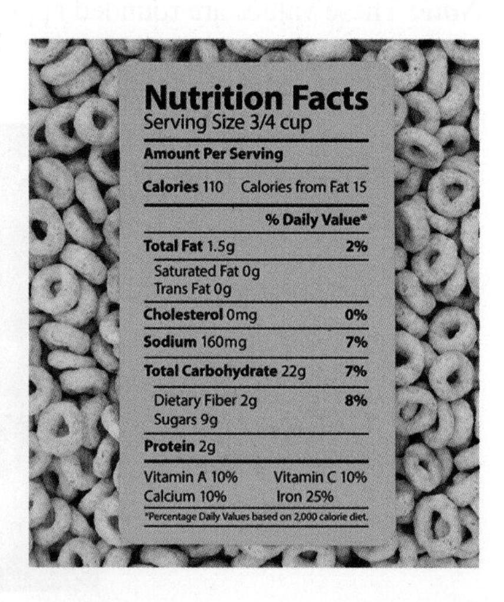

tlvd-11427

WORKED EXAMPLE 4 Calculating the number of kilojoules consumed

You eat 150 g of baked beans from a 210-g can.
Calculate how many kilojoules of baked beans you have consumed if the nutrition information providing the energy content from the label of a can of baked beans shows that:

a. **a serving size of 210 g of beans is 810 kJ**
b. **a 100-g serve of beans is 385 kJ.**

Cheesy Bean Fajitas **Makes 8**
1. Combine 400g beef strips, 1 tablespoon Mexican Fajita Seasoning and 1 tablespoon oil in a bowl. Mix until well coated. Heat a non-stick frying pan, add beef strips and cook for 2 minutes or until browned. Set aside.
2. Add 1 sliced red onion and 1 sliced green capsicum and cook for 3 minutes, or until softened. Add **420g can HEINZ Baked Beans - Cheesy Tomato Sauce**, cook until warmed through.
3. Top 8 flour tortillas with lettuce leaves, bean mixture and beef strips. Wrap to enclose filling. Serve.

NUTRITION INFORMATION
SERVINGS PER PACKAGE: 2 SERVING SIZE: 210g

AVG QUANTITY	PER SERVING	PER 100g
ENERGY	810kJ	385kJ
PROTEIN	11.1g	5.3g
FAT, TOTAL	1.7g	0.8g
- SATURATED	0.6g	0.3g
CARBOHYDRATE	27.5g	13.1g
- SUGARS	8.6g	4.1g
DIETARY FIBRE	11.1g	5.3g
SODIUM	670mg	320mg
POTASSIUM	565mg	270mg
FOLATE	56µg (28% RDI*)	27µg
IRON	2.7mg (22% RDI*)	1.3mg

* Recommended Dietary Intake (Average Adult)

INGREDIENTS
Navy Beans (51%), Cheesy Tomato Sauce (49%) [Tomatoes (25%), Water, Sugar, Maize Thickener (1422), Cheese (1.1%), Salt, Yeast Extract, Food Acids (Acetic Acid, Citric Acid), Flavours].

‡ Based on the Australian Guide to Healthy Eating. One serve of Vegetables (including Legumes) = 75g. Aim for a variety of vegetables each day

THINK / **WRITE**

a. 1. Calculate the number of manufacturer's serves.

a. Number of serves is $\frac{150}{210} = 0.714$.

2. The kJ consumed is the number of serves multiplied by the number of kJ per serve.

$0.714 \times 810 = 578.6$
578.6 kJ is consumed from eating 150 g of baked beans.

b. 1. Divide the weight of the amount you have consumed by 100.

b. Number of serves per 100 g is $\frac{150}{100} = 1.5$.

2. Multiply the number of kJ by the number of serves per 100 g.

$1.5 \times 385 = 577.5$
577.5 kJ is consumed from eating 150 g of baked beans.

Note: The 1.07 kJ difference between the two calculations is most likely due to the manufacturer rounding the values in either the per-serve or the 100-g value.

5.3.3 Fat, protein and carbohydrates

- Most of the energy in food comes from fat, protein and carbohydrates. Protein, fats and carbohydrates are converted into energy in different quantities.
- Based on the Atwater general system:
  1 gram of carbohydrates = 17 kJ
  1 gram of protein = 17 kJ
  1 gram of fat = 37 kJ
  1 gram of dietary fibre = 8 kJ
  1 gram of alcohol = 29 kJ
  Note: The Atwater general system is a system for allocating energy values to foods. It was developed in the 19th century by the American chemist W. O. Atwater. The system uses a single energy value (factor) for each of the main groups of energy nutrients (protein, fat, and carbohydrate), regardless of the food in which it is found.

- To calculate the energy available from a food, multiply the number of grams of carbohydrate, protein and fat by 17, 17 and 37, respectively. Then add the results together.

Calculating the energy content of food

- **The formula for calculating the energy content of food from the amount of carbohydrate, protein, and fat can be summarised as:**

$$E = 17c + 17p + 37f$$

where p is the protein in grams, f is the fat in grams and c is the carbohydrates in grams.

WORKED EXAMPLE 5 Calculating the energy available from food items

Calculate the energy available, in both calories and kilojoules, from a slice of bread with a tablespoon of peanut butter on it containing 16 g carbohydrates, 7 g protein, and 9 g fat.

THINK	WRITE
1. Define the number of carbohydrates, protein and fat in grams.	$c = 16$ $p = 7$ $f = 9$
2. To calculate the energy content, multiply the number of grams of carbohydrate, protein and fat by 17, 17 and 37, respectively. Then add the results together using the formula.	$E = 17 \times 16 + 17 \times 7 + 37 \times 9$ $= 272 + 119 + 333$ $= 724$ The energy available from a slice of bread with a tablespoon of peanut butter on it containing 16 g carbohydrates, 7 g protein and 9 g fat is 724 kJ.
3. To calculate the energy content in kilocalories, divide the amount in kilojoules by 4.184.	$\frac{724}{4.184} \approx 173.0$ The energy available from a slice of bread with a tablespoon of peanut butter on it containing 16 g carbohydrates, 7 g protein and 9 g fat is approximately 173 kilocalories.

5.3.4 Recommended daily energy needs

- The average daily energy requirement for an adult is approximately 8700 kJ.
- The recommended daily intake of fat for an average adult is 70 g.
- Knowing the number of kilojoules in the food you eat can help ensure you get the recommended amount of energy for your needs.
- Daily energy requirements depend on your age, health, job, level of exercise and metabolism.

WORKED EXAMPLE 6 Calculating the recommended daily intake of fat for a child

A 3-year-old child has daily energy needs of 4900 kJ. Calculate the recommended daily intake, to 2 decimal places, of fat for a child with daily energy needs of 4900 kJ.

THINK	WRITE
1. Determine the ratio of the child's daily energy needs against the recommended daily intake of energy for an average adult.	$\frac{4900}{8700} \approx 0.563$
2. To calculate the recommended daily intake of fat for the child, multiply the recommended daily intake of fat of an average adult by the ratio of the child's energy needs against an adult's energy needs.	$\frac{4900}{8700} \times 70 \approx 39.43$ The recommended daily intake of fat for a child with daily energy needs of 4900 kJ is 39.43 g.

Exercise 5.3 Energy in food

learnon

5.3 Quick quiz on | 5.3 Exercise

Simple familiar	Complex familiar	Complex unfamiliar
1, 2, 3, 4, 5, 6, 7, 8, 9, 10, 11, 12, 13, 14, 15, 16, 17, 18, 19, 20, 21	N/A	N/A

Simple familiar

1. For breakfast, Tyler ate an egg muffin (2200 kJ) and an orange juice (525 kJ). Calculate, in kilojoules, how much energy Tyler consumed.
2. Given that 1 kilocalorie is equal to 4.184 kilojoules, complete the following table listing some different types of snacks. Give the values in kilocalories to the nearest whole number.

Food item	Kilocalories	Kilojoules
265 g T-bone steak grilled		2001
140 g chopped grilled chicken		966
100 g lamb ribs		669
425 g tuna in brine		1849
100 g crumbed flounder		940

3. **WE3** Consider the following nutritional information.

Food item	Energy per serve
2 slices supreme pan pizza	2474 kJ
1 hamburger	563 kcal
6 sticks chicken satay	216 kcal
1 wedge of fruit cake	743 kJ
1 iced doughnut	306 kcal
1 slice of chocolate cake	1054 kJ

a. Calculate how many kilocalories, to the nearest whole number, there are in 1 slice of pizza and 2 iced doughnuts.
b. Calculate how many kilocalories, to the nearest whole number, there are in 12 sticks of chicken satay and 1 wedge of fruit cake.

4. Mark consumed 2 macaroons (848 kJ) and a milkshake (1187 kJ) for morning tea. Calculate the energy content of the meal.

5. A large birthday cream cake contains 6112 kJ. Jo cuts the cake into 8 equal slices. Calculate how many kilojoules there are in each slice of cake.

6. **WE4** You eat 125 g of yoghurt from a 175 g tub. Calculate how many kilojoules of yoghurt you have consumed, to the nearest whole number, if the nutrition information providing the energy content from the label shows that 1 serving size of 175 g yoghurt is 659 kJ.

7. You eat 2 chocolate Easter eggs from a pack of 10. Calculate how many kilojoules of chocolate you have consumed if the nutrition information from the label shows that the energy content for the pack of 10 Easter eggs is 2240 kJ.

8. **WE5** According to the Atwater general system:
1 gram of carbohydrates = 17 kJ
1 gram of protein = 17 kJ
1 gram of fat = 37 kJ
1 gram of dietary fibre = 8 kJ
1 gram of alcohol = 29 kJ
Calculate the energy available, in kilojoules, from a slice of sourdough bread which contains 13 g carbohydrates, 2.2 g protein, and 0.8 g fat.

9. Calculate the energy available, in kilojoules, from a 20-g cookie which contains 11.8 g carbohydrates, 1.8 g protein, and 4.8 g fat.

10. A suggested breakdown of our daily energy intake is:

Breakfast	20%
Morning snack	10%
Lunch	20%
Afternoon snack	10%
Dinner	40%

If an inactive person requires on average 8000 kJ energy during the day, calculate how much energy should be in each of their meals.

11. WE6 The ideal daily fat intake is 20%. Since 1 gram of fat is equal to 37 kJ, calculate the ideal fat intake (in grams) for Heidi and her diet of 7400 kJ a day.

12. The average serving of 100 g of chicken and noodles is 622 kJ. Oiyee ate 350 g of chicken and noodles. Calculate the energy content of the meal in kilocalories to the nearest whole number.

13. The percentage daily intake (%DI) is based on the recommended amounts of energy and nutrients needed to meet daily nutritional needs of an average adult male. The percentages are based on the values below.

	Energy	Protein	Fat	Carbohydrates
Reference value used in %DI	8700 kJ	50 g	70 g	310

a. Calculate the recommended daily amount of fat for a person with daily energy needs of 7200 kJ.
b. Calculate the recommended daily amount of protein for a person with daily energy needs of 9300 kJ.
   Give your answers correct to 2 decimal places.

14. The following information is printed on the takeaway container of a 200-g deluxe hamburger.

	Energy	Protein	Fat	Carbohydrate
Adult quantity per day	8700 kJ	50 g	70 g	310 g
Hamburger	2010 kJ	25 g	25.5 g	35 g
%DI				

Calculate each %DI nutrient value for the hamburger based on a daily energy level of 8700 kJ. Give your answers correct to 1 decimal place where necessary.

15. Bridgette is 17 years old. On a typical day she eats:

Breakfast	Lunch	Dinner
1 bowl cereal 480 kJ 1 glass milk 400 kJ 1 slice of toast 447 kJ	1 banana 374 kJ 1 chicken-and-mayo sandwich 1672 kJ 1 bottle of water 0 kJ	1 soft drink 600 kJ 1 grilled chicken breast 1264 kJ 1 serving of potatoes 365 kJ 1 serving of carrots 135 kJ

A suggested breakdown of our daily energy intake for the main meals is:

Breakfast 20%
Lunch 20%
Dinner 40%

An average 16–18-year-old girl requires 9200 kJ of energy during the day. Determine if Bridgette's energy intake for the main meals is enough.

16. At a 10-year-old boy's birthday, each child eats:
- serving of popcorn 1351 kJ
- chocolate bar 1332 kJ
- ice cream 633 kJ
- cheeseburger 2209 kJ
- serving of French fries 2579 kJ
- soft drink 600 kJ.

a. Calculate the energy content of each child's birthday party meal in calories.
b. An average 10–12-year-old boy should intake 8800 kJ each day. Determine what effect each birthday meal will have on a child's energy intake for the day.

17. A balanced diet for an average adult is made up of the following nutrients each day:

	Energy	Protein	Fat	Carbohydrate
Quantity per day	8700 kJ	50 g	70 g	310 g

Determine the difference in maximum protein intake between a 7000 kJ daily diet and a 9000 kJ daily diet. Give your answer to 1 decimal place.

18. A balanced diet for an average adult is made up of the following nutrients each day:

	Energy	Protein	Fat	Carbohydrate
Quantity per day	8700 kJ	50 g	70 g	310 g

Determine the difference in maximum carbohydrate intake between a 7200 kJ daily diet and a 9800 kJ daily diet. Give your answer to 1 decimal place.

19. Below is a list of food items and their energy values.

Food item	Energy (kJ)	Food item	Energy (kJ)
White bread (slice)	306	Hamburger	1800
Wholegrain bread (slice)	252	Sushi	155
Vegemite (1 tsp)	30	White rice (1 serve)	930
Egg	330	Chicken breast	1040
Bacon (2 rashers)	1060	Steak (200 g)	1560
Cornflakes (bowl)	497	Sausages (2)	1400
Coco Pops (bowl)	728	Mixed vegetables (1 serve)	330
Apple	284	Garden salad (1 serve)	300
Banana	432	Chocolate mousse (1 serve)	770
Doughnut	770	Ice-cream (1 serve)	450
Yoghurt (1 serve)	596	Tuna sandwich	350

a. Choose your favourite five food items from the list and calculate the total energy intake in calories.
b. Plan a breakfast meal. Calculate the total energy intake.

20. At a childcare centre at snack time, the 4-year-old children can choose from a platter of various fruits.
The suggested breakdown for snack time is 10%.

- 100-g serving of apple 202 kJ
- 100-g serving of blueberries 239 kJ
- 100-g serving of grapes 290 kJ
- 100-g serving of raisins 1256 kJ
- 100-g serving of watermelon 126 kJ

A child (4–6 years old) requires 6300 kJ each day.

a. Ruby eats a 50-g serving of grapes and a 20-g serving of raisins.
   i. Calculate how many kilojoules were in Ruby's snack.
   ii. Calculate what percentage of Ruby's daily energy requirement, to 2 decimal places, she has at snack time.

b. Dean is hungry and eats a 100-g serving of blueberries, a 200-g serving of watermelon and a 25-g serving of apple.
   i. Calculate how many kilojoules were in Dean's snack.
   ii. Calculate the percentage of Dean's daily energy requirement, to 2 decimal places, that he has at snack time.

21. The table shown contains incomplete data for the number of grams of fat, protein and carbohydrates and the number of kilojoules for some selected foods.

Food	Fat (g)	Protein (g)	Carbohydrates (g)	Kilojoules
1 hot cross bun	3	3		485
100 g of chocolate cake	16	4	56	
100 g of roast chicken	14	26	0	
70 g of bacon	8	21	0	655
2 grilled sausages		13	15	1113
1 piece of fish (flake), no batter	1	21	0	
1 banana	0	1	20	
1 apple	0	0	17	
250 mL of milk	10	8	12	
210 g of tinned tomato soup	0.8		14.9	324

Determine which food in the list has the highest number of kilojoules. Explain whether this is what you expected.

Fully worked solutions for this chapter are available online.

LESSON
5.4 Energy in activities

SYLLABUS LINKS

- Use units of energy to describe the amount of energy in activities.

5.4.1 Energy inputs and outputs

- The human body gets energy from food it consumes. We need this energy to be active.
- When you are active, your body burns more energy.
- People's daily energy requirements differ based on their gender, how much exercise they do, their height and weight, and whether or not they suffer from particular illnesses or disorders.
- The following table shows the estimated amount of energy the human body of an average body weight requires each day to maintain good health, based on age group, gender and level of activity.

Age (years)	Males (kJ)	Females (kJ)
1–3	5200	4900
4–6	6300	6300
7–9	7500	7500
10–14	9300	7800
15–18	11 600	8900
18–70 Light work Medium work Heavy work	 10 000 12 500 14 250	 8000 9200 10 750
Pregnant women Breastfeeding women		10 150 11 800

- A way to expend more energy is to increase the time spent on an activity.
  For example, a cappuccino and muffin have a total energy content of 1745 kJ.
  The equivalent activity of 1745 kJ is to walk for 81 minutes, jog for 35 minutes or run for 20 minutes.
- Below is a table of activities and the energy used per hour for an average person weighing 70 kg.

Activity	Kilojoules burnt per hour
Computing	500
Walking at 3.5 km/h	1000
Aerobics	7000
Sitting on a train	500
Watching TV	350
Reading	500
Sleeping	250
Housework	600
Running	5300
Swimming	1100

WORKED EXAMPLE 7 Calculating the energy used in activities in a day

Jo, aged 45, eats 8200 kJ per day. Every morning she walks on her treadmill for 30 minutes. She uses 1000 kJ/h. She catches the train to and from work for 45 minutes each way and uses 500 kJ/h. She works as an administrator and does computing work for 6 hours a day using 500 kJ/h. Calculate the amount of kilojoules she has left to use.

THINK	WRITE
1. Calculate the kilojoules Jo uses for each activity.	Walking on treadmill = 1000 kJ for 1 hour = 500 kJ for 30 mins Catching train = 500 kJ for 1 hour = 750 kJ for 1.5 hours Computer work = 500 kJ for 1 hour = 3000 kJ for 6 hours
2. Total the energy output (kilojoules) used for activities.	Energy output = 500 + 750 + 3000 = 4250 kJ
3. Calculate the energy difference.	Energy difference = energy input − energy output = 8200 − 4250 = 3950 kJ Jo still needs to burn 3950 kJ.

5.4.2 Basal energy expenditure

- You can now calculate the food energy in kilojoules, but how much energy do you need to meet the demands of your **basal metabolic rate (BMR)** and **physical activity level (PAL)**?
- The basal metabolic rate (BMR) or basal energy expenditure is the calculation of how much energy it takes to maintain basic body functions on a daily basis, regardless of activity level.
- The amount of energy that the body needs to function while resting for 24 hours is known as BMR. This number of kilojoules reflects how much energy your body requires to support vital bodily functions.
- The BMR is the minimum amount of energy that a body requires to fuel its normal metabolic activity at rest.

Calculating BMR

- **BMR can be calculated using the Mifflin St. Jeor formulas:**

Male: BMR = 10 × weight (kg) + 6.25 × height (cm) − 5 × age (years) + 5
Female: BMR = 10 × weight (kg) + 6.25 × height (cm) − 5 × age (years) − 161

This gives a result in kilocalories, so to convert to kilojoules, multiply by 4.184.

tlvd-11428

WORKED EXAMPLE 8 Calculating BMR

A 16-year-old boy weighs 60 kg and is 180 cm tall. Calculate his BMR to the nearest whole number.

THINK	WRITE
1. List the information (age, height and weight) for the boy.	Weight = 60 kg Height = 180 cm Age = 16 years
2. Use the male BMR equation to calculate the BMR in kilocalories.	$\text{BMR} = (10 \times \text{weight}) + (6.25 \times \text{height}) - (5 \times \text{age}) + 5$ $= (10 \times 60) + (6.25 \times 180) - (5 \times 16) + 5$ $= 600 + 1125 - 80 + 5$ $= 1650 \text{ kcal}$
3. Convert BMR from kilocalories to kilojoules.	$\text{BMR} = 1650 \times 4.184$ $\approx 6931 \text{ kJ}$ A 16-year-old boy who weighs 60 kg and is 180 cm tall has a BMR of 6931 kJ.

5.4.3 Physical activity level

- The physical activity level (PAL) is a way to express a person's daily physical activity as a number, and is used to estimate a person's total energy expenditure in combination with the basal metabolic rate.
- It can be used to compute the amount of food energy a person needs to consume in order to maintain a particular lifestyle.

Calculating PAL

$$\text{PAL} = \frac{\textbf{total energy expenditure}}{\textbf{BMR}}$$

- The physical activity level can be estimated using a list of the physical activities a person performs within a 24-hour period and the amount of time spent on each activity, e.g. walking to work, light housework, swimming, carrying bricks at work, or whatever applies to an individual person.
- PAL is estimated at 1.4 for inactive men and women, 1.6 for moderately active women and 1.7 for moderately active men.
- There are five activity 'levels' ranging from sedentary to athletic.

- These factors are based most often upon the rigor of your lifestyle and exercise routine.

Activity	Level description	Activity factor
Sedentary	Little or no exercise/desk job	1.2
Lightly active	Light exercise/sports 1–3 days per week	1.375
Moderately active	Moderate exercise/sports 3–5 days per week	1.55
Very active	Hard exercise/sports 3–5 days per week	1.725
Extremely active/athletic	Hard exercise/sports daily and/or a physical job	1.9

Calculating total energy expenditure

- **To calculate your total energy expenditure in a 24-hour period, use the formula:**

$$\text{Total energy expenditure} = \text{PAL} \times \text{BMR}$$

WORKED EXAMPLE 9 Calculating daily energy needs

A 15-year-old girl weighs 60 kg, is 165 cm tall and has an activity level that is classified as lightly active. Calculate her daily energy needs to the nearest whole number.

THINK	WRITE
1. List the information (age, height and weight) for the girl.	Weight = 60 kg Height = 165 cm Age = 15 years
2. Use the female BMR equation to calculate the BMR in kilocalories.	$\text{BMR} = (10 \times \text{weight}) + (6.25 \times \text{height}) - (5 \times \text{age}) - 161$ $= (10 \times 60) + (6.25 \times 165) - (5 \times 15) - 161$ $= 600 + 1031.25 - 75 - 161$ $= 1395.25\text{ kcal}$
3. Convert BMR from kilocalories to kilojoules.	$\text{BMR} = 1395.25 \times 4.184$ $= 5837.726\text{ kJ}$
4. Determine the activity factor (PAL).	For lightly active people, activity factor is 1.375.
5. Calculate the daily energy needs.	$\text{Total energy expenditure} = \text{PAL} \times \text{BMR}$ $= 1.375 \times 5837.726$ ≈ 8027 A 15-year-old girl who weighs 60 kg and is 165 cm tall has daily energy needs of 8027 kJ.

on Resources

Interactivity Basal metabolic rate (BMR) (int-6909)

Exercise 5.4 Energy in activities

5.4 Quick quiz	5.4 Exercise

Simple familiar	Complex familiar	Complex unfamiliar
1, 2, 3, 4, 5, 6, 7, 8, 9, 10, 11, 12, 13, 14, 15, 16, 17, 18, 19, 20	N/A	N/A

Simple familiar

1. **WE8** Calculate the BMR for the following people.
   a. A 128-cm, 8-year-old girl weighing 28 kg
   b. A 97-cm, 3-year-old boy weighing 12 kg
2. Identify whether the following statement is True or False.
   The BMR for a 180-cm 20-year-old woman weighing 55 kg is 1414 kJ.
3. **WE9** Use the following PAL activity table to calculate the daily energy needs to the nearest whole number for:

Activity	Level description	Activity factor
Sedentary	Little or no exercise/desk job	1.2
Lightly active	Light exercise/sports 1–3 days per week	1.375
Moderately active	Moderate exercise/sports 3–5 days per week	1.55
Very active	Hard exercise/sports 3–5 days per week	1.725
Extremely active/athletic	Hard exercise/sports daily and/or a physical job	1.9

   a. a 165-cm, sedentary 68-year-old woman weighing 88 kg
   b. a 178-cm, very active 17-year-old boy weighing 68 kg.
4. Megan has a BMR of 6400 kJ. Her body's energy expenditure breakdown is:

Energy expenditure breakdown	
Liver	27%
Brain	19%
Other organs	19%
Skeletal muscle	18%
Kidneys	10%
Heart	7%

   Determine how much energy her brain uses of her BMR.
5. Show that a 165-cm, 15-year-old male weighing 60 kg has an approximate BMR of 1561.25 kilocalories.
6. Luke is 24 years old, works in an office and requires 12 500 kJ for his active lifestyle. Every day on his way to work he stops to buy a bacon-and-egg roll that contains 2194 kJ for breakfast.
   Calculate the percentage of his daily energy needs that Luke consumes in the morning for breakfast. Give your answer to 1 decimal place.

7. WE7 Emily, aged 25, eats 7600 kJ per day. Each day she walks to and from work for 20 minutes (per direction), burning 600 kJ/h. At work she serves customers for 7 hours, burning 500 kJ/h. When she gets home she dances for 30 minutes, burning 1200 kJ/h.
Calculate the amount of energy in kilojoules she has left to burn.

8. Avril ate a bucket of hot chips with an energy content of 5700 kJ. Calculate how long it will take her to burn the energy from the hot chips if she runs at a rate of 75 kJ/min.

9. Calculate the number of kJ an average person uses when gardening for 1.2 hours if they use 23 kJ/min.

10. Calculate the number of kJ an average person uses when circuit-training for 30 minutes at 53 kJ/min.

11. Calculate the number of kJ an average person uses when ironing for 10 minutes at 17 kJ/min.

12. Calculate the number of kJ an average person uses when playing tennis for two and a half hours at 31 kJ/min.

13. Every afternoon, Eddie works out at the gym for 2 hours, burning 4000 kJ/h. He lives a very active lifestyle and requires 13 500 kJ of energy per day.
Determine how much energy he has left for the remaining hours in the day.
Consider whether this is enough for someone who lives a very active lifestyle.

14. Michelle is a sales representative and drives her car two hours a day. If she uses 6.4 kJ per hour per kg of body weight and she weighs 78 kg, calculate how many kilojoules she burns driving.

15. Jordan has a very active lifestyle as a 20-year-old bricklayer who requires 14 800 kJ of energy per day. For lunch Jordan eats 2 chicken-and-mayo sandwiches, each containing 1672 kJ.
Calculate the percentage of his daily energy requirement that Jordan has for lunch. Give your answer to 1 decimal place.

16. Melinda is 50 years old. During the week she is inactive and only requires 8000 kJ of energy; during the weekend she is very active, requiring 10 400 kJ per day.
Calculate how many more kJ she requires per day on the weekend than during the week.

17. Sarah swims every morning as a member of a swim squad. She uses 29 kJ per hour per kg of her body weight. She swims for 90 minutes and uses approximately 2500 kJ.
Calculate Sarah's approximate body weight correct to 1 decimal place.

18. Matt weighs 65 kg and plays basketball every Saturday. He uses 40 kJ per hour per kg of his body weight. His game lasts for 40 minutes.
Calculate how many kilojoules, to the nearest whole number, he burns playing basketball.

19. John weighs 75 kg and plays lawn bowls every weekday. He uses 5 kJ per 30 minutes per kg of his body weight. His lawn bowl games last for 3 hours each day.
Calculate how many kilojoules John burns each week.

20. The table below shows the average length of time it takes to burn 1000 kJ for different activities.

Activity	Time required to use 1000 kJ
Computing	2 hours
Walking at 3.5 km/h	1 hour
Aerobics	10 minutes
Sitting on a train	2 hours
Watching TV	2 hours and 30 minutes
Bike riding	50 minutes
Working in class, studying	2 hours and 30 minutes
Eating	3 hours
Sleeping	4 hours

James, aged 18, leads an active life. He consumes 12 500 kJ of energy per day. On an average day, he spends 6 hours in class studying, 8 hours sleeping, 1 hour and 30 minutes eating, 1 hour watching TV and 30 minutes walking to and from school, and the rest of his time he spends playing sport.
Determine whether James eats enough to meet his energy requirements so that he can play sport.

Fully worked solutions for this chapter are available online.

LESSON
5.5 Energy consumption

SYLLABUS LINKS

- Use units of energy to describe consumption of electricity, e.g. kilowatt hours.

5.5.1 Energy use

- Electrical appliances use energy.
- The unit of energy is the joule or kilojoule, and the unit for the amount of energy used in a second is the **watt**. The watt is the unit of power.

Watts and milliwatts

One watt is defined as the energy consumption rate of one joule per second.

$$1\,\text{W} = 1\,\text{J/s}$$

The energy in milliwatts (mW) is equal to the energy in watts multiplied by 1000.

$$\text{Milliwatt} = \text{watt} \times 1000$$

- The faster you use energy, the more watts you use.
  For example, a 1500 W hair dryer uses energy 60 times faster than a 25 W globe.
- The kilowatt hour or kWh is a unit of energy.
- Most energy rating labels provide the energy consumption value in kilowatt hours.

Kilowatt hours (kWh)

To calculate the energy used in one kilowatt hour (kWh), the number of watts used per hour divided by 1000.

$$\text{Kilowatt hour} = \frac{\text{number of watts used per hour}}{1000}$$

For example, if a hair dryer uses 1500 W consistently for one hour, then the energy used is 1.5 kWh.

tlvd-11429

WORKED EXAMPLE 10 Solving energy conversion problems

An oven uses 11 000 W in an hour. Convert this to:

a. **kilowatt hours**

b. **milliwatts.**

THINK	WRITE
a. To convert from W per hour to kWh, divide by 1000.	a. $\frac{11\,000}{1000} = 11$ A 11 000 W oven is equal to 11 kWh.
b. To convert from W to mW, multiply by 1000.	b. $11\,000 \times 1000 = 11\,000\,000$ A 11 000 W oven is equal to 11 000 000 mW.

- Energy-supply companies charge you for the wattage of the appliance that you are using and multiply it by the number of hours that it is used.
- A 100 W ceiling fan that is used for 1 hour is equal to:

$$100 \times 1 = 100 \text{ watts/hour}$$

- If the ceiling fan was used for half an hour, it uses:

$$100 \times 0.5 = 50 \text{ watts/hour}$$

- Most energy rating labels provide the energy consumption value in kilowatts per hour. A kilowatt hour is equal to 1 kW delivered continuously for 1 hour (3600 seconds).

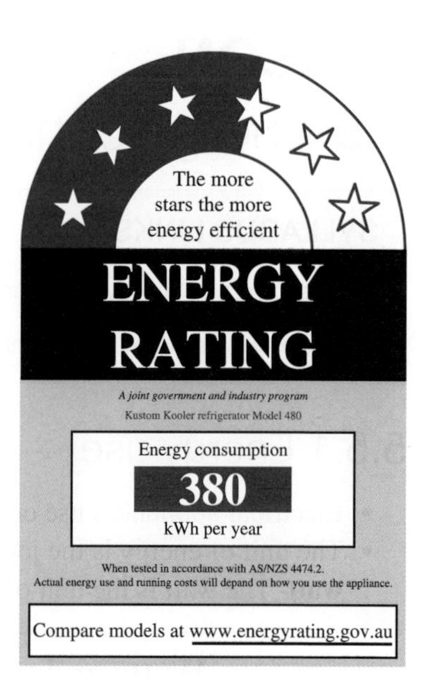

The more
stars the more
energy efficient
ENERGY
RATING
A joint government and industry program
Kustom Kooler refrigerator Model 480
Energy consumption
380
kWh per year
When tested in accordance with AS/NZS 4474.2.
Actual energy use and running costs will depand on how you use the appliance.
Compare models at www.energyrating.gov.au

Energy consumption

We can calculate the energy consumption (E) of an appliance by multiplying the number of power units (P) by the time (t).

$$E = P \times t$$

where E is in joules, P is in watts and t is in seconds.

WORKED EXAMPLE 11 Determining energy consumption

Determine the energy consumption of a 11 000 W hair dryer for the time duration of 30 seconds.

THINK	WRITE
1. Define the power in watts and the time in seconds.	$P = 11\,000$, $t = 30$
2. To calculate energy consumption, use the formula $E = P \times t$.	$E = P \times t$ $= 11\,000 \times 30$ $= 330\,000$
3. Write the answer.	The energy consumption is 330 000 J.

- The power in watts is equal to the energy in joules divided by the time in seconds.

$P = \dfrac{E}{t}$, so watts $= \dfrac{\text{joules}}{\text{seconds}}$ or $W = \dfrac{J}{s}$.

WORKED EXAMPLE 12 Determining power consumption

Calculate the power consumption of a light globe that has an energy consumption of 90 J for the time duration of 3 seconds.

THINK	WRITE
1. Define the energy in joules and time in seconds.	$E = 90$, $t = 3$
2. To calculate power consumption, use the formula $P = \dfrac{E}{t}$.	$P = \dfrac{90}{3} = 30$
3. Write the answer.	The power consumption of the light globe is 30 W.

- Some appliances do not give the wattage, but only the values for current and voltage.

Calculating wattage using voltage and current

$$P = V \times I$$

where power (P) is in watts, voltage (V) is in volts and current (I) is in amperes.

WORKED EXAMPLE 13 Calculating wattage

A clock radio has 240 V and 25 mA. Calculate the clock radio's wattage.

THINK	WRITE
1. Convert the current mA to A, by dividing by 1000.	$25\text{ mA} = \frac{25}{1000}\text{ A}$ $= 0.025\text{ A}$
2. Define the voltage in volts and the current in amps.	240 V, 0.025 A
3. To calculate the wattage, use the formula power = voltage × current.	Power = voltage × current $P = 240 \times 0.025$ $= 6$ The clock radio has a wattage of 6 W.

Exercise 5.5 Energy consumption

learnon

5.5 Quick quiz on | 5.5 Exercise

Simple familiar	Complex familiar	Complex unfamiliar
1, 2, 3, 4, 5, 6, 7, 8, 9, 10, 11, 12, 13, 14, 15, 16	N/A	N/A

These questions are even better in jacPLUS!

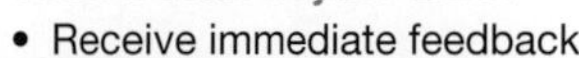
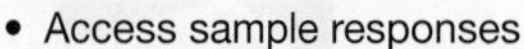
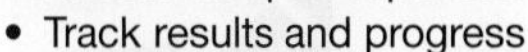

- Receive immediate feedback
- Access sample responses
- Track results and progress

Find all this and MORE in jacPLUS

Simple familiar

1. WE10 A computer uses 350 W in an hour. Convert the measurement to kilowatt hours.
2. Identify whether the following statement is True or False.
   A 400 000 000 W generator is the same as a 400 MW generator.
3. A light sensor uses 0.002 W. Convert the measurement to milliwatts.
4. WE11 Calculate the energy consumption of a 250 W games console for the time duration of 1 hour.
5. WE12 Calculate the power consumption of a microwave that has an energy consumption of 144 000 J for the time duration of 2 minutes.

6. WE13 A toaster has 240 V and 7 A. Calculate the toaster's wattage.
7. A woman's average energy requirement is 8000 kJ. Calculate how many watt hours, to 1 decimal place, this is.
8. Determine how long it will take to power a 2400 W clothes dryer so that it is the same as an adult using 12 000 kJ.
9. Calculate the length of time that the average energy used by an adult in a day (12 000 kJ) is the same as powering a 1.5 W mobile phone charger.
10. Calculate how much energy, in kJ, a 75-watt light bulb uses when it is turned on for 25 minutes.

11. Assume your electrical bill shows that you used 1250 kWh over a 30-day period.
    a. Calculate the energy used in kJ for the 30-day period.
    b. Calculate the energy in joules per day.
    c. At the rate of \$0.12/kWh, calculate your electricity bill for the month.
12. A 100-watt light bulb is 20% efficient (the bulb converts 20% of electrical energy into light and 80% is wasted by being transformed into heat).
    a. Calculate how much energy it uses in 12 hours of operation.
    b. Calculate how much energy the bulb converts to light during 12 hours.
    c. Convert the total energy use to kWh.
13. An electric clothes dryer has a power rating of 3000 W. A family does 5 loads of laundry each week for 4 weeks. It takes the dryer 1 hour to complete each load.
    a. Calculate the energy used in kWh.
    b. The electricity company charges \$0.08/kWh. Calculate the operating cost for 4 weeks.

14. The electric lights in a household use 400 watts per hour and average four hours per day, every day for one year.
   a. Determine how many kWh per year this represents.
   b. If replacing the lights with a fluorescent bulb would save 60 W per night, calculate the savings in kWh that this represents in one year.
   c. The cost per kWh is $0.0825. If the fluorescent bulb costs $18 but lasts for 10 years, consider whether it would be a wiser investment than incandescent bulbs. Explain your answer.
15. A single-door, manual-defrost refrigerator uses 650 kWh/year. A large two-door automatic-defrost refrigerator uses 1870 kWh/year. If 1 kWh = 860 kcal, calculate how many kcal/year each refrigerator uses.

16. An air conditioner is used for a total of 125 days, 24 hours per day at a rate of 7.25 kWh per hour. The cost per kWh is $0.0875.
   a. Calculate the total number of kWh used per year.
   b. Determine the cost of air-conditioning for one year.
   c. Calculate how many kcal are used per year. (1 kWh = 860 kcal)

Fully worked solutions for this chapter are available online.

LESSON
5.6 Review

5.6.1 Summary

doc-41755

Hey students! Now that it's time to revise this chapter, go online to:

Access the chapter summary

Review your results

Watch teacher-led videos

Practise questions with immediate feedback

Find all this and MORE in jacPLUS

5.6 Exercise

learnon

5.6 Exercise

Simple familiar	Complex familiar	Complex unfamiliar
1, 2, 3, 4, 5, 6, 7, 8, 9, 10, 11, 12, 13, 14, 15, 16, 17, 18, 19, 20	N/A	N/A

These questions are even better in jacPLUS!
- Receive immediate feedback
- Access sample responses
- Track results and progress

Find all this and MORE in jacPLUS

Simple familiar

1. The luggage limit on an international flight is 2 suitcases of 25 kg each. A passenger checks in for a flight to Singapore and her luggage has a mass of 42.3 kg.
   Calculate the extra mass she is entitled to take in her luggage.
2. Consider the following nutritional information.

Food item	Energy per serve
2 slices supreme pan pizza	2474 kJ
1 hamburger	563 kcal
6 sticks chicken satay	216 kcal
1 wedge of fruit cake	743 kJ
1 iced doughnut	306 kcal
1 slice of chocolate cake	1054 kJ

   a. Calculate how many kilojoules there are in 2 slices of pizza and 1 iced doughnut.
   b. Calculate how many kilojoules there are in 1 hamburger and 1 slice of chocolate cake. Give your answer correct to 1 decimal place.
3. You eat 125 g of yoghurt from a 175 g tub of yoghurt. Calculate how many kilojoules of yoghurt you have consumed if the nutrition information providing the energy content from the label shows that a 100 g serve of yoghurt is 376 kJ.

4. Calculate the energy available from a 30-g serving of porridge that contains 23.1 g carbohydrates, 2.7 g protein, and 1.5 g fat.

5. Calculate the energy available in kilocalories from a 75-g serving of lemon meringue pie that contains 29.25 g carbohydrates, 3 g protein, and 9.8 g fat. Give your answer to 1 decimal place.

6. The percentage daily intake (%DI) is based on the recommended amounts of energy and nutrients needed for an average adult male to meet his daily nutritional needs.
The percentages are based on the values below.

Nutrient	Energy	Protein	Fat	Carbohydrate
Reference value used in %DI	8700 kJ	50 g	70 g	310 g

Calculate the recommended daily amount of fat for a person with daily energy needs of 8300 kJ. Give your answer to 1 decimal place.

7. Calculate the BMR for a 155-cm, 13-year-old boy weighing 43.5 kg.

8. Calculate the daily energy needs, to the nearest whole number, for:
   a. a 171-cm, lightly active 78-year-old man weighing 72 kg
   b. a 165-cm, moderately active 14-year-old girl weighing 45 kg.

9. Calculate the number of kilojoules an average person uses for the following activities and give your answers to 2 decimal places.
   a. Swimming breast stroke for 30 minutes at 7.2 kcal/min
   b. Mopping a floor for 18 minutes at 2.79 kcal/min

10. A coffee machine uses 1500 W in an hour. Convert the measurement to kilowatt hours.

11. Calculate the energy consumption of a 3500 W oven for 20 minutes.

12. A juice extractor has 230 V and 25 A. Calculate its wattage.

13. Kimberley is 42 years old, works in a shop and requires 8500 kJ for her lifestyle. Every day on her way to work she stops to buy a cafe latte and croissant that contain 1286 kJ.
Calculate the percentage of her daily energy needs, to 1 decimal place, that Kimberley consumes in the morning.

14. The table shown contains data for the number of grams of fat, protein and carbohydrates for some selected foods.

Food (100-g serving)	Carbohydrate (g)	Protein (g)	Fat (g)
Bananas	23	1	0
Raspberries	12	1	1
Strawberries	8	1	0
Grapes	18	1	0
Pears	15	0	0
Raisins	79	3	0
Apricots	11	1	0
Cheese spread	4	7	29
Cheddar cheese	1	25	33
Cottage cheese	3	12	5

a. Calculate the energy content of 100 g of raspberries, 100 g of strawberries, 50 g of raisins and 50 g of the cheese spread.
b. According to the Australia New Zealand Food Standards Code, a balanced diet for an average adult is made up of the following nutrients each day:

	Energy	Protein	Fat	Carbohydrate
Quantity per day	8700 kJ	50 g	70 g	310 g

i. If 50 g of grapes is consumed, calculate the percentage of carbohydrates that have been eaten for the day.
ii. If 200 g of cheddar cheese is consumed, calculate the percentage of fat that has been eaten for the day. Give your answers to 1 decimal place.

15. The table below shows the calories per kilo per hour used for different activities lasting for an hour.

Activity (1 hour)	Calories per kilo per hour
Surfing	3
Playing golf	5
Playing drums	4
Playing rugby	10
Ballet	6
Cycling	7
Waterskiing	6
Playing squash	12
Water aerobics	4

a. Calculate how many kilojoules, to the nearest whole number, a person weighing 68 kg uses to play squash for 20 minutes.
b. Calculate how many kilojoules, to the nearest whole number, a person weighing 55 kg uses to do a ballet class for 2 hours.

16. A balanced diet for an average adult is made up of the following nutrients each day:

Nutrient	Energy	Protein	Fat	Carbohydrate
Quantity per day	8700 kJ	50 g	70 g	310 g

Determine the difference in maximum protein intake between a 6500 kJ daily diet and a 10 500 kJ daily diet. Give your answer to the nearest whole number.

17. A family eat Mexican food for lunch. Each family member eats:
    - 100 g beef mince 1033 KJ
    - 50 g baked taco shells 982 KJ
    - 50 g tortillas 655 KJ
    - 50 g cheese 844 KJ
    - 200 g chili con carne 848 KJ
    - 100 g tomato 97 KJ
    - 100 g onions 185 kJ

    a. Calculate the energy content of each family member's meal.

    b. An average person should intake 8700 kJ each day. If 20% of the daily energy intake should be for lunch, calculate the effect the lunch meal will have on a family member's energy intake for the day.

18. Your electrical bill shows that you used 4520 kWh over a 30-day period. At the rate of \$0.16/kWh, calculate your electricity bill for the month.

19. A washing machine has a power rating of 500 W. A family does 8 loads of laundry each week for 4 weeks. It takes the washing machine 50 minutes to complete each load. The electricity company charges \$0.11/kWh. Calculate the operating cost for 4 weeks.

20. A 42-inch LED television is used for a total of 82 days, 8 hours per day at a rate of 0.346 kWh per hour. The cost per kWh is \$0.0925.
    When the 42-inch LED television is not used, it is on standby at a rate of 0.021 kWh per hour.
    Calculate the total cost (both operational and standby) of using the television for the 82 days.

Fully worked solutions for this chapter are available online.

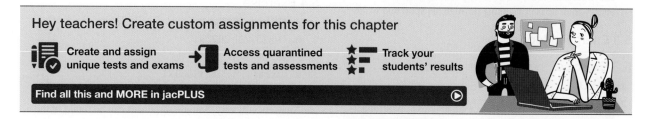

Answers

Chapter 5 Units of energy

5.2 Units of energy

5.2 Exercise

1. a. 2000 mg b. 15 100 kg c. 1.2 g
   d. 0.022 kg e. 40 g f. 0.013 kg
2. a. 500 000 mg b. 0.02 t c. 109 020 g
   d. 3800 kg e. 800 mg f. 5 kg
3. 1.8 kg
4. a. False b. True
5. a. 2000 J b. 3 500 000 J
6. a. 52 MJ b. 370 000 MJ
7. a. 418.4 kJ b. 9204.8 kJ
8. 834 kcal
9. $k = 4.184x$
10. a. 1841 kJ b. 229 kcal
11. 864.5 kg
12. 171.915 t
13. a. 824 kJ b. 2734 kJ
14. a.

Food item	Kilocalories	Kilojoules
Can of Coke	154	$154 \times 4.184 = 644.336$ kJ
Can of Fanta	195	$195 \times 4.184 = 815.88$ kJ
Apple juice	102	$102 \times 4.184 = 426.768$ kJ
Glass of milk	164	$164 \times 4.184 = 686.176$ kJ
Cappuccino	53	$53 \times 4.184 = 221.752$ kJ

   b. 6267 kJ
   c. 23.57%
15. Answers will vary. Sample responses can be found in the worked solutions in the online resources.
16. Since Harry prefers not to exercise and eats a lot of snacks and fast food, he is eating more than he is burning and will likely gain weight as a result. It would be a good recommendation for Harry to eat healthier food with a lower calorie count.
    Ben is a healthy eater and very active, although he probably could eat a bit more, since he is potentially burning more energy than he is consuming through his sport and work.

5.3 Energy in food

5.3 Exercise

1. 2725 kJ
2.

Food item	Kilocalories	Kilojoules
265 g T-bone steak grilled	478	2001
140 g chopped grilled chicken	231	966
100 g lamb ribs	160	669
425 g tuna in brine	442	1849
100 g crumbed flounder	225	940

3. a. 908 kcal b. 610 kcal
4. 2035 kJ
5. 764 kJ
6. 471 kJ
7. 448 kJ
8. 288 kJ
9. 408.8 kJ
10. Breakfast: 1600 kJ
    Morning snack: 800 kJ
    Lunch: 1600 kJ
    Afternoon snack: 800 kJ
    Dinner: 3200 kJ
11. 40 g
12. 520 kcal
13. a. 57.93 g b. 53.45 g
14.

	Energy	Protein	Fat	Carbohydrate
Adult quantity per day	8700 kJ	50 g	70 g	310 g
Hamburger	2010 kJ	25 g	25.5 g	35 g
%DI	23.1%	50%	36.4%	11.3%

15. Bridgette is not eating enough for breakfast or dinner, but she is eating more for lunch.
16. a. 8704 kJ
    b. Nearly all the children's recommended energy intake is consumed in 1 meal. The children have not eaten balanced meals throughout the day and have consumed a lot of unhealthy foods.
17. 11.5 g
18. 92.6 g
19. Answers will vary. Sample responses can be found in the worked solutions in the online resources.
20. a. i. 396.2 kJ ii. 6.29%
    b. i. 541.5 kJ ii. 8.60%
21. The 100-g piece of chocolate cake has the most kilojoules. This is to be expected as it is high in fat and carbohydrates.

5.4 Energy in activities

5.4 Exercise

1. a. 879 kcal; 3677 kJ
   b. 716.25 kcal; 2997 kJ
2. False
3. a. 7080 kJ b. 12360 kJ
4. 1216 kJ
5. Answers will vary. Sample responses can be found in the worked solutions in the online resources.
6. 17.6%
7. 3100 kJ
8. 76 minutes
9. 1656 kJ

10. 1590 kJ
11. 170 kJ
12. 4650 kJ
13. 5500 kJ
    This is probably not enough for someone who lives a very active lifestyle.
14. 998.4 kJ
15. 22.6%
16. 2400 kJ
17. 57.5 kg
18. 1733 kJ
19. 11 250 kJ
20. James has 6700 kJ to play sport. This should be sufficient.

5.5 Energy consumption

5.5 Exercise

1. 0.35 kW
2. True
3. 2 mW
4. 900 000 J
5. 1200 W
6. 1680 W
7. 2222.2 Wh
8. $83\frac{1}{3}$ minutes
9. 92.6 days
10. 112.5 kJ
11. a. 4 500 000 kJ
    b. 1.5×10^8 J/day
    c. $150
12. a. 4.32×10^6 J
    b. 864 000 J
    c. 1.2 kWh
13. a. 60 kWh
    b. $4.80
14. a. 584 kWh/year
    b. 87.6 kWh/year
    c. Yes, the bulb would be paid for in just under 2.5 years.
15. Single-door: 559 000 kcal/year
    Two-door: 1 608 200 kcal/year
16. a. 21 750 kWh/year
    b. $1903.13
    c. 18 705 000 kcal/year

5.6 Review

5.6 Exercise

1. 7.7 kg
2. a. 3754.3 kJ b. 3409.6 kJ
3. 470 kJ
4. 494.1 kJ
5. 217.7 kcal
6. 66.8 g
7. 5622 kcal
8. a. 8075 kcal b. 8108 kcal
9. a. 903.74 kJ b. 210.12 kJ
10. 1.5 kWh
11. 4 200 000 J
12. 5750 W
13. 15.1%
14. a. 1738 kJ
    b. i. 2.9% ii. 94.3%
15. a. 1138 kJ b. 2761 kJ
16. 23 g
17. a. 4644 kJ
    b. Each family member is consuming approximately 53.4% of their recommended daily intake, which is way above the 20% recommended for lunch.
18. $723.20
19. $1.47
20. $23.55

6 Data classification, presentation and interpretation

LESSON SEQUENCE

Fully worked solutions for this chapter are available online.

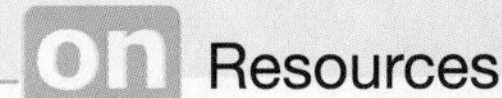

Solutions Solutions — Chapter 6 (sol-1259)

Digital documents Learning matrix — Chapter 6 (doc-41756)
Quick quizzes — Chapter 6 (doc-41757)
Chapter summary — Chapter 6 (doc-41758)

LESSON
6.1 Overview

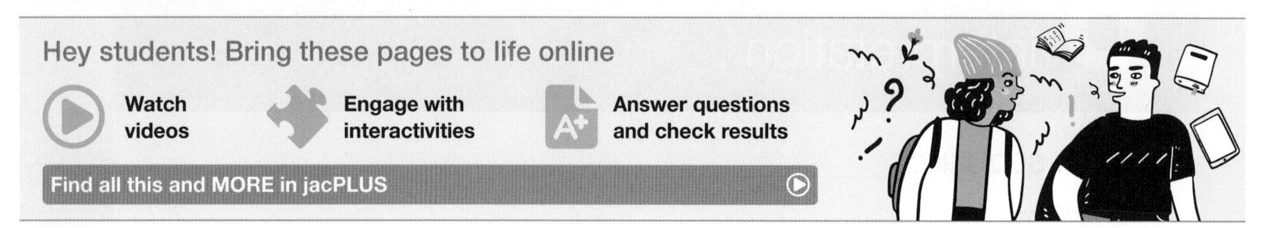

6.1.1 Introduction

The world we live in produces an ever-increasing amount of data. Being able to understand and interpret these large sets of data is a valuable skill in the 21st century. To be able to interpret data, you need to first know how to collect it. The Australian Bureau of Meteorology collects data about the weather, analyses it, and then uses that knowledge to predict what the weather will be like tomorrow, next week and even next year. The Australian Bureau of Statistics runs a census night every five years to collect data about the population. This data is then analysed so future decisions can be made by the government.

Social media companies collect data about how you use their platforms. This data is analysed and used to adapt these platforms to the way you use them (and to help these companies maximise profits). This demonstrates the importance of understanding how to collect and then interpret data to make important decisions.

6.1.2 Syllabus links

Lesson	Lesson title	Syllabus links
6.2	**Classifying data**	○ Identify examples of categorical data. ○ Identify examples of numerical data.
6.3	**Categorical data**	○ Display categorical data in tables and column graphs.
6.4	**Numerical data**	○ Display numerical data as frequency distribution tables, dot plots, stem plots and histograms.
6.5	**Recognising and identifying outliers**	○ Recognise and identify outliers from a dataset.
6.6	**Comparing different data [complex]**	○ Compare the suitability of different methods of data presentation in real-world contexts [complex].

LESSON
6.2 Classifying data

SYLLABUS LINKS

- Identify examples of categorical data.
- Identify examples of numerical data.

6.2.1 Types of data

- Statistics is the science of collecting, organising, presenting, analysing and interpreting data. The information collected is called **data**.
- The information in the flowchart shown can be used to determine the type of data being considered.

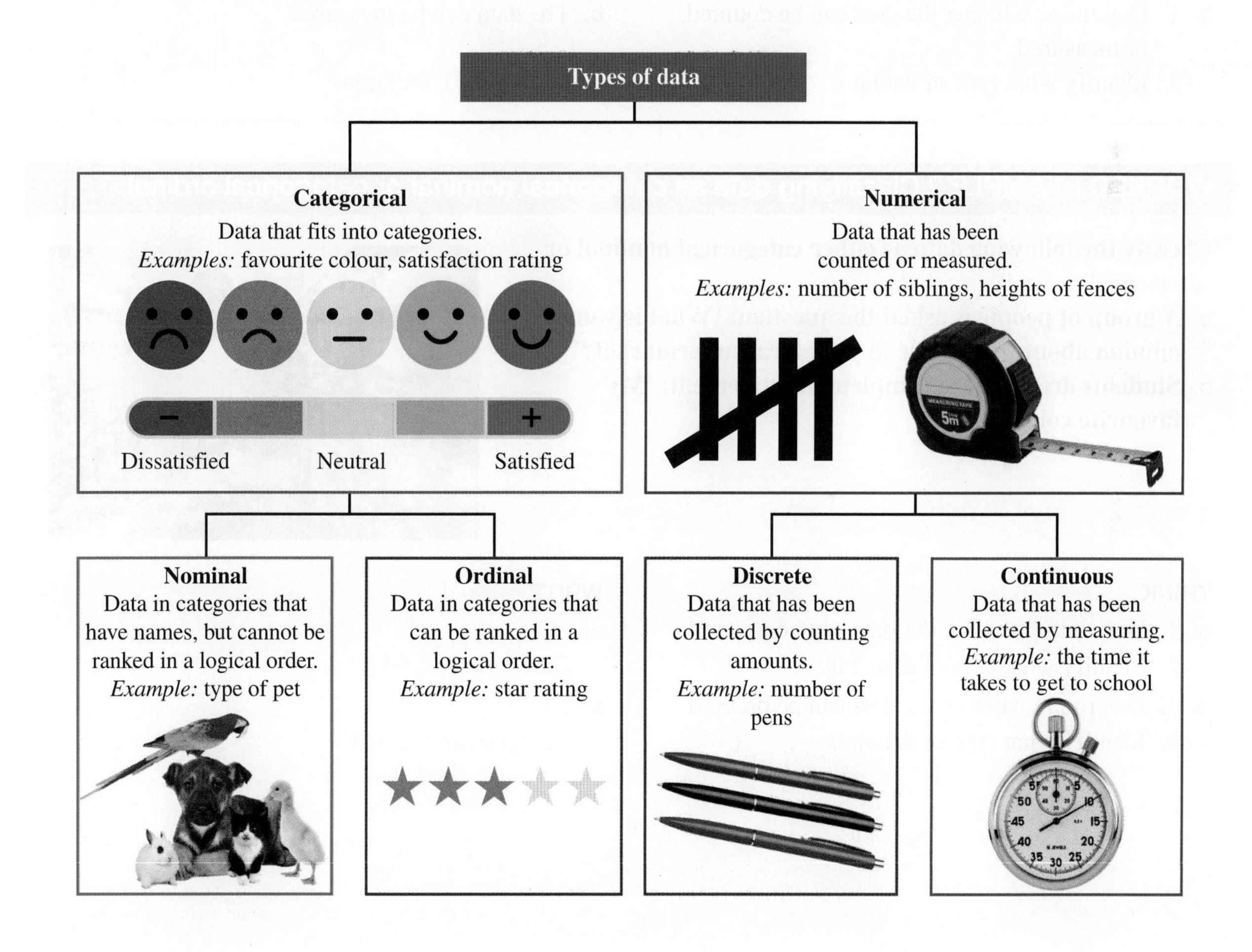

WORKED EXAMPLE 1 Classifying data as numerical discrete or numerical continuous

Classify the following data as either numerical discrete or numerical continuous.

a. A group of people is asked the question: 'How many cars does your family have?'

b. Students are asked to complete the statement: 'My height is ...'

THINK	WRITE
a. 1. Determine whether the data can be counted or measured.	a. The data can be counted.
2. Identify what type of data it is.	Numerical discrete
b. 1. Determine whether the data can be counted or measured.	b. The data can be measured.
2. Identify what type of data it is.	Numerical continuous

WORKED EXAMPLE 2 Classifying data as categorical nominal or categorical ordinal

Classify the following data as either categorical nominal or categorical ordinal.

a. A group of people is asked the question: 'What is your opinion about the service in your local supermarket?'

b. Students are asked to complete the statement: 'My favourite colour is ...'

THINK	WRITE
a. 1. Determine whether the data can be ordered.	a. Yes
2. Identify what type of data it is.	Categorical ordinal
b. 1. Determine whether the data can be ordered.	b. No
2. Identify what type of data it is.	Categorical nominal

6.2.2 Categorical data with numerical values

- There are instances when categorical data is expressed by numerical values.
- 'Phone numbers ending in 5' and 'postcodes' are examples of categorical data represented by numerical values.

tlvd-4691

WORKED EXAMPLE 3 Classifying data as categorical or numerical

Classify the following data as either categorical or numerical.
a. A set of data is collected about bank account numbers.
b. A group of people are asked: 'How many minutes do you exercise per day?'
c. A group of people are asked: 'What is the area of your backyard?'
d. Students are asked to complete the statement: 'The subject I like the most is ...'

THINK	WRITE
a. 1. Determine whether the data can be counted or measured.	a. Neither
2. Identify what type of data it is.	Categorical
b. 1. Determine whether the data can be counted or measured.	b. The data can be measured.
2. Identify what type of data it is.	Numerical
c. 1. Determine whether the data can be counted or measured.	c. The data can be measured.
2. Identify what type of data it is.	Numerical
d. 1. Determine whether the data can be counted or measured.	d. Neither
2. Identify what type of data it is.	Categorical

Exercise 6.2 Classifying data

learnon

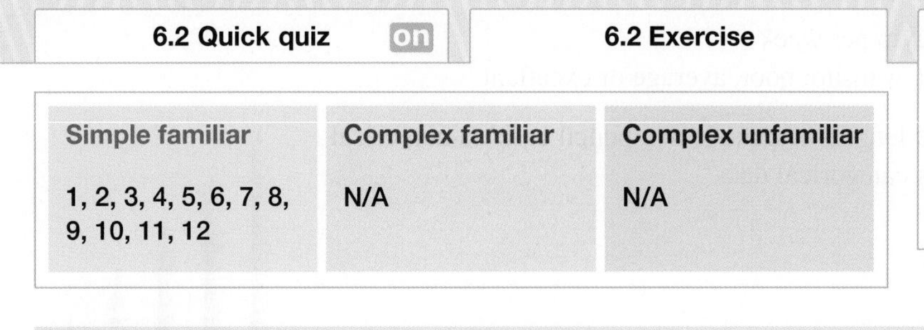

6.2 Quick quiz on	6.2 Exercise	
Simple familiar	Complex familiar	Complex unfamiliar
1, 2, 3, 4, 5, 6, 7, 8, 9, 10, 11, 12	N/A	N/A

These questions are even better in jacPLUS!
- Receive immediate feedback
- Access sample responses
- Track results and progress

Find all this and MORE in jacPLUS

Simple familiar

1. WE1 Classify the following data as either numerical discrete or numerical continuous.
   a. The amount of time spent on homework per night
   b. The marks in a test

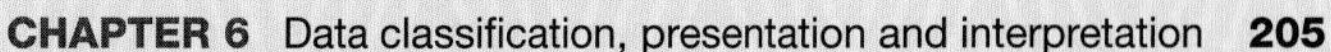

2. Explain in your own words why 'the number of pieces of fruit sold' is numerical discrete data while 'the quantity (in kg) of fruit sold' is numerical continuous data.

3. Classify the following data as either discrete or continuous.

   a. Data is collected about the water levels in a dam over a period of one month.
   b. An insurance company is collecting data about the number of work injuries in ten different businesses.
   c. 350 students are selected by a phone company to answer the question: 'How many pictures do you take per week using your mobile phone?'
   d. Customers in a shopping centre are surveyed about the number of times they visited the shopping centre in the last 12 months.
   e. 126 people participated in a 5-km charity marathon. Their individual times are recorded.
   f. A group of students are asked to measure the perimeter of their bedrooms. This data is collected and used for a new project.

4. WE2 Classify the following data as either categorical nominal or categorical ordinal.
   a. The brands of cars sold by a car yard over a period of time
   b. The opinion about the quality of a website: poor, average or excellent

5. Explain in your own words why 'favourite movie' is categorical nominal data while 'the rating for a movie' is categorical ordinal data.

6. Classify the following data as either ordinal or nominal.
   a. The members of a football club are asked to rate the quality of the training grounds as 'poor', 'average' or 'very good'.
   b. A group of students are asked to state the brand of their mobile phone.
   c. A fruit and vegetable shop is conducting a survey, asking the customers, 'Do you find our products fresh?' Possible answers are 'Never', 'Sometimes', 'Often', 'Always'.
   d. A chocolate company conducts a survey with the question: 'Do you like sweets?'. The possible answers are 'Yes' and 'No'.

7. WE3 Classify the following data as either numerical or categorical.
   a. Data collected on 'whether the subject outcomes 1, 2 and 3 have been completed'
   b. The number of learning outcomes per subject

8. Explain in your own words why 'country phone codes' is categorical data while 'the number of international calls' is numerical data.

9. Classify the following data as either numerical or categorical.
   a. The amount of time spent at the gym per week
   b. The opinion about the quality of a website: poor, average or excellent

10. Explain in your own words why 'the lengths of pencils in a pencil case' is numerical data while 'the colours of pencils' is categorical data.

11. A train company recorded the time by which trains were late over a period of one month. Classify if this data is numerical discrete, numerical continuous, categorical nominal or categorical ordinal.

12. Classify the following data as either numerical or categorical.
    a. The average time spent by customers in a homeware shop is recorded.
    b. Mothers are asked: 'What brand of baby food do you prefer?'
    c. An online games website asked the players to state their gender.
    d. A teacher recorded the lengths of his students' long jumps.

Fully worked solutions for this chapter are available online.

LESSON
6.3 Categorical data

SYLLABUS LINKS

- Display categorical data in tables and column graphs.

6.3.1 One-way or frequency tables

- Categorical data can be displayed in tables or graphs.
- **One-way tables** or **frequency distribution tables** are used to organise and display data in the form of frequency counts.
- The **frequency** of a score is the number of times the score occurs in the set of data.
- Frequency distribution tables display one variable only.
- For large amounts of data, a **tally** column helps make accurate counts of the data.
- The frequency distribution table shown displays the tally and the frequency for a given data set.

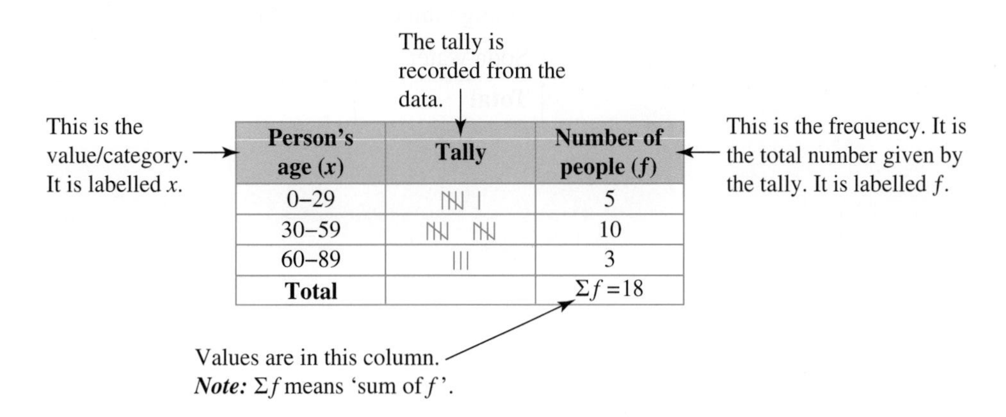

Person's age (x)	Tally	Number of people (f)
0–29	𝍸 |	5
30–59	𝍸 𝍸	10
60–89	|||	3
Total		$\Sigma f = 18$

The tally is recorded from the data.
This is the value/category. It is labelled x.
This is the frequency. It is the total number given by the tally. It is labelled f.
Values are in this column.
Note: Σf means 'sum of f'.

tlvd-4692

WORKED EXAMPLE 4 Constructing a frequency distribution table

Alana is organising the Year 11 formal and has to choose three types of drinks out of the five available: cola, lemonade, apple juice, orange juice and soda water. She decided to survey her classmates to determine their favourite drinks. The data collected is shown below.

orange juice,	**cola,**	**cola,**	**cola,**	**apple juice,**	**lemonade,**
lemonade,	**cola,**	**apple juice,**	**lemonade,**	**orange juice,**	**orange juice,**
lemonade,	**cola,**	**apple juice,**	**orange juice,**	**cola,**	**soda water,**
orange juice,	**cola,**	**soda water,**	**orange juice,**	**cola,**	**cola**

By constructing a frequency table, determine the three drinks Alana should choose for the formal.

THINK

1. Draw a frequency table with three columns and the headings 'Drink', 'Tally' and 'Frequency'.

WRITE

Drink	Tally	Frequency

2. Write all the possible outcomes in the first column and 'Total' in the last row.

Drink	Tally	Frequency
Apple juice		
Cola		
Lemonade		
Orange juice		
Soda water		
Total		

3. Fill in the 'Tally' column.

Drink	Tally	Frequency
Apple juice	|||	
Cola	~~||||~~ ||||	
Lemonade	||||	
Orange juice	~~||||~~|	
Soda water	||	
Total		

4. Fill in the 'Frequency' column. Ensure the table has a title also.

Favourite drinks for a Year 11 class		
Drink	**Tally**	**Frequency**
Apple juice	|||	3
Cola	~~||||~~ ||||	9
Lemonade	||||	4
Orange juice	~~||||~~ |	6
Soda water	||	2
Total		24

5. Write your answer as a sentence.

Alana should choose cola, orange juice and lemonade for the Year 11 formal.

6.3.2 Two-way frequency tables

- **Two-way tables** are used to examine the relationship between two related categorical variables.
- Because the entries in these tables are frequency counts, they are also called **two-way frequency tables**.
- The table shown is a two-way table that relates two variables.
- It displays the relationship between age groups and party preference.

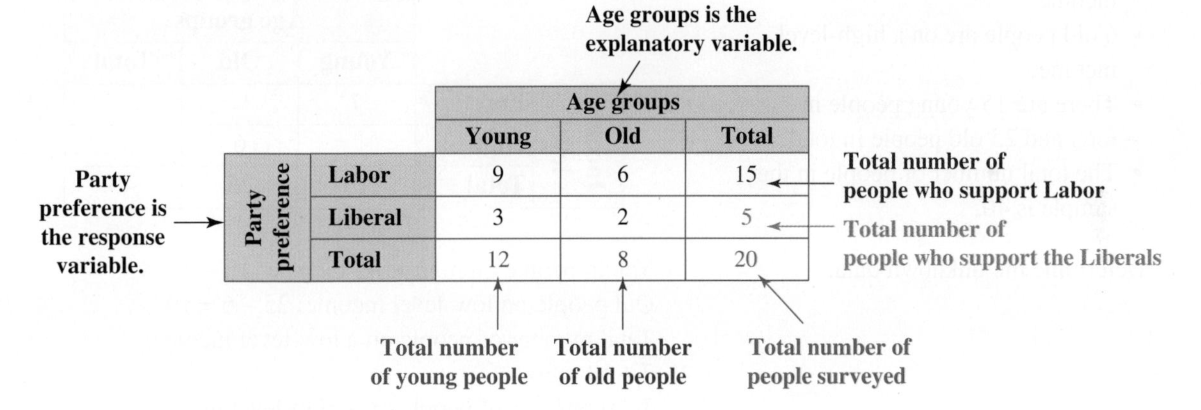

		Age groups		
		Young	Old	Total
Party preference	Labor	9	6	15
	Liberal	3	2	5
	Total	12	8	20

- One variable is **explanatory** (age groups) and the other variable is **response** (party preference).
- The explanatory variable is the variable that can be manipulated to produce changes in the response variable.
- The response variable changes as the explanatory variable changes.
- Its outcome is a response to the input of the explanatory variable.

WORKED EXAMPLE 5 Constructing a two-way frequency table to display data

A sample of 40 people was surveyed about their income level (low or high). The sample consisted of 15 young people (25–45 years old) and 25 old people (above 65 years old). 7 young people were on a low-level income while 6 old people were on a high-level income.
Display this data in a two-way table and determine the number of young people on a high-level income.

THINK	WRITE
1. Identify the two categorical variables.	Age groups: young and old people Income levels: low and high
2. Decide where the two variables are placed.	The explanatory variable is 'Age groups' — columns. The response variable is 'Income level' — rows.
3. Construct the two-way table and label both the columns and the rows.	(see table below)

		Income levels for young and old people		
		Age groups		
		Young	Old	Total
Income level	Low			
	High			
	Total			

4. Fill in the known data:
   - 7 young people are on a low-level income.
   - 6 old people are on a high-level income.
   - There are 15 young people in total and 25 old people in total.
   - The total number of people in the sample is 40.

		Income levels for young and old people		
		Age groups		
		Young	Old	Total
Income level	Low	7		
	High		6	
	Total	15	25	40

5. Determine the unknown data.

Young people on high-level income: $15 - 7 = 8$
Old people on low-level income: $25 - 6 = 19$
Total number of people on a low-level income:
$7 + 19 = 26$
Total number of people on a high-level income:
$8 + 6 = 14$

6. Fill in the table.

		Income levels for young and old people		
		Age groups		
		Young	Old	Total
Income level	Low	7	19	26
	High	8	6	14
	Total	15	25	40

7. Write the answer as a sentence.

The total number of young people on a high-level income is 8.

6.3.3 Two-way relative frequency tables

- Consider the party preference table shown. As the number of young and old people is different, it is difficult to compare the data.

		Party preference for young and old people		
		Age groups		
		Young	Old	Total
Party preference	Labor	9	6	15
	Liberal	3	2	5
	Total	12	8	20

- Nine young people and six old people prefer Labor; however, there are more young people surveyed than old.
- In order to accurately compare the data, we need to convert these values in a more meaningful way as proportions or percentages.

Formula for relative frequency

The frequency given as a proportion is called relative frequency.

$$\textbf{Relative frequency} = \frac{\textbf{score}}{\textbf{total number of scores}}$$

- The relative frequency table for the party preference data is:

		Party preference for young and old people		
		Age groups		
		Young	Old	Total
Party preference	Labor	$\frac{9}{20}$	$\frac{6}{20}$	$\frac{15}{20}$
	Liberal	$\frac{3}{20}$	$\frac{2}{20}$	$\frac{5}{20}$
	Total	$\frac{12}{20}$	$\frac{8}{20}$	$\frac{20}{20}$

- In simplified form using decimal numbers, the relative frequency table for party preference becomes:

		Party preference for young and old people		
		Age groups		
		Young	Old	Total
Party preference	Labor	0.45	0.30	0.75
	Liberal	0.15	0.10	0.25
	Total	0.60	0.40	1.00

Note: In a relative frequency table, the total value is always 1.

Formula for percentage relative frequency

The frequency written as a percentage is called percentage relative frequency (% relative frequency). We can use the following formula to calculate it:

$$\text{Percentage relative frequency} = \frac{\text{score}}{\text{total number of scores}} \times 100\%$$

or

$$\text{Percentage relative frequency} = \text{relative frequency} \times 100\%$$

tlvd-4693

WORKED EXAMPLE 6 Calculating the percentage relative frequency

Consider the party preference data discussed previously, with the relative frequencies shown.

		Party preference for young and old people		
		Age groups		
		Young	Old	Total
Party preference	Labor	0.45	0.30	0.75
	Liberal	0.15	0.10	0.25
	Total	0.60	0.40	1.00

Calculate the percentage relative frequency for all entries in the table out of the total number of people. Discuss your findings.

THINK

1. Write the formula for % relative frequency.

WRITE

Percentage relative frequency = relative frequency × 100%

THINK

2. Substitute all scores into the percentage relative frequency formula.

WRITE

		Party preference for young and old people		
		Age groups		
		Young	Old	Total
Party preference	Labor	0.45×100	0.30×100	0.75×100
	Liberal	0.15×100	0.10×100	0.25×100
	Total	0.60×100	0.40×100	1.00×100

3. Simplify.

		Party preference for young and old people		
		Age groups		
		Young	Old	Total
Party preference	Labor	45	30	75
	Liberal	15	10	25
	Total	60	40	100

4. Discuss your findings.

Of the total number of people, 75% prefer Labor and 25% prefer the Liberals. 60% of people are young and 40% of people are old. 45% of people who prefer Labor are young and 30% of people who prefer the Liberals are old. 15% of people who prefer the Liberals are young and 10% of people who prefer the Liberals are old.

6.3.4 Grouped column graphs

- **Grouped column graphs** are used to display the data for two or more categories.
- They are visual tools that allow for easy comparison between various categories of data sets.
- The frequency is measured by the height of the column.
- Both axes have to be clearly labelled, including scales and units, if used.
- The title should explicitly state what the grouped column graph represents.
- Consider the data recorded in the two-way distribution table shown.

		Income levels for young and old people		
		Age groups		
		Young	Old	Total
Income level	Low	7	19	26
	High	8	6	14
	Total	15	25	40

- The grouped column graph for this data is shown. Notice that all rectangles have the same width.

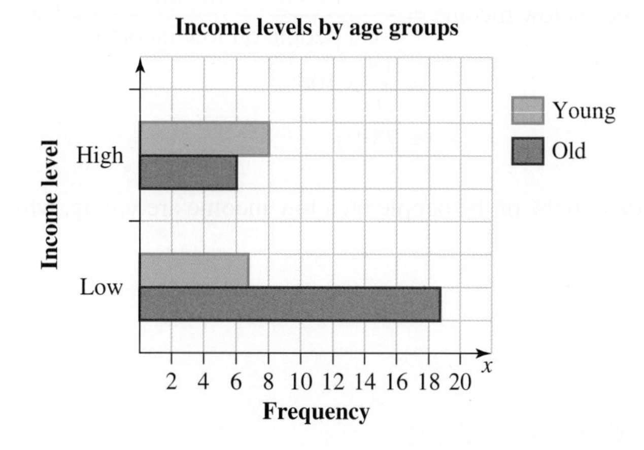

- The colour or pattern of the rectangles differentiates between the two categories, 'Young' and 'Old'.

Making comparisons

- It is helpful to change the table to a percentage frequency table to compare the income levels of old and young people.
- We first calculate the percentage relative frequency for all the entries in the two-way table out of the total number of people.

		Income levels for young and old people		
		Age groups		
		Young %	Old %	Total %
Income level	Low	$\frac{7}{40} \times 100 = 17.5$	$\frac{19}{40} \times 100 = 47.5$	$\frac{26}{40} \times 100 = 65.0$
	High	$\frac{8}{40} \times 100 = 20.0$	$\frac{6}{40} \times 100 = 15.0$	$\frac{14}{40} \times 100 = 35.0$
	Total %	$\frac{15}{40} \times 100 = 37.5$	$\frac{25}{40} \times 100 = 62.5$	$\frac{40}{40} \times 100 = 100$

- From these percentages we can conclude that:
  - 17.5% of the people surveyed are young on a low income and 47.5% of the people are old on a low income.
  - 20% of the total number of people surveyed are young on a high income while 15% are old on a high income.
  - 37.5% of the people surveyed are young while 62.5% are old.
  - 65% of the people surveyed are on a low income while 35% are on a high income.
- Another way of comparing the results of this survey is to calculate percentages out of the total numbers in columns and in rows.

Note: This method of comparison is more applicable when data has even or close to even numbers of each age group surveyed.

Row 1: Percentages of young people and old people on low incomes out of the total number of people on a low income

$$\begin{aligned}\text{Young on low income} &= \frac{\text{young on low income}}{\text{total people on low income}} \times 100\% \\ &= \frac{7}{26} \times 100\% \\ &= 26.9\%\end{aligned}$$

$$\begin{aligned}\text{Old on low income} &= \frac{\text{old on low income}}{\text{total people on low income}} \times 100\% \\ &= \frac{19}{26} \times 100\% \\ &= 73.1\%\end{aligned}$$

These percentages indicate that 26.9% of the people on a low income are young, while 73.1% are old.

Row 2: Percentages of young and old people on high incomes out of the total number of people on a high income

$$\begin{aligned}\text{Young on high income} &= \frac{\text{young on high income}}{\text{total people on high income}} \times 100\% \\ &= \frac{8}{14} \times 100\% \\ &= 57.1\%\end{aligned}$$

$$\begin{aligned}\text{Old on high income} &= \frac{\text{old on high income}}{\text{total people on high income}} \times 100\% \\ &= \frac{6}{14} \times 100\% \\ &= 42.9\%\end{aligned}$$

These percentages indicate that 57.1% of the people on a high income are young while 42.9% are old.

Column 1: Percentages of young people on low and high income out of the total number of young people surveyed

$$\begin{aligned}\text{Young on low income} &= \frac{\text{young on low income}}{\text{total number of young}} \times 100\% \\ &= \frac{7}{15} \times 100\% \\ &= 46.7\%\end{aligned}$$

$$\begin{aligned}\text{Young on high income} &= \frac{\text{young on high income}}{\text{total number of young}} \times 100\% \\ &= \frac{8}{15} \times 100\% \\ &= 53.3\%\end{aligned}$$

These percentages indicate that 46.7% of the young people surveyed are on a low income while 53.3% are on a high income.

Column 2: Percentages of old people on low and high income out of the total number of old people surveyed

$$\begin{aligned}\text{Old on low income} &= \frac{\text{old on low income}}{\text{total number of old}} \times 100\% \\ &= \frac{19}{25} \times 100\% \\ &= 76\%\end{aligned}$$

$$\begin{aligned}\text{Old on high income} &= \frac{\text{old on high income}}{\text{total number of old}} \times 100\% \\ &= \frac{6}{25} \times 100\% \\ &= 24\%\end{aligned}$$

These percentages indicate that 76% of the old people surveyed are on a low income while 24% of old people are on a high income.

WORKED EXAMPLE 7 Constructing a grouped column graph to display the percentage relative frequencies of the given data

The data shown is part of a random sample of 25 students who participated in a questionnaire run by the Australian Bureau of Statistics through Census@School. The aim of the questionnaire was to determine how often the students use the internet for schoolwork.
Draw a grouped column graph displaying the percentage relative frequencies of the data set given.

		Use of internet for schoolwork		
		Place		
		School	Home	Total
Time spent	Rarely	2	8	10
	Sometimes	4	4	8
	Often	6	1	7
	Total	12	13	25

THINK

WRITE/DRAW

1. Calculate the % relative frequency for the data needed.

		Use of the internet for schoolwork		
		Place		
		School	Home	Total
Time spent	Rarely	$\frac{2}{25} \times 100 = 8$	$\frac{8}{25} \times 100 = 32$	$\frac{10}{25} \times 100 = 40$
	Sometimes	$\frac{4}{25} \times 100 = 16$	$\frac{4}{25} \times 100 = 16$	$\frac{8}{25} \times 100 = 32$
	Often	$\frac{6}{25} \times 100 = 24$	$\frac{1}{25} \times 100 = 4$	$\frac{7}{25} \times 100 = 28$
	Total	$\frac{12}{25} \times 100 = 48$	$\frac{13}{25} \times 100 = 52$	100

2. Draw a labelled set of axes with accurate scales. Ensure the graph has a title. The maximum percentage is 32%. Therefore, go up by 5% increments from 0 to 35%.

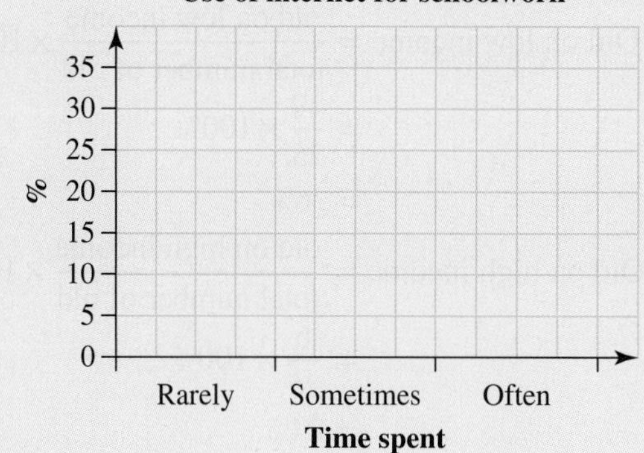

3. Draw the sets of corresponding rectangles. The graph can be drawn vertically.

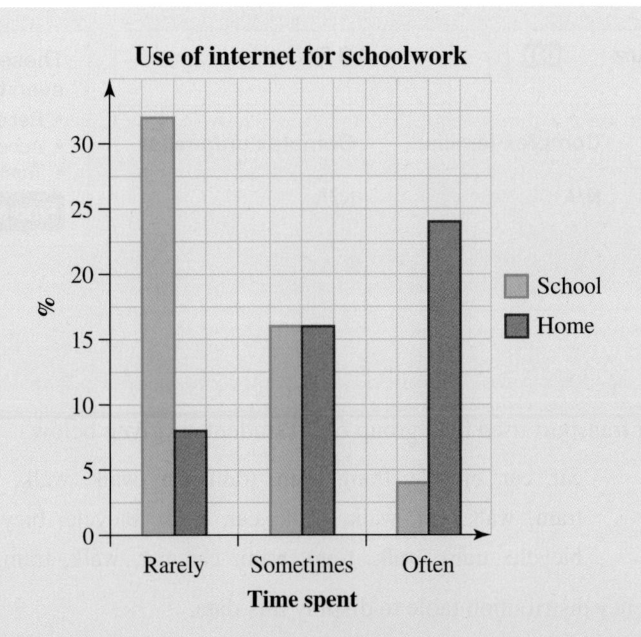

4. The graph can also be drawn horizontally.

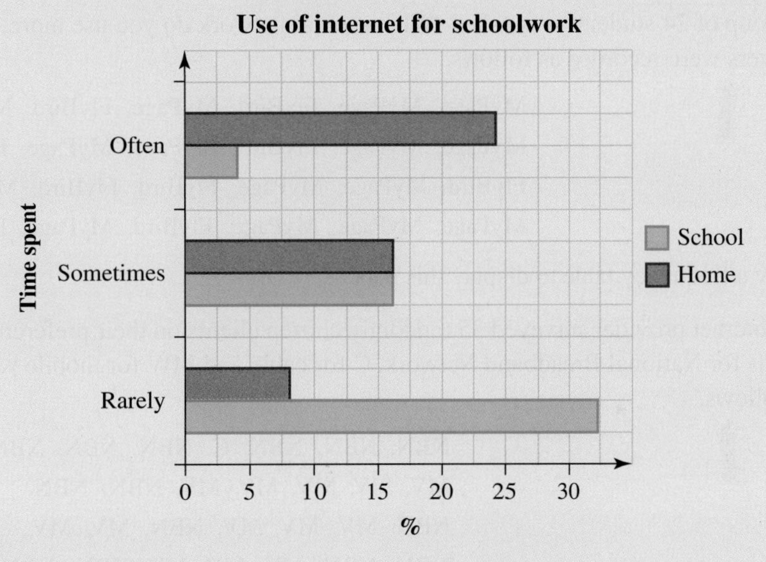

on Resources

Interactivity Create a bar graph (int-6493)

Exercise 6.3 Categorical data

learn on

6.3 Quick quiz	6.3 Exercise

Simple familiar	Complex familiar	Complex unfamiliar
1, 2, 3, 4, 5, 6, 7, 8, 9, 10, 11, 12, 13, 14, 15, 16, 17, 18	N/A	N/A

Simple familiar

1. WE4 The form of transport used by a group of 30 students is given below.

   car, car, bicycle, tram, tram, tram, car, walk, walk, walk,
   tram, walk, car, walk, walk, car, walk, bicycle, bicycle, walk,
   bicycle, tram, walk, tram, tram, car, car, walk, tram, tram

   Construct a frequency distribution table to display this data.

2. A group of 24 students was asked 'Which social network do you use more, MyPage or FlyBird?' Their answers were recorded as follows.

   MyPage, MyPage, FlyBird, MyPage, FlyBird, MyPage,
   MyPage, MyPage, FlyBird, MyPage, MyPage, FlyBird,
   FlyBird, MyPage, MyPage, FlyBird, FlyBird, MyPage,
   MyPage, MyPage, MyPage, FlyBird, MyPage, FlyBird

   Draw a frequency table to display this data.

3. An internet provider surveyed 35 randomly chosen clients on their preference of internet connection. NBN stands for National Broadband Network, C for cable and MW for mobile wireless. Their answers were as follows.

   NBN, NBN, NBN, C, NBN, NBN, NBN,
   MV, MV, MV, MV, MV, NBN, NBN,
   NBN, MV, MV, MV, NBN, MV, MV,
   NBN, NBN, MV, MV, MV, NBN, NBN,
   NBN, C, C, MV, MV, MV, MV

   Display the data shown in a frequency distribution table.

4. The school canteen surveyed a group of 500 students about the quality of the sandwiches offered at lunch. 298 students were of the opinion that the sandwiches offered at the school canteen were very good, 15 students thought they were of poor quality, 163 students thought they were excellent, while the rest said they were average.
   Construct a frequency distribution table for this set of data.

5. **WE5** Determine the missing values in the two-way table shown.

		Parking place and car theft		
		Parking type		
		In driveway	On the street	Total
Theft	Car theft		37	
	No car theft	16		
	Total		482	500

6. A survey of 263 students found that 58 students who owned a mobile phone also owned a tablet, while 11 students owned neither. The total number of students who owned a tablet was 174.
Construct a complete two-way table to display this data and calculate all the unknown entries.

7. The numbers of the Year 11 Maths students at Senior Secondary College who passed and failed their exams are displayed in the table shown.

		Year 11 students' Mathematics results		
		Result		
		Pass	Fail	Total
Subject	Essential Mathematics		23	
	General Mathematics		49	74
	Mathematical Methods	36	62	
	Specialist Mathematics	7		25
	Total	80		

a. Complete the two-way table.
b. Determine the number of Year 11 students who study Mathematics at Senior Secondary College.
c. Calculate the total number of students enrolled in Essential Mathematics.

8. George is a hairdresser who keeps a record of all his customers and their hairdressing requirements. He made a two-way table for last month and recorded the numbers of customers who wanted a haircut only, and those who wanted a haircut and colour.
a. Design a two-way table that George would need to record the numbers of young and old customers who wanted a haircut only, and those who wanted a haircut and colour.
b. George had 56 old customers in total and 2 young customers who wanted a haircut and colour. Place these values in the table from part a.
c. George had 63 customers during last month, with 30 old customers requiring a haircut and colour. Fill in the rest of the two-way table.

9. The manager at the local cinema recorded the movie genres and the ages of the audience at the Monday matinee movies. The total audience was 156 people. A quarter of the audience were children under the age of 10, 21 were teenagers and the rest were adults. There were 82 people who watched the drama movie, 16 teenagers who watched comedy, and 11 people under the age of 10 who watched the drama movie.
   a. Construct a complete two-way table.
   b. Determine the total number of adults who watched the comedy movie.
   c. Calculate the number of children.

10. **WE6** Consider the data given in the two-way table shown.

		Cat and dog owners		
		Dog owners		
		Own	**Do not own**	**Total**
Cat owners	**Own**	12	78	90
	Do not own	151	64	215
	Total	163	142	305

   a. Calculate the relative frequencies for all the entries in the two-way table given. Give your answers correct to 3 decimal places or as fractions.
   b. Calculate the percentage relative frequencies for all the entries in the two-way table out of the total number of pet owners.

11. Using the two-way table from question 7, construct:
   a. the relative frequency table, correct to 4 decimal places
   b. the percentage relative frequency table.

12. Consider the data given in the two-way table shown.

		Swimming attendance		
		People		
		Adults	**Children**	**Total**
Day of the week	**Monday**	14	31	45
	Tuesday	9	17	26
	Wednesday	10	22	32
	Thursday	15	38	53
	Friday	18	45	63
	Saturday	27	59	86
	Sunday	32	63	95
	Total	125	275	400

   a. Calculate the relative frequencies for all the entries for adults out of the total number of adults.
   b. Calculate the percentage relative frequencies for all the entries for adults out of the total number of adults.

13. A ballroom studio runs classes for adults, teenagers and children in Latin, modern and contemporary dances. The enrolments in these classes are displayed in the two-way table shown.

		Dance class enrolments			
		People			
		Adults	Teenagers	Children	Total
Dance	Latin			23	75
	Modern		34	49	96
	Contemporary	36		62	
	Total	80			318

a. Complete the two-way table.
b. Construct a relative frequency table for this data set.
c. Construct a percentage relative frequency table for this data set.

14. WE7 Bianca manages a bookstore. She recorded the number of fiction and non-fiction books she sold over two consecutive weeks in the two-way table below.

		Types of books sold in a fortnight		
		Week		
		Week 1	Week 2	Total
Type of book	Fiction	51	73	124
	Non-fiction	37	28	65
	Total	88	101	189

Construct a grouped column graph for this data.

15. Construct a grouped column graph for the data displayed in the following two-way table.

		Type of exercise enjoyed by different age groups		
		People		
		Teenagers	Adults	Total
Type of exercise	Running	12	10	22
	Walking	8	15	23
	Total	20	25	45

16. The horizontal column graph below displays data about school sector enrolments in each of the Australian states collected by the ABC in 2017.

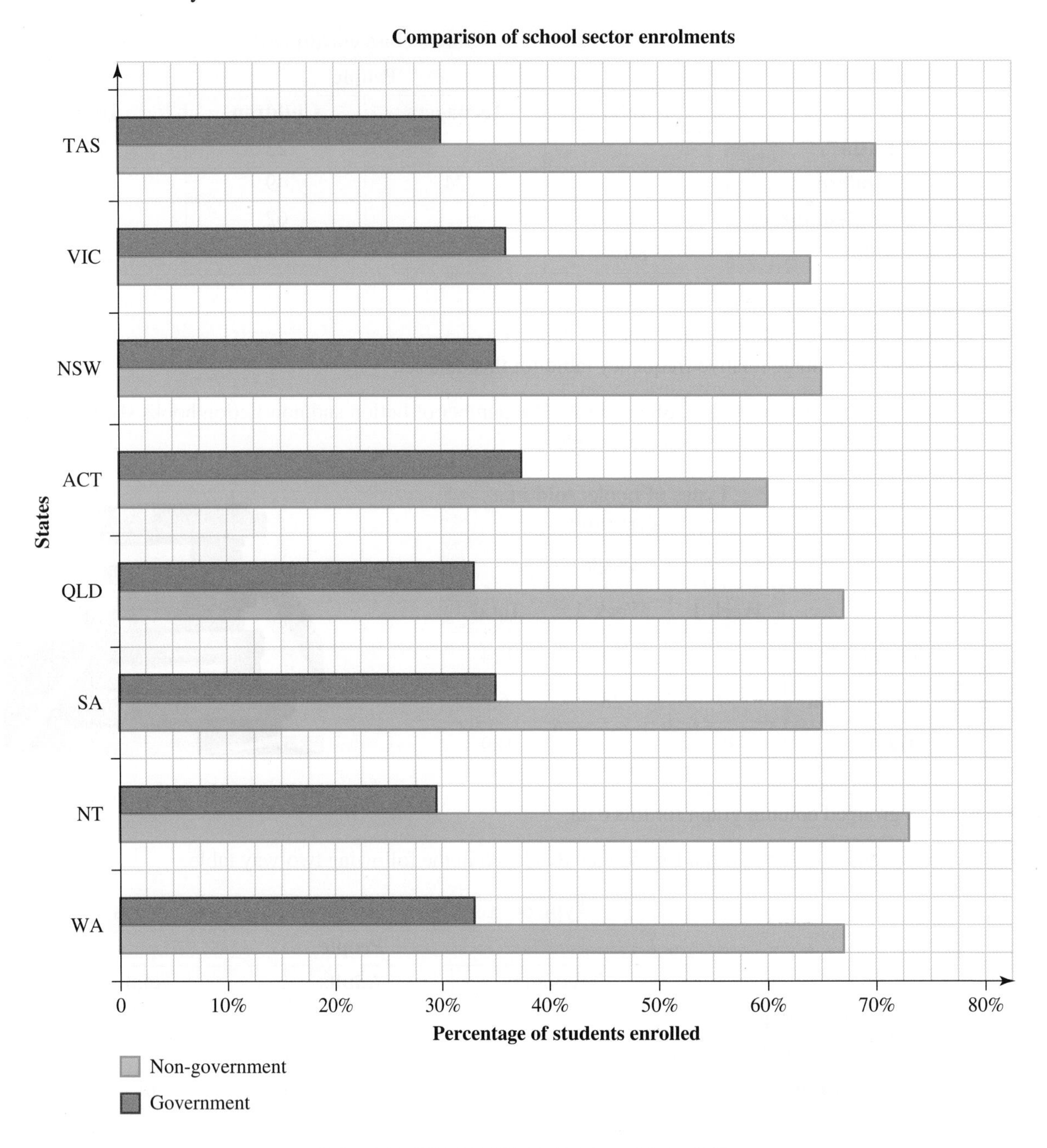

Identify which two states and territories have the highest percentage of students enrolled in government schools.

17. Human blood is grouped in 8 types: O+, O−, A+, A−, B+, B−, AB+ and AB−. The following % relative frequency table shows the percentage of Australians and Chinese who have a particular blood type.

Blood type	Percentage of Australia's population	Percentage of China's population
O+	40	47.7
O–	9	0.3
A+	31	27.8
A–	7	0.2
B+	8	18.9
B–	2	0.1
AB+	2	5.0
AB–	1	0.01

Source: Wikipedia; http://en.wikipedia.org/wiki/ABO_blood_group_system

Use a spreadsheet or otherwise to construct a grouped column graph to display the data for both Australia's and China's populations.

18. a. Car insurance companies calculate premiums according to the perceived risk of insuring a person and their vehicle. From historical data, age is a perceived 'risk factor'. In order to determine the premium of their car insurance policies, Safe Drive insurance company surveyed 300 of its customers.
    They determined three age categories: category A for young drivers, category B for mature drivers, and category C for older drivers. They then recorded whether the customer had a traffic accident within the last year. Their records showed that 113 drivers from category B and 17 drivers out of 47 in category C did not have a traffic accident within the last year.
    If 75 drivers had a traffic accident last year and 35 of the drivers were in category A, construct a complete two-way table for this data set.
    b. Identify how many mature drivers had a traffic accident last year.
    c. Determine the total number of drivers who did not have a traffic accident last year.
    d. Use a spreadsheet or otherwise to construct a grouped column graph for this data.

Fully worked solutions for this chapter are available online.

LESSON
6.4 Numerical data

SYLLABUS LINKS

- Display numerical data as frequency distribution tables, dot plots, stem plots and histograms.

6.4.1 Displaying numerical data

- The visual display of data is an important tool in communicating the information collected.
- Numerical data can be displayed in tables or graphical representations.

Frequency distributions

- **Frequency distributions** (also known as **frequency distribution tables** or **frequency tables**) for numerical data are constructed in the same way as for categorical data, given that they display frequency counts in both cases.

Ungrouped numerical data

- For **ungrouped data**, each score is recorded separately. For large sets of data, it is helpful to use tally marks.

tlvd-4694

WORKED EXAMPLE 8 Constructing a frequency table for data

For his statistics project, Radu surveyed 50 students on the number of books they had read over the past six months. The data recorded is listed below.

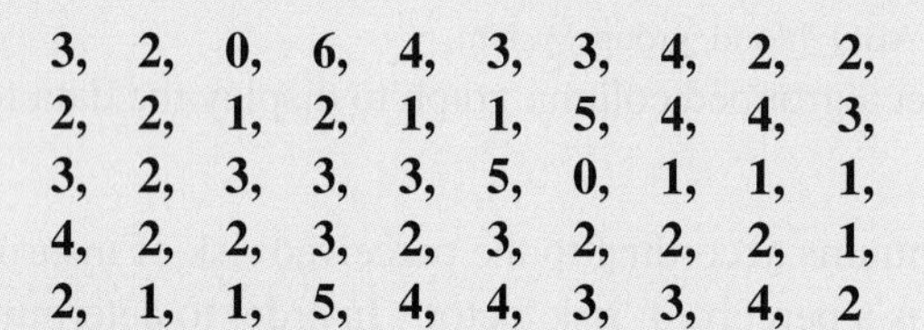

3, 2, 0, 6, 4, 3, 3, 4, 2, 2,
2, 2, 1, 2, 1, 1, 5, 4, 4, 3,
3, 2, 3, 3, 3, 5, 0, 1, 1, 1,
4, 2, 2, 3, 2, 3, 2, 2, 2, 1,
2, 1, 1, 5, 4, 4, 3, 3, 4, 2

Construct a frequency table for this data.

THINK / **WRITE**

1. Draw a frequency table with three columns to display the variable (number of books), tally and frequency.
   - The data has scores between 0 as the lowest score and 6 as the highest score.
   - The last cell in the first column is labelled 'Total'.

Number of books	Tally	Frequency
0		
1		
2		
3		
4		
5		
6		
Total		

2. Complete the 'Tally' and 'Frequency' columns.

Number of books	Tally	Frequency
0	||	2
1	𝍸 ||||	9
2	𝍸 𝍸 𝍸	15
3	𝍸 𝍸 ||	12
4	𝍸 |||	8
5	|||	3
6	|	1
Total		50

3. Check the frequency total with the total number of data. 50

Grouped data

- **Grouped data** is data that has been organised into groups or class intervals.
- It is used for very large amounts of data and for continuous data.
- Care has to be taken when choosing the class sizes for the data collected.
- The most common class sizes are 5 or 10 units.
- A class interval is usually written as 5– < 10, meaning that all scores between 5 and 10 are part of this group, including 5 but excluding 10.

WORKED EXAMPLE 9 Constructing a frequency table for grouped data

The data below displays the maximum daily temperature in Melbourne for every day of a month.

Date	Maximum temp (°C)	Date	Maximum temp (°C)	Date	Maximum temp (°C)
1	27.4	11	29.1	21	22.4
2	20.7	12	34.3	22	28.3
3	21.4	13	35.1	23	38.3
4	18.9	14	22.9	24	27.6
5	18.2	15	24.8	25	21.9
6	21.2	16	23.1	26	22.4
7	30.6	17	22.0	27	31.2
8	37.0	18	27.8	28	21.7
9	21.1	19	32.6	29	23.3
10	21.4	20	23.5	30	21.9
				31	24.7

Source: Copyright Commonwealth Bureau of Meterology

Construct a frequency table for this data.

THINK

1. Draw a frequency table with three columns to display the variable (temperature (°C)), tally and frequency.
   - The minimum score is 18.2 °C and the maximum score is 38.3 °C, so the range is 20.1 °C. Create five groups of class size 5 °C between 18 °C and 43 °C.
   - The last cell in the first column is labelled 'Total'.

WRITE

Temperature (°C)	Tally	Frequency
18– < 23		
23– < 28		
28– < 33		
33– < 38		
38– < 43		
Total		

2. Complete the 'Tally' and 'Frequency' columns.

Temperature (°C)	Tally	Frequency
18– < 23	𝍸𝍸𝍷𝍷𝍷𝍷	14
23– < 28	𝍸𝍷𝍷𝍷	8
28– < 33	𝍸	5
33– < 38	𝍷𝍷𝍷	3
38– < 43	𝍷	1
Total		31

3. Check the frequency total with the total number of data.

31

6.4.2 Dot plots

- A dot plot is a graphical representation of numerical data made up of dots.
- A dot plot can also be used to represent categorical data, with dots being stacked in a column above each category.
- All data are displayed using identical equally spaced dots, where each dot represents one score of the variable.
- The height of the column of dots represents the frequency for that score.

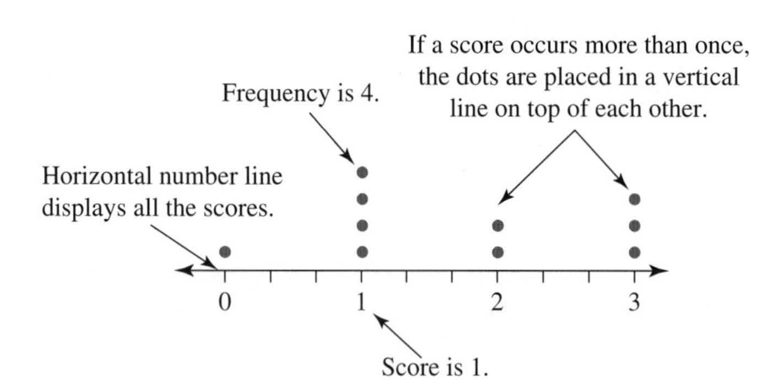

tlvd-4695

WORKED EXAMPLE 10 Constructing a dot plot for a data set

At an electronics store, customers were surveyed about the number of TV sets they have in their household. The data collected from 20 customers is displayed in the following frequency table.

Number of TV sets	Frequency
0	1
1	3
2	5
3	8
4	2
5	1
Total	20

Construct a dot plot for this set of data.

THINK

1. Draw a labelled horizontal line for the variable 'Number of TV sets'.
   - Start from the minimum score, 0, and continue until you've placed the maximum score on the line (5 in this example).
   - The spaces between scores have to be equal.

WRITE/DRAW

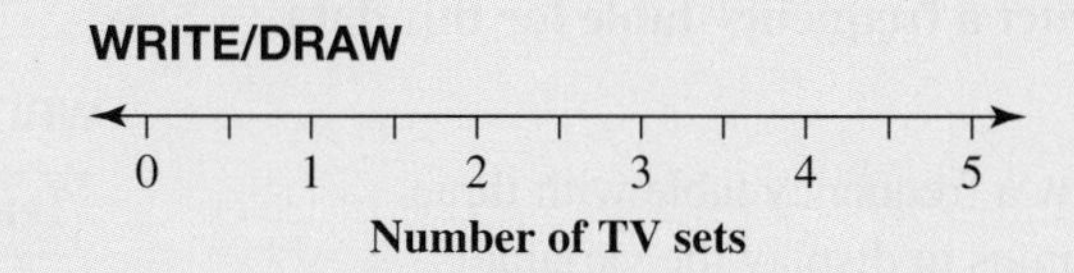

2. Place the dots for each score.
   - The frequency for 0 is 1. Place one dot above score 0.
   - The frequency for 1 is 3. Place three dots above score 1.
   - Continue until you've placed all data above the line.
   - Include a title.

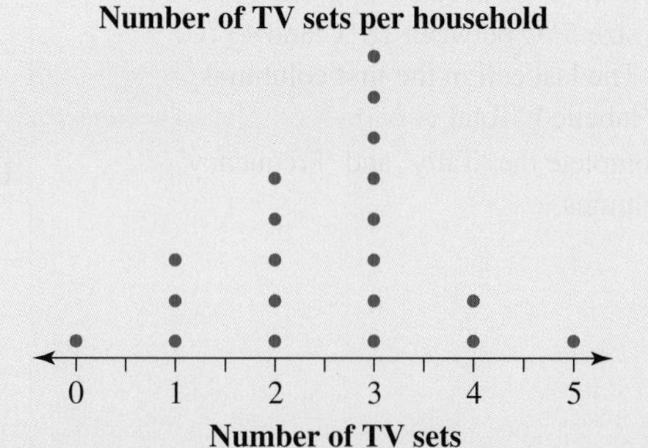

6.4.3 Histograms

- A histogram is a graphical representation of numerical data.
- It is very similar to the column graph used to represent categorical data, because the data is displayed using rectangles of equal width.
- However, although there is a space before the first column, there are no spaces between columns in a histogram.
- Each column represents a different score of the same variable.
- The vertical axis always displays the values of the frequency.

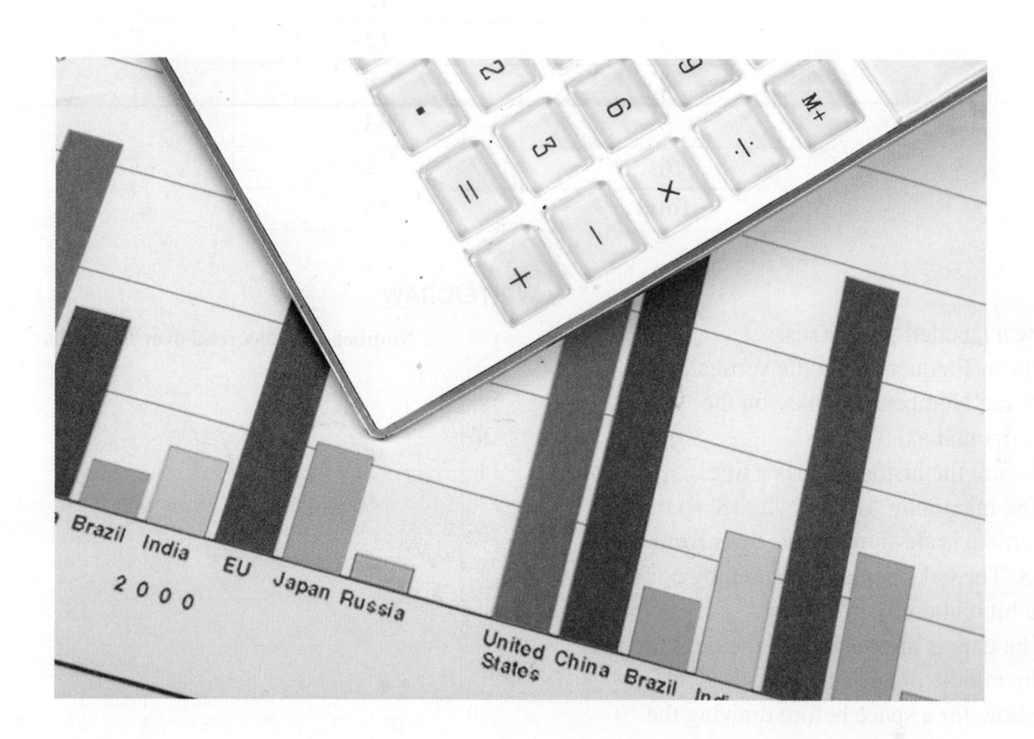

Features

Ungrouped data	Grouped data
The value of the variable is written in the middle of the column under the horizontal axis.	The ends of the class interval are written under the ends of the corresponding columns.

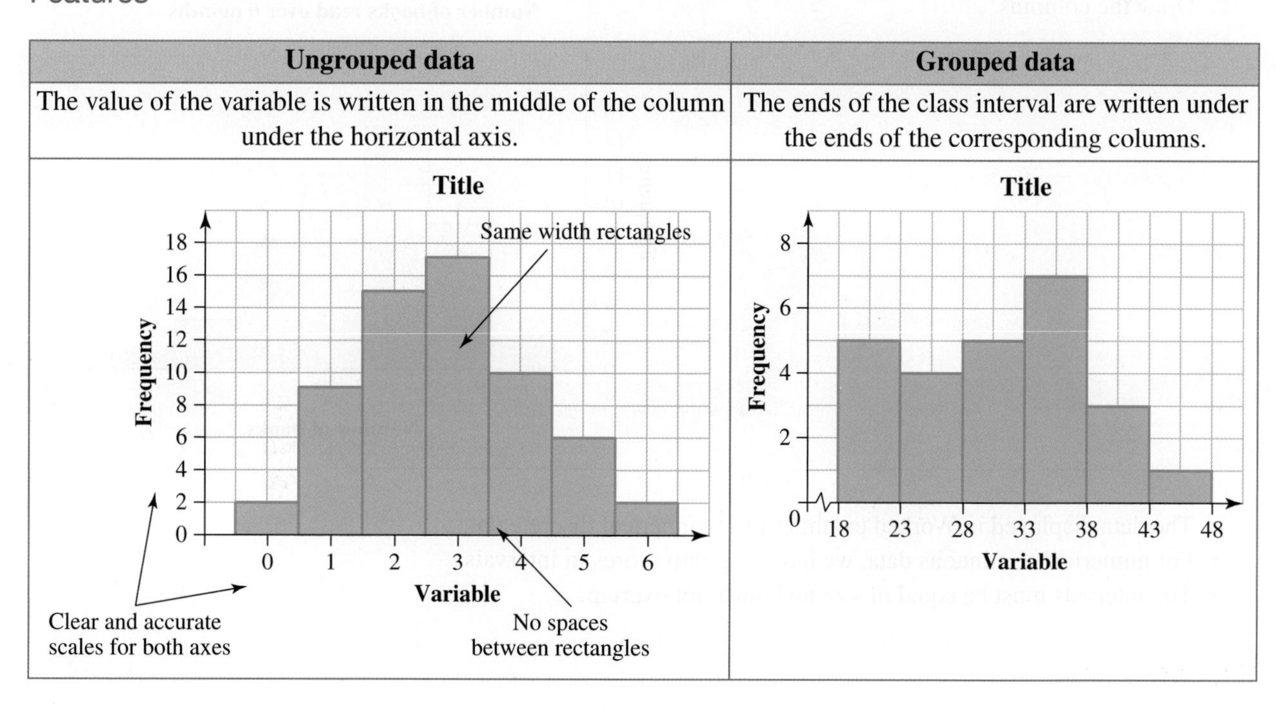

WORKED EXAMPLE 11 Constructing a histogram for numerical discrete data

Draw a histogram for the data given in the frequency table below.

Number of books read over six months	Frequency
0	2
1	9
2	15
3	12
4	8
5	3
6	1
Total	50

THINK

1. Draw a labelled set of axes:
   - Place 'Frequency' on the vertical axis.
   - Place 'Number of books' on the horizontal axis.
   - Ensure the histogram has a title.
   - The maximum frequency is 18, so the vertical scale should have ticks from 0 to 18. The ticks could be written by ones (a bit cluttered) or by twos.
   - The data is ungrouped, so the ticks are in the middle of each column.
   - Allow for a space before drawing the columns of the histogram.

WRITE/DRAW

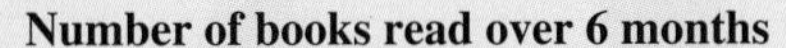

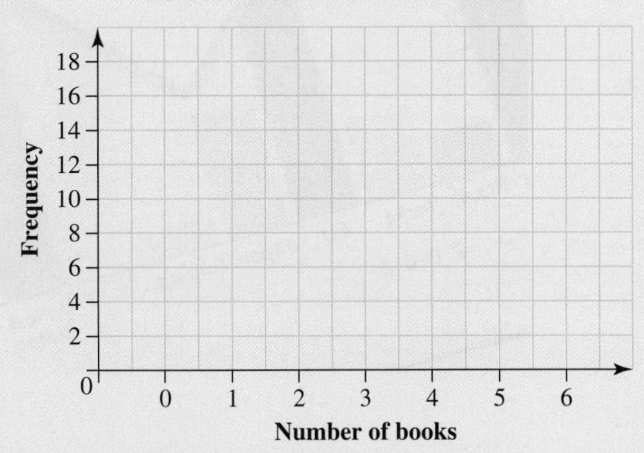

2. Draw the columns.

Number of books read over 6 months

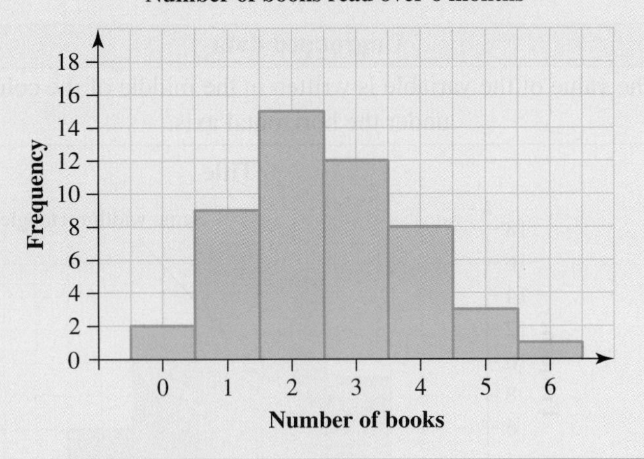

- The data displayed in Worked example 11 is numerical discrete data.
- For numerical continuous data, we have to group scores in intervals.
- The intervals must be equal in size and must not overlap.

tlvd-4696

WORKED EXAMPLE 12 Constructing a histogram for numerical continuous data

Using the frequency table of maximum daily temperatures in Melbourne in December 2021, construct a histogram to represent this set of data.

Temperature (°C)	Frequency
18– < 23	14
23– < 28	8
28– < 33	5
33– < 38	3
38– < 43	1
Total	31

THINK

1. Draw a labelled set of axes:
   - Place 'Frequency' on the vertical axis.
   - Place 'Temperature (°C)' on the horizontal axis.
   - Ensure the histogram has a title.
   - The maximum frequency is 14, so the vertical scale should have ticks from 0 to 16. The ticks could be written by ones (a bit cluttered) or by twos.
   - The data is grouped so the ticks should be at the edges of each column.
   - Allow for a space before drawing the columns of the histogram.

   Note: The -^- symbol shows the graph does not start from zero.

WRITE/DRAW

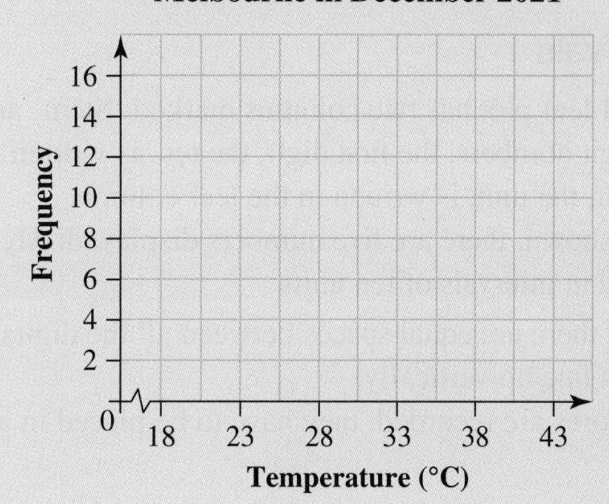

2. Draw the columns.

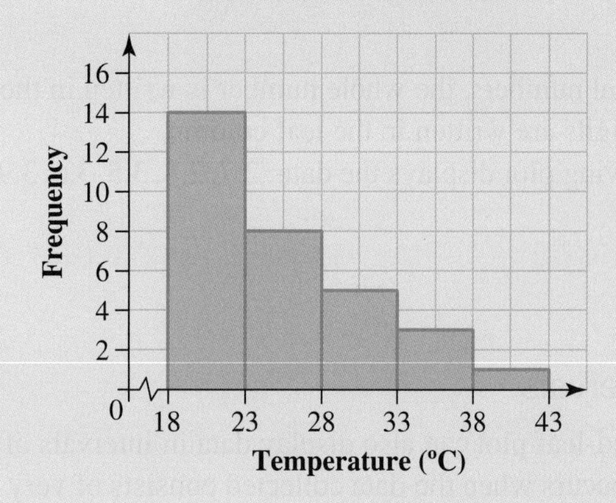

6.4.4 Stem-and-leaf plots

- A stem-and-leaf plot is a graphical representation of grouped numerical data.
- The stem-and-leaf plot and the histogram shown are graphical representations of the same set of data.
- The two graphical representations have similar shapes.
- However, a stem-and-leaf plot displays all the data collected, while a histogram loses the individual scores.

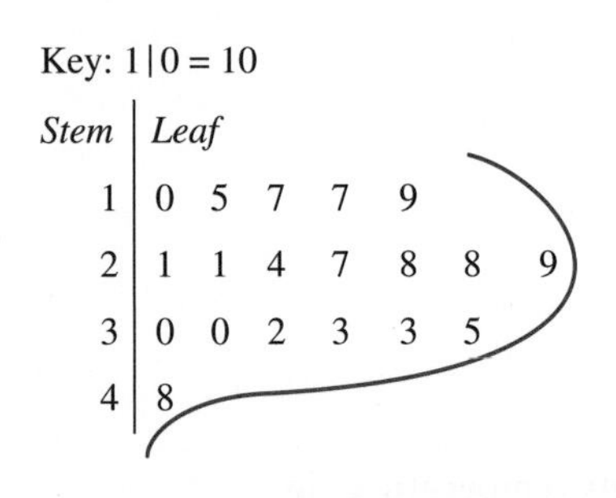

Key: 1 | 0 = 10

Stem	Leaf
1	0 5 7 7 9
2	1 1 4 7 8 8 9
3	0 0 2 3 3 5
4	8

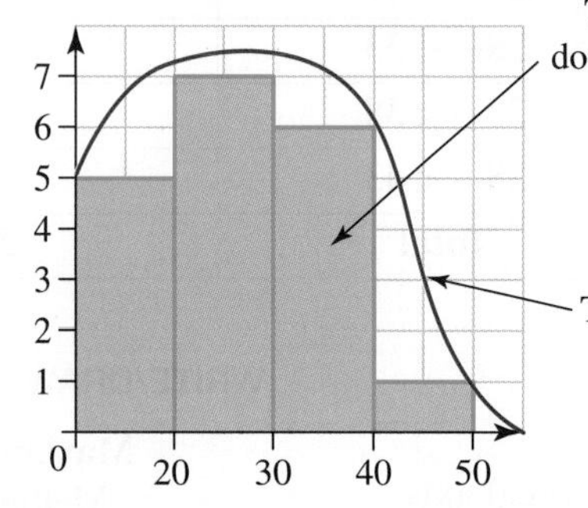

This column shows six scores but does not display the individual scores.

The histogram has the same shape as the stem-and-leaf plot.

Ten-unit intervals

- A stem-and-leaf plot has two columns marked 'Stem' and 'Leaf'.
- For two-digit numbers, the first digit, the ten, is written in the stem column; the second digit, the unit, is written in the leaf column.
- In the plot shown, there are five numbers displayed: 31, 37, 35, 48 and 42. The data is displayed in intervals of ten units.

Key: 3|1 = 31

Stem	Leaf
3	1 7 5
4	8 2

- Ensure that there are equal spaces between all the digits in the leaf column and that all numbers line up vertically.
- Once all scores are recorded, they have to be placed in an increasing order.

Key: 3|1 = 31

Stem	Leaf
3	1 5 7
4	8 8

- For three-digit numbers, the first two digits are written in the first column and the third digit in the second column.
- In the following plot there are four numbers: 120, 125, 126 and 134.

Key: 3|1 = 31

Stem	Leaf
12	0 5 6
13	4

- For decimal numbers, the whole number is written in the stem column and the decimal digits are written in the leaf column.
- The following plot displays the data: 2.7, 2.8, 3.5, 3.6, 3.9, 4.1.

Key: 2|7 = 2.7

Stem	Leaf
2	7 8
3	5 6 9
4	1

Five-unit intervals

- A stem-and-leaf plot can also display data in intervals of five units. This situation occurs when the data collected consists of very close values.
  For example, the set of data 11, 12, 12, 13, 14, 15, 17, 18, 21, 22, 26, 26, 27, 29 is better represented in five-unit intervals. In this case, the stem will consist of the values 1, 1*, 2 and 2*.
- The first 1 represents the data values from 10 to 14 inclusive, while 1* represents the data values from 15 to 19 inclusive.
- The first 2 represents the data values from 20 to 24 inclusive, while 2* represents the data values from 25 to 29 inclusive.

Key: 2|3 = 23
2*|6 = 26

Stem	Leaf
1	1 2 2 3 4
1*	5 7 8
2	1 2
2*	6 6 7 9

tlvd-4697

WORKED EXAMPLE 13 Constructing a stem-and-leaf plot using ten- and five-unit intervals

The heights of 20 students are listed below. All measurements are given in centimetres. Construct a stem-and-leaf plot for this set of data:
a. using ten-unit intervals
b. using five-unit intervals.

156, 172, 162, 164, 174, 151, 150, 169, 171, 169,
167, 161, 153, 155, 165, 172, 148, 166, 169, 158

THINK	WRITE/DRAW
a. 1. Draw a horizontal line and a vertical line, as shown, and label the two columns *Stem* and *Leaf*.	**a.** *Stem* | *Leaf*
2. Write the values in the stem part of the graph using intervals of 10 units. As the values of the data are three-digit numbers, we write the first two digits in the stem column in increasing order. The lowest value is 148 and the highest value is 174. The stem will have the numbers 14, 15, 16 and 17.	*Stem* | *Leaf* 14 | 15 | 16 | 17 |
3. Write the values in the leaf part of the graph. Start with the first score, 156, and place 6 to the right of 15. The next value is 172. Place 2 to the right of 17.	*Stem* | *Leaf* 14 | 15 | 6 16 | 17 | 2
4. Continue until all data is displayed. Count data points to check there are 20.	*Stem* | *Leaf* 14 | 8 15 | 6 1 0 3 5 8 16 | 2 4 9 9 7 1 5 6 9 17 | 2 4 1 2
5. Rewrite all data in increasing order and add a key.	Key: 14|8 = 148 *Stem* | *Leaf* 14 | 8 15 | 0 1 3 5 6 8 16 | 1 2 4 5 6 7 9 9 9 17 | 1 1 2 2 4

b. 1. Rewrite the stem part of the graph using intervals of five units.

b.

Stem	*Leaf*
14	
14*	
15	
15*	
16	
16*	
17	
17*	

2. Write the values in the leaf part of the graph and add a key.

Key: $15|1 = 151$

$15^*|8 = 158$

Stem	*Leaf*
14*	8
15	0 1 3
15*	5 6 8
16	1 2 4
16*	5 6 7 9 9 9
17	1 2 2 4

on Resources

Interactivities Create a histogram (int-6494)

Stem plots (int-6242)

Create stem plots (int-6495)

Dot plots, frequency tables and histogram and bar charts (int-6243)

Exercise 6.4 Numerical data

learn on

6.4 Quick quiz on | **6.4 Exercise**

Simple familiar	Complex familiar	Complex unfamiliar
1, 2, 3, 4, 5, 6, 7, 8, 9, 10, 11, 12, 13, 14, 15, 16, 17, 18, 19, 20	N/A	N/A

Simple familiar

1. **WE8** A group of 25 Year 11 students was surveyed by a psychologist about how many hours of sleep they each have per night. The set of scores below shows this data.

6, 6, 7, 8, 5, 10, 8, 8, 9, 6, 4, 7, 7,
6, 5, 10, 7, 8, 6, 8, 9, 7, 5, 9, 7

Display this data in a frequency distribution table.

2. Using the random sampler of 30 students from the Census@School (Australian Bureau of Statistics website), Eva downloaded the responses of 30 Year 11 students across Australia to the question, ‘What is the length, in cm, of your foot without a shoe?’ Measurements are given to the nearest centimetre.

28, 27, 28, 23, 27, 29, 24, 26, 29, 26,
23, 23, 27, 25, 28, 24, 26, 28, 24, 28,
22, 26, 23, 26, 29, 21, 23, 25, 24, 26

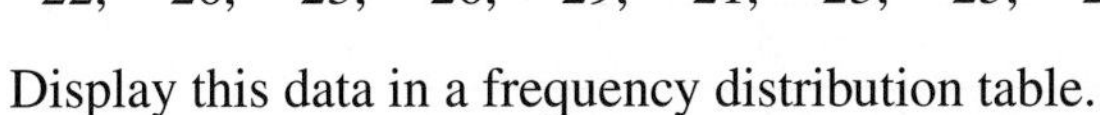

Display this data in a frequency distribution table.

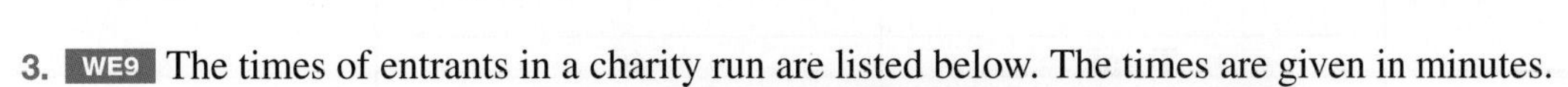

3. **WE9** The times of entrants in a charity run are listed below. The times are given in minutes.

34, 29, 57, 45, 26, 40, 19, 28, 33, 37, 39, 46,
18, 52, 19, 36, 28, 19, 20, 19, 54, 38, 38, 51,
19, 21, 53, 36, 37, 25, 22, 30, 25, 34, 18, 17

Display this data in a frequency distribution table.

4. The data set shown represents the number of times last month a randomly chosen group of 30 people spent their weekends away from home.

4, 3, 2, 0, 0, 0, 1, 2, 0, 1, 0, 0, 2, 1, 1,
4, 1, 1, 0, 0, 0, 2, 1, 1, 0, 2, 1, 3, 3, 0

Construct a frequency distribution table for this set of data, including a column for the tally.

5. For research purposes, 36 Tasmanian giant crabs were measured and the widths of their carapaces are given below.

35.3, 34.9, 36.0, 35.4, 37.0, 34.2,
35.9, 35.5, 35.0, 36.2, 35.7, 33.1,
36.5, 36.4, 37.2, 35.4, 34.9, 35.1,
34.8, 35.8, 35.2, 35.6, 37.1, 36.7,
35.0, 34.2, 36.3, 33.9, 35.2, 34.1,
36.5, 36.4, 36.6, 34.3, 35.8, 35.7

Construct a frequency table for this set of data.

6. WE10 The following frequency distribution table displays the data collected by MyMusic website about the number of songs downloaded by 40 registered customers per week.

Number of songs	Frequency
2	12
3	8
4	7
5	6
6	3
7	1
Total	40

Construct a dot plot for this data.

7. Daniel asked his classmates 'How many pairs of shoes do you have?'
He collected the data and listed it in the following frequency table.

Pairs of shoes	Frequency
3	9
4	5
5	3
6	2
7	3
8	2
9	1
Total	25

Construct a dot plot for this set of data.

8. The frequency table shown displays the ages of 30 people who participated in a spelling contest.

Age	Frequency
11	3
12	6
13	8
14	7
15	3
16	2
17	1

Construct a dot plot for this set of data.

9. The data given represents the results in a typing test, in characters per minute, of a group of 24 students in a beginners class.

36, 39, 40, 38, 41, 39, 37, 40,
41, 38, 34, 35, 39, 38, 37, 36,
39, 41, 39, 38, 42, 40, 39, 37

Construct a dot plot for this set of data.

10. WE11 The following frequency distribution table displays the data collected by a Year 11 student about the number of emails his friends send per day.

Number of emails	Frequency
0	1
1	12
2	17
3	5
4	2
5	3
Total	40

Construct a histogram for this data.

11. 'In how many languages can you hold an everyday conversation?' is one of the questions in the Census@School questionnaire. Using the random sampler from the Census@School (Australian Bureau of Statistics) website, Brennan downloaded the responses of 200 randomly chosen students. He displayed the data in the frequency distribution table shown.

Number of languages	Frequency
1	127
2	50
3	13
4	4
5	3
6	2
7	1
Total	200

Construct a histogram for this data.

12. A group of 24 students was given a task that involved searching the internet to find how to construct an ungrouped histogram. The times, in minutes, they spent searching for the answer was recorded.

10, 15, 18, 12, 19, 13, 14, 10,
16, 18, 12, 15, 18, 12, 13, 12,
15, 12, 12, 10, 15, 11, 10, 16

Draw an ungrouped histogram for this set of data.

13. WE12 A communications provider surveyed 50 randomly chosen customers about the number of times they check their email account on weekends.

5, 10, 2, 17, 8, 8, 13, 6, 3, 0,
11, 1, 6, 10, 1, 4, 0, 11, 5, 12,
0, 3, 4, 5, 6, 5, 14, 9, 4, 10,
14, 9, 2, 16, 3, 9, 8, 10, 5, 11,
7, 2, 5, 6, 1, 2, 2, 12, 9, 17

a. Display the data collected in a frequency distribution table using class intervals of 3.
b. Construct a histogram for this data.

14. The following frequency distribution table shows the number of road fatalities in Victoria over a period of 12 months per age group, involving people between 30 and 70 years old.

Age group	Frequency
30– < 40	47
40– < 50	27
50– < 60	41
60– < 70	35
Total	150

Source: TAC road safety statistical summary, page B

Construct a histogram for this data.

15. The data below displays the responses of 50 Year 11 students who participated in a questionnaire related to the amount of money they earned from their part-time jobs over the previous week.

30,	103,	50,	30,	0,	90,	0,	80,	68,	157,
0,	350,	123,	70,	80,	50,	0,	330,	210,	26,
25,	0,	0,	50,	230,	50,	20,	60,	0,	305,
177,	126,	90,	0,	0,	12,	200,	120,	15,	45,
90,	30,	60,	0,	0,	80,	0,	260,	25,	150

a. Display this data in a frequency table.
b. Draw a grouped histogram to represent this data.

16. WE13 The BMI or body mass index is the number that shows the proportion between a person's mass and height. The formula used to calculate the BMI is:

$$\text{BMI} = \frac{\text{weight (kg)}}{\text{height (m)}}$$

The BMIs of 24 students are listed below.

25, 24, 18, 16, 29, 23, 20, 21,
21, 17, 19, 23, 25, 24, 32, 22,
22, 23, 30, 18, 23, 20, 19, 24

Display this data in a stem-and-leaf plot.

17. The marks out of 50 obtained by a Year 11 Essential Mathematics group of students on their exam are given below.

36, 41, 29, 50, 45, 23, 48, 56,
20, 12, 43, 33, 35, 44, 32, 49,
39, 48, 50, 18, 43, 20, 38, 29

Display this data in a stem-and-leaf plot.

18. The manager of a timber yard started an inventory for the upcoming stocktake sale. The lengths of the first 30 timber planks recorded are given.

250, 220, 245, 229, 260, 210, 250, 261, 244, 218,
250, 251, 216, 243, 232, 210, 212, 227, 219, 207,
231, 243, 204, 230, 265, 220, 206, 253, 229, 225

Display this data set in an ordered stem-and-leaf plot.

19. Consider the following set of data, which represents the heights, in cm, of 40 children under the age of 3.

58.9, 54.6, 62.9, 45.7, 49.6, 62.8, 43.2, 56.7, 69.0, 65.3,
57.8, 59.4, 66.7, 49.6, 72.7, 35.2, 72.8, 75.6, 32.9, 56.7,
54.8, 45.2, 69.5, 47.3, 68.6, 67.4, 63.4, 61.5, 52.6, 48.0,
56.3, 78.4, 39.9, 75.3, 45.2, 56.9, 66.3, 55.4, 63.8, 72.3

Using a spreadsheet or otherwise, construct a grouped histogram for this set of data.

20. The reaction times using their non-dominant hand of a group of people are listed below. The times are given in deciseconds (a tenth of a second).

48, 39, 40, 36, 35, 37, 46, 41, 43, 39, 35,
50, 59, 37, 32, 39, 42, 43, 46, 33, 38, 30,
41, 29, 36, 38, 51, 42, 34, 39, 36, 37, 41

Using a spreadsheet or otherwise, construct a grouped histogram for this data set.

Fully worked solutions for this chapter are available online.

6.5 Recognising and identifying outliers

SYLLABUS LINKS

- Recognise and identify outliers from a dataset.

6.5.1 Identifying possible outliers

- Outliers are extreme values on either end of a data set that appear very different from the rest of the data.
- An outlier can affect the summary statistics of your data.

WORKED EXAMPLE 14 Identifying any possible outliers in the data set

The number of hours a casual worker works each week are shown for the past 10 weeks.

12, 8, 13, 9, 10, 24, 13, 10, 9, 11

a. Identify any possible outliers in the data set.
b. Explain a possible reason for the outlier.

THINK	WRITE
a. 1. Organise the values in ascending order.	a. 8, 9, 9, 10, 10, 11, 12, 13, 13, 24
2. Outliers are extreme values at either end of the data set.	8, 9, 9, 10, 10, 11, 12, 13, 13, **24** 24 is a possible outlier.
b. 1. Think of reasons a casual worker would work 24 hours in a week.	b. There are many possibilities. For example, it could be a student who is on school holidays that week and could do more hours, or it could be that it was a busy time at work and they needed more staff.

tlvd-11430

WORKED EXAMPLE 15 Determining a possible outlier and its value

Referring to the dot plot shown, determine if there is a possible outlier and, if so, identify its value.

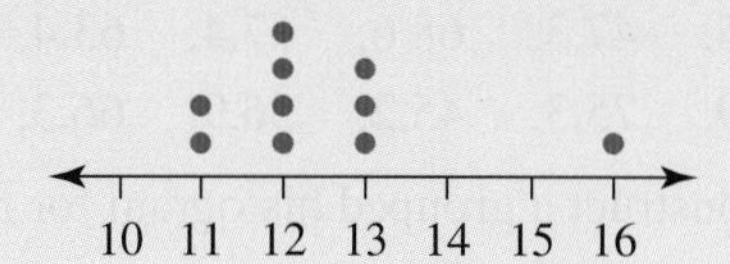

THINK	WRITE
1. Look at the range of the data from smallest to largest.	Smallest data point = 11 Largest data point = 16
2. Outliers are extreme values at either end of the data set.	Since most of the data is clustered around 11–13, then 16 is a possible outlier.

Exercise 6.5 Recognising and identifying outliers

learn on

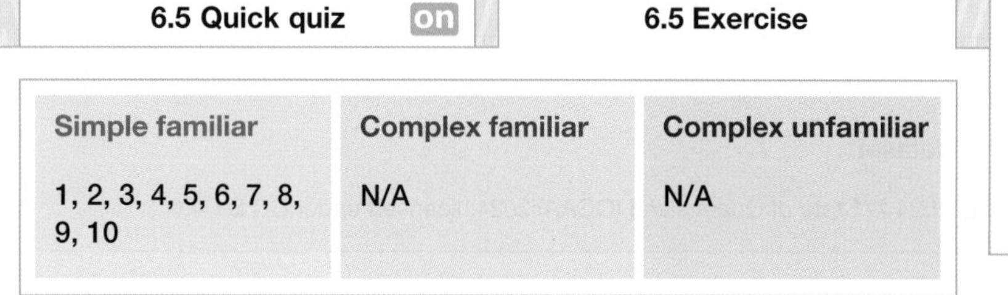

6.5 Quick quiz
on
6.5 Exercise
Simple familiar
1, 2, 3, 4, 5, 6, 7, 8, 9, 10
Complex familiar
N/A
Complex unfamiliar
N/A

These questions are even better in jacPLUS!
- Receive immediate feedback
- Access sample responses
- Track results and progress

Find all this and MORE in jacPLUS

Simple familiar

1. WE15 The number of hours of homework completed by a student each night over two weeks is shown.

   1, 1.5, 1, 1.5, 2, 2, 1.5, 1, 0, 2, 5, 1, 2, 2.5

   a. Identify any possible outliers in the data set.
   b. Explain a possible reason for the outlier.

2. The number of people playing golf at a golf course over a week is shown.

   75, 82, 78, 63, 77, 125, 90

   a. Identify any possible outliers in the data set.
   b. Explain a possible reason for the outlier.

3. WE16 Referring to the dot plot shown, determine if there is a possible outlier and, if so, identify its value.

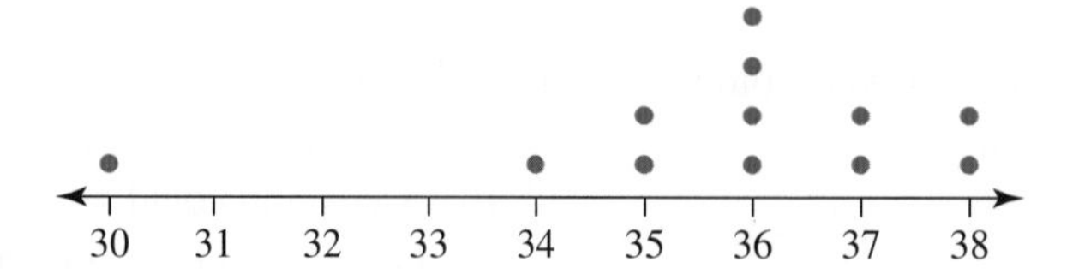

4. Referring to the dot plot shown, determine if there is a possible outlier and, if so, identify its value.

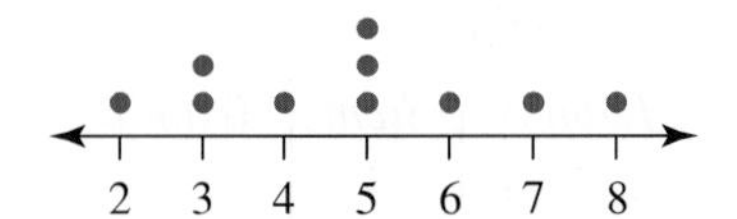

5. The graph shows the ages of customers in a music shop on a Saturday morning. Determine if there is a possible outlier group and, if so, identify this group.

Ages of customers in a music shop on a Saturday morning

Number of customers: 0, 2, 4, 6, 8

Age (years): 8, 12, 16, 20, 24, 28, 32, 36

6. The following data set shows the number of students attending the school canteen over a two-week period.

30, 28, 25, 30, 29, 33, 27, 51, 33, 27

a. Organise the data in order of smallest to largest.
b. Represent the data as a dot plot.
c. Identify any possible outliers.
d. Explain a possible reason for the outlier.

7. The following data was recorded during the high jump competition at a school athletics carnival.

Student	1	2	3	4	5	6	7	8
Height (m)	1.25	1.35	1.45	1.40	1.30	1.50	130	1.45

a. Identify any possible outliers.
b. Provide a possible explanation for the outlier found.

8. Data was collected during a survey of the number of pets people own.

2, 1, 0, 3, 1, 1, 2, 1, 2, 3, 3, 2, 0, 1, 4, 3, 2, 3, 3, 12, 3, 0

a. Determine if there are any outliers.
b. Provide a possible explanation for this outlier.

9. The following stem-and-leaf plot shows Team A and Team B basketball scores.

Key: $5|5 = 55$

Team A	*Stem*	*Team B*
1	4	
	5	5 7 8 9
	6	0 2 3 5 7
6 5 4 3	7	0 1 2
8 6 5 4 3 2 1	8	8
	9	0

a. Determine which team has the most obvious outlier.
b. Identify the outlier determined in part a.

10. The following dot plots compare Mathematics test scores out of 20 for Class 1 and Class 2.

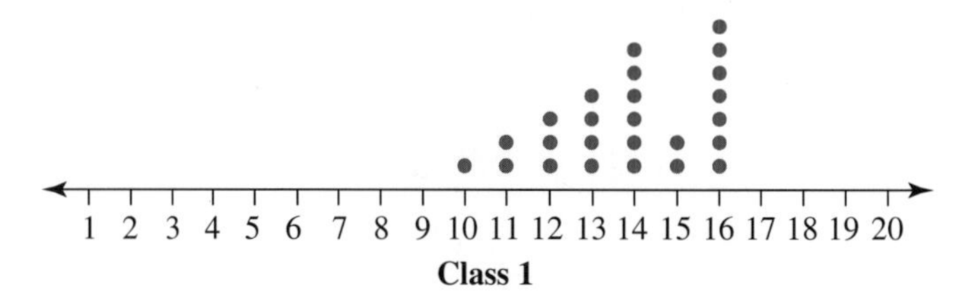

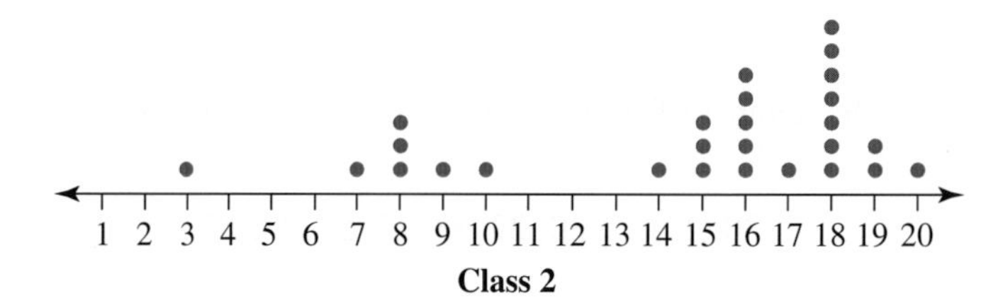

a. Determine which class *does not* have an obvious outlier.
b. Identify any outliers.

Fully worked solutions for this chapter are available online.

6.6 Comparing different data [complex]

SYLLABUS LINKS

- Compare the suitability of different methods of data presentation in real-world contexts [complex].

Source: Essential Mathematics Senior Syllabus 2024 © State of Queensland (QCAA) 2024; licensed under CC BY 4.0.

6.6.1 Comparing methods of presenting data

- When data is collected, it is a set of numbers that is often difficult to interpret.
- When data is presented in a table or graph, it is easier to interpret the data.
- When choosing a presentation method for your data, you want it to be easy to understand visually and to clearly demonstrate the point you are trying to make.

tlvd-11431

WORKED EXAMPLE 16 Comparing and interpreting the data presentation

Use the NRLW 2018–2020 table shown to answer the following questions.
a. Determine the technique used to emphasise the point in the table.
b. Describe what the table is trying to emphasise.
c. Comment on the effectiveness of the representation of the data.

Category	2018	2019	2020
Goals	24	24	27
Line break assists	28	28	37
Line breaks	55	48	66
Points	212	192	222
Tries	41	36	42
Try assists	25	20	23

THINK	WRITE
a. Look at the table to see what method has been used to emphasise key points.	a. The highest values in each category have been highlighted to make them stand out.
b. Looking at the highlighted values, identify what they show about the data.	b. The highlighted values show that 5 of the 6 categories are higher in 2020 than they were in 2018 and 2019. This suggests there has been an improvement in the game over this time.
c. Ask yourself: does the method of representing the data clearly demonstrate the points NRLW is trying to make?	c. This table clearly demonstrates the points they are trying to make so it is effective. A column graph could have also been used to demonstrate these same points.

WORKED EXAMPLE 17 Interpreting the graph

Use the graph shown to answer the following questions.
a. Propose why a column graph was used to represent this data.
b. Determine what this graph is trying to represent.
c. Explain why some of the years don't have any columns.

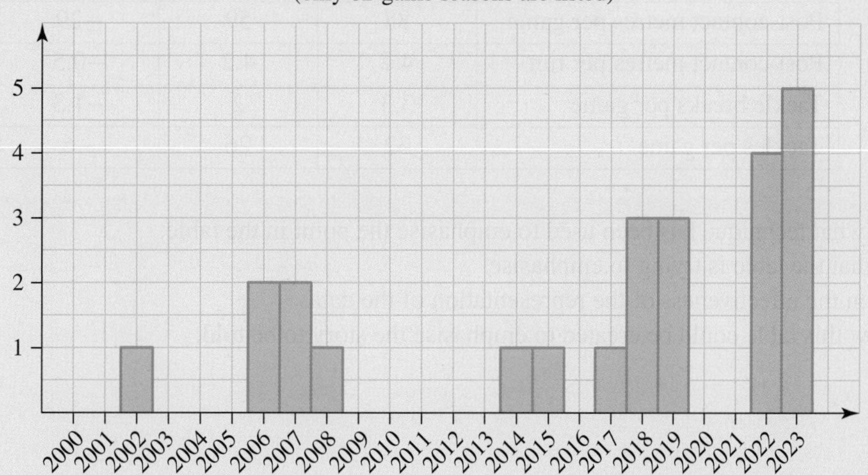

THINK	WRITE
a. Look at the data that is being graphed to decide what type of graph to use.	a. A column graph can clearly represent the number of players who have played in the All-NBA Team before playing 65 NBA games. It can show how many players fit this category each year.
b. Look for what this graph is representing or what story it is trying to tell.	b. With taller and more columns to the right of the graph, it suggests more players are playing in the All-NBA Team in recent years.
c. Read the labelling and heading of the graph to understand what the graph is showing.	c. The vertical axis represents the number of players to play in the All-NBA Team before they have played 65 games. Where there are no columns, this suggests that during these years there were no players playing in the All-NBA Team.

Exercise 6.6 Comparing different data [complex]

learn on

6.6 Quick quiz on | 6.6 Exercise

Simple familiar	Complex familiar	Complex unfamiliar
N/A	1, 2, 3, 4, 5, 6, 7	8, 9, 10

Complex familiar

1. WE16 Use the NRL 2020–2021 table shown to answer the following questions.

	Jason Taumalolo		
Stat attack	**2020**	**2021**	**DIFF**
Games	16	3	—
Metres per match	207	147	−60
Metres per run	11.2	10.5	−0.7
Post-contact metres per game	88	59	−29
Post-contact metres per run	4.7	4.2	−0.5
Tackle breaks per game	3.3	2	−1.3
Tackles per game	32	26	−6

a. Determine what technique has been used to emphasise the point in the table.
b. Describe what the table is trying to emphasise.
c. Comment on the effectiveness of the representation of the data.
d. Explain how this table could be created to emphasise the story to be told.

2. WE17 Using the AFL age breakdown table shown, answer the following questions.

Age at RD 1, 2018			Age breakdown						
				Under 21s		21–24		25–29	
Rank	**Club**	**Ave**	**Club**	**Total**	**Rank**	**Total**	**Rank**	**Total**	**Rank**
1	**Hawthorn**	24.66	**Adelaide**	11	12	11	17	19	①
2	**Adelaide**	24.53	**Brisbane**	15	3	15	7	12	10
3	**GWS Giants**	24.36	**Carlton**	16	2	12	15	11	14
4	**Geelong**	24.36	**Collingwood**	11	12	15	7	14	8
5	**Collingwood**	24.26	**Essendon**	13	5	14	11	15	5
6	**Port Adelaide**	24.15	**Fremantle**	13	5	17	4	8	18
7	**Sydney**	24.15	**Geelong**	12	8	11	17	19	①
8	**Essendon**	24.03	**Gold Coast**	17	1	12	15	12	10
9	**Melbourne**	23.93	**GWS Giants**	9	17	15	7	14	8
10	**West Coast**	23.86	**Hawthorn**	10	15	14	11	15	5
11	**Fremantle**	23.82	**Melbourne**	7	18	24	1	9	17
12	**Richmond**	23.79	**North Melbourne**	14	4	17	4	10	16
13	**Carlton**	23.74	**Port Adelaide**	12	8	13	13	16	③
14	**Western Bulldogs**	23.72	**Richmond**	11	12	16	6	16	③
15	**St Kilda**	23.55	**St Kilda**	10	15	21	2	12	10
16	**North Melbourne**	23.42	**Sydney**	12	8	15	7	12	10
17	**Brisbane Lions**	23.28	**West Coast**	13	5	13	13	15	5
18	**Gold Coast**	23.17	**Western Bulldogs**	12	8	18	3	11	14

a. Identify the technique that has been used to represent this data.
b. Construct a column graph for the teams and the number of players they have 'in the zone'.
c. Decide whether the frequency table or the column graph is the best to represent your data. Explain.

3. Twenty students in a Year 11 class were asked how many siblings there are in their family. The results are shown.

1, 3, 5, 3, 2, 3, 4, 3, 4, 1, 3, 2, 3, 2, 1, 1, 1, 7, 5, 4

a. Create a frequency table without grouping.
b. Create a dot plot.
c. Explain whether the frequency table or the dot plot is more effective.

4. The graph shows the completion rate for each team in the NRL.

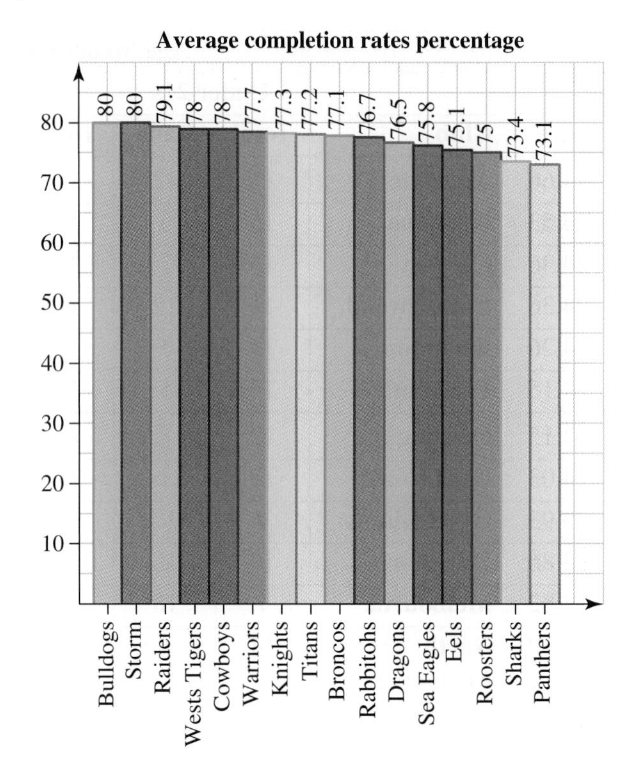

a. Identify the team with the lowest completion rate.
b. Identify the teams that have the same completion rate.
c. Comment on the techniques used to make this graph clear to read and understand.

5. The data shown compares Michael Jordan to LeBron James in a number of areas.

	Michael Jordan	**LeBron James**
Points	32 292	31 927
Rebounds	6672	8671
Assists	5633	8439
Steals	2514	1907
Blocks	893	911
Total minutes	41 011	45 419
Total games	1072	1175
Field goal%	0.497%	0.504%
3_{pts} field goal%	0.327%	0.344%
Championships	6	3

a. Determine which player has the greater field goal percentage.
b. Determine which player has more assists.
c. Explain why it is difficult to represent all this data on a column graph.

6. The graph shows the percentage of netball injuries.

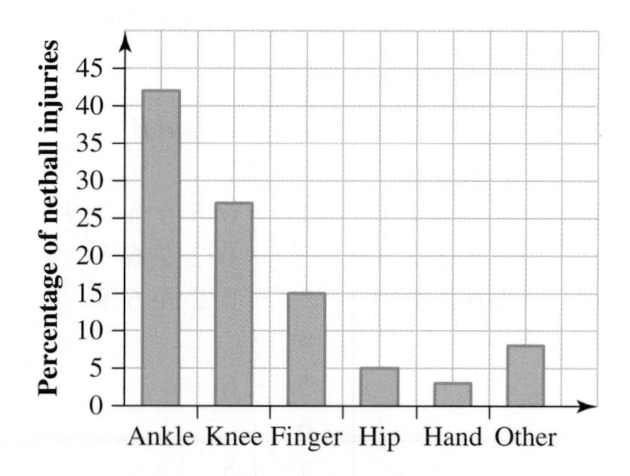

a. Determine the type of graph that was used to represent the data.
b. Explain why this type of graph was chosen.
c. From the graph, determine the most common injury in netball.

7. The daily number of hits a fashion blogger gets on her new website over three weeks is:

126,	356,	408,	404,	420,	425,	176,
167,	398,	433,	446,	419,	431,	189,
120,	431,	390,	495,	454,	215,	117

Display the data set using an appropriate graphical display.

Complex unfamiliar

8. A side-by-side bar chart shows the distribution of road fatalities over a year.

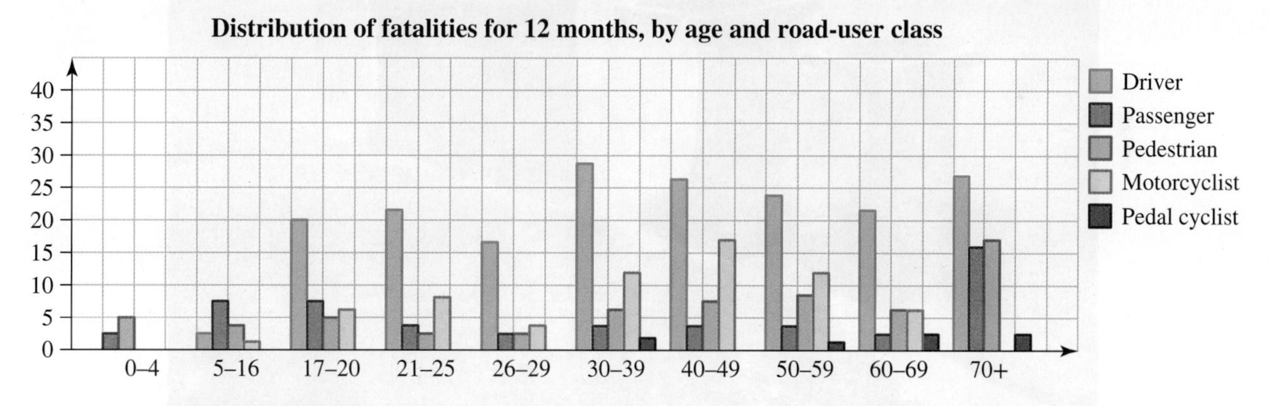

The government wants to introduce a road campaign. Determine which age groups the government should focus on. Explain your answer.

9. The comparisons between the battery lives of two mobile phone brands are shown in the back-to-back stem-and-leaf plot below. Identify the benefits of representing the data using a steam-and-leaf plot.

Key: $6 \mid 1 = 61$ hours

Brand A	*Stem*	*Brand B*
8 8 7 5	0	7
9 7 4 1 0	1	0 5 5 5 7 9
2 2 2 1	2	0 2 2 6 7
8 6 4 2 0	3	0 2 4 6 8
	4	
	5	6
1	6	
	7	5

10. A sample of Queensland residents were asked to state the environmental issue that was most important to them. The results were sorted by the age of the people surveyed. The results were as follows.

	Environmental issue		
Age	**Reducing pollution**	**Conserving water**	**Recycling rubbish**
Under 30	52	63	41
Over 30	32	45	23

Identify the type of data display that you would use to compare the data sets and construct this graph.

Fully worked solutions for this chapter are available online.

LESSON

6.7 Review

6.7.1 Summary

doc-41758

Hey students! Now that it's time to revise this chapter, go online to:

Access the chapter summary **Review your results** **Watch teacher-led videos** **Practise questions with immediate feedback**

Find all this and MORE in jacPLUS

6.7 Exercise

learnon

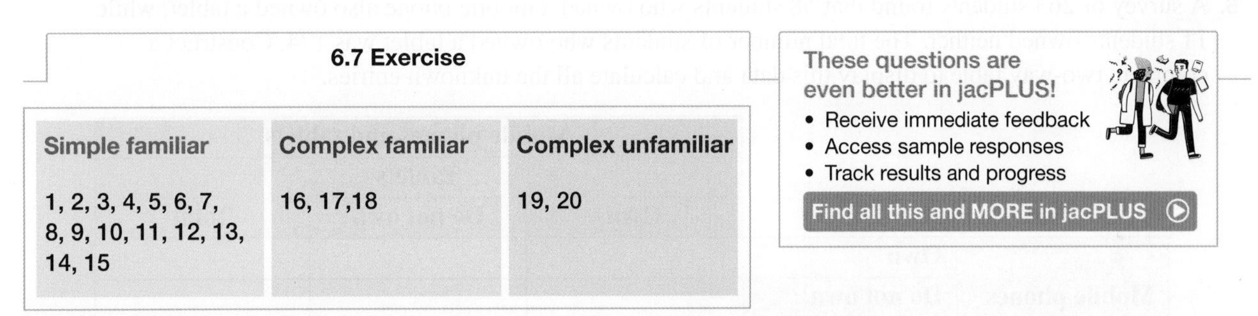

6.7 Exercise

Simple familiar	Complex familiar	Complex unfamiliar
1, 2, 3, 4, 5, 6, 7, 8, 9, 10, 11, 12, 13, 14, 15	16, 17,18	19, 20

These questions are even better in jacPLUS!
• Receive immediate feedback
• Access sample responses
• Track results and progress
Find all this and MORE in jacPLUS

Simple familiar

1. Classify the following fruit shop data as numerical, discrete, continuous, categorical, nominal or ordinal.
   a. A list of all fruit types sold at the fruit shop
   b. The number of pieces of fruit sold in a day
   c. The quantity of fruit (in kg) sold in a day
   d. A ranking of the most popular to least popular fruits sold in summer months
2. The Australian Bureau of Statistics collects data on the price of unleaded petrol. Identify whether this data is nominal, ordinal, discrete or continuous.
3. Classify the following data as either categorical ordinal or numerical discrete.
   a. The number of children enrolled in 20 high schools
   b. Someone's opinion on how well their car was serviced
4. At a hospital nursing station, the following information is available about a patient.
   - Temperature: 36.7 °C
   - Blood type: A
   - Blood pressure: 120/80 mm Hg
   - Response to treatment: excellent

   Determine which information is ordinal.

5. For each of the following two-way frequency tables, complete the missing entries.

a.

		Grades		
		Year 12	Year 11	Total
Transport preference	Tram	25	i	47
	Train	ii	iii	iv
	Total	51	v	92

b.

		Voters' ages	
		18–45	Over 45
Party preference	Labor	i	42%
	Liberal	53%	ii
	Total	iii	iv

6. A survey of 263 students found that 58 students who owned a mobile phone also owned a tablet, while 11 students owned neither. The total number of students who owned a tablet was 174. Construct a complete two-way table to display this data and calculate all the unknown entries.

		Mobile phones and tablets		
		Tablets		
		Own	Do not own	Total
Mobile phones	Own			
	Do not own			
	Total			

7. Consider the stem plot shown.
Write a data set that matches the stem plot shown.

Key: 7|6 = 7.9

Stem	*Leaf*
7	8
8	0 8 9
9	1 6 7 8
10	3 5 8
11	2

8. The dot plot shows the number of goals scored by a soccer team over a season.

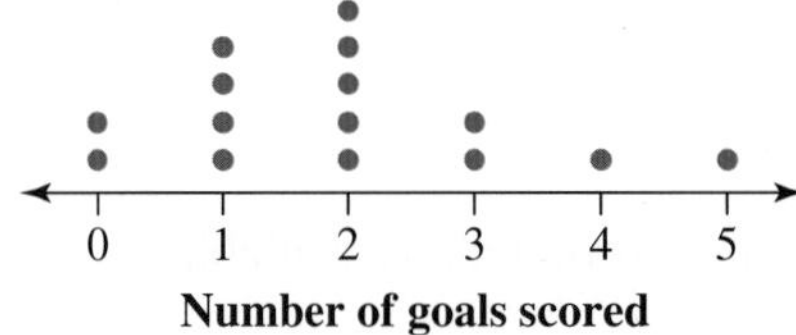

Number of goals scored by soccer team over a season

a. Determine the greatest number of goals scored in a match.
Goals = □

b. Determine in how many matches the team scored no goals.
Matches = □

9. As part of his preparation for the Essential Mathematics exam, Andrew completes a weekly test. The results, in percentages, for his last 9 tests are:

67, 61, 69, 38, 68, 66, 71, 69, 69

Identify any possible outliers.

10. In a group of 165 people aged 16–30 years, 28 people aged 23–30 years have a learner driver licence while 15 people aged 16–22 years have a probationary driver licence. Out of the 117 people who have a licence, 59 are learners.

		Type of driver licence for people aged 16–30 years		
		Age groups		
		16–22 years	23–30 years	Total
Driver licence	L	31	28	59
	P	15	43	58
	None	27	21	48
	Total	73	92	165

If the total number of people aged 16–22 years surveyed was 73, calculate the number of people aged 23–30 who have a probationary driver licence.

11. One hundred teenagers were surveyed about their favourite type of music genre. The data was organised into a frequency table.

Music genre	Frequency
Hip hop/rap	28
Pop/R&B/soul	27
Rock	26
Country	3
Blues/jazz	4
Classical	2
Alternative	10

a. Determine whether the data is numerical or categorical.
b. Identify the data type of 'music genre'.
c. Calculate what percentage of teenagers preferred rock music.

12. A road worker recorded the number of vehicles that passed through an intersection in the morning and in the afternoon and summarised the data in a two-way table. A total of 47 heavy vehicles passed through the intersection during the day and 126 light vehicles in the afternoon.

Of the 238 vehicles that passed through the intersection in the morning, 199 were light vehicles.

a. Construct a two-way table for this set of data.
b. Determine the number of vehicles that passed through the intersection during the whole day.
c. Determine the number of heavy vehicles that passed through the intersection in the afternoon.

13. The numbers of the immediate family members of the teachers in a small school are given.

5, 6, 3, 8, 2, 2, 3, 4, 3, 4, 6, 7,
2, 5, 4, 5, 4, 6, 1, 3, 4, 5, 5, 4

a. Present this set of data in a frequency table.
b. Present this data as a dot plot.

14. Students from a Year 11 class were asked about their favourite subjects and the following data was recorded.

Maths	English	PE	Science	PE
Art	Maths	Science	English	Science
Cooking	PE	English	Cooking	Maths
PE	Art	Science	Maths	Art
Science	Cooking	Art	PE	PE

a. Construct a frequency table using the data and record the number of students who were surveyed.
b. Calculate the percentage of students who preferred Maths.
c. Identify the most popular subject. Calculate the percentage of students who preferred this subject.
d. Determine what type of data this is.

15. Calculate the percentage relative frequency, correct to 1 decimal place, for all the entries in the two-way table given.

		Eye colour and hair colour			
		Eye colour			
		Adults	**Teenagers**	**Children**	**Total**
Hair colour	**Light**	49	38	26	113
	Dark	106	12	29	147
	Total	155	50	55	260

Complex familiar

16. A car company surveyed a group of 250 people about the number of cars they had in their household. The data is displayed in the frequency distribution table below.

Number of cars	Frequency
1	36
2	107
3	89
4	13
5	5
Total	250

Determine the best way to represent this data and sketch the graph.

17. The following graph shows the number of people in selected occupations in Queensland.

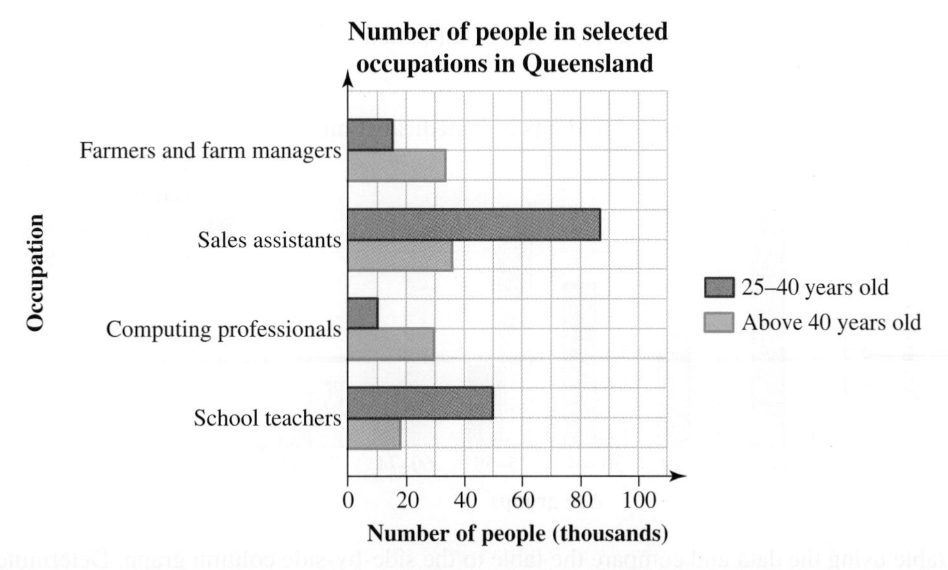

a. Identify the type of graph that is used to represent the data and explain why it is used.
b. Use the graph to estimate the number of people aged 25–40 who were sales assistants in Queensland at the time the data was collected.
c. Estimate the number of school teachers in Queensland.
d. Identify in which occupation there are about 30 000 people above the age of 40.
e. Identify in which occupation the total number of workers is about 40 000.

18. A coffee bar serves either skim, reduced-fat or whole milk in coffees. The coffees sold on a particular day are shown in the table below, sorted by the type of milk and the weight range of the customers.

		Weight range	
		Underweight	**Overweight**
Type of milk	**Skim**	87	124
	Reduced fat	55	73
	Whole	112	49

a. Identify the number of coffees that were sold on this day.
b. Propose the best way to represent this data graphically and draw the graph.
c. Determine the percentage of underweight customers who asked for skim milk in their coffees out of the total number of underweight customers. Give your answer to the nearest whole number.
d. Determine the percentage of coffees sold that contained reduced-fat milk out of the total number of coffees sold. Give your answer to the nearest whole number.
e. If this was the daily trend of sales for the coffee bar, determine the percentage of the coffee bar's customers you would expect to be overweight. Give your answer to the nearest whole number.

Complex unfamiliar

19. The following graphical display summarises the ages of patients seen by two doctors in a medical surgery during one particular day.

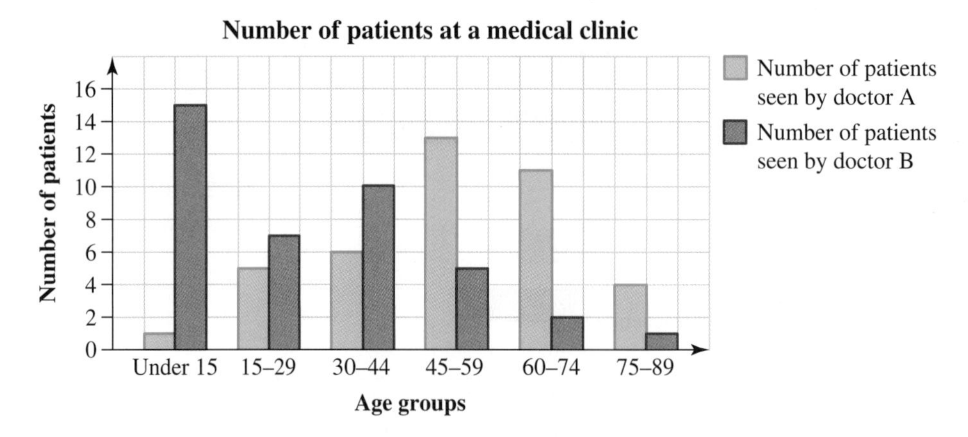

Create a table using the data and compare the table to the side-by-side column graph. Determine the most effective way to represent the data and explain why this is the most effective way.

20. The local childcare centre, HappyTodd, is attended by 37 children aged 2–4 years. In order to provide an appropriate diet for these toddlers, their weights, in kilograms, were recorded at enrolment.

14.5,	12.4,	16.3,	13.0,	13.6,	12.9,	15.4,	14.7,
14.3,	14.9,	13.8,	15.2,	15.4,	16.1,	13.2,	13.8,
13.5,	13.4,	15.1,	12.7,	14.2,	13.6,	14.4,	14.1,
14.7,	14.8,	14.2,	14.3,	13.5,	13.1,	14.6,	15.3,
14.2,	15.5,	15.7,	15.8,	16.5			

Determine the best way to represent this data and draw the graph.

Fully worked solutions for this chapter are available online.

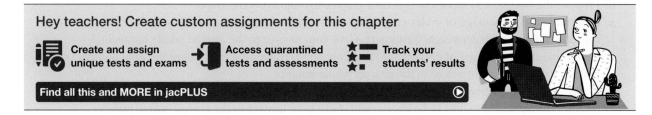

Answers

Chapter 6 Data classification, presentation and interpretation

6.2 Classifying data

6.2 Exercise

1. a. Numerical continuous
   b. Numerical discrete
2. The number of pieces of fruit is discrete data as it is represented by counting numbers. The quantity of fruit sold is continuous data because it represents a measurement; there is always another value between any two values.
3. a. Continuous b. Discrete
   c. Discrete d. Discrete
   e. Continuous f. Continuous
4. a. Categorical nominal
   b. Categorical ordinal
5. 'Favourite movies' identifies a category; it cannot be ordered. 'The rating for a movie' can be ordered in categories like poor, average, very good and excellent.
6. a. Ordinal b. Nominal
   c. Ordinal d. Nominal
7. a. Categorical
   b. Numerical
8. Although country phone codes are numerical, they are neither countable nor measurable. They represent categories. The number of international calls is numerical data as it is countable.
9. a. Numerical
   b. Categorical
10. Data collected on the lengths of pencils is a measurement, while data collected on the colours of pencils can be represented in colour categories.
11. Numerical continuous
12. a. Numerical b. Categorical
    c. Categorical d. Numerical

6.3 Categorical data

6.3 Exercise

1.

Mode of transport to school		
Mode of transport	**Tally**	**Frequency**
Car	卌 ||	7
Tram	卌 ||||	9
Bicycle	||||	4
Walk	卌 卌	10
Total		30

2.

Favourite social network		
Social network	**Tally**	**Frequency**
MyPage	卌 卌 卌	15
FlyBird	卌 ||||	9
Total		24

3.

Internet connection		
Type of internet connection	**Tally**	**Frequency**
NBN	卌 卌 卌	15
C	|||	3
MW	卌 卌 卌 ||	17
Total		35

4.

Opinion	Frequency
Excellent	163
Very good	298
Average	24
Poor	15
Total	500

5.

		Parking place and car theft		
		Parking type		
		In driveway	**On the street**	**Total**
Theft	**Car theft**	2	37	39
	No car theft	16	445	461
	Total	18	482	500

6.

		Mobile phones and tablets		
		Tablets		
		Own	**Do not own**	**Total**
Mobile phones	**Own**	58	78	136
	Do not own	116	11	127
	Total	174	89	263

7. a.

		Year 11 students' Mathematics results		
		Result		
		Pass	Fail	Total
Subject	Essential Mathematics	12	23	35
	General Mathematics	25	49	74
	Mathematical Methods	36	62	98
	Specialist Mathematics	7	18	25
	Total	80	152	232

b. 232 students

c. 35 students

8. a.

		Hairdressing requirements per age group		
		Age group		
		Young	Old	Total
Hairdressing requirement	Haircut			
	Haircut and colour			
	Total			

b.

		Hairdressing requirements per age group		
		Age group		
		Young	Old	Total
Hairdressing requirement	Haircut			
	Haircut and colour	2		
	Total		56	

c.

		Hairdressing requirements per age group		
		Age group		
		Young	Old	Total
Hairdressing requirement	Haircut only	5	26	31
	Haircut and colour	2	30	32
	Total	7	56	63

9. a.

		Monday matinee audience per movie genre		
		Movie genres		
		Drama	Comedy	Total
Audience type	Adults	66	30	96
	Teenagers	5	16	21
	Children	11	28	39
	Total	82	74	156

b. 30

c. 39

10. a. Relative frequencies

		Cat and dog owners		
		Dog owners		
		Own	Do not own	Total
Cat owners	Own	0.039	0.256	0.295
	Do not own	0.495	0.210	0.705
	Total	0.534	0.466	1.000

b. Percentage relative frequencies

		Cat and dog owners (%)		
		Dog owners		
		Own (%)	Do not own (%)	Total
Cat owners	Own (%)	3.9	25.6	29.5
	Do not own (%)	49.5	21.0	70.5
	Total	53.4	46.6	100.0

11. a. Relative frequency table

		Year 11 students' Mathematics results		
		Result		
		Pass	Fail	Total
Subject	Essential Mathematics	0.0517	0.0991	0.1509
	General Mathematics	0.1078	0.2112	0.3190
	Mathematical Methods	0.1552	0.2672	0.4224
	Specialist Mathematics	0.0302	0.0776	0.1078
	Total	0.3448	0.6552	1.0000

b. Percentage relative frequency table

		Year 11 students' Mathematics results		
		Result		
		Pass (%)	Fail (%)	Total%)
Subject	Essential Mathematics (%)	5.17	9.91	15.09
	General Mathematics (%)	10.78	21.12	31.90
	Mathematical Methods (%)	15.52	26.72	42.24
	Specialist Mathematics (%)	3.02	7.76	10.78
	Total (%)	34.48	65.52	100.00

Note: Due to rounding, some of the percentages written do not add up to the actual value in some of the total columns. To have the correct values in the total columns, ensure that the actual values (no rounding) are summed up.

12. a. Relative frequencies

Swimming attendance	
Day of the week	**Adults**
Monday	0.112
Tuesday	0.072
Wednesday	0.080
Thursday	0.120
Friday	0.144
Saturday	0.216
Sunday	0.256
Total	1.000

b. Percentage relative frequencies

Swimming attendance	
Day of the week	**Adults (%)**
Monday	11.2
Tuesday	7.2
Wednesday	8.0
Thursday	12.0
Friday	14.4
Saturday	21.6
Sunday	25.6
Total	100.0

13. a.

		Dance class enrolments			
		People			
		Adults	Teenagers	Children	Total
Dance	Latin	31	21	23	75
	Modern	13	34	49	96
	Contemporary	36	49	62	147
	Total	80	104	134	318

b. Relative frequency table

		Dance class enrolments			
		People			
		Adults	Teenagers	Children	Total
Dance	Latin	0.10	0.07	0.07	0.24
	Modern	0.04	0.11	0.15	0.30
	Contemporary	0.11	0.15	0.19	0.46
	Total	0.25	0.33	0.42	1.00

Note: Due to rounding, some of the percentages written do not add up to the actual value in some of the total columns. To have the correct values in the total columns, ensure that the actual values (no rounding) are summed up.

c. Percentage relative frequency table

		Dance class enrolments (%)			
		People			
		Adults (%)	Teenagers (%)	Children (%)	Total
Dance	Latin (%)	10	7	7	24
	Modern (%)	4	11	15	30
	Contemporary (%)	11	15	19	46
	Total (%)	25	33	42	100

14.

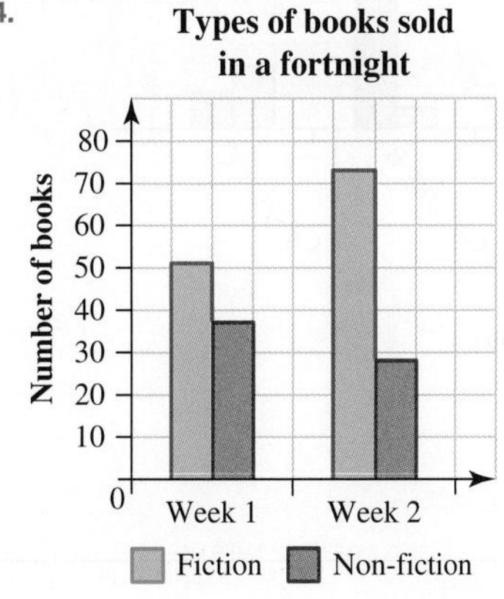

15. **Types of fitness exercise for people**

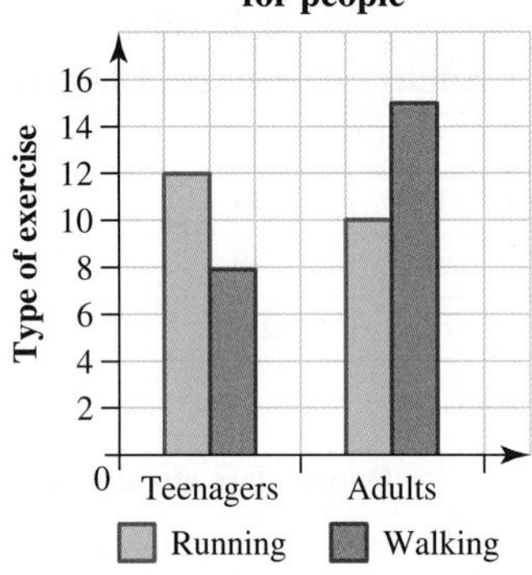

Running Walking

16. VIC and ACT

17. See the figure at the bottom of the page.*

18. a.

		Drivers and traffic accidents			
		Drivers			
		Category A	Category B	Category C	
		Young drivers	Mature drivers	Older drivers	Total
Event	Traffic accident	35	10	30	75
	No traffic accident	95	113	17	225
	Total	130	123	47	300

b. 10

c. 225

d.

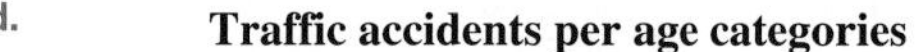

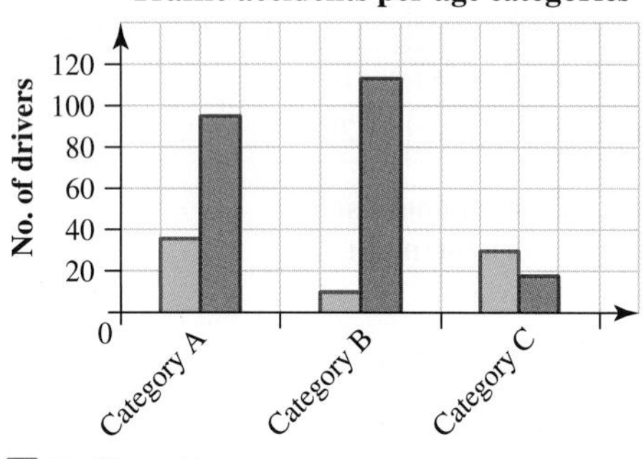

Traffic accidents
No traffic accidents

6.4 Numerical data

6.4 Exercise

1.

Hours of sleep	Tally	Frequency
4	|	1
5	|||	3
6	𝍸	5
7	𝍸 |	6
8	𝍸	5
9	|||	3
10	||	2
Total		25

2.

Foot length	Tally	Frequency
21	|	1
22	|	1
23	𝍸	5
24	||||	4
25	||	2
26	𝍸 |	6
27	|||	3
28	𝍸	5
29	|||	3
Total		30

3.

Time	Tally	Frequency
17– < 22	𝍸 𝍸	10
22– < 27	||||	4
27– < 32	||||	4
32– < 37	𝍸	5
37– < 42	𝍸 |	6
42– < 47	||	2
47– < 52	|	1
52– < 57	|||	3
57– < 62	|	1
Total		36

*17.

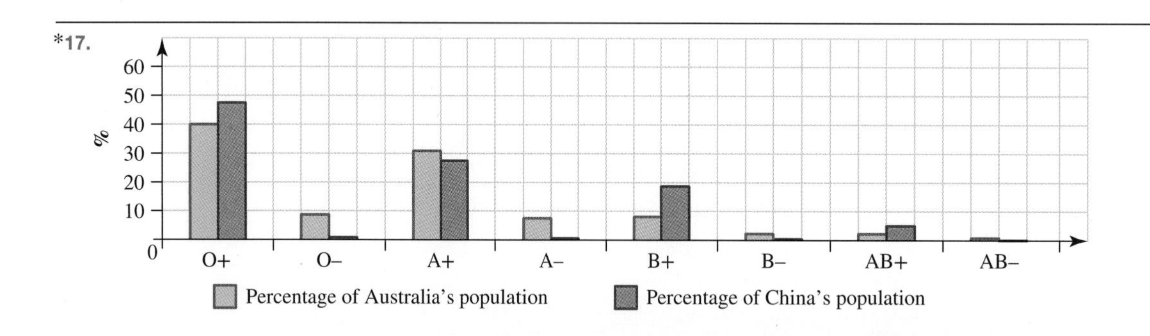

4.

Weekends	Tally	Frequency
0	卌 卌 |	11
1	卌 ||||	9
2	卌	5
3	|||	3
4	||	2
Total		30

5.

Width	Tally	Frequency
33.0– < 34.0	||	2
34.0– < 35.0	卌 ||	7
35.0– < 36.0	卌 卌 卌	15
36.0– < 37.0	卌 ||||	9
37.0– < 38.0	|||	3
Total		36

6.

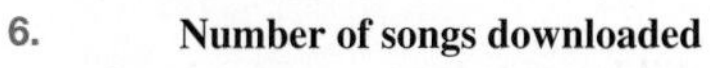

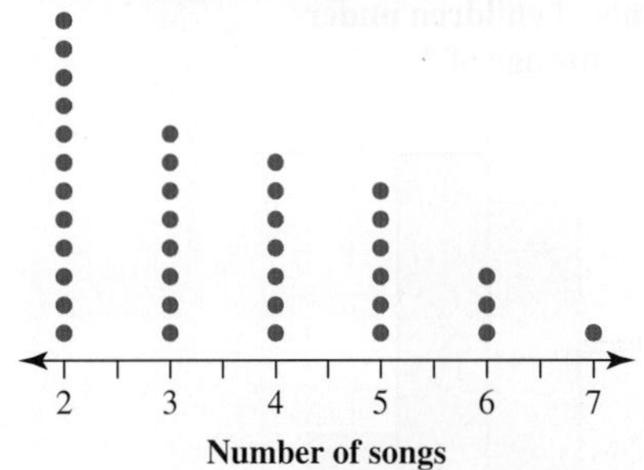

7.

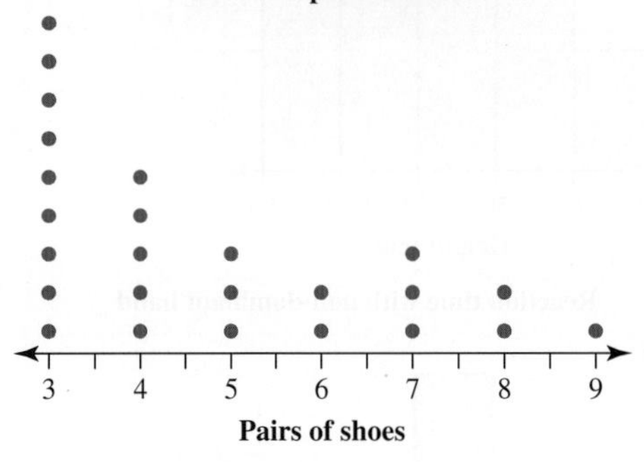

8.

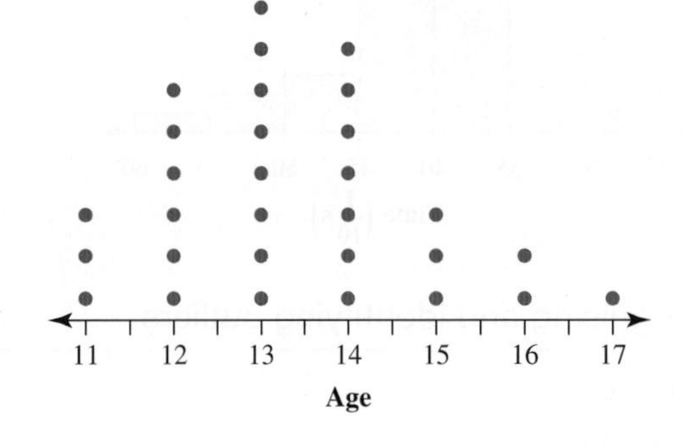

9.

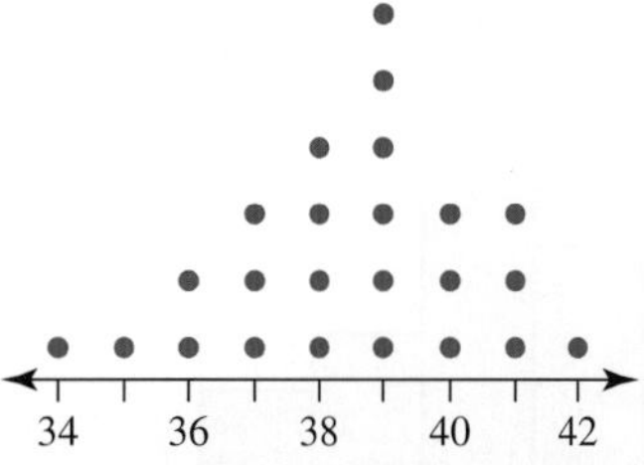

10.

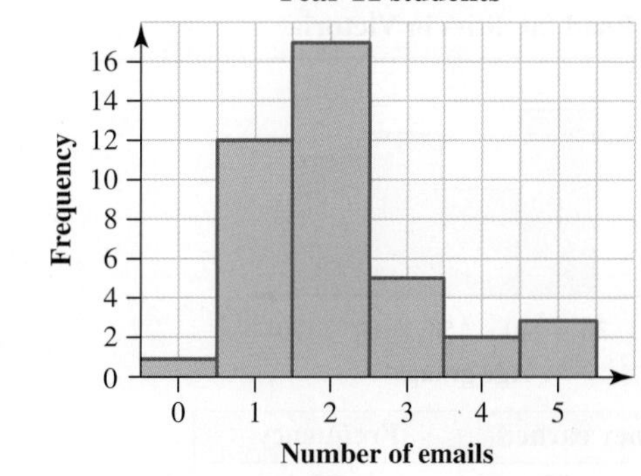

11.

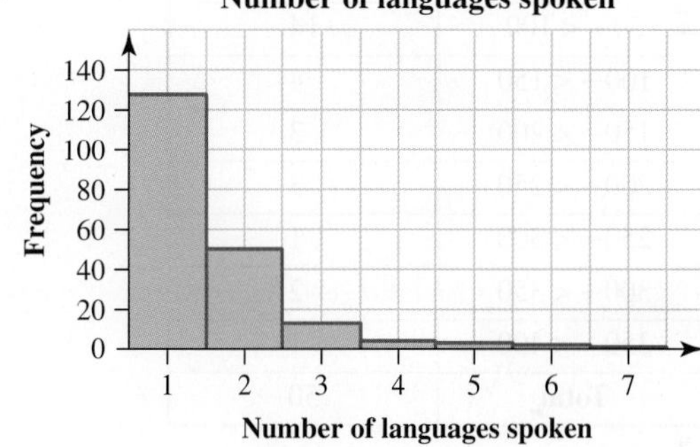

12.

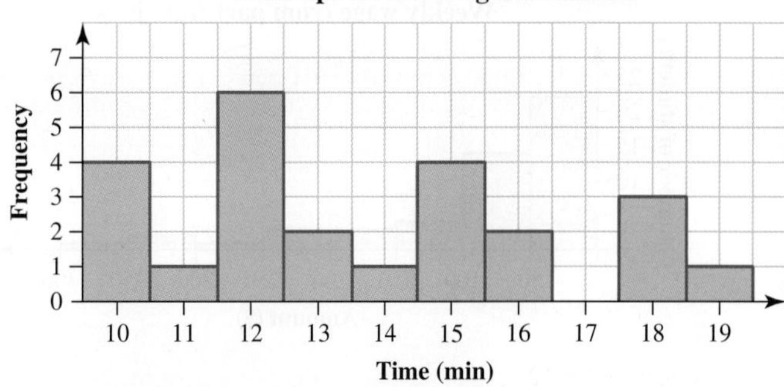

13. a.

Email check	Frequency
0– < 3	11
3– < 6	12
6– < 9	8
9– < 12	11
12– < 15	5
15– < 18	3

b.

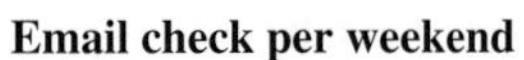

Email check per weekend

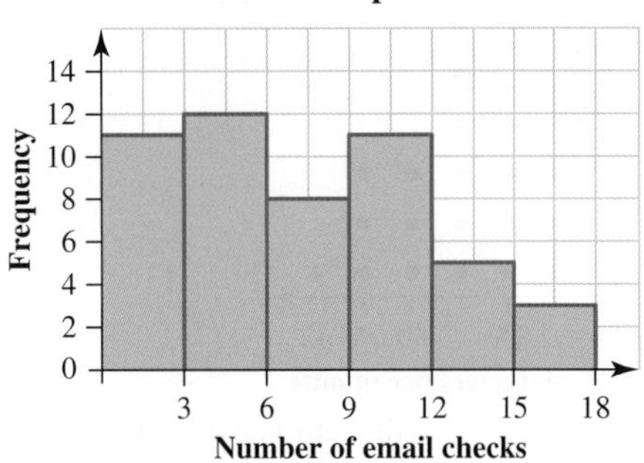

14.

Road fatalities in Victoria

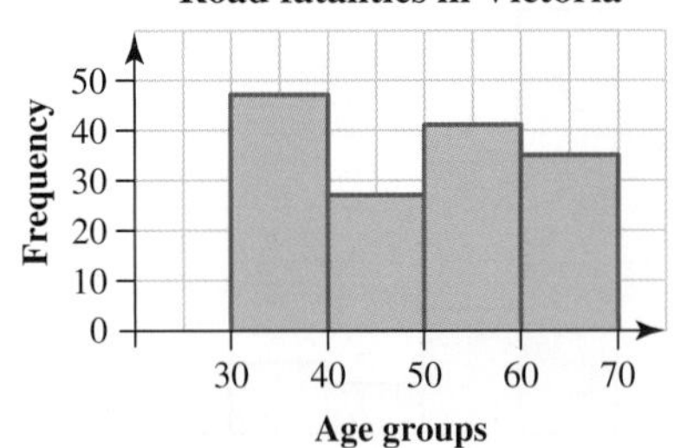

15. a.

Money earned	Frequency
0– < 50	22
50– < 100	14
100– < 150	4
150– < 200	3
200– < 250	3
250– < 300	1
300– < 350	2
350– < 400	1
Total	50

b.

Weekly wage from part-time jobs

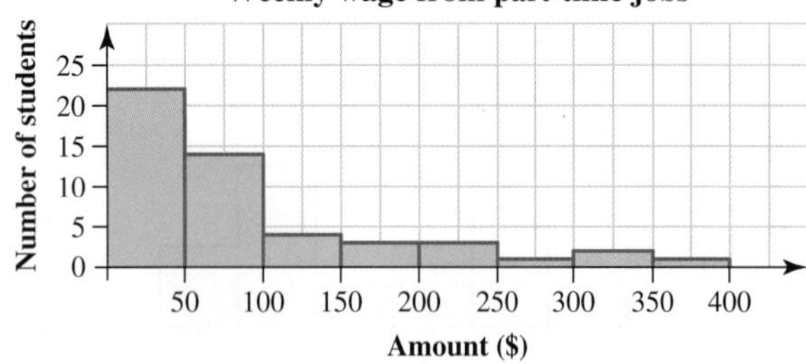

16. Key: 1|2 = 12
1*|6 = 16

Stem	*Leaf*
1*	6 7 8 8 9 9
2	0 0 1 1 2 2 3 3 3 3 4 4 4
2*	5 5 9
3	0 2

17. Key: 1|2 = 12

Stem	*Leaf*
1	2 8
2	0 0 3 9 9
3	2 3 5 6 8 9
4	1 3 3 4 5 8 8 9
5	0 0 6

18. Key: 20|4 = 204

Stem	*Leaf*
20	4 6 7
21	0 0 2 6 8 9
22	0 0 5 7 9 9
23	0 1 2
24	3 3 4 5
25	0 0 0 1 3
26	0 1 5

19.

Heights of children under the age of 3

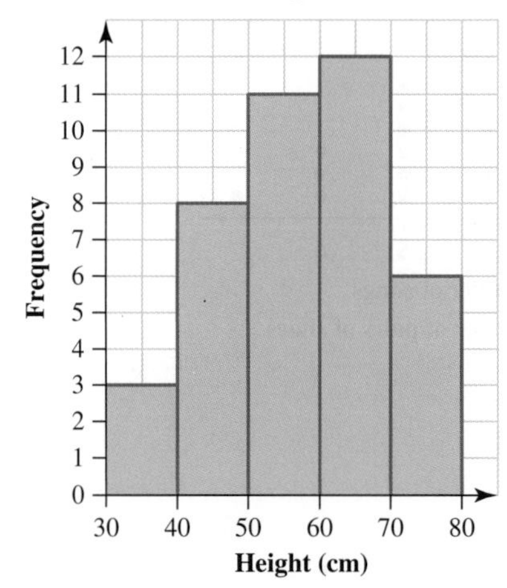

20.

Reaction time with non-dominant hand

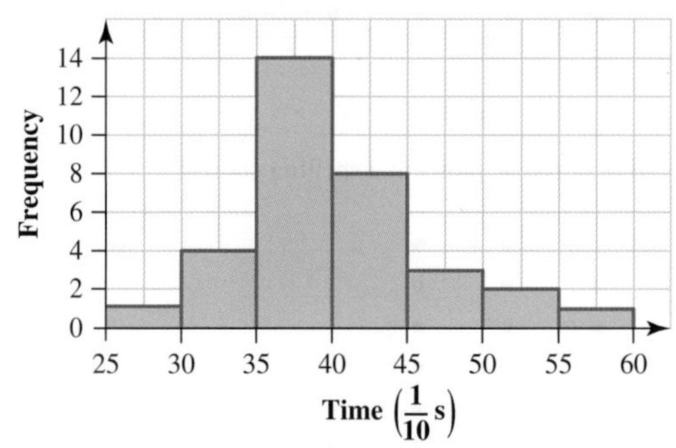

6.5 Recognising and identifying outliers

6.5 Exercise

1. a. 5 is a possible outlier.
   b. Sample responses can be found in the worked solutions in the online resources.

2. a. 125 is a possible outlier.
   b. Sample responses can be found in the worked solutions in the online resources.
3. A possible outlier is 30.
4. There is no obvious outlier.
5. A possible outlier is the age group 32–36.
6. a. 25, 27, 27, 28, 29, 30, 30, 33, 33, 51
   b. See figure at the bottom of the page*
   c. 51 is a possible outlier.
   d. Sample responses can be found in the worked solutions in the online resources.
7. a. Student 7 at 130 m
   b. Sample responses can be found in the worked solutions in the online resources.
8. a. 12 is a possible outlier.
   b. Sample responses can be found in the worked solutions in the online resources.
9. a. Team A
   b. 41 appears to be a possible outlier.
10. a. Class 2
    b. 3 is the obvious outlier.

6.6 Comparing different data [complex]

6.6 Exercise

1. Sample responses can be found in the worked solutions in the online resources.
2. Sample responses can be found in the worked solutions in the online resources.
3. a.

Number of siblings	Frequency
1	5
2	3
3	6
4	3
5	2
6	0
7	1
Total	20

   b.

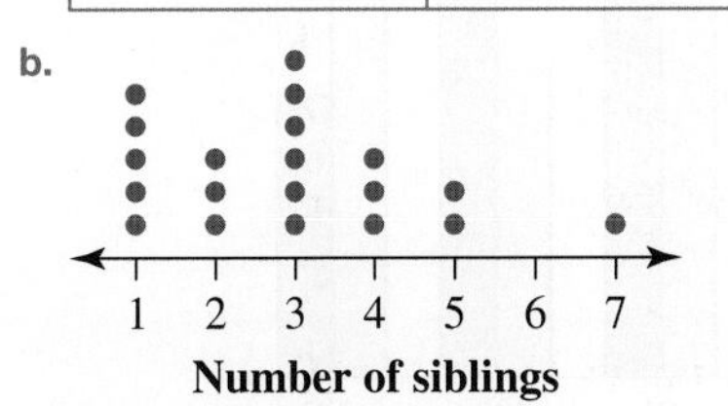

   c. Sample responses can be found in the worked solutions in the online resources.
4. a. Panthers
   b. Bulldogs and Storm, and West Tigers and Cowboys
   c. Sample responses can be found in the worked solutions in the online resources.
5. a. LeBron James
   b. LeBron James
   c. Due to some of the data being discrete (countable) and some being percentages, it is difficult to have all on the same axis.
6. a. Column graph.
   b. Sample responses can be found in the worked solutions in the online resources.
   c. Ankle injury

7–9. Sample responses can be found in the worked solutions in the online resources.

10. A side-by-side bar chart or back-to-back bar chart would be a suitable data display.

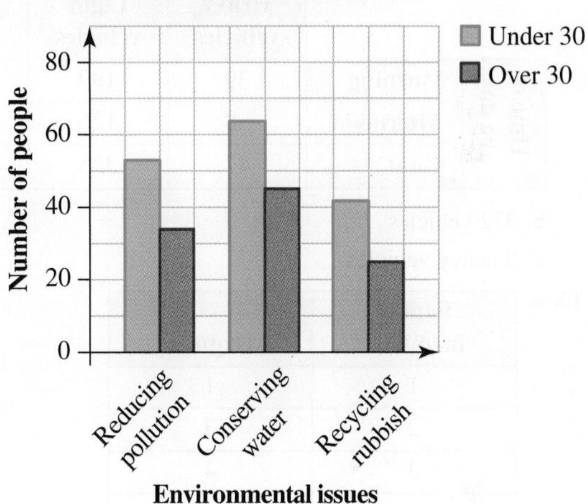

6.7 Review

6.7 Exercise

1. a. Categorical nominal
   b. Numerical discrete
   c. Numerical continuous
   d. Categorical ordinal
2. Continuous
3. a. Numerical discrete
   b. Categorical ordinal
4. Response to treatment: excellent
5. a. i. 22 ii. 26 iii. 19 iv. 45 v. 41
   b. i. 47% ii. 58% iii. 100% iv. 100%

*6 b

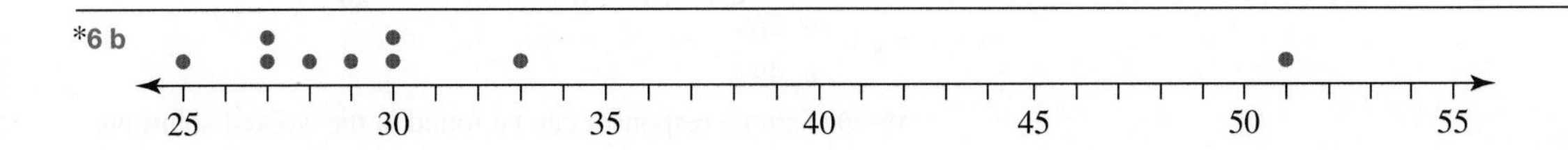

6.

		Mobile phones and tablets		
		Tablets		
		Own	Do not own	Total
Mobile phones	Own	58	78	136
	Do not own	116	11	127
	Total	174	89	263

7. 7.8, 8.0, 8.8, 8.9, 9.1, 9.6, 9.7, 9.8, 10.3, 10.5, 10.8, 11.2

8. a. 5 b. 2

9. Outlier = 38

10. 43

11. a. Categorical
 b. Categorical, nominal
 c. 26%

12. a.

		Number of vehicles		
		Type of vehicle		
		Heavy vehicles	Light vehicles	Total
Time of the day	Morning	39	199	238
	Afternoon	8	126	134
	Total	47	325	372

 b. 372 vehicles
 c. 8 heavy vehicles

13. a.

Family members	Frequency
1	1
2	3
3	4
4	6
5	5
6	3
7	1
8	1

 b. Draw a horizontal line for the number of family members. Place the numbers 1–8 evenly spaced along the line. Place the dots for each score.

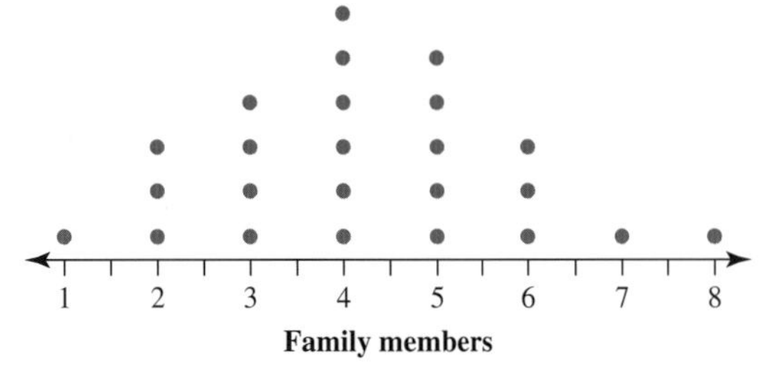

14. a.

Subject	Tally	Frequency (f)
Maths	||||	4
Art	||||	4
Cooking	|||	3
PE	~~||||~~ |	6
Science	~~||||~~	5
English	|||	3
Total		25

 b. 16%
 c. PE; 24%
 d. Nominal (categorical)

15.

		Eye colour and hair colour			
		Eye colour			
		Adults	Teenagers	Children	Total
Hair colour	Light	18.8%	14.6%	10.0%	43.4%
	Dark	40.8%	4.6%	11.2%	56.6%
	Total	59.6%	19.2%	21.2%	100%

16. Sample responses can be found in the worked solutions in the online resources.

17. a. Sample responses can be found in the worked solutions in the online resources.
 b. 87 000
 c. 68 000
 d. Computing professionals
 e. Computing professionals

18. a. 500
 b. Side-by-side column graph

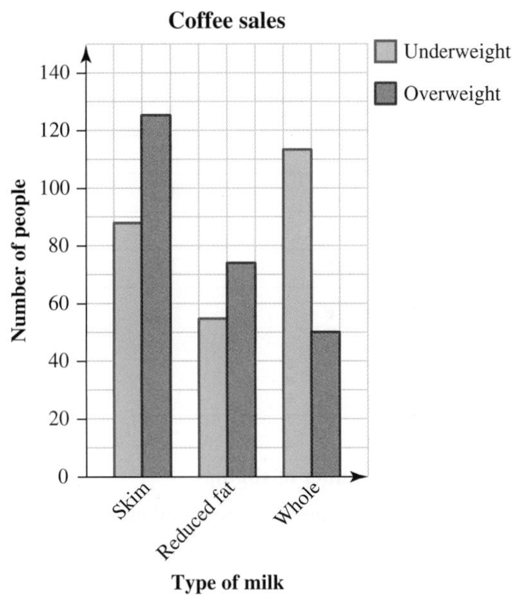

 c. 34%
 d. 26%
 e. 49%

19–20. Sample responses can be found in the worked solutions in the online resources.

7 Earning money

LESSON SEQUENCE

Fully worked solutions for this chapter are available online.

Resources

 **Solutions** Solutions — Chapter 7 (sol-1260)

 **Digital documents** Learning matrix — Chapter 7 (doc-41759)
Quick quizzes — Chapter 7 (doc-41760)
Chapter summary — Chapter 7 (doc-41761)

LESSON
7.1 Overview

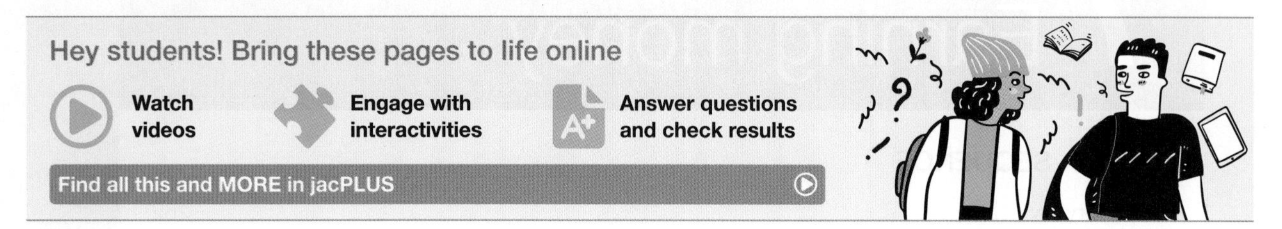

7.1.1 Introduction

In this chapter, you will investigate different kinds of employment as well as the different investment options for growing your savings. Every business has to pay its employees either an annual salary or a wage based on an hourly rate. Financial investments are another way for businesses and individuals to make money — it is important to understand how investments work to be able to decide whether an investment is a good idea or not.

A sound knowledge of financial mathematics is essential in a range of careers, including financial consultancy, accountancy, business management and pay administration.

7.1.2 Syllabus links

Lesson	Lesson title	Syllabus links
7.2	**Salary, wages and overtime**	◯ Find earnings, including salary, wages, overtime, piecework and commission. ◯ Interpret entries on a selection of wage or salary pay slips and timesheets.
7.3	**Income payments — commission, piecework and royalties**	◯ Find earnings, including salary, wages, overtime, piecework and commission.
7.4	**Convert between different pay periods [complex]**	◯ Convert between annual, monthly, fortnightly, weekly and hourly rates of earning [complex].
7.5	**Personal taxation, PAYG and employee superannuation**	◯ Understand the purpose of taxation and the use of tax file numbers. ◯ Understand the purpose of superannuation. ◯ Interpret entries on a simple PAYG summary.
7.6	**Determining PAYG tax using tax tables [complex]**	◯ Use tax tables to determine PAYG tax for periodic (weekly/fortnightly/monthly) earnings [complex].
7.7	**Determining taxable income by applying tax deductions [complex]**	◯ Apply the concepts of taxable income, gross income, allowable deductions and levies in simple contexts [complex].
7.8	**Calculating simple tax returns [complex]**	◯ Calculate a simple income tax return and net income using current income tax rates [complex].

Source: Essential Mathematics Senior Syllabus 2024 © State of Queensland (QCAA) 2024; licensed under CC BY 4.0.

LESSON
7.2 Salary, wages and overtime

SYLLABUS LINKS

- Find earnings, including salary, wages, overtime, piecework and commission.
- Interpret entries on a selection of wage or salary pay slips and timesheets.

Note: Concepts shown in grey are covered in other lessons.

7.2.1 Salary and wages

- Employees may be paid for their work in a variety of ways. Most receive either an **hourly wage** or a **salary**.

Ways of paying employees

Salary	**Wages**
A fixed amount of money paid per year (annually), usually paid fortnightly or monthly regardless of the number of hours worked.	A fixed amount of money per hour worked (hours worked outside the normal work period are paid at a higher amount).
Examples of salaried jobs include: architect, company director, data analyst, teacher, doctor, accountant, federal or state government minister.	Examples of waged jobs include: waiter, kitchen hand, bar attendant, receptionist, technician, retail assistant, fruit picker or packer.

Key points

- **Normal working hours in Australia are *38 hours* per week.**
- **There are *52 weeks* in a year.**
- **There are *26 fortnights* in a year (this value will be slightly different for a leap year).**
- **There are *12 months* in a year.**

7.2.2 Overtime

- **Overtime** is paid when a wage earner works longer than the regular hours of work.
- These additional payments are often referred to as penalty rates.
- Penalty rates are usually paid for working on weekends, public holidays or at night.
- The extra hours are paid at a higher hourly rate, normally calculated at either time and a half or double time.

Calculating overtime

The overtime hourly rate is usually a multiple of the regular hourly rate. Some examples of overtime rates include:

- **1.5 × regular hourly wage (*time and a half*)**
- **2 × regular hourly wage (*double time*)**
- **2.5 × regular hourly wage (*double time and a half*)**

tlvd-4810

WORKED EXAMPLE 1 Calculating wage earned over a weekend

A hospitality worker is paid time and a half for working on Saturday nights and double time for working on Sundays or public holidays. The normal hourly pay rate is $15.50. If the worker worked 5 hours on Saturday and 6 hours on Sunday, calculate the amount of money they earned over the weekend.

THINK	WRITE
1. Calculate the hourly rate of pay for Saturday. Time and a half means × 1.5.	$15.50 \times 1.5 = \$23.25$ per hour
2. Calculate the amount they were paid for Saturday.	$23.25 \times 5 = \$116.25$
3. Calculate the hourly rate of pay for Sunday. Double time means × 2.	$15.50 \times 2 = \$31$ per hour
4. Calculate the amount paid for Sunday.	$31 \times 6 = \$186$
5. Write the answer as a sentence. *Note:* This is the amount they earned before tax was taken out.	The amount earned for the weekend is $\$116.25 + \$186 = \$302.25$.

7.2.3 Timesheets and pay slips

- A **pay slip** summarises your work during your particular pay period. This includes hours worked, gross and net income, superannuation, and the amount of tax withheld.
- A **timesheet** records the number of hours worked over a particular pay period.

WORKED EXAMPLE 2 Interpreting entries on a timesheet

Review the following weekly employee timesheet and answer the questions.

Weekly Employee Timesheet

Company Name
Address 1
Address 2
City, State Postcode
(00) 0000 0000
www.company-name.com

Employee Name: ____________

Supervisor Name: ____________

Week of: 6/24/2019

Day of Week	Regular [h]:mm	Overtime [h]:mm	Sick [h]:mm	Annual Leave [h]:mm	Public Holiday [h]:mm	Unpaid Leave	Other [h]:mm	TOTAL [h]:mm
Mon 6/24								0:00
Tue 6/25	8:00	2:15						10:15
Wed 6/26								0.00
Thu 6/27								0.00
Fri 6/28								0:00
Sat 6/29								0.00
Sun 6/30								0:00
Total Hrs:	**8:00**	**2:15**	**0:00**	**0:00**	**0:00**	**0:00**	**0:00**	**10:15**
Rate/Hour:	15.00	23.00	15.00	15.00	15.00	0.00	0.00	
Total Pay:	120.00	51.75	0.00	0.00	0.00	0.00	0.00	**$171.75**

Total Hours Reported (h:mm): 10:15
Total Pay: $171.75

a. **Determine how many hours the employee worked during the week.**
b. **Calculate their total pay.**
c. **Determine how many hours overtime they worked.**
d. **Calculate their regular hourly rate.**

THINK	WRITE
a. Look for the total hours row.	a. 10 h 15 min
b. Look at the bottom row, which calculates the total pay.	b. $171.75
c. Look at the overtime column.	c. 2 h 15 min
d. Look at their rate/hour under the regular column.	d. $15 per hour

WORKED EXAMPLE 3 Interpreting entries on a pay slip

Review the following weekly employee pay slip and answer the questions.

Pay slip

Leasa Butcher 45 Commercial Dve Shailer Park QLD 4128	Period: 14/07/2024 to 21/07/2024 Process Date: 21/07/2024 Process No: 57 Emp. No: 3 Status: Full Time Permanent Award: Clerical Award 2010 Classif.: Level 543 Super Fund.:

Description	Rate	Hours/Qty	Amount	Hours/Qty	Amount
Base Pay	45.0000	32.0000	1,440.00	32.0000	1,440.00
Time and a half	67.5000				
Double Time	90.0000				
Annual Leave	45.0000				
Leave Loading	7.8750				
Public Holiday	45.0000				
Personal Leave	45.0000				
RDO	45.0000				
RDO	(45.0000)			2.0000	
		Gross Pay:	**1,440.00**		
TAX (Tax Free Threshold + W/Wo Leave Loading)			(335.00)		(335.00)
		Net Pay:	**1,105.00**		
Personal Leave		1.4615		17.2529	
Annual Leave		2.9232		(27.5609)	
Super Casual			136.10		136.80

a. **Determine their regular rate.**
b. **Calculate their gross pay.**
c. **Determine their net pay.**
d. **Calculate how much tax they paid.**
e. **Determine how often they are paid.**

THINK	WRITE
a. Look for the base pay rate.	**a.** \$32 per hour
b. Look for the gross pay line in bold.	**b.** \$1440.00
c. Look for the net pay line in bold.	**c.** \$1105.00
d. Look at the tax amount in brackets.	**d.** \$335.00
e. The pay period at the top right is from 14/7/24 to 21/7/24.	**e.** The employee is paid weekly.

Exercise 7.2 Salary, wages and overtime

7.2 Quick quiz	7.2 Exercise

These questions are even better in jacPLUS!
- Receive immediate feedback
- Access sample responses
- Track results and progress

Find all this and MORE in jacPLUS

Simple familiar	Complex familiar	Complex unfamiliar
1, 2, 3, 4, 5, 6, 7, 8, 9, 10, 11, 12, 13, 14, 15, 16, 17, 18, 19, 20	N/A	N/A

Simple familiar

1. If an employee earns \$2000 each week, explain whether this is a salary or a wage.
2. If an employee earns a salary of \$90 000 a year and gets paid monthly, determine how much they get paid each month.
3. **WE1** Calculate the time and a half rate if the normal rate is:
   a. \$15
   b. \$21.50.
4. Calculate the double time rate if the normal rate is:
   a. \$25.50
   b. \$18.25.
5. Calculate the double time and a half rate if the normal rate is:
   a. \$26
   b. \$32.50.
6. A lawyer is offered a job with a salary of either \$74 000 per year or \$40 per hour. Assuming that they work 80 hours every fortnight, determine which pay is greater.
7. Calculate the fortnightly pay of an hourly wage of \$32.32 for 77.5 hours per fortnight.
8. Calculate the wage for the following hourly rates and hours worked.
   a. \$19.75 for 74.75 hours
   b. \$24.85 for 45.75 hours
   c. \$45.30 for 35.25 hours
9. Calculate the fortnightly wage of a person who earns \$22.25 per hour and works:
   a. 26 hours for both weeks
   b. 32 hours for both weeks.
10. Calculate the fortnightly wage (to the nearest cent) of a person who earns \$26.50 per hour and works:
    a. 32 hours in the first week and 27.5 hours in the second week
    b. 25.5 hours in the first week and 33.25 hours in the second week.
11. If a person is paid \$32.50 per hour, determine the following weekly wage:
    a. 20 hours at normal rate and 8 hours at double time
    b. 23 hours at normal rate and 7 hours at double time
12. If a person is paid \$18.25 per hour, determine the weekly wage to the nearest cent if they work:
    a. 23.5 hours at normal rate and 10.5 hours at double time
    b. 27.25 hours at normal rate and 8.5 hours at double time.

13. WE2 Answer the following questions using the weekly timesheet shown.

Weekly Employee Timesheet

Company Name
Address 1
Address 2
City, State Postcode
(00) 0000 0000
www.company-name.com

Employee Name: ____________
Supervisor Name: ____________
Week of: 24/6/2024

Day of Week	Regular [h]:mm	Overtime [h]:mm	Sick [h]:mm	Annual Leave [h]:mm	Public Holiday [h]:mm	Unpaid Leave	Other [h]:mm	TOTAL [h]:mm
Mon 6/24								0:00
Tue 6/25	6:00	3:15						9:15
Wed 6/26								0.00
Thu 6/27								0.00
Fri 6/28	4:00							4.00
Sat 6/29								0.00
Sun 6/30		3:30						3:30
Total Hrs:	**10:00**	**6:45**	**0:00**	**0:00**	**0:00**	**0:00**	**0:00**	**16:45**
Rate/Hour:	15.00	23.00	15.00	15.00	15.00	0.00	0.00	
Total Pay:	150.00	155.25	0.00	0.00	0.00	0.00	0.00	**$305.25**

Total Hours Reported (h:mm): 16:45
Total Pay: 305.25

a. Determine how many hours the employee worked during the week.
b. Calculate their total pay.
c. Determine how many hours overtime they worked.
d. Calculate their regular hourly rate.

14. Answer the following questions using the weekly timesheet shown.

Weekly Employee Timesheet

Company Name
Address 1
Address 2
City, State Postcode
(00) 0000 0000
www.company-name.com

Employee Name: ____________
Supervisor Name: ____________
Week of: 15/3/2024

Day of Week	Regular [h]:mm	Overtime [h]:mm	Sick [h]:mm	Annual Leave [h]:mm	Public Holiday [h]:mm	Unpaid Leave	Other [h]:mm	TOTAL [h]:mm
Mon 6/24								0:00
Tue 6/25								0:00
Wed 6/26	5:00							5.00
Thu 6/27	3:30							3.30
Fri 6/28								0:00
Sat 6/29	3:30	2:30						6.00
Sun 6/30		3:00						3:00
Total Hrs:	**12:00**	**5:30**	**0:00**	**0:00**	**0:00**	**0:00**	**0:00**	**17:30**
Rate/Hour:	22.00	33.00	22.10	22.00	22.00	0.00	0.00	
Total Pay:	264.00	181.50	0.00	0.00	0.00	0.00	0.00	**$445.50**

Total Hours Reported (h:mm): 17:30
Total Pay: $

a. Determine how many hours the employee worked during the week.
b. Calculate their total pay.
c. Determine if they earn a salary or a wage.
d. Determine how much they earned in overtime.
e. Calculate their overtime hourly rate.

15. WE3 For the following questions, refer to the pay slip shown.

Pay slip

John Citizen 45 Commercial Dr Shailer Park QLD 4128	Period: 18/01/2024 to 25/01/2024 Process Date: 25/07/2024 Process No: 57 Emp. No: 3 Status: Full Time Permanent Award: Clerical Award 2010 Classif.: Level 543 Super Fund.:

Description	Rate	Hours/Qty	Amount	Hours/Qty	Amount
Base Pay	45.0000	36.00	1620.00	36.0000	1620.00
Time and a half	67.5000				
Double Time	90.0000				
Annual Leave	45.0000				
Leave Loading	7.8750				
Public Holiday	45.0000				
Personal Leave	45.0000				
RDO	45.0000				
RDO	(45.0000)			2.0000	
		Gross Pay:	**1620.00**		
TAX (Tax Free Threshold + W/Wo Leave Loading)			(412.00)		(412.00)
		Net Pay:	**1208.00**		
Personal Leave		1.4615		17.2529	
Annual Leave		2.9232		(27.5609)	
Super Casual			170.10		170.10

a. Determine the employee's regular rate.
b. Calculate their gross pay.
c. Determine their net pay.
d. Calculate how much tax they paid.
e. Determine how often they are paid.

16. For the following questions, refer to the pay slip shown.

Pay slip

John Citizen 45 Commercial Dr Shailer Park QLD 4128	Period: 03/04/2024 to 11/04/2024 Process Date: 11/04/2024 Process No: 57 Emp. No: 3 Status: Full Time Permanent Award: Clerical Award 2010 Classif.: Level 543 Super Fund.:

Description	Rate	Hours/Qty	Amount	Hours/Qty	Amount
Base Pay	65.0000	32.0000	2080	32.0000	2080
Time and a half	97.5000				
Double Time	130.0000				
Annual Leave	65.0000				
Leave Loading	11.3750				
Public Holiday	65.0000				
Personal Leave	65.0000				
RDO	65.0000				
RDO	(45.0000)			2.0000	
		Gross Pay:	**2080.00**		
TAX (Tax Free Threshold + W/Wo Leave Loading)			(438.50)		
		Net Pay:	**1641.50**		
Personal Leave		1.4615		17.2529	
Annual Leave		2.9232		(27.5609)	
Super Casual			218.40		136.80

a. Determine the employee's time and a half rate.
b. Calculate their net pay.
c. Determine if they earn a wage or a salary.
d. Calculate how much tax they paid.

17. A retail worker is paid time and a half for working on Saturday and double time for working on Sundays or public holidays. The normal pay rate is $18.75. The worker worked over a long weekend, with the Monday being a public holiday. The hours worked for Saturday, Sunday and Monday (public holiday) were 8, 5 and 6 hours respectively.
Calculate the amount of money they earned over the weekend.

18. If a person is paid $975 for working a 25-hour week and all their hours are time and a half, determine their normal hourly rate.

19. If a person is paid $1457 for the week, their normal rate is $23.50 per hour and all their hours are double time, determine how many hours they worked that week.

20. If a person is paid $500 for working 20 hours in a week and half of their hours were at their regular rate and the rest at time and a half, determine their regular rate.

Fully worked solutions for this chapter are available online.

LESSON
7.3 Income payments — commission, piecework and royalties

SYLLABUS LINKS

- Find earnings, including salary, wages, overtime, piecework and commission.

7.3.1 Commission

- A **commission** is paid to a salesperson when an item is sold. For example, when real estate agents sell houses, they are paid a commission.
- Commissions are calculated as a percentage of the sale price.
- If no sales are made, no commission is received.
- A commission table is often used to determine the value of a commission.

tlvd-4813

WORKED EXAMPLE 4 Calculating the commission to be paid

When selling cars, a salesperson is paid according to the table shown.

Sale price	Commission	Plus
Between 0 and $80 000	2% of sale price	0
Between $80 001 and $140 000	1.5% of the amount over $80 000	$1600
$140 000 and over	1.1% of the amount over $140 000	$2500

If a car is sold for $200 000, calculate the commission paid to the salesperson.

THINK	WRITE
1. Since the amount of commission varies according to the sale price, determine which of the three ranges the sale falls into. 200 000 > 140 000, so the price falls into the third range ($140 000 and over). The commission for this range is based on the amount of the sale *over* $140 000. Calculate the amount over $140 000.	$200 000 −$140 000 $60 000
2. The commission is 1.1% of $60 000; write 1.1% as a decimal.	$1.1\% = \frac{1.1}{100} = 0.011$
3. Multiply 0.011 by the amount over $140 000 — that is, $60 000.	1.1% of $60 000 = 0.011 × $60 000 = $660
4. The table also specifies an additional commission of $2500, due to the sale being over $140 000. Calculate the total of the commission.	$660 + $2500 = $3160
5. Write the answer as a sentence.	The total commission is $3160.

7.3.2 Salary and commissions

- In some industries, the rate of sales varies widely, so it is not practical to live on commissions only.
  For example, car salespeople are given a base salary in addition to a commission.

tlvd-11432

WORKED EXAMPLE 5 Calculating the total wage including salary and commission

A car salesperson earns a salary of \$1200 per month plus a commission of 3% of the total sales that they make that month. In February, they sold cars worth \$198 500. Calculate their total wage (salary + commission) for that month.

THINK	WRITE
1. Their commission is 3% of sales. Write 3% as a decimal.	$3\% = \frac{3}{100} = 0.03$
2. Calculate 3% of the total sales in order to determine the commission.	3% of $\$198\,500 = 0.03 \times \$198\,500$ $= \$5955$
3. Their total wage is the commission plus their base salary. Add the commission of \$5955 to the base salary of \$1200.	$\$5955 + \$1200 = \$7155$
4. Write the answer as a sentence.	The total wage paid in February is \$7155.

7.3.3 Piecework

- A person paid by **piecework** is paid a fixed amount for each item they produce.
- There may be bonuses for faster workers.

WORKED EXAMPLE 6 Calculating the amount earned by piecework

A dressmaker gets paid \$4.75 for every dress they sew. If they sewed 124 dresses last week, calculate the amount of money they earned.

THINK	WRITE
1. The dressmaker gets \$4.75 for each dress, so multiply this amount by the number of dresses sewn.	$\$4.75 \times \$124 = \$589$
2. Write the answer as a sentence.	Last week the dressmaker was paid \$589.00.

7.3.4 Royalties

- A **royalty** is a payment made to an author, composer or creator for each copy of the work or invention sold. For example, if a pop star sold $2 million worth of music last year, they are entitled to some of the profits.
- Other people, such as managers, also get a share.
- Royalties are usually calculated as a percentage of the total sales (not total profit).

WORKED EXAMPLE 7 Calculating royalties

Last year, a new rock star sold music worth $2 156 320. Their recording contract specifies that they receive 2.4% royalty on the total sales.
Determine the amount of money they earned last year.

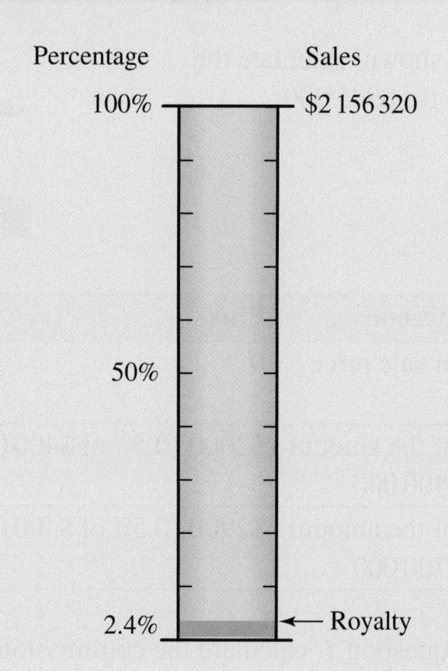

THINK	WRITE
1. They earned 2.4% of the sales ($2 156 320). Write 2.4% as a decimal.	$2.4\% = \frac{24}{100}$ $= 0.024$
2. Calculate 2.4% of the total sales; multiply 0.024 by the total sales.	2.4% of \$2 156 320 $= 0.024 \times \$2\,156\,320$ $= \$51\,751.68$
3. Write the answer as a sentence.	The rock star earned $51 751.68 in royalties.

on Resources

Interactivities Piecework (int-6069)
Commission and royalties (int-6070)

Exercise 7.3 Income payments — commission, piecework and royalties

7.3 Quick quiz	7.3 Exercise

These questions are even better in jacPLUS!
- Receive immediate feedback
- Access sample responses
- Track results and progress

Find all this and MORE in jacPLUS

Simple familiar	Complex familiar	Complex unfamiliar
1, 2, 3, 4, 5, 6, 7, 8, 9, 10, 11, 12, 13, 14, 15, 16	N/A	N/A

Simple familiar

1. WE4 Using the commission table shown, calculate the commission on the sale of a house for $945 000.

Sale price	Commission	Plus
Between $0 and $400 000	0.5% of sale price	0
Between $400 001 and $700 000	0.3% of the amount over $400 000	$2000 (0.5% of $400 000)
$700 001 and over	0.2% of the amount over $700 000	$2900 (0.5% of $400 000 + 0.3% of $300 000)

2. Using the commission table from question 1, calculate the commission for the following sales.
   a. $330 000 b. $525 000 c. $710 000 d. $1 330 000
3. Using the commission table from question 1, calculate the commission on the sale of a $500 000 house.
4. WE5 A car salesperson earns a salary of $1400 per month plus a commission of 3.5% of the total sales they make that month. In April, they sold cars worth $155 000. Calculate their total wage (salary + commission) for that month.
5. WE6 A shoemaker is paid $5.95 for each pair of running shoes they make. If the shoemaker made 235 pairs of shoes last week, calculate the amount they were paid.
6. If a software engineer gets paid $3.40 for every line of computer code they write, determine how much they will make if they write 865 lines in a fortnight.
7. A dressmaker gets paid per item they sew. If they received $545 for sewing 45 items, determine the amount they were paid per item.
8. WE7 Last year, a pop star sold $5 342 562 worth of music on their website and via music streaming services. Their contract calls for a 3.5% royalty on all sales. Determine the amount they earned last year.
9. If an author earned $25 560 on the basis of sales of $568 000, determine the royalty payment.

10. An actor is paid a 2.5% royalty plus $300 000 cash to act in the latest blockbuster movie.
   a. Complete the following table of royalties.

Time	Jan–Mar	Apr–Jun	Jul–Aug	Sep–Dec
Sales	$123 400	$2 403 556	$432 442	$84 562
Royalty payment				

   b. Calculate the total amount the actor received, including the cash.
11. To keep matters simple, the winner of a television singing competition receives a royalty of 1% on sales of their first album. If this year's winner sells $4 563 453 worth of music, determine the royalty payment.
12. A new children's book author is offered a choice of $100 000 cash or a 4% royalty on sales. Calculate the sales needed for the royalty offer to match the cash offer.
13. A hot-shot used-car salesperson earns $1500 per month plus 6.2% commission on all sales. If they sold $243 540 worth of cars last month, calculate their total wage.
14. A songwriter receives royalties on the sales of their songs. They received $11 473.75 in royalties for the sale of $458 950 in songs. Explain how the royalty percentage they receive can be determined, and hence calculate the royalty percentage they receive.

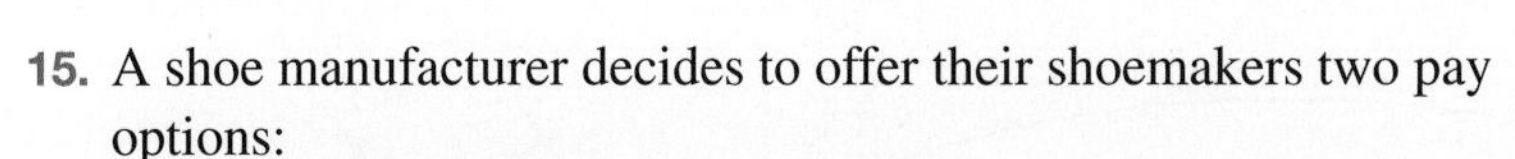

15. A shoe manufacturer decides to offer their shoemakers two pay options:
   Option 1: $8.25 per pair of shoes made
   Option 2: receive a commission of 3.5% on the shoes sold
   A shoemaker can make 15 pairs of shoes each day (over 5 days) and each pair of shoes sells for $75.95. In one week, the manufacturer will sell 250 pairs of shoes.
   Determine which option will earn the shoemaker the most money. Justify your answer using calculations.
16. A telecommunications salesperson receives a base wage of $450 per week plus 2.5% commission on all new telephone plans sold during the week. In one week, their earnings (before tax) were $502.25.

   a. Calculate the total amount (in dollars) of telephone plans they sold during the week.
   b. If each plan was $95, determine how many plans they sold.
   c. The salesperson is offered the choice to remain on a weekly base wage of $530 with an increase of 3.5% in commission on all sales, or to receive an annual salary of $37 500. Determine which option you would recommend they choose. Justify your answers using calculations.

Fully worked solutions for this chapter are available online.

LESSON
7.4 Convert between different pay periods [complex]

SYLLABUS LINKS

- Convert between annual, monthly, fortnightly, weekly and hourly rates of earning [complex].

7.4.1 Different pay periods

- Employees may get paid at different times; for example, they may be paid weekly, fortnightly or monthly.
- This has to be taken into account if you want to know your annual salary.

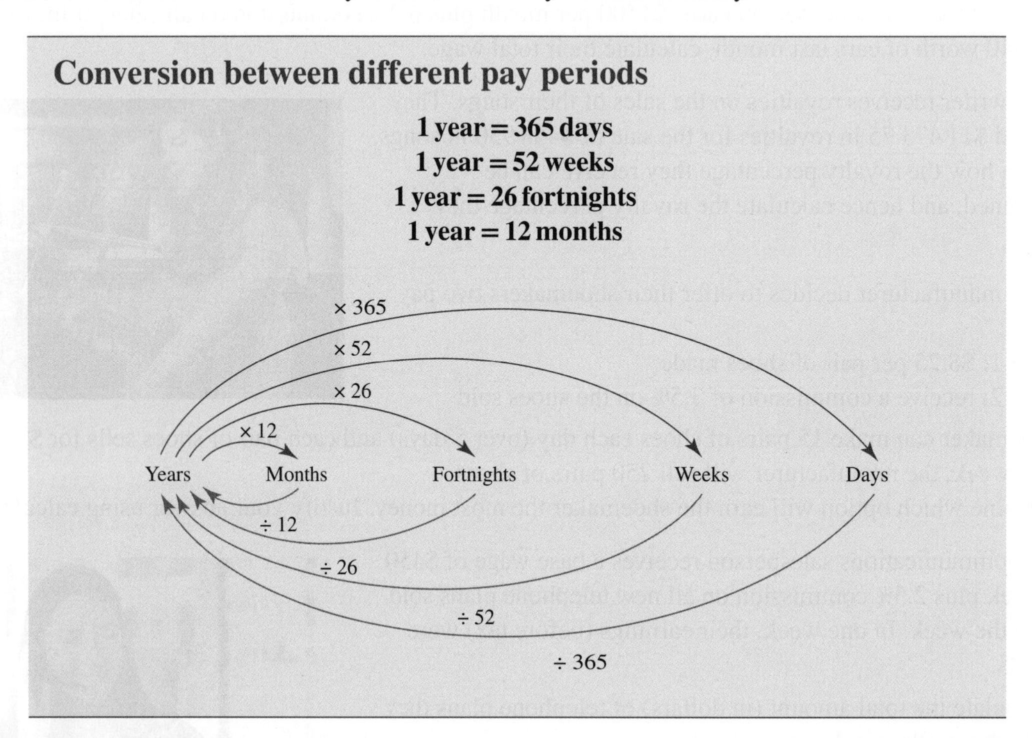

WORKED EXAMPLE 8 Solving problems of converting annual income

Phoenix earns an annual income of $64 500. To the nearest cent, convert this income to:

a. weekly income

b. fortnightly income

c. monthly income.

THINK	WRITE
a. There are 52 weeks in a year, so divide the annual income by 52.	a. Weekly income $= \frac{\$64\,500}{52} = \$1240.38$
Write the answer.	Phoenix has a weekly income of $1240.38.

b. There are 26 fortnights in a year, so divide the annual income by 26.

Write the answer.

b. Fortnightly income $= \frac{\$64\,500}{26} = \$2480.77$

Phoenix has a fortnightly income of $2480.77.

c. There are 12 months in a year, so divide the annual income by 12.

Write the answer.

c. Monthly income $= \frac{\$64\,500}{12} = \$5375.00$

Phoenix has a monthly income of $5375.00.

WORKED EXAMPLE 9 Solving problems of converting weekly income

Amy works in a pharmacy earning $554.25 a week. Calculate to the nearest cent their:
a. fortnightly income
b. annual income.

THINK	WRITE
a. There are 2 weeks in a fortnight, so multiply the weekly income by 2.	a. Fortnightly income $= \$554.25 \times 2$ $= \$1108.50$
Write the answer.	Amy has a fortnightly income of $1108.50.
b. There are 52 weeks in a year, so multiply the weekly income by 52.	b. Annual income $= \$554.25 \times 52$ $= \$28\,821.00$
Write the answer.	Amy has a fortnightly income of $28 821.00.

tlvd-11433

WORKED EXAMPLE 10 Solving problems of converting monthly income

Kyle works as a data analyst earning $7565.50 a month. Calculate to the nearest cent their:
a. annual income
b. fortnightly income
c. weekly income.

THINK	WRITE
a. There are 12 months in a year, so multiply the monthly income by 12.	a. Annual income $= \$7565.50 \times 12$ $= \$90786.00$
Write the answer.	Kyle has an annual income of $90 786.00.
b. Using the annual income, there are 26 fortnights in a year, so divide the annual income by 26.	b. Fortnightly income $= \$90\,786 \div 26$ $= \$3491.77$
Write the answer.	Kyle has a fortnightly income of $3491.77.
c. Using the annual income, there are 52 weeks in a year, so divide the annual income by 52.	c. Weekly income $= \$90\,786 \div 52$ $= \$1745.88$
Write the answer.	Kyle has a weekly income of $1745.88.

tlvd-11434

WORKED EXAMPLE 11 Calculating salary paid per fortnight [complex]

a. A bank employee earns a salary of $61 000 per year. Calculate their pay per fortnight.
b. A fast-food employee is paid $15.50 per hour. If they worked 72 hours last fortnight, calculate the amount they were paid.

THINK	WRITE
a. 1. Divide the annual salary by 26.07 and round to the nearest cent.	a. $\$61\,000 \div 26 \approx \$2346.15$
2. Write the answer as a sentence.	The bank employee's fortnightly pay is $2346.15.
b. 1. The worker is paid $15.50 for each hour of work, so multiply this amount by the number of hours worked.	b. $\$15.50 \times 72 = \$1116.00$
2. Write the answer as a sentence.	Last fortnight, the fast-food employee was paid $1116.00.

Exercise 7.4 Convert between different pay periods [complex] learnon

7.4 Quick quiz on | 7.4 Exercise

Simple familiar	Complex familiar	Complex unfamiliar
N/A	1, 2, 3, 4, 5, 6, 7, 8, 9, 10, 11, 12	13, 14

Complex familiar

1. Complete the following sentences:
   a. There are ____________ weeks in a year.
   b. There are ____________ fortnights in a year.
   c. There are ____________ days in a year.
2. Complete the following sentences:
   a. To convert an annual income to a monthly income, ____________ the annual income by ____________.
   b. To convert a weekly income to a fortnightly income, ____________ the weekly income by ____________.
3. WE8 Patricia earns an annual income of $86 250. To the nearest cent, convert this income to:
   a. weekly income b. fortnightly income c. monthly income.
4. Ryan earns an annual income of $62 660. To the nearest cent, convert this income to:
   a. weekly income b. fortnightly income c. daily income.
5. WE9 Aaron works in a bike shop earning $759.20 a week. Calculate to the nearest cent their:
   a. fortnightly income b. annual income.

6. Hung works part-time while studying at university. They earn $356.50 a week. Calculate to the nearest cent their:

   a. fortnightly income b. annual income c. monthly income.

7. **WE10** Claire works as a waitress. They earn $3568.50 per month. Calculate to the nearest cent their:

   a. annual income b. fortnightly income c. daily income.

8. Steve earns money playing NBL basketball. They earn $12 550 a month. Calculate to the nearest cent their:

   a. annual income
   b. fortnightly income
   c. weekly income
   d. earnings for each point, to the nearest cent, if on average they shot 20 points per game and they played one game each week.

9. Colleen works as a bookkeeper; they earn $6795 per month. Calculate their weekly income to the nearest cent.

10. Scott earns $80 per hour as a concreter. If he works 40 hours per week for 45 weeks of the year, calculate his monthly income.

11. A worker earns $20.40 an hour. They need to earn a minimum of $700 each week before tax to buy food, pay rent and bills, and have some money for entertainment. Calculate the minimum number of hours they have to work each week.
    Give your answer correct to 2 decimal places.

12. **WE11** Over the last four weeks, a person has worked 35, 36, 34 and 41 hours. If she earns $24.45 per hour, calculate the amount she earns for each of the two fortnights.

Complex unfamiliar

13. A factory worker receives an hourly rate of $28.40 to work a standard 38-hour week. They receive overtime of time and a half for any hour worked above 38 hours. On average, they work 42.5 hours each week.
    If the worker were offered a salary of $68 000, determine if they will be better off remaining on a wage or taking the salary offer.
    Justify your answer using calculations.

14. There is a proposal to have the same penalty rates for both Saturday and Sunday of time and a half. Currently workers receive time and a half to work on Saturdays and double time to work on Sundays.
    A waiter receives penalty rates and their hourly rate is $17.75.
    Calculate the percentage, to 2 decimal places, by which their hourly rate will need to increase to ensure they receive the same amount they currently earn for working both Saturdays and Sundays, assuming they work the same number of hours each day.

Fully worked solutions for this chapter are available online.

LESSON
7.5 Personal taxation, PAYG and employee superannuation

SYLLABUS LINKS

- Understand the purpose of taxation and the use of tax file numbers.
- Understand the purpose of superannuation.
- Interpret entries on a simple PAYG summary.

7.5.1 The purpose of taxation

- Taxation is a means by which the federal and state governments raise revenue for public services, welfare and community needs by imposing charges on citizens, organisations and businesses.
- Services include education, health, pensions for the elderly, unemployment benefits, public transport and much more.

Tax file numbers

- A tax file number (TFN) is a personal reference number for every tax-paying individual, company, fund and trust. Tax file numbers are valid for life and are issued by the Australian Taxation Office (ATO).

Pay as you go (PAYG) tax

- Pay as you go (PAYG) is a system whereby your employer withholds a certain amount of tax from your weekly, fortnightly, or monthly pay. This tax is then sent to the ATO.
- Tax is only required to be paid once you earn over $18 200 per annum.

WORKED EXAMPLE 12 Determining the tax bracket and tax on an income

Using the tax table shown, determine which tax bracket the following taxable income fits in and identify the tax on this taxable income.

Taxable income	Tax on this income
0–$18 200	Nil
$18 201–$45 000	19c for each $1 over $18 200
$45 001–$120 000	$5092 plus 32.5c for each $1 over $45 000
$120 001–$180 000	$29 467 plus 37c for each $1 over $120 000
$180 001 and over	$51 667 plus 45c for each $1 over $180 000

a. **$16 500** **b.** **$33 875** **c.** **$127 500**

THINK	WRITE
a. $16 500 fits in the 0–$18 200 bracket.	**a.** Tax is nil.
b. $33 875 fits in the $18 201–$45 000 bracket.	**b.** Tax is 19c for each dollar over $18 200.
c. $127 500 fits in the $120 001–$180 000 bracket.	**c.** Tax is $29 467 plus 37c for each $1 over $120 000.

- At the end of the financial year, your employer will provide you with a PAYG summary that shows the amount of tax you have paid in that financial year.

Calculating net pay

Net pay = gross pay − tax withheld

WORKED EXAMPLE 13 Interpreting entries on a simple PAYG summary

Referring to the PAYG summary shown, answer the following.

PAYG payment summary — individual non-business

Payment summary for year ending 30 June 2023

Payee details

Citizen, John
45 Commercial Dr
Shailer Park QLD 4128

NOTICE TO PAYEE

If this payment summary shows an amount in the total tax withheld box you must lodge a tax return. If no tax was withheld, you may still have to lodge a tax return.

For more information on whether you have to lodge, or about this payment and how it is taxed, you can:

- visit **www.ato.gov.au**
- refer to *Individual tax return Instructions*
- phone **13 28 61** between 8.00am and 6.00pm (EST), Monday to Friday

	Date/Month/Year		Date/Month/Year
Period of Payment	30/07/2022	to	30/06/2023

Payee's tax file number 123456782

TOTAL TAX WITHHELD $ 6,899

			Lump sum payments		Type
Gross payments	$	43,280	A $	0	
CDEP payments	$	0	B $	0	
Reportable fringe benefits amount FBT year 1 April to 21 March	$		D $	0	
Reportable employer superannuation contributions	$	13,100	E $	0	
Total allowances	$				

Total allowances are not included in the gross payments above. The amount needs to be shown separately in your tax return.

a. **Identify the TFN.**
b. **Calculate the tax withheld.**
c. **Calculate the net pay.**

THINK	WRITE
a. On the PAYG summary, look for 'Payee's tax file number'.	a. TFN is 123456782.
b. On the PAYG summary, look for the 'Total tax withheld' box.	b. Tax withheld is \$6899.
c. Use the formula net pay = gross pay − tax withheld, where gross pay = \$43 280.	c. Net pay = \$43 280 − \$6899 = \$36 381

7.5.2 Superannuation

- All workers in Australia get an additional sum of money for their retirement. This sum is called **superannuation**.
- In 2024, the law requires employers to pay an additional 11.5% of annual salary into a recognised superannuation fund.
- The amount is based on usual earnings before tax and is calculated at each pay cycle.
- Some workers choose to contribute additional funds from their wages to increase their superannuation.
- There are tax incentives, such as paying lower tax rates on superannuation lump sums, to encourage workers to save for their retirement.

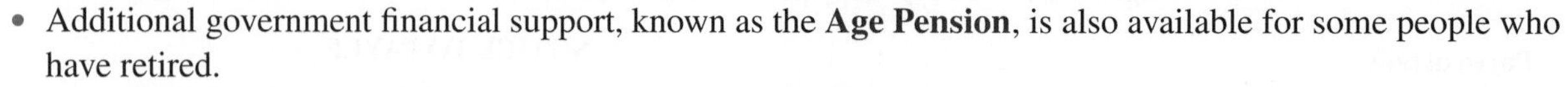

- Additional government financial support, known as the **Age Pension**, is also available for some people who have retired.
- The amount a retired person receives in the pension depends on their personal financial security and wealth.

tlvd-11435

WORKED EXAMPLE 14 Calculating the amount of superannuation paid by the employer each fortnight

A worker's hourly rate is $28.75 and he works 38 hours each week. The worker is paid fortnightly. Calculate the amount of superannuation the employer pays on his behalf each fortnight.

THINK	WRITE
1. Calculate the fortnightly wage. Note that if the employee works 38 hours per week, he works 76 (38 × 2) hours per fortnight.	$76 \times \$28.75 = \$2185.00$
2. Calculate 11.5% of his fortnightly wage. Remember to convert 11.5% to a decimal by dividing by 100.	$11.5\% \times \$2185 = 0.115 \times \$2185$ $\approx \$251.28$
3. Write the answer as a sentence.	The employer pays $251.28 into the worker's superannuation fund each fortnight.

7.5.3 Annual leave loading

- Some workers will receive an extra payment on top of the 4-week annual leave pay.
- It is usually 17.5% of their normal pay for 4 weeks.
- It will depend on the working agreement workers have with their employers if they receive an **annual leave loading**.
- The purpose of annual leave loading is to compensate workers who are unable to earn additional money through overtime while on leave.

tlvd-4811

WORKED EXAMPLE 15 Calculating the amount paid for annual leave loading before tax

A worker's annual salary is $58 056. Calculate the amount, in dollars, paid for annual leave loading before tax.

THINK	WRITE
1. Calculate the weekly salary by dividing the annual salary by 52.14.	$\frac{\$58\,056}{52.14} \approx \$1113.46$
2. Calculate the wage for 4 weeks.	$\$1113.46 \times 4 = \$4453.84$
3. Calculate 17.5% of the 4-week wage. Remember to convert 17.5% to a decimal by dividing by 100.	$17.5\% \times \$4453.84 = 0.175 \times \$4453.84$ $\approx \$779.42$
4. Write the answer as a sentence.	The amount paid for annual leave loading, before tax, is $779.42.

Resources

Interactivity Special rates (int-6068)

Exercise 7.5 Personal taxation, PAYG and employee superannuation

learnon

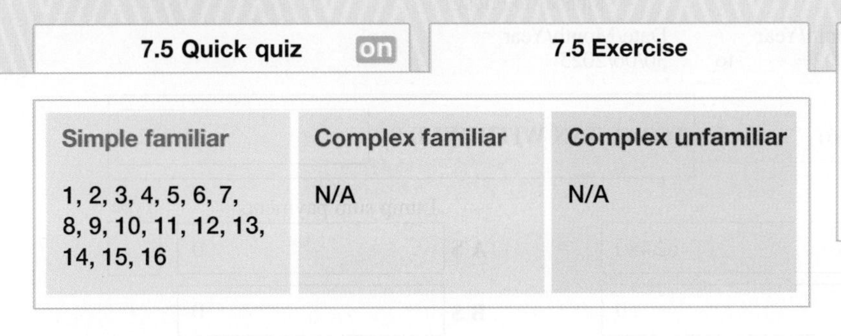

7.5 Quick quiz on | 7.5 Exercise

Simple familiar	Complex familiar	Complex unfamiliar
1, 2, 3, 4, 5, 6, 7, 8, 9, 10, 11, 12, 13, 14, 15, 16	N/A	N/A

These questions are even better in jacPLUS!

- Receive immediate feedback
- Access sample responses
- Track results and progress

Find all this and MORE in jacPLUS

Simple familiar

1. WE12 Using the tax table shown, determine which tax bracket the following taxable income fits in and identify the tax on this taxable income.

Taxable income	Tax on this income
0–$18 200	Nil
$18 201–$45 000	19c for each $1 over $18 200
$45 001–$120 000	$5092 plus 32.5c for each $1 over $45 000
$120 001–$180 000	$29 467 plus 37c for each $1 over $120 000
$180 001 and over	$51 667 plus 45c for each $1 over $180 000

a. $14 895

b. $42 159

c. $133 500

2. Using the tax table shown, determine which tax bracket the following taxable income fits in and identify the tax on this taxable income.

Taxable income	Tax on this income
0–$18 200	Nil
$18 201–$45 000	19c for each $1 over $18 200
$45 001–$120 000	$5092 plus 32.5c for each $1 over $45 000
$120 001–$180 000	$29 467 plus 37c for each $1 over $120 000
$180 001 and over	$51 667 plus 45c for each $1 over $180 000

a. $18 000
b. $99 550
c. $220 845

3. WE13 Referring to the PAYG summary shown, answer the following.

PAYG payment summary — individual non-business

Payment summary for year ending 30 June 2023

Payee details

Citizen, John
45 Commercial Dr
Shailer Park QLD 4128

NOTICE TO PAYEE

If this payment summary shows an amount in the total tax withheld box you must lodge a tax return. If no tax was withheld, you may still have to lodge a tax return.

For more information on whether you have to lodge, or about this payment and how it is taxed, you can:

- visit **www.ato.gov.au**
- refer to *Individual tax return Instructions*
- phone **13 28 61** between 8.00am and 6.00pm (EST), Monday to Friday

	Date/Month/Year		Date/Month/Year
Period of Payment	30/07/2022	to	30/06/2023

Payee's tax file number 562434961

TOTAL TAX WITHHELD $ 9 272

			Lump sum payments	Type
Gross payments	$ 62 480	A $	0	
CDEP payments	$ 0	B $	0	
Reportable fringe benefits amount FBT year 1 April to 21 March	$	D $	0	
Reportable employer superannuation contributions	$ 656040	E $	0	
Total allowances	$			

Total allowances are not included in the gross payments above. The amount needs to be shown separately in your tax return.

a. Identify the TFN.
b. Calculate the tax withheld.
c. Calculate the net pay.

4. Referring to the PAYG summary shown, answer the following.

PAYG payment summary — individual non-business

Payment summary for year ending 30 June 2023

Payee details

Citizen, John
45 Commercial Dr
Shailer Park QLD 4128

NOTICE TO PAYEE

If this payment summary shows an amount in the total tax withheld box you must lodge a tax return. If no tax was withheld, you may still have to lodge a tax return.

For more information on whether you have to lodge, or about this payment and how it is taxed, you can:

- visit **www.ato.gov.au**
- refer to *Individual tax return Instructions*
- phone **13 28 61** between 8.00am and 6.00pm (EST), Monday to Friday

	Date/Month/Year		Date/Month/Year
Period of Payment	30/07/2022	to	30/06/2023

Payee's tax file number 123456782 — **TOTAL TAX WITHHELD $** 19 787

			Lump sum payments	Type
Gross payments	$ 92 748	A $	0	
CDEP payments	$ 0	B $	0	
Reportable fringe benefits amount FBT year 1 April to 21 March	$	D $	0	
Reportable employer superannuation contributions	$	E $	0	
Total allowances	$		Total allowances are not included in the gross payments above. The amount needs to be shown separately in your tax return.	

a. Calculate the tax withheld.
b. Calculate the net pay.
c. Determine how much the employer would pay in compulsory superannuation.

5. Referring to the PAYG summary shown, answer the following.

PAYG payment summary — individual non-business

Payment summary for year ending 30 June 2023

Payee details

Citizen, John
45 Commercial Dr
Shailer Park QLD 4128

NOTICE TO PAYEE

If this payment summary shows an amount in the total tax withheld box you must lodge a tax return. If no tax was withheld, you may still have to lodge a tax return.

For more information on whether you have to lodge, or about this payment and how it is taxed, you can:

- visit **www.ato.gov.au**
- refer to *Individual tax return Instructions*
- phone **13 28 61** between 8.00am and 6.00pm (EST), Monday to Friday

	Date/Month/Year		Date/Month/Year
Period of Payment	30/07/2022	to	30/06/2023

Payee's tax file number 987543211

TOTAL TAX WITHHELD $ 24 823

		Lump sum payments		Type
Gross payments	$ 105246	A $	0	
CDEP payments	$ 0	B $	0	
Reportable fringe benefits amount FBT year 1 April to 21 March	$	D $	0	
Reportable employer superannuation contributions	$ 14 500	E $	0	
Total allowances	$			

Total allowances are not included in the gross payments above. The amount needs to be shown separately in your tax return.

a. Identify the TFN.
b. Calculate the tax withheld.
c. Calculate the net pay.
d. Determine how much over the compulsory superannuation was contributed to the superannuation in this financial year.

6. Calculate 11.5% of the following yearly salaries.

a. $15 650
b. $25 902
c. $46 890
d. $78 465
e. $99 560
f. $104 240

7. If Xavier earns $76 500 p.a. calculate how much superannuation they earn in that year.

8. An employee is paid $85 500 before tax. Determine how much superannuation they earned that year.

9. Fredrick earns $9560.50 per month. Determine how much superannuation they earn that year.

10. WE14 A worker's hourly rate is $29.45 and they work 25.75 hours each week. The worker is paid fortnightly. Calculate the amount of superannuation the employer pays on her behalf each fortnight.

11. A school principal earns an annual salary of $155 750.
    a. Calculate the amount they earn each month.
    b. Calculate their superannuation fund payment if the fund is paid 11.5% of their annual salary.
12. A salary earner makes $62 000 per year.
    a. Calculate the amount they earned each month.
    b. Calculate their superannuation fund payment each month if they receive 11.5% superannuation.
    c. Calculate the total amount deposited into the superannuation fund in a year.
13. **WE15** A worker's annual salary is $85 980. Calculate the amount, in dollars, paid for annual leave loading before tax.
14. An employee earns $425 per week. Determine how much leave loading they would earn that year.
15. Leah earns annual leave loading. If they are paid $64 000 p.a. determine the amount of leave loading they earn that year to the nearest cent.
16. Some employers offer a superannuation bonus scheme. They pay you 11.5%, then for every additional 1% of your salary that you contribute, they match it with a further 1%. For example, if you contribute 2% of your salary to superannuation, you will receive a total of 11.5% + 2% + 2% = 15.5% into your fund.
    A software salesperson makes $70 000 per year and decides to contribute 3% of their salary into their superannuation.
    a. Calculate the salesperson's contribution from their own salary into their superannuation each year.
    b. The salesperson's employer matches their superannuation contribution. In total, calculate the amount they receive into their superannuation fund each year.

Fully worked solutions for this chapter are available online.

LESSON
7.6 Determining PAYG tax using tax tables [complex]

SYLLABUS LINKS

- Use tax tables to determine PAYG tax for periodic (weekly/fortnightly/monthly) earnings [complex].

7.6.1 Tax tables

- Tax tables are produced by the Australian Taxation Office (ATO) to help employers determine how much tax needs to be withheld each pay.
- The ATO provide tables for weekly, fortnightly, and monthly pay periods.

WORKED EXAMPLE 16 Interpreting a tax table (weekly)

Using the table shown, determine the amount of tax that must be withheld if an individual is paid $825 per week.

Weekly earnings ($)	Amount to be withheld ($)
816.00	119.00
817.00	119.00
818.00	119.00
819.00	120.00
820.00	120.00
821.00	120.00
822.00	121.00
823.00	121.00
824.00	121.00
825.00	122.00
826.00	122.00
827.00	122.00
828.00	123.00
829.00	123.00
830.00	123.00

Weekly earnings ($)	Amount to be withheld ($)
831.00	124.00
832.00	124.00
833.00	125.00
834.00	125.00
835.00	125.00
836.00	126.00
837.00	126.00
838.00	126.00
839.00	127.00
840.00	127.00
841.00	127.00
842.00	128.00
843.00	128.00
844.00	128.00
845.00	129.00

THINK	WRITE
Using the tax table, look for the row where weekly earnings are $825.	The tenth line of the left side of the table is for weekly earnings of $825.
Read the table to see how much weekly tax needs to be withheld.	$122.00

WORKED EXAMPLE 17 Calculating net pay by subtracting the tax withheld

A factory worker is paid $19.94 per hour. Each week they work 38 hours, with an additional 4 hours overtime paid at time and a half. Using the following taxation table, calculate their net pay (the amount they receive each week).

Weekly earnings ($)	Tax withheld ($)
856.00	133.00
857.00	133.00
858.00	133.00
859.00	134.00
860.00	134.00
861.00	134.00
862.00	135.00
863.00	135.00
864.00	135.00
865.00	136.00
866.00	136.00
867.00	136.00
868.00	137.00
869.00	137.00
870.00	137.00
871.00	138.00
872.00	138.00
873.00	138.00
874.00	139.00
875.00	139.00
876.00	139.00
877.00	140.00
878.00	140.00

Weekly earnings ($)	Tax withheld ($)
879.00	141.00
880.00	141.00
881.00	141.00
882.00	142.00
883.00	142.00
884.00	142.00
885.00	143.00
886.00	143.00
887.00	143.00
888.00	144.00
889.00	144.00
890.00	144.00
891.00	145.00
892.00	145.00
893.00	145.00
894.00	146.00
895.00	146.00
896.00	146.00
897.00	147.00
898.00	147.00
899.00	147.00
900.00	148.00

THINK	WRITE
1. Calculate the weekly wage for normal hours.	$38 \times \$19.94 = \$757.72$
2. Calculate the overtime.	$\$19.94 \times 1.5 \times 4 = \$119.64$
3. Calculate the total weekly wage.	$\$757.72 + \$119.64 = \$877.36$
4. Using the table, locate the weekly wage and read the amount of tax withheld.	Weekly wage is $877; tax withheld is $140.
5. Calculate the net pay by subtracting the tax withheld from the total wages.	Weekly income $= \$877.36 - \$140$ $= \$737.36$

7.6 Quick quiz on	7.6 Exercise

Simple familiar	Complex familiar	Complex unfamiliar
N/A	1, 2, 3, 4, 5, 6	N/A

Complex familiar

1. WE16 Using the tax table shown, determine the required amount of tax to be withheld from an employee earning the following weekly pay.

Weekly earnings ($)	Amount to be withheld ($)
816.00	119.00
817.00	119.00
818.00	119.00
819.00	120.00
820.00	120.00
821.00	120.00
822.00	121.00
823.00	121.00
824.00	121.00
825.00	122.00
826.00	122.00
827.00	122.00
828.00	123.00
829.00	123.00
830.00	123.00

Weekly earnings ($)	Amount to be withheld ($)
831.00	124.00
832.00	124.00
833.00	125.00
834.00	125.00
835.00	125.00
836.00	126.00
837.00	126.00
838.00	126.00
839.00	127.00
840.00	127.00
841.00	127.00
842.00	128.00
843.00	128.00
844.00	128.00
845.00	129.00

a. $835　　b. $844　　c. $837　　d. $833

2. Using the table shown, determine the amount of tax withheld from the following fortnightly pay.

Fortnightly earnings ($)	Tax withheld ($)	Fortnightly earnings ($)	Tax withheld ($)
2682.00	602.00	2710.00	612.00
2684.00	602.00	2712.00	612.00
2686.00	604.00	2714.00	614.00
2688.00	604.00	2716.00	614.00
2690.00	604.00	2718.00	614.00
2692.00	606.00	2720.00	616.00
2694.00	606.00	2722.00	616.00
2696.00	606.00	2724.00	616.00
2698.00	608.00	2726.00	618.00
2700.00	608.00	2728.00	618.00
2702.00	608.00	2730.00	618.00
2704.00	610.00	2732.00	620.00
2706.00	610.00	2734.00	620.00
2708.00	610.00	2736.00	620.00

a. $2710 b. $2720 c. $2698 d. $2684

3. Consider the table shown.

Monthly earnings ($)	Tax withheld ($)	Monthly earnings ($)	Tax withheld ($)
4792.67	949.00	4866.33	975.00
4797.00	953.00	4870.67	979.00
4801.33	953.00	4875.00	979.00
4805.67	958.00	4879.33	979.00
4810.00	958.00	4883.67	984.00
4814.33	958.00	4888.00	984.00
4818.67	962.00	4892.33	984.00
4823.00	962.00	4896.67	988.00
4827.33	962.00	4901.00	988.00
4831.67	966.00	4905.33	988.00
4836.00	966.00	4909.67	992.00
4840.33	966.00	4914.00	992.00
4844.67	971.00	4918.33	997.00
4849.00	971.00	4922.67	997.00
4853.33	971.00	4927.00	997.00
4857.67	975.00	4931.33	1001.00
4862.00	975.00	4935.67	1001.00

i. Calculate the amount of tax to be withheld from the following monthly pay.

a. $4901 b. $4875 c. $4823

ii. If an employee had the following monthly tax withheld, determine their approximate monthly earnings.

a. $962 b. $975 c. $1001

4. WE17 Consider the following table.

Weekly earnings ($)	Amount to be withheld ($)
816.00	119.00
817.00	119.00
818.00	119.00
819.00	120.00
820.00	120.00
821.00	120.00
822.00	121.00
823.00	121.00
824.00	121.00
825.00	122.00
826.00	122.00
827.00	122.00
828.00	123.00
829.00	123.00
830.00	123.00

Weekly earnings ($)	Amount to be withheld ($)
831.00	124.00
832.00	124.00
833.00	125.00
834.00	125.00
835.00	125.00
836.00	126.00
837.00	126.00
838.00	126.00
839.00	127.00
840.00	127.00
841.00	127.00
842.00	128.00
843.00	128.00
844.00	128.00
845.00	129.00

If Chloe earned $822 a week working as a waitress, use the table to calculate:

a. Chloe's annual earnings

b. the yearly tax withheld by her employer

c. Chloe's annual net earnings.

5. Using the table shown, determine the amount of tax withheld from the following weekly pay.

Weekly earnings ($)	Amount to be withheld ($)
816.00	119.00
817.00	119.00
818.00	119.00
819.00	120.00
820.00	120.00
821.00	120.00
822.00	121.00
823.00	121.00
824.00	121.00
825.00	122.00
826.00	122.00
827.00	122.00
828.00	123.00
829.00	123.00
830.00	123.00

Weekly earnings ($)	Amount to be withheld ($)
831.00	124.00
832.00	124.00
833.00	125.00
834.00	125.00
835.00	125.00
836.00	126.00
837.00	126.00
838.00	126.00
839.00	127.00
840.00	127.00
841.00	127.00
842.00	128.00
843.00	128.00
844.00	128.00
845.00	129.00

a. $1650 b. $1670 c. $1690

6. Using the table, calculate the amount of tax withheld from the following monthly pay.

Monthly earnings ($)	Tax withheld ($)
4792.67	949.00
4797.00	953.00
4801.33	953.00
4805.67	958.00
4810.00	958.00
4814.33	958.00
4818.67	962.00
4823.00	962.00
4827.33	962.00
4831.67	966.00
4836.00	966.00
4840.33	966.00
4844.67	971.00
4849.00	971.00
4853.33	971.00
4857.67	975.00
4862.00	975.00
4866.33	975.00

Monthly earnings ($)	Tax withheld ($)
4870.67	979.00
4875.00	979.00
4879.33	979.00
4883.67	984.00
4888.00	984.00
4892.33	984.00
4896.67	988.00
4901.00	988.00
4905.33	988.00
4909.67	992.00
4914.00	992.00
4918.33	997.00
4922.67	997.00
4927.00	997.00
4931.33	1001.00
4935.67	1001.00
4940.00	1001.00

a. $58 760.04 b. $59 124 c. $57 615.96

Fully worked solutions for this chapter are available online.

LESSON
7.7 Determining taxable income by applying tax deductions [complex]

SYLLABUS LINKS

- Apply the concepts of taxable income, gross income, allowable deductions and levies in simple contexts [complex].

7.7.1 Income tax

- **Income tax** is a tax levied on people's financial income. It is deducted from each fortnightly or monthly pay.
- The amount of income tax is based on **total income** and **tax deductions**, which determine a worker's **taxable income**.

7.7.2 Tax deductions

- Workers who spend their own money for work-related expenses are entitled to claim the amount spent as tax deductions.
- The deductions are subtracted from the taxable income, which lowers the amount of money earned and hence reduces the amount of tax to be paid.
- What can be claimed as tax deductions is determined by the Australian Taxation Office.
- Some examples of deductions that can be claimed include using your car to travel to work-related events, purchasing materials, union fees, donations made to charities and using a home office.
- Deductions must be work-related expenses and evidence, such as receipts, must be provided.

Calculating taxable income

Taxable income = total income − tax deductions

WORKED EXAMPLE 18 Determining taxable income

An employee has a total income of $65 650 p.a. They are able to claim $2430 worth of deductions. Determine their taxable income.

THINK	WRITE
Taxable income = total income − deductions Substitute the relevant values in the formula.	Taxable income = total income − deductions = $65 650 − $2430 = $63 220
Write the answer.	The taxable income is $63 220.

7.7.3 The Medicare levy

- Australian residents have access to health care through Medicare, which is partly funded by taxpayers through the payment of a **Medicare levy**.
- The rate of Medicare levy is dependent on taxable income.
- The standard rate is 2%. This is reduced for people with a taxable income below a certain amount, and increased (via a surcharge) for high-income earners who do not have private health insurance.

Calculating total tax payable

Total tax payable = tax paid + Medicare levy

WORKED EXAMPLE 19 Calculating the payable amount of Medicare levy

A plumber's taxable income for the financial year is \$65 850. They claim \$5680 in work-related expenses (deductions) and they also have private health insurance.
Determine the amount they have to pay for the Medicare levy.

THINK	WRITE
1. Calculate the plumber's taxable income.	Taxable income = \$65 850 − \$5680 = \$60 170
2. As the plumber has private health insurance, no surcharge is applicable. Calculate 2% of the taxable income. Remember to convert 2% to a decimal.	2% × \$60 170 = 0.02 × \$60 170 = \$1203.40
3. Write the answer as a sentence.	They will pay \$1203.40 for the Medicare levy.

Exercise 7.7 Determining taxable income by applying tax deductions [complex]

7.7 Quick quiz on | 7.7 Exercise

Simple familiar	Complex familiar	Complex unfamiliar
N/A	1, 2, 3, 4, 5, 6, 7, 8, 9, 10, 11, 12	13, 14

Complex familiar

1. **WE18** An employee has a total income of \$72 000 p.a. They will claim \$1850 worth of deductions that year. Determine their taxable income.
2. James has a yearly income of \$89 250. They are able to claim \$3100 worth of deductions that year. Determine their taxable income.
3. Ronda earned \$98 450 this year and was able to claim travel deductions of \$850 and work-related expenses of \$350. Calculate their taxable income.
4. Evanka earned \$120 000 this year and was able to claim travel deductions of \$520 and work-related expenses of \$220. Calculate their taxable income.
5. Jack was paid \$105 670 this financial year and was able to claim a deduction of \$720 for travel and \$520 for their uniform. Calculate their taxable income.
6. If an employee was earning a salary of \$2200 per week and for that financial year they were able to claim \$5200 in deductions, calculate their taxable income.
7. If Jacky was earning \$13 250 per month during the financial year and was able to claim \$375 per week of deductions, calculate their taxable income in that financial year.
8. **WE19** Calculate the Medicare levy for the following taxable incomes. Assume there is no surcharge applicable.
   a. \$60 400 b. \$77 300 c. \$89 400 d. \$108 423
9. A shop worker earns \$18.95 per hour. They work 52 hours in a fortnight. Calculate the amount of Medicare levy (no surcharge) they will be expected to pay at the end of the financial year.
   Note: Assume there are 26.07 fortnights in a year.
10. If an employee has private health insurance and earns \$72 450 p.a. with no deductions, calculate their Medicare levy.
11. Frances was paid \$99 725 this financial year and was able to claim a deduction of \$835 for travel and \$325 for their uniform. Calculate their Medicare levy assuming they have private health insurance.
12. If an employee was earning a salary of \$2125 per week and, for that financial year, they were able to claim \$7300 in deductions, calculate their Medicare levy assuming they had private health insurance.

Complex unfamiliar

13. Linda earns \$2350 per week and has private health insurance. Determine the amount of tax deductions she had in the financial year if she had to pay \$2340 in Medicare levy.
14. Stephen pays a Medicare levy of \$1924 and has private health insurance. If he claimed \$3800 tax deduction the same year, calculate his total income for this financial year.

Fully worked solutions for this chapter are available online.

LESSON
7.8 Calculating simple tax returns [complex]

SYLLABUS LINKS

- Calculate a simple income tax return and net income using current income tax rates [complex].

7.8.1 Tax paid on taxable income

- The calculation of income tax is based upon an income tax table. The income tax table at the time of writing is:

Taxable income	Tax on this income
0–$18 200	Nil
$18 201–$45 000	19c for each $1 over $18 200
$45 001–$120 000	$5092 plus 32.5c for each $1 over $45 000
$120 001–$180 000	$29 467 plus 37c for each $1 over $120 000
$180 001 and over	$51 667 plus 45c for each $1 over $180 000

Note: The income tax table is subject to change.

WORKED EXAMPLE 20 Calculating income tax when there are no deductions

A person earns $76 500. Determine the amount of tax they should pay if they have no deductions.

THINK	WRITE
1. Taxable income = total income − deductions Substitute the relevant values into the formula.	Taxable income = total income − deductions = $76 500 − 0 = $76 500
2. Determine the tax bracket they fit into.	$76 500 fits within the $45 001–$120 000 bracket.
3. Determine the difference between $76 500 and $45 000.	Difference = $76 500 − $45 000 = $31 500
4. Determine the tax payable on $76 500, using the tax table.	Tax payable = $5092 + (0.325 × $31 500) = $5092 + $10 237.50 = $15 329.50
5. Write the answer.	The amount of tax payable is $15 329.50.

tlvd-4812

WORKED EXAMPLE 21 Calculating the income tax payable

Calculate the income tax payable by a teacher who earns a salary of $67 400. Tax deductions for the teacher are $4240 and they earned $1240 in bank interest.

THINK	WRITE
1. Determine the total income by adding the salary and bank interest together.	$67 400 + $1240 = $68 640
2. Subtract the tax deductions from the total income.	Taxable income = $68 640 − $4240 = $64 400
3. Determine the tax bracket based on the taxable income of $64 400.	Taxable income is between $45 000 and $120 000.
4. Determine the amount of tax to be paid.	$5092 plus 32.5c for every $1 over $45 000
5. Determine the amount of taxable income over $45 000 by subtracting $45 000 from the teacher's taxable income.	$64 400 − $45 000 = $19 400
6. Determine the tax rate amount as 32.5% of $19 400. Remember to convert 32.5% to a decimal.	0.325 × $19 400 = $6305
7. Add the 'plus amount' to calculate the payable tax.	$5092 + $6305 = $11 397
8. Write the answer as a sentence.	The total tax payable by the teacher is $11 397.

- At the end of the financial year, workers submit a tax return which lists their deductions and all money earned.
  - If the total amount of tax paid over the year is less than is required, the worker will have to pay the difference.
  - If the amount of tax paid over the year is more than is required, the worker receives a refund.
- Australian workers who provide their tax file number and do not work another job claim the tax-free threshold. This means that they do not pay tax on the first $18 200 earned in the financial year.

Calculating tax refund or debt

Tax refund or debt = PAYG tax − total tax payable

7.8.2 Calculating tax payable

- Calculating the tax payable on gross income can be done using a spreadsheet and inserting formulas that determine the amount of tax payable for each tax tier.

WORKED EXAMPLE 22 Calculating the amount of tax payable on gross income

A law clerk's annual salary is $45 675. She claims $450 in tax deductions and receives $250 in interest from investments. Using a spreadsheet, calculate the amount of tax payable for the financial year by using the tax table provided.

THINK / **WRITE**

1. Create a spreadsheet with headings 'Gross income', 'Interest', 'Deductions' and 'Taxable income'. Enter the values.

	A	B	C	D
1	Gross income	Interest	Deductions	Taxable income
2	45 675	250	450	
3				
4				
5				
6				

2. In cell D2 calculate the taxable income: add the interest ($250) to the annual salary ($45 675) and subtract the deductions ($450) by writing the formula $= A2 + B2 - C2$.

D2 *fx* = A2 + B2 − C2

	A	B	C	D
1	Gross income	Interest	Deductions	Taxable income
2	45 675	250	450	45 475
3				
4				
5				
6				

3. Insert a column to represent the tax bracket that the law clerk falls into.

	A	B	C	D	E
1	Gross income	Interest	Deductions	Taxable income	Tax tier (45 001–120 000)
2	45 675	250	450	45 475	
3					
4					
5					
6					

4. Insert a formula to calculate the tax payable for this tax tier.

Type the following formula into E2:
$= 5092 + (D2 - 45\,000)^{*}\ 0.325$

5. Calculate the tax payable.

	A	B	C	D	E
1	Gross income	Interest	Deductions	Taxable income	Tax tier (45 001–120 000)
2	45 675	250	450	45 475	5246.375
3					
4					
5					
6					

6. Write the answer to the nearest cent.

The tax payable by the law clerk is $5246.38.

7.8.3 Preparing a wage sheet for tax purposes

- A wage sheet shows a list of workers and details of their earnings.
- These include net wages, pay rates, gross deductions and allowances.
- A wage sheet may look similar to the one shown.

Employee	Pay rate ($)	Normal hours worked	Overtime	Penalty rate		Allowance	Gross pay ($)	Tax withheld ($)	Net pay ($)
			1.5	1.5	2				
Dave	19.50	38	2	5		25.5	945.75	163.00	782.75
Jose	21.80	32	4	3.5	5		1253.50	271.00	982.50
Neeve	25.70	25		4		15	1002.30	183.00	819.30
Niha	22.45	38				10	853.10	131.00	722.10
Yen	29.85	36			6		1074.60	208.00	866.60

- A pay slip (or pay sheet) shows individual workers' details, including the hours worked, pay rate, net wage, tax deduction, leave days and superannuation paid.
- A wage sheet can be set up in a spreadsheet containing formulas that perform the necessary calculations.

WORKED EXAMPLE 23 Using a wage sheet

Using a spreadsheet, complete the following wage sheet.

Employee	Pay rate ($)	Normal hours worked	Overtime	Penalty rate		Allowance ($)	Gross pay ($)	Tax withheld ($)	Net pay ($)
			1.5	1.5	2				
Honura	**38.95**	**38**	**4**		**6**	**15**		**620**	
Skye	**28.40**	**38**	**3**	**5**				**289**	
Beau	**19.15**	**35**		**5**				**234**	

THINK | **WRITE**

1. Set up a spreadsheet.

	A	B	C	D	E	F	G	H	I	J
1			Normal						Tax	
2		Pay	hours	Overtime	Penalty rate		Allowance	Gross	with-	Net
3	Employee	rate ($)	worked	1.5	1.5	2	($)	pay ($)	held ($)	pay ($)
4	Honura	38.95	38	4		6	15		620	
5	Skye	28.40	38	3	5				289	
6	Beau	19.15	35		5				234	

2. Calculate the pay for hours worked:
Pay for normal hours formula (e.g. 38*38.95)
B4*C4

Overtime pay calculation (e.g. 4*1.5*38.95)
D4*D3*B4

Pay for working on penalty rates (e.g. 0*1.5*38.95 + 6*2*38.95).
E4*E3*B4 + F4*F3*B4

3. Calculate gross pay (= hours worked plus allowances).

In cell H4 enter the following formula:
H4 = B4*C4 + D4*D3*B4 + E4*E3*B4 + F4*F3*B4 + G4

4. Complete the table by filling the formula in column H down for all workers.

	A	B	C	D	E	F	G	H	I	J
1										
2				Overtime	Penalty rate					
3	Employee	Pay rate ($)	Normal hours worked	1.5	1.5	2	Allowance ($)	Gross pay ($)	Tax with-held ($)	Net pay ($)
4	Honura	38.95	38	4		6	15	2196.20	620	
5	Skye	28.40	38	3	5			1420	289	
6	Beau	19.15	35		5			813.88	234	

5. Calculate the net pay by deducting tax withheld.

In cell J4 type: = H4 – I4

6. Write the answer.

	A	B	C	D	E	F	G	H	I	J
1										
2				Overtime	Penalty rate					
3	Employee	Pay rate ($)	Normal hours worked	1.5	1.5	2	Allowance ($)	Gross pay ($)	Tax with-held ($)	Net pay ($)
4	Honura	38.95	38	4		6	15	2196.20	620	1576.20
5	Skye	28.40	38	3	5			1420	289	1131
6	Beau	19.15	35		5			813.88	234	579.88

Exercise 7.8 Calculating simple tax returns [complex]

learnon

7.8 Quick quiz on | 7.8 Exercise

Simple familiar	Complex familiar	Complex unfamiliar
N/A	1, 2, 3, 4, 5, 6, 7, 8, 9, 10, 11, 12, 13, 14, 15	16, 17, 18, 19, 20

Complex familiar

1. WE20 A person earns $83 000. Determine the amount of tax they should pay if they have no deductions.
2. An employee earns $122 000. Determine the amount of tax they should pay if they have no deductions.
3. WE21 Calculate the income tax payable by a baker whose salary is $51 260. The baker's tax deductions are $2120. The baker earned $1850 in bank interest.

4. Using the taxation table provided:

Taxable income	Tax on this income
0–$18 200	Nil
$18 201–$45 000	19c for each $1 over $18 200
$45 001–$120 000	$5092 plus 32.5c for each $1 over $45 000
$120 001–$180 000	$29 467 plus 37c for each $1 over $120 000
$180 001 and over	$51 667 plus 45c for each $1 over $180 000

i. determine the tax bracket the following annual salaries fall into
ii. hence, calculate the amount of tax payable.

a. $37 500 b. $15 879 c. $85 670 d. $131 000

5. A nurse's gross annual salary is $58 284 and they are paid monthly.

a. Calculate their gross monthly salary.
b. Using the tax table shown, determine the amount of tax withheld from their monthly salary.

Monthly earnings ($)	Tax withheld ($)
4792.67	949.00
4797.00	953.00
4801.33	953.00
4805.67	958.00
4810.00	958.00
4814.33	958.00
4818.67	962.00
4823.00	962.00
4827.33	962.00
4831.67	966.00
4836.00	966.00
4840.33	966.00
4844.67	971.00
4849.00	971.00
4853.33	971.00
4857.67	975.00
4862.00	975.00
4866.33	975.00

Monthly earnings ($)	Tax withheld ($)
4870.67	979.00
4875.00	979.00
4879.33	979.00
4883.67	984.00
4888.00	984.00
4892.33	984.00
4896.67	988.00
4901.00	988.00
4905.33	988.00
4909.67	992.00
4914.00	992.00
4918.33	997.00
4922.67	997.00
4927.00	997.00
4931.33	1001.00
4935.67	1001.00
4940.00	1001.00

c. Determine the percentage of tax, to 2 decimal places, they pay each month.
d. Calculate the Medicare levy they pay.

6. An accountant pays $1719.40 in Medicare levy.

a. If they have private health insurance, calculate their taxable income for the year.
b. Calculate their tax payable for the financial year. Use the tax table provided in lesson 7.5.1.

7. A hairdresser receives a weekly wage of \$930.60, which includes 38 hours of normal pay plus 4 hours of overtime, paid at time and a half.

a. Explain how their hourly rate can be determined, and hence calculate the hourly rate.

b. Determine their annual salary if they work an average of 40 hours a week (38 hours at normal pay and 2 hours overtime). Assume 52 weeks in 1 year.

c. Using your answer from part b, determine the amount of tax payable at the end of the financial year.

d. The hairdresser forgot to claim \$1980 in deductions and is looking forward to receiving \$1980 in a tax refund. By recalculating the amount of tax payable at the end of the financial year based on their new taxable income, explain why they won't receive \$1980 as tax refund.

8. A truck driver earns \$24.07 per hour for working a 76-hour fortnight. An overtime rate of time and a half is paid for additional hours worked over the 76 hours during Monday to Friday, and double time is paid for hours worked on weekends (Saturday and Sunday). Assume 52 weeks in 1 year.
Over a fortnight, the truck driver worked 80 hours Monday–Friday and 15 hours over the weekend.

a. Calculate the gross fortnightly wage of the truck driver.

Fortnightly earnings (\$)	Tax withheld (\$)
2682.00	602.00
2684.00	602.00
2686.00	604.00
2688.00	604.00
2690.00	604.00
2692.00	606.00
2694.00	606.00
2696.00	606.00
2698.00	608.00
2700.00	608.00
2702.00	608.00
2704.00	610.00
2706.00	610.00
2708.00	610.00
2710.00	612.00

Fortnightly earnings (\$)	Tax withheld (\$)
2712.00	612.00
2714.00	614.00
2716.00	614.00
2718.00	614.00
2720.00	616.00
2722.00	616.00
2724.00	616.00
2726.00	618.00
2728.00	618.00
2730.00	618.00
2732.00	620.00
2734.00	620.00
2736.00	620.00
2738.00	622.00
2740.00	622.00

b. Show that the truck driver's net fortnightly pay was \$2089.84 by using the PAYG tax table shown above.

c. If the truck driver claims \$2486 in deductions, calculate their taxable income, and hence calculate the amount of payable tax for the year by using the tax table provided in section 7.5.1.

d. Calculate the Medicare levy the driver is expected to pay, given that they have private health insurance.

9. A computer technician's annual salary is $67 374. They claim $1580 in deductions and receive $225 in dividends from shares.
   a. The technician is paid monthly. Calculate their gross monthly salary.
   b. Their employer withholds $1200 PAYG tax each month. Calculate the total amount of tax withheld for the year.
   c. Determine the tax payable by the computer technician for the financial year. Hence, determine whether they receive a tax refund or are required to pay more tax, and calculate the amount.

10. WE22 A fast-food manager's annual salary is $48 131. They claim $980 in tax deductions and receive $500 in interest from investments. Using a spreadsheet, calculate the amount of tax payable for the financial year.

11. Using a spreadsheet, calculate the tax payable by the following workers.

Employee	Gross salary ($)	Interest ($)	Deductions ($)	Taxable income	Tax free (0–18 200) 0c	Tax tier (18 201–45 000) 19c	Tax tier (45 001–120 000) 32.5c	Tax tier (120 001–180 000) 37c	Tax tier (180 000+) 45c
Stan	52 895	350							
Bett	65 845		650						
Xiao	36 080	125							
Mohamed	75 040	870	1500						

12. WE23 Using a spreadsheet, complete the following wage sheet.

Employee	Pay rate ($)	Normal hours worked	Overtime	Penalty rate		Allowance	Gross pay ($)	Tax withheld ($)	Net pay ($)
			1.5	1.5	2				
Rex	15.85	28		7	7			124	
Tank	22.15	35		5		15		167	
Gert	30.10	32	1.5		8	19		367	

13. A pay sheet for an individual employee is shown. Complete the pay sheet using a spreadsheet, and hence determine the employee's net weekly pay and amount of superannuation paid into their superannuation fund.

Entitlements	Unit	Rate	Total ($)
Wages for ordinary hours worked	30 hours	$35.05	
Total ordinary hours = 30 hours			
Penalty (double time)	5 hours	$70.10	
	Gross payment		

Deductions	
Taxation	$325
Total deductions	
Net payments	

Employer superannuation contribution	
Contribution	$

14. The taxation table is shown.

Taxable income	Tax on this income
0–$18 200	Nil
$18 201–$45 000	19c for each $1 over $18 200
$45 001–$120 000	$5092 plus 32.5c for each $1 over $45 000
$120 001–$180 000	$29 467 plus 37c for each $1 over $120 000
$180 001 and over	$51 667 plus 45c for each $1 over $180 000

Using a spreadsheet, explain how the 'plus' calculations of $5092, $29 467 and $51 667 are found for the different tax levels in the table.

15. A waiter is paid an hourly rate of $21.50. They receive penalty rates of time and a half for working Saturdays and double time for working Sundays and public holidays.
Over the Christmas period, they worked 5 hours on Friday night, 8 hours on Saturday, 9 hours on Sunday and 7.5 hours on Boxing Day (a public holiday). Their employer withholds $185 in tax. Prepare a pay sheet using a spreadsheet to represent the waiter's weekly pay.

Complex unfamiliar

16. The Australian tax system is a tiered system, as shown in the table.

Taxable income	Tax on this income
0–$18 200	Nil
$18 201–$45 000	19c for each $1 over $18 200
$45 001–$120 000	$5092 plus 32.5c for each $1 over $45 000
$120 001–$180 000	$29 467 plus 37c for each $1 over $120 000
$180 001 and over	$51 667 plus 45c for each $1 over $180 000

A dentist's annual salary is $97 605.
Explain why the tax payable is not found by calculating 32.5% of $97 605.

17. A mechanical engineer paid $14 812 in tax over the course of the year. At the end of the financial year, they have $985 worth of deductions to claim. Determine their taxable income and the tax refund they should expect.

18. A teacher's annual salary is $94 961. They decide to reduce their time to 0.8 (this means they receive 0.8 of their salary and work 4 days out of 5), claiming that they will be on the lower tax level and pay less in tax overall, so their net fortnightly pay will not be much lower.
By calculating the tax payable for both the full-time salary and the 0.8 salary, determine how accurate their claim is.

19. Each year, an IT technician moves up one level on a pay scale. The current pay scale is shown.

Level	Annual salary
1–1	$51 758
1–2	$53 258
1–3	$55 333
1–4	$57 508
1–5	$60 088

The technician is currently on level 1–3. At the start of the following year, they move up the pay scale to level 1–4.

a. Determine the percentage change in their wage correct to 2 decimal places.

A new working agreement is reached, which increases the pay level of technicians on levels 3 and 4 by 3.75%.

b. Determine the percentage change in the technician's net salary from their original 1–3 wage, correct to 2 decimal places.

c. After the pay increase of 3.75%, another technician moves to level 2–3 with an annual salary of $65 880. The overall percentage change in their wage is 8.5%. Determine the salary at level 2–2 before and after the pay increase.

20. A new taxation system is being proposed.

Taxable income	Tax on this income
0–$19 500	Nil
$19 501–$65 000	21c for each $1 over $19 500
$65 001–$125 000	$9555 plus 35.5c for each $1 over $65 000
$125 001–$180 000	$30 855 plus 37c for each $1 over $125 000
$180 001 and over	$51 205 plus 45c for each $1 over $180 000

By comparing this proposed taxation table with the current table, determine whether this would be a fairer taxation system. Justify your answer using calculations.

Fully worked solutions for this chapter are available online.

LESSON
7.9 Review

7.9.1 Summary

doc-41761

Hey students! Now that it's time to revise this chapter, go online to:

Access the chapter summary | Review your results | Watch teacher-led videos | Practise questions with immediate feedback

Find all this and MORE in jacPLUS

7.9 Exercise

learnon

7.9 Exercise

Simple familiar	Complex familiar	Complex unfamiliar
1, 2, 3, 4, 5, 6, 7, 8, 9, 10, 11, 12, 13	14, 15, 16, 17, 18	19, 20

These questions are even better in jacPLUS!
- Receive immediate feedback
- Access sample responses
- Track results and progress

Find all this and MORE in jacPLUS

Simple familiar

1. An employee has a total income of $87 500 p.a. They will claim $2350 worth of deductions that year. Calculate the taxable income.
2. Meredith walks dogs on the weekend. She charges $14.00 per dog plus $6.00 an hour. She offers her clients a 5% discount for paying in cash. Calculate the amount she would charge for someone paying cash to walk 3 dogs for 2 hours.
3. The energy usage and rates for a household are as shown. There is a 25% discount for paying the bill before the due date. If the bill was paid before the due date, determine the amount paid. (Remember to include GST.)

Electricity

Tariff	Bill days	Current reading	Previous reading	Total usage (kWh)	Charge/rate (c/kWh)
All day	90	78 250	77 682		25.08
Service to property	90			**$1.105/day**	

Gas

Tariff	Bill days	Current reading	Previous reading	Total usage (kWh)	Charge/rate (c/MJ)
All day	90	8920	8138		2.452
Service to property	90			**$0.985/day**	

4. If Raffaele earns $56 420 p.a., calculate the superannuation they earn that year.

5. Determine the term that represents receipt of payment per item constructed.

6. Determine the amount of fortnightly pay from an annual salary of $56 200.

7. A bookkeeper receives a pay rise. Their original weekly pay was $758 and their new weekly pay is $775. Calculate the approximate percentage change of their pay.

8. A librarian pays $1137.60 in Medicare levy. Calculate their annual salary.

9. A real estate agent takes a commission of 2.2% on the sale of a house that sold for $945 000. Determine how much commission they made.

10. If Robbie earns an annual income of $92 550, calculate their weekly income.

11. A nurse is paid $28.50 per hour for an 80-hour fortnight. Assuming 26.07 fortnights per year, calculate:
    a. the nurse's annual salary
    b. the nurse's annual superannuation (assuming 9.5% superannuation)
    c. the nurse's weekly pay.

12. A novelist gets a royalty payment of 2.9% of gross sales of their book. Complete the following table.

	January	February	March	April
Gross sales	$45 000	$125 000	$320 000	
Royalty				$1508

13. A person earns $77 300. Determine the amount of tax they should pay if they have no deductions.

Complex familiar

14. An employee earns $133 000. Determine the amount of tax they should pay if they can claim $8450 in deductions.

15. Calculate the amount of superannuation that is paid each fortnight (assume 11.5% superannuation) for the following annual salaries.
    a. $67 899 b. $98 765 c. $101 010 d. $123 456

16. Employees at a cafe are paid penalty rates for working on weekends. The normal hourly rate is $18.50. Penalty rates for Saturday are time and a half, and double time for Sundays and public holidays. Four employees worked on the long weekend. The hours worked are shown in the table.

Employee	Saturday	Sunday	Monday (public holiday)
Tran	5	8	
Gus		6	
Meg	3.5		7
Warren	4		6

Calculate the earnings before tax for each employee.

17. A factory worker's hourly rate is $21.50. They work 38 hours per week, with 4 hours overtime paid at time and a half. The tax withheld each week is $195. They claim $25 a week for the uniform and earn $120 in interest for the year. (Assume 52.14 weeks in the year.)
    a. Calculate the taxable income and Medicare levy.
    b. Explain why the worker should expect to receive a tax refund. Support your explanation with calculations.
18. The fortnightly pay slip for an employee is shown.

Entitlements	Unit	Rate	Total ($)
Wages for ordinary hours worked	76 hours	$20.95	
Total ordinary hours = 76 hours			
Overtime	15 hours	$31.40	
Travel allowance	350 km	$0.66/km	
		Gross payments	

Deductions	
Taxation	$430
Total deductions	
Net payment	

    a. Using a spreadsheet, calculate the employee's fortnightly pay.
    b. Assuming the employee receives the same fortnightly pay for the year, calculate their annual taxable income. Assume 26.07 fortnights in the year.
    c. Determine if the employee will receive a tax refund or have to pay more in tax.

Complex unfamiliar

19. A window cleaner has the opportunity to move from casual to full-time employment. Workers employed on a casual basis do not receive annual or sick leave. All employees in the cleaning business receive 11.5% superannuation.
    The window cleaner's current casual hourly rate is $25.76. A worker employed full-time earns $900.60 per week.
    Determine if they should move to full-time employment.
    Justify your answer by calculating their annual salary based on working 38 hours each week at the normal hourly rate and their annual leave loading.

20. Business groups are seeking a change to the taxation system. Their proposal is as follows:

Taxable income	Tax on this income
0–$29 000	Nil
$29 001–$85 000	35c for each $1 over $29 000
$85 001–$175 000	$19 600 plus 40c for each $1 over $85 000
$175 001	$55 600 plus 58c for each $1 over $175 000

Compare the tax payable for the following annual salaries under the new proposed system and the current taxation table.

i. $25 890 ii. $45 870 iii. $67 950 iv. $81 940 v. $195 870

Evaluate, using reasoning, how workers will be better or worse off under the proposed new taxation system.

Fully worked solutions for this chapter are available online.

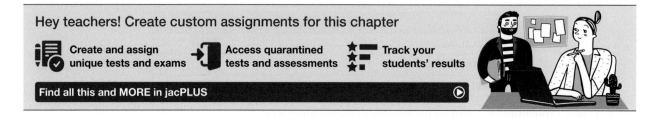

Answers

Chapter 7 Earning money

7.2 Salary, wages and overtime

7.2 Exercise

1. Salary
2. $7500
3. a. $22.50 b. $32.25
4. a. $51 b. $36.50
5. a. $65 b. $81.25
6. The wage (hourly rate) is the greater pay ($3200 per fortnight compared to $2838.51).
7. $2504.80
8. a. $1476.31 b. $1136.89 c. $1596.83
9. a. $1157 b. $1424
10. a. $1576.75 b. $1556.88
11. a. $1170 b. $1202.50
12. a. $812.13 b. $807.56
13. a. 16 h 45 min b. $305.25 c. 6 h 45 min d. $15.00 per hour
14. a. 17 h 30 min b. $445.50 c. Wage d. $181.50 e. $33.00
15. a. $45.00 per hour b. $1620.00 c. $1208.00 d. $412.00 e. Weekly
16. a. $97.50 per hour b. $1641.50 c. Salary d. $438.5
17. $637.50
18. $26
19. 31 hours
20. $20 per hour

7.3 Income payments — commission, piecework and royalties

7.3 Exercise

1. $3390
2. a. $1650
   b. $2375
   c. $2920
   d. $4160
3. $2300
4. $6825
5. $1398.25
6. $2941.00
7. $12.11
8. $186 989.67
9. 4.5%
10. a. See the table at the bottom of the page.*
    b. $376 099
11. $45 634.53
12. $2 500 000
13. $16 599.48
14. The royalty percentage can be found by dividing the royalty amount by the total sale amount and multiplying by 100. The songwriter's royalty percentage is 2.5%.
15. Option 1:
    $15 \times 5 \times 8.25 = \$618.75$
    Option 2:
    $250 \times 75.95 \times 0.035 = \$664.56$
    Option 2 is the better option.
16. a. $2090
    b. 22
    c. If they sell 22 new telephone plans each week, taking the salary option of $37 500 per annum would give them a higher weekly wage.

7.4 Convert between different pay periods [complex]

7.4 Exercise

1. a. 52 weeks b. 26 fortnights c. 365 days
2. a. Divide by 12. b. Multiply by 2.
3. a. $1658.65 b. $3317.31 c. $7287.50
4. a. $1205.00 b. $2410.00 c. $171.67
5. a. $1518.40 b. $39 478.40
6. a. $1205.00 b. $2410.00 c. $171.67
7. a. $42 822.00 b. $1647.00 c. $117.32
8. a. $150 600 b. $5792.31 c. $2896.15 d. $289.62
9. $1568.08
10. $12 000
11. 34.31 hours
12. $1833.75
13. They should take the salary offer.
14. 16.65%

*10. a.

	Jan–Mar	Apr–Jun	Jul–Aug	Sept–Dec
Sales	$123 400	$2 403 556	$432 442	$84 562
Royalty payment	$3085.00	$60 088.90	$10 811.05	$2114.05

7.5 Personal taxation, PAYG and employee superannuation

7.5 Exercise

1. a. $0–$18 200, nil.
   b. $18 201–$45 000, 19c for each $1 over $18 200
   c. $120 001–$180 000, $29 467 plus 37c for each $1 over $120 000
2. a. $0–$18 200
   b. $45 001–$120 000
   c. $180 001 and over
3. a. TFN: 562434961
   b. $9272
   c. $53 208
4. a. $19 787 b. $72 961 c. $10 666.02
5. a. TFN: 987543211
   b. $24 823
   c. $80 423
   d. $2396.71
6. a. $1799.75 b. $2978.73 c. $5392.35
   d. $9023.48 e. $11 449.40 f. $11 987.60
7. $8797.50
8. $9832.50
9. $13 193.49
10. $174.42
11. a. $12 979.17
    b. $17 911.25 annually, $1492.60 monthly
12. a. $5166.67 b. $594.17 c. $7130.04
13. $1157.42
14. $297.50
15. $861.54
16. $12 250

7.6 Determining PAYG tax using tax tables [complex]

7.6 Exercise

1. a. $125 b. $128 c. $126
   d. $125
2. a. $612 b. $616 c. $608
   d. $602
3. i. a. $988
      b. $979
      c. $962
   ii. a. $4818.67 – $4827.33
      b. $4857.67 – $4866.33
      c. $4931.33 – $4940.00
4. a. $42 744 b. $6292 c. 36 452
5. a. Fortnightly pay = $1650

   $$\text{Weekly pay} = \frac{1650}{2} = \$825$$

   Tax withheld = $122.00

   b. Fortnightly pay = $1670

   $$\text{Weekly pay} = \frac{1670}{2} = \$835$$

   Tax withheld = $125.00

   c. Fortnightly pay = $1690

   $$\text{Weekly pay} = \frac{1690}{2} = \$845$$

   Tax withheld = $129.00

6. a. Annual pay = $58 760.04

   $$\text{Monthly pay} = \frac{\$58\,760.04}{12} = \$4896.67$$

   Tax withheld = $988.00

   b. Annual pay = $59 124

   $$\text{Monthly pay} = \frac{\$59\,124}{12} = \$4927$$

   Tax withheld = $997.00

   c. Annual pay = $57 615.96

   $$\text{Monthly pay} = \frac{\$57\,615.96}{12} = \$4801.33$$

   Tax withheld = $953.00

7.7 Determining taxable income by applying tax deductions [complex]

7.7 Exercise

1. $70 150
2. $86 150
3. $97 250
4. $119 260
5. $104 430
6. $109 200
7. $139 500
8. a. $1208 b. $1546 c. $1788 d. $2168.46
9. $513.79
10. $1449
11. $1971.30
12. $2064
13. $5200
14. $100 000

7.8 Calculating simple tax returns [complex]

7.8 Exercise

1. $17 442
2. $30 207
3. $7038.75
4. a. i. $18 201–$45 000
      ii. $3667
   b. i. 0–$18 200
      ii. $0

c. i. $45 001–$120 000

ii. $18 309.75

d. i. $120 001–$180 000

ii. $33 537

5. a. $4857 b. $975

c. 20.07% d. $1165.68

6. a. $85 970 b. $18 407.25

7. a. $21.15

Working 4 hours overtime equates to the same pay as working 6 hours normal time ($4 \times 1.5 = 6$). Calculate the number of 'normal' hours worked ($38 + 6 = 44$) and divide into wage.

b. $45 091.80

c. $5121.84

d. $4733.24

They won't receive $1980 as a tax refund because the tax payable is a percentage of their taxable income. Their taxable income was reduced by $1980 but the tax payable was reduced by 32.5% of 1980 (0.325 × 1980).

8. a. $2695.84

b. $2695.84 – $606 = $2089.84

c. $12 438.90

d. $1352.12

9. a. $5614.50

b. $14 400

c. They will receive a refund of $2476.82.

10. $5953.58

11. Stan = $7771.63, Bett = $11 655.38, Xiao = $3420.95, Mohamed = $14 650.25

12. See the table at the bottom of the page.*

13.

Entitlements	Unit	Rate	Total
Wages for ordinary hours worked	30 hours	$35.05	$1051.50
Total ordinary hours = 30 hours			
Penalty (double time)	5 hours	$70.10	$350.50
		Gross payment	**$1402.00**

Deductions	
Taxation	$325
Total deductions	**$325**
Net payment	**$1077.00**

Employer superannuation contribution	
Contribution	$161.23

14. The 'plus' calculations are found by adding the tax payable in the previous brackets.

15. See the table at the bottom of the page.*

16. Sample responses can be found in the worked solutions in the online resources.

17. Taxable income = $73 922.69

Refund = $320.13

18. The claim is partly correct. Yes, the teacher will pay $6172.47 less in tax overall; however, they're incorrect about being on a lower tax level, as they're still in the taxable income bracket of $45 001– $120 000. Reducing their salary by 20% has an overall reduction of $12 819.73 after tax. This means their fortnightly pay will be reduced by $493.07 or 17.4%.

19. a. 3.93%

b. 7.83%

c. Before pay increase = $60 718.89

After pay increase = $62 995.85

20. The middle earners will pay less but low-income earners will pay more and high-income earners will pay less; therefore, it does not seem to be a fairer system.

7.9 Review

7.9 Exercise

1. $85 150
2. $51.30
3. $288.52
4. $6488.30
5. Piecework
6. $2149.73
7. 2.24%

*12.

Employee	Pay rate ($)	Normal hours worked	Overtime	Penalty rate		Allowance	Gross pay ($)	Tax withheld ($)	Net pay ($)
			1.5	1.5	2				
Rex	15.85	28		7	7		832.13	124	708.13
Tank	22.15	35		5		15	956.38	167	789.38
Gert	30.10	32	1.5		8	19	1531.53	367	1164.53

*15.

Hourly rate	Normal hours	Saturday	Sunday	Public holiday	Gross pay	PAYG tax	Net pay
		1.5	2	2			
21.5	5	8	9	7.5	$1075.00	$185.00	$890.00

8. $56 880

9. $20 790

10. $1779.81

11. a. $59 439.60 b. $5646.76 c. $1140

12. See the table at the bottom of the page.[†]

13. $15 589.50

14. $31 150.5

15. a. $299.52 b. $435.67

c. $445.58 d. $544.59

16. See the table at the bottom of the page.*

17. a. $962.82

b. They have paid $3091.60 more in tax than they are required to. Therefore, they will receive a tax refund.

18. a.

Entitlements	Unit	Rate	Total ($)
Wages for ordinary hours worked	76	$20.95	$76 \times 20.95 = 1592.20$
Total ordinary hours = 78 hours			
Overtime	15	$31.40	$31.40 \times 15 = 471$
Travel allowance	350	$0.66	$350 \times 0.66 = 231.00$
		Gross payment	$1592.20 + 471 + 231 = 2294.20$
Deductions ($)			
Taxation			430
Total deductions			**430**
Net payment			$2294.20 - 430 = 1864.20$

b. $59 809.79

c. Their PAYG tax is higher than the required tax. Therefore, they will get a return of $11 210.10 – $9905.11 = $1304.99.

19. Casual employment will give them more money overall. Considering taking annual leave, working full-time would be the better deal. They will also have sick leave and more job security.

20. Under the proposed taxation system, lower income earners will pay less in tax. Higher income earners will pay more due to a higher percentage tax rate. Middle income workers will pay slightly more under the proposed system.

[†]12.

	January	February	March	April
Gross sales	$45 000	$125 000	$320 000	$\frac{1508}{0.029} = \$52\,000$
Royalty	$2.9\% \times \$45\,000$ $= 0.029 \times \$45\,000$ $= \$1305$	$2.9\% \times \$125\,000$ $= 0.029 \times \$125\,000$ $= \$3625$	$2.9\% \times \$320\,000$ $= 0.029 \times \$320\,000$ $= \$9280$	$1508

*16.

Employee	Saturday ($27.75)	Sunday ($37)	Monday (public holiday) ($37)	Gross wage
Tran	5	8		$434.75
Gus		6		$222
Meg	3.5		7	$356.13
Warren	4		6	$333

8 Budgeting [complex]

LESSON SEQUENCE

Fully worked solutions for this chapter are available online.

on Resources

Solutions Solutions — Chapter 8 (sol-1261)

Digital documents Learning matrix — Chapter 8 (doc-41762)
Quick quizzes — Chapter 8 (doc-41763)
Chapter summary — Chapter 8 (doc-41764)

LESSON
8.1 Overview

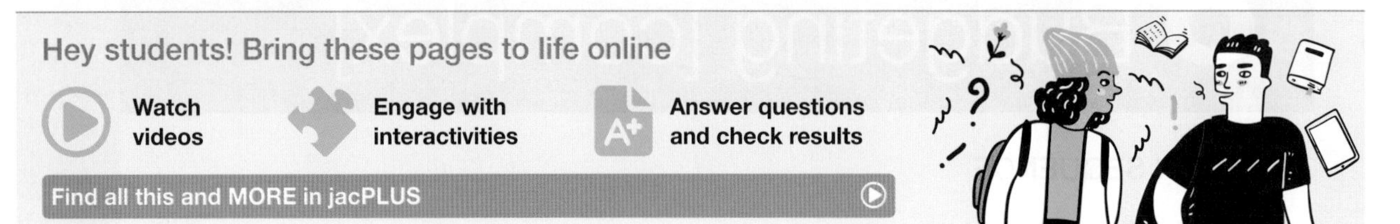

8.1.1 Introduction

It is important to understand how to manage financial matters, as money is needed to pay for basic personal essentials: food, shelter and clothing. Being informed about your spending choices will help you make smart financial decisions, provide those essentials, and also allow you to successfully manage your personal finances as well as engage in community financial matters. This chapter will help you understand some of these concepts.

Before you take the first steps to becoming an independent adult, it's a good idea to be aware of some of the costs you'll be expected to pay. By carefully tracking income and expenditures, you can gain insight into your financial health, identify areas for improvement, and make informed decisions to achieve financial stability and prosperity.

Basic expenses, such as internet services, utilities bills and car insurance, can't be avoided, so they should be included in your budget. Many companies offer discounts for their services, so research your options. By managing your finances, you can have an understanding of your income and expenses and can make good decisions now and in the future. Whether you're aiming to save for a major purchase, pay off debt, or build wealth for the future, mastering the art of budgeting is essential for taking control of your financial destiny.

8.1.2 Syllabus links

Lesson	Lesson title	Syllabus links
8.2	**Products and services**	◯ Investigate the costs involved in independent living [complex].
8.3	**Rent and utility costs**	◯ Investigate the costs involved in independent living [complex].
8.4	**Car expenses**	◯ Investigate the costs involved in independent living [complex].
8.5	**Personal budget plan**	◯ Prepare a personal budget plan [complex].

LESSON
8.2 Products and services

SYLLABUS LINKS

- Investigate the costs involved in independent living [complex].

8.2.1 Internet services

- Finding a plan that suits you will depend on your individual needs and circumstances.
- Most plans offer unlimited data with a fixed cost structure, but there are still some that have a monthly data limit.
- This type of plan would have variable costs each month if the data limit is exceeded, and can get expensive.

tlvd-5080

WORKED EXAMPLE 1 Determining the most cost-effective option for products

Byron is trying to decide between two internet plans, as shown.

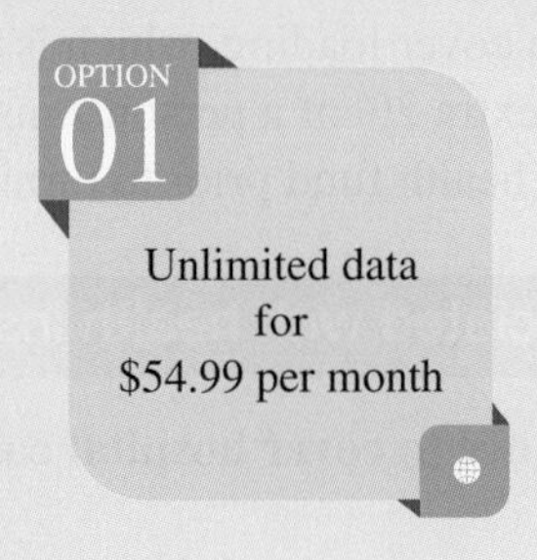

OPTION
01
Unlimited data
for
$54.99 per month

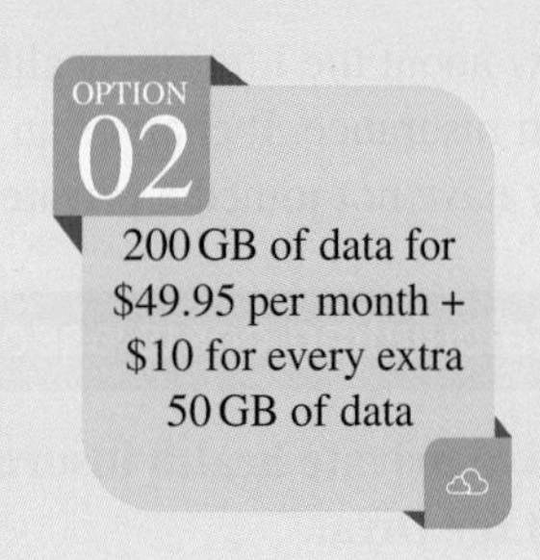

OPTION
02
200 GB of data for
$49.95 per month +
$10 for every extra
50 GB of data

Determine the most cost-effective option if Byron uses:

a. 100 GB of data per month

b. 250 GB of data per month.

THINK	WRITE
a. 1. Option 1 has unlimited data, and option 2 will not exceed the 200 GB in the plan.	**a.** Option 1 = \$54.99 Option 2 = \$49.95
2. Write the answer.	Option 1 costs \$54.99 and option 2 costs \$49.95, so option 2 is more cost-effective.
b. 1. Option 2 will exceed the data limit by 50 GB, which is an additional \$10.	**b.** Option 1 = \$54.99 Option 2 = \$49.95 + \$10 = \$59.95
2. Write the answer.	Option 1 costs \$54.99 and option 2 costs \$59.95, so option 1 is more cost-effective.

8.2.2 Health insurance

- To be covered by health insurance, a regular fee is paid to an insurer weekly, fortnightly, monthly or annually.
- The amount of the fee will depend on the type of cover you require (hospital or extras), and your age.
- All private health insurers can be compared at the Australian government website www.privatehealth.gov.au.
- In Australia, having appropriate private hospital health insurance may mean that you do not have to pay the Medicare levy surcharge, which can be up to an extra 1.5% of your taxable income on top of the Medicare levy that is paid by everyone.

Medicare levy surcharge

The Medicare levy surcharge is calculated on your taxable income from the table below.

Threshold	Base tier	Tier 1	Tier 2	Tier 3
Single threshold	$97 000 or less	$97 001–$113 000	$113 001–$151 000	$151 001 or more
Family threshold	$194 000 or less	$194 001–$226 000	$226 001–$302 000	$302 001 or more
Medicare levy surcharge	0%	1%	1.25%	1.5%

- It is also important to know about the lifetime health cover loading, which is used to encourage younger people to get private health insurance. It charges an extra 2% of a person's insurance fee for every year after they have turned 30 if they have not joined a private health fund prior to turning 30.

WORKED EXAMPLE 2 Calculating costs associated with health insurance

HealthCo Hospital offers Diya private health insurance to cover hospital costs for $79 per month.

a. Calculate the annual cost for Diya.

b. Diya's taxable income is $98 000 per annum. Determine if it is more cost-effective for Diya to join HealthCo or pay the Medicare levy surcharge.

c. Diya doesn't join HealthCo until she turns 32. Determine the cost per month for Diya, taking into account the lifetime health cover loading.

THINK	WRITE
a. 1. Multiply the monthly cost by 12 to calculate the annual cost.	a. $\$79 \times 12 = \$948$
2. Write the answer.	The annual cost for Diya is $948.
b. 1. Calculate the 1% Medicare levy surcharge on Diya's taxable income of $95 000.	b. $1\%\text{ of }\$98\,000 = \frac{1}{100} \times \$98\,000 = \$980$
2. Write the answer.	The Medicare levy surcharge on Diya's taxable income is $980, which is more than the private health insurance cost. Therefore, it is more cost-effective for Diya to join HealthCo.

c. 1. Diya will be charged an extra 2% for every year after she turned 30. | c. $2 \times 2\% = 4\%$

2. Calculate 4% of 79. | $\frac{4}{100} \times \$79 = \$3.16$

3. Add the lifetime health cover loading to the monthly cost. | $\$79 + \$3.16 = \$82.16$

4. Write the answer. | Diya will pay $82.16 per month.

8.2.3 Car insurance

Car insurance covers the costs associated with vehicle accidents, such as repairs, replacement vehicles and medical expenses.

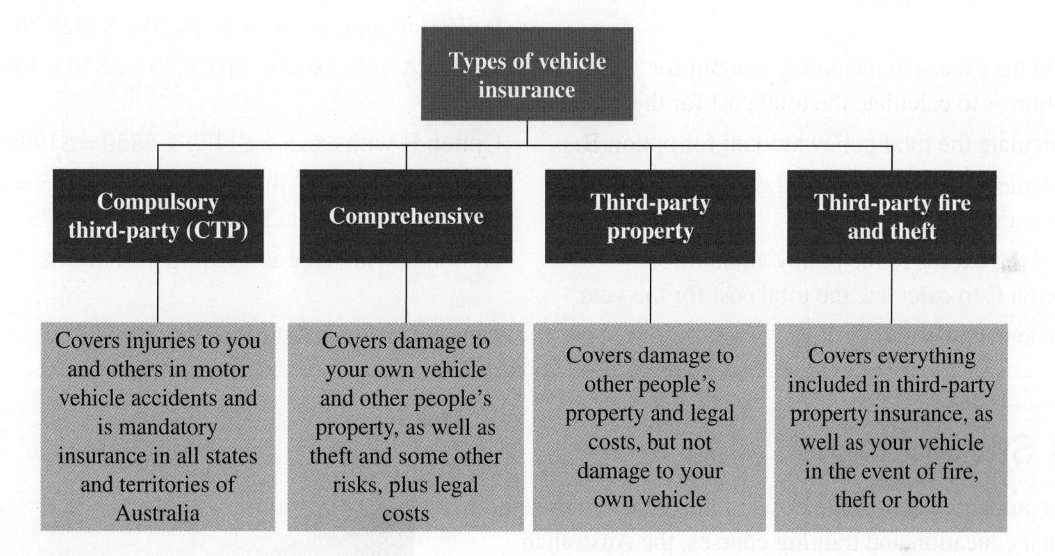

- Car insurance costs depend on how long you've been driving, your vehicle type, age, parking spot, and coverage choice.
- With comprehensive insurance, you can pick between your car's market value or an agreed value, but there's an extra cost, called excess, you pay before the insurance company sorts your claim. If you're under a certain age, that excess cost is usually higher.

tlvd-5081

WORKED EXAMPLE 3 Determining the most cost-effective car insurance

A 19-year-old driver has three options for yearly comprehensive car insurance for a small car that will be parked in a locked garage in a suburban property. All options include a replacement vehicle, provision of transportation if the car is unable to be driven, and costs for injuries for all passengers.
Option A: $1143, 10% online discount, $850 excess
Option B: $1149, $850 excess
Option C: $1155, 15% online discount, $800 excess
The excess fee is the amount that is paid to the insurer if a claim is made.
If the driver had an accident in the first year of their policy, determine which option is the most cost-effective.

THINK	WRITE
1. Calculate the policy amount with the 10% discount from option A.	Option A with 10% discount: $10\% \times \$1143 = \$114.30$ Policy amount: $\$1143 - \$114.30 = \$1028.70$
2. Add the excess to the policy amount for option A to calculate the total cost for the year.	Option A with excess: $\$1028.70 + \$850 = \$1878.70$
3. Calculate the total policy amount for option B.	Option B with excess: $\$1149 + \$850 = \$1999$
4. Calculate the policy amount with the 15% discount for option C.	Option C with 15% discount: $15\% \times \$1155 = \$173.25$ Policy amount: $\$1155 - \$173.25 = \$981.75$
5. Add the excess to the policy amount for option C to calculate the total cost for the year.	Option C with excess: $\$981.75 + \$800 = \$1781.75$
6. Write the answer.	Option C is the most cost-effective.

8.2.4 Student loans

To assist students to cover the costs of university study and/or vocational education and training courses, the Australian government offers Commonwealth assistance loans known as VET Student Loans and HECS-HELP.

VET Student Loans

- If studying an approved higher-level (diploma and above) vocational education and training (VET) course, you can access the Australian government VET Student Loans program, which will assist you in paying tuition fees.
- The limit, or loan cap, for a VET Student Loan differs based on the course being studied. If you undertake a course that costs more than the loan cap, you will need to pay the difference up-front to your course provider.

HECS-HELP

- Australian universities offer Commonwealth Supported Places (CSP) to Australian residents. These are university enrolments subsidised by the Australian government.
- If you enroll in a CSP, the Australian government will pay some of your fees. You have to pay the rest. This is known as your *student contribution amount*, and can be quite costly.
- To cover the student contribution amount, a government loan, known as HECS-HELP, is offered.
- The limit to a HECS-HELP loan is usually $121 844 and you need to repay it through the Australian tax system once you earn above the compulsory repayment threshold, which is $51 550 in 2023–24.

- The following table shows the repayment interest rates for taxable annual incomes in the 2023–24 financial year.

2023–2024 repayment income (RI)	Repayment % rate
Below $51 550	Nil
$51 550–559 518	1.0%
$59 519–$63 089	2.0%
$63 090–$66 875	2.5%
$66 876–$70 888	3.0%
$70 889–$75 140	3.5%
$75 141–$79 649	4.0%
$79 650–$84 429	4.5%
$84 430–$89 494	5.0%
$89 495–594 865	5.5%
$94 866–$100 557	6.0%
$100 558–$106 590	6.5%
$106 591–$112 985	7.0%
$112 986–$119 764	7.5%
$119 765–$126 950	8.0%
$126 951–$134 568	8.5%
$134 569–$142 642	9.0%
$142 643–$151 200	9.5%
$151 201 and above	10%

tlvd-11436

WORKED EXAMPLE 4 Calculating the monthly HECS-HELP repayment

Courtney studied a Bachelor of Psychological Science at university, accruing a total HECS-HELP debt of $35 124. She is now a practising psychologist and is expected to repay her HECS-HELP loan from her monthly gross income.

a. If Courtney earns $120 000 per annum, calculate her monthly HECS-HELP repayment.

b. Determine how long, to the nearest month, it will take Courtney to pay off her HECS-HELP debt if she remains on this annual salary.

THINK	WRITE
a. 1. Calculate the repayment percentage rate for an annual income of $120 000.	a. 8.5%
2. Calculate the annual amount of HECS-HELP owing by finding 8.0% of 120 000.	$\frac{8.0}{100} \times \$120\,000 = \$9600$
3. Divide by 12 to calculate the monthly repayments of HECS-HELP.	$\$9600 \div 12 = \$800$
4. Write the answer.	Courtney's monthly HECS-HELP repayment is $800.
b. Divide the total HECS-HELP debt by the monthly payment amount. Round up to calculate the total number of months and write the answer.	b. $\$35\,124 \div \$800 = 43.905$ It will take Courtney 44 months to repay her HECS-HELP debt.

Exercise 8.2 Products and services

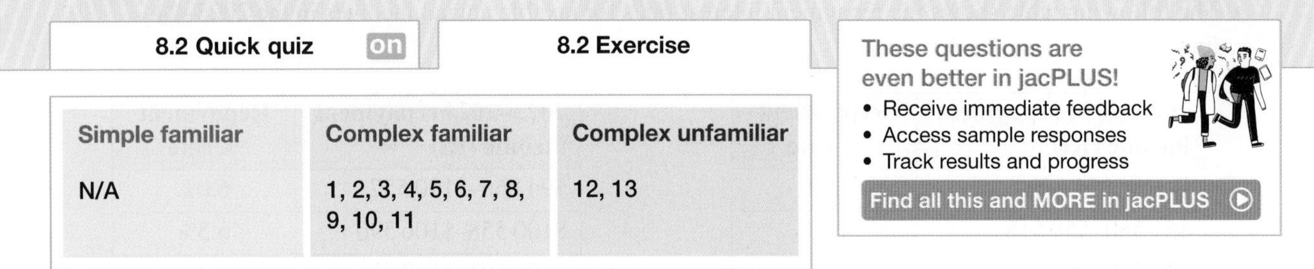

8.2 Quick quiz	8.2 Exercise	
Simple familiar	**Complex familiar**	**Complex unfamiliar**
N/A	1, 2, 3, 4, 5, 6, 7, 8, 9, 10, 11	12, 13

These questions are even better in jacPLUS!
- Receive immediate feedback
- Access sample responses
- Track results and progress

Find all this and MORE in jacPLUS

Complex familiar

1. **WE1** Connie is trying to decide between the two internet plans shown.

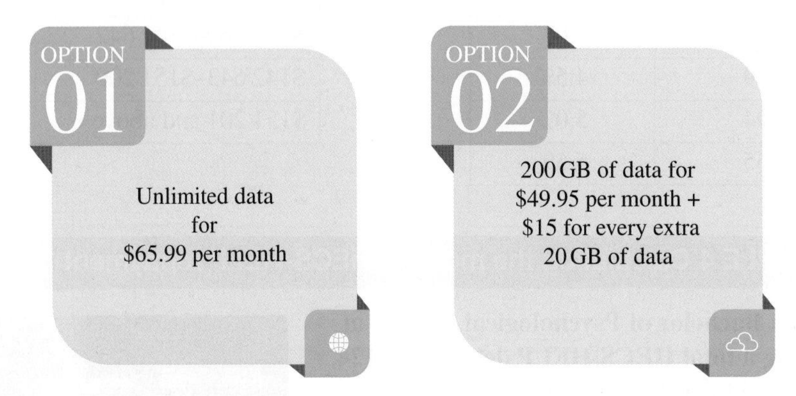

OPTION
01
Unlimited data
for
$65.99 per month
OPTION
02
200 GB of data for
$49.95 per month +
$15 for every extra
20 GB of data

Determine the most cost-effective option if Connie uses:

a. 150 GB of data per month
b. 220 GB of data per month
c. 260 GB of data per month.

2. Aaron is trying to decide between the two internet plans shown.

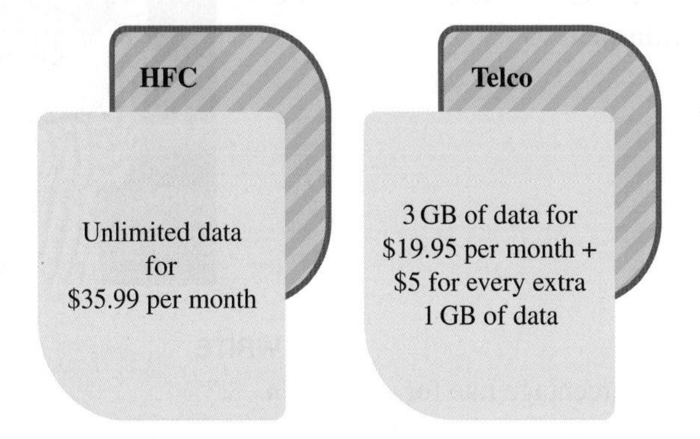

HFC
Unlimited data
for
$35.99 per month
Telco
3 GB of data for
$19.95 per month +
$5 for every extra
1 GB of data

a. Determine the most cost-effective option if Aaron uses:
   i. 3 GB of data per month
   ii. 5 GB of data per month
   iii. 7 GB of data per month.

b. HFC offers Aaron a deal whereby he pays $1 for the first month and then normal rates for the remainder of the year. If Aaron regularly uses 7 GB of data per month, calculate the yearly cost for each plan and determine the best plan for Aaron.

3. WE2 HIB Hospital + Extras offers John private health insurance to cover hospital costs for $125 per month.
   a. Calculate the annual cost for John.
   b. John's taxable income is $110 000 per annum. Determine if it is more cost-effective for John to join HIB or pay the Medicare levy surcharge.
   c. John doesn't buy private health insurance until he turns 37. If he joins HIB, determine the cost per month for John, taking into account the lifetime health cover loading.

4. Visit the Australian government website www.privatehealth.gov.au and determine the most appropriate private health insurance for yourself.
5. Private health insurance costs for Marco are provided below:

Type	Cost per month ($)
Hospital	83
Hospital + extras	131

   a. Calculate the difference in cost per year for hospital cover and hospital + extras cover.
   b. Determine the cost of each type of insurance per month if Marco does not join until he is 43 years old.
6. WE3 A 21-year-old driver has three options to purchase comprehensive car insurance for their car. All options include a replacement vehicle and costs for injuries for all passengers.

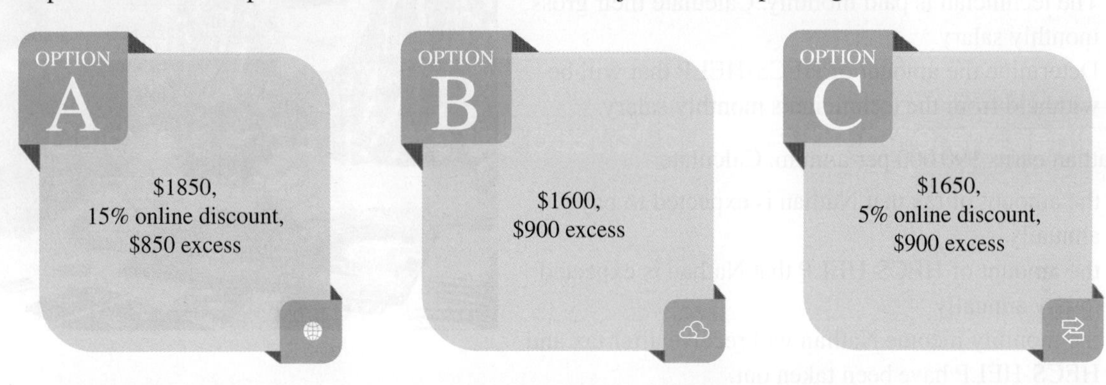

OPTION
A
$1850,
15% online discount,
$850 excess
OPTION
B
$1600,
$900 excess
OPTION
C
$1650,
5% online discount,
$900 excess

   The excess fee is the amount that is to be paid to the insurer if a claim is made.
   If the driver had an accident in the first year of their policy, determine which option is the most cost-effective.
7. Burke is 19 years old and has just had a car accident in which she was at fault. The following excesses apply to claims for at-fault motor vehicle accidents for her comprehensive car insurance:
   - Basic excess of $750 for each claim
   - An additional age excess of $1500 for drivers under 25 years of age
   - An additional age excess of $300 for drivers 25 years of age or over with no more than 2 years of driving experience

   Determine how much excess Burke is required to pay her insurance company.

8. An insurance company offers customers the following discounts on the annual cost of car insurance.

Type of discount	Discount	Conditions
Multi-policy discount	15%	Owner has 2 or more insurance policies with the company.
No claim bonus	20%	Owner has not had an insurance claim in at least 5 years.

Calculate the discounted annual cost for car insurance for:

a. a driver whose basic cost is \$870 per year, who has three insurance policies and has never made a claim in their 10 years with this insurance company
b. a driver whose basic cost is \$1100 per year, who has no other insurance policies and has never made a claim in their 6 years with this insurance company
c. a driver whose basic cost is \$85 per month, who has two insurance policies and has made a claim last year after a small accident.

9. WE4 Anoush studied a Bachelor of Arts at university, accruing a total HECS-HELP debt of \$42 400. She is now working and has to repay her HECS-HELP loan from her monthly gross income.
a. If Anoush earns \$70 000 per annum, calculate her monthly HECS-HELP repayment.
b. Determine how long, to the nearest month, it will take Anoush to pay off her HECS-HELP debt if she remains on this annual salary.

10. A computer technician's annual salary is \$67 374.
a. The technician is paid monthly. Calculate their gross monthly salary.
b. Determine the amount of HECS-HELP that will be withheld from the technician's monthly salary.

11. Nathan earns \$90 000 per annum. Calculate:
a. the amount of tax that Nathan is expected to pay annually
b. the amount of HECS-HELP that Nathan is expected to pay annually
c. the monthly income Nathan will receive after tax and HECS-HELP have been taken out.

Complex unfamiliar

12. Josie is 22 years old and has been with the same car insurance company since obtaining her driving licence at 18 years of age. She has three other types of insurance with this company. Her car insurance costs \$1035 per year prior to any discounts being applied.

Using the information from questions 7 and 8, calculate the annual amount that Josie is required to pay after having her first accident.

13. Use an online car insurance calculator to determine the cost of annual insurance for your family car. Consider two other companies and calculate the most cost-effective insurance option for your family car.

Fully worked solutions for this chapter are available online.

LESSON
8.3 Rent and utility costs

SYLLABUS LINKS

- Investigate the costs involved in independent living [complex].

8.3.1 Rent

- When renting a property, a tenancy agreement or lease must state how much rent is to be paid, how frequently and for how long.
- All rent payments need to be paid in advance.
- A deposit, known as a rental bond, is also usually part of a tenancy agreement. It is often equivalent to four weeks' worth of rent and is kept as a form of insurance in case of damage to the property.

tlvd-5082

WORKED EXAMPLE 5 Calculating costs associated with rental property

A tenancy agreement is shown below.

General tenancy agreement (Form 18a)
Residential Tenancies and Rooming Accommodation Act 2008

Item 7	**Rent**	$ 380 per ☒ week ☐ fortnight ☐ month See clause 8(1)
Item 8	**Rent must be paid on the**	Friday (Insert day. See clause 8(2)) day of each 14 Dec 2024 (Insert week, fortnight or month)
Item 9	**Method of rent payment**	Insert the way the rent must be paid. See clause 8(3) Details for direct credit BSB no. ______ Bank/building society/credit union ______ Account no. ______ Account name ______ Payment reference ______
Item 10	**Place of rent payment**	Insert where the rent must be paid. See clause 8(4) to 8(6)
Item 11	**Rental bond amount**	$ 2000 See clause 13

Calculate:

a. the bond amount payable

b. the total amount payable on 14 December 2024

c. the amount of rent to be paid each month, to the nearest dollar.

THINK	WRITE
a. Find the bond amount in the tenancy agreement.	a. Bond amount = $2000
b. 1. Calculate the sum of the bond and the weekly rent.	b. $\$2000 + \$380 = \$2380$
2. Write the answer.	The total amount payable on 14 December is $2380.
c. 1. Calculate the amount of rent to be paid per year.	c. $\$380 \times 52 = \$19\,760$
2. Divide the yearly amount of rent by 12.	$\frac{\$19\,760}{12} = \$1647$
3. Write the answer.	To the nearest dollar, the amount of rent to be paid each month is $1647.

8.3.2 Electricity and gas bills

- There are three main parts to our electricity and gas supply systems: wholesalers (or generators), distributors and retailers.
  - Wholesalers produce electricity and extract gas.
  - Distributors own and maintain infrastructure such as power poles, wiring and gas lines. When there are disruptions to the electricity or gas supply, the distributor is contacted to fix the problem.
  - Retailers purchase electricity and gas from wholesalers and then sell them to customers.
  - There are many electricity and gas options available to customers, and customers can select whichever deal best suits their needs. Therefore, it is important to know how much you will be charged per unit of energy used and what discounts are available before you sign any contracts.

The standard measurement unit of electricity consumption

The standard measurement unit of electricity consumption is the kilowatt hour (kWh). One kWh is the amount of energy produced by an appliance outputting 1 kW (1000 W) of power for an hour. Gas consumption is measured in megajoules (MJ).

- To understand electricity and gas bills, it is important to know some of the terms that are used.

Energy account

Mary Sample
123 Main Road
Sampleton, 0000

Inquiries:	555 888 000
Customer No.:	123 456 789
Due date:	12 Jan 2025
Total due:	$286.00

Your average daily use at:
123 Main Road, Sampleton, 0000

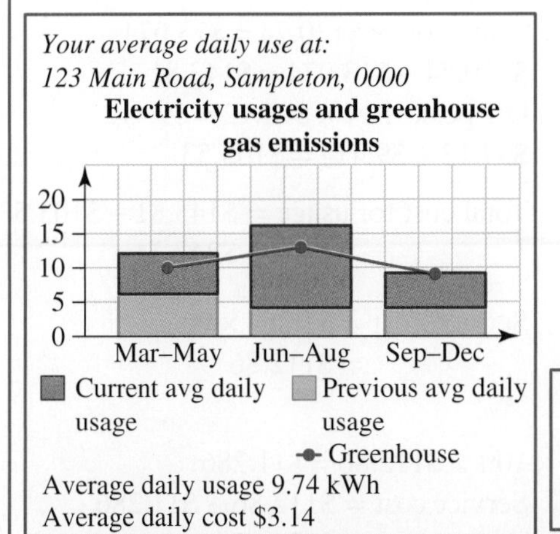

Average daily usage 9.74 kWh
Average daily cost $3.14

Average daily usage is how much energy was used each day on average.
Average cost per day is how much you pay each day on average for the billing period.

Electricity charges (charges based on estimated meter reading)								
Billing period: 1 Sep–1 Dec								
Tariff	**Meter no.**	**Bill days**	**Current reading**	**Previous reading**	**Total usage (kWh)**	**Charge/rate (c/kWh)**	**$ exc. GST**	**$ inc. GST**
Peak	111111	91	793	459	334	28.9	96.53	106.18
Off-peak	111111	91	4861	4309	552	12.8	70.66	77.72
Service to property		91			**$1.020/day**		92.82	102.10

Bill days is the number of days for the billing period.
Charge/kWh is how much you are being charged, in cents, per kilowatt hour.
Peak and off-peak are the different time periods for electricity usage. They will have different charges per kWh depending on the energy plan.
Service to property is a fixed charge that is also called the 'daily supply charge'.

Note: GST is a tax imposed by the Australian federal government on goods and services. A tax of 10% is added to an amount.

tlvd-5083

WORKED EXAMPLE 6 Calculating the total electricity bill amount

The meter readings and charge/rates for a household are shown.

Electricity charges (charges based on meter reading)								
Billing period: 1 Jan–31 March								
Tariff	**Meter no.**	**Bill days**	**Current reading**	**Previous reading**	**Total usage (kWh)**	**Charge/rate (c/kWh)**	**$ exc. GST**	**$ inc. GST**
Peak		**90**	**12 791**	**12 353**	**438**	**29.85**		
Off-peak		**90**	**2 587**	**1 883**	**704**	**13.37**		
Service to property		**90**			**$1.254/day**			

Calculate the total bill amount.

THINK	WRITE
1. Looking at the table, determine the peak and off-peak electricity usage.	Peak: 438 Off-peak: 704
2. Calculate the cost for each time period by multiplying the usage by the charge/rate.	Peak: $438 \times \$0.2985 \approx \$130.74$ Off-peak: $704 \times \$0.1337 \approx \$94.12$
3. Add GST by adding 10% of each amount. (See chapter 4, lesson 4.3.3. The amount of GST on an item can be determined by dividing by 10 if the amount is pre-GST.)	Peak: $10\% \times \$130.74 = \$13.074$ $\$130.74 + \$13.074 = \$143.81$ Off-peak: $10\% \times \$94.12 = \$9.412$ $\$94.12 + \$9.412 = \$103.53$
4. Calculate the usage cost.	Total cost for usage $= \$143.81 + \$103.53$ $= \$247.34$
5. Calculate the cost of service to the property by multiplying the cost per day by the number of billing days.	Service cost $= \$1.254 \times 90$ $= \$112.86$
6. Add 10% GST to the service cost.	$10\% \times \$112.86 = \$11.286$ Service cost $= \$112.86 + \$11.286$ $= \$124.15$
7. Calculate the total cost by adding the usage cost to the service cost.	Total cost $= \$247.34 + \$124.15$ $= \$371.49$
8. Write the answer as a sentence.	The bill total is \$371.49.

8.3.3 Water bills

- Homeowners are charged a fixed fee for the supply of water and removal of wastewater.
- They are also charged a fee for the management of stormwater services, and some households have to pay a recycled-water usage fee, depending on the property.
- The amount of water used in the home over a fixed period, usually three months, is measured through a water meter fixed to the main water supply in the home.

Water usage

Homeowners are charged for their water usage at a rate in dollars per kilolitre (kL).

1000 litres = 1 kilolitre

***Note:* GST does not apply to water bills.**

WORKED EXAMPLE 7 Calculating water rates with usage and fixed charges

Calculate the water rates for a home with the following usage and fixed charges.

Bill details as at 31 March	Billing period	Value	Price
Water service charge	1 Jan–31 Mar	22.51	
Sewerage service charge	1 Jan–31 Mar	145.90	
Stormwater service	1 Jan–31 Mar	18.70	
Usage	**Volume**	**Charge/kL**	**Price**
Water usage	55	\$2.00	
Recycled water usage	35	\$1.79	

THINK	WRITE
1. Calculate the cost of the water usage by multiplying the volume by the cost.	$\text{Water usage} = 55 \times \$2$ $= \$110$
2. Calculate the cost of the recycled water usage by multiplying the volume by the cost.	$\text{Recycled water} = 35 \times \$1.79$ $= \$62.65$
3. Calculate the fixed service cost.	$\$22.51 + \$145.90 + \$18.70 = \$187.11$
4. Add the fixed service cost to the water usage cost.	$\$187.11 + \$110 + \$62.65 = \$359.76$
5. Write the answer as a sentence.	The water rates amount to \$359.76.

8.3.4 Council rates

- Councils provide services and infrastructure to people living in communities, such as parklands, libraries, rubbish collection and road maintenance.
- To support councils in providing these services, homeowners pay **council rates**.
- Council rates are calculated as a percentage of the capital improved value of the land and buildings within the council municipality.

Calculating council rates

$$\text{Rate in the dollar} = \frac{\text{revenue}}{\text{council capital improved value}}$$

Then multiply the rate in the dollar by the capital improved value of the property.

Annual council rates = rate in the dollar × property capital improved value

For example, if the council plans to raise $15 million for the annual budget and there is $2 billion in capital improved value within the municipality, then the rate in the dollar is $\frac{\$15\,000\,000}{\$2\,000\,000\,000} = 0.0075$.

If the capital improvement on a ratepayer's house and land is $750 000, then the payable council rates for that property are $0.0075 \times \$750\,000 = \$5625$.

Note: GST does not apply to council rates.

WORKED EXAMPLE 8 Calculating the payable council rates for a property

A council municipality plans to raise $15 million. There is $3.25 billion in capital improved value in the municipality. Calculate the payable council rates for a property with $275 000 capital improved value. Write your answer correct to the nearest cent.

THINK	WRITE
1. Calculate the rate in the dollar. Divide the revenue to be raised by the capital improved value.	$\frac{\$15\,000\,000}{\$3\,250\,000\,000} = \frac{15}{3250}$ $= 0.004\,62...$
2. Multiply the rate in the dollar by the capital improved value.	$0.004\,62... \times \$275\,000 = \$1269.23$
3. Write the answer as a sentence.	The council rates payable are $1269.23.

 Resources

 **Interactivity** Electricity bills (int-6912)

Exercise 8.3 Rent and utility costs

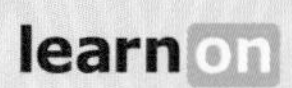

8.3 Quick quiz	8.3 Exercise

Simple familiar	Complex familiar	Complex unfamiliar
N/A	1, 2, 3, 4, 5, 6, 7, 8, 9, 10, 11, 12, 13, 14	15, 16

Complex familiar

1. WE5 A tenancy agreement is shown.

General tenancy agreement (Form 18a)
Residential Tenancies and Rooming Accommodation Act 2008

Item 7 **Rent** $ 550 per ☒ week ☐ fortnight ☐ month See clause 8(1)

Item 8 **Rent must be paid on the** Wednesday day of each 21 August 2024
Insert day. See clause 8(2) Insert week, fortnight or month

Item 9 **Method of rent payment** Insert the way the rent must be paid. See clause 8(3)

Details for direct credit
BSB no. Bank/building society/credit union
Account no. Account name
Payment reference

Item 10 **Place of rent payment** Insert where the rent must be paid. See clause 8(4) to 8(6)

Item 11 **Rental bond amount** $ See clause 13

Calculate:

a. the bond amount payable
b. the total amount payable on 21 August 2024
c. the amount of rent to be paid each month, to the nearest dollar.

2. A tenancy agreement is shown.

General tenancy agreement (Form 18a)
Residential Tenancies and Rooming Accommodation Act 2008

Item 7 **Rent** $ 2250 per ☐ week ☐ fortnight ☒ month See clause 8(1)

Item 8 **Rent must be paid on the** 16th of each month day of each 16 January 2025
Insert day. See clause 8(2) Insert week, fortnight or month

Item 9 **Method of rent payment** Insert the way the rent must be paid. See clause 8(3)

Details for direct credit
BSB no. Bank/building society/credit union
Account no. Account name
Payment reference

Item 10 **Place of rent payment** Insert where the rent must be paid. See clause 8(4) to 8(6)

Item 11 **Rental bond amount** $ See clause 13

a. Calculate:

i. the amount of rent to be paid each week, to the nearest dollar
ii. the total amount payable on 16 January 2025.

b. A family has the choice between this rental property and the rental property given in question 1. Discuss which of the two options is more cost-effective for the family.

3. **WE6** The meter readings and charges/rates for a household are as shown.

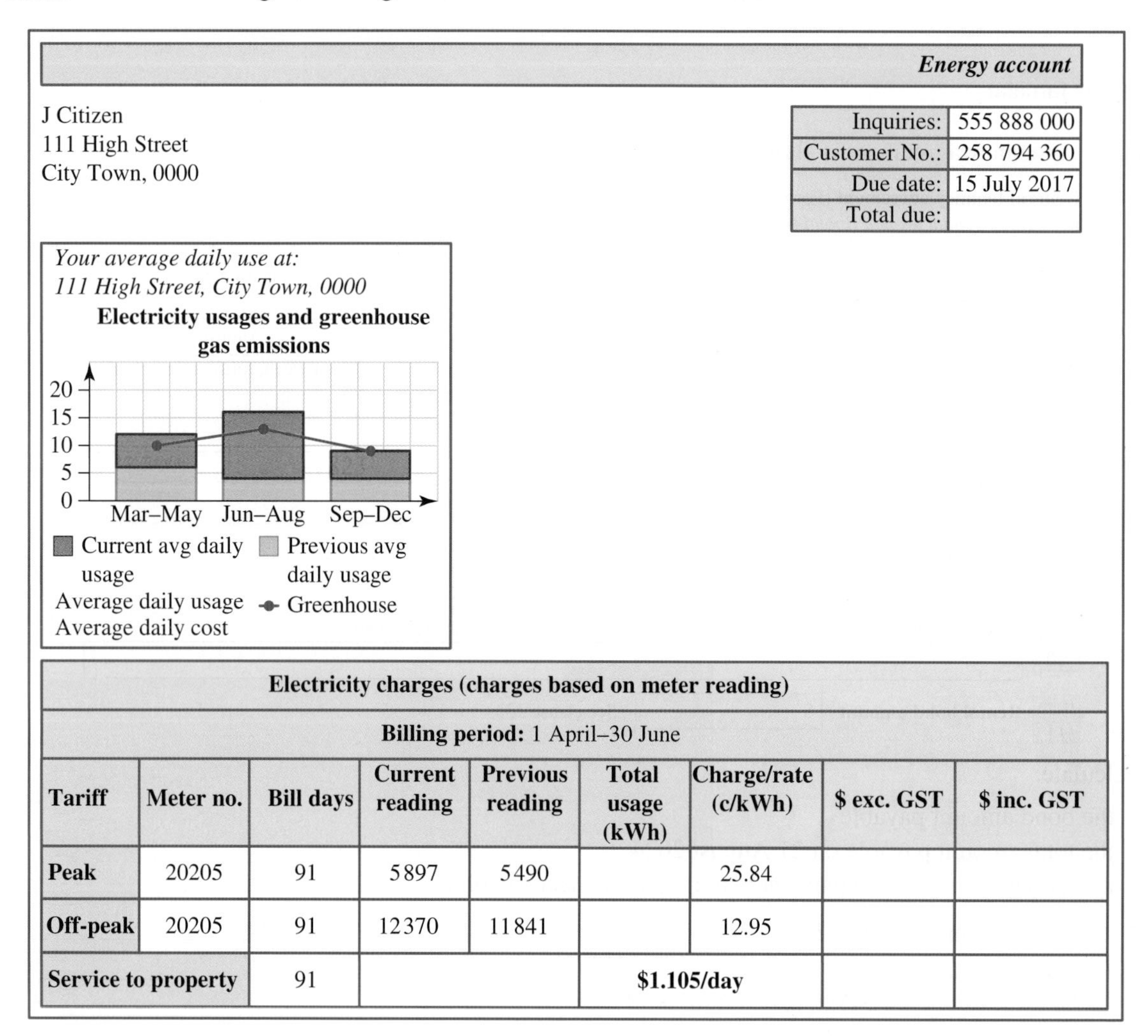

Energy account

J Citizen
111 High Street
City Town, 0000

Inquiries:	555 888 000
Customer No.:	258 794 360
Due date:	15 July 2017
Total due:	

Your average daily use at:
111 High Street, City Town, 0000
Electricity usages and greenhouse gas emissions

20
15
10
5
0
Mar–May Jun–Aug Sep–Dec
Current avg daily usage
Previous avg daily usage
Average daily usage
Greenhouse
Average daily cost

Electricity charges (charges based on meter reading)								
Billing period: 1 April–30 June								
Tariff	**Meter no.**	**Bill days**	**Current reading**	**Previous reading**	**Total usage (kWh)**	**Charge/rate (c/kWh)**	**\$ exc. GST**	**\$ inc. GST**
Peak	20205	91	5897	5490		25.84		
Off-peak	20205	91	12370	11841		12.95		
Service to property		91			**\$1.105/day**			

Calculate the total bill amount. (Remember to include GST.)

4. Calculate the energy charges (excluding GST) for each of the following, giving your answers correct to the nearest cent.

	Energy usage	Charge/rate
a.	563 kWh	10.56c/kWh
b.	1070 MJ	2.629c/MJ
c.	895 kWh	14.64c/kWh
d.	6208 MJ	1.827c/MJ

5. The usage component of a family's energy bill for a quarter is \$545 excluding GST. The charge/rate is 11.43c/kWh. Calculate the family's energy usage for the quarter. Give your answer to the nearest whole number.

6. Part of an electricity bill is shown below.

Electricity charges (charges based on meter reading)								
Billing period: 1 Sep–31 Dec								
Tariff	**Meter no.**	**Bill days**	**Current reading**	**Previous reading**	**Total usage (kWh)**	**Charge/rate (c/kWh)**	**$ exc. GST**	**$ inc. GST**
All day	28 710	91	47 890	46 995		30.54		
Service to property		91			**$1.017/day**			

a. Write:

i. the charge/rate
ii. the daily service-to-property fee
iii. the usage amount in kWh.

b. Calculate:

i. the amount charged for electricity usage excluding GST
ii. the total amount for the bill, excluding GST
iii. the total amount for the bill, including GST.

7. There is a 15% discount off the total bill for customers who pay their energy bill before the due date. The energy usage of a customer is:
- electricity: 456 kWh at 12.46c/kWh
- gas: 967 MJ at 1.956c/MJ
- $95 fixed fee for service to the property.

If the customer pays before the due date, calculate the amount paid including GST.

8. A family's electricity usage is shown in the following table.

Tariff	Bill days	Current reading	Previous reading	Total usage	Charge/rate (c/kWh)
Peak	90	58 973	58 615		29.05
Off-peak	90	8569	8028		15.59
Service to property	90			**$1.08/day**	

Calculate the total cost of the family's electricity usage, including GST and correct to the nearest 5 cents.

9. **WE7** Calculate the water rates for a home with the following usage and fixed charges.

Bill details as at 31 March	Billing period	Value	Price
Water service charge	1 Jan–31 Mar	22.51	
Sewerage service charge	1 Jan–31 Mar	145.90	
Stormwater	1 Jan–31 Mar	18.70	
Usage	**Volume (L)**	**Charge/kL**	**Price**
Water usage	58 000	$2.00	
Recycled water usage	39 000	$1.79	

10. A customer pays $245.85 for their water bill for three months. They are charged a fixed service fee of $180.30. The charge/kL is $1.95. The customer does not have recycled-water service. Calculate the amount of water, in kilolitres, used by the customer for the three months. Give your answer correct to 2 decimal places.

11. **WE8** A council municipality plans to raise $12 million. There is $1.8 billion in capital improved value in the municipality. Calculate the payable council rates for a property with $275 000 capital improved value.

12. The council rates for a home that has a capital improved value of $325 000 are $1950.
   a. Calculate the rate in the dollar charged by the council.
   b. Calculate the amount of revenue the council intends to collect if the total amount of capital improved value for the municipality is $2.67 billion.
13. Here is an incomplete water bill received by a household.

Bill details as at 30 June	Billing period	Value	Price
Water service charge	1 Apr–30 Jun	23.80	
Sewerage service charge	1 Apr–30 Jun	150.20	
Stormwater	1 Apr–30 Jun	17.95	
Usage	**Volume (L)**	**Charge/kL**	**Price**
Water usage	64 000		130.56
Recycled water usage	38 000		
		Total	**$393.57**

   a. Complete the water bill by filling in the missing values.

   As at 1 July, the charges per kL will increase by 4%, the water service charge will increase by $2.50 and the sewerage service charge will increase by $1.50.

   b. Calculate the new monthly bill.
   c. Determine the percentage by which the average water bill will increase (assuming water usage remains the same). Give your answer correct to 2 decimal places.
14. A household's bills for three months (one quarter) are shown.

Electricity bill

Tariff	Bill days	Current reading	Previous reading	Total usage	Charge/rate (c/kWh)
Peak	90	12 589	11 902		31.08
Off-peak	90	2 058	1 511		14.05
Service to property	90				$1.075/day

Water bill

Bill details as at 31 March	Billing period	Value	Price
Water service charge	1 Jan–31 Mar	22.51	
Sewerage service charge	1 Jan–31 Mar	145.90	
Stormwater	1 Jan–31 Mar	18.70	
Usage	**Volume (L)**	**Charge/kL**	**Price**
Water usage	55	$2.00	
Recycled water usage	35	$1.79	

Council rates

Capital improved value $280 000; rate in the dollar 0.008

   a. Show that the quarterly council rate instalment is $560.
   b. Calculate the amount that the household spends on electricity, water and council rates for the quarter. (Remember to include GST where relevant.)
   c. The family's monthly net income is $6019.35 (after tax). Calculate the percentage of the income that goes towards paying the household bills. Give your answer correct to 2 decimal places.

Complex unfamiliar

15. A customer has a choice between two electricity plans, A and B. The two plans are shown below.

Plan A		Plan B	
Charge/rate		**Charge/rate**	
All day	12.03c/kWh	Off-peak (11 pm–7 am)	9.08c/kWh
		Peak	13.01c/kWh
Service to property	$1.01/day	Service to property	$0.99/day
Pay-on-time discount	15%	Pay-on-time discount	12%

The customer's energy usage over 90 billing days is 534 (peak time) and 757 (off-peak time).

Determine which plan is more cost-effective for the customer. Justify your answer by calculating the electricity costs for each plan if the customer pays on time and if they don't pay on time.

16. A family has both electricity and gas connected to their house. Their average electricity usage for 90 billing days is 987 kWh and their average gas usage is 1650 MJ per month (30 days). The family decides to change their provider. They have the following choices.

Choice 1: Separate electricity and gas providers

- Electricity: all-day rate 10.03c/kWh
- Pay-on-time discount (electricity only): 12.5%
- Gas: all-day rate 1.89c/MJ

Choice 2: Bundle both electricity and gas with one provider and receive the following discounts

- 17% discount on gas
- 10% discount on electricity
- Electricity daily rate: 12.01c/kWh
- Gas daily rate: 2.03c/MJ

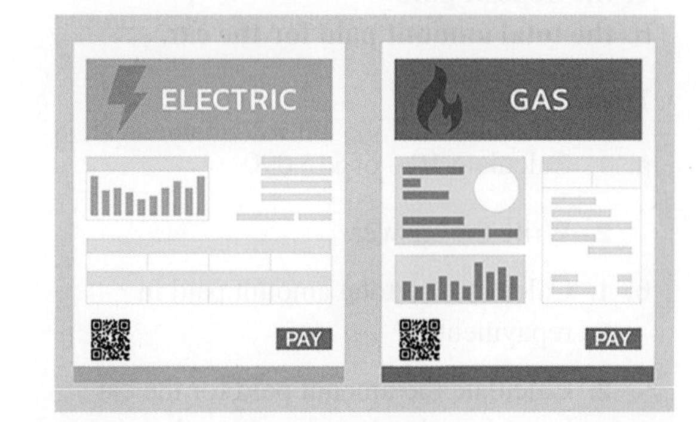

Service-to-property charges are the same for both providers.
Determine which choice will be more cost-effective for the family, assuming they always pay on time.

Fully worked solutions for this chapter are available online.

LESSON
8.4 Car expenses

SYLLABUS LINKS

- Investigate the costs involved in independent living [complex].

Source: Essential Mathematics Senior Syllabus 2024 © State of Queensland (QCAA) 2024; licensed under CC BY 4.0.

8.4.1 Planning for the purchase of a vehicle

- Many costs are associated with buying a vehicle, but the primary cost is the purchase price or sale price.
- It is important to know how much the vehicle is worth and then shop around for the best price.

Paying for the vehicle

- Many financial institutions offer loans to purchase vehicles.
- As with all loans, it is important to know exactly how much interest is charged and what additional costs apply, such as administration fees and loan servicing fees.

tlvd-5087

WORKED EXAMPLE 9 Calculating costs involved in buying a vehicle

A customer takes out a loan to purchase a car for \$15 090. The monthly repayments are \$294.26 and the loan is for 5 years. The customer pays a 10% deposit.
Calculate:
a. the deposit paid
b. the total amount paid for the car.

THINK	WRITE
a. 1. Calculate 10% of \$15 090.	a. $\frac{10}{100} \times \$15\,090 = \$1509$
2. Write the answer.	The deposit paid is \$1509.
b. 1. Calculate the total amount paid in repayments.	b. Monthly repayments = \$294.26 for 5 years $\$294.26 \times 12 \times 5 = \$17\,655.60$
2. Calculate the amount paid for the car by adding the deposit paid to the total amount in monthly repayments.	Amount paid for the car $= \$1509 + \$17\,655.60$ $= \$19\,164.60$
3. Write the answer.	The total amount paid for the car is \$19 164.60.

8.4.2 On-road costs for new and used vehicles

- In addition to car insurance, there are other on-road costs such as registration and stamp duty.
- A **registration fee** is a combination of administration fees, taxes and charges paid to legally drive a vehicle on the roads.

- Stamp duty is a government tax on certain transactions, including the purchase of a motor vehicle. It is charged based on:
  - the date the vehicle was registered or the registration transferred
  - whether it is a passenger car or a non-passenger car
  - the dutiable value of the vehicle.

WORKED EXAMPLE 10 Calculating the amount of stamp duty for vehicle purchases

Calculate the amount of stamp duty to be paid for the following vehicle purchases, given the rates in the following table.

a. A 4-cylinder car purchased for $14 560

b. A 4-cylinder car purchased for $104 900

Type of vehicle	Rate
• hybrid — any number of cylinders • electric	Up to $100 000: $2 for each $100, or part of $100 More than $100 000: $4 for each $100, or part of $100
• 1 to 4 cylinders • 2 rotors • steam powered	Up to $100 000: $3 for each $100, or part of $100 More than $100 000: $5 for each $100, or part of $100

THINK	WRITE
a. 1. Look at the table and determine the percentage to be paid.	a. The vehicle costs less than $100 000 and has four cylinders. Therefore, the percentage is 3%.
2. Calculate 3% of the vehicle purchase price.	$\frac{3}{100} \times \$14\,560 = \$436.80$
3. Write the answer as a sentence.	The stamp duty to be paid on a four-cylinder vehicle purchased for $14 560 is $436.80.
b. 1. Determine the percentage to be paid.	b. The vehicle costs more than $100 000 and has four cylinders. Therefore, the percentage is 5%.
2. Calculate 5% of the vehicle purchase price.	$\frac{5}{100} \times \$104\,900 = \$5245$
3. Write the answer as a sentence.	The stamp duty to be paid on a 4-cylinder vehicle purchased for $104 900 is $5245.

8.4.3 Sustainability, fuel consumption rates, servicing and tyres

- The running costs of a vehicle include fuel consumption and maintenance.
- The latter includes the cost to service the vehicle and replace the tyres.
- **Fuel consumption** rates show the amount of fuel (petrol or diesel) used per 100 km. For example, 7.8 L/100 km means that the vehicle uses an average of 7.8 litres of fuel for every 100 km travelled.

WORKED EXAMPLE 11 Calculating the fuel consumption rate

A vehicle travels 580 km and uses 62 litres of fuel. Calculate the fuel consumption rate in L/100 km. Give your answer correct to 2 decimal places.

THINK	WRITE
1. Divide the total distance travelled by 100 km (this gives the distance in 100-km lots).	$\frac{580}{100} = 5.8$
2. Divide the amount of fuel in litres by the answer in step 1. Round to 2 decimal places.	$\frac{62}{5.8} = 10.6896$ ≈ 10.69
3. Write the answer as a sentence.	The fuel consumption rate is 10.69 L/100 km.

- Vehicles require servicing by experienced mechanics to maintain performance.
- Services are generally required every 10 000–15 000 km, depending on the manufacturer's guidelines.
- The service cost will include labour, plus the cost to replace parts.

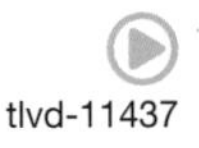
tlvd-11437

WORKED EXAMPLE 12 Calculating the cost of servicing a vehicle

A driver travels an average of 450 km every week.

a. If the vehicle requires servicing every 15 000 km, determine how many times this vehicle will need servicing over three years.

b. At the 30 000-km service, the vehicle needs the following parts. (The prices for the parts include GST.)
- **Oil filter $30**
- **Fuel filter $25**
- **Oil $35**
- **Windscreen wiper blades $15**

The service will take 3 hours of labour at $105 per hour excluding GST.
Calculate the total cost for the 30 000-km service.

THINK	WRITE
a. 1. Calculate the number of weeks the driver takes to travel 15 000 km.	a. $\frac{15\,000}{450} = 33.33$
2. Calculate the number of weeks in 3 years.	$52 \times 3 = 156$
3. Divide the number in step 2 by the number in step 1.	$\frac{156}{33.33} = 4.68$
4. Write the answer as a sentence.	There are four 15 000-km services in 3 years.
b. 1. Calculate the labour costs.	b. $3 \times \$105 = \$315$ Add 10%. GST: 10% of $\$315 = \$31.50$ Labour costs $= \$315 + \$31.50$ $= \$346.50$
2. Calculate the cost of parts.	$\$30 + \$25 + \$35 + \$15 = \$105$
3. Add the labour costs to the cost of the parts.	$\$105 + \$346.50 = \$451.50$
4. Write the answer as a sentence.	The total cost for the 30 000-km service is $451.50.

- Vehicles without the regulated amount of tyre tread are deemed unsafe and the owner is required to have the tyres replaced.
- The tyres are usually checked during routine vehicle servicing to ensure that they are safe to use and are wearing evenly.

WORKED EXAMPLE 13 Determining how many times tyres need to be replaced

For a particular brand of tyres, the manufacturer recommends replacement after 40 000 km. A driver using these tyres travels an average of 550 km each week. Determine how many times they should expect to replace the tyres in 5 years.

THINK	WRITE
1. To calculate the number of weeks it takes to travel 40 000 km, divide 40 000 km by the average weekly distance.	$\frac{40\,000}{550} = 72.73$
2. To calculate the number of years it takes to travel 40 000 km, divide the answer by 52.	$\frac{72.73}{52} = 1.3986$
3. To calculate the number of times the tyres should be replaced in 5 years, divide 5 by the answer from step 2.	$\frac{5}{1.399} = 3.574$
4. Write the answer as a sentence.	In 5 years the driver should expect to replace the tyres 3 times.

8.4.4 Calculate and compare the cost of purchasing different vehicles using a spreadsheet

- A spreadsheet can be used to calculate and compare the cost of purchasing different vehicles.
- Many resources that compare vehicle performance and costs are available.
- To compare vehicles, it is important to compare similar vehicles and common elements, such as fuel consumption and servicing costs.
- Registration and insurance costs will vary due to other factors, such as location of the vehicle and state of registration.

WORKED EXAMPLE 14 Determining the most effective purchase using a spreadsheet

Using a spreadsheet, determine which one of the following three cars is the most cost-effective to buy. Assume that the average distance travelled each year is 15 000 km, the average price of fuel per litre is $1.20, and all of the cars have the same safety rating.

Four-door sedan 1.6-litre engine				
Model	**Sale price**	**Fuel consumption (L/100 km)**	**Servicing (km)**	**Average service costs**
Car A	**$15 480**	**8.2**	**15 000**	**$450**
Car B	**$16 250**	**7.5**	**20 000**	**$350**
Car C	**$14 999**	**9.8**	**10 000**	**$475**

THINK | **WRITE**

1. Create a spreadsheet.

	A	B	C	D	E
1	Vehicle	Sales price	Fuel consumption (L/100 km)	Servicing (km)	Average service costs
2	Car A	$15 480.00	8.2	15 000	$450.00
3	Car B	$16 250.00	7.5	20 000	$350.00
4	Car C	$14 999.00	9.8	10 000	$475.00

2. Calculate the number of litres used (on average) for one year of driving.

Average distance travelled = 15 000 km

Number of 100-km units: $\frac{15\,000}{100} = 150$

Insert the formula '= 150*C2' into the cell F2.
Copy the formula into cells F3 and F4.

3. Calculate the cost to fuel the car for one year at $1.20/L.

Enter the formula '= F2*1.2' into cell G2.
Copy the formula into cells G3 and G4.

	A	B	C	D	E	F	G
1	Vehicle	Sales price	Fuel consumption (L/100 km)	Servicing (km)	Average service costs	Litres/ years	Fuel cost
2	Car A	$15 480.00	8.2	15 000	$450.00	1230	$1476.00
3	Car B	$16 250.00	7.5	20 000	$350.00	1125	$1350.00
4	Car C	$14 999.00	9.8	10 000	$475.00	1470	$1764.00

4. Calculate the servicing costs. To compare the costs, calculate the proportional service costs by dividing the average distance travelled (15 000 km) by the service distance for each car, then multiply by the average service costs.

Enter the formula '= 15 000/D2*E2' into cell H2.
Copy the formula into cells H3 and H4.

	A	B	C	D	E	F	G	H
1	Vehicle	Sales price	Fuel consumption (L/100 km)	Servicing (km)	Average service costs	Litres/ years	Fuel cost	Service cost
2	Car A	$15 480.00	8.2	15 000	$450.00	1230	$1476.00	$450.00
3	Car B	$16 250.00	7.5	20 000	$350.00	1125	$1350.00	$262.50
4	Car C	$14 999.00	9.8	10 000	$475.00	1470	$1764.00	$712.50

5. Calculate the total cost for each car over the year.

Insert the formula '= B2 + G2 + H2' into cell I2.
Copy the formula into cells I3 and I4.

	A	B	C	D	E	F	G	H	I
1	Vehicle	Sales price	Fuel consumption (L/100 km)	Servicing (km)	Average service costs	Litres/ years	Fuel cost	Service cost	Total cost
2	Car A	$15 480.00	8.2	15 000	$450.00	1230	$1476.00	$450.00	$17 406.00
3	Car B	$16 250.00	7.5	20 000	$350.00	1125	$1350.00	$262.00	$17 862.50
4	Car C	$14 999.00	9.8	10 000	$475.00	1470	$1764.00	$712.00	$17 475.50

6. Write the answer as a sentence.

The most cost-effective car to buy is car A, with an overall annual cost in the first year of $17 406.

Resources

Interactivity Car loans (int-6913)

Exercise 8.4 Car expenses

learnon

8.4 Quick quiz	8.4 Exercise

Simple familiar	Complex familiar	Complex unfamiliar
N/A	1, 2, 3, 4, 5, 6, 7, 8, 9, 10, 11, 12	13, 14

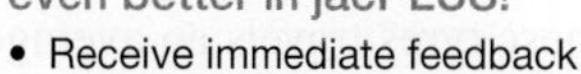
These questions are even better in jacPLUS!
- Receive immediate feedback
- Access sample responses
- Track results and progress

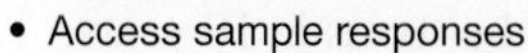
Find all this and MORE in jacPLUS

Complex familiar

1. WE9 A customer takes out a loan to purchase a car for $17 995. The monthly repayments are $313.38 and the loan is for 5 years. The customer pays a 10% deposit. Calculate:
   a. the deposit paid
   b. the total amount paid for the car.

2. A $24 995 car is purchased through finance. A 10% deposit is paid, with monthly repayments of $487.22 for 5 years. Calculate:
   a. the deposit paid
   b. the amount borrowed for the car (principal)
   c. the total amount paid for the car.
3. WE10 Stamp duty payable on the purchase of four-cylinder vehicles is:
   - for a vehicle that cost $100 000 or less: 3% ($3 per $100 or part thereof)
   - for a vehicle that cost more than $100 000: 5% ($5 per $100 or part thereof).

   Calculate the amount of stamp duty to be paid for the following vehicle purchases.

   a. $22 995
   b. $125 000

4. The stamp duty paid for a four-cylinder vehicle was \$250. Determine the price paid for the vehicle.

5. WE11 A vehicle travels 640 km and uses 58 litres of fuel. Calculate the fuel consumption rate in L/100 km. Give your answer correct to 2 decimal places.

6. A vehicle's fuel consumption rate for a journey was calculated at 8.9 L/100 km. If the vehicle used a total of 71 L of fuel for the journey, determine how far it travelled. Give your answer to the nearest whole kilometre.

7. A family is on a road trip. The vehicle they are travelling in averages 765 km for 99 L of fuel.
   a. Calculate the fuel consumption rate correct to 2 decimal places.
   b. If the average fuel price is \$1.34/L, calculate the cost per 100 km.

8. WE12 A driver travels on average 350 km every week. Assume 52 weeks in a year.
   a. If the vehicle requires servicing every 10 000 km, calculate how often it will need servicing over 2 years.
   b. At the 30 000 km service, the vehicle needs the following parts and labour. (The prices for the parts include GST.)
      - Oil filter \$20
      - Fuel filter \$35
      - Oil \$40
      - Windscreen wiper blades \$10
      - 2.5 hours labour at \$85 per hour excluding GST

      Calculate the cost for the 30 000 km service.

9. WE13 For a particular brand of tyres, the manufacturer recommends replacement after 35 000 km. A driver using these tyres travels an average of 2050 km each month. Determine how many times she should expect to replace her tyres in 10 years.

10. A vehicle's average fuel consumption is 9.8 L/100 km. The driver fills the tank with 65 litres of fuel. Determine the average number of kilometres the vehicle can travel before the tank is empty.

11. WE14 Using a spreadsheet, work out which one of the following three cars is the most cost-effective to buy. Assume that the average distance travelled each year is 15 000 km and the cost of fuel is \$1.15.

Four-door sedan (medium-sized car)				
Model	**Sale price**	**Fuel consumption (L/100 km)**	**Servicing (km)**	**Average service costs**
Car A	\$37 990	12.7	15 000	\$255
Car B	\$36 748	8.6	20 000	\$595
Car C	\$38 490	10.5	10 000	\$235

12. The fuel consumption for a vehicle travelling in city traffic is recorded as 9.8 L/100 km. When travelling along highways, the fuel consumption is recorded as 7.5 L/100 km.
    a. Demonstrate that the average fuel consumption is 8.65 L/100 km.
    b. The vehicle travels 126 km along city roads and 458 km along the highway. Calculate the total number of litres of fuel used, correct to 1 decimal place.
    c. The cost of fuel is \$1.25 per litre. Calculate the cost of the fuel for this trip, correct to the nearest cent.
    d. When calculating the fuel costs using the average fuel consumption, the value is different to the actual cost. Explain why.

Complex unfamiliar

13. A vehicle is purchased for \$24 980 through finance that attracts 6% interest for 5 years. The owner pays a 10% deposit and then pays \$496.28 monthly. This value for the monthly repayments includes a \$250 one-off administration fee to set up the loan and a monthly \$5 loan-servicing fee.
Demonstrate that the comparison rate for the loan is 6.489% per annum.

14. A person wants to buy a new car. They drive to work in the city, a return distance of 45 km, Monday to Friday. On the weekend, they drive along a major highway to visit a relative who lives in a regional town 112.5 km from the city.
They want to choose one of the following cars.

Car	Fuel consumption for city driving (L/100 km)	Fuel consumption for highway driving (L/100 km)
A	12.5	7.8
B	11.2	8.5

Based on the weekly distances driven by the person, determine which car would be more cost-effective to purchase. Justify your answer using calculations.

Fully worked solutions for this chapter are available online.

LESSON
8.5 Personal budget plan

SYLLABUS LINKS

- Prepare a personal budget plan [complex].

8.5.1 Monthly income

- When managing money, it can be helpful to create a monthly budget.
- A **budget** is a list of all planned income and costs.
- A budget can be prepared for long periods of time, but for individuals it is usually a monthly plan.
- In order to create a monthly budget, monthly income and expenses need to be known.
  Monthly income is $\frac{1}{12}$ of the total annual income.

WORKED EXAMPLE 15 Calculating monthly income

Calculate the monthly income of a worker who has a net annual salary of $71 400 and earns $1590 each year in bank interest.

THINK

1. Calculate the total annual income as the sum of salary and bank interest.
2. Calculate the total monthly income after tax by dividing the answer in step **1** by 12.
3. Write the answer as a sentence.

WRITE

1. $\text{Total annual income after tax} = \$71\,400 + \$1590 = \$72\,990$
2. $\text{Total monthly income after tax} = \dfrac{\$72\,990}{12} = \$6082.50$
3. The worker's monthly income after tax is $6082.50.

8.5.2 Fixed and variable expenses

- **Fixed expenses** are expenses that occur every month (or week, fortnight, year or other time period) and are always the same amount.
  - These may include rent or mortgage, school fees, health and car insurance.
- **Variable expenses** occur from time to time and are not a fixed amount.
  - These may include food, entertainment, clothing and other items that can be controlled or varied.
- To reduce your expenses in order to save more money, you would look at reducing your variable expenses.

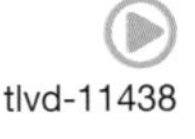
tlvd-11438

WORKED EXAMPLE 16 Calculating total monthly expenses

Determine the total monthly expenses from the following list of fixed and variable expenses. The variable expenses represent the amounts spent in a month.

	Fixed expenses		Variable expenses	
Item	**Frequency**	**Amount**	**Item**	**Amount**
Rent	**Weekly**	**$230**	**Food**	**$423**
Health insurance	**Monthly**	**$78**	**Clothing**	**$107**
Vehicle registration	**Yearly**	**$620**	**Entertainment**	**$85**
Vehicle insurance	**Yearly**	**$389**	**Car repairs**	**$325**

THINK

1. Convert any weekly expenses to monthly expenses by multiplying by $\frac{52}{12}$. The only weekly cost is the rent of $230.
2. Determine the total of all of the annual expenses and divide by 12 to convert it to monthly expenses.
3. Calculate the total fixed expenses by adding the numbers from steps **1** and **2** to the monthly fixed cost of health insurance ($78).

WRITE

1. $\text{Monthly rent} = \$230 \times \dfrac{52}{12} = \$996.67$ (correct to 2 decimal places)
2. $\text{Annual expenses} = \$620 + \$389 = \$1009$

   $\text{Monthly cost} = \dfrac{\$1009}{12} = \$84.08$ (correct to 2 decimal places)
3. $\text{Total monthly fixed expenses} = \$996.67 + \$84.08 + \$78 = \$1158.75$

4. Calculate the total variable expenses by adding the numbers in the last column of the table.

$$\text{Total monthly variable expenses} = \$423 + \$107 + \$85 + \$325 = \$940$$

5. To calculate the total monthly expenses, add the total fixed expenses to the total variable expenses obtained in step 4.

$$\begin{aligned}\text{Total monthly expenses} &= \text{fixed expenses} + \text{variable expenses} \\ &= \$940 + \$1158.75 \\ &= \$2098.75\end{aligned}$$

6. Write the answer as a sentence.

The total expenses are $2098.75 per month.

8.5.3 Financial planning

- To help in planning your finances, you can prepare a budget to estimate income and expenditure.
- A personal or family budget can be prepared using a spreadsheet to calculate the total fixed and variable expenses for each month and then subtracting this from the net monthly income.

Profit and loss

A *profit* is made when the net monthly income is more than the total monthly expenses.

A *loss* is made when the net monthly income is less than the total monthly expenses.

- A personal or family budget can help to:
  - ensure that you do not spend more money than you earn
  - decide what you can and can't afford
  - estimate the amount of money that you can save
  - control and decrease your expenses in order to save more money.
- It should be noted that a budget gives only an approximation of the real-life situation as it is based on estimates and does not include unexpected expenses.

WORKED EXAMPLE 17 Using a spreadsheet to determine the profit or loss at the end of a month

A nurse's net annual salary is $59 500. Her expenses are shown in the following table.

Spending description	Amount
Rent (monthly)	$658
Electricity (quarterly)	$295
Water usage (quarterly)	$150
Health insurance (monthly)	$205
Car finance (fortnightly)	$172.50
Vehicle registration (yearly)	$595
Vehicle insurance (half-yearly)	$680
Food (weekly)	$75
Entertainment (monthly)	$200
Fuel (weekly)	$68

Using a spreadsheet, determine whether the nurse has made a profit or loss at the end of each month.

THINK

1. Calculate the monthly costs.

For yearly, divide by 12.
For half-yearly, divide by 6.
For weekly, multiply by 52, then divide by 12.
For quarterly, divide by 3.
For fortnightly, multiply by 26, then divide by 12.

WRITE

Insert the needed formulas into the relevant cells in column C:
= (cell in column B) /12
= (cell in column B) /6
= (cell in column B)*52/12
= (cell in column B) /3
= (cell in column B)*26/12

	A	B	C
1	**Spending description**	**Amount**	**Monthly expenses**
2	Rent (monthly)	$658.00	$658.00
3	Electricity (quarterly)	$295.00	$98.33
4	Water usage (quarterly)	$150.00	$50.00
5	Health insurance (monthly)	$205.00	$205.00
6	Car finance (fortnightly)	$172.50	$373.75
7	Vehicle registration (yearly)	$595.00	$49.58
8	Vehicle insurance (half-yearly)	$680.00	$113.33
9	Food (weekly)	$75.00	$325.00
10	Entertainment (monthly)	$200.00	$200.00
11	Fuel (weekly)	$68.00	$294.67

2. Calculate the total monthly spending.

Create a new row: Monthly expenses
Insert '= sum(C2 : C11)' into cell C12.

3. Calculate the net annual income.

Create a new row: Net monthly income
Insert '59500' into B13.
Insert '= B13/12' into C13.

4. Calculate the profit/loss for the month.

Create new row: Profit/loss
Insert '= C13 − C12' into cell C14.

	A	B	C
1	**Spending description**	**Amount**	**Monthly expenses**
2	Rent (monthly)	$658.00	$658.00
3	Electricity (quarterly)	$295.00	$98.33
4	Water usage (quarterly)	$150.00	$50.00
5	Health insurance (monthly)	$205.00	$205.00
6	Car finance (fortnightly)	$172.00	$373.75
7	Vehicle registration (yearly)	$595.00	$49.58
8	Vehicle insurance (half-yearly)	$680.00	$113.33
9	Food (weekly)	$75.00	$325.00
10	Entertainment (monthly)	$200.00	$200.00
11	Fuel (weekly)	$68.00	$294.67
12	**Monthly expenses**		**$2 367.66**
13	**Net monthly income**		**$4 958.33**
14	**Profit/loss**		**$2 590.67**

5. Write the answer as a sentence.

The nurse makes a profit of $2590.67.

Resources

Interactivity Budgeting (int-6914)

Exercise 8.5 Personal budget plan

8.5 Quick quiz | 8.5 Exercise

Simple familiar	Complex familiar	Complex unfamiliar
N/A	1, 2, 3, 4, 5, 6, 7, 8, 9, 10, 11, 12	13, 14

These questions are even better in jacPLUS!
- Receive immediate feedback
- Access sample responses
- Track results and progress

Find all this and MORE in jacPLUS

Complex familiar

1. **WE15** Calculate the net monthly income of a worker who has a net annual salary of $63 200 and earns $3745 each year in bank interest.
2. A bank teller earns $1100 per week (after tax) plus $2200 per year in bank interest (after tax). Calculate their net monthly income.
3. Complete the following table.

	Net annual salary	Total net annual bank interest	Total net monthly income
a.	$70 500	$4812	
b.	$81 234	$10 387	
c.	$90 349	$9885	
d.	$24 500	$42 500	

4. **WE16** Determine the total monthly expenses from the following list of fixed and variable expenses. The variable expenses represent the amounts spent in a month.

	Fixed expenses		Variable expenses	
Item	**Frequency**	**Amount**	**Item**	**Amount**
Rent	Weekly	$205	Food	$494
Health insurance	Monthly	$89	Clothing	$205
Vehicle registration	Yearly	$540	Entertainment	$123
Vehicle insurance	Yearly	$499	Car repairs	$72

5. A baker earns $18.50 an hour after tax. She works an average of 38 hours a week plus 7 hours overtime, paid at time and a half. Her monthly expenses are 47% of her net monthly income.
   Calculate how much money she spends per month.

6. Determine the total monthly cost from the following list of fixed and variable expenses. Variable expenses represent the amounts spent in a month.

Fixed expenses			Variable expenses	
Item	**Frequency**	**Amount**	**Item**	**Amount**
Home mortgage	Monthly	$1254	Food	$563
Health insurance	Monthly	$124	Clothing	$321
Vehicle registration	Yearly	$702	Entertainment	$389
Vehicle insurance	Yearly	$899	Car repairs	$396
School fees	Half-yearly	$5300	Travel	$1379

7. Complete this table of income and costs for a family of five over 4 months.

	Net annual salary	Net monthly bank interest	Monthly fixed expenses	Monthly variable expenses	Monthly profit/loss
a.	$73 700	$784	$3623	$2563	
b.	$73 700	$792	$3623	$639	
c.	$73 700	$804	$3623	$4456	
d.	$73 700	$823	$3623	$1065	

8. WE17 The net annual salary for a childcare worker is $43 470. Their expenses are shown in the following table.

Expenses	Amount
Rent (weekly)	$225
Electricity (quarterly)	$150
Water usage (quarterly)	$75
Travel to work (weekly)	$60
Food (weekly)	$45
Entertainment (monthly)	$250

Using a spreadsheet, determine the amount of profit or loss the childcare worker makes at the end of each month.

9. A university student who lives with her parents has the following expenses:
   - She pays her parents $70 per week for board and food.
   - A monthly ticket for public transport costs her $80.
   - She spends on average $45 a month on books and stationery.
   - Her single health insurance premium is $68.55 a month.
   - Entertainment and snacks cost her about $90 a month.
   - The university enrolment fee is $900 a year.
   - She pays approximately $80 per month for clothes and accessories.

Prepare a monthly budget if the student's income consists of Youth Allowance ($355.40 per fortnight) and birthday and Christmas presents ($250 a year), and calculate the amount of money that she can save per month.

10. The following information shows the expenses and income for a couple.

Expenses	Mortgage	$672 per fortnight
	Food	$123 per week
	Clothing	$78 per week
	Transport	$35 per day
	Entertainment	$120 per week
	Health insurance	$200 per month
	Car (insurance, registration, fuel, servicing)	$600 per month
Net income	Partner A	$1680 per fortnight
	Partner B	$1100 per week

a. Using a spreadsheet, calculate the total monthly spending.
b. Demonstrate that the couple make a monthly profit.

The couple would like to save to buy a block of land for $350 000. To buy the block of land, they will need to make a 10% deposit.

c. If the couple's spending does not change, determine how many months it will take to save the 10% deposit. Give your answer correct to 1 decimal place.
d. The couple's financial adviser recommends they reduce their spending so that it's less than 50% of their net monthly income. Suggest where they should adjust their spending and justify your answer by using calculations.

11. The following shows a personal budget prepared by a graduate IT technician.

Item	Frequency	Amount	Monthly
Rent	Weekly	$253.85	$1100
Bills	Monthly	$145	$145
Travel	Monthly	$350	$350
Entertainment	Monthly	$330	$330
Clothes	Fortnightly	$203.08	$440
Insurance	Yearly	$1150	$95.83
Monthly spending			$2460.83
Net monthly income			
Profit			**$2355.92**

a. Determine the technician's net monthly income and hence calculate their net annual income.
b. Calculate the percentage amount they spend of their net income. Give your answer correct to 2 decimal places.
c. The technician would like to save up for an overseas holiday and has budgeted $5500 for the trip. If they can save 50% of their monthly net income, determine how many months it will take to have saved enough money for the trip.

12. A family of four are going on an overseas holiday for 6 weeks. They have saved $20 000 for the trip and recorded the costs for the essentials but don't know how much to spend on food.

Item	Cost
Passports	$832
Flights	$5668
Travel insurance	$658
Accommodation	$245 per night
Food	

a. They estimate food costs to be about $45 per day. Calculate the amount of money remaining from the $20 000 that they can spend on their trip.

b. The family reassess and decide to adjust their spending to save more money for their trip. Their monthly spending is $2585 and net monthly income is $4580.
If they can save half of their monthly profit for 4 months, calculate the amount of money they will have for their overseas trip.

Complex unfamiliar

13. A family's net annual income is $69 480. Their spending is shown in the following table.

Item	Frequency	Amount
Holiday/entertainment	Yearly	$4500
Mortgage	Weekly	$245
Insurance	Yearly	$1980
Car costs	Monthly	$950
Food	Weekly	$145
Electricity, water	Quarterly	$580
School fees	Yearly	$8500
Health insurance	Monthly	$368

Prepare a monthly budget and determine whether the family are living within their means (i.e. making a profit).
The family need to adjust their spending. By identifying the fixed and variable spending items, explain which items the family should consider reducing.

14. Prepare your personal monthly budget (or your family budget if you do not have any income).

Determine the possibilities for cutting some of the expenses.

Fully worked solutions for this chapter are available online.

LESSON
8.6 Review

8.6.1 Summary

doc-41764

Hey students! Now that it's time to revise this chapter, go online to:

- Access the chapter summary
- Review your results
- Watch teacher-led videos
- Practise questions with immediate feedback

Find all this and MORE in jacPLUS

8.6 Exercise

learnon

8.6 Exercise

Simple familiar	Complex familiar	Complex unfamiliar
N/A	1, 2, 3, 4, 5, 6, 7, 8, 9, 10, 11, 12, 13, 14, 15, 16	17, 18, 19, 20

These questions are even better in jacPLUS!
- Receive immediate feedback
- Access sample responses
- Track results and progress

Find all this and MORE in jacPLUS

Complex familiar

1. An insurance company offers customers the following discounts on the annual cost for car insurance.

Type of discount	Discount	Conditions
Multi-policy discount	7%	Owner has 2 or more insurance policies with the company.
No claim bonus	15%	Owner has not made an insurance claim for at least 5 years.

The following excesses apply to claims for at-fault motor vehicle accidents for the comprehensive car insurance:
- Basic excess of $550 for each claim
- An additional age excess of $1100 for drivers under 25 years of age

Fred has been with the same insurance company since obtaining his licence at the age of 18, has never had an accident, and he has 4 other insurance policies with this company. His car insurance costs $1450 per year prior to any discounts being applied.
Determine the annual amount that Fred is required to pay after having his first accident at the age of 24.

2. In its annual budget, the local council plans to raise $25 million in revenue to repair local infrastructure. The capital improvement for the area is valued at $1.5 billion. Determine the council rate quarterly instalment for a ratepayer whose house has a capital improved value of $475 000.

3. A person spends 65% of their net annual income. If their net annual income is $58 480, calculate the monthly profit they make.

4. A person earns $74 600 net annual salary and also receives a bonus payment from their employer of $3500. Calculate their fortnightly salary.

5. In calculating the budget for a small business, determine whether each of the following can be considered a variable expense.
   a. Rent
   b. Employee wages
   c. Public liability insurance
   d. Electricity bills

6. The energy usage and rates for a household are as shown.

Electricity bill

Tariff	Bill days	Current reading	Previous reading	Total usage (kWh)	Charge/rate (c/kWh)
All day	90	78 250	77 682		25.08
Service to property	90			**$1.105/day**	

Gas

Tariff	Bill days	Current reading	Previous reading	Total usage (kWh)	Charge/rate (c/MJ)
All day	90	8920	8138		2.452
Service to property	90			**$0.985/day**	

There is a 25% discount for paying the bill before the due date. Determine the amount paid if the bill was paid before the due date.

7. A manager's annual salary is $65 850. Calculate the amount of HECS-HELP they pay per month.

8. A worker's net annual income is $34 105. The worker makes a monthly loss of $145. Calculate their monthly spending.

9. A car can travel 485 km on 68 litres of fuel. Determine the fuel consumption in L/100 km correct to 2 decimal places.

10. In its annual budget, the local council plans to raise $25 million in revenue to repair local infrastructure. The capital improvement for the area is valued at $1.5 billion. Determine the council rate quarterly instalment for a ratepayer whose house has a capital improved value of $475 000.

11. Complete the following table.

	Energy usage	Charge/rate	Cost
a.	987 kWh	19.02c/kWh	
b.	2058 MJ		$53.28
c.		15.48c/kWh	$98.61
d.	5439 MJ	1.534c/MJ	

12. Calculate the monthly net income for each of the following.
    a. A net annual income of $63 480
    b. A weekly net income of $945.35
    c. A fortnightly net income of $1450

13. A household's daily water usage is 1700 litres. The total water bill for 60 billing days is $392.12, which includes $187.10 for water services. Calculate the charge per kilolitre.

14. The following table shows the net annual incomes and monthly spending of four people. Calculate the profit or loss for each person.

	Net annual income	Monthly spending	Profit/loss
a.	$51 890	$3950	
b.	$84 755	$8640	
c.	$75 120	$6580	
d.	$62 308	$4980	

15. There is a 12% discount for customers who pay their energy bill before the due date. The energy usage of a customer is:
    - electricity: 875 kWh at 19.02c/kWh
    - gas: 720 MJ at 2.034c/MJ.

    The customer is also charged a $112 fixed fee for service to the property.
    If the customer pays the bill before the due date, calculate the amount they paid, including GST.

16. The quarterly council rates for a property with capital improved value of $385 000 are $721.88.
    a. Calculate the amount in the dollar charged annually by the council.
    b. If the total capital improved value for all properties within the council municipality is $1.5 billion, calculate the amount of revenue the council plans to raise this year. Give your answer correct to the nearest thousand dollars.

Complex unfamiliar

17. The cost of bread has increased each year due to inflation. Over the last 2 years, the average annual inflation rates were 1.5% and 2% respectively. If the cost of bread today is $3.04, explain how you would determine the cost of bread 2 years ago.
    Justify your answer using calculations.

18. The monthly credit card spending of a family of four is $8650, which includes all bills and household expenses. Partner A's annual net salary is $74 486 and Partner B's fortnightly wage is $1685.
    Show that the family spends approximately 88% of their net monthly income.

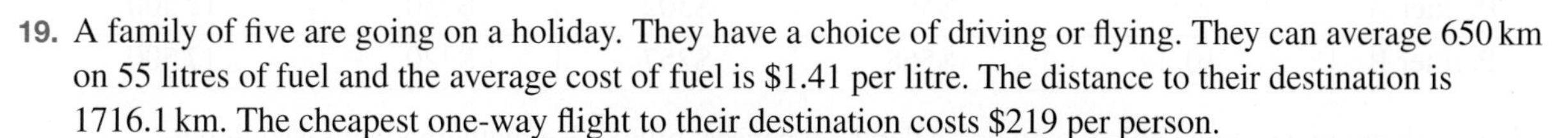

19. A family of five are going on a holiday. They have a choice of driving or flying. They can average 650 km on 55 litres of fuel and the average cost of fuel is $1.41 per litre. The distance to their destination is 1716.1 km. The cheapest one-way flight to their destination costs $219 per person.

Determine, using mathematical reasoning, which option is more cost-effective. Explain your answer.

20. A family of 6 live in a 5-bedroom house in the outer suburbs of a major city. They own two cars. Partner A uses one car to drive to the local train station; partner B uses the other car to drive their children to school, sports training and holidays.
They spend \$165 on food each week. Their health insurance costs \$368 per month, their house insurance costs \$1580 a year, and their annual council rates are \$3580. The children's school fees total \$2850 a year. Partner A's net annual salary is \$74 805.
Partner B works part-time Monday–Friday as an administrator at a local law firm. Their hourly rate is \$27.50 and their hours are 9 am–3 pm with a 45-minute unpaid lunch break. Their employer withholds \$85 in PAYG tax each week.
Their bills and other expenses are shown in the following tables.

Electricity bill

Billing period: 1 Jul–30 Sep					
Tariff	**Bill days**	**Current reading**	**Previous reading**	**Total usage**	**Charge/rate (c/kWh)**
Peak	91	23 890	23 150		33.58
Off-peak	91	48 901	48 334		15.60
Service to property	91			**\$1.148/day**	

Water bill

Bill details as at 30 September	Billing period	Value	Price
Water service charge	1 Jul–30 Sep	25.75	
Sewerage service charge	1 Jul–30 Sep	149.80	
Stormwater	1 Jul–30 Sep	22.05	
Usage	**Volume (kL)**	**Charge/kL**	**Price**
Water usage	65	\$2.05/kL	
Recycled water usage	45	\$1.82/kL	

Cars

Average fuel price: \$1.31/litre					
Vehicle	**Fuel consumption (L/100 km)**	**Registration (annual)**	**Insurance (6-monthly)**	**Servicing (every 15 000 km)**	**Average distance travelled each year**
Partner A	8.8	\$568	\$302	\$250	12 500
Partner B	10.2	\$568	\$257	\$320	17 500

To be solvent means to be able to pay one's debts (expenses, bills). Use mathematical reasoning to determine whether the family is solvent.

Fully worked solutions for this chapter are available online.

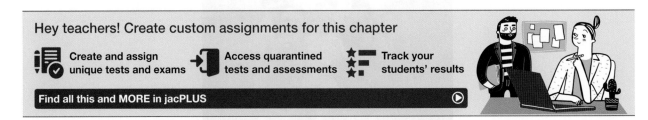

Answers

Chapter 8 Budgeting [complex]

8.2 Products and services

8.2 Exercise

1. a. Option 2 b. Option 2 c. Option 1
2. a. i. Telco ii. Telco iii. HFC
   b. HFC
3. a. $1500
   b. Medicare levy surcharge
   c. $142.50
4. Sample responses can be found in the worked solutions in the online resources.
5. a. $576
   b. Hospital = $104.58
   Extras = $165.06
6. Option A
7. $2250
8. a. $565.50 b. $880 c. $867
9. a. $175 b. 243 months
10. a. $5614.50 b. $168.44
11. a. $19 717 b. $4950 c. $5444.42
12. $3129.75
13. Sample responses can be found in the worked solutions in the online resources.

8.3 Rent and utility costs

8.3 Exercise

1. a. $2200 b. $2750 c. $2383
2. a. i. $519 ii. $4326
   b. The rental property in question 2 is more cost-effective as it costs $519 per week, whereas the rental property in question 1 costs $550 per week.
3. $301.67
4. a. $59.45 b. $28.13
   c. $131.03 d. $113.42
5. 4768 kWh
6. a. i. 30.54c/kWh ii. $1.017 iii. 895 kWh
   b. i. $273.33 ii. $365.88 iii. $402.47
7. $159.63
8. $314.10
9. $372.92
10. 33.62 kL
11. $1833.33
12. a. 0.006 b. $16 020 000
13. a. See the table at the bottom of the page.*
    b. $405.63
    c. 3.06%
14. a. $\dfrac{0.008 \times \$280\,000}{4} = \$560$
    b. $1345.60
    c. 7.45%
15. Plan B is more cost-effective for both paying on time and not paying on time.
16. Choice 1 is more cost-effective for the family, as it costs $60.06 and choice 2 costs $63.36.

8.4 Car expenses

8.4 Exercise

1. a. $1799.50 b. $20 602.30
2. a. $22 495.50 b. $22 495.50 c. $31 732.70
3. a. $689.85 b. $6250
4. $8333.33
5. 9.06 L/100 km
6. 798 km
7. a. 12.94 L/100 km
   b. $17.34
8. a. 3 times b. $338.75
9. 7 times
10. 663 km
11. Car B
12. a. $\dfrac{9.8 + 7.5}{2} = 8.65\text{ L}/100\text{ km}$
    b. 46.7 L
    c. $58.38

*13. a.

Bill details as at 30 June	Billing period	Value	Price
Water service charge	1 Apr–30 Jun	23.80	$23.80
Sewerage service charge	1 Apr–30 Jun	150.20	$150.20
Stormwater	1 Apr–30 Jun	17.95	$17.95
Usage	**Volume (L)**	**Charge/kL**	**Price**
Water usage	64 000	$\dfrac{130.56}{64} = \$2.04$	$130.56
Recycled water usage	38 000	$\dfrac{71.06}{38} = \$1.87$	$71.06
		Total	$393.57

d. The average fuel consumption is a value that assumes equal distance (1 : 1) of driving under the two conditions (city and highway). In this case, the ratio of city to highway driving was 126 : 458 (1 : 3.6). This means the overall fuel consumption is lower than the average value would predict, because 3.6 times more driving took place at the lower fuel consumption.

13. Total monthly payments: $496.28 × 60 = $29 776.80
Deposit: 10% × $24 980 = $2498
Principal borrowed: $24 980 − $2498 = $22 482
Interest paid: $29 776.80 − $22 482 = $7294.80

$$I = Pin$$

$$\$7294.80 = \$22\,482 \times i \times 5$$

$$\frac{\$7294.80}{\$22\,482 \times 5} = r$$

$$r \approx 6.489\%$$

14. Car B has a lower average fuel consumption (9.85 L/100 km) and will therefore be more cost-effective.

8.5 Personal budget plan

8.5 Exercise

1. $5578.75
2. $4950
3. a. $6276
b. $7635.08
c. $8352.83
d. $5583.33
4. $1957.91
5. $1827.40
6. $5442.75
7. a. $739.67 profit
b. $2671.67 profit
c. $1133.33 loss
d. $2276.67 profit
8. $1867.50 profit
9. $48.98
10. a. $4711.58
b. The net monthly income of $8406.67 exceeds the monthly spending, for a monthly profit of $3695.09.
c. 9.5 months
d. Areas in which spending can be reduced (adjusted) include clothing, entertainment, transport and food. Mortgage, health insurance and car registration/insurance are fixed spending.
11. a. Net monthly income: $4816.75
Net annual income: $57 801
b. 51.09%
c. 2.28 months
12. a. $662
b. $3990
13. See the table at the bottom of the page.*

Fixed	Variable
Mortgage	Holiday/entertainment
Insurance	Car costs
Health insurance	Food
School fees	Electricity, water

Variable items can be reduced, particularly non-essential items such as holidays/entertainment. These items do not have fixed costs, which means that their costs can be adjusted based on usage or need.

14. Sample responses can be found in the worked solutions in the online resources.

*13.

Item	Frequency	Amount	Monthly cost
Holiday/entertainment	Yearly	$4500	$\frac{\$4500}{12} = \$375$
Mortgage	Weekly	$245	$\$245 \times \frac{52}{12} = \$1061.67$
Insurance	Yearly	$1980	$\frac{\$1980}{12} = \$165$
Car costs	Monthly	$950	$950.00
Food	Weekly	$145	$\$145 \times \frac{52}{12} = \$628.33$
Electricity, water	Quarterly	$580	$\frac{\$580}{3} = \$193.33$
School fees	Yearly	$8500	$\frac{\$8500}{12} = \$708.33$
Health insurance	Monthly	$368	$368.00
		Total	$375 + $1061.67 + $165 + $950 + $628.33 + $193.33 + $708.33 + $368 = **$4449.66**
Income	Yearly	**$69 480**	$\frac{\$69\,480}{12} = \mathbf{\$5790.00}$
		Profit	$5790 − $4449.66 = **$1340.34**

8.6 Review

8.6 Exercise

1. 2796.23
2. $1979.17
3. $1705.67
4. $3003.85
5. a. No b. No c. No d. Yes
6. $288.52
7. $192
8. $2987.08
9. 14.02 L/100 km
10. $1979.17
11.

	Energy usage	Charge/rate	Cost
a.	987 kWh	19.02c/kWh	$187.73
b.	2058 MJ	2.589c/MJ	$53.28
c.	637 kWh	15.48c/kWh	$98.61
d.	5439 MJ	1.534c/MJ	$83.43

12. a. $5290 b. $4096.52 c. $3141.67
13. $2.01/kL
14. a. $374.17 profit b. $1577.08 loss c. $320 loss d. $212.33 profit
15. $283.69
16. a. 0.0075 b. $11 250 000
17. Divide the current price of bread by 1.015×1.02.
18. Sample responses can be found in the worked solutions in the online resources.
19. Sample responses can be found in the worked solutions in the online resources.
20.

Item	Cost	Frequency	Monthly cost
Food	$165.00	Weekly	$715.00
Health insurance	$368.00	Monthly	$368.00
House insurance	$1580.00	Yearly	$131.67
School fees	$2850.00	Yearly	$237.50
Council rates	$3580.00	Yearly	$298.33
Electricity	$485.56	Quarterly	$161.85
Water	$412.75	Quarterly	$137.58
Car insurance	$559.00	Half-yearly	$93.17
Car registration	$1136.00	Yearly	$94.67
Fuel	$3779.35	Yearly	$314.95
Servicing	$581.67	Yearly	$48.47
		Total	**$2601.19**
		Net monthly income	**$8993.56**
		Profit	**$6392.37**

Yes, the family are solvent. Their net monthly income is greater than their spending.

UNIT

2 Data and travel

UNIT 2 Data and travel
TOPIC 1 Data collection

9 Census, surveys and simple survey procedures

LESSON SEQUENCE

Fully worked solutions for this chapter are available online.

on Resources

Solutions Solutions — Chapter 9 (sol-1262)

Digital documents Learning matrix — Chapter 9 (doc-41765)
Quick quizzes — Chapter 9 (doc-41766)
Chapter summary — Chapter 9 (doc-41767)

LESSON
9.1 Overview

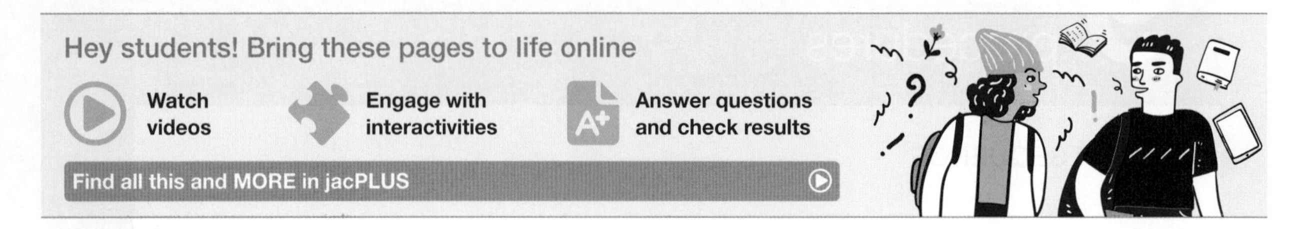

9.1.1 Introduction

The world we live in produces an ever-increasing amount of data. Being able to understand and interpret these large sets of data is a valuable skill in the 21st century. To be able to interpret data, you need to first know how to collect it. The Australian Bureau of Meteorology collects data about the weather, analyses it, and then uses that knowledge to predict what the weather will be like tomorrow, next week and even next year. The Australian Bureau of Statistics runs a census night every five years to collect data about the population. This data is then analysed so future decisions can be made by the government.

Social media companies collect data about how you use their platforms. This data is analysed and used to adapt these platforms to the way you use them (and to help these companies maximise profits). This demonstrates the importance of understanding how to collect and then interpret data to make important decisions.

9.1.2 Syllabus links

Lesson	Lesson title	Syllabus links
9.2	**Investigating a census**	◯ Investigate the procedure for conducting a census. ◯ Understand the purpose of sampling to provide an estimate of population values when a census is not used.
9.3	**Sampling methods — advantages and disadvantages [complex]**	◯ Investigate the advantages and disadvantages of conducting a census [complex]. ◯ Investigate the different kinds of samples [complex]. ◯ Investigate the advantages and disadvantages of these kinds of samples [complex].
9.4	**Target population**	◯ Identify the target population to be surveyed.
9.5	**Investigating questionnaire design principles [complex]**	◯ Investigate questionnaire design principles, including simple language, unambiguous questions, consideration of number of choices, issues of privacy and ethics, and freedom from bias [complex].

LESSON
9.2 Investigating a census

SYLLABUS LINKS

- Investigate the procedure for conducting a census.
- Understand the purpose of sampling to provide an estimate of population values when a census is not used.

9.2.1 Collecting data by census

- **Data collection** is a process in which data is collected to obtain information and draw conclusions about issues of concern regarding a given population.
- A **population** consists of a complete group of people, objects, events, etc. with at least one common characteristic.
- Any subset of a population is called a **sample**.
- Data is collected using either a **census** of the entire population or a **survey** of a sample of the population.

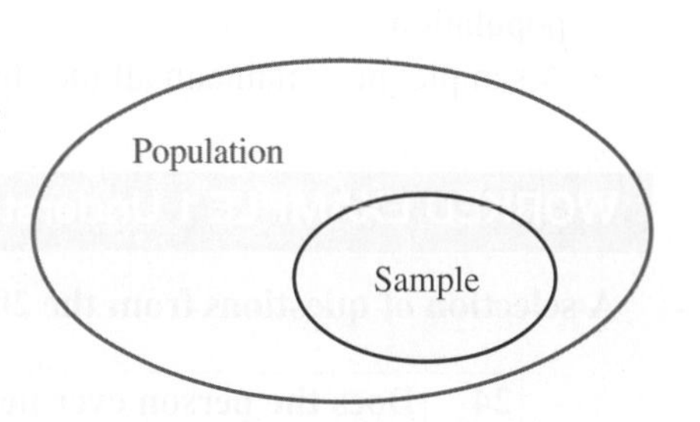

Census

- A census is conducted by official bodies at regular time intervals on a set date. In Australia, a census of the entire population and housing is conducted by the **Australian Bureau of Statistics (ABS)** every five years on **census night**.
- The last census in Australia occurred in August 2021.

Procedure for conducting a census

A census consists of three main stages:

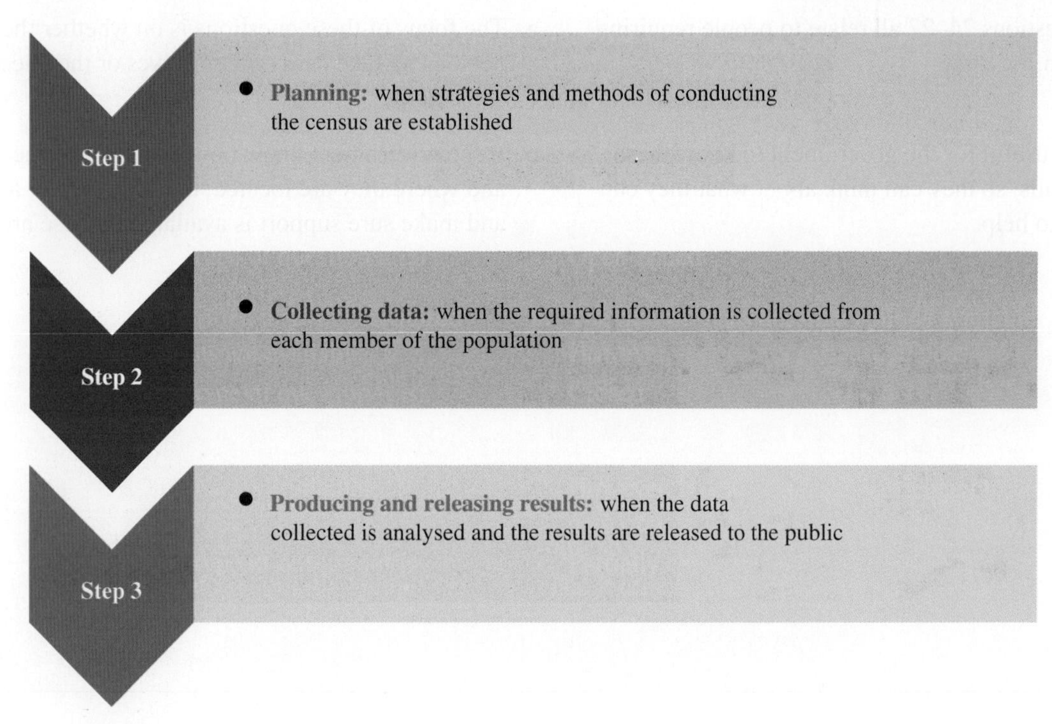

Step 1
Planning: when strategies and methods of conducting the census are established
Step 2
Collecting data: when the required information is collected from each member of the population
Step 3
Producing and releasing results: when the data collected is analysed and the results are released to the public

9.2.2 Survey and sampling

Survey

- A survey is conducted using a sample of the population.
- A survey is a method of collecting data when a census cannot be used.
- Using a sample of the population to conduct a survey can give a good estimate of the data of the entire population.
- It is more expensive and time-consuming to conduct a census, so a survey is often the preferred option of collecting data.

Sampling

- Sampling is the process of selecting a sample of a population to provide an estimate of the entire population.
- A sample must maintain all the characteristics of the population it represents.

WORKED EXAMPLE 1 Understanding census information

A selection of questions from the 2021 ABS Census are shown.

24	**Does the person ever need someone to help with, or be with them for, self-care activities?**
25	**Does the person ever need someone to help with, or be with them for, body movement activities?**
26	**Does the person ever need someone to help with, or be with them for, communication activities?**
27	**What are the reasons for the need for assistance or supervision shown in Questions 24, 25 and 26?**

a. **Explain what the focus of these questions is.**
b. **Explain why this information is useful for the government to know.**

THINK	WRITE
a. Questions 24–27 all relate to people requiring help.	a. The focus of these questions is on whether the person can take care of themselves or they need some help.
b. It's useful for the government to know these details, so they can think about what they can do to help.	b. If the government know how many people need help and where they are located, they can budget for this and make sure support is available in these areas.

WORKED EXAMPLE 2 Deciding whether to use a census or a survey

For each of the following situations, identify whether the data should be collected using a census or a survey. Give reasons for your choice.

a. **All students in a school are asked to state their method of travel to and from school.**

b. **Data is collected from every third person leaving a shopping centre about the time they spent shopping.**

c. **Data is collected from 500 households regarding their average monthly water usage.**

THINK	WRITE
a. 1. Determine whether the group considered represents an entire population or a sample.	a. All students in a school represent the whole student population of the school.
2. Identify whether a census or a survey is required.	As this represents the entire student population for the given school, a census is required.
b. 1. Determine whether the group considered represents an entire population or a survey.	b. As only every third person is involved in the data collection, this represents a sample of the whole population of people leaving the shopping centre.
2. Identify whether a census or a survey is required.	As this represents a sample, a survey is required.
c. 1. Determine whether the group considered represents an entire population or a survey.	c. Only 500 households are involved in the data collection, so this represents a sample of all the households.
2. Identify whether a census or a survey is required.	As this represents a sample, a survey is required.

Exercise 9.2 Investigating a census

learnon

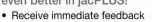

9.2 Quick quiz on	9.2 Exercise

Simple familiar	Complex familiar	Complex unfamiliar
1, 2, 3, 4, 5, 6, 7, 8	N/A	N/A

Simple familiar

1. To conduct a census, the three main stages are:
   a. ____________
   b. ____________
   c. ____________

2. When selecting a sample, explain an important characteristic.

3. If a sample of 300 students was selected, determine whether a census or survey was being conducted, and explain your answer.

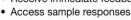

4. WE1 A selection of questions from the Census is shown.

49	How did the person get to work on Tuesday 10 August 2021?
50	*Last week,* how many hours did the person work in all jobs?
51	Did the person *actively* look for work at any time in the *last four weeks*?
52	If the person had found a job, could the person have started work *last week*?

a. Explain what the focus of these questions is.
b. Explain why this information is useful for the government to know.

5. A selection of questions from the Census is shown.

32	What is the highest year of primary or secondary school the person has *completed*?
33	Has the person *completed* any educational qualification?
34	What is the level of the *highest* qualification the person has *completed*?
35	What is the main field of study for the person's *highest* qualification *completed*?

a. Explain what is the focus of these questions.
b. Explain why this information is useful for the government to know.

6. WE2 For each of the following situations, determine whether the data should be collected using a census or a survey. Give reasons for your choice.
   a. An online business recorded the time that its customers spent searching for a product.
   b. Fifty per cent of the Year 12 students at a secondary college were asked to state their preference for either online tutorials or face-to-face tutorials.
   c. The heights of all the patients at a local hospital were recorded.

7. For each of the following, identify which situation represents a sample and which represents a population. Give reasons for your choice.
   a. The number of absences of all Year 12 students in a school and the number of absences of all students in the school
   b. The number of passenger airplanes landing at an airport and the total number of airplanes landing at the same airport

8. Identify whether a census or a survey is required to collect the data in the following statistical investigations. Give reasons for your answer.
   a. Water savings in 150 suburbs across the country
   b. Highest educational level of all people in Australia
   c. Roll marking at homeroom in a secondary college
   d. Asking every fifth person leaving the theatre whether they liked the play
   e. Asking every customer in a car dealership to fill in a questionnaire

Fully worked solutions for this chapter are available online.

LESSON

9.3 Sampling methods — advantages and disadvantages [complex]

SYLLABUS LINKS

- Investigate the advantages and disadvantages of conducting a census [complex].
- Investigate the different kinds of samples [complex].
- Investigate the advantages and disadvantages of these kinds of samples [complex].

9.3.1 Census and survey

- When considering whether to conduct a census, the following advantages and disadvantages should be considered.

Advantages and disadvantages of conducting a census

Advantages

The advantages of the information collected through a census is that it:

- represents a true measure of the entire population
- can be used for further studies
- provides detailed information about minority groups within the entire population.

Disadvantages

The disadvantages of the information collected through a census include the:

- high costs involved
- long period of time required to collect, analyse, produce and release the results
- fact that data can become out of date quite quickly.

For example, a company manufacturing school shoes wanted to collect data on the shoe size of every student in the country. This process would be very costly and it would take a long time to collect, analyse and produce the results. The company should investigate a sample of the student population for a fast and cost-effective estimate.

9.3.2 Collecting data by surveys and sampling

- A survey of a sample of a given population is considered when a census cannot be used.
- **Sampling** is the process of selecting a sample of a population to provide an estimate of the entire population.

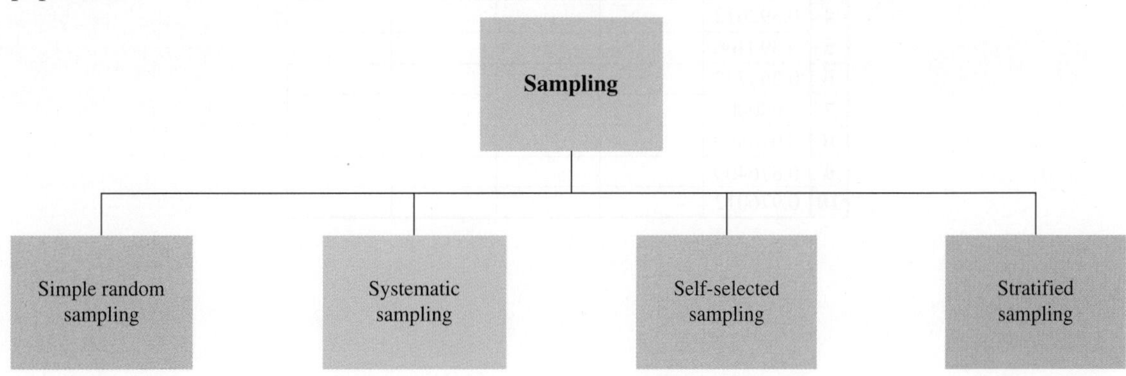

Simple random sampling

- **Simple random sampling** is the basic method of selecting a sample.
- This type of sampling ensures that each individual in a population has an equal chance of being selected for the sample.

Formula to calculate the sample size

The sample size can be calculated using the formula:

$$S = \sqrt{N}, \text{ where } N \text{ is the size of the population}$$

- Consider a population of 25 Year 11 Mathematics students and a sample of 5 students.
- A basic method of choosing the sample is by assigning each Year 11 student a unique number from 1 to 25, writing each number on a piece of paper, placing all the papers in a box or a bowl, shaking it well, and then choosing 5 pieces of paper.
- The students who correspond to the numbers drawn will be a sample of the Year 11 student population.

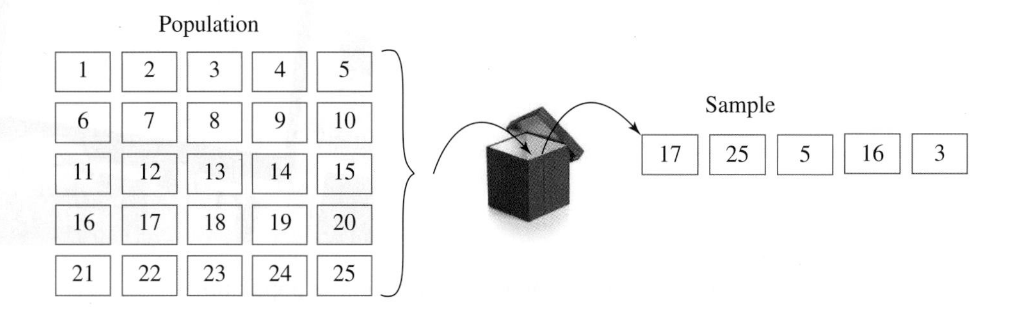

Random number generator using calculators or computer software

Calculators and computer software have a random number generator that makes the process of selecting a sample a lot easier. An Excel worksheet generates random numbers using the RAND( ) or RANDBETWEEN(a, b) command.

The RAND( ) command generates a random number between 0 and 1. Depending on the size of the sample, these generated numbers have to be multiplied by *n*, where *n* is the size of the population.

	A	B	C	D	E
1	0.433821				
2	0.78391				
3	0.547901				
4	0.892612				
5	0.390466				
6	0.264742				
7	0.690003				
8	0.070899				
9	0.876409				
10	0.976012				

RANDBETWEEN(a, b) generates a random number from a to b. For the population of Year 11 Mathematics students, the sample of five students can be generated by using RANDBETWEEN(1, 25).

	A	B	C	D	E	F
1	9					
2	3					
3	25					
4	3					
5	17					

If the same number appears twice, a new number will need to be generated in its place to achieve five different numbers.

tlvd-4690

WORKED EXAMPLE 3 Selecting a sample

Select a sample of 12 days between 1 December and 31 January:

a. by hand

b. using a random number generator.

THINK	WRITE
a. 1. Write every member of the population on a piece of paper.	a. There are 31 days in December and 31 days in January. 62 pieces of paper will be required to write down each day. 1/12, 2/12, … , 30/01, 31/01 *Note:* Alternatively, each day could be assigned a number, in order, from 1 to 62.
2. Fold all papers and put them in a box.	Ensure that the papers are folded properly so no number can be seen.
3. Select the required sample. A sample of 12 days is required, so randomly choose 12 papers from the box.	Sample: 15, 9, 41, 12, 1, 7, 36, 13, 26, 5, 50, 48
4. Convert the numbers into the data represented.	15 represents 15/12, 9 represents 9/12, 41 represents 10/01, 12 represents 12/12, 1 represents 1/12, 7 represents 7/12, 36 represents 5/01, 13 represents 13/12, 26 represents 26/12, 5 represents 5/12, 50 represents 19/01, 48 represents 17/01.
5. Write the sample selected.	Sample: 15/12, 9/12, 10/01, 12/12, 1/12, 7/12, 5/01, 13/12, 26/12, 5/12, 19/01, 17/01
b. 1. Assign each member of the population a unique number.	b. Assign each day a number, in order, from 1 to 62: $1/12 = 1$, $2/12 = 2$, … , $31/01 = 62$

2. Open a new Excel worksheet and use the RANDBETWEEN(1, 62) command to generate the random numbers required.

	A	B	C	D	E	F
1	29					
2	27					
3	17					
4	26					
5	4					
6	45					
7	41					
8	25					
9	10					
10	28					
11	16					
12	8					

Note: Some of the numbers may be repeated. For this reason, more than 12 numbers should be generated. Select the first 12 unique numbers.

3. Convert the numbers into the data represented.

29 represents 29/12, 27 represents 27/12, 17 represents 17/12, 26 represents 26/12, 4 represents 4/12, 45 represents 14/01, 41 represents 10/01, 25 represents 25/12, 10 represents 10/12, 28 represents 28/12, 16 represents 16/12, 8 represents 8/12.

4. Write the sample selected.

Sample: 29/12, 27/12, 17/12, 26/12, 4/12, 14/01, 10/01, 25/12, 10/12, 28/12, 16/12, 8/12

Systematic sampling

- **Systematic sampling** or **systematic random sampling** requires a starting point chosen at random, with members of the population chosen at regular intervals.

  For example, if a sample of 5 students out of 30 is needed for a survey, to ensure that each member of the population has an equal chance of being chosen, divide the whole population by the size of the sample (in this case, $30 \div 5 = 6$). So, the starting point will be a random number from 1 to 6 and then every 6th student will be chosen.

Starting point interval formula

The starting point interval is chosen using the formula:

$$\textbf{Maximum starting point} = N - I(s - 1)$$

where N is the size of the population

s is the size of the sample

I is the sampling interval and is the whole number of $\frac{N}{s}$.

WORKED EXAMPLE 4 Using a systematic random sampling method to determine a sample

Using a systematic random sampling method, a researcher wants to choose 8 people from a population of 59 people. Identify one such sample.

THINK	WRITE
1. Write a list of all the people in the population.	For the purpose of this example, we are going to use the initials of the people. GF, AS, TG, YH, ID, BK, CD, YT, UE, OM, LP, HI, BT, SJ, FR, IV, BX, MN, IT, UM, WK, FA, FP, ST, VP, AA, AR, BT, OD, PM, LC, RP, PO, GV, BF, TP, AD, IP, CR, AN, OO, FC, LM, LC, AF, AK, PY, SF, GH, VS, MR, CB, FJ, RM, CS, KF, GS, WM, FX
2. Using the systematic random sampling method $\frac{N}{s}$, choose a sampling interval.	$I = \frac{N}{s}$ $= \frac{59}{8}$ $= 7.37$ The researcher could choose every 7th or 8th person.
3. Choose the starting point.	Let $I = 7$. Maximum starting point $= N - I(s-1)$ $= 59 - 7 \times (8-1)$ $= 59 - 49$ $= 10$ The starting point will be a number from 1 to 10. Let this number be 4.
4. Form the sample.	The sample will be formed by the 4th, 11th, 18th, 25th, 32nd, 39th, 46th and 53rd person. GF, AS, TG, **YH**, ID, BK, CD, YT, UE, OM, **LP**, HI, BT, SJ, FR, IV, BX, **MN**, IT, UM, WK, FA, FP, ST, **VP**, AA, AR, BT, OD, PM, LC, **MR**, PO, GV, BF, TP, AD, IP, **CR**, AN, OO, FC, LM, LC, AF, **AK**, PY, SF, GH, VS, RP, CB, **FJ**, RM, CS, KF, GS, WM, FX
5. Write the sample selected.	YH, LP, MN, VP, MR, CR, AK, FJ

Self-selected sampling

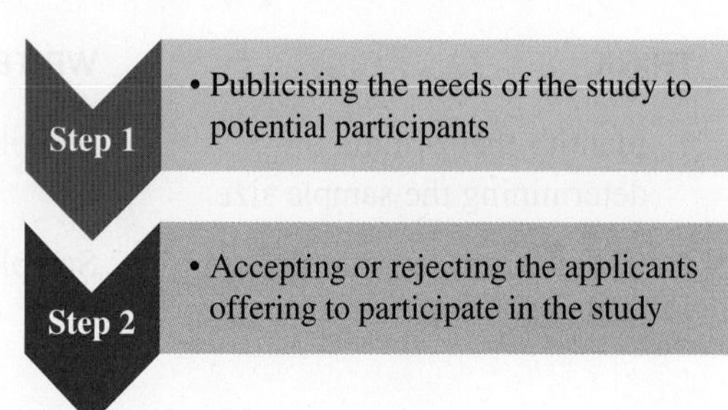

Step 1
• Publicising the needs of the study to potential participants
Step 2
• Accepting or rejecting the applicants offering to participate in the study

- **Self-selected sampling** is used when the members of a population are given the choice to participate in research.
- In this type of sampling, the researcher chooses the **sampling strategy**, such as:
  - asking for volunteers to participate in the trial of a drug
  - a website asking customers to answer a short questionnaire.
- Self-selected sampling requires two steps:
  For example, if you were asked to form a sample of 20 students out of 200 Year 11 students, the two steps would be:
  - advertising your study around the school
  - accepting the Year 11 students who showed an interest in participating in your study and rejecting any students who are not in Year 11.

WORKED EXAMPLE 5 Identifying steps to form a self-selected sample

A pharmaceutical company wants to trial a new flu vaccine. The study requires adults aged between 20 and 80 years. Identify the steps required to form a self-selected sample of 100 participants.

THINK	WRITE
1. Determine the ways to publicise the study.	Possible ways to publicise the study are: • online (social media or email) • on TV • on radio • at hospitals.
2. Identify the characteristics required for accepting or rejecting the applicants.	Accept 100 adults aged between 20 and 80 years old. Reject any others.

Stratified sampling

- **Stratified sampling** is used when there are variations in the characteristics of a population.
- This method requires the population to be divided into subgroups called **strata**.
  For example, if the Year 12 students of a secondary college were asked about their favourite movie, the preferences could be different for boys and girls. For this reason, the sample would have to reflect the same proportion of boys and girls as the actual population.

> **Calculating sample size for subgroups**
>
> $$\textbf{Sample size for each subgroup} = \frac{\textbf{sample size}}{\textbf{population size}} \times \textbf{subgroup size}$$

- Once the sample size of each subgroup has been determined, random sampling or systematic sampling can be used to form the sample required.

WORKED EXAMPLE 6 Using sample size for a subgroup to determine the sample

Calculate the number of female students and male students required to be part of a sample of 25 students if the student population is 652, with 317 male students and 335 female students.

THINK	WRITE
1. Identify the formula for determining the sample size.	$\text{Sample size for each subgroup} = \frac{\text{sample size}}{\text{population size}} \times \text{subgroup size}$
2. Calculate the sample size for each subgroup.	$\text{Sample size for female students} = \frac{25}{652} \times 335 = 13$ $\text{Sample size for male students} = \frac{25}{652} \times 317 = 12$
3. Write the answer.	The sample should contain 13 female students and 12 male students.

Exercise 9.3 Sampling methods — advantages and disadvantages [complex]

9.3 Quick quiz on | 9.3 Exercise

Simple familiar	Complex familiar	Complex unfamiliar
N/A	1, 2, 3, 4, 5, 6, 7, 8, 9, 10, 11, 12, 13, 14, 15, 16	17, 18

Complex familiar

1. WE3 Select a sample of 5 students to participate in a relay from a group of 26 students:
   a. 'by hand'
   b. using a random number generator.
2. A small business has 29 employees. The owner of the business has decided to survey 8 employees on their opinion about working conditions. Select the required sample:
   a. 'by hand'
   b. using a random number generator.
3. Calculate the approximate sample size of a population of 23 500 people.
4. Generate a sample of 10 members from a population of 100 people using the Excel worksheet random number generator and the following command:
   a. RAND( )
   b. RANDBETWEEN(0, 100).
5. WE4 A car manufacturer wants to test 5 cars from a lot of 32 cars. Create one possible sample by using a systematic random sampling method.

6. A quality control officer is conducting a survey of 46 products. She decides to sample 9 products using a systematic random sampling method. Create two possible samples by using systematic random sampling methods.
7. Stratified sampling is used in the following surveys. Calculate the number of members needed from each subgroup.
   a. A school has a population of 127 Year 12 students where 61 students are boys. A sample of 11 students is chosen to represent all the Year 12 students in an interstate competition.
   b. The local cinema is running a promotional movie screening. A sample of 10 people from those attending the screening will be chosen to receive a free pass to the next movie screening. If the total number of people to attend the promotional screening is 138, with 112 adults and 26 children, calculate the number of children and adults who will receive a free movie pass in order to keep a proportional sampling.
8. WE5 The local council is seeking volunteers to participate in a study on how to improve the local public library. Determine the steps required to form a self-selected sample of 50 participants.

9. A gym instructor is conducting a study on the effects of regular physical exercise on the wellbeing of teenagers. Determine the steps required to form a self-selected sample of 30 participants.

10. Census At School Australia is a project that collects information from Australian students using a questionnaire developed by the Australian Bureau of Statistics. It is not compulsory for students to participate in this census; however, students can volunteer to participate.
Explain what type of sampling is used in this project.

11. A farmer has 46 black sheep and 29 white sheep. He wants to try a new diet on a sample of 10 sheep.
    a. Determine what type of sampling the farmer is using.
    b. Calculate the number of black sheep and white sheep required for this trial in order to give a proportional representation of both types of sheep.

12. A manufacturing company of mobile phones is testing two batches of 100 phones each for defects. In the first batch, every 5th phone is checked, while in the second batch, every 7th phone is checked.
    a. Determine the method of sampling that is used for the testing.
    b. Determine the number of phones tested for defects in the two batches.
    c. Calculate the maximum starting point for each batch.

13. WE6 Calculate the sample size of the subgroups in a sample of 33 people from a population of 568 males and 432 females.

14. Calculate the sample size of the subgroups in a sample of 15 cars from a population of 179 white cars and 215 black cars.

15. At Guntawang Secondary College, a survey is conducted to collect information from the 230 students enrolled.
    a. Calculate the appropriate sample size for this population.
    b. Explain how the sample could be chosen.

16. Generate a sample of 12 members from a population of 140 people using the Excel worksheet random number generator and the following command:
    a. RAND( )
    b. RANDBETWEEN(1, 140).

Complex unfamiliar

17. If a sample size of 35 was taken using simple random sampling, determine the size of the population it was taken from.

18. If simple random sampling is used and the population size doubles, identify the effect this has on the size of the sample. Justify your response.

Fully worked solutions for this chapter are available online.

LESSON
9.4 Target population

SYLLABUS LINKS

- Identify the target population to be surveyed.

9.4.1 Target population for a survey

- The members of a population could be any entity, such as people, animals, organisations and businesses.
- A survey is concerned with two populations: the target population and the survey population.
- A target population is the entire population while the survey population includes those members of the target population.
- Suppose you wish to know how many games Brisbane Broncos supporters went to in the past season.
- You would want to survey Brisbane Broncos supporters and not Penrith Panthers supporters.

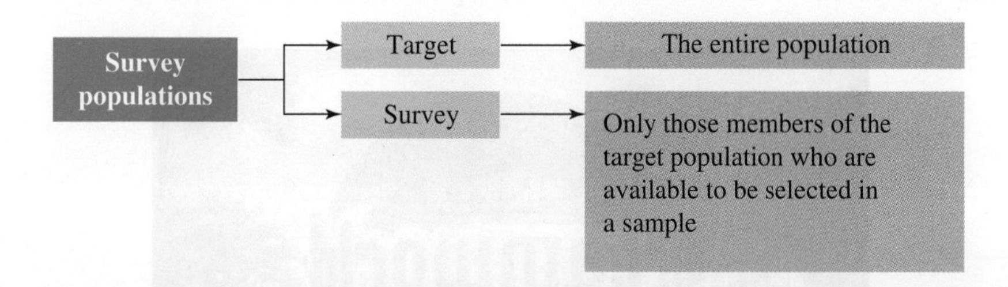

WORKED EXAMPLE 7 Identifying the target population

A mechanic wants to know if their clients are satisfied with the work performed on their cars. Describe what the target population is.

THINK	WRITE
The mechanic wants to understand what the clients think about the work done on their cars.	The target population is the people that have had their car serviced by the mechanic.

WORKED EXAMPLE 8 Explaining the importance of the target population

An art exhibition surveyed people about their enjoyment of the exhibition.

a. Describe the target population.

b. Explain the importance of identifying the target population.

THINK	WRITE
a. The aim of the survey is to understand the level of enjoyment from the art exhibition.	a. The target population is the people who attended the art exhibition.
b. To make assumptions about the population, the survey participants must come from that population.	b. Any conclusions would not be accurate if the participants completing the survey did not attend the art exhibition.

Exercise 9.4 Target population

9.4 Quick quiz	9.4 Exercise

Simple familiar	Complex familiar	Complex unfamiliar
1, 2, 3, 4, 5, 6, 7, 8	N/A	N/A

These questions are even better in jacPLUS!
- Receive immediate feedback
- Access sample responses
- Track results and progress

Find all this and MORE in jacPLUS

Simple familiar

1. WE7 A mathematics tutoring company wants to know if their students are satisfied with the tutoring they receive. Describe what the target population is.
2. The clothing store Country Road want to know the quality of the service their staff provide to their customers. Describe what the target population is.
3. WE8 A school surveyed students to find the most popular subject in Year 11.
   a. Describe the target population.
   b. Explain the importance of identifying the target population.
4. Dreamworld want to know the level of satisfaction of people who visit their theme park.

   a. Describe the target population.
   b. Explain the importance of identifying the target population.
5. To make sure an orchard was selling fresh apples, a worker randomly selects apples that have been picked to see if they are fresh. Describe the target population.
6. A school surveys its staff to see who wants to go to school camp. Describe the target populations.
7. A company is conducting a survey about time spent by employees to travel to work. Define the target population.
8. A company is conducting a survey in order to obtain information required to improve its employees' work conditions. Define the target population in the following surveys.
   a. A survey about the amount of money spent weekly on childcare
   b. A survey about the amount of time spent using a computer

Fully worked solutions for this chapter are available online.

LESSON
9.5 Investigating questionnaire design principles [complex]

SYLLABUS LINKS

- Investigate questionnaire design principles, including simple language, unambiguous questions, consideration of number of choices, issues of privacy and ethics, and freedom from bias [complex].

9.5.1 Questionnaires

- Surveys can be administered using paper or electronic questionnaires, face-to-face interviews or by telephone.
- A **questionnaire** is a list of questions used to collect data from the survey's participants.
- Designing questionnaires is probably one of the most important tasks when administering a survey.

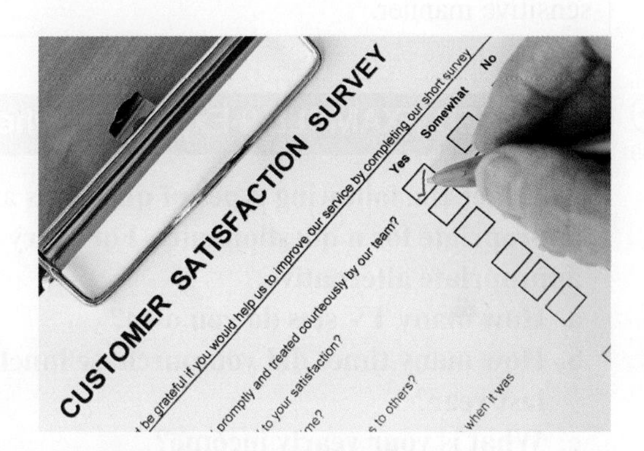

Some dos and don'ts for wording questions in a survey are shown in the table below.

Use simple and easy-to-understand language.	Avoid abbreviations and jargon. E.g. ABS (Australian Bureau of Statistics), QCE (Queensland Certificate of Education), etc.
Use short sentences. E.g. 'There are many types of movies available in cinemas that people watch. Which type do you think most represents your preference?' could be replaced by a simple, straightforward question such as 'What type of movie do you prefer?'	Avoid vague or ambiguous questions by asking precise questions.
Ask for one piece of information at a time and avoid double-barrelled questions. E.g. 'How satisfied are you with your job and your boss?' is a double-barrelled question, putting two pieces of information together.	Avoid double negatives. E.g. 'Would you say that your Mathematics teacher is not unqualified?' is an affirmation; however, some people might not be aware of this and the question might look confusing.

(continued)

(continued)

Ask questions that are impartial and do not suggest the 'correct' answer.	Avoid leading questions. E.g. 'Do you agree with the new school council structure? Yes or No.' could be replaced with a question such as 'Do you agree or disagree with the new school council structure? 1. Strongly agree, 2. Agree, 3. Disagree, 4. Strongly disagree.'
Give consideration to sensitive issues like privacy and ethics. E.g. Questions about issues such as income, religious beliefs, political views and gender orientation are sensitive topics and they have to be stated in a sensitive manner.	Avoid or eliminate bias. E.g. 'How would you rate your Mathematics textbook: Excellent, Good or Average?' The bias is in the absence of a negative option. Questions with 'always' and 'never' can also bias participants.

tlvd-11439

WORKED EXAMPLE 9 Explaining whether questions are appropriate for a questionnaire

Consider the following types of questions and explain why they are not appropriate for a questionnaire. For every question, identify a more appropriate alternative.

a. **How many TV sets do you own?**

b. **How many times did you purchase lunch from the school's cafeteria last year?**

c. **What is your yearly income?**

THINK	WRITE
a. 1. Identify any concerns about the question.	a. The question assumes that all respondents own a TV set. Some respondents might have a TV set in their household but not own it.
2. Identify a possible replacement question or questions.	'Do you own a TV set? If yes, how many TV sets do you own?'
b. 1. Identify any concerns about the question.	b. It is hard to understand what period 'last year' covers. The question assumes that all students buy lunch from the school's cafeteria. It might be hard to recall the exact number of times.
2. Identify a possible replacement question or questions.	'Did you buy lunch at the school's cafeteria last school year (or over the last 12 months)? If yes, state the approximate number of times.' Alternatively, provide categories like 0–5, 6–10, 10–15, etc.
c. 1. Identify any concerns about the question.	c. People usually don't like to state their exact income. Broad categories are more likely to be answered honestly.
2. Identify a possible replacement question or questions.	'What category best describes your annual income? Less than \$25 000, \$25 000–\$50 000, \$50 000–\$75 000, etc.'

9.5.2 Types of questions

- Questions have to be designed keeping in mind the relevance to the survey.
- The data collected has to be clear and easy to analyse.

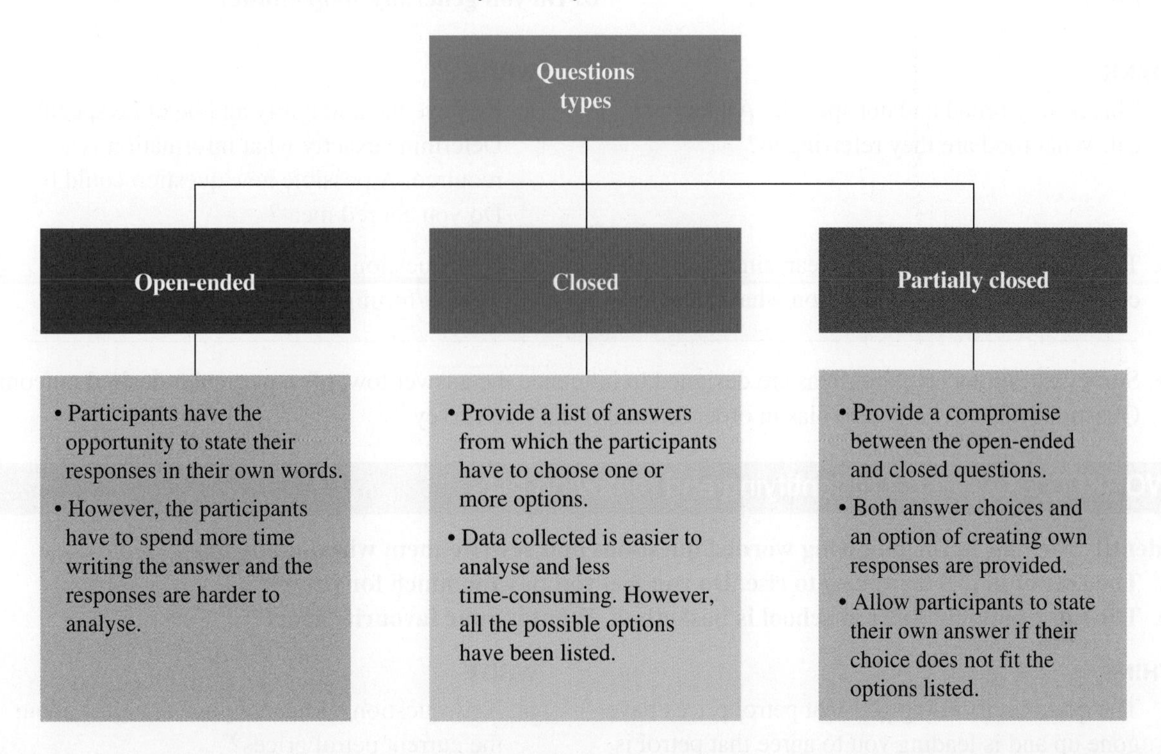

WORKED EXAMPLE 10 Identifying types of questions

Determine whether the following questions are open-ended, partially closed or closed.

a. **What mode of transport do you use when travelling to school?**

b. **What mode of transport do you use to travel to school? Multiple options may be selected.**
1. Walk 2. Bus 3. Tram 4. Bike 5. Car

c. **What mode of transport do you use to travel to school?**
1. Walk 2. Bus 3. Tram 4. Bike 5. Car 6. Other ☐

THINK	WRITE
a. 1. Determine if a written answer is required or possible options are available.	a. The question requires the participant to write their own answer.
2. Determine what type of question it is.	Open-ended
b. 1. Determine if a written answer is required or possible options are available.	b. The question has possible options available. However, it does not exhaust all possible options.
2. Determine what type of question it is.	Closed
c. 1. Determine if a written answer is required or possible options are available.	c. The question has possible options available. However, it does allow the participant to write their own option.
2. Determine what type of question it is.	Partially closed

- Sometimes survey questions can be written vaguely or can be interpreted differently by different people. To avoid getting a varied response, questions should avoid any ambiguity.

WORKED EXAMPLE 11 Identifying and removing ambiguities

Identify the ambiguities in the following worded questions and rewrite them without any ambiguity.
a. **Do you eat?**
b. **Do you generally shop online?**

THINK	WRITE
a. This is very broad and not specific. All humans eat. What food are they referring to?	a. Remove the ambiguity and be more specific. Determine exactly what information is required. A possible new question could be: Do you eat red meat?
b. The word 'generally' is not clear, since people could have different opinions on what it means.	b. New question: Do you shop online weekly/fortnightly/monthly/yearly/never?

- Survey questions involving bias are designed to influence the answer towards a particular desired outcome. Questions need to have zero bias in order to comprise a true survey.

WORKED EXAMPLE 12 Identifying and removing bias

Identify the bias in the following worded questions and rewrite them without any bias.
a. **The cost of petrol continues to rise. Do you feel you pay too much for petrol?**
b. **The most popular sport at school is basketball. What is your favourite sport?**

THINK	WRITE
a. The question is telling you that petrol prices have gone up and is leading you to agree that petrol is getting expensive.	a. New question: What are your thoughts about the current petrol prices?
b. The question gives you information about basketball's popularity, which could lead you to claim that basketball is also your favourite sport.	b. New question: What is your favourite sport?

Exercise 9.5 Investigating questionnaire design principles [complex]

learnon

9.5 Quick quiz on | 9.5 Exercise

Simple familiar	Complex familiar	Complex unfamiliar
N/A	1, 2, 3, 4, 5, 6, 7, 8, 9, 10, 11, 12, 13	14

These questions are even better in jacPLUS!
- Receive immediate feedback
- Access sample responses
- Track results and progress

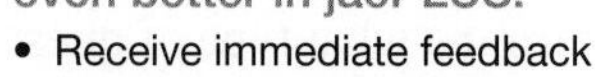
Find all this and MORE in jacPLUS

Complex familiar

1. WE9 Consider the following types of questions and explain why they are not appropriate for a questionnaire. For every question, identify a possible acceptable replacement.
   a. What brand of mobile phone do you have?
   b. Do you agree or disagree that drinking alcohol on the beach is not permitted but smoking is permitted?
   c. What church do you attend on Sundays?

2. Consider the following types of questions and explain why they are not appropriate for a questionnaire. For every question, identify a possible acceptable replacement.

a. Do you like pizza, pasta or both?
b. Are your parents not unclear in their expectations of you?

3. WE10 Determine whether the following questions are open-ended, partially closed or closed.
a. How would you rate the quality of the meat at our supermarket?
1. Excellent 2. Good 3. Average 4. Poor
b. What is your favourite brand of cereal?
c. Can you improve your grades in Mathematics? Yes or No.
d. What steps have you taken to improve your Mathematics grades?
1. Completing more homework
2. Asking the teacher for assistance more often
3. Spending more time understanding the concepts
4. Other — insert your answer here: ________________

4. Write one question of each of the following types.
a. Open-ended
b. Closed
c. Partially closed

5. Design an open-ended question, a partially closed question and a closed question for a questionnaire surveying the students in your year level about the internet search engine they use.

6. The following questions are not properly worded for a questionnaire. Explain why, and rewrite them using the guidelines in this lesson.
a. Do you catch a bus to work, or a tram?
b. Which of these is the most important issue facing teenagers today? Circle one answer.
The environment, binge drinking, teenage pregnancy
c. Do you feel better now that you have stopped smoking?
d. How much money do you earn per week?

7. Determine if the following questions are open-ended, partially closed or closed.
a. What is your country of birth?
☐ Australia ☐ Other: ________________________
b. Do you have an account on a social networking site?
☐ Yes ☐ No
c. What is your preferred way of spending weekends?

8. WE11 Identify the ambiguities in the following worded questions and rewrite them without any ambiguity.
a. Do you exercise?
b. Do you do homework?

9. Identify the ambiguities in the following worded questions and rewrite them without any ambiguity.
a. Do you support a team?
b. Do you sleep?

10. Rewrite the following questions without any ambiguity.
   a. Do you usually purchase your groceries online?
   b. Do you travel to school?
11. WE12 Identify the bias in the following worded questions and rewrite them without any bias.
   a. With interest rates going up, do you feel the cost-of-living pressures?
   b. Statistics say the crime rate on the Gold Coast is at its highest. Do you feel the police have crime under control on the Gold Coast? Explain.
12. Identify the bias in the following worded questions and rewrite them without any bias.
   a. The price of an iPhone continues to grow. Do you feel the latest iPhone is good value?
   b. Do you prefer a new and improved computer or a slower, older model?
13. The ABS conducts a census every five years. To monitor changes that might occur between these times, surveys are conducted on samples of the population. The ABS selects a representative sample of the population and interviewers are allocated particular households. It is important that no substitutes occur in the sampling. The interviewer must persevere until the selected household supplies the information requested. It is a legal requirement that selected households cooperate.

   The following questionnaire is reproduced from the ABS website www.abs.gov.au. It illustrates the format and types of questions asked by an interviewer collecting data regarding employment from a sample.

Q.1.	**I WOULD LIKE TO ASK ABOUT LAST WEEK, THAT IS, THE WEEK STARTING MONDAY THE ... AND ENDING (LAST SUNDAY THE .../YESTERDAY).**	
Q.2.	LAST WEEK DID ... DO ANY WORK AT ALL IN A JOB, BUSINESS OR FARM?	
	Yes	☐ **Go to Q.5**
	No	☐
	Permanently unable to work	☐ **No More Questions**
	Permanently not intending to work (if aged 65+ only)	☐ **No More Questions**
Q.3.	LAST WEEK DID ... DO ANY WORK WITHOUT PAY IN A FAMILY BUSINESS?	
	Yes	☐ **Go to Q.5**
	No	☐
	Permanently not intending to work (if aged 65+ only)	☐ **No More Questions**
Q.4.	DID ... HAVE A JOB, BUSINESS OR FARM THAT ... WAS AWAY FROM BECAUSE OF HOLIDAYS, SICKNESS OR ANY OTHER REASON?	
	Yes	☐
	No	☐ **Go to Q.13**
	Permanently not intending to work (if aged 65+ only)	☐ **No More Questions**
Q.5.	DID … HAVE MORE THAN ONE JOB OR BUSINESS LAST WEEK?	
	Yes	☐
	No	☐ **Go to Q.7**
Q.6.	THE NEXT FEW **QUESTIONS** ARE ABOUT THE JOB OR BUSINESS IN WHICH ... USUALLY WORKS THE MOST HOURS.	
Q.7.	DOES ... WORK FOR AN EMPLOYER, OR IN ... OWN BUSINESS?	
	Employer	☐
	Own business	☐ **Go to Q.10**
	Other/Uncertain	☐ **Go to Q.9**

Q.8.	IS ... PAID A WAGE OR SALARY, OR SOME OTHER FORM OF PAYMENT?	
	Wage/Salary	☐ **Go to Q.12**
	Other/Uncertain	☐
Q.9.	WHAT ARE ... (WORKING/PAYMENT) ARRANGEMENTS?	
	Unpaid voluntary work	☐ **Go to Q.13**
	Contractor/Subcontractor	☐
	Own business/Partnership	☐
	Commission only	☐
	Commission with retainer	☐ **Go to Q.12**
	In a family business without pay	☐ **Go to Q.12**
	Payment in kind	☐ **Go to Q.12**
	Paid by the piece/item produced	☐ **Go to Q.12**
	Wage/salary earner	☐ **Go to Q.12**
	Other	☐ **Go to Q.12**
Q.10.	DOES ... HAVE EMPLOYEES (IN THAT BUSINESS)?	
	Yes	☐
	No	☐
Q.11.	IS THAT BUSINESS INCORPORATED?	
	Yes	☐
	No	☐
Q.12.	HOW MANY HOURS DOES … USUALLY WORK EACH WEEK IN (THAT JOB/THAT BUSINESS/ALL … JOBS)?	
	1 hour or more	☐ **No More Questions**
	Less than 1 hour/no hours	☐
	Insert occupation questions if required	
	Insert industry questions if required	
Q.13.	AT ANY TIME DURING THE LAST 4 WEEKS HAS … BEEN LOOKING FOR FULL-TIME OR PART-TIME WORK?	
	Yes, full-time work	☐
	Yes, part-time work	☐
	No	☐ **No More Questions**
Q.14.	AT ANY TIME IN THE LAST 4 WEEKS HAS …	
	Written, phoned or applied in person to an employer for work?	☐
	Answered an advertisement for a job?	☐
	Looked in newspapers?	☐
	Checked factory notice boards, or used the touchscreens at Centrelink offices?	
	AT ANY TIME IN THE LAST 4 WEEKS HAS …	
	Been registered with Centrelink as a jobseeker?	☐
	Checked or registered with an employment agency?	☐
	Done anything else to find a job?	☐

	Advertised or tendered for work	☐
	Contacted friends/relatives	☐
	Other	☐ **No More Questions**
	Only looked in newspapers	☐ **No More Questions**
	None of these	☐ **No More Questions**
Q.15.	IF … HAD FOUND A JOB COULD … HAVE STARTED WORK LAST WEEK?	
	Yes	☐
	No	☐
	Don't know	☐
	Remaining questions are only required if Duration of Unemployment is needed for output or to derive the long-term unemployed.	
Q.16.	WHEN DID … BEGIN LOOKING FOR WORK?	
	Enter Date	
	Less than 2 years ago	…../…../….. **DD MM YY**
	2 years or more ago	…../…../….. **DD MM YY**
	5 years or more ago	…../…../….. **DD MM YY**
	Did not look for work	☐
Q.17.	WHEN DID … LAST WORK FOR TWO WEEKS OR MORE?	
	Enter Date	
	Less than 2 years ago	…../…../….. **DD MM YY**
	2 years or more ago	…../…../….. **DD MM YY**
	5 years or more ago	…../…../….. **DD MM YY**
	Has never worked (for two weeks or more)	☐ **No More Questions**

Reading the questionnaire carefully, you will note that, although the questions are labelled 1–17, there are only 15 questions requiring answers (two are introductory statements to be read by the interviewer). Because of directions to forward questions, no individual would be asked all 15 questions.

a. Determine how many questions could be asked of those who have a job.
b. Determine how many questions could be answered by unemployed individuals.
c. Determine how many questions apply to those not in the labour force.

Complex unfamiliar

14. Choose a topic of interest to you and conduct a survey.
Design an interview questionaire of a similar format to the ABS survey, using directions to forward questions. Decide on a technique to select a representative sample of the students in your class and administer your questionaire to this sample. Collect and draw conclusions from your results.

Fully worked solutions for this chapter are available online.

LESSON
9.6 Review

9.6.1 Summary

doc-41767

Hey students! Now that it's time to revise this chapter, go online to:

Access the chapter summary | Review your results | Watch teacher-led videos | Practise questions with immediate feedback

Find all this and MORE in jacPLUS

9.6 Exercise

learnon

9.6 Exercise

Simple familiar	Complex familiar	Complex unfamiliar
1, 2, 3, 4, 5, 6, 7, 8, 9, 10, 11	12, 13, 14, 15, 16	17, 18

These questions are even better in jacPLUS!
- Receive immediate feedback
- Access sample responses
- Track results and progress

Find all this and MORE in jacPLUS

Simple familiar

1. Determine which of the following data collection situations requires a census or survey.
   a. The employment status of a group of 1200 people over 18
   b. The employment status of all people living in Australia at a given time
   c. The employment status of a group of females aged between 30 and 40 years
   d. Temperatures in Australia from June 2023 to June 2024
2. Determine which of the following represents a population or sample.
   a. A group of 25 members of a fitness club
   b. Every 10th cereal box from a batch of 500 boxes
   c. Every second male listener of a radio show
   d. All the girls in Year 11
3. A publishing company of mathematics textbooks is checking an Essential Mathematics textbook for printing errors. Every 5th page is checked in one book and every 23rd page is checked in a second book. Determine the type of sampling used.
4. An animal farm is selecting a sample of 20 animals to try a new vaccine on. The farm has 140 cows and 60 horses. Determine the number of cows and horses required to be part of this sample.
5. Identify one advantage of basic random sampling over self-selected sampling.
6. A random number generator is used to generate 40 numbers between 0 and 1 to select a sample of 40 from a population of 100. Determine how this could be generated.

7. Determine if the following statements are True or False.
   a. Closed questions are questions that only some people can answer.
   b. A bias is when a person has a different opinion than the rest of the sample.
   c. The target population is the part of the population to be surveyed.
   d. Open-ended questions are designed to allow people to write their own answer.

8. Blanca, the local veterinarian, records the name of every dog owner. Determine and explain what type of data collection this is.

9. A random number generator is used to select 5 students from a class of 30. The first 10 numbers generated in a list are:
   87, 49, 28, 07, 16, 58, 10, 21, 19, 45, ...
   Reading the random numbers from left to right, determine the first student number selected.

10. Maya is conducting a survey on the preferred way of doing research for statistical investigations. Identify the target population if she decides to survey:
   a. a sample of 20 Year 11 students
   b. a sample of 30 students from Years 11 and 12
   c. a sample of 50 students from the whole school.

11. When we obtain data from the whole population, we conduct a _______________; on the other hand, a survey obtains data from a _______________ of the population.

Complex familiar

12. Mathsisfun Secondary College conducted a survey to collect information from 540 students enrolled.
   a. Calculate the appropriate sample size for this population.
   b. Explain how the sample could be chosen.

13. For each of the following, determine whether a census or a survey has been used.
   a. Two hundred people in a shopping centre are asked to nominate the supermarket where they do most of their grocery shopping.
   b. To find the most popular new car on the road, 500 new car buyers are asked what make and model of car they purchased.
   c. To find the most popular new car on the road, the make and model of every new car registered is recorded.
   d. To find the average mark in the Mathematics half-yearly examination, every student's mark is recorded.
   e. To test the quality of tyres on a production line, every 100th tyre is road tested.

14. Determine whether the following questions are open-ended, closed or partially closed.
   a. What is your preferred Mathematics topic?
   b. What is your preferred Mathematics topic?
      1. Algebra 2. Geometry 3. Statistics 4. Trigonometry
   c. What is your preferred Mathematics topic?
      1. Algebra 2. Geometry 3. Statistics 4. Trigonometry 5. Other _____________.

15. Select a sample of 5 students from your class.
 a. Select the sample 'by hand'.
 b. Select the sample using a random number generator.
 c. Compare your sample with the samples of other classmates.

16. A company manufacturing tablet devices is testing two batches of 150 tablet devices for defects. In the first batch every 6th tablet is checked, while in the second batch every 10th tablet is checked.
 a. Determine how many tablets are being tested for defects in the two batches.
 b. Determine the maximum starting point for each batch.

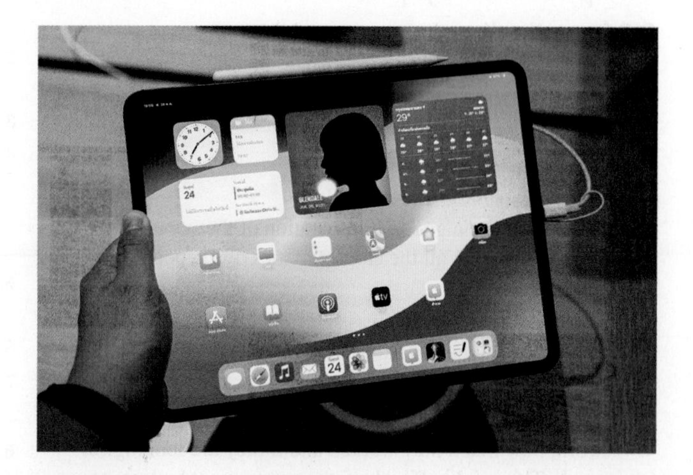

Complex unfamiliar

17. Surveys are conducted on samples to determine the characteristics of the population.

	Sample	Population
a.	Year 11 students	Student drivers
b.	Year 12 students	Students with part-time jobs
c.	Residents attending a Neighbourhood Watch meeting	Residents of a suburb
d.	Students in the school choir	Music students in the school
e.	Cars in a shopping centre car park	Cars on the road
f.	Males at a football match	People who watch TV
g.	Users of the local library	Teenagers

Discuss whether the samples selected would provide a reliable indication of the population's characteristics.

18. If simple random sampling is used and the population size triples, identify the effect this has on the size of the sample. Justify your response.

Fully worked solutions for this chapter are available online.

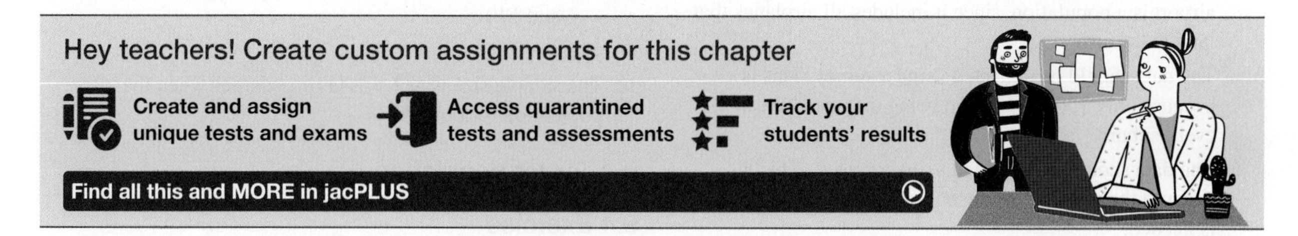

Hey teachers! Create custom assignments for this chapter
Create and assign unique tests and exams
Access quarantined tests and assessments
Track your students' results
Find all this and MORE in jacPLUS

Answers

Chapter 9 Census, surveys and simple survey procedures

9.2 Investigating a census

9.2 Exercise

1. a. Planning
   b. Collecting data
   c. Producing and releasing results
2. A sample is to be a good representation of the population, so it must maintain all the characteristics of the population it represents.
3. Since it was a sample that was selected, it was a sample of the population, so most likely a survey was conducted.
4. a. The focus is work-related.
   b. It is useful for the government to know if people had work, how much they worked and if they are looking for work. This might affect how the government budgets to support people out of work.
5. a. The focus is on the level of education obtained.
   b. It is important to the government to know if there are enough people qualified to work in certain areas or if the government needs to encourage more people to study in areas where there might be a shortage of workers in the future.
6. a. The data should be collected using a census. This is because the response of the entire population of customers is needed.
   b. The data should be collected using a survey. This is because only 50% of the Year 11 population is needed, which is a sample.
   c. The data should be collected using a census. This is because the response of the entire population of the patients is needed.
7. a. The number of absences of all Year 12 students is a sample, since it is only one part of the school. The number of absences of all students in the school is a population, since all students are included.
   b. The number of passenger airplanes landing at an airport is a sample, since it is only one group of airplanes that land at an airport. The total number of airplanes landing at an airport is a population, since it includes all airplanes that land at the airport.
8. a. The data should be collected using a survey. This is because only 150 suburbs are being included, not every suburb.
   b. The data should be collected using a census. This is because the response of the entire population of Australia is needed.
   c. The data should be collected using a census. This is because the response of the entire homeroom is needed.
   d. The data should be collected using a survey. This is because only every fifth person is being asked, not every person.
   e. The data should be collected using a census. This is because a response is being sought from every customer.

9.3 Sampling methods — advantages and disadvantages [complex]

9.3 Exercise

1. a., b. Any sample with 5 different numbers between 1 and 26
2. a., b. Any sample with 8 different numbers between 1 and 29
3. 153 people
4. a. Any sample with 10 different numbers between 1 and 100
   b. Any sample with 10 different numbers between 1 and 100
5. 1–8 if selecting every sixth car, or 1–4 if selecting every seventh car
6. From 1 to 6, selecting every fifth product
7. a. 5 boys and 6 girls
   b. 8 adults and 2 children
8. Publicising includes advertising at the local library, local radio station, letter drop in the surrounding households and advertisements at the local shopping centres.
   All applicants could be accepted as there are no restrictions set on the study.
9. Publicising using advertisements in the local paper, or at the local gym
10. Self-selected sampling
11. a. Stratified sampling
    b. 6 black sheep and 4 white sheep
12. a. Systematic sampling
    b. 20 phones in the first batch; 14 phones in the second batch
    c. Maximum starting point is 5 in the first batch when selecting every fifth phone, and 9 in the second batch when selecting every seventh phone.
13. 19 males and 14 females
14. 7 white cars and 8 black cars
15. a. 15 students
    b. Either using a random number generator or using systematic sampling
16. a. Any sample with 12 different numbers between 1 and 140
    b. Any sample with 12 different numbers between 1 and 140
17. 1225
18. The sample size is $\sqrt{2} \approx 1.41$ times larger when the population is doubled.

9.4 Target population

9.4 Exercise

1. The target population is the students who are tutored by the mathematics tutoring company.
2. Since it is Country Road who want to know about their staff, the target population is Country Road customers.
3. a. Since it is the most popular subject in Year 11, the target population is Year 11 students at that school.

b. Any conclusions would not be accurate if the participants completing the survey were not Year 11 students from that school.

4. a. The target population is the people who visit the Dreamworld theme park.

   b. For the survey to be accurate, the people taking the survey need to be from the target population.

5. The target population is the apples from the orchard.

6. The target population is the staff from the school.

7. The target population is all the company employees.

8. a. The target population is all employees with children.

   b. The target population is all employees required to use a computer for work.

9.5 Investigating questionnaire design principles [complex]

9.5 Exercise

1. a. It is assumed all people have mobile phones. A possible replacement question is 'Do you own a mobile phone? If yes, what brand do you have?'

   b. This is a double-barrelled question. It should be separated into two questions: 'On a scale of 1 (strongly agree) to 5 (strongly disagree), should drinking alcohol be permitted at the beach? Should smoking be banned at the beach?'

   c. The question assumes that all people go to church on Sundays and that all people are of a certain religion. A possible replacement question is 'Do you attend a place of prayer?
   - Never
   - Rarely
   - Sometimes
   - Often
   - Regularly'

2. a. This is a leading question. 'Do you like pizza, pasta, both or neither?'

   b. This is a double negative question.
   'Are your parents' expectations of you clear?'

3. a. Closed b. Open-ended
   c. Closed d. Partially closed

4. a. Sample responses can be found in the worked solutions in the online resources.

   b. Sample responses can be found in the worked solutions in the online resources.

   c. Sample responses can be found in the worked solutions in the online resources.

5. Possible answers:
   Open-ended question: 'What is your preferred search engine when surfing the internet?'
   Partially closed question: 'The internet search engine I prefer to use is:
   ☐ Firefox ☐ Google ☐ Safari ☐ Other ....'
   Closed question: 'The internet search engine I prefer to use is:
   ☐ Firefox ☐ Google ☐ Safari ☐ None of these
   ☐ I don't use the internet'

6. a. This is a double-barrelled question. Possible replacement: 'What mode of transport do you use to go to work?
   ☐ Bus ☐ Tram ☐ Other ....'

   b. Some people might have other opinions, such as smoking or lack of social skills or too much technology.
   Possible replacement: 'In your opinion, what is the most important issue facing teenagers today?'

   c. Leading question suggesting the desired answer. Possible replacement: 'How do you feel now after stopping smoking?'

   d. Too personal. Possible replacement: Your annual income is:
   ☐ less than $30 000
   ☐ $30 000–$50 000
   ☐ $50 000–$70 000
   ☐ $70 000–$90 000
   ☐ more than $90 000

7. a. Partially closed

   b. Closed

   c. Open-ended

8. a. The ambiguity is how one defines exercise and how regularly it needs to be done to be considered exercise.
   New question: Do you run weekly?

   b. The ambiguity is the regularity of homework that needs to be done to be considered 'doing homework'.
   New question: Do you complete more than 10 hours per week of homework?

9. a. The ambiguity is the type of team and sport.
   New question: Do you support an NRL team?

   b. The ambiguity is that all people sleep, but we don't know how much they sleep.
   New question: Do you sleep on average 8 hours per night?

10. a. How many times per month do you purchase your groceries online?

    b. Do you travel to school by bus?

11. a. This is a loaded question that states that interest rates are going up.
    New question: This year have you felt there is a higher cost of living?

    b. Giving the statistics leads to the response.
    New question: Do you feel the police have crime under control on the Gold Coast?

12. a. This is loaded question, since it states the cost is growing.
    New question: Will you purchase the latest iPhone?

    b. The question suggests that a new computer performs far better than an old one, so it is leading to a certain response.
    New question: Do you like the new computer model?

13. Sample responses can be found in the worked solutions in the online resources.

14. Sample responses can be found in the worked solutions in the online resources.

9.6 Review

9.6 Exercise

1. Survey: a, c, d
   Census: b
2. Sample: a, b, c
   Population: d
3. Systematic sampling
4. Cows = 14; horses = 6
5. It removes the possibility of bias.
6. To get the randomly generated number to fit in the population size, they will all have to be multiplied by the population size, which is 100.
7. True: d
   False: a, b, c
8. Survey
9. 28
10. a. Year 11 students
    b. Year 11 and 12 students
    c. All students at the school
11. When we obtain data from the whole population, we conduct a census; on the other hand, a survey obtains data from a sample of the population.
12. a. 23
    b. Randomly or systematically
13. a. Survey b. Survey c. Census
    d. Census e. Survey
14. a. Open-ended
    b. Closed
    c. Partially closed
15. a. Any set of 5 students
    b. Any set of 5 students
    c. The sets would most likely be different.
16. a. 25 tablets; 15 tablets
    b. 6 and 10
17. a. There would be many more student drivers in Year 12 than in Year 11 — perhaps also some in Year 10.
    b. Students with part-time jobs are in lower year levels as well.
    c. Residents not at the Neighbourhood Watch meeting have been ignored.
    d. Other music students who play instruments and don't belong to the choir have been excluded.
    e. The composition of cars in a shopping centre car park is not representative of the cars on the road.
    f. Females have been excluded.
    g. Users of the local library would not reflect the views of teenagers.
18. The sample size is $\sqrt{3} \approx 1.73$ times larger when the population is tripled.

10 Sources of bias

LESSON SEQUENCE

Fully worked solutions for this chapter are available online.

on Resources

Solutions Solutions — Chapter 10 (sol-1263)

Digital documents Learning matrix — Chapter 10 (doc-41768)
Quick quizzes — Chapter 10 (doc-41769)
Chapter summary — Chapter 10 (doc-41770)

LESSON
10.1 Overview

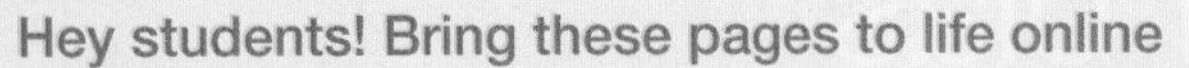

Hey students! Bring these pages to life online

 Watch videos

 Engage with interactivities

 Answer questions and check results

Find all this and MORE in jacPLUS

10.1.1 Introduction

In the modern world of technology and information, data collection is important in order to understand, analyse and make decisions across various fields. Whether in scientific research, business operations, public policy, or everyday life, gathering and interpreting data helps us understand the world and drives progress.

Sources of bias can make data and survey results inaccurate and unreliable, leading to misrepresentation of the phenomena being studied. Bias can arise from various sources, including the design of the study, the selection of participants, the formulation of questions, and the interpretation of results.

Overall, it's very important for researchers, analysts, and practitioners to be aware of the sources of bias, as this helps them to lessen its effects and ensure that data and survey results accurately reflect the reality being studied.

Strategies such as random sampling, careful questionnaire design, transparency in data collection methods, and critical evaluation of findings can help minimise bias and improve the validity and reliability of research outcomes.

10.1.2 Syllabus links

Lesson	Lesson title	Syllabus links
10.2	**Faults in collecting data**	◯ Describe the faults in the process of collecting data. ◯ Describe sources of error in surveys, including sampling error and measurement error.
10.3	**Investigating possible misrepresentation of results [complex]**	◯ Investigate the possible misrepresentation of the results of a survey due to misunderstanding the procedure or the reliability of generalising the survey findings to the entire population [complex].
10.4	**Investigating errors and misrepresentation in surveys [complex]**	◯ Investigate errors and misrepresentation in surveys, including examples of media misrepresentations of surveys [complex].

LESSON
10.2 Faults in collecting data

SYLLABUS LINKS

- Describe the faults in the process of collecting data.
- Describe sources of error in surveys, including sampling error and measurement error.

10.2.1 Bias

- **Bias** is a distortion that occurs during the data collection process that produces results that are not representative of the population.
- Care has to be taken when planning for a survey to avoid and, if possible, eliminate bias.
- Accurate data collection is important in ensuring the quality of any statistical study, because its results are often used in decision-making.
- Sources of bias can be that:
  - the sample is not representative of the population being studied
  - questions do not respect the guidelines required.

WORKED EXAMPLE 1 Identifying and eliminating bias in questions

The following questions were asked of theatre-goers.

1. **Rate your level of enjoyment of the show:**
   Highly enjoyable / very enjoyable / enjoyable
2. **How good were your seats at the show?**
   Outstanding / exceptional / fantastic

For the above questions:

a. **identify whether they are biased and explain your reasoning**
b. **modify the questions to eliminate the bias.**

THINK	WRITE
1. a. Determine whether the possible responses cover all possible options.	The question is biased because all the responses are positive and it doesn't allow for a negative opinion.
b. Eliminate the bias by having a good range of options.	Rate your level of enjoyment of the show: Very enjoyable / enjoyable / not enjoyable
2. a. Determine whether the possible responses cover all possible options.	All the responses are positive and the question doesn't allow for a negative or neutral opinion.
b. Eliminate the bias by having a good range of options.	How good were your seats at the show? Outstanding / average / poor

10.2.2 Sources of error

- When forming a sample and recording data from the population, there are two main errors to look for: sampling error and measurement error.

Sampling error

- This type of error is probably the most common type of error that happens in a survey.
- **Sampling errors** occur when surveys are conducted on one sample out of a large number of possible survey samples.

- Some sampling errors occur when:
  - the sample is too small
  - subgroups of the population are unrepresented or overrepresented
  - the sample was not selected at random.

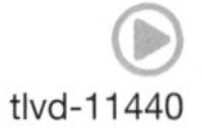
tlvd-11440

WORKED EXAMPLE 2 Identifying sampling errors in surveys

The school principal wants to get the students' perspective about an issue at the school. They think of the following three ways to select the sample of students.

1. **Select the two school captains.**
2. **Select the 15 school prefects.**
3. **Select students who are playing on the basketball court.**

Answer the following questions about the above samples.

a. **Determine if the samples are biased.**

b. **Explain how the principal can select an unbiased sample of students.**

THINK	WRITE
a. Identify whether the sample is large enough, if it is a good representation of the population, and whether it is selected randomly.	a. 1. The sample is biased, because it is small and not a representation of the whole school. 2. The sample is biased because the school prefects are not a representation of the whole school. 3. The sample is biased because all basketballers are not a representation of the whole school.
b. To represent the view of all students, the sample needs to include students from all year levels.	b. To ensure the sample is not biased, the principal should conduct a random stratified sample.

Measurement error

- When measuring data, you want to be accurate and precise.
- Errors and faults in measurements may occur due to the following:

Systematic errors

- **Systematic errors** are errors that affect the accuracy of a measurement and cannot be improved by repeating the experiment. This can be due to a faulty measuring device.

Random errors

- **Random errors** are chance variations in measurements that affect the precision of measurement. This could include incorrect reading of the scale of a measuring instrument (e.g. ruler).

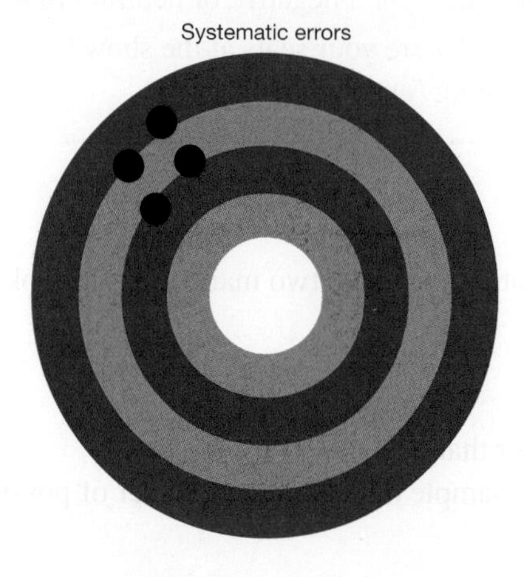

WORKED EXAMPLE 3 Calculating measurement errors

Consider the diagram shown.

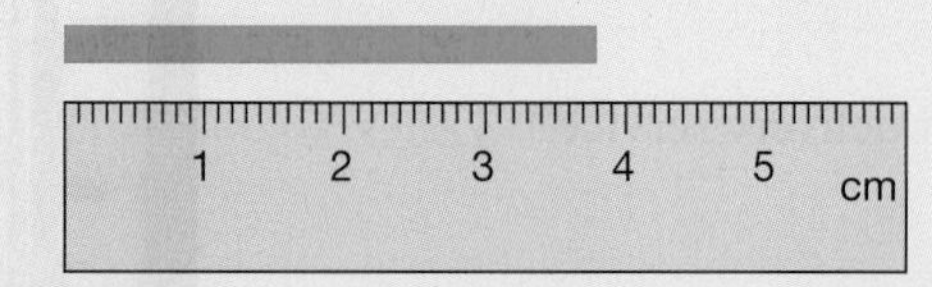

a. **Determine the length of the blue object.**
b. **If the ruler shown below were used to measure the blue object, explain the accuracy of the measurement.**

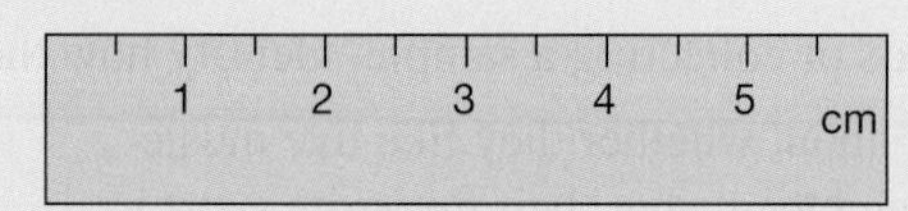

THINK	WRITE
a. Read the scale on the ruler. Measure to the nearest cm less than the length of the blue object, and then count the increments to the end of the blue object.	a. Each increment represents 0.1 cm. The nearest cm less is 3 cm and then there are 8 increments, with each increment representing 0.1 cm. Length $= 3 + 8 \times 0.1 = 3.8$ cm
b. Determine the accuracy to which the ruler can measure.	b. The scale on the ruler measures in 0.5-cm intervals. The ruler in part **a** measures in 0.1-cm intervals. Since the ruler in part **a** measures using smaller increments, it would give a more accurate measurement.

Exercise 10.2 Faults in collecting data

learnon

10.2 Quick quiz on | 10.2 Exercise

Simple familiar	Complex familiar	Complex unfamiliar
1, 2, 3, 4, 5, 6, 7, 8, 9, 10, 11, 12	N/A	N/A

These questions are even better in jacPLUS!
- Receive immediate feedback
- Access sample responses
- Track results and progress

Find all this and MORE in jacPLUS

Simple familiar

1. WE1 The following questions were asked of people after they ate at a restaurant.
   i. Rate your meal:
   Excellent / very enjoyable / enjoyable
   ii. How good was the service you received?
   Outstanding / very good / good

   For the above questions:
   a. describe if they are biased and, if so, explain why
   b. modify the questions to eliminate the bias.

2. The following questions were asked of people purchasing items from an online electrical store.

   i. Rate your purchase:
   Very happy / happy / satisfied
   ii. How would you describe your delivery service?
   Speedy / very quick / quick

   For the above questions:

   a. describe if they are biased and, if so, explain why
   b. modify the questions to eliminate the bias.

3. For each of the following methods of collecting a sample, identify how they might result in a biased sample.
   a. Surveying people at a concert about whether they like live music
   b. Asking people at the Australian Open what their favourite sport is
   c. Asking school students if they think the Age Pension is too high

4. WE2 The management of the Brisbane Bullets basketball team want to survey the general public on the impact the team has on the Brisbane community. To conduct the survey, they consider selecting the sample in one of the following three ways:

   i. Survey people as they leave one of their home games.
   ii. Select 5 people at random from a Brisbane shopping centre.
   iii. Select 100 people at random from a Brisbane shopping centre.

   Determine if these samples are biased and explain why.

5. A travel company wants to conduct a survey to get feedback on the quality of their service. Describe how they could conduct an unbiased survey.

6. An internet service provider wants to conduct a survey on how their customers rate their internet service. Describe how they could conduct an unbiased survey.

7. WE3 Use the diagram shown to answer the following questions.

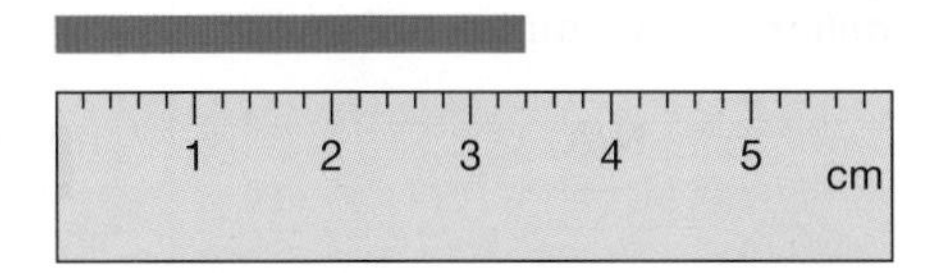

   a. Determine the length of the green object.
   b. If the ruler shown was used to measure the green object, discuss the accuracy of the measurement.

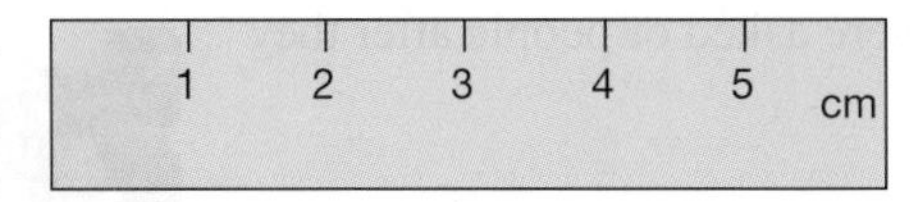

8. Decide if the following could produce a measurement error and explain why.
   a. Using a 30-cm ruler to measure student heights in a Year 11 class
   b. Timing students' reaction times with a stopwatch by having them catch a falling object
   c. Recording a mass of 0.076 g of an object that actually has a mass of 0.76 g

9. Modify each of the following survey questions to eliminate bias.
   a. I don't like the colour of our uniform. Do you like our school uniform?
   b. Do you prefer to watch the mighty Brisbane Broncos, or the poorly performing Brisbane Bullets?
   c. I always see students on their phones. Do you think they spend too much time on their phones?
   d. Would you prefer to watch the Brisbane Lions at the home of AFL, the MCG, or at the SCG with its flat atmosphere?
10. When a measurement is made, there is always a level of error. For the following, determine the lowest and highest possible measure.
   a. 1.5 ± 0.1 cm b. 15.6 ± 0.2 g c. 28.25 ± 0.05 kg d. 5.93 ± 0.03 s
11. An experiment was conducted to measure the mass of a ball bearing with scales that have an error of ± 0.2 g. If the scales measured 25.9 g:
   a. determine the maximum possible mass of the ball bearing
   b. determine the minimum possible mass of the ball bearing.
12. A mobile phone provider conducted a survey on 750 customers about what customers mainly used their mobile phone for. The following results were found.
   a. Calculate how many customers use their phones mainly for calls and photos.
   b. Determine the type of sampling that was likely used for this survey.

Mobile phone usage	
28%	SMS
6%	Photos
46%	Social media
4%	Calls
16%	Apps

Fully worked solutions for this chapter are available online.

LESSON 10.3 Investigating possible misrepresentation of results [complex]

SYLLABUS LINKS

- Investigate the possible misrepresentation of the results of a survey due to misunderstanding the procedure or the reliability of generalising the survey findings to the entire population [complex].

10.3.1 Misrepresentation of results

- In Chapter 9 and lesson 9.2, we have looked at the importance of sampling and good questioning in surveys. If this is not done correctly, we can get biased results.
- It is not just important to correctly set up the survey, it is also important how the results of the survey are interpreted.

WORKED EXAMPLE 4 Commenting on misrepresentation of results

A travel company surveyed 20 people who went on one of their day trips to Uluru. From the 20 people surveyed, 15 of them said they would recommend the tour to their friends. Using the results of the survey, it was reported back to the head of the travel company that 75% of customers love the tours and would recommend them to their friends.

a. Comment on the sampling technique.
b. Comment on the survey procedure.
c. Comment on the interpretation of results.

THINK	WRITE
a. The sample taken should be a representation of the population.	a. The sample is from only one tour group, which is not a representation of all the tours the company offers.
b. The question should ask clearly what the company wants to find out.	b. Even though recommending friends to do the tour is a positive, the question could be more specific about the parts of the tour customers enjoyed most, or give them a scale to represent their level of enjoyment of the tour.
c. It is important to have clear questions with an appropriate sampling technique to make generalisation using the obtained data.	c. This data is collected from one tour group, so to say that 75% of customers like the tours is not accurate. The sample would need to be from people randomly selected from all their tours.

10.3.2 Allowing for errors in results

- When you use a sample as a representation of a population, a margin of error should be taken into account.
- The **margin of error** is the range of values that result when using a sample of a population.
  For example, if 60% of the people in a sample answered 'Yes' to a question, then it is assumed that, for a 5% margin of error, between 57.5% and 62.5% of the population answered 'Yes' to that question.

Calculating the margin of error

The margins of error are calculated using the formula:

Lower margin $= x\% - \dfrac{e\%}{2}$ **and upper margin** $= x\% + \dfrac{e\%}{2}$

where $x\%$ **is the percentage of the sample considered,**

$e\%$ **is the percentage error.**

- The **confidence level** is used to determine the percentage of results that will not fall within the margin of error stated.
- If the confidence level for the situation described above is 95%, it means that there is a 5% risk that the results will not fall between 57.5% and 62.5%.

Calculating the margin of error

Risk = 100% − confidence level

This table provides recommended sample sizes for two different margins of error.

Population size	Sample size		Population size	Sample size	
	5%	10%		5%	10%
10	10		275	163	74
15	14		300	172	76
20	19		325	180	77
25	24		350	187	78
30	28		375	194	80
35	32		400	201	81
40	36		425	207	82
45	40		450	212	82
50	44		475	218	83
55	48		500	222	83
60	52		1000	286	91
65	56		2000	333	95
70	59		3000	353	97
75	63		4000	364	98
80	66		5000	370	98
85	70		6000	375	98
90	73		7000	378	99
95	76		8000	381	99
100	81	51	9000	383	99
125	96	56	10 000	385	99
150	110	61	15 000	390	99
175	122	64	20 000	392	100
200	134	67	25 000	394	100
225	144	70	50 000	397	100
250	154	72	100 000	398	100

Source: Isaac and Michael, 1981; Smith, M. F, 1983.

tlvd-11441

WORKED EXAMPLE 5 Calculating the margin of error and confidence level

a. Calculate the lower and upper margins for a 10% margin of error if 24% of the members of a sample were found to be blue eyed.

b. If the confidence level is 85%, determine the risk of the results falling outside the margins calculated in part a.

THINK	WRITE
a. 1. Write the formula for the lower margin.	**a.** Lower margin $= x\% - \dfrac{e\%}{2}$
2. Substitute the known values and simplify.	$= 24\% - \dfrac{10\%}{2}$ $= 24\% - 5\%$ $= 19\%$
3. Write the formula for the upper margin.	Upper margin $= x\% + \dfrac{e\%}{2}$
4. Substitute the known values and simplify.	$= 24\% + \dfrac{10\%}{2}$ $= 24\% + 5\%$ $= 29\%$
5. Write the answer.	Between 19% and 29% of the population is blue eyed.

b. 1. Calculate the risk.

b. Risk = 100% − confidence level
= 100% − 85%
= 15%

2. Write the answer.

There is a 15% risk that the blue-eyed members will fall outside the interval 19%–29%.

Exercise 10.3 Investigating possible misrepresentation of results [complex]

learnon

10.3 Quick quiz on | 10.3 Exercise

Simple familiar	Complex familiar	Complex unfamiliar
N/A	1, 2, 3, 4, 5, 6, 7, 8, 9, 10	11, 12

These questions are even better in jacPLUS!
- Receive immediate feedback
- Access sample responses
- Track results and progress

Find all this and MORE in jacPLUS

Complex familiar

1. Comment on the following sampling techniques.
   a. Selecting 10 students from the library, to see if they think the library is a good place to study
   b. Selecting 30 students in the canteen line to see if they like the canteen food

2. Forty rugby players were surveyed, and 30 of them said that rugby players were the fittest sportspeople. Rugby promoters claim that 75% of people believe rugby players are the fittest sportspeople. Comment on the interpretation of the results.

3. WE4 A survey was conducted on 150 people in Cavil Avenue (Gold Coast). They were asked what the best beach in Australia was, and 120 of them said the best beach was Gold Coast Beach. The survey was used to claim that 80% of people believe the Gold Coast Beach is the best beach in Australia.
   a. Comment on the sampling technique.
   b. Comment on the survey procedure.
   c. Comment on the interpretation of results.

4. The five school leaders were asked if they thought the process for selecting leaders in the school was a fair process. Comment on the reliability of this data.

5. WE5 a. Calculate the lower and upper margins for a 5% margin of error if 11% of the batteries in a sample were found to be batteries with defects.
   b. If the confidence level is 95%, determine the risk of the results falling outside the margins calculated in part a.

6. a. Calculate the lower and upper margins for a 4% margin of error if 96% of the members of a sample of school students were found to be Year 12 students.
   b. If the confidence level is 92%, determine the risk of the results falling outside the margins calculated in part a.

7. The manager of a local supermarket asked 10 customers 'Do you like to shop at our supermarket?'. Eight customers replied 'Yes'. Once he collected this data, the manager placed a poster stating that '80% of people like to do their shopping in our supermarket.' Explain whether this interpretation is correct.
8. Calculate the confidence level of a survey if 2.5% of its results do not fall within a given margin of error.
9. Calculate the lower and upper margins for:
   a. a 3% margin of error if 20% of the members of a sample were found to be people over 40 years old
   b. a 5% margin of error if 70% of the cars in a sample were found to be sedans.
10. If the confidence level for the samples in question 6 is 96%, calculate the percentage of results that will not fall within the margins of error calculated.

Complex unfamiliar

11. Determine the margin of error that has a lower and upper margin of 25.5% and 32.5% if 29% of the members in the sample had blonde hair.
12. Calculate the lower and upper number of eggs for a 4% margin of error if 8% of the eggs in the sample of 50 eggs were found to be cracked.

Fully worked solutions for this chapter are available online.

LESSON
10.4 Investigating errors and misrepresentation in surveys [complex]

SYLLABUS LINKS

- Investigate errors and misrepresentation in surveys, including examples of media misrepresentations of surveys [complex].

10.4.1 Statistical interpretation bias

- Once the data have been collected, collated and subjected to statistical calculations, bias may still occur in the interpretation of the results.
- *Misleading graphs* can be drawn, leading to a biased interpretation of the data. Graphical representations of a set of data can give a visual impression of 'little change' or 'major change' depending on the scales used on the axes.
- *The use of terms* such as 'majority', 'almost all' and 'most' is open to interpretation.
- When we consider that 50.1% 'for' and 49.9% 'against' represents a 'majority for' an issue, it's clear that the true figures have been hidden behind words with very broad meanings.
- Although we would probably not learn the real facts, we should be wary of statistical issues quoted in such terms.

WORKED EXAMPLE 6 Interpreting statistical bias in the collected data

Discuss why the following selected samples could provide bias in the statistics collected.

a. In order to determine the extent of unemployment in a community, a committee phoned two households (randomly selected) from each page of the local telephone directory during the day.

b. A newspaper ran a feature article on the use of animals to test cosmetics. A form beneath the article invited responses to the article.

THINK	WRITE
a. 1. Consider phone book selection.	a. Phoning two randomly selected households per page of the telephone directory is possibly a representative sample.
2. Consider those with no phone contact.	However, those without a home phone and those with unlisted numbers could not form part of the sample.
3. Consider the hours of contact.	An unanswered call during the day would not necessarily imply that the resident was at work.
b. 1. Consider the newspaper circulation.	b. Selecting a sample from a circulated newspaper excludes those who do not have access to the paper.
2. Consider the urge to respond.	In emotive issues such as these, only those with strong views will bother to respond, so the sample will likely overrepresent 'extreme' points of view.

10.4.2 Misrepresentation of graphs

- Misleading graphs are sometimes deliberately misleading.
- There are numerous ways of misrepresenting data. Examples of misleading graphs are those where:
  - the vertical scale is too big or too small, skips numbers, or doesn't start at zero
  - the graph isn't labelled properly
  - data points are left out.
- The graph shown displays the horizontal axis that jumps from 2015 to 2018 while the rest of the scale represents equal intervals of one year. This gives the impression that the increase from between the first two columns is a lot bigger than the increases between the rest of the columns.

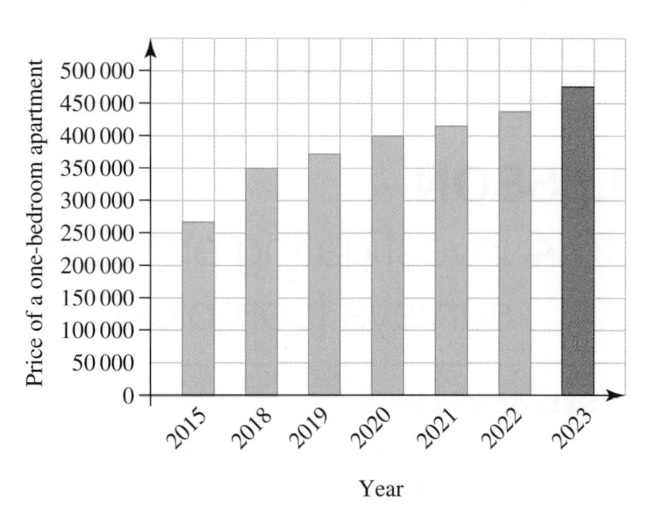

- A more accurate representation of this data is revealed in the corresponding histogram shown.

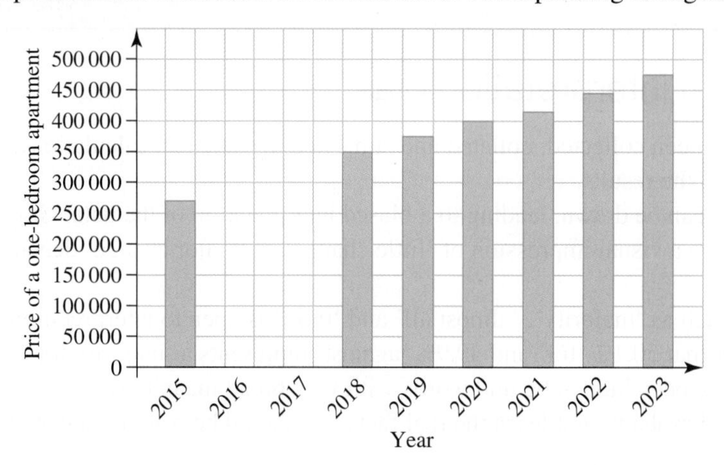

tlvd-11442

WORKED EXAMPLE 7 Understanding whether a graph is misleading

Identify in what way is the graph shown misleading or deceiving.

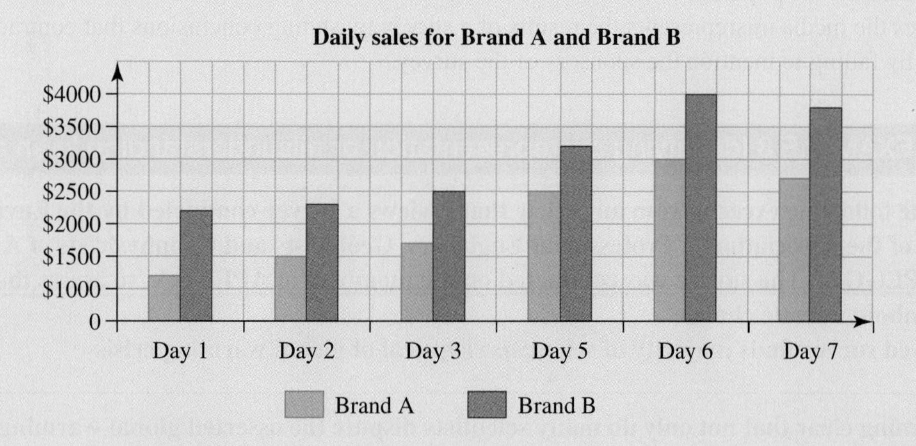

THINK	WRITE
1. Check the starting point of the vertical axis.	The vertical axis does not have an even scale. The first incremental increase is $1000 (from $0 to $1000) whereas following incremental increases are all of $500 (e.g. $1000 to $1500). This gives the impression that the sales for Brand B are a lot higher than the ones for Brand A. If we compare the first two columns, it looks like the sales for Brand B are more than double the sales for Brand A. This assumption is incorrect. Here is the graph with the scale showing even incremental increases of $500.

In reality, the sales for Brand B are roughly double the sales of Brand A on Day 1.

2. Check the accuracy of the scales on both axes.	The vertical axis has equally spaced scales. The horizontal axis does not have a correct scale. Notice that Day 4 sales are missing. This makes the jump from the sales on Day 3 to the sales on Day 5 look a lot higher.

10.4.3 Misrepresentation and misunderstanding

- A **misrepresentation** of the results of a survey is an untrue statement that does not reflect the characteristics of the population.
- Sometimes the media misrepresents the results of a survey by stating conclusions that contradict a survey's report or by failing to mention the sponsors of the survey.

tlvd-11443

WORKED EXAMPLE 8 Identifying misrepresentation or misunderstanding in an article

Consider the following excerpt from an article that reviews a survey conducted by the Environment Committee of the Association of Professional Engineers, Geologists and Geophysicists of Alberta, Canada (APEGGA). The survey was conducted on the members of APEGGA 'to assess their beliefs and values about climate change'.
Peer-reviewed survey finds majority of scientists skeptical of global warming crisis

It is becoming clear that not only do many scientists dispute the asserted global warming crisis, but these skeptical scientists may indeed form a scientific consensus.

Don't look now, but maybe a scientific consensus exists concerning global warming after all. Only 36 per cent of geoscientists and engineers believe that humans are creating a global warming crisis, according to a survey reported in the peer-reviewed *Organization Studies*. By contrast, a strong majority of the 1077 respondents believe that nature is the primary cause of recent global warming and/or that future global warming will not be a very serious problem.[1]

The Actual Survey, Science or science fiction? professionals' discursive construction of climate chage'[2], states that 'Given the similarity of the survey respondents to APEGGA's general membership, this suggests [a 3.1% error] that responses may be generalizable to the membership as a whole' and '27.4% believe it is caused by primarily natural factors (natural variation, volcanoes, sunspots, lithosphere motions, etc.), 25.7% believe it is caused by primarily human factors (burning fossil fuels, changing land use, enhanced water evaporation due to irrigation), and 45.2% believe that climate change is caused by both human and natural factors'.[3]

1. ***Source: Forbes*, http://www.forbes.com/sites/jamestaylor/2013/02/13/peer-reviewed-survey-find s-majority-of-scientists-skeptical-of-global-warming-crisis/**
2. **Lianne M. Lefsrud, University of Alberta, Canada; Renate E. Meyer, Vienna University of Economics and Business, Austria and Copenhagen Business School, Denmark.**
3. ***Source:* SAGE journals, *Organization Studies*, http://oss.sagepub.com/content/33/11/1477.full**

Identify any misinterpretations or misunderstandings found in the article when discussing the survey.

THINK	WRITE
1. Start with the title of the article and check the survey to check its reliability.	'Peer-reviewed survey finds majority of scientists skeptical of global warming crisis' The title generalises the findings of the survey to all scientists. The survey states '… generalizable to the membership as a whole.'
2. Comment on the misrepresentation of the results.	As only members of an organisation were surveyed, the results can be generalised only to the organisation's scientific population and not all scientists.

3. Check each sentence and compare with the results stated in the survey.	Article: 'By contrast, a strong majority of the 1077 respondents believe that nature is the primary cause of recent global warming and/or that future global warming will not be a very serious problem.' Survey: '27.4% believe it is caused by primarily natural factors (natural variation, volcanoes, sunspots, lithosphere motions, etc.), 25.7% believe it is caused by primarily human factors (burning fossil fuels, changing land use, enhanced water evaporation due to irrigation), and 45.2% believe that climate change is caused by both human and natural factors.'
4. Comment on the misrepresentation or misunderstanding of the results.	According to the survey, only 27.4% of the respondents believe that global warming is caused primarily by natural factors, while the article states that '... a strong majority of the 1077 respondents believe that nature is the primary cause of recent global warming ...' It is clear this article is presenting a biased view.

Exercise 10.4 Investigating errors and misrepresentation in surveys [complex]

learn on

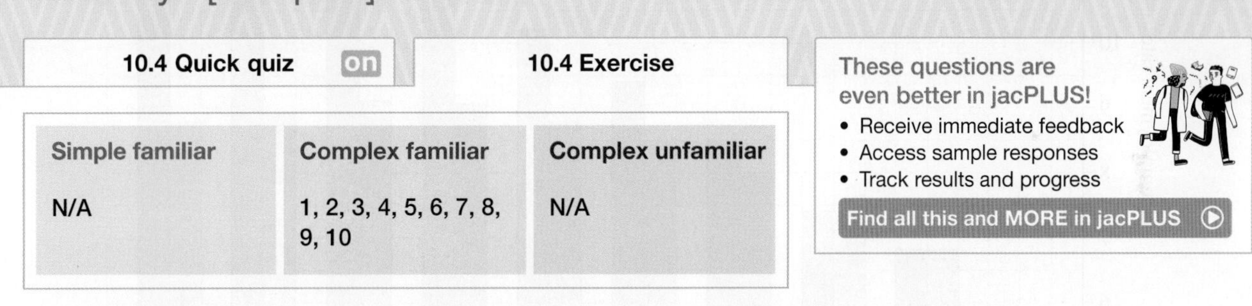

10.4 Quick quiz on

10.4 Exercise

Simple familiar	Complex familiar	Complex unfamiliar
N/A	1, 2, 3, 4, 5, 6, 7, 8, 9, 10	N/A

These questions are even better in jacPLUS!
• Receive immediate feedback
• Access sample responses
• Track results and progress
Find all this and MORE in jacPLUS

Complex familiar

1. **WE6** Discuss why the following selected sample could provide bias in the statistics collected.
   To determine the popularity of the latest Audi car, a survey was conducted in front of the local Audi car yard.
2. Discuss why the following selected sample could provide bias in the statistics collected.
   To determine who people were going to vote for in an upcoming election, five people were called and asked who they were going to vote for.
3. Discuss why the following selected samples could provide bias in the statistics collected.
   A billboard was posted on a freeway with an advertisement against caged chickens. The billboard also invited people to comment on the advertisement.

4. WE7 The graph shown represents the relationship between job satisfaction of a group of people surveyed and the years of schooling before conditioning on income (blue columns) and after conditioning on income (pink rectangles).
Determine in what way this graph is misleading or deceiving.

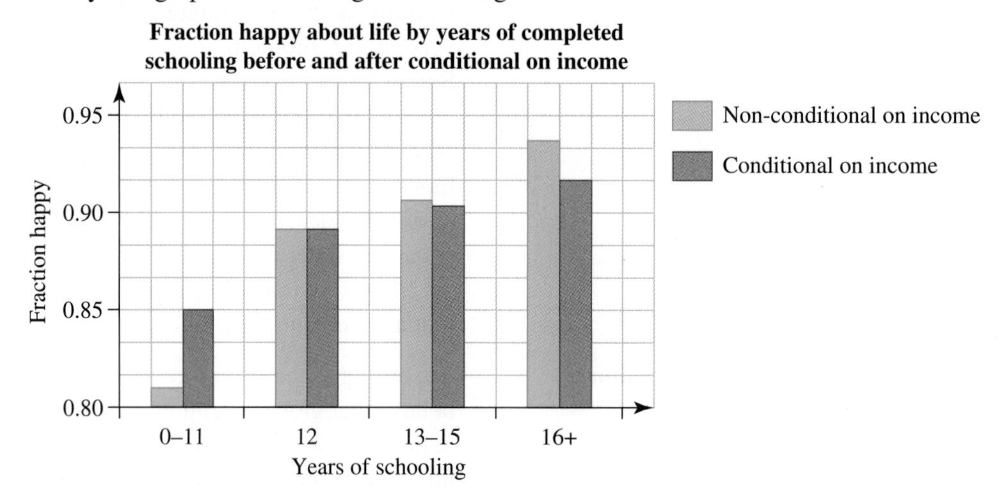

Source: Oreopoulos, P., Salvanes, K. G., 'Priceless: the nonpecuniary benefits of schooling', *Journal of Economic Perspectives*, vol. 25, no. 1, Winter 2011, pp. 159–84, http://pubs.aeaweb.org/doi/pdfplus/10.1257/jep.25.1.159

5. Describe how this graph is misleading or deceiving.

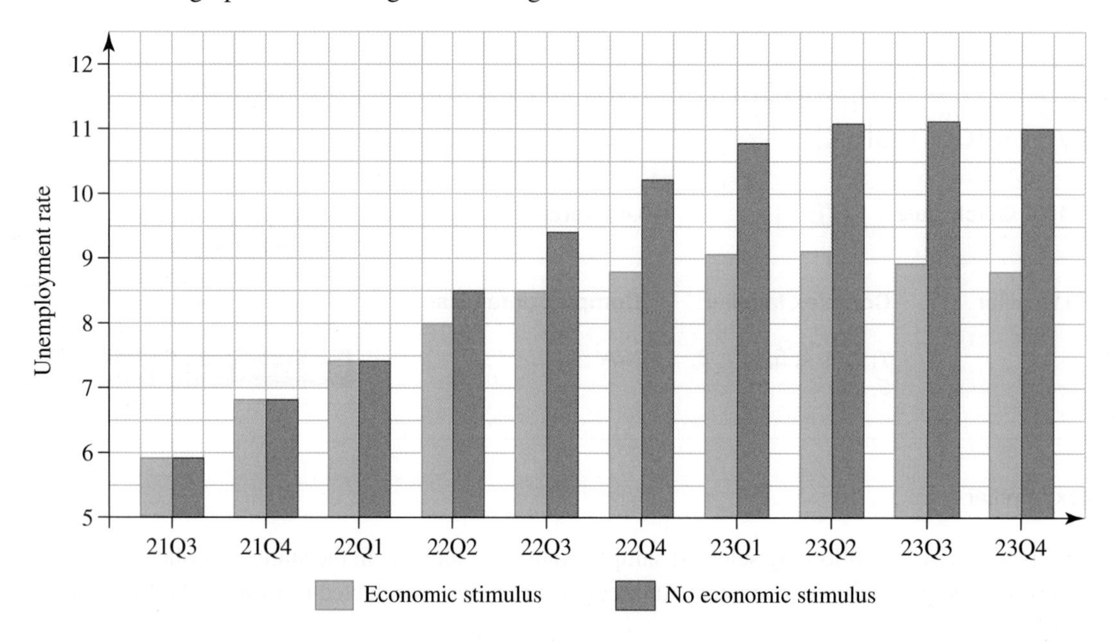

6. WE8 In the same survey discussed in Worked example 8, the authors of the study state that 'The petroleum industry — through oil and gas companies, related industrial services, and consulting services — is the largest employer, either directly or indirectly, of professional engineers and geoscientists in Alberta... These professionals and their organizations are regulated by a single professional self-regulatory authority — APEGGA.'
This fact is not stated in the article. Explain why the author of the article did not mention this fact.

7. In the lead-up to an election, various polls are taken to see who is likely to win the election. One poll says Party A is leading by 2.4% and another poll says Party B is leading by 1.1%. Party A says, 'The polls say we are in front by 2.4%, so we are very confident of winning the election.'
Explain how this misrepresents the survey results.

8. The following graph was plotted to demonstrate the effect of climate change in Sydney.

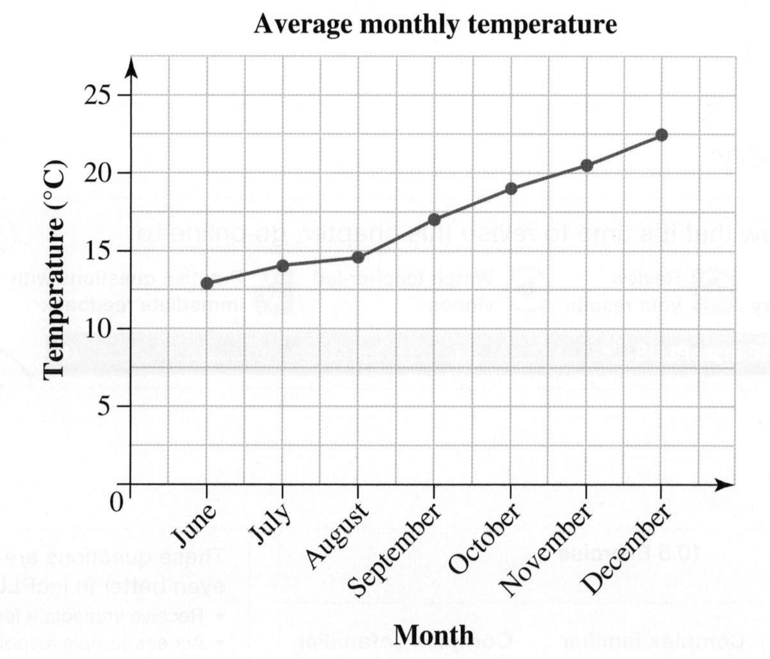

Describe why this graph is misleading.

9. If you investigate different data, you may find two sets of data that are similar if graphed a certain way. From the graph shown, comment on the similarity of these two data sets.

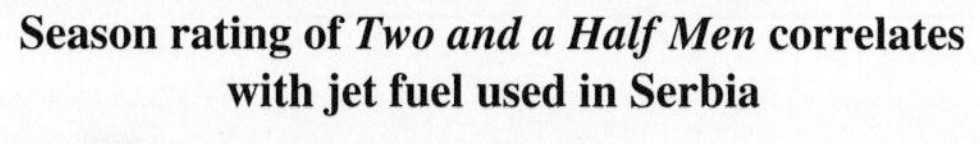

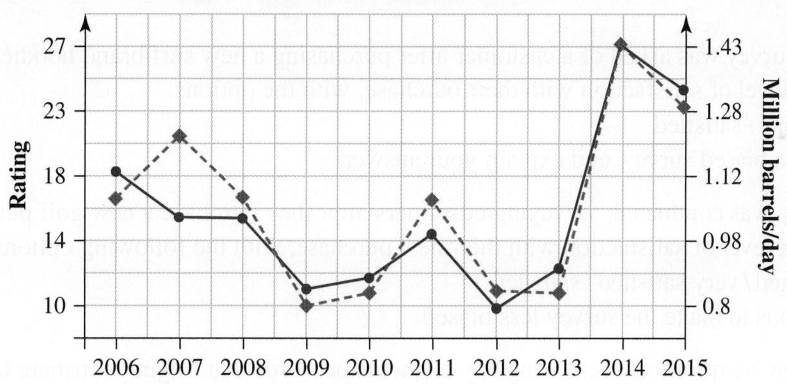

◆ - Season rating of *Two and a Half Men*

●— Volume of jet fuel used in Serbia in millions of barrels per day

10. A person conducted a survey in the cereal isle of a supermarket. They surveyed each person who picked up Brand A cereal and asked what their favourite cereal is. From the data collected, 64% of people chose Brand A as their favourite cereal. The Brand A company then wants to advertise that 64% of people choose Brand A as their favourite cereal.
Discuss whether this statement is a misrepresentation and explain your answer.

Fully worked solutions for this chapter are available online.

LESSON
10.5 Review

10.5.1 Summary

doc-41770

Hey students! Now that it's time to revise this chapter, go online to:

Access the chapter summary | Review your results | Watch teacher-led videos | Practise questions with immediate feedback

Find all this and MORE in jacPLUS

10.5 Exercise

learnon

10.5 Exercise

Simple familiar	Complex familiar	Complex unfamiliar
1, 2, 3, 4, 5, 6, 7, 8	9, 10, 11, 12, 13, 14, 15, 16, 17, 18, 19	20

These questions are even better in jacPLUS!
- Receive immediate feedback
- Access sample responses
- Track results and progress

Find all this and MORE in jacPLUS

Simple familiar

1. A quick online survey was asked of a customer after purchasing a new surf brand hoodie. The survey asked about the level of satisfaction with their purchase, with the options:
   Very happy / happy / satisfied
   Decide if this is a biased survey, and explain your answer.
2. An online survey was conducted, surveying customers after they purchase a new golf putter. The survey asked about their level of satisfaction with their new purchase, with the following options:
   Extremely satisfied / very satisfied / satisfied
   Modify the options to make the survey less biased.
3. Modify the following questions, removing any elements or words that might contribute to bias in responses.
   a. The poor homeless people, through no fault of their own, experience great hardship during the freezing winter months. Would you contribute to a fund to build a shelter to house our homeless?
   b. Most people think that, since we've developed as a nation in our own right and broken many ties with Great Britain, we should adopt a national flag without the Union Jack. You'd agree with this, wouldn't you?
4. Modify each of the following questions to eliminate any bias.
   a. You'd know that our Australian 50 cent coin is in the shape of a dodecagon, wouldn't you?
   b. Many in the workforce toil long hours for low wages. By comparison, politicians seem to get life pretty easy when you take into account that they only work for part of the year and they receive all those perks and allowances. You'd agree, wouldn't you?
5. Modify the questions in questions 3 and 4 so the expected responses are reversed.

6. When a measurement is made, there is always a level of error. For the following, determine the lowest and highest possible measure.
   a. 7.7 ± 0.1 cm
   b. 22.5 ± 0.2 g
   c. 36.55 ± 0.05 kg
   d. 7.94 ± 0.04 s
7. Decide if the following could produce a measurement error and explain why.
   a. You use a 1-m ruler to measure the height of a two-storey building.
   b. You record the mass of an object as 2.42 g, when it was 2.042 g.
8. Use the diagram shown to answer the following questions.

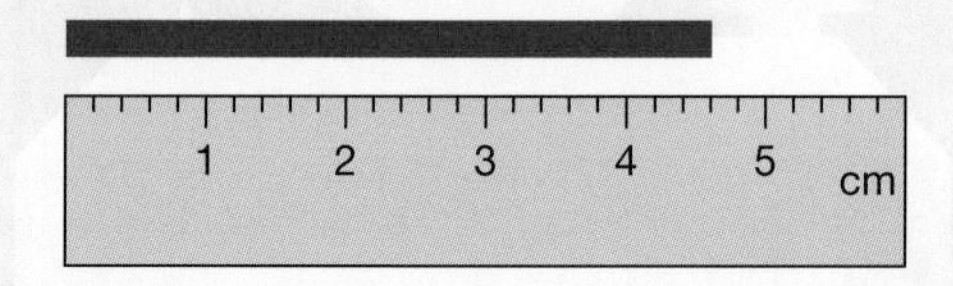

   a. Calculate the length of the red object.
   b. If the ruler shown below was used to measure the red object, explain the accuracy of the measurement.

1 2 3 4 5 cm

Complex familiar

9. Comment on the following sampling techniques.
   a. Surveying students from the school's netball team to determine their favourite sport
   b. Selecting 5 students from Year 10 to ask if they think all students are given too much homework
10. A manager of a bike store asked 20 customers if they liked shopping at the store. Of the 20 customers, 14 said that they did. The manager then publicised that 70% of customers like shopping in the store. Comment on the validity of this statement.
11. Calculate the lower and upper margins for a 7% margin of error if 12% of the batteries of a sample were found to be defective.
12. In question 11, if the confidence level is 90%, calculate the risk that the results will fall outside the margins.
13. Calculate the lower and upper margins for a:
   a. 7% margin of error if 10% of the members of a sample were found to be small dogs
   b. 4% margin of error if 85% of the cars in a sample were found to be red cars
   c. 8% margin of error if 92% of the electronic devices of a sample were found to be laptops.
14. Calculate the confidence level of a survey if 3.5% of its results do not fall within a given margin of error.

15. In the lead-up to a state election, polls are taken to see who is most likely to win the election. One poll says Party X is leading by 1.7% and another poll says Party Y is leading by 0.8%. Party X says, ‘The polls say we are in front by 1.7%, so we expect we will win the election.’
Explain how this misrepresents the survey results.

16. The following graph represents the recruitment of nurses over a five-year period.

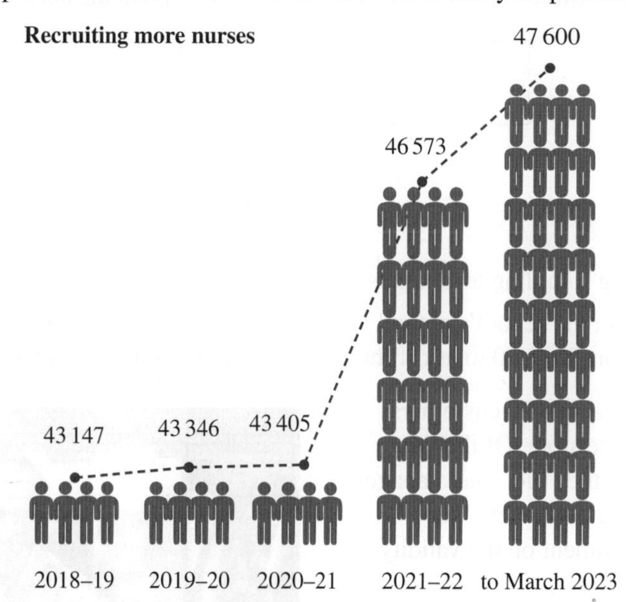

Explain why this graph is misleading.

17. The following graph demonstrates global cooling.

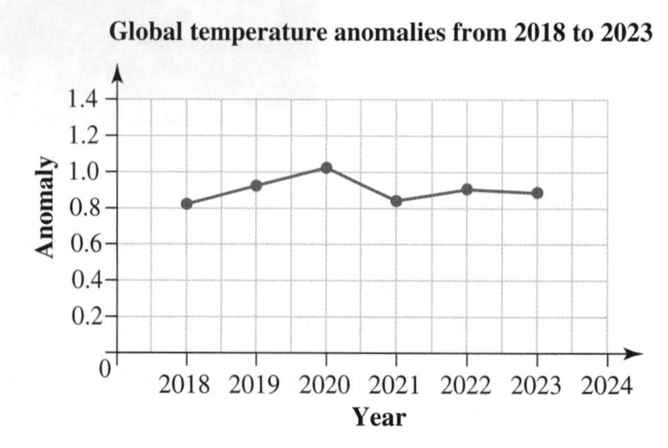

Explain how this graph misleads the reader.

18. Discuss why the following selected sample could provide bias in the statistics collected.
Tourists in a national park in Africa were selected to answer if they preferred to view animals in a zoo or in the wild. From the people surveyed, 90% said they preferred to view animals in the wild.

19. During an experiment, the time it took for a ball to hit the ground was measured to be 2.25 seconds.
The stopwatch used to measure this time had an error of ± 0.2 s.
a. Determine the maximum possible time it took for the ball to hit the ground.
b. Determine the minimum possible time it took for the ball to hit the ground.

Complex unfamiliar

20. The government wants to improve sporting facilities in Brisbane. They decide to survey 1000 people about what facilities they would like to see improved. To do this, they choose the first 1000 people going through the gate at a football match at Suncorp Stadium. Evaluate, using reasoning, how the government could conduct a survey to identify how to improve the facilities at the ground.

Fully worked solutions for this chapter are available online.

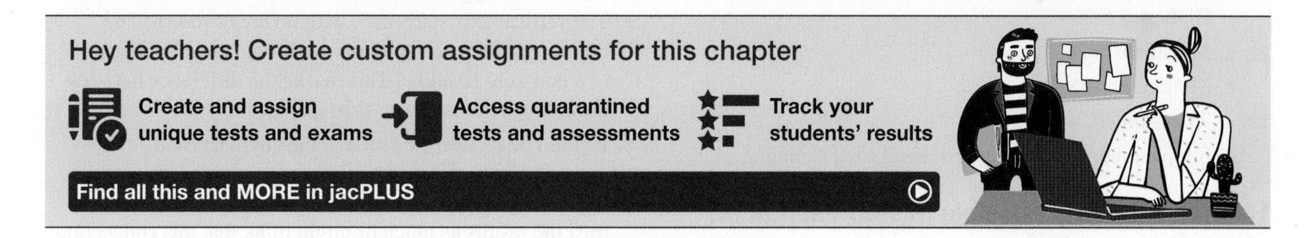

Hey teachers! Create custom assignments for this chapter
Create and assign unique tests and exams
Access quarantined tests and assessments
Track your students' results
Find all this and MORE in jacPLUS

Answers

Chapter 10 Sources of bias

10.2 Faults in collecting data

10.2 Exercise

1. Sample responses can be found in the worked solutions in the online resources.
2. Sample responses can be found in the worked solutions in the online resources.
3. Sample responses can be found in the worked solutions in the online resources.
4. Sample responses can be found in the worked solutions in the online resources.
5. Sample responses can be found in the worked solutions in the online resources.
6. Sample responses can be found in the worked solutions in the online resources.
7. a. 3.4 cm
   b. Sample responses can be found in the worked solutions in the online resources.
8. Sample responses can be found in the worked solutions in the online resources.
9. Sample responses can be found in the worked solutions in the online resources.
10. a. Lowest = 1.4 cm
    Highest = 1.6 cm
    b. Lowest = 15.4 g
    Highest = 15.8 g
    c. Lowest = 28.20 kg
    Highest = 28.30 kg
    d. Lowest = 5.90 s
    Highest = 5.96 s
11. a. Maximum = 26.1 g
    b. Minimum = 25.7 g
12. a. 75 customers
    b. Simple random sampling or self-selected sampling

10.3 Investigating possible misrepresentation of results [complex]

10.3 Exercise

1. Sample responses can be found in the worked solutions in the online resources.
2. Sample responses can be found in the worked solutions in the online resources.
3. Sample responses can be found in the worked solutions in the online resources.
4. Sample responses can be found in the worked solutions in the online resources.
5. a. Lower margin = 8.5%; upper margin = 13.5%
   b. 5%
6. a. Lower margin = 94%; upper margin = 98%
   b. 8%
7. This is a biased sample as the people surveyed are already shopping at the given supermarket.
8. 97.5%
9. a. Lower margin = 18.5%; upper margin = 21.5%
   b. Lower margin = 67.5%; upper margin = 72.5%
10. 4%
11. 7%
12. Between 3 and 5 eggs

10.4 Investigating errors and misrepresentation in surveys [complex]

10.4 Exercise

1. Sample responses can be found in the worked solutions in the online resources.
2. Sample responses can be found in the worked solutions in the online resources.
3. Sample responses can be found in the worked solutions in the online resources.
4. The vertical axis starts at 0.80 with no break shown. The differences between columns look exaggerated because a non-zero baseline has been used. The percentage difference is quite minimal, approximately 7%, between the people who never finished high school and the people who have a postgraduate degree. Here is the graph with a baseline of zero:

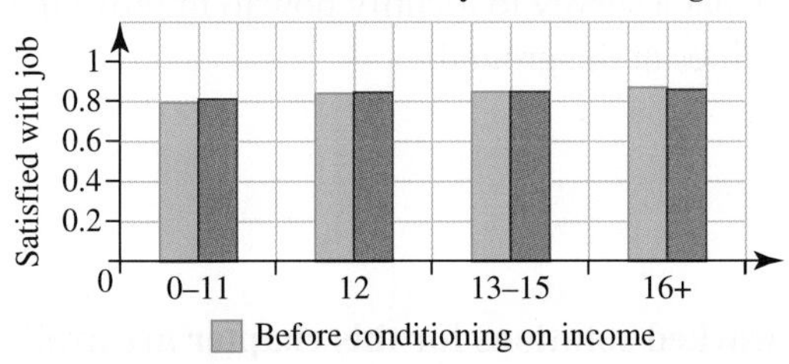

5. The vertical axis starts at 5, with no break shown. The differences between columns are exaggerated by the use of a non-zero baseline. This makes the difference between the columns look greater than they actually are.
6. It could be to hide the fact that the engineers and geoscientists surveyed are, in some way, dependent on or paid by the petroleum industry. The reader might not trust the results as much or might think that this connection makes the geoscientists and the engineers somewhat biased.
7. Sample responses can be found in the worked solutions in the online resources.
8. Sample responses can be found in the worked solutions in the online resources.
9. Sample responses can be found in the worked solutions in the online resources.
10. Sample responses can be found in the worked solutions in the online resources.

10.5 Review

10.5 Exercise

1. Sample responses can be found in the worked solutions in the online resources.
2. Sample responses can be found in the worked solutions in the online resources.

3. Sample responses can be found in the worked solutions in the online resources.
4. Sample responses can be found in the worked solutions in the online resources.
5. Sample responses can be found in the worked solutions in the online resources.
6. a. Lowest = 7.6 cm
   Highest = 7.8 cm
   b. Lowest = 22.3 g
   Highest = 22.7 g
   c. Lowest = 36.50 kg
   Highest = 36.60 kg
   d. Lowest = 7.90 s
   Highest = 7.98 s
7. Sample responses can be found in the worked solutions in the online resources.
8. a. 4.6 cm
   b. Sample responses can be found in the worked solutions in the online resources.
9. Sample responses can be found in the worked solutions in the online resources.
10. Sample responses can be found in the worked solutions in the online resources.
11. Lower margin = 8.5%
    Upper margin = 15.5%
12. 10%
13. a. Lower margin = 6.5%; upper margin = 13.5%
    b. Lower margin = 83%; upper margin = 87%
    c. Lower margin = 88%; upper margin = 96%
14. Confidence level = 96.5%
15. Sample responses can be found in the worked solutions in the online resources.
16. Sample responses can be found in the worked solutions in the online resources.
17. Sample responses can be found in the worked solutions in the online resources.
18. Sample responses can be found in the worked solutions in the online resources.
19. a. Maximum = 2.45 s b. Minimum = 2.05 s
20. Sample responses can be found in the worked solutions in the online resources.

UNIT 2 Data and travel
TOPIC 2 Graphs

11 Reading, interpreting and using graphs

LESSON SEQUENCE

Fully worked solutions for this chapter are available online.

Resources

Solutions Soutions — Chapter 11 (sol-1264)

Digital documents Learning matrix — Chapter 11 (doc-41771)
Quick quizzes — Chapter 11 (doc-41772)
Chapter summary — Chapter 11 (doc-41773)

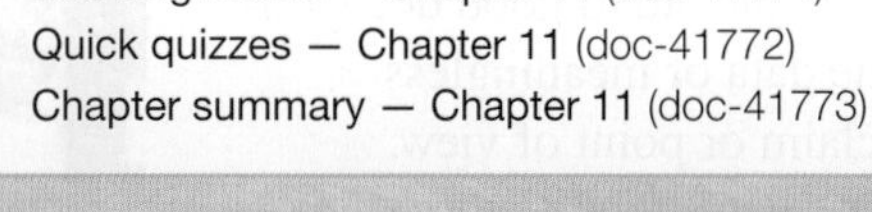

LESSON
11.1 Overview

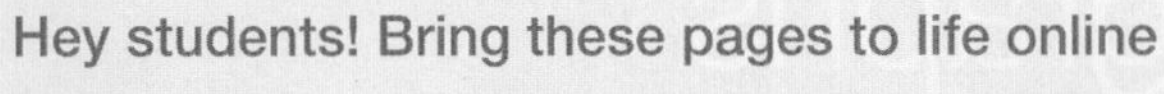

Find all this and MORE in jacPLUS

11.1.1 Introduction

Data collection and analysis have become a major focus of governments and businesses in the 21st century. How to effectively handle the data collected is one of the biggest challenges to overcome in the modern world. You may have seen graphs or charts of housing prices in the news or online. One house price report stated that in December 2021 the median house price was $1.6 million in Sydney, $1.12 million in Melbourne and $1.4 million in Brisbane. This chapter will help you to understand the answers to these questions.

When analysed properly, data can be an invaluable aid to good decision-making. However, deliberate distortion of the data or meaningless pictures can be used to support almost any claim or point of view. Whenever you read an advertisement or hear a news report, you need to have a healthy degree of skepticism about the reliability of the source and nature of the data presented.

In 2020, when the COVID-19 pandemic started, news and all forms of media were flooded with data. These data were used to inform governments worldwide about infection rates, recovery rates and all sorts of other important information. They guided the decision-making process in determining the restrictions that were imposed or relaxed to maintain a safe community. A solid understanding of data analysis is crucially important, as it is very easy to fall prey to statistics that are designed to confuse and mislead.

11.1.2 Syllabus links

Lesson	Lesson title	Syllabus links
11.2	**Two-way tables**	◯ Interpret information in two-way tables.
11.3	**Line graphs, pie graphs and step graphs**	◯ Interpret information presented in graphs, e.g. step graphs, column graphs, pie graphs, picture graphs, line graphs. ◯ Use graphs in practical situations.
11.4	**Column graphs and picture graphs**	◯ Interpret information presented in graphs, e.g. step graphs, column graphs, pie graphs, picture graphs, line graphs. ◯ Use graphs in practical situations.
11.5	**Misleading graphs [complex]**	◯ Discuss and interpret tables and graphs, including misleading graphs found in the media and in factual texts [complex]. ◯ Interpret graphs in practical situations [complex].

LESSON
11.2 Two-way tables

SYLLABUS LINKS

- Interpret information in two-way tables.

11.2.1 Reading and interpreting two-way tables

- In this section, we consider in more detail how to read and interpret two-way tables.
- The table shown is a two-way table that relates two variables: the ability to swim and age group.

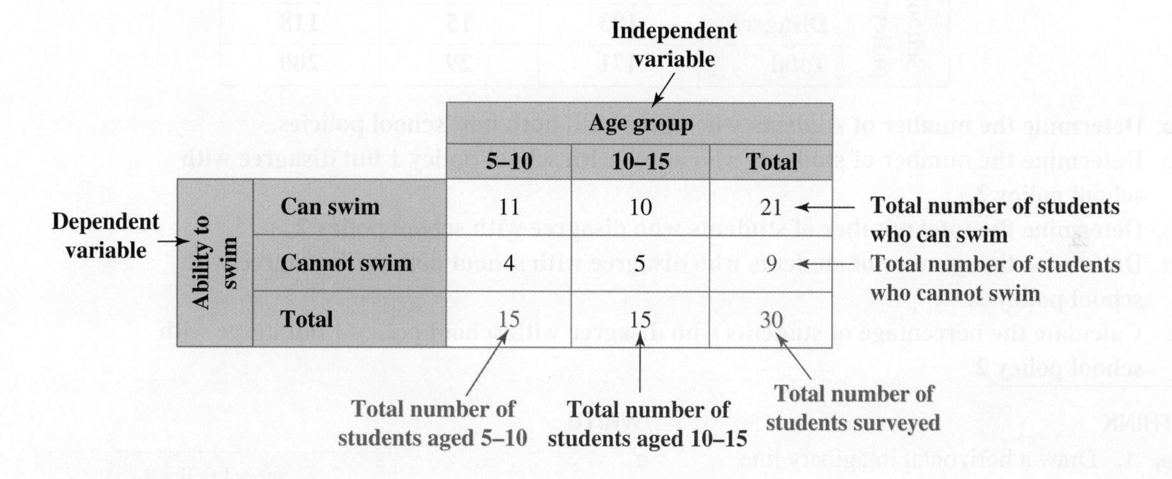

Ability to swim		Age group 5–10	10–15	Total
	Can swim	11	10	21
	Cannot swim	4	5	9
	Total	15	15	30

- If we want to determine the number of students aged 5–10 who can swim, draw an imaginary line through the row 'Can swim' and an imaginary line through the column '5–10'.
- The cell where the two imaginary lines meet has the answer: 11 students aged 5–10 can swim.

Ability to swim		Age group 5–10	10–15	Total
	Can swim	11	10	21
	Cannot swim	4	5	9
	Total	15	15	30

- The other cells of the table represent:

Ability to swim		Age group 5–10	10–15	Total
	Can swim	11 students 5–10 years old who can swim	10 students 10–15 years old who can swim	21 students who can swim
	Cannot swim	4 students 5–10 years old who cannot swim	5 students 10–15 years old who cannot swim	9 students who cannot swim
	Total	15 students 5–10 years old surveyed	15 students 10–15 years old surveyed	30 students surveyed

tlvd-4958

WORKED EXAMPLE 1 Analysing and interpreting a two-way table

Two new school policies were introduced at Burumin High School.
School policy 1: The school uniform has to be worn on the way to school as well as on the way home from school.
School policy 2: Mobile phones are not to be used at school between 8:30 am and 3:30 pm.
The two-way table shown displays the responses to a survey conducted on 200 students.

		School policy 1		
		Agree	Disagree	Total
School policy 2	Agree	68	14	82
	Disagree	103	15	118
	Total	171	29	200

a. **Determine the number of students who agree with both new school policies.**
b. **Determine the number of students who agree with school policy 1 but disagree with school policy 2.**
c. **Determine the total number of students who disagree with school policy 2.**
d. **Determine the number of students who disagree with school policy 1 but agree with school policy 2.**
e. **Calculate the percentage of students who disagree with school policy 1 but agree with school policy 2.**

THINK

a. 1. Draw a horizontal imaginary line through the row 'Agree' with school policy 2 and a vertical imaginary line through the column 'Agree' with school policy 1.

WRITE

a.

		School policy 1		
		Agree	Disagree	Total
School policy 2	Agree	68	14	82
	Disagree	103	15	118
	Total	171	29	200

2. Read the number of students in the box where the two lines meet.

68 students agree with both new school policies.

b. 1. Draw a horizontal imaginary line through the row 'Disagree' with school policy 2 and a vertical imaginary line through the column 'Agree' with school policy 1.

b.

		School policy 1		
		Agree	Disagree	Total
School policy 2	Agree	68	14	82
	Disagree	103	15	118
	Total	171	29	200

2. Read the number of students in the box where the two lines meet.

103 students agree with school policy 1 but disagree with school policy 2.

c. 1. Draw a horizontal imaginary line through the row 'Disagree' with school policy 2 and a vertical imaginary line through the column 'Total'.

c.

		School policy 1		
		Agree	**Disagree**	**Total**
School policy 2	**Agree**	68	14	82
	Disagree	103	15	118
	Total	171	29	200

2. Read the number of students in the box where the two lines meet.

118 students disagree with school policy 2.

d. 1. Draw a horizontal imaginary line through the row 'Agree' with school policy 2 and a vertical imaginary line through the column 'Disagree' with school policy 1.

d.

		School policy 1		
		Agree	**Disagree**	**Total**
School policy 2	**Agree**	68	14	82
	Disagree	103	15	118
	Total	171	29	200

2. Read the number of students in the box where the two lines meet.

14 students disagree with school policy 1 but agree with school policy 2.

e. 1. Using your answer from part **d**, divide the number of students by the total number of students, and multiply by 100.

$$\text{Percentage} = \frac{14}{200} \times 100$$
$$= 7\%$$

2. Write your answer.

Seven per cent of students disagree with school policy 1 but agree with school policy 2.

Exercise 11.2 Two-way tables

learnon

11.2 Quick quiz on | **11.2 Exercise**

Simple familiar	Complex familiar	Complex unfamiliar
1, 2, 3, 4, 5, 6, 7, 8, 9, 10	N/A	N/A

These questions are even better in jacPLUS!
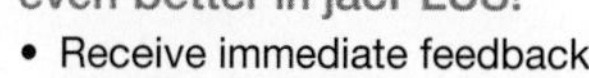
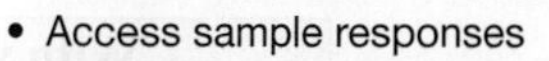

- Receive immediate feedback
- Access sample responses
- Track results and progress

Find all this and MORE in jacPLUS

Simple familiar

1. **WE1** The students in a school were surveyed on whether they wear glasses for reading or not. The data is displayed in the two-way table shown.

		Year level		
		Year 11	**Year 12**	**Total**
Reading glasses	**Wear glasses for reading**	26	23	49
	Do not wear glasses for reading	291	313	603
	Total	317	335	652

a. Determine the total number of students who were surveyed.
b. Determine the number of students in Year 11 who wear glasses for reading.
c. Determine the number of students in both years who do not wear glasses for reading.
d. Determine the number of students in Year 12 who were surveyed.
e. Calculate the percentage of students in Year 11 who wear glasses, correct to 1 decimal place.

2. The two-way table shown displays the way a group of 150 students check the weather.

		Method of checking the weather		
		Weather app	TV news	Total
Students	Junior (Years 7–10)	21	47	68
	Senior (Years 11 and 12)	75	7	82
	Total	96	54	150

a. Determine the number of students who check the weather using a weather app.
b. Determine the number of senior students who check the weather by watching the TV news.
c. Determine the number of junior students who check the weather using a weather app.
d. Determine the number of junior students who were surveyed.
e. Determine what percentage of students who were surveyed are senior students, correct to 1 decimal place.

3. The contingency table shown below displays the information gained from a medical test screening for a virus. A positive test indicates that the patient has the virus.

		Test results		
		Accurate	Not accurate	Total
Virus	With virus	45	3	48
	Without virus	922	30	952
	Total	967	33	1000

a. Determine the number of patients who were screened for the virus.
b. Determine the number of positive tests that were recorded (that is, the number of tests in which the virus was detected).

4. For the two-way table shown, calculate the number of nurses in the regional area.

		Location		
		Metro	Regional	Total
Career	Teachers	58	26	84
	Nurses	77	39	116
	Total	135	65	200

5. The two-way table shown displays the relationship between year level and pierced ears.

Pierced ears		Year level: Year 12	Year 11	Total
	Pierced	2	35	37
	Not pierced	41	7	48
	Total	43	42	85

a. Determine the number of students in Year 11 who have had their ears pierced.
b. Calculate the total number of students surveyed.
c. Determine the number of students in Year 12 who have had their ears pierced.
d. Calculate the total number of students who have not had their ears pierced.

6. The two-way table below indicates the results of a radar surveillance system. If the system detects an intruder, an alarm is activated.

Detection		Test results: Alarm activated	Alarm not activated	Total
	Intruders	40	8	48
	No intruders	4	148	152
	Total	44	156	200

a. If the alarm is activated, determine the percentage chance that there actually is an intruder.
b. If the alarm is not activated, determine the percentage chance that there is an intruder.
c. Calculate the percentage of accurate results over the test period.
d. Comment on the overall performance of the radar detection system.

The information below is to be used in questions 7–9.

A test for a disease is not always accurate. The table shown indicates the results of a trial on a number of patients who were known to either have the disease or not have the disease.

Disease status		Test results: Accurate	Not accurate	Total
	With disease	57	3	60
	Without disease	486	54	540
	Total	543	57	600

7. Calculate the overall accuracy of the test.

8. Calculate the percentage of patients with the disease who have had it detected by the test.

9. Determine whether the accuracy is greater for positive (with disease) or negative (without disease) results.

10. Airport scanning equipment was tested by scanning 200 pieces of luggage. Prohibited items were placed in 50 bags and the scanning equipment detected 48 of them. The equipment detected prohibited items in five bags that did not have any forbidden items in them.
Determine the overall percentage accuracy of the scanning equipment.

Fully worked solutions for this chapter are available online.

LESSON
11.3 Line graphs, pie graphs and step graphs

SYLLABUS LINKS

- Interpret information presented in graphs, e.g. step graphs, column graphs, pie graphs, picture graphs, line graphs.
- Use graphs in practical situations.

11.3.1 Line graphs

- Data is often presented in the form of graphs. Graphical representations of data are easier to read and interpret.
- **Line graphs** are used to represent the relationship between two numerical continuous data sets.
- These graphs consist of individually plotted points joined with a straight line.
- The graph shown represents the relationship between two numerical continuous data sets: the temperature readings (°C) in Canberra recorded hourly on one day in August.

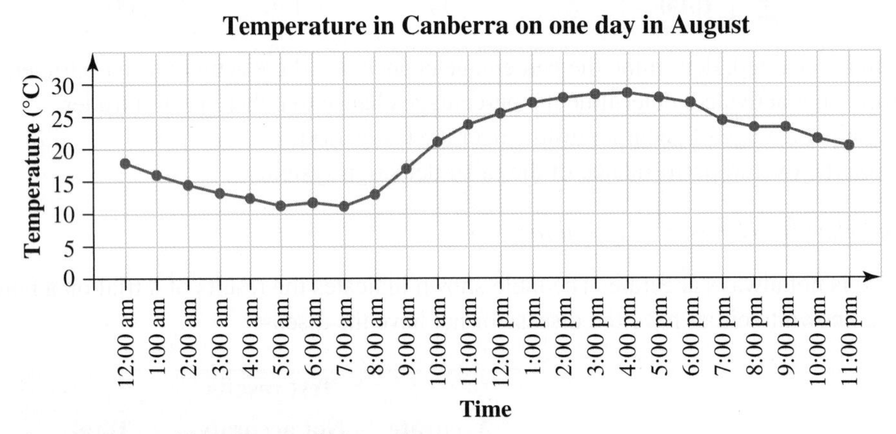

Source: Adapted from Bureau of Meteorology

- The information we can gather from this graph is:
  - the minimum (lowest) temperature of the day and the time when it occurred (10.9 °C at 7:00 am)
  - the maximum (highest) temperature of the day and the time when it occurred (28.5 °C at 4:00 pm)
  - the temperature at any hour of the day
  - the time of the day when a certain temperature occurred
  - the times between which the temperature was increasing (from 7:00 am to 4:00 pm)
  - the times between which the temperature was decreasing (from 4:00 pm to 8:00 pm).

WORKED EXAMPLE 2 Analysing and interpreting a line graph

The graph shown is a representation of the temperature readings (°C) in Sydney on one day in March, recorded hourly.

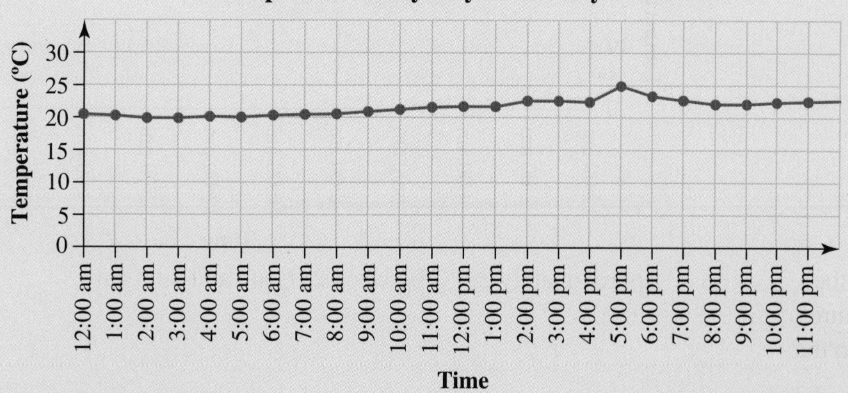

Source: Adapted from Bureau of Meteorology, http://www.bom.gov.au/products/IDN60901/IDN60901.94768.shtml

a. **Determine the maximum temperature in Sydney on this day.**
b. **Determine the minimum temperature in Sydney on this day.**
c. **Determine the times between which the temperature in Sydney was increasing.**
d. **Determine the temperature in Sydney at 12:00 pm.**
e. **Determine the times when the temperature in Sydney was 24 °C.**

THINK	WRITE
a. 1. The maximum temperature occurs at the highest point on the graph.	a. 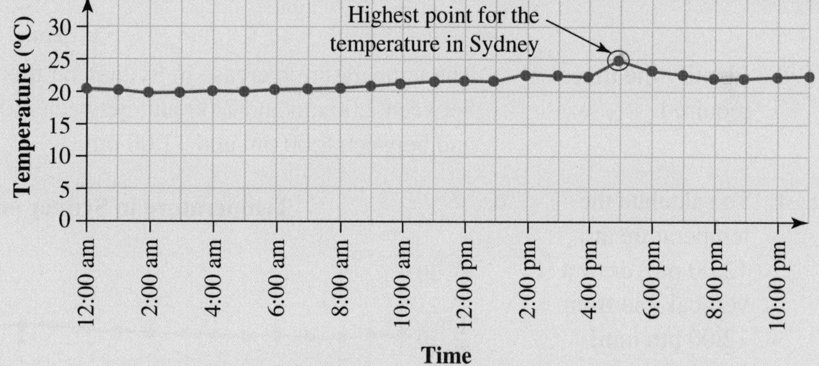
2. Identify the temperature shown on the graph.	Approximately 25 °C at 5:00 pm

b. 1. The minimum temperature occurs at the lowest point on the graph.

b.

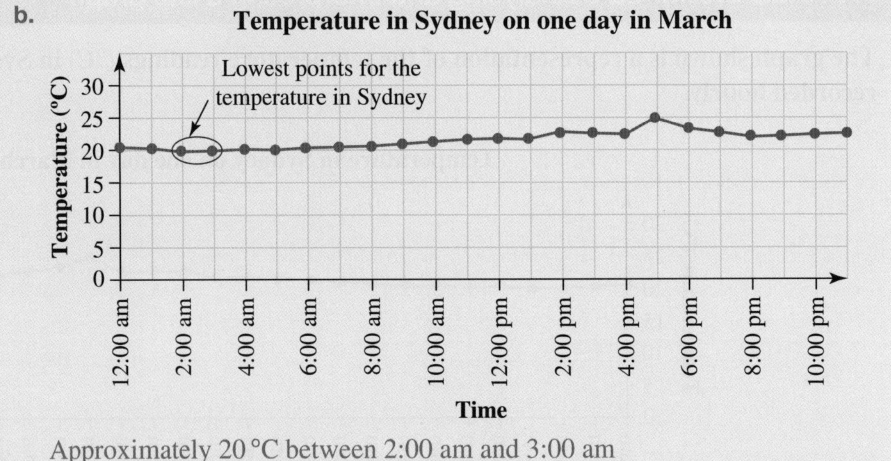

2. Identify the temperature shown on the graph.

Approximately 20 °C between 2:00 am and 3:00 am

c. 1. The increase in temperature is shown by a line that slopes upwards from left to right.

c.

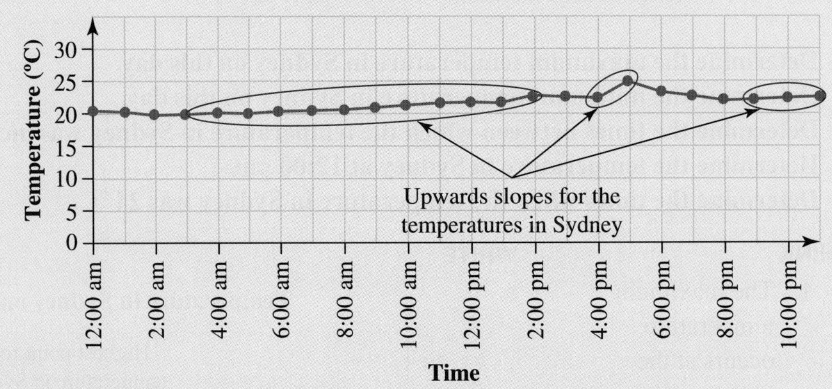

2. Identify the times required.

The temperature increase in Sydney on this day occurred between 2:00 am and 2:00 pm, between 4:00 pm and 5:00 pm and between 8:00 pm and 11:00 pm.

d. 1. To calculate the temperature at 12:00 pm, draw a vertical line from 12:00 pm until it meets the line graph.

d.

Temperature in Sydney on one day in March

Temperature (°C)

Time

2. From the point of intersection, draw a horizontal line to the left until it crosses the vertical axis.

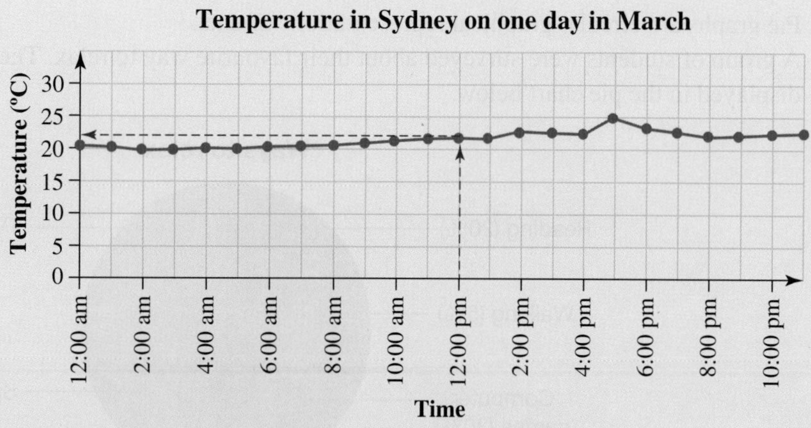

3. Read the temperature on the vertical axis.

The temperature at 12:00 pm was approximately 22 °C.

e. 1. To calculate the time when the temperature is 24 °C, draw a horizontal line starting at 24 °C until it meets the line graph.

e.

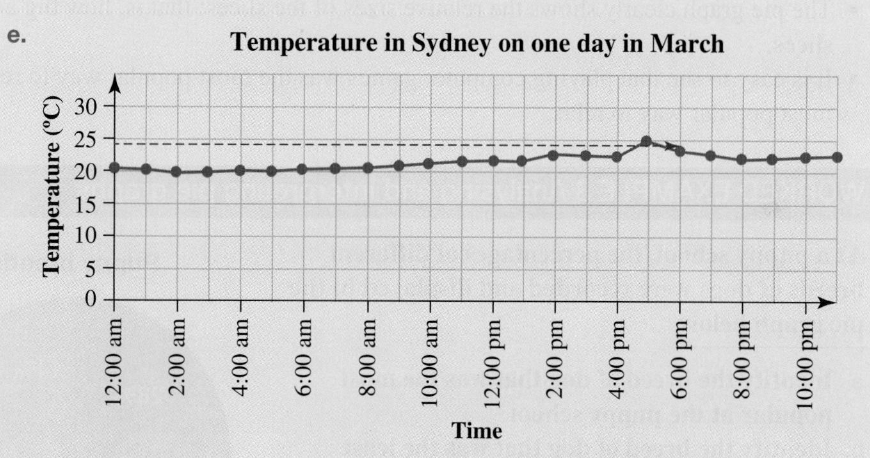

2. From the points of intersection, draw vertical lines down until they cross the horizontal axis.

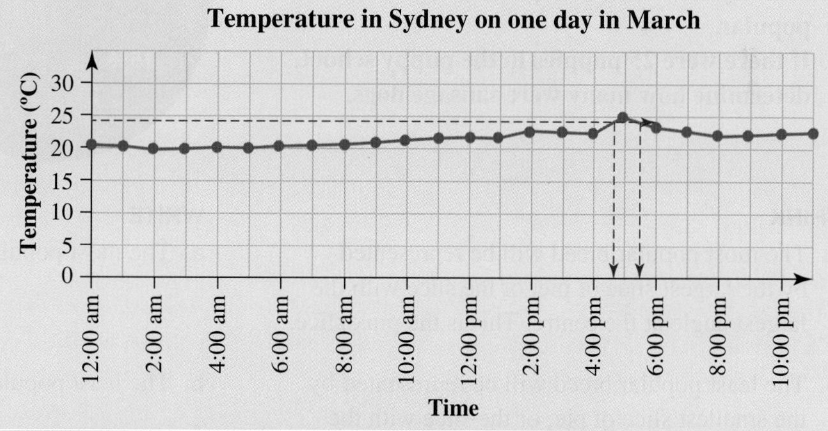

3. Read the times on the horizontal axis.

Notice that this line crosses the line graph in two points.
The temperature was 24 °C at around 4:45 pm and 5:30 pm.

11.3.2 Pie graphs

- **Pie graphs** are circles divided into sectors.
- Each sector represents one item of the data and has to be coloured differently.
- All labels and other writing have to be horizontal and equally distanced from the circle.
- Pie graphs can be used to represent categorical data.

- A pie chart uses pie slices to show the relative sizes of data.
- Pie graphs are circular graphical representations of data.
- A group of students were surveyed about their favourite way to relax. The results of the survey were displayed in the pie chart below.

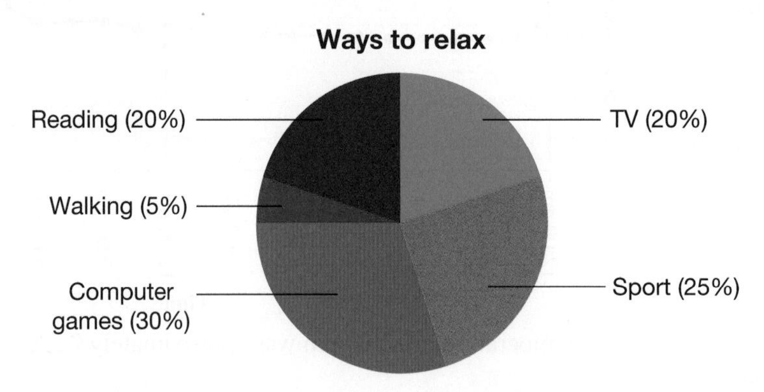

- The pie graph clearly shows the relative sizes of the slices; that is, how big a slice is compared to the other slices.
- It is easy to see that playing computer games was the most popular way to relax, and sport was the second most popular way to relax.

WORKED EXAMPLE 3 Analysing and interpreting pie graphs

At a puppy school, the percentages of different breeds of dogs were recorded and displayed in the pie graph below.

a. **Identify the breed of dog that was the most popular at the puppy school.**
b. **Identify the breed of dog that was the least popular.**
c. **If there were 25 puppies in the puppy school, determine how many were sausage dogs.**

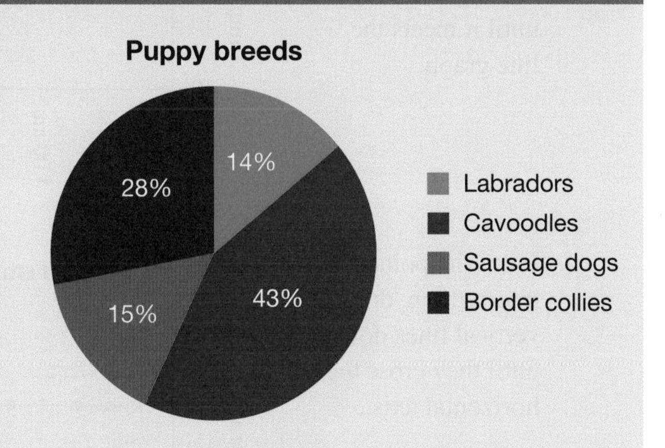

THINK	WRITE
a. The most popular breed will be represented by the largest slice of pie, or the slice with the largest angle at the centre. This is the pink slice.	a. The most popular breed is the cavoodle.
b. The least popular breed will be represented by the smallest slice of pie, or the slice with the smallest angle at the centre. This is the blue slice.	b. The least popular breed is the labrador.
c. 15% of the puppies are sausage dogs, so calculate 15% of 25.	c. 15% of 25 $\frac{15}{100} \times 25$ $= 0.15 \times 25$ $= 3.75$ 4 puppies are sausage dogs (rounded to the nearest whole number).

11.3.3 Step graphs

- **Step graphs** are made up of horizontal straight lines.
- These graphs are used when the value remains constant over intervals.
- Postage and parking costs are examples of step graphs.
- The first step represents a cost of \$0.60 for envelopes with a weight of less than 20 g. This means that regardless of whether the envelope weighs 10 g or 19 g, the postage cost is \$0.60.
- The open circle at the end of the interval means that the postage cost for 20 g is not \$0.60; it jumps to the next step, which is \$1.20.
- This is why the next step has a closed dot at the beginning of the interval.
  - The closed dot, •, means that the interval includes that value.
  - The empty circle, ∘, means that the interval does not include that value.

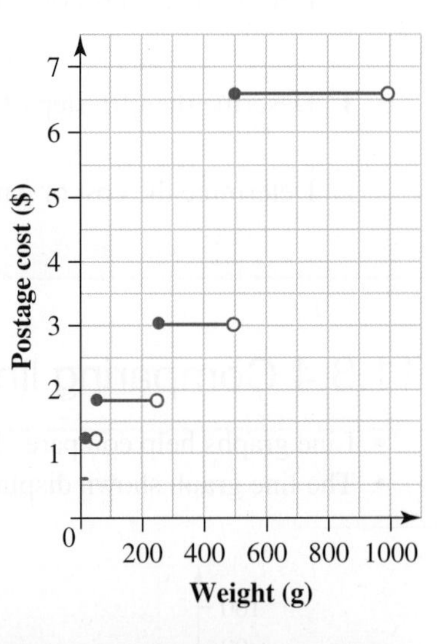

WORKED EXAMPLE 4 Analysing and interpreting a step graph

The step graph shown displays the cost of parking in a car park. Describe the costs involved if you were to use this car park.

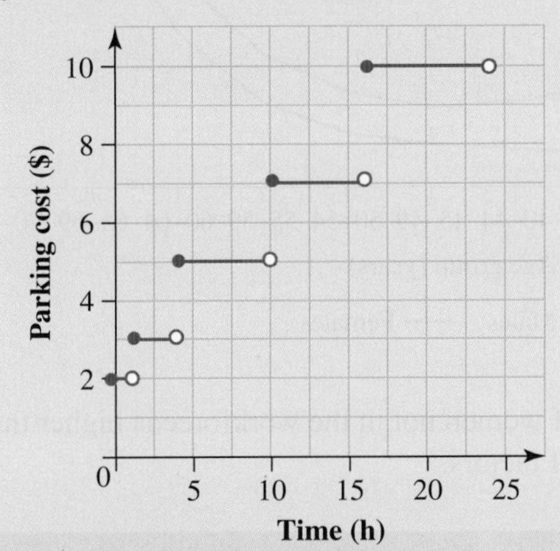

a. Calculate the cost of parking for 5 hours.
b. Calculate the cost of parking for 15 hours.
c. Calculate the cost of parking for 23 hours.

THINK	WRITE
a. 1. Describe the first step of the graph.	**a.** The cost for parking for one hour or less is \$2.
2. Describe the second step of the graph.	As soon as an hour has passed, the cost for between 1 hour and up to 4 hours of parking is \$3.
3. Determine the cost of parking.	The five-hour mark is on the third step, so the cost will be \$5.
b. 1. Describe the third step of the graph.	**b.** The third step shows a cost of \$5 for between 4 hours and up to 10 hours.
2. Describe the fourth step of the graph.	The fourth step shows a cost of \$7 for parking between 10 hours and up to 16 hours.

3. Determine the cost of parking.	The 15-hour mark is on the fourth step, so the cost will be $7.
c. 1. Describe the fifth step of the graph.	c. The fifth step shows a cost of $10 for parking between 16 hours and up to 24 hours.
2. Determine the cost of parking.	The 23-hour mark is on the fifth step, so the cost will be $10.

11.3.4 Comparing line graphs

- Line graphs help compare similar types of data.
- The line graph shown displays people not in the workforce, by age and sex, in 2019–2020 in Australia.

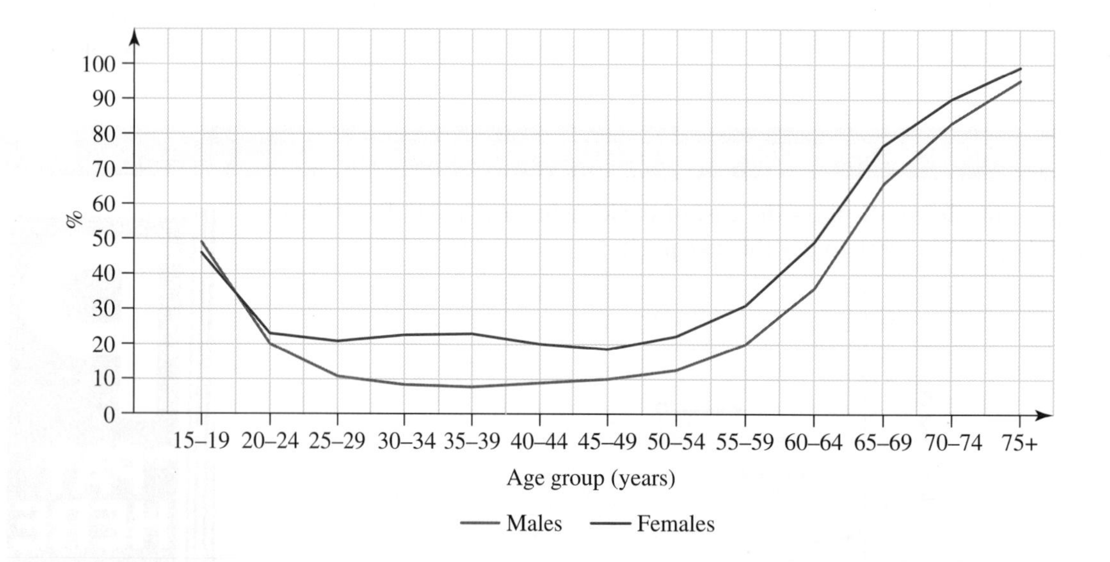

- This graph clearly shows that the percentage of women not in the workforce is higher throughout the years than it is for men. This could be for a variety of factors.

tlvd-4959

WORKED EXAMPLE 5 Comparing line graphs

The line graph shown displays the total number of employees by age for the period 2010–2024.

a. **Determine when the number of employed people aged 25–35 was equal to the number of employed people aged 35 and above.**

b. **Determine the number of employed people aged 25–35 and the number of employed people aged 35 and above in 2020.**

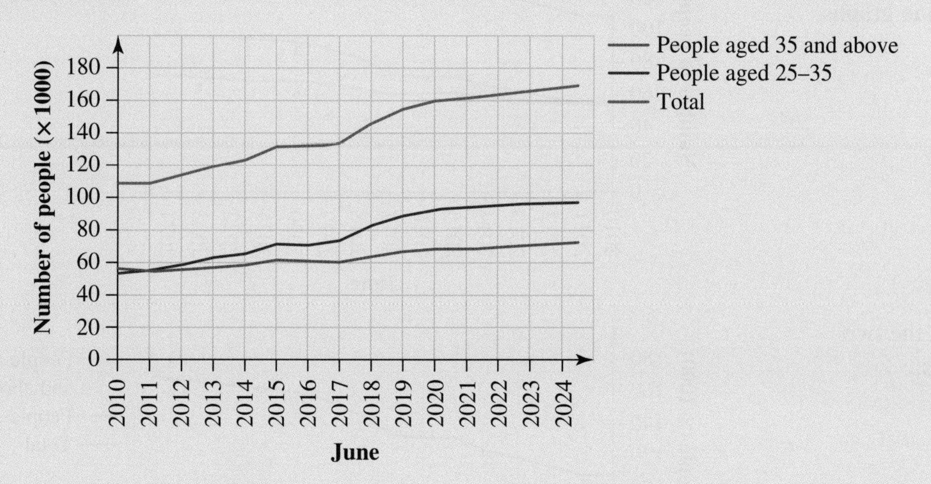

THINK	WRITE
a. 1. Mark the point where the two line graphs intersect.	a. 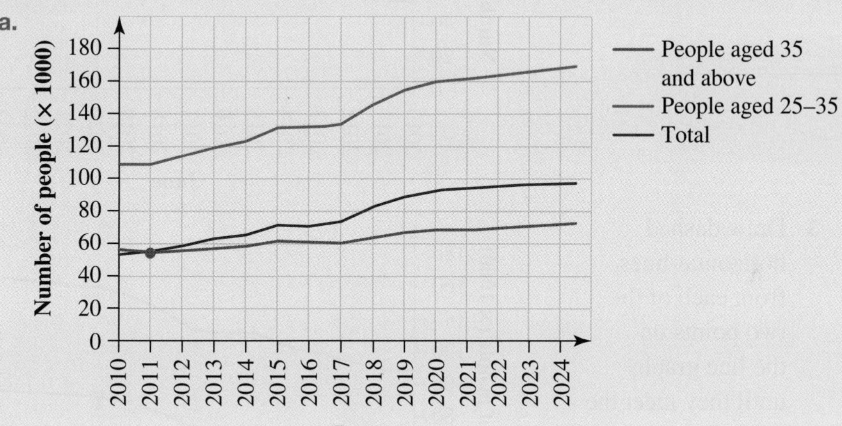
2. Draw a dotted vertical line from this point until it meets the horizontal axis.	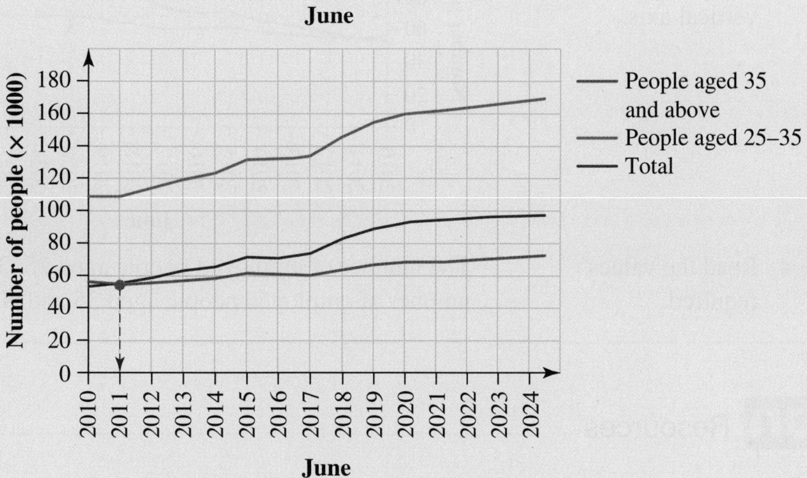

3. Write the answer. In 2011, the numbers of employed people aged 25–35 and employed people aged 35 and above are equal.

b. 1. Draw a dashed vertical line starting at 2020 until it touches the two line graphs.

b.

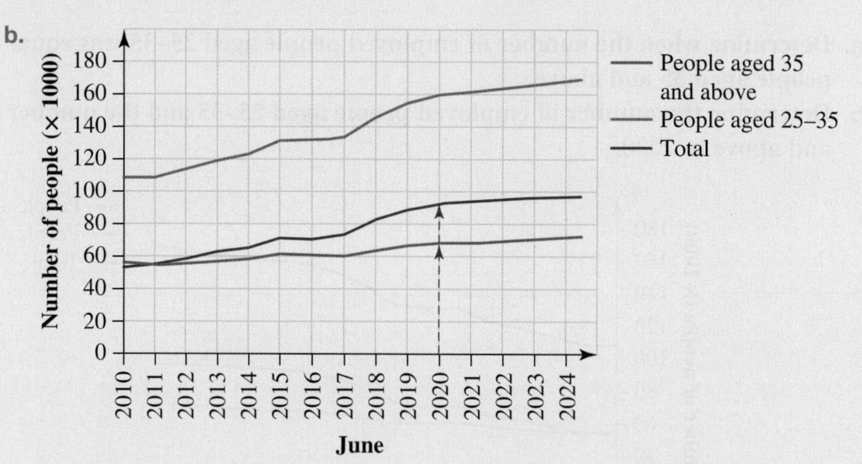

2. Mark the two points.

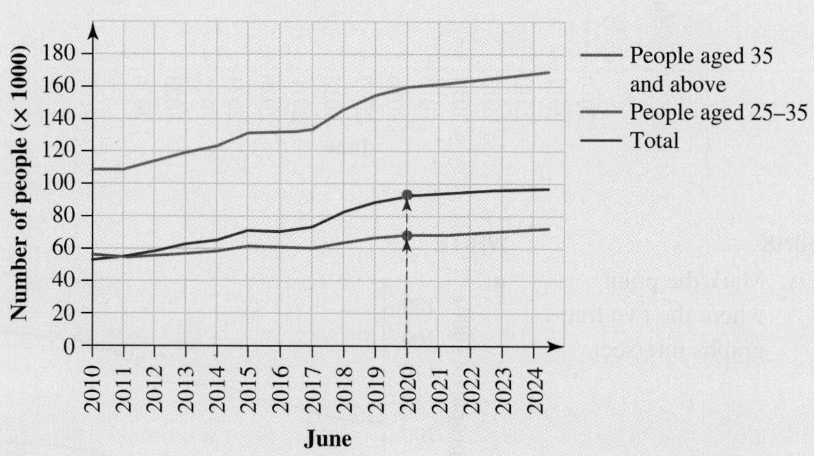

3. Draw dashed horizontal lines from each of the two points on the line graphs until they meet the vertical axis.

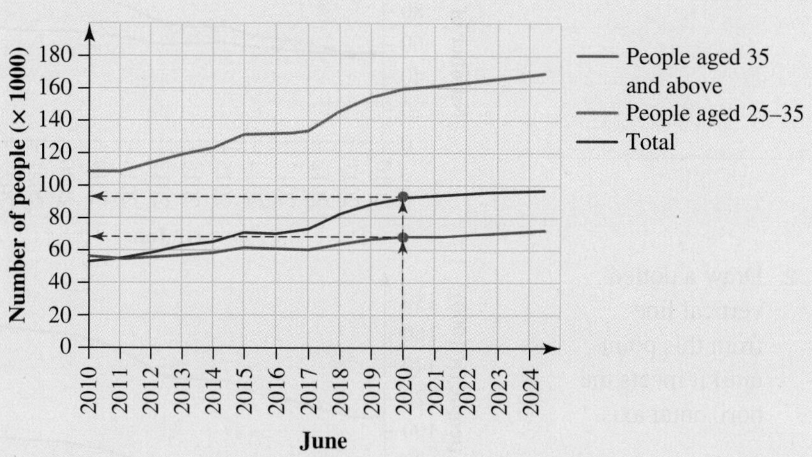

4. Read the values required. The number of employed people aged 25–35 in 2020 was 95 000 and the number of employed people aged 35 and above was 70 000.

on Resources

Digital document Investigation: Graphical displays of data (doc-15910)

Exercise 11.3 Line graphs, pie graphs and step graphs

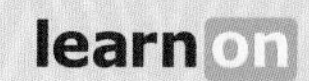

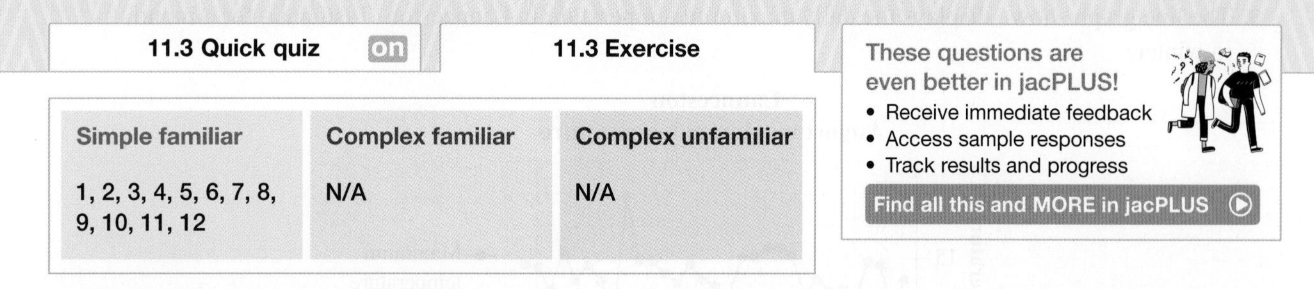

Simple familiar

1. WE2 The graph shows the temperature readings (°C) in Melbourne on a day in March, recorded hourly.

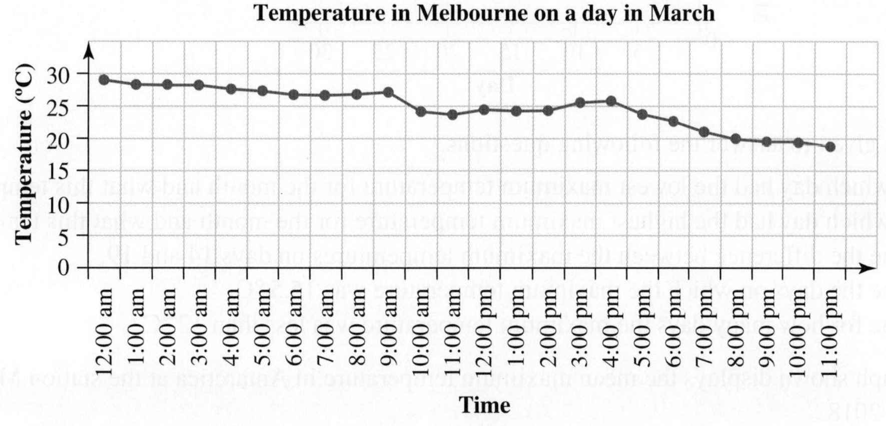

Source: Exchange-Rates.com http://www.exchange-rates.org/history/AUD/USD/G/30.

a. Identify the maximum temperature in Melbourne on this day.
b. Identify the minimum temperature in Melbourne on this day.
c. Identify the times between which the temperature in Melbourne was decreasing.
d. Determine the temperature in Melbourne at 9:00 am on this day.
e. Determine the times at which the temperature in Melbourne was 21 °C.

2. The line graph shown displays the average rainfall levels (mm) in Australia for a 13-year period.

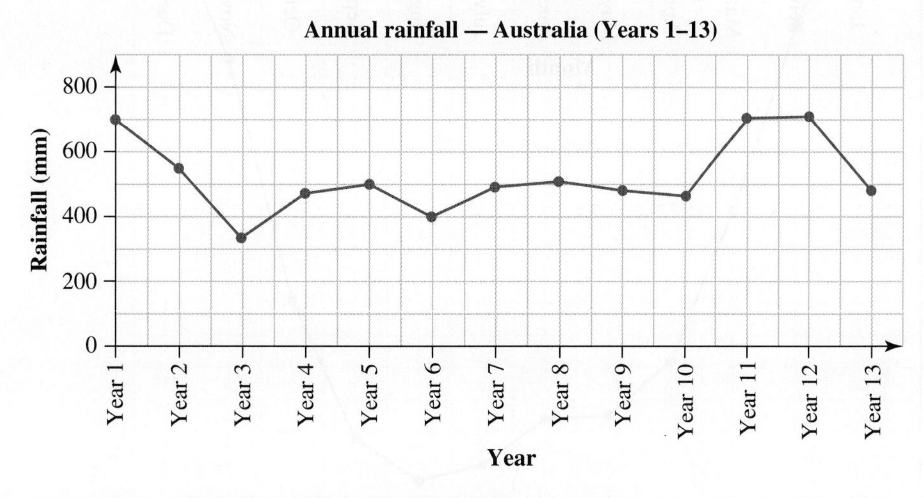

a. Determine when the maximum average annual rainfall occurred.
b. Determine the average annual rainfall in Australia for the sixth year.
c. Identify the periods during which the average annual rainfall in Australia was increasing.
d. Identify the years in which the average annual rainfall in Australia was 500 mm.
e. Determine when the minimum average annual rainfall in Australia occurred and what it was.

3. The line graph shown displays the daily maximum temperature in Launceston, Tasmania, for a month in winter.

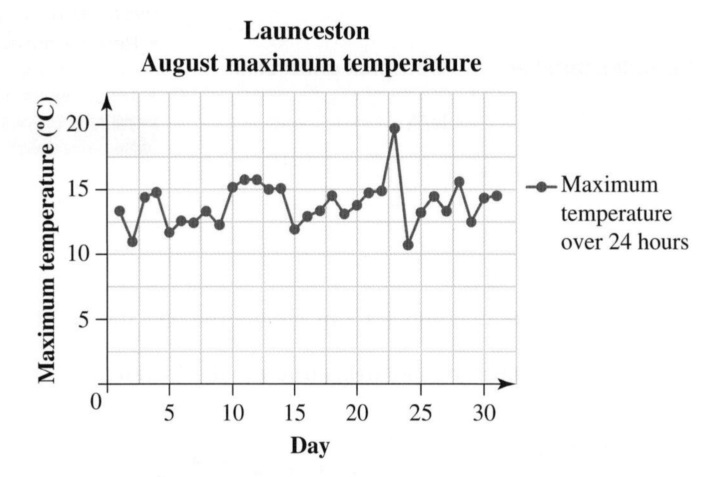

Use the line given to answer the following questions.

a. Identify which day had the lowest maximum temperature for the month and what this temperature was.
b. Identify which day had the highest maximum temperature for the month and what this temperature was.
c. Determine the difference between the maximum temperatures on days 14 and 19.
d. Determine the days on which the maximum temperature was 15.5 °C.
e. Determine for how many days the maximum temperature was less than 12 °C.

4. The line graph shown displays the mean maximum temperature in Antarctica at the station Mawson for the years 1954–2018.

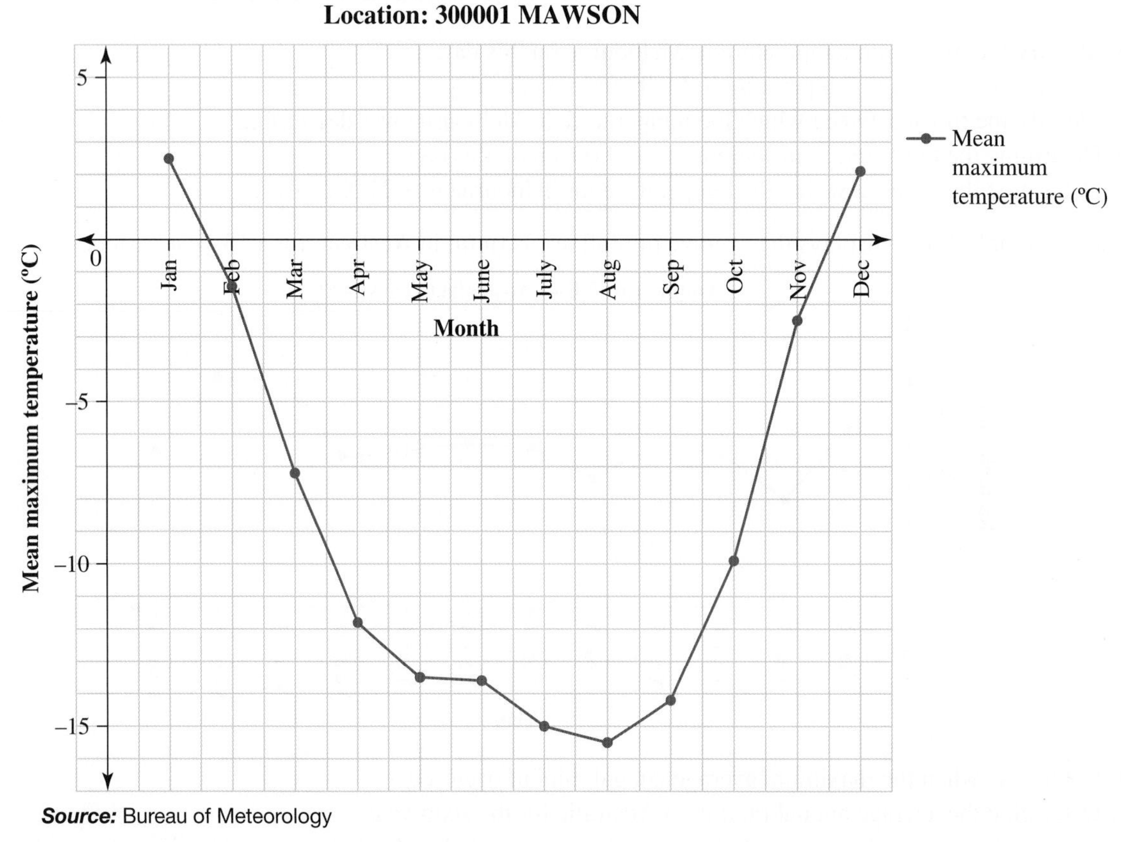

Source: Bureau of Meteorology

Use the line graph given to answer the following questions.

a. Identify which month had the lowest mean maximum temperature and what this temperature was.
b. Identify which month had the highest mean maximum temperature and what this temperature was.
c. Determine the difference between the two maximum temperatures found in parts **a** and **b**.
d. Identify the month in which the mean maximum temperature was −2.5 °C.
e. Determine for how many months the mean maximum temperature was greater than −8 °C.

5. **WE3** A suburban florist took an inventory of the most popular flowers they sold over a number of days. The percentages of the most popular flowers were recorded and displayed in the pie graph below.

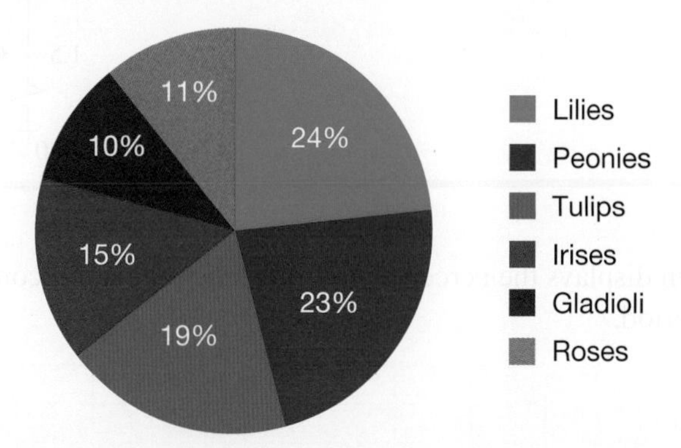

a. Identify which flower was the most popular at the florist during this time.
b. Identify which flower was the least popular.
c. If there were 220 flowers sold during this time, determine how many were irises.

6. A survey was conducted to determine the usual mode of transport to school for a group of students. The results are displayed in the pie graph below.

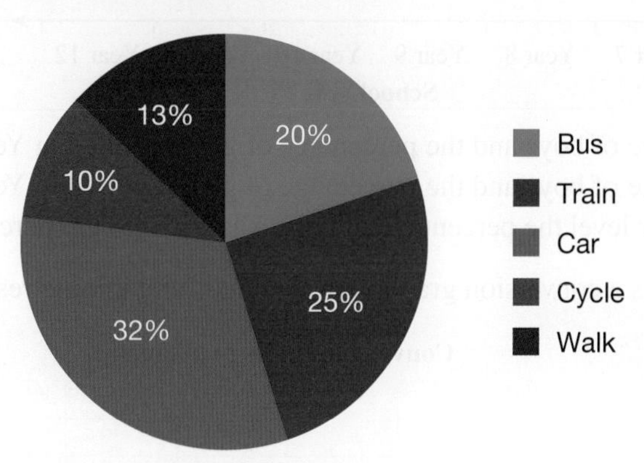

a. Identify the mode of transport most students use to get to school.
b. Identify the mode of transport the students use least to get to school.
c. If 1500 students were surveyed, determine how many used public transport to get to school.

7. **WE4** The step graph shown represents the cost of parking at the domestic terminal at Brisbane Airport.

a. Calculate the cost of parking for 45 minutes.
b. Calculate the cost of parking for 2 hours.
c. Calculate the cost of parking for 3.5 hours.

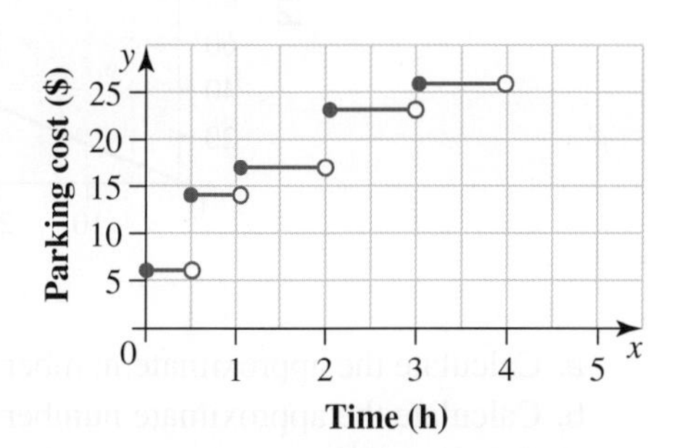

8. The cost of an international call is displayed in the step graph shown.

   a. Determine the cost of a 5-minute call.
   b. Determine the cost of a 50-minute call.
   c. Determine the length of a call if the cost of the call is $3.50.

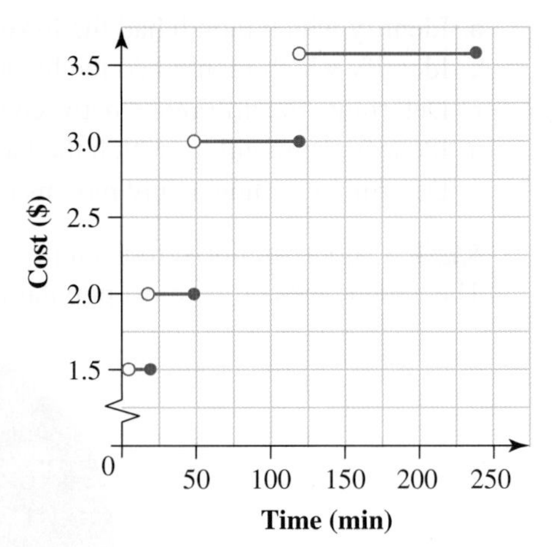

9. WE5 The line graph shown displays the percentage of full-time regional secondary school students by gender over a 12-month period.

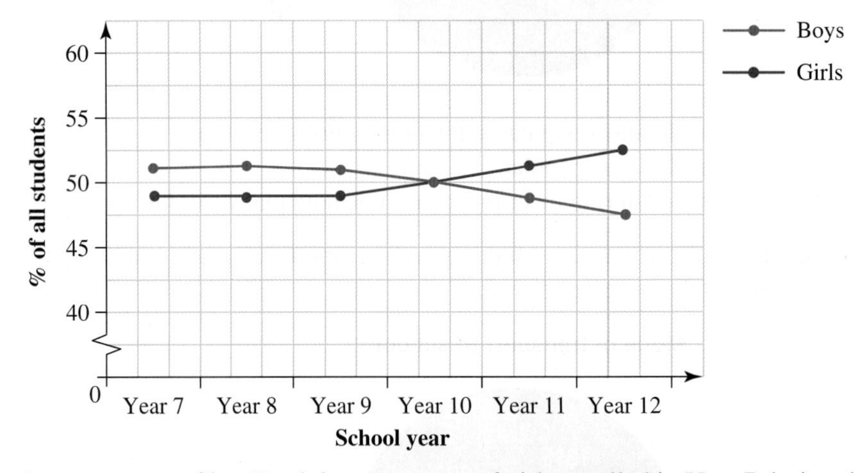

   a. Determine the percentage of boys and the percentage of girls enrolled in Year 7 during the year.
   b. Determine the percentage of boys and the percentage of girls enrolled in Year 12 during the year.
   c. Determine in which year level the percentage of boys was equal to the percentage of girls.

10. The graph shown represents a conversion graph between miles and kilometres.

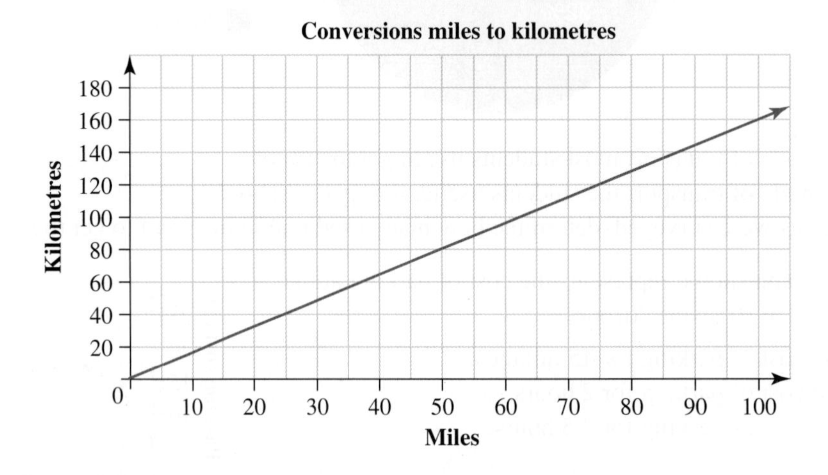

   a. Calculate the approximate number of kilometres equal to 80 miles.
   b. Calculate the approximate number of kilometres equal to 50 miles.
   c. Calculate the approximate number of miles equal to 30 kilometres.
   d. Calculate the approximate number of miles equal to 140 kilometres.

11. Use the temperature conversion graph given to answer the following questions.

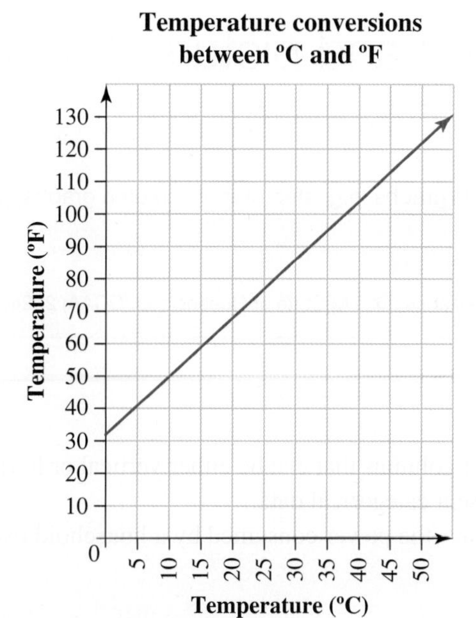

a. Determine the temperature in °F equivalent to 35 °C.
b. Determine the temperature in °F equivalent to 10 °C.
c. Determine the temperature in °C equivalent to 41 °F.
d. Determine the temperature in °C equivalent to 104 °F.

12. The line graph shown displays the gender wage gap in Australia over a 20-year period.

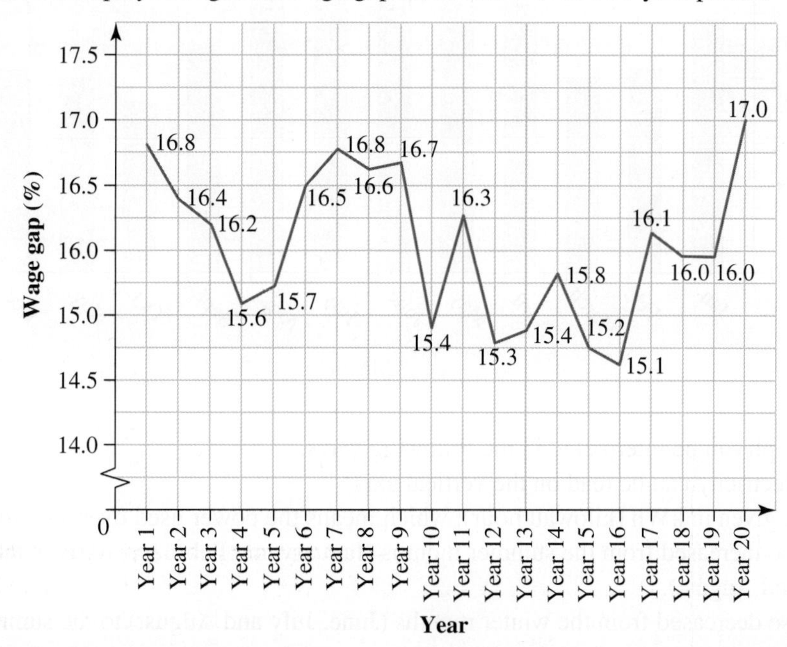

a. Determine in which year the wage gap was at its lowest percentage and what that value was.
b. Determine in which year the wage gap was at its highest percentage and what that value was.
c. Calculate the percentage increase in the wage gap in years 1–20.
d. Determine between which years the highest decrease in the wage gap occurred.
e. Identify the year in which the wage gap was 16.5%.

Fully worked solutions for this chapter are available online.

LESSON
11.4 Column graphs and picture graphs

SYLLABUS LINKS

- Interpret information presented in graphs, e.g. step graphs, column graphs, pie graphs, picture graphs, line graphs.
- Use graphs in practical situations.

Source: Essential Mathematics Senior Syllabus 2024 © State of Queensland (QCAA) 2024; licensed under CC BY 4.0.

11.4.1 Column graphs

- **Column graphs** are made up of columns that can be either vertical or horizontal.
- These graphs are used to represent categorical data.
- The column graph shown displays the power consumed by a household over a period of one year.

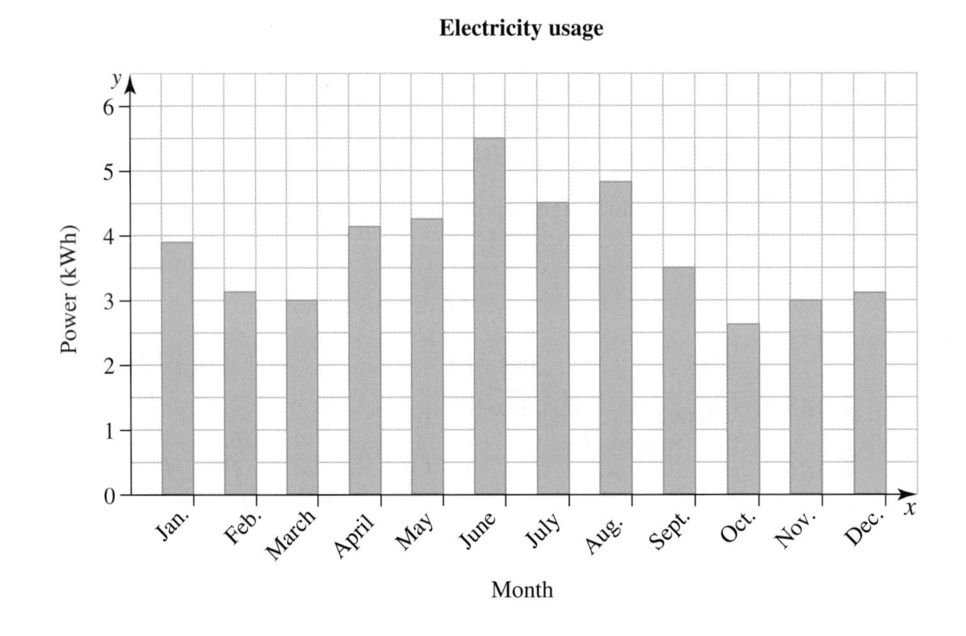

- This column graph can be interpreted in the following ways.
  - The use of electricity can be read on the vertical axis.
  - The units are given in kWh (kilowatt hour), which means the power used over a period of one hour.
  - Electricity use increased from the summer months (January and February) to the winter months (June, July and August).
  - Electricity use decreased from the winter months (June, July and August) to the summer month (December).

WORKED EXAMPLE 6 Analysing and interpreting column graphs

Water bills display the water use in a household using column graphs. Describe the water consumption for the household represented by the column graph shown.

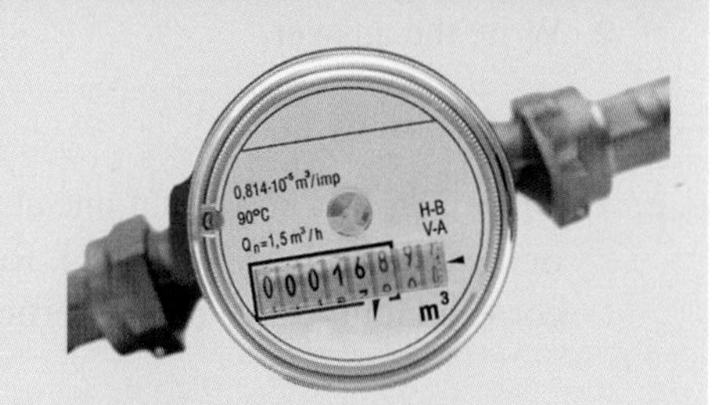

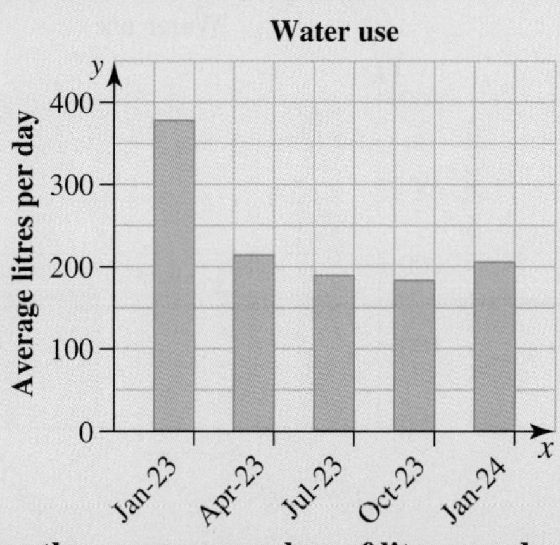

a. **Determine the average number of litres used per day in the January 2023 bill.**
b. **Determine the average number of litres used per day in the July 2023 bill.**
c. **Determine the average number of litres used per day in the January 2024 bill.**
d. **Use the information provided in the graph to comment on the household usage over the five months.**

THINK	WRITE
a. 1. Draw a dashed horizontal line at the top of the column labelled Jan-23. Read the corresponding value on the vertical axis.	**a.** 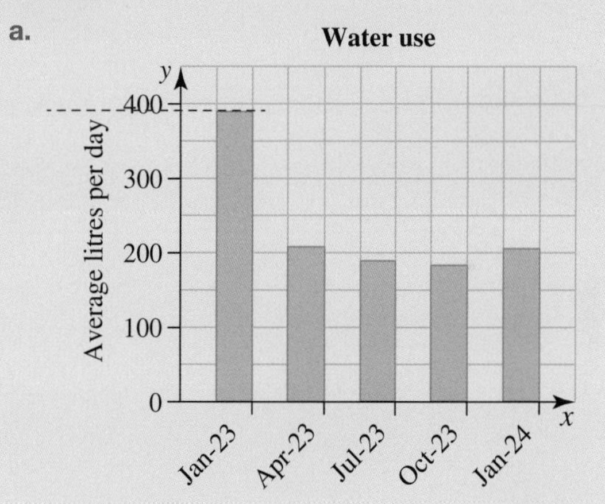
2. Write the answer.	The average number of litres per day used in the first quarter of 2023 (January, February and March) is approximately 390 L.
b. 1. Draw a dashed horizontal line at the top of the column labelled Jul-23. Read the corresponding value on the vertical axis.	**b.** 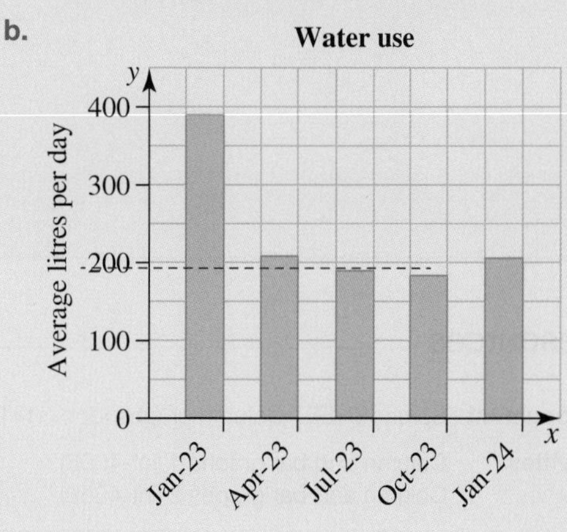

2. Write the answer.	The average number of litres per day used in the third quarter of 2023 (July, August and September) is approximately 190 L.
c. 1. Draw a dashed horizontal line at the top of the column labelled Jan-24. Read the corresponding value on the vertical axis.	**c.** 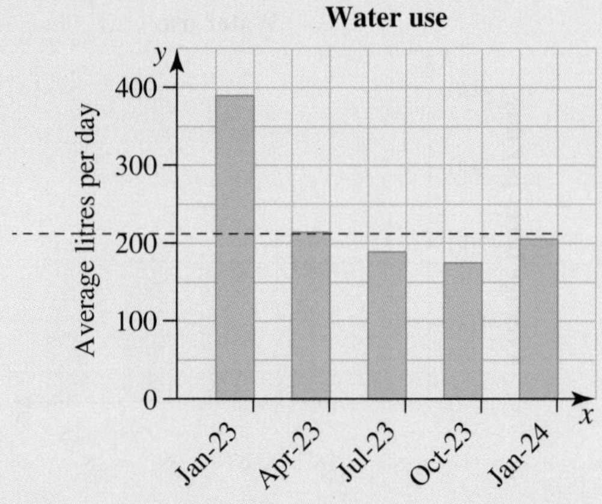
2. Write the answer.	The average number of litres per day used in the first quarter of 2024 (January, February and March) is approximately 210 L.
d. 1. Interpret the graph.	**d.** This household decreased its usage of water from the first quarter 2023 to the last quarter of 2023. More water is used in the quarter starting in January 2024 compared to the second half of 2023. However, the average number of litres used per day in the quarter starting in January 2024 is a lot lower than in the quarter starting in January 2023. Possible reasons include: • There are fewer people in the household. • The garden was watered less because of water restrictions. • There is higher rainfall than average.

Resources

Digital document SpreadSHEET Column graphs (doc-3441)

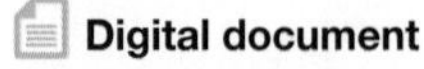
Interactivities Column and bar graphs 1 (int-4058)
Column and bar graphs 2 (int-4059)

11.4.2 Picture graphs

- **Picture graphs** or **pictographs** are statistical graphs that use pictures to display categorical data.
- The picture graph shown displays the average sales of four categories of fruit by a retailer on Mondays and Saturdays.

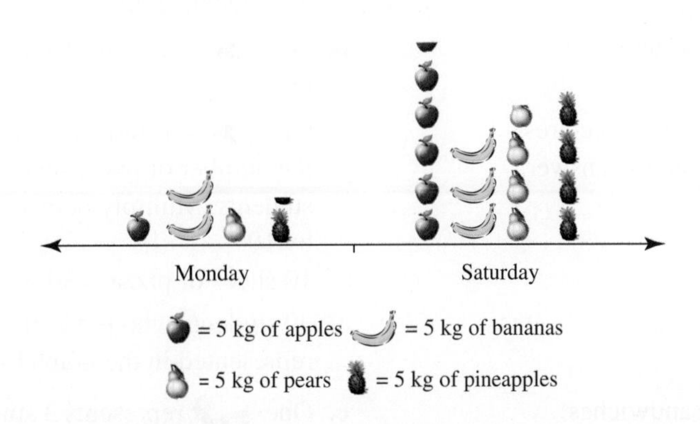

- In order to be able to read the sales of the fruits, read the key underneath the graph.
- The fruit shop sells an average of 5 kg of apples on Mondays, and 26 kg of apples on Saturdays.
- Parts of a picture represent parts of the full quantity represented by the full picture.

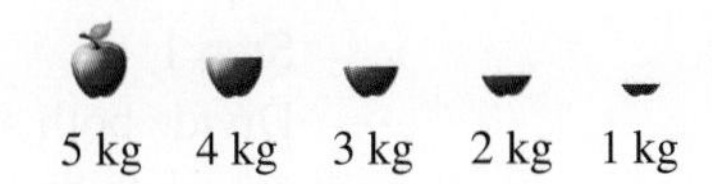

- Pictographs are typically easy to read and show trends in data clearly. However, sometimes they are hard to read, especially when parts of the picture are shown.

tlvd-11446

WORKED EXAMPLE 7 Analysing and interpreting picture graphs

A group of 50 students was asked to state which of the following foods they liked the most: pizza, hamburgers or sandwiches. The results are displayed in the picture graph shown.

a. Determine the number of students who prefer hamburgers.

b. If 30 students preferred pizza, determine how many slices of pizza would have to be drawn.

c. If 16 students preferred sandwiches, determine how many sandwiches would have to be drawn.

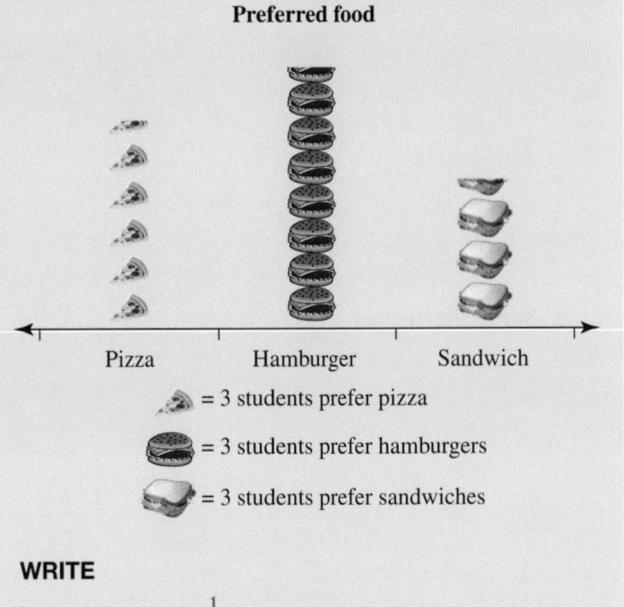

THINK	WRITE
a. 1. Count the number of pictures that represent hamburgers.	a. There are $7 + \frac{1}{3}$ hamburgers.

2. Multiply the number of pictures by the corresponding number given in the key.

Each [hamburger] represents 3 students.

$$\left(7+\frac{1}{3}\right)\times 3 = 7\times 3+\frac{1}{3}\times 3$$
$$= 21+1$$
$$= 22 \text{ students}$$

3. Write the answer.

22 students prefer hamburgers.

b. 1. Check the key for pizzas.

b. One [pizza] represents 3 students who prefer pizza.

2. Determine the number of pictures required and work out the answer.

One [pizza] = 3 students. We need to determine the number of pizza slices required for 30 students. Multiply both sides of the equation by 10.

10 slices of pizza = 30 students

3. Write the answer.

30 students who prefer pizza would be represented in the graph by 10 slices of pizza.

c. 1. Check the key for sandwiches.

c. One [sandwich] represents 3 students who prefer sandwiches.

2. Determine the number of pictures required and work out the answer.

One [sandwich] = 3 students

We need to determine the number of [sandwich] required for 16 students.

Step 1

Divide both sides of the equation by 3 to calculate the number of [sandwich] required to represent one student.

$\frac{1}{3}$ [sandwich] = 1

Step 2

Multiply both sides of the equation by the number of students required.

$16\times\frac{1}{3}$ [sandwich] = 1 student $\times 16$

$\frac{16}{3}$ [sandwich] = 16 students

Step 3

Convert the improper fraction into a mixed number.

$\frac{16}{3}$ [sandwich] $=5\frac{1}{3}$ [sandwich]

3. Write the answer.

16 students who prefer sandwiches would be represented by $5\frac{1}{3}$ [sandwich].

Exercise 11.4 Column graphs and picture graphs

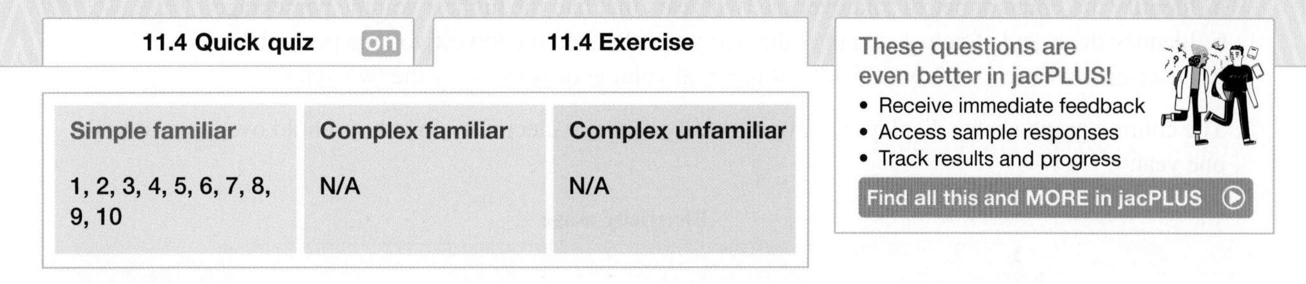

11.4 Quick quiz
on
11.4 Exercise
Simple familiar
1, 2, 3, 4, 5, 6, 7, 8, 9, 10
Complex familiar
N/A
Complex unfamiliar
N/A
These questions are even better in jacPLUS!
• Receive immediate feedback
• Access sample responses
• Track results and progress
Find all this and MORE in jacPLUS

Simple familiar

1. **WE6** The column graph shown displays the average daily usage of gas for a household over a period of one year.

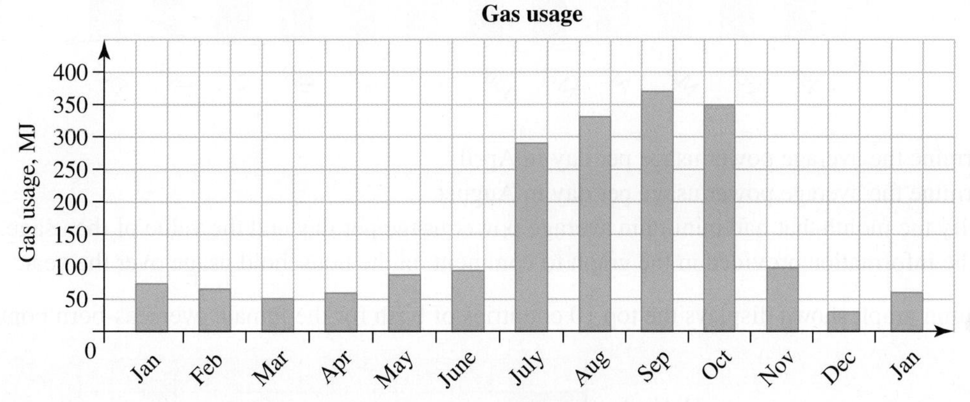

   a. Determine the average gas usage per day in June of this year.
   b. Determine the average gas usage per day in September of this year.
   c. Determine what months the average gas usage was 60 MJ.
   d. Use the information provided in the graph to comment on the household usage over the year.

2. The horizontal column graph shown displays the percentage of petrol volume sold, by brand, over a two-year period.

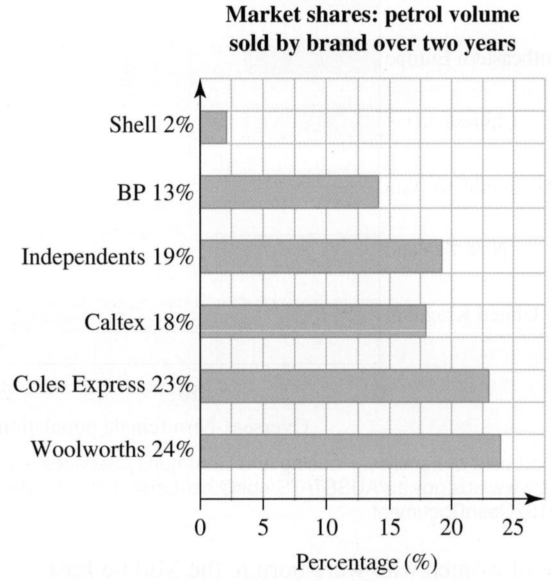

Source: Australian Institute of Petroleum (AIP), http://www.aip.com.au/pricing/facts/Facts_About_the_Australian_Retail_Fuels_Market_and_Prices.htm

a. Identify the brand of petrol sold over the two years that had the highest volume percentage.
b. Identify the brand of petrol sold over the two years that had the lowest volume percentage.
c. Determine which company sold 18% of the total volume of petrol over the two years.

3. The column graph shown displays the average daily usage of electricity for a household over a period of one year.

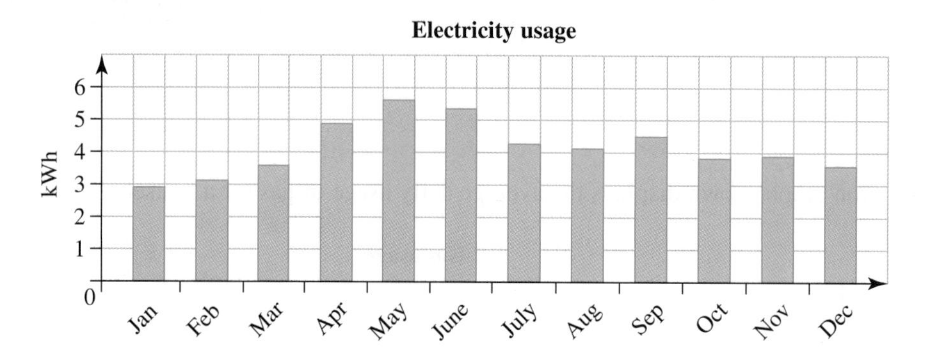

a. Determine the average power usage per day in April.
b. Determine the average power usage per day in August.
c. Identify the month that had minimum average power usage per day and the value of the usage.
d. Use the information provided in the graph to comment on the household usage over the year.

4. The column graph shown displays the top 10 countries of birth for the female overseas-born population.

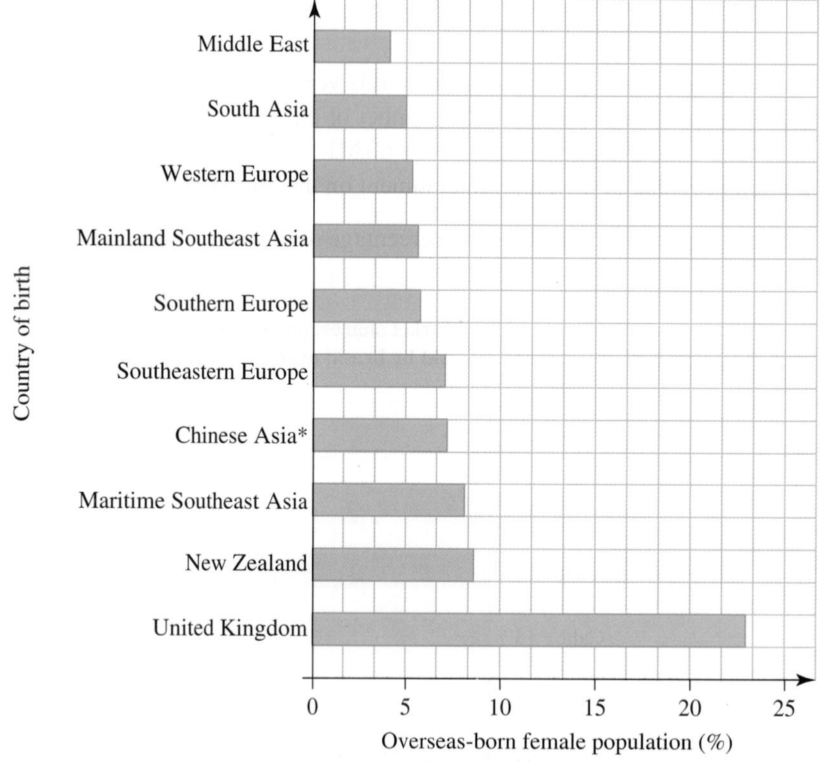

Source: Department of Families, Housing and Community Services and Indigenous Affairs, http://www.abs.gov.au/AUSSTATS/abs@.nsf/Lookup/3201.0Main+Features 1Jun%202010?OpenDocument

a. Determine the percentage of women who were born in the Middle East.
b. Determine which country the highest percentage of women come from.
c. Determine the percentage of women who came from Southeastern Europe.

5. The graph shown represents the population estimate for states and territories in Australia.

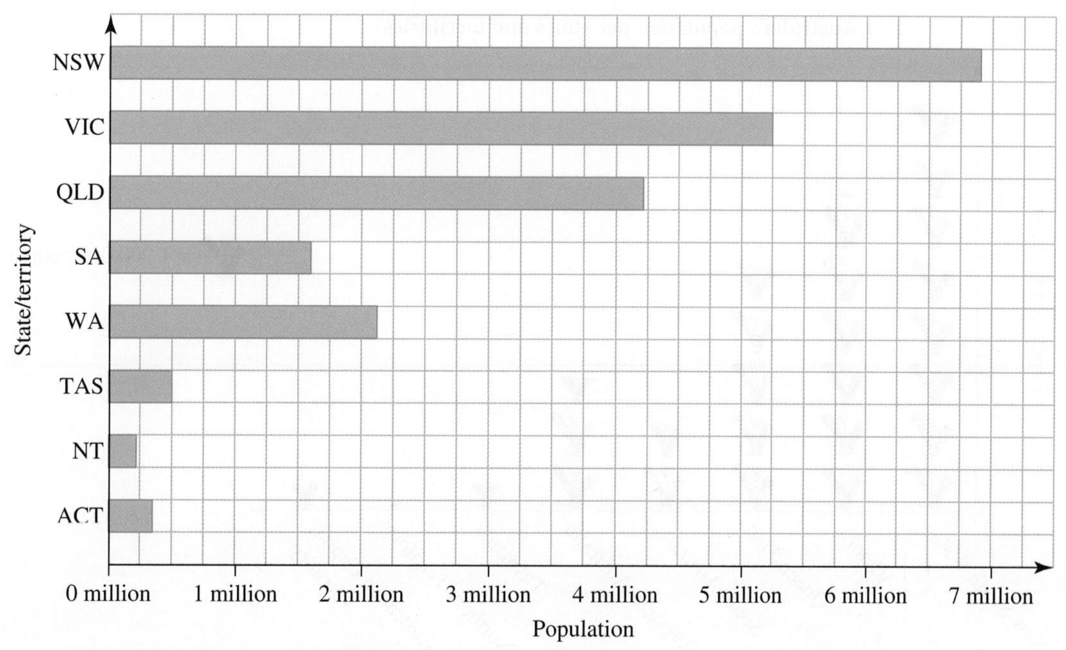

Source: ABS http://www.abs.gov.au/websitedbs/d3310114.nsf/Home/Animated+Historical+Population+Chart

Determine the approximate population of each of the Australian states and territories.

6. The column graph shown displays the average daily usage of gas for a household over a 31-day month.

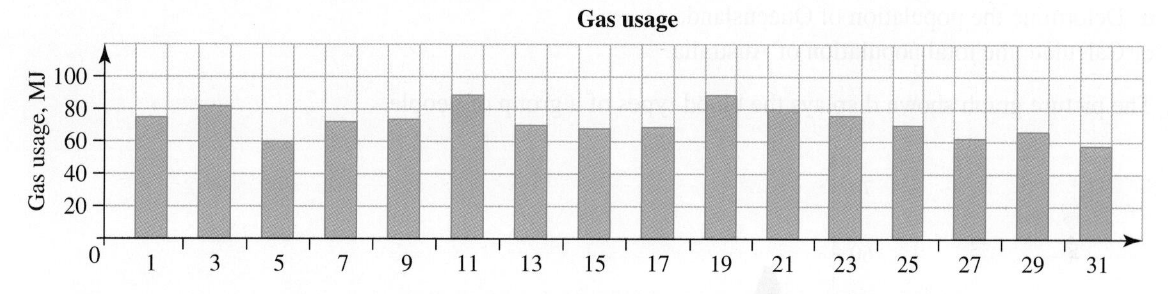

a. Determine the gas usage on the thirteenth of the month.
b. Identify the day that had the maximum gas usage for the month and state the gas usage.
c. Identify the day that had the minimum gas usage for the month and state the gas usage.

7. WE7 The average prices of houses in three suburbs are represented in the picture graph shown.

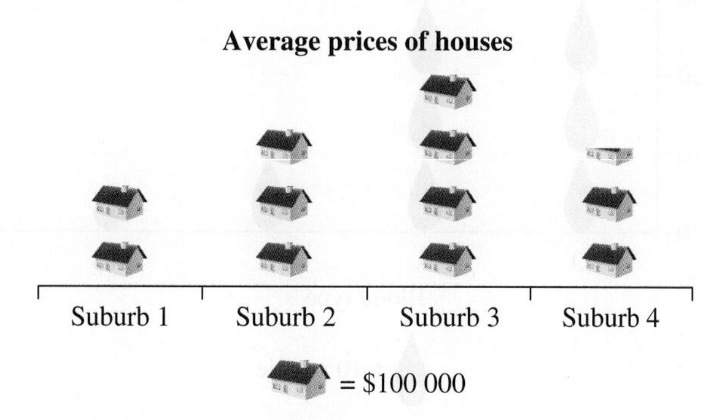

a. Determine the average price of houses in Suburb 3.
b. Determine the average price of houses in Suburb 1.
c. Determine by how much the houses in Suburb 2 are more expensive than the houses in Suburb 4.

8. The population of each Australian state and territory is represented in the picture graph shown.

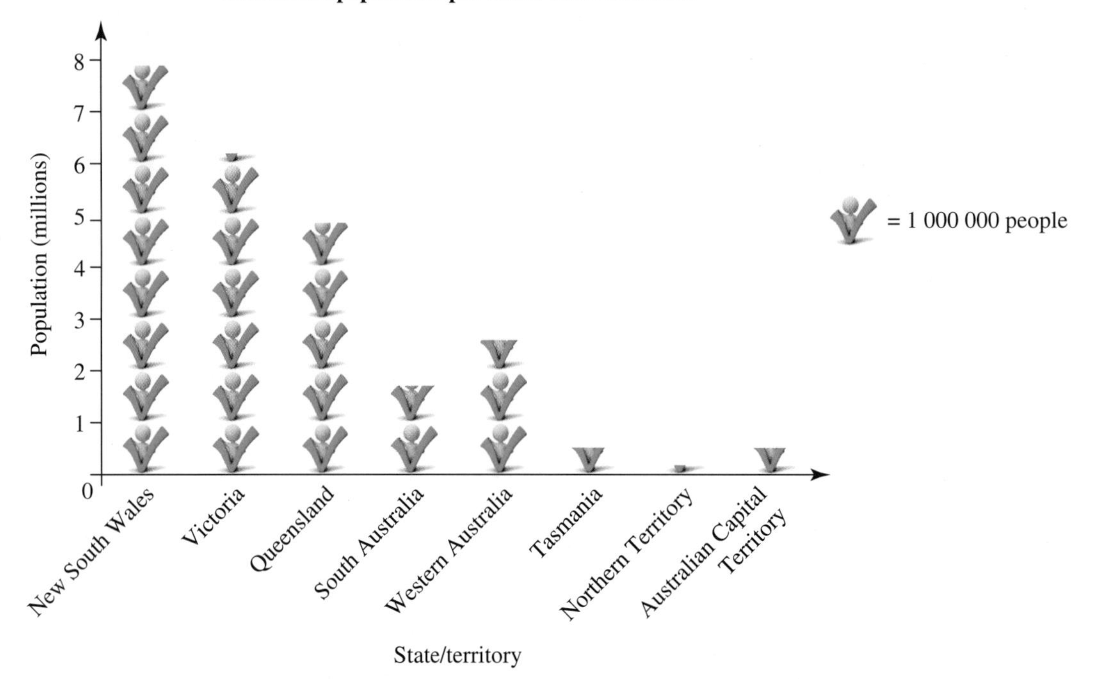

Source: Adapted from Australian Bureau of Statistics, http://www.abs.gov.au/ausstats/abs@.nsf/mf/3101.0

a. Determine the population of Tasmania.
b. Determine the population of Queensland.
c. Calculate the total population of Australia.

9. The picture graph shown displays the blood types of a group of people.

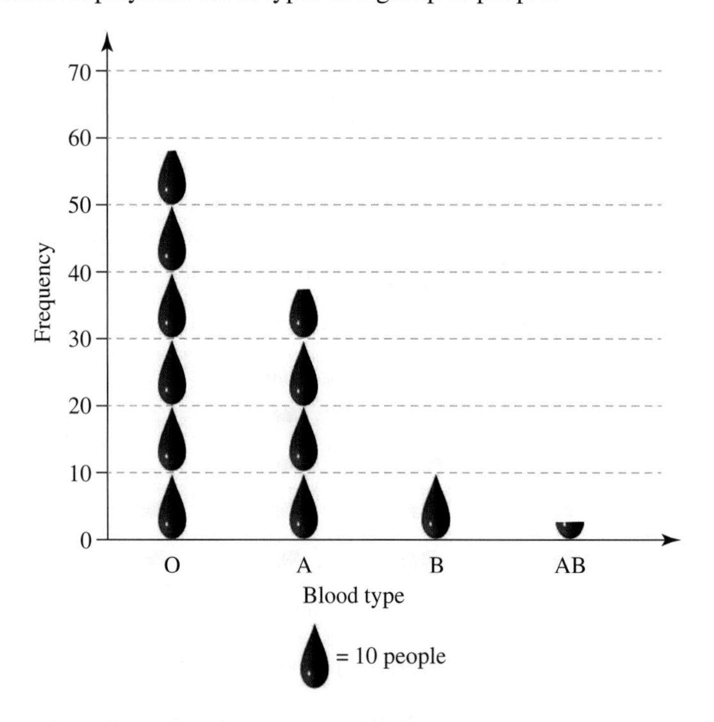

Determine the total number of people who were surveyed.

10. A group of 48 students were asked whether they liked sports at school.
Their answers are displayed in the picture graphs shown.

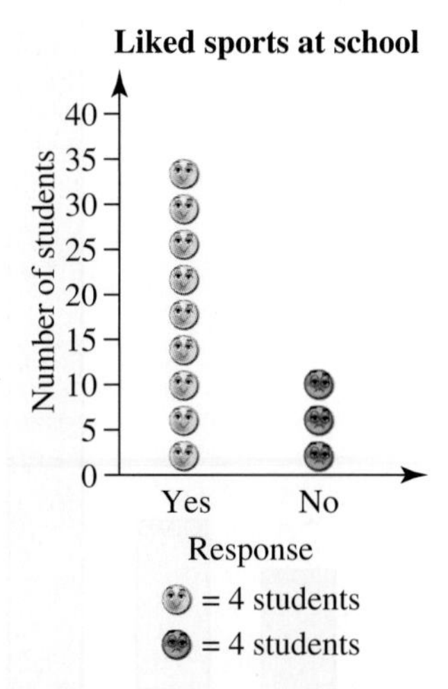

a. Determine the number of students who liked sports at school.
b. Determine the total number of students who were surveyed.

Fully worked solutions for this chapter are available online.

LESSON
11.5 Misleading graphs [complex]

SYLLABUS LINKS

- Discuss and interpret tables and graphs, including misleading graphs found in the media and in factual texts [complex].
- Interpret graphs in practical situations [complex].

11.5.1 Methods of misrepresenting data

- Many people have reasons for misrepresenting data.
- For example, politicians may wish to magnify the progress achieved during their term, or businesspeople may wish to accentuate their reported profits.
- There are numerous ways of misrepresenting data. In this section, only graphical methods of misrepresentation are considered.

The vertical and horizontal axes

- It is obvious that the steeper the graph, the better the growth looks.
- A 'rule of thumb' for statisticians is that, for the sake of appearances, the vertical axis should be two-thirds to three-quarters the length of the horizontal axis.
- This rule was established in order to compare graphs more easily.
- The figure shown illustrates how distorted a graph appears when the vertical axis is disproportionately large.

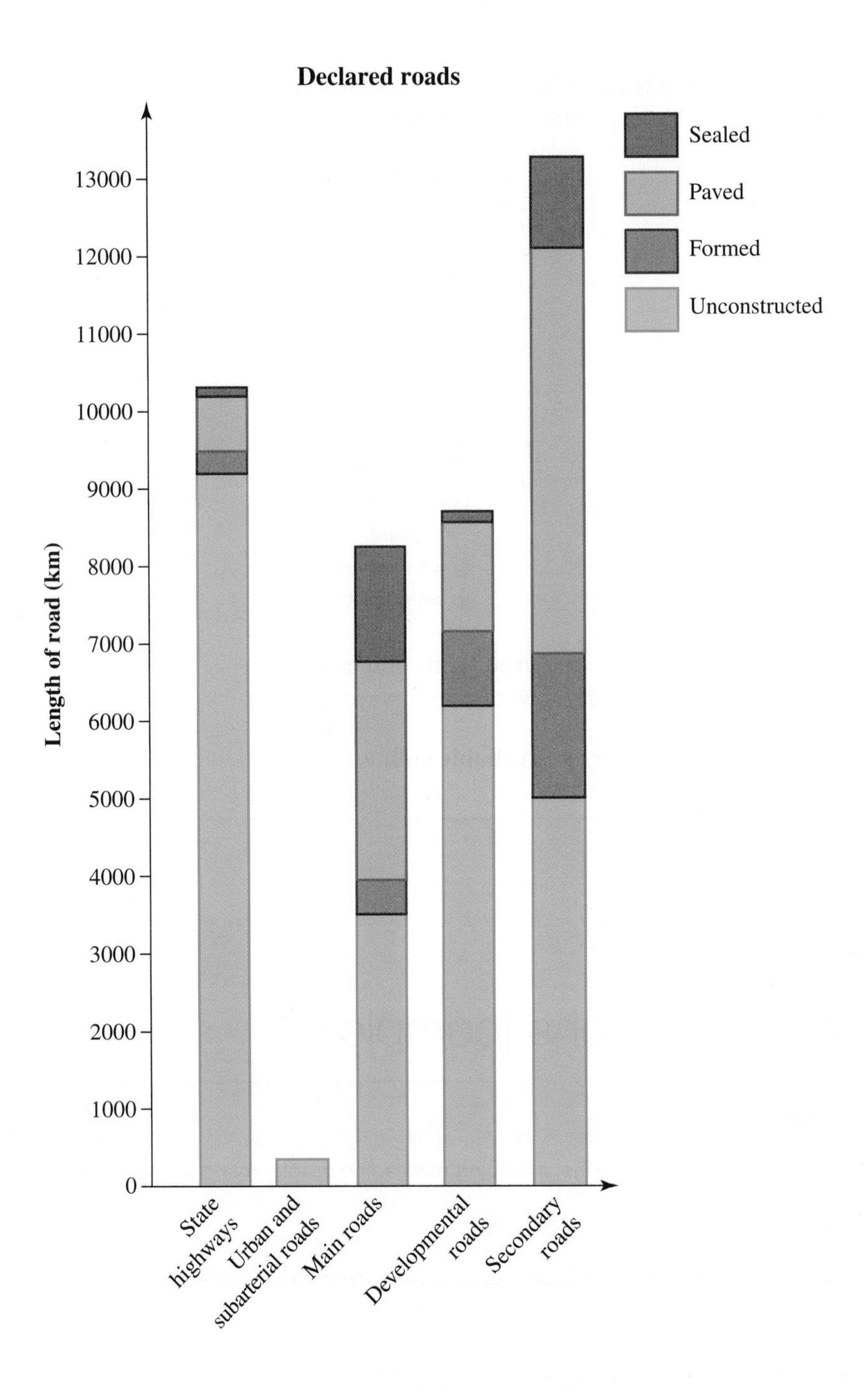

Changing the scale on the vertical axis

- The following table shows the holdings of a corporation during a particular year.

Quarter	Holdings (× $1 000 000)
Jan–Mar	200
Apr–June	200
July–Sept	201
Oct–Dec	202

- Here is one way of representing the data.

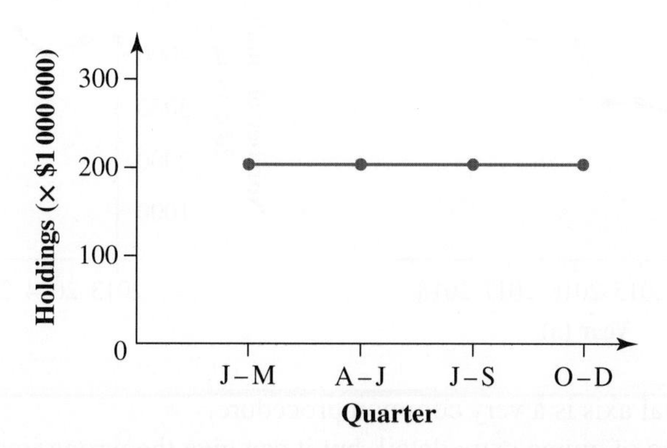

- Now look at the following graph showing the same information with a modified scale on the *y*-axis.

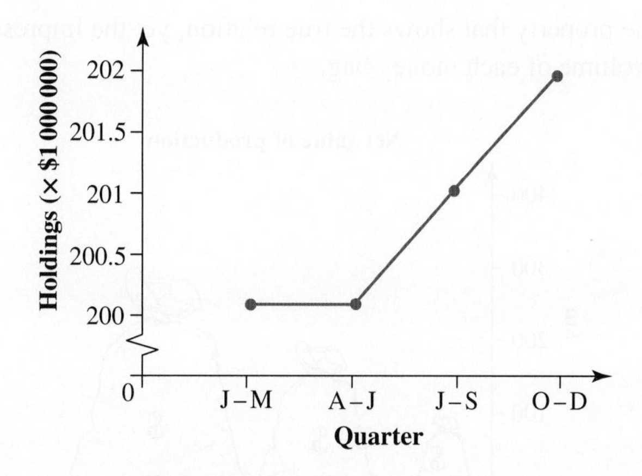

Omitting certain values

- If you ignore the second quarter's value, which shows no increase, then the graph looks even better.

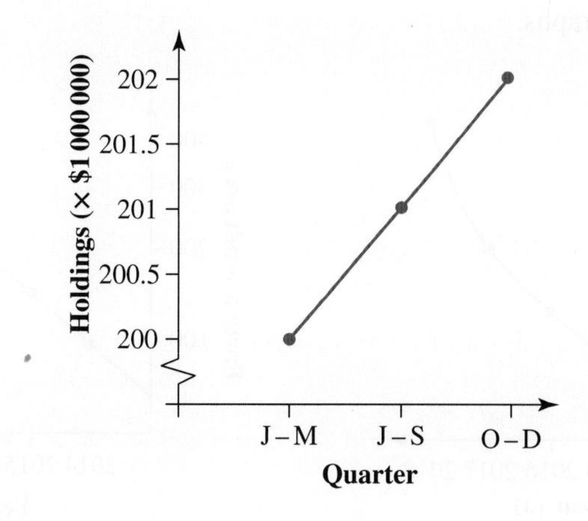

Foreshortening the vertical axis

- Look at the following figures. Notice in graph (a) that the numbers from 0 to 4000 have been omitted.
- In graph (b), these numbers have been inserted. The rate of growth of the company looks far less spectacular in graph (b) than in graph (a). This is known as foreshortening the vertical axis.

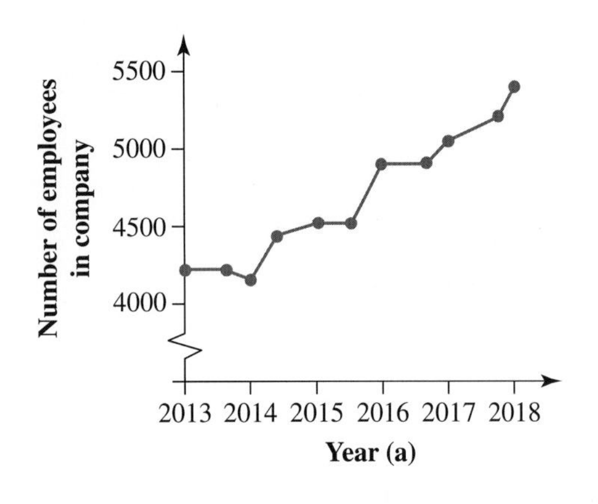

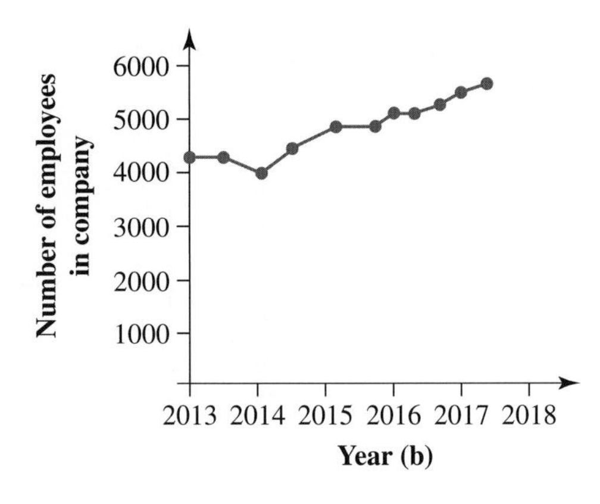

- Foreshortening the vertical axis is a very common procedure.
- It does have the advantage of giving extra detail, but it can give the wrong impression about growth rates.

Visual impression

- In this graph, height is the property that shows the true relation, yet the impression of a much greater increase is given by the volume of each money bag.

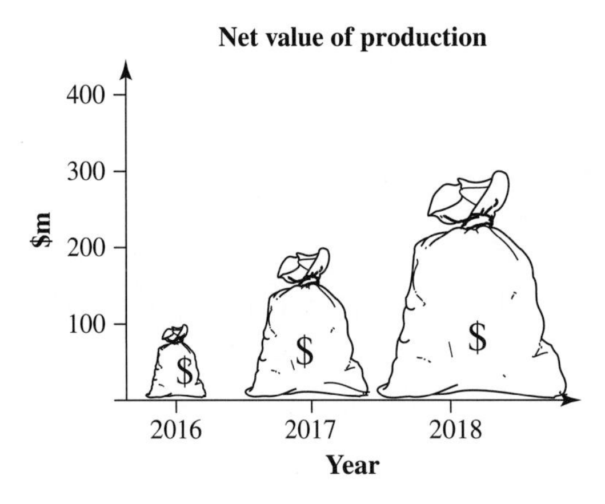

A non-linear scale on an axis or on both axes

- Consider the following two graphs.

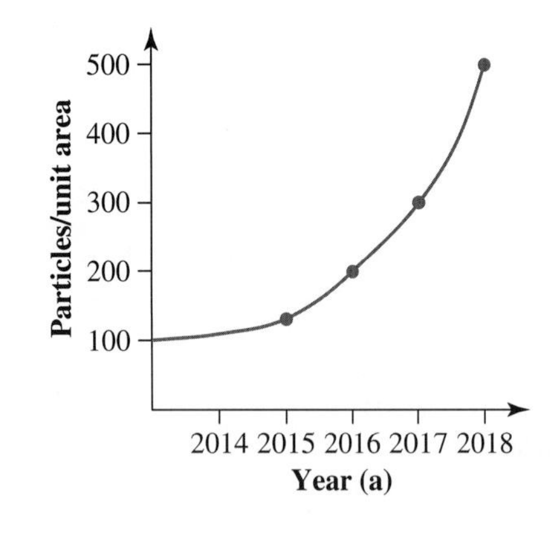

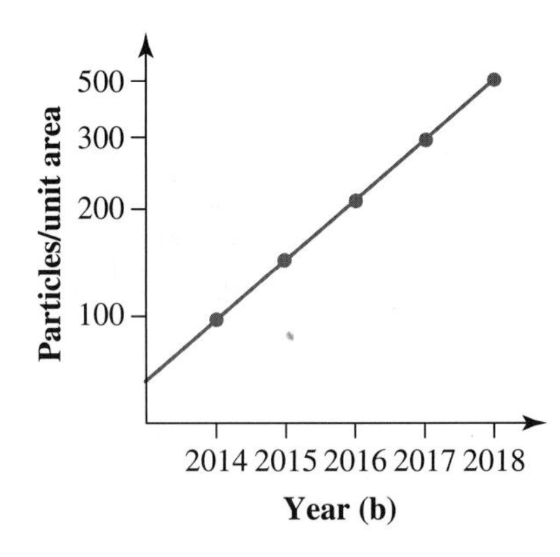

- Both of these graphs show the same numerical information, but graph (a) has a linear scale on the vertical axis and graph (b) does not.
- Graph (a) emphasises the ever-increasing rate of growth of pollutants, whereas graph (b) suggests a slower, linear growth.

tlvd-11447

WORKED EXAMPLE 8 Determining whether a graph is misleading

The following data shows wages and profits for a certain company. All figures are in millions of dollars.

Year	2015	2016	2017	2018
Wages	6	9	13	20
% increase in wages	25	50	44	54
Profits	1	1.5	2.5	5
% increase in profits	20	50	66	100

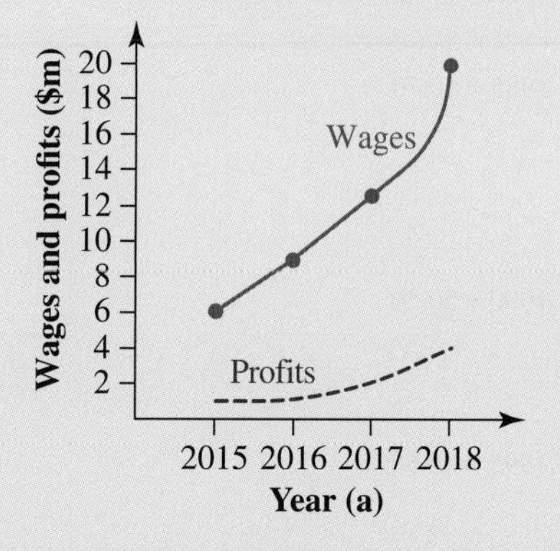

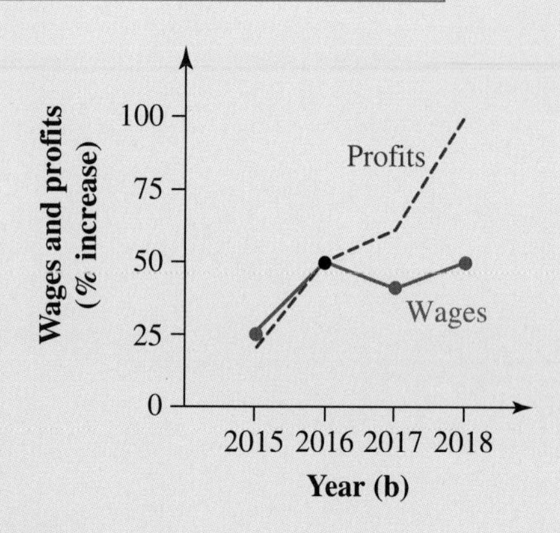

a. Determine whether the graphs accurately reflect the data.

b. Decide which graph you would rather have published if you were:

i. an employer dealing with employees requesting pay increases

ii. an employee negotiating with an employer for a pay increase.

THINK	WRITE
a. 1. Look at the scales on both axes. Both scales are linear.	**a.** The graphs do represent the data accurately. However, they use different units on the y-axis, so they look quite different and therefore suggest different conclusions.
2. Look at the units on both axes. Graph (a) has its y-axis in \$, whereas graph (b) has its y-axis in %.	
b. i. 1. Compare wage increases with profit increases.	**b. i.** The employer would prefer graph (a) because they could argue that employees' wages were increasing at a greater rate than profits.
2. The employer wants a graph that shows wages are already increasing exponentially to argue against pay increases.	
ii. 1. Consider again the increases in wages and profits.	**ii.** The employee would choose graph (b), arguing that profits were increasing at a great rate while wage increases clearly lagged behind.
2. The employee sees that profits are increasing at a much greater rate than wages, and thinks this is unfair.	

Misleading picture graphs

- There are times when picture graphs can be misleading. This can happen either on purpose or by mistake.

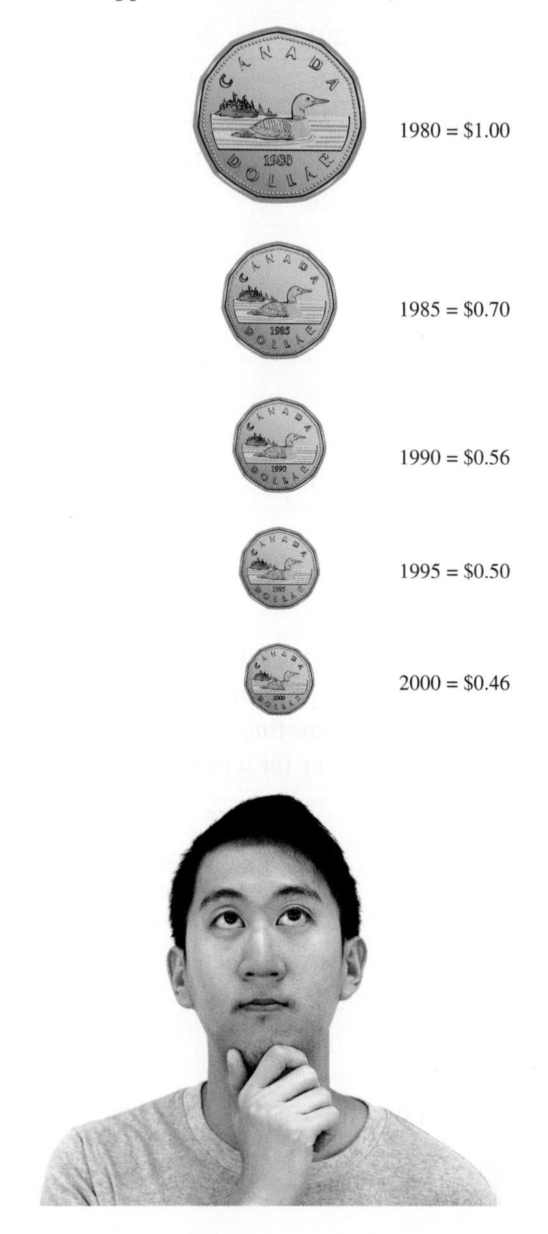

- The purpose of the picture graph shown is to express the idea that the value of the Canadian dollar decreased from $1.00 to less than a half, $0.46, over the 20 years from 1980 to 2000, due to inflation.
- However, the visual impact is a lot stronger because the difference between the area of the dollar that represents $1.00 in 1980 is about three times larger than the area that represents $0.46 in 2000.
- The picture graph shown below is a more accurate representation of this situation.

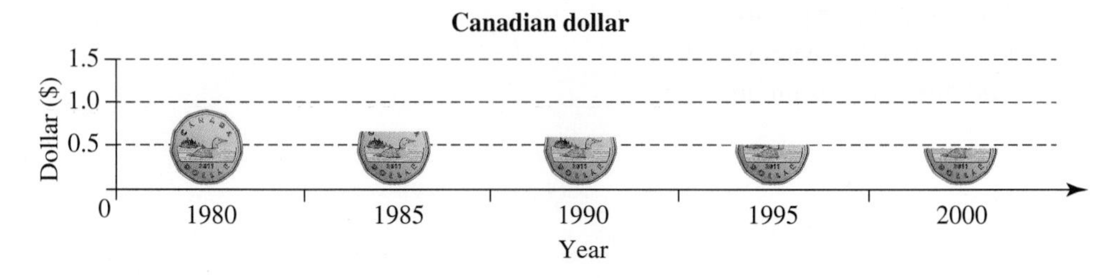

Exercise 11.5 Misleading graphs [complex]

learnon

11.5 Quick quiz	11.5 Exercise

Simple familiar	Complex familiar	Complex unfamiliar
N/A	1, 2, 3, 4, 5, 6, 7, 8, 9, 10	N/A

Complex familiar

1. WE8 The graph shows the money spent on research in a company for 2022, 2023 and 2024.

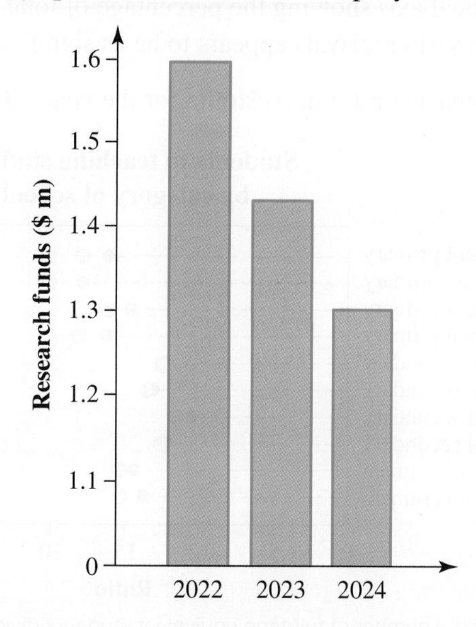

Sketch another bar graph that minimises the appearance of the fall in research funds.

2. Examine this graph of employment growth in a company.

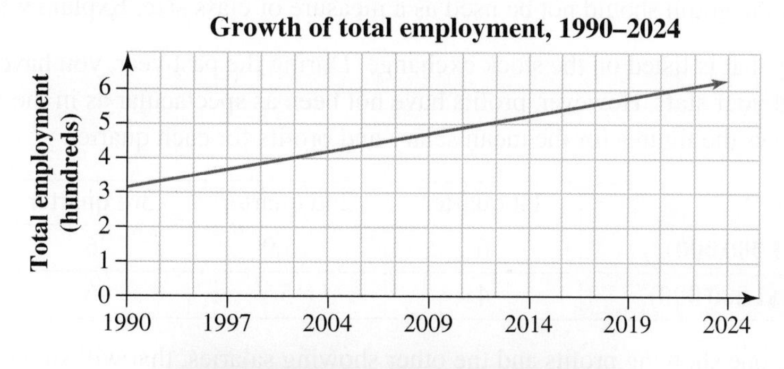

Explain why this graph is misleading.

3. Examine this graph.

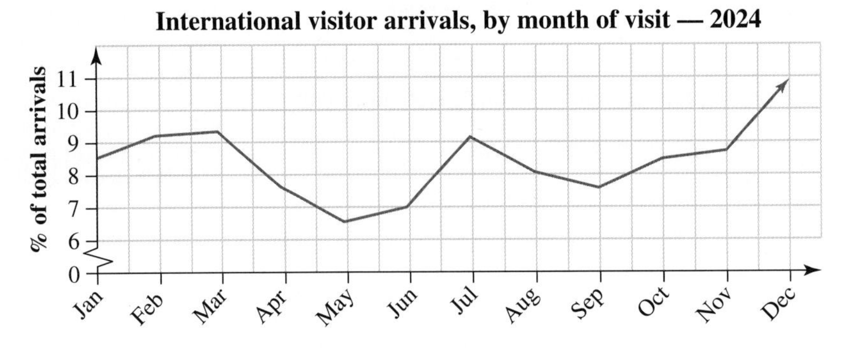

a. Sketch the graph with the vertical axis showing the percentage of total arrivals starting at 0.
b. Explain whether the change in visitor arrivals appears to be as significant as the original graph suggests.

4. This graph shows the student-to-teacher ratio in Australia for the years 2008 and 2018.

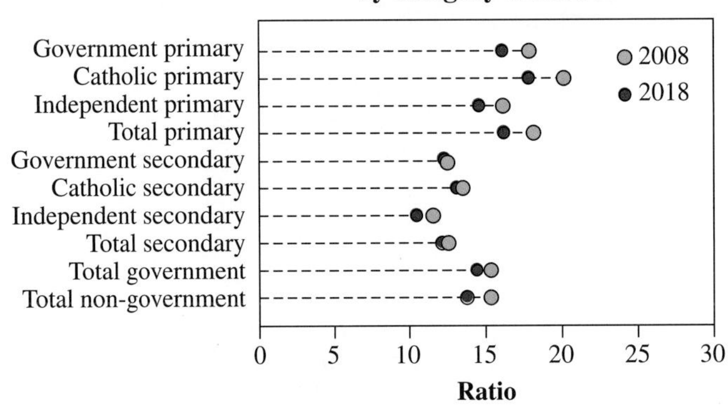

Note: Ratio = number of full-time equivalent students divided by the number of full-time equivalent teaching staff

a. Describe what has generally happened to the ratio of students to teaching staff over the 10-year period.
b. A note says that the graph should not be used as a measure of class size. Explain why.

5. You run a company that is listed on the stock exchange. During the past year, you have given substantial rises in salary to all your staff. However, profits have not been as spectacular as in the year before. The following table shows the figures for the mean salary and profits for each quarter.

Quarter	1st quarter	2nd quarter	3rd quarter	4th quarter
Profits (× $1 000 000)	6	5.9	6	6.5
Salaries (× $1 000 000)	4	5	6	7

Sketch two graphs, one showing profits and the other showing salaries, that will show you in the best possible light to your shareholders.

6. You are a manufacturer and your plant is discharging heavy metals into a waterway. Your own chemists do tests every 3 months, and the following table shows the results for a period of 2 years.

Year	2023				2024			
Month	Jan.	Apr.	Jul.	Oct.	Jan.	Apr.	Jul.	Oct.
Concentration (parts per million)	7	9	18	25	30	40	49	57

Sketch a graph that will show your company in the best light.

7. This pie graph shows the break-up of national health expenditure in 2023–24 from three sources: the Australian government, state and local government, and non-government. (This expenditure relates to private health insurance, injury compensation insurers and individuals.)

Source	Expenditure ($m)	%
Australian government	37 229	45
State and local government	21 646	25
Non-government	28 004	30

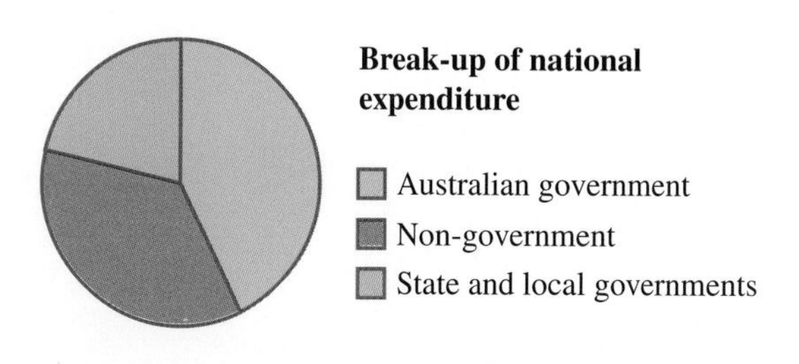

a. Comment on the claim that $87 000 m was spent on health from these three sources.
b. Identify which area contributes least to national health expenditure. Comment on its quoted percentage.
c. Identify which area contributes the next greatest amount to national health expenditure. Comment on its quoted percentage.
d. The Australian government contributes the greatest amount. Comment on its quoted percentage.
e. Consider the pie graph.
   i. Based on the percentages shown in the table, determine what the angles should be.
   ii. Based on the actual expenditures, determine what the angles should be.
   iii. Measure the angles in the pie graph and comment on their values.

8. This graph shows how the $27 that a buyer pays for a new album by their favourite singer-songwriter is distributed among the departments of a major recording company involved in its production and marketing.

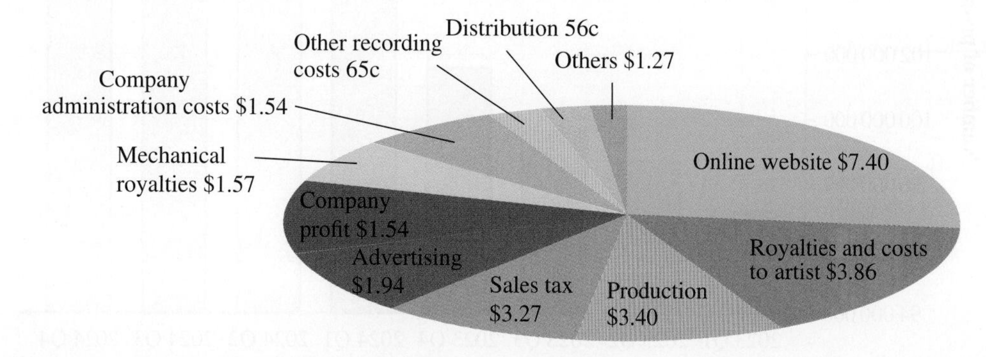

Determine if the graph is misleading and justify your answer.

9. The pictograph shows how much pizza, fries and ice creams were sold in a day. Determine whether the graph is misleading or not and justify your answer.

Food sold in one day

Pizzas

Fries

Ice-cream cones

10. The graph shown demonstrates the number of people in the US receiving welfare. Determine whether the graph is misleading or not and justify your answer.

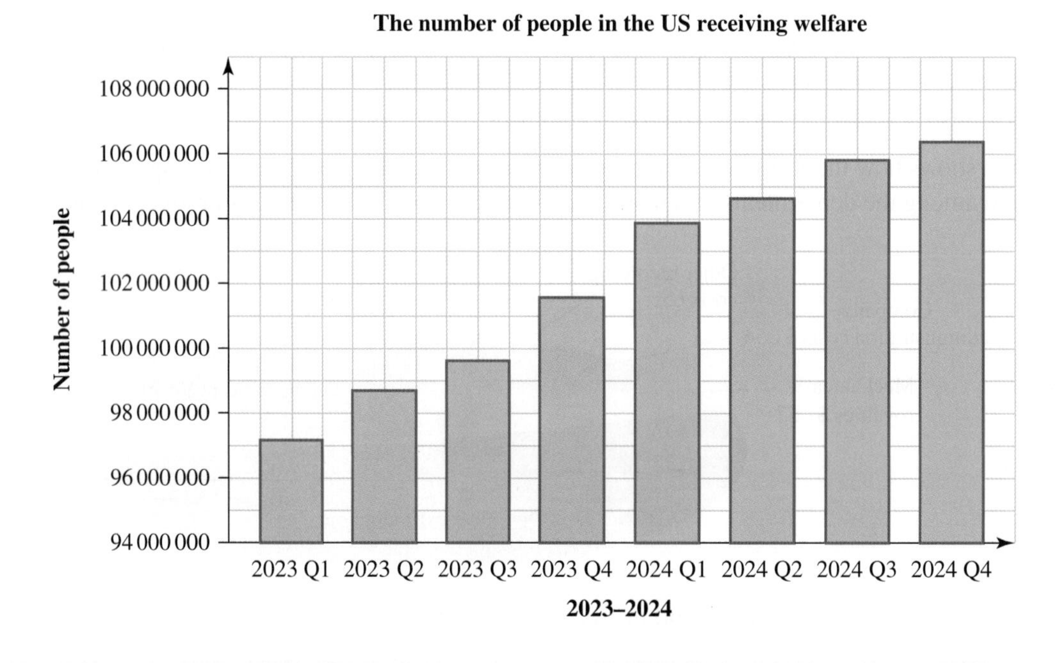

Fully worked solutions for this chapter are available online.

LESSON
11.6 Review

11.6.1 Summary

doc-41773

Hey students! Now that it's time to revise this chapter, go online to:

Access the chapter summary | Review your results | Watch teacher-led videos | Practise questions with immediate feedback

Find all this and MORE in jacPLUS

11.6 Exercise

learnon

11.6 Exercise

Simple familiar	Complex familiar	Complex unfamiliar
1, 2, 3, 4, 5, 6, 7, 8, 9, 10, 11, 12, 13, 14, 15	16, 17	18

These questions are even better in jacPLUS!

- Receive immediate feedback
- Access sample responses
- Track results and progress

Find all this and MORE in jacPLUS

Simple familiar

1. The literacy and numeracy skills of 142 primary school students are recorded in the table shown.

		Literacy skills		
		Good	Poor	Total
Numeracy skills	Good	101	12	113
	Poor	26	3	29
	Total	127	15	142

Determine the number of students who have good numeracy skills but poor literacy skills.

2. A skip bin can be hired for \$120 for the first 2 days, \$160 for more than 2 days and up to and including 4 days, \$210 for more than 4 days and up to and including 7 days, and \$280 for more than 7 days and up to and including 10 days.

   Sketch a step graph that represents the relationship between the cost of hiring the skip bin and the number of days hired.

3. The daily minimum temperatures for the month of June recorded at Mount Wellington station are displayed in the graph shown.

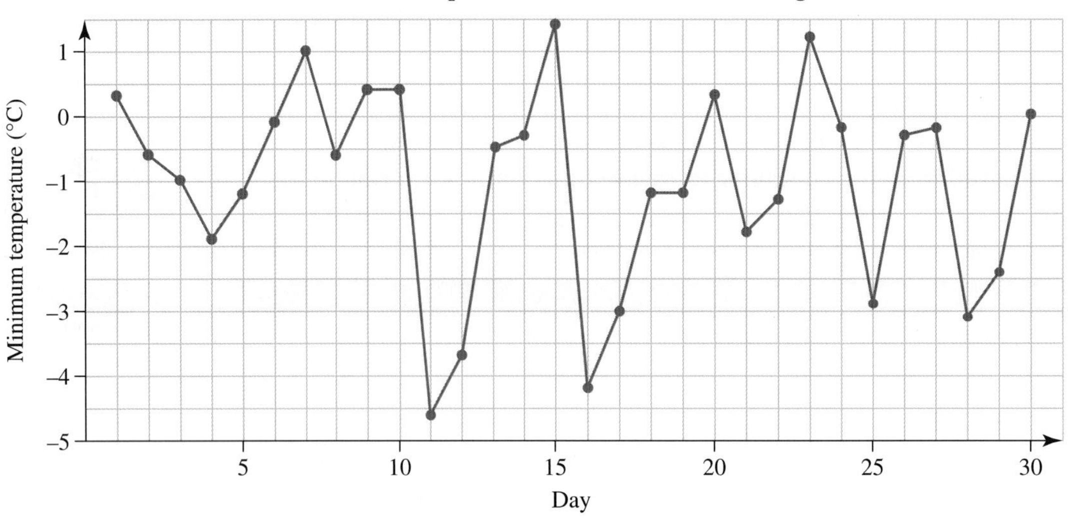

Minimum temperature over 24h — Minimum temperature period uncertain • No data

Source: Adapted from http://www.bom.gov.au/jsp/ncc/cdio/weatherData/av?p_display_type=dataDGraph&p_stn_num=094087&p_nccObsCode=123&p_month=06&p_startYear=2006.

Determine the highest minimum temperature in June that was recorded at Mount Wellington station.

4. The step graph shown displays the parking fees at a shopping centre.

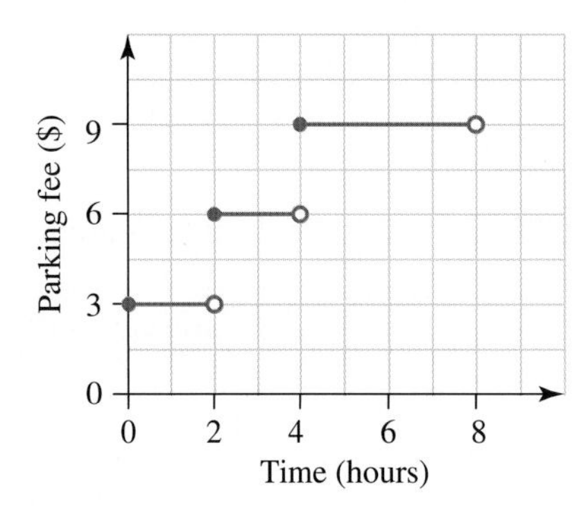

Determine the cost of parking for 2.5 hours.

5. The column graph shown displays the labour force participation rate for mothers aged 20–54 years by age of their youngest child.

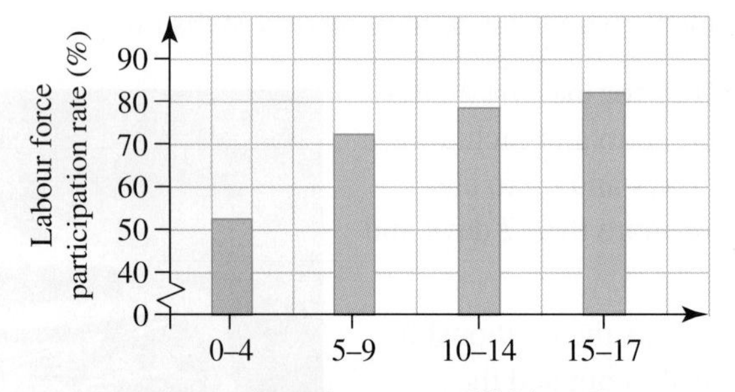

Source: Adapted from Department of Families, Housing and Community Services and Indigenous Affairs, http://www.abs.gov.au/AUSSTATS/abs@.nsf/Lookup/3201.0Main+Features1Jun%202010?OpenDocument

Explain what the labour force participation rate of approximately 78% represents.

6. Determine the total number of people used to draw the following column graph.

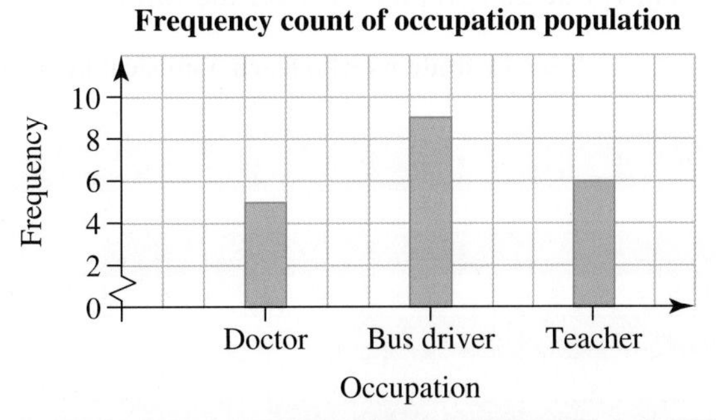

Source: ABS http://www.abs.gov.au/websitedbs/a3121120.nsf/89a5f3d8684682b6ca256de4002c809b/e200e8e572a2ae52ca25794900127f4f!OpenDocument

7. A group of students collected clothing and shoes to donate to a charity shop.

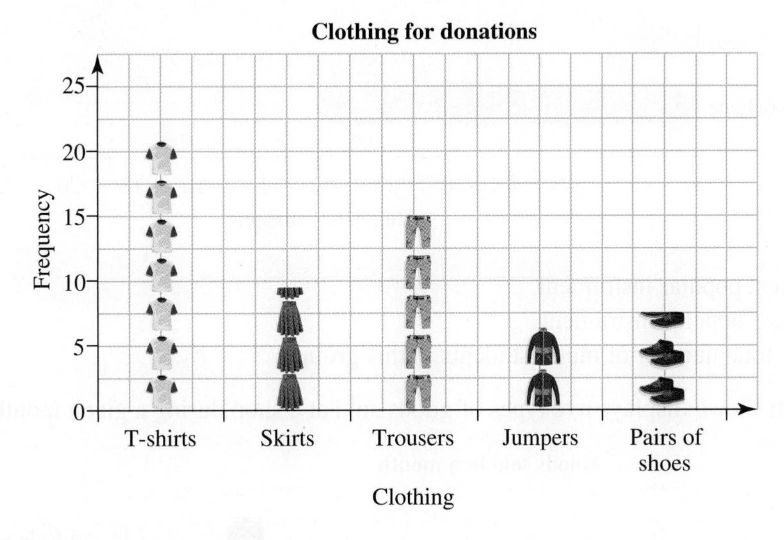

If each symbol represents 3 items, determine the number of T-shirts, skirts, trousers, jumpers and pairs of shoes respectively from the given picture graph.

8. A transport company charges different fees depending on the weight of the luggage. Luggage less than or equal to 5 kg is free of charge. Luggage over 5 kg and less than or equal to 10 kg has a fee of \$15, and luggage over 10 kg and less than or equal to 20 kg has a fee of \$25.
Luggage heavier than 20 kg and less than or equal to 30 kg incurs a charge of \$30. Construct a step graph to represent this data.

9. A group of Year 11 music students were asked which musical instrument they were learning. The results are summarised in the graph below. Use this graph to answer the following questions.

Year 11 students who learn a musical instrument

Musical instrument
Piano
Guitar
Violin
Drums
Flute
Trumpet
Clarinet
Saxophone
0 2 4 6 8 10 12 14 16 18 20 22
Number of students

a. Identify the most popular instrument.
b. Identify the least popular instrument.
c. Determine the total number of music students in this group.

10. The picture graph shown displays five types of goods sold at a shop during a given month.

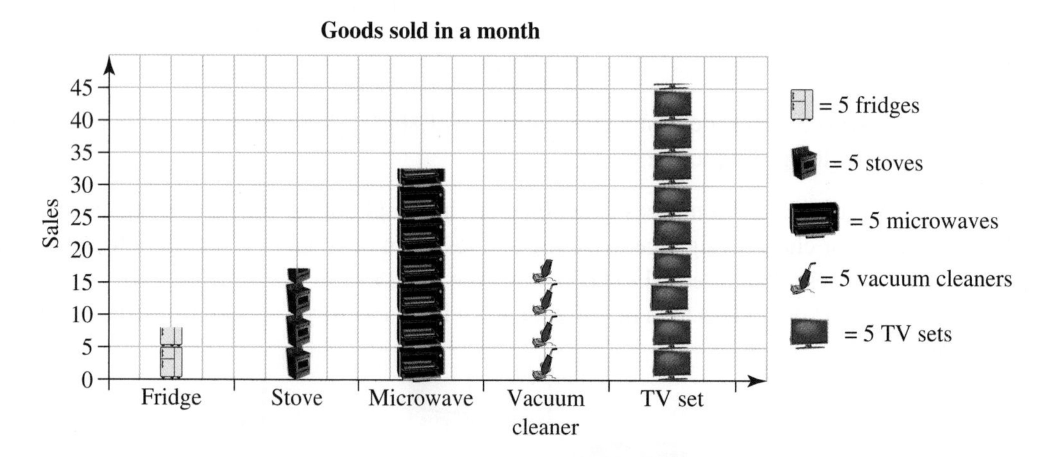

The sales team earn $50 commission on the sale of every fridge, stove, microwave and vacuum, and $25 commission on the sale of every TV.
Determine the total monthly commission.

11. The two-way table shown displays the extracurricular activities that 90 Year 11 students undertake outside school hours.

		Other		
		Singing	Playing an instrument	Total
Sport played	Soccer	4	12	16
	Netball	23	17	40
	Basketball	8	26	34
	Total	35	55	90

a. Identify the number of students who play soccer and are learning to sing.
b. Identify the number of students who play basketball.
c. Calculate the percentage of students who play both an instrument and basketball, correct to 1 decimal place.

12. Construct the two-way table from question 11, adding 30 more students who are learning to dance in their spare time. Two of these students play soccer, 15 play netball and 13 play basketball.

13. The line graphs shown display the percentages of water stored at Samson Brook during 2015, 2016, 2017 and 2018.

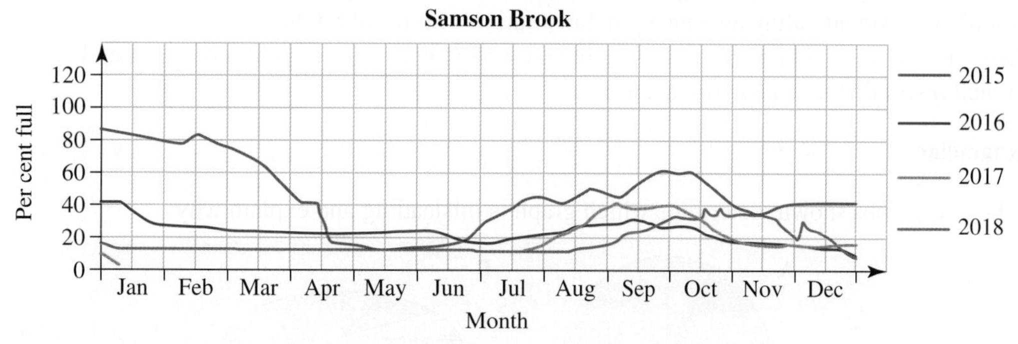

Source: Adapted from Australian Government Bureau of Meteorology http://water.bom.gov.au/waterstorage/awris/#urn:bom.gov.au:awris:common:codelist:feature:samsonbrook

a. Identify the month with the highest percentage of stored water in the four years.
b. Determine in what months the amount of water stored in 2015 was the same as the amount of water stored in 2018.
c. Determine between what months the water stored in 2017 was greater than the water stored in 2018.
d. Identify which year had the highest variation in the percentage of the water stored.

Use the following data collected from the census for questions 14 and 15.

The table shown details age groups and education details for residents of Townsville.

	Males	Females	Persons
Total persons	48 396	47 068	95 464
Age groups			
0–4 years	3 003	2 850	5 853
5–14 years	6 415	5 978	12 393
15–19 years	3 749	3 819	7 568
20–24 years	4 783	4 266	9 049
25–34 years	7 306	6 900	14 206
35–44 years	6 873	6 846	13 719
45–54 years	6 630	6 295	12 925
55–64 years	4 919	4 422	9 341
65–74 years	2 723	2 797	5 520
75–84 years	1 576	2 073	3 649
85 years and over	419	822	1 241
Age of persons attending an educational institution			
0–4 years	357	344	701
5–14 years	5 662	5 293	10 955
15–19 years	2 171	2 623	4 794
20–24 years	1 054	1 487	2 541
25 years and over	1 351	2 219	3 570

Source: Australian Bureau of Statistics, Census of Population and Housing: Townsville.

14. Construct a two-way table displaying the number of male and female 15–19-year-olds in Townsville compared with all other age groups there. Then, determine if the percentage of males who are 15–19 years of age is greater than the percentage of females who are 15–19 years old. Provide calculations to support your answers to each of these.

15. Construct a two-way table displaying males and females of 15–19 years 'Attending an educational institution' and 'Not attending an educational institution'. Show all totals.
Consider whether it would be correct to say that only 55% of 15–19-year-old females attend an educational institution. Explain your answer.

Complex familiar

16. From the two graphs shown, determine which graph is misleading and explain why.

A

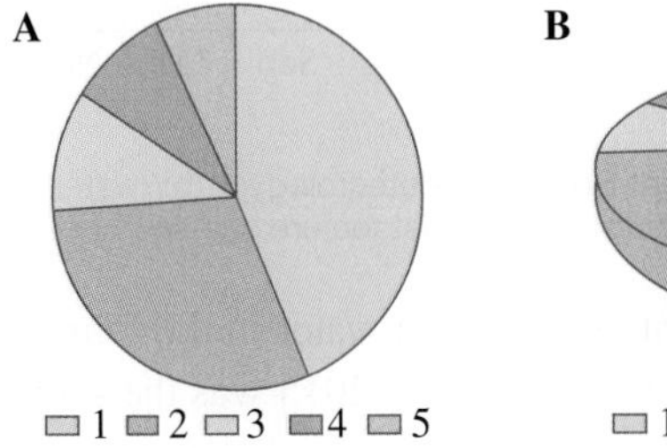

B

1 2 3 4 5

17. Compare the two graphs shown. Identify which graph is misleading and which graph is not. Justify your response.

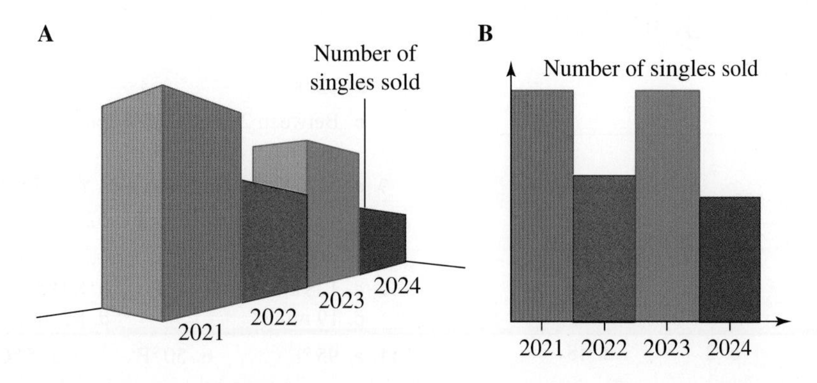

Complex unfamiliar

18. Determine if the graph shown is misleading, and if so, explain why and sketch the graph so it's no longer misleading.

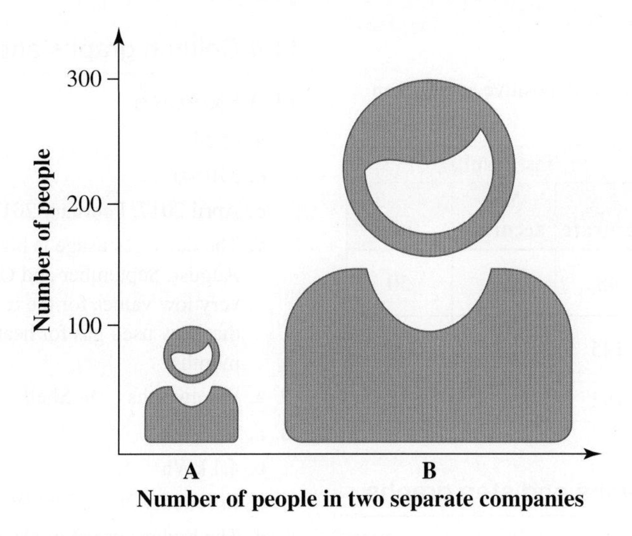

Fully worked solutions for this chapter are available online.

Answers

Chapter 11 Reading, interpreting and using graphs

11.2 Two-way tables

11.2 Exercise

1. a. 652 b. 26 c. 603 d. 335 e. 8.2%
2. a. 96 b. 7 c. 21 d. 68 e. 54.7%
3. a. 1000 b. 75
4. 39
5. a. 35 b. 85 c. 2 d. 48
6. a. 90.9%
   b. 5.1%
   c. 94%
   d. Sample responses are available in the worked solutions in the online resources.
7. 90.5%
8. 95%
9. The test has a greater accuracy with positive results than with negative results.
10.

		Test results		
		Accurate	Not accurate	Total
Bags	With prohibited items	48	2	50
	With no prohibited items	145	5	150
	Total	193	7	200

11.3 Line graphs, pie graphs and step graphs

11.3 Exercise

1. a. 29 °C
   b. 19 °C
   c. The temperature is decreasing during 12:00–7:00 am, 9:00–11:00 am, 12:00–2:00 pm and 4:00–11:00 pm.
   d. 27 °C
   e. 7 pm
2. a. Year 12
   b. 400 mm
   c. Years 3–5, years 6–8 and years 10–12
   d. Years 5 and 8
   e. Year 3, 330 mm
3. a. 24th, 10.6 °C d. 11th, 12th and 28th
   b. 23rd, 19.6 °C e. 4 days
   c. 2 °C
4. a. August, −15.5 °C b. January, 2.5 °C
   c. 18 °C d. November
   e. 5
5. a. Lilies b. Gladioli c. 33
6. a. Car b. Cycle c. 675
7. a. $14 b. $24 c. $26
8. a. $1.50
   b. $2.00
   c. Between 2 hours and 4 hours, not including 2 hours exactly
9. a. 51% Year 7 boys and 49% Year 7 girls
   b. 47.5% Year 12 boys and 52.5% Year 12 girls
   c. Year 10
10. a. 128 km b. 80 km
    c. 19 miles d. 87 miles
11. a. 95 °F b. 50 °F c. 5 °C d. 40 °C
12. a. Year 16, 15.1%
    b. Year 20, 17.0%
    c. 0.2%
    d. From Year 9 to Year 10
    e. Year 6

11.4 Column graphs and picture graphs

11.4 Exercise

1. a. 95 MJ
   b. 370 MJ
   c. April 2017, February 2017, or January 2018
   d. The daily gas usage is higher in the winter months, July, August, September and October. This usage decreases to very low values for the rest of the year. This could mean that they used gas for heating the house during the winter months.
2. a. Woolworths b. Shell c. Caltex
3. a. 4.9 kWh
   b. 4.1 kWh
   c. In January 2018, 2.9 kWh
   d. The highest usage of electricity is in the period April–June 2018. The lowest usage of electricity is in January 2018.
4. a. 4%
   b. United Kingdom
   c. 7%
5. NSW 6 900 000; VIC 5 200 000; QLD 4 180 000; SA 1 700 000; WA 2 100 000; TAS 400 000; NT 190 000; ACT 350 000.
6. a. 70 MJ b. 11th, 90 MJ c. 31st, 58 MJ
7. a. $400 000 b. $200 000 c. $50 000
8. a. 500 000 approx.
   b. 4 800 000 approx.
   c. 24 200 000
9. 110
10. a. 12 b. 48

11.5 Misleading graphs [complex]

11.5 Exercise

1. Sample responses can be found in the worked solutions in the online resources.
2. The horizontal axis uses the same division for 5- and 7-year periods.
3. a. Sample responses can be found in the worked solutions in the online resources.
   b. No
4. a. Student-to-teacher ratios have improved slightly.
   b. Country schools have smaller class sizes.
5. Sample responses can be found in the worked solutions in the online resources.
6. Sample responses can be found in the worked solutions in the online resources.
7. a. The claim is accurate enough in the context ($86 879 m actually).
   b. State and local governments. The stated 25% is correct (rounded up from 24.9%).
   c. Non-government organisations. The stated 30% is rounded down from 32.2%. The percentages being quoted seem to be rounded to the nearest 5%.
   d. The quoted percentage (45%) has been rounded up from 42.9%. This could be considered misleading in some contexts.
   e. i. 162°, 90°, 108°
      ii. 154°, 90°, 116°
      iii. 154°, 78°, 128°. Even though the pie graph gives a rough picture of the relative contributions of the three sectors, it has not been carefully drawn.
8. Sample responses can be found in the worked solutions in the online resources.
9. The graph is misleading since the ice-cream picture is larger than the picture of the fries. More fries were sold than ice-cream cones.
10. The graph is misleading due to the vertical axis not starting at zero.

11.6 Review

11.6 Exercise

1. 12
2. 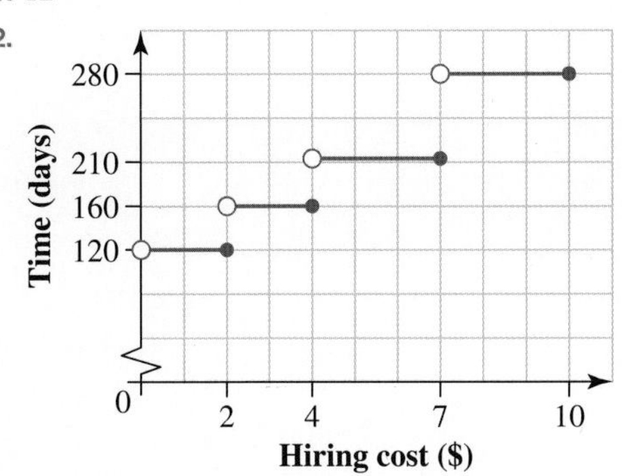

3. 15 June
4. $6
5. Women aged 20–54 with the youngest child aged 10–14
6. 20 people
7. 21, 10, 15, 6, 8
8. 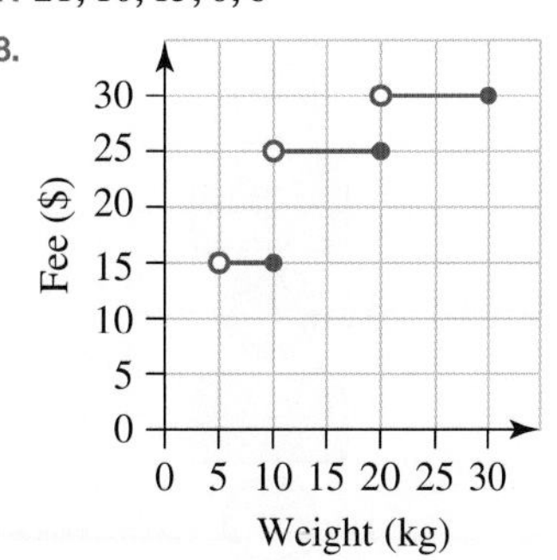

9. a. Piano
   b. Clarinet
   c. 87
10. $5000
11. a. 4
    b. 34
    c. 28.9%
12.

		Other			
		Singing	Playing an instrument	Learning to dance	Total
Sport played	Soccer	4	12	2	18
	Netball	23	17	15	55
	Basketball	8	26	13	47
	Total	35	55	30	120

13. a. January
    b. May and November
    c. Between July and October
    d. 2015
14.

	Male	Female	Total
15–19 years	3749	3819	7568
All other ages	44 647	43 249	87 896
Total	48 396	47 068	95 464

No — 7.7% for males and 8.1% for females

15.

		Gender		
		Male	Female	Total
Educational institution	Attending	2171	2623	4794
	Not attending	1578	1196	2774
	Total	3749	3819	7568

No — 69% attend

16. Graph B is misleading since it is a 3D graph.
17. Graph A is misleading.

18. This graph is misleading since you may look at the area of the two shapes and not just the height of the shapes.

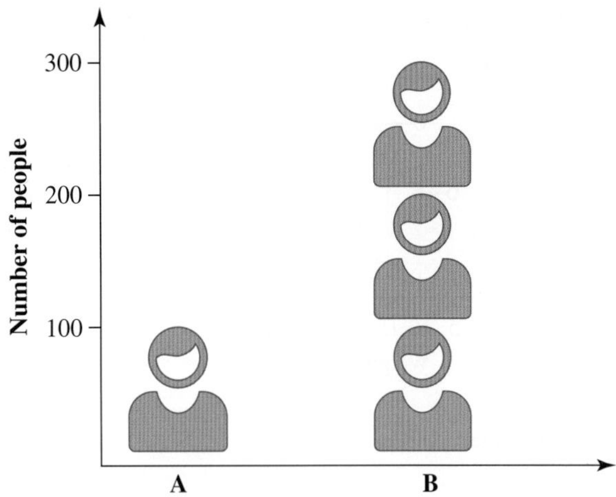

UNIT 2 Data and travel
TOPIC 2 Graphs

12 Drawing and using graphs

LESSON SEQUENCE

Fully worked solutions for this chapter are available online.

on Resources

 **Solutions** Solutions — Chapter 12 (sol-1265)

Digital documents Learning matrix — Chapter 12 (doc-41774)
Quick quizzes — Chapter 12 (doc-41775)
Chapter summary — Chapter 12 (doc-41776)

LESSON
12.1 Overview

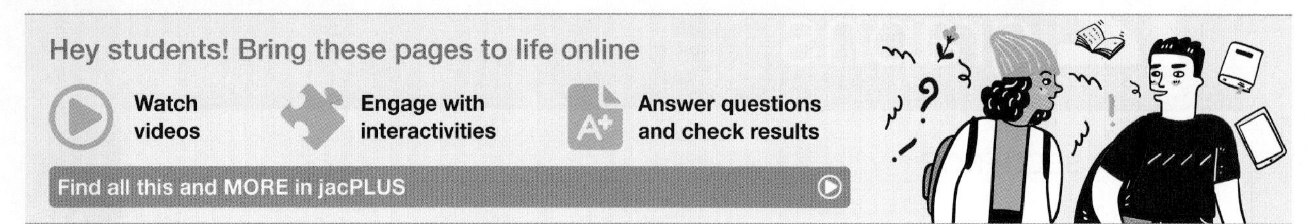

12.1.1 Introduction

In today's data-driven world, it's important to be able to visually represent information effectively. Knowing how to draw and use various types of graphs, such as line graphs, column graphs and pie charts, helps us clearly and precisely show complex data. These graphical representations offer a succinct way to analyse trends, patterns and relationships within datasets.

Moreover, advances in technology have improved the process of graphing, providing sophisticated tools and software for the creation and manipulation of graphs.

Using technology to graph not only improves efficiency but also enables individuals and organisations to extract actionable insights from data, which makes decision-making easier and drives innovation.

12.1.2 Syllabus links

Lesson	Lesson title	Syllabus links
12.2	**Drawing and using line graphs [complex]**	◯ Draw a line graph to represent any data that demonstrates a continuous change [complex]. ◯ Interpret graphs in practical situations [complex]. ◯ Draw graphs from given data to represent practical situations [complex]. ◯ Interpret the point of intersection and other important features (x- and y-intercepts) of given graphs of two linear functions drawn from practical contexts [complex].
12.3	**Drawing and using column graphs and pie graphs**	◯ Determine which type of graph is best used to display a dataset. ◯ Use graphs in practical situations.
12.4	**Graphing with technology [complex]**	◯ Use spreadsheeting software to tabulate and graph data [complex]. ◯ Use graphs in practical situations.

LESSON
12.2 Drawing and using line graphs [complex]

SYLLABUS LINKS

- Draw a line graph to represent any data that demonstrates a continuous change [complex].
- Interpret graphs in practical situations [complex].
- Draw graphs from given data to represent practical situations [complex].
- Interpret the point of intersection and other important features (x- and y-intercepts) of given graphs of two linear functions drawn from practical contexts [complex].

12.2.1 Drawing line graphs with numerical continuous data

- In Chapter 6, data was classified into categorical data and numerical data.
- Numerical data is classified as discrete or continuous.
- Line graphs can display the relationship between two numerical continuous variables.
  - The values of the independent variable are represented on the horizontal axis (the x-axis).
  - The values of the dependent variable are represented on the vertical axis (the y-axis).
- Line graphs that represent numerical continuous data are drawn with solid lines between the data points.
- Line graphs clearly show trends in the data.
- Consider the two graphs below.

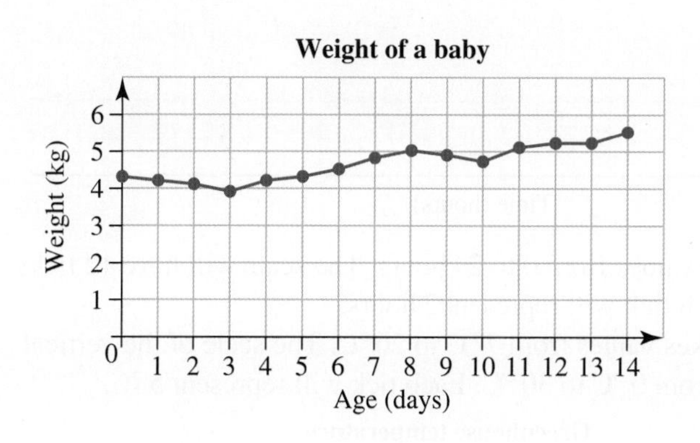

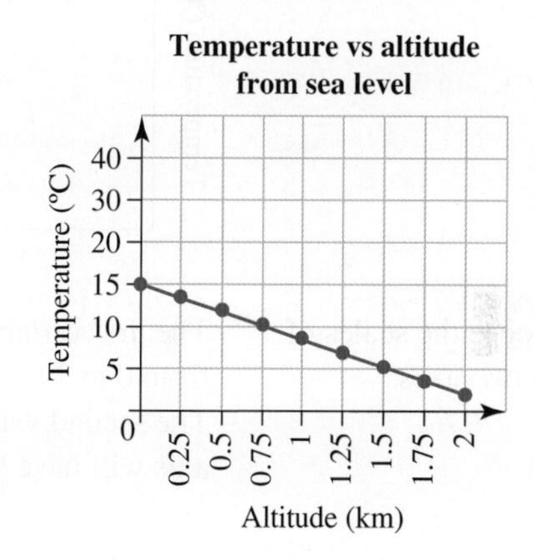

- The first line graph displays the change of the weight of a baby over a period of 14 days. There is an overall upward trend. As age increases, weight increases.
  'Age' is the independent variable and 'weight' is the dependent variable.
- The second line graph displays the change in the temperature of the atmosphere in relationship to altitude (the vertical distance from sea level). There is a downward trend. As the altitude increases, the temperature decreases.
  'Altitude' is the independent variable and 'temperature' is the dependent variable.

WORKED EXAMPLE 1 Sketching a continuous line graph

A greenhouse worker recorded the temperature in the greenhouse every hour for a period of 24 hours. The data were recorded in the table of values shown.
Sketch a continuous line graph for this dataset.

Time, hours	0	1	2	3	4	5	6	7	8	9	10	11	12
Temperature, °C	9	8	7	9	10	12	15	17	20	21	23	25	26

Time, hours	13	14	15	16	17	18	19	20	21	22	23	24
Temperature, °C	24	22	20	17	13	12	10	9	9	11	13	16

THINK	WRITE
1. Draw the two axes and label them with the independent variable and the dependent variable.	The independent variable is the time and the dependent variable is the temperature in the greenhouse. The temperature is the dependent variable because its values depend on the time of reading. The temperature does not affect the time.

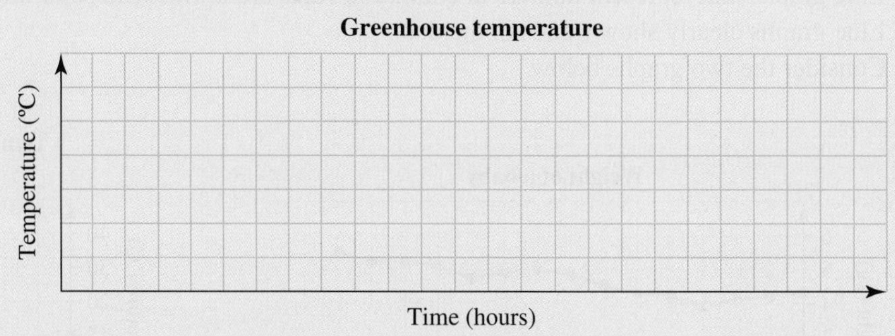

THINK	WRITE
2. Choose the scales of the two axes.	The first variable takes values from 0 to 24 hours. The scale will have 12 ticks from 0 to 24 hours. Each tick will represent 2 hours. The second variable takes values from 7 °C to 26 °C. The scale of the vertical axis will have 6 ticks from 0 °C to 30 °C. Each tick will represent 5 °C.

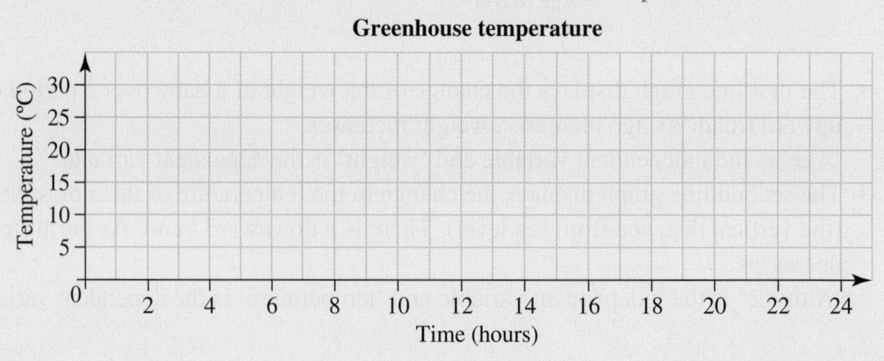

3. Plot the first data point.

The first data point is at time 0 hours, when the temperature is 9 °C.

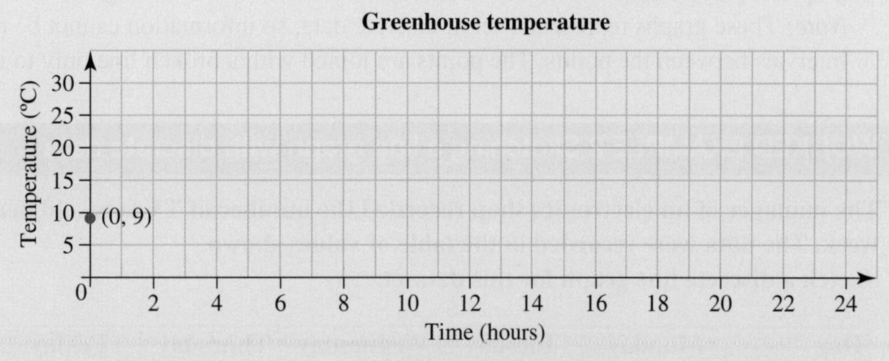

4. Plot the remaining values.

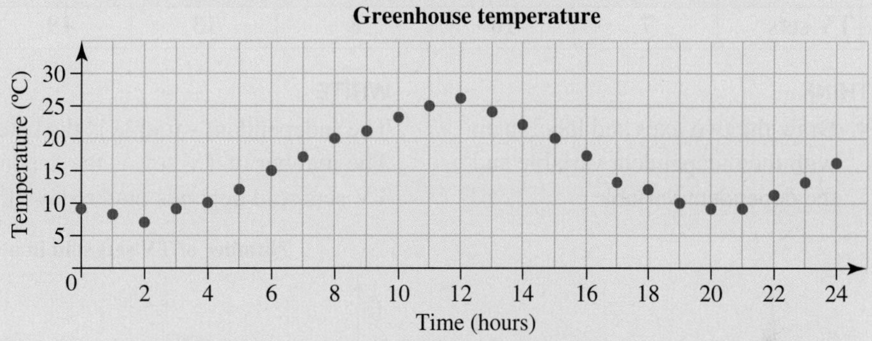

5. Connect all the points with a continuous line.

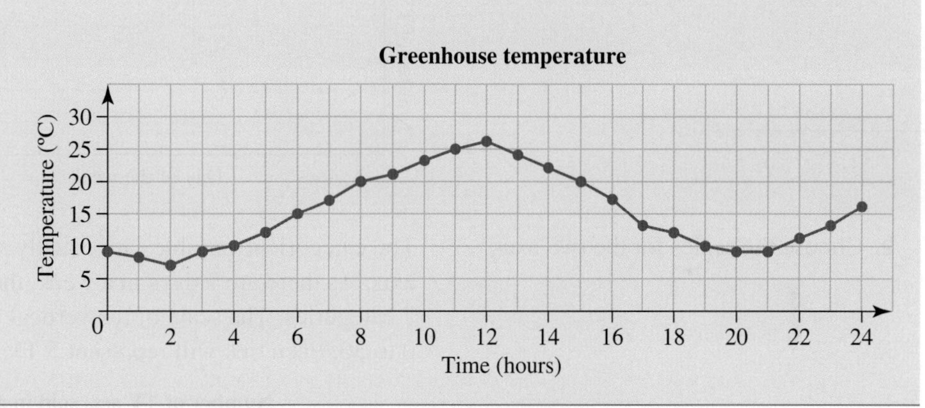

12.2.2 Drawing line graphs with discrete numerical data

- Numerical discrete data, such as the number of marks in a test or the final score in a footy game, can be represented by line graphs showing trends over time.
  Consider the two graphs below.

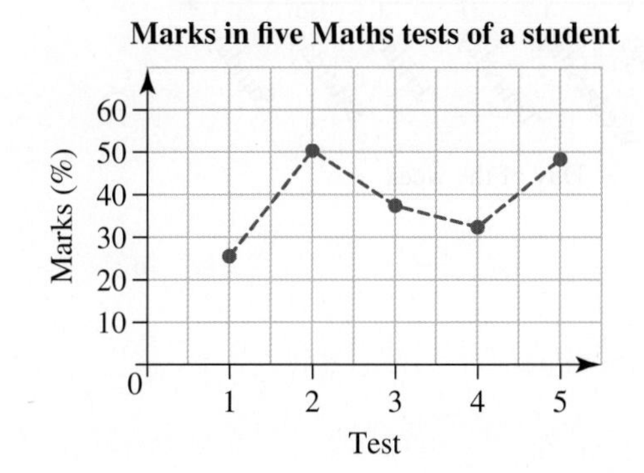

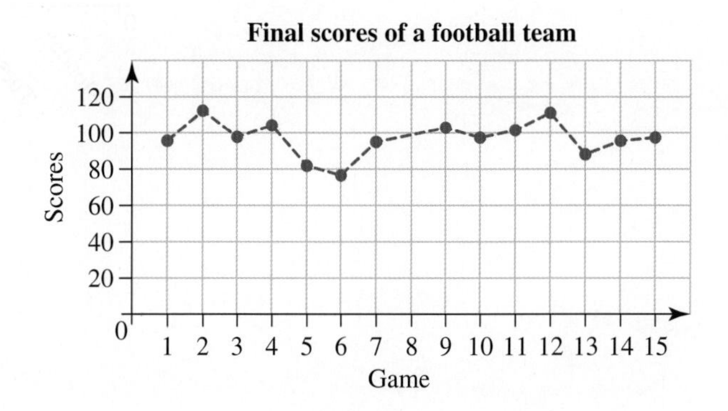

- The first line graph displays the results of the first five Mathematics tests of a student. The second line graph displays the final scores of a football team in 15 consecutive games.
  Note: These graphs represent a set of discrete data, so information cannot be read from the graphs at intervals between the points. The points are joined with a broken line only to illustrate any trend.

WORKED EXAMPLE 2 Sketching a line graph with discrete data

The manager of an electronics shop recorded the number of TV sets sold per day over a period of one week. The data were recorded in the table of values shown.
Sketch a discrete line graph for this dataset.

Day	Monday	Tuesday	Wednesday	Thursday	Friday	Saturday	Sunday
TV sets	7	10	8	13	19	25	18

THINK

1. Draw the two axes and label them with the independent variable and the dependent variable.

WRITE

The independent variable is the categorical data day of the week. The number of TV sets is the dependent variable. The number of TV sets sold depends on the day of the week.

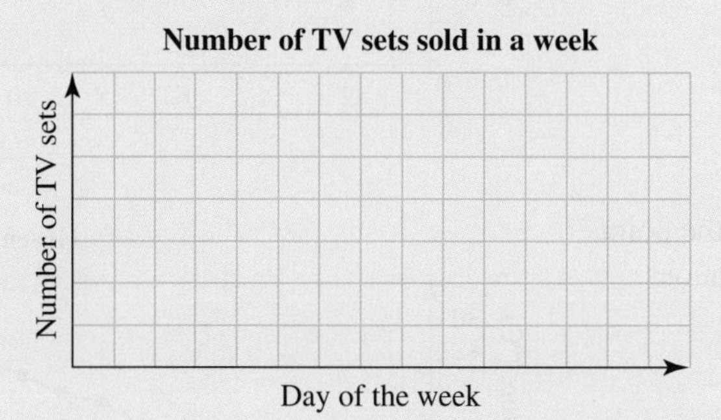

2. Choose the scales for the two axes.

The categorical variables are equally spaced on the horizontal axis. As there are 7 days in a week, the horizontal axis will have 7 categories. The scale of the vertical axis will have 6 ticks from 0 to 30. Each tick will represent 5 TV sets.

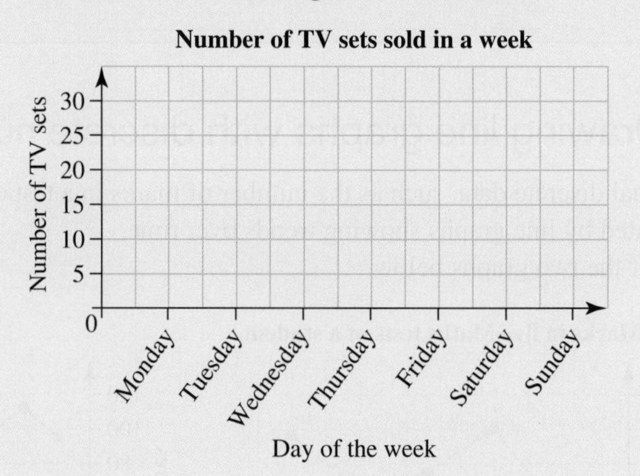

3. Plot the first data point.

The first data point is on Monday, when the store sold 7 TV sets.

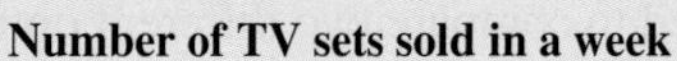

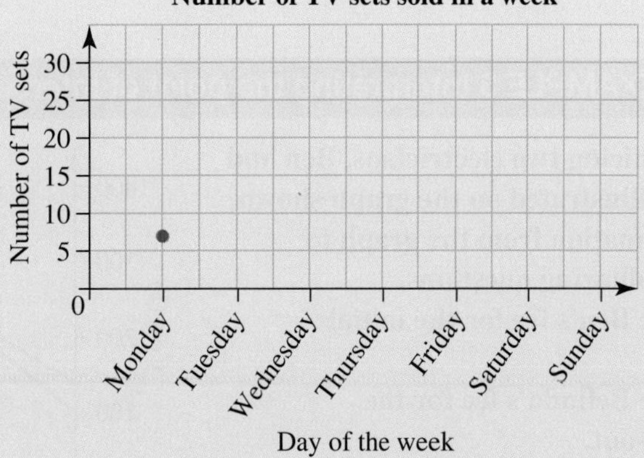

4. Plot the remaining values.

Number of TV sets sold in a week

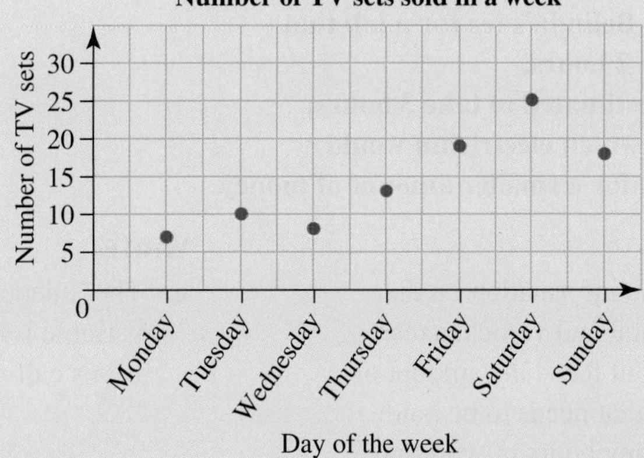

5. Connect all the points with a dotted line because the data is discrete.

Number of TV sets sold in a week

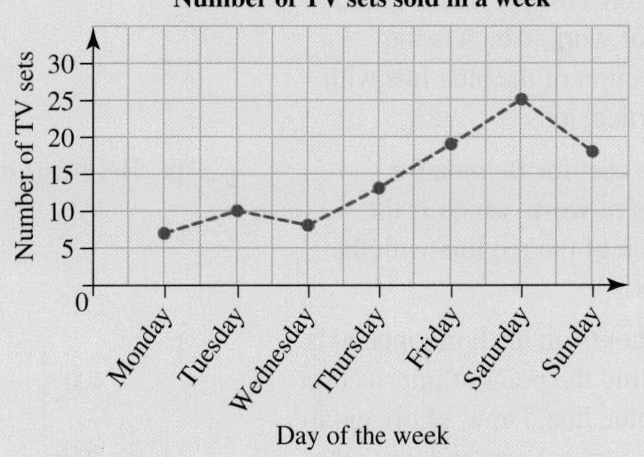

12.2.3 Interpreting the point of intersection of two line graphs

Line graphs can be used to compare two different options related to the same situation.

tlvd-11448

WORKED EXAMPLE 3 Interpreting the point of intersection of two line graphs

The costs of hiring two electricians, Ben and Belinda, are illustrated on the graph shown. Use the information from the graph to answer the following questions.

a. Determine Ben's fee for the initial call-out.

b. Determine Belinda's fee for the initial call-out.

c. Determine Ben's fee for a job that would take 2 hours.

d. Determine Belinda's fee for a job that would take 2 hours.

e. If a job is estimated to take 3 hours, determine which electrician would complete it for a smaller amount of money.

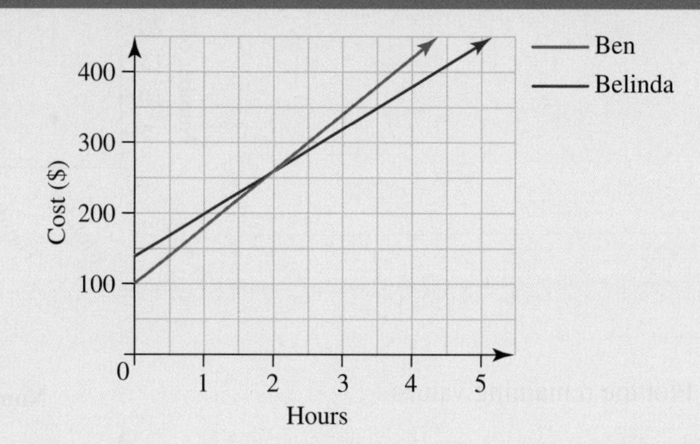

THINK	WRITE
a. 1. Read the information on the horizontal and vertical axes.	a. The independent variable is hours. The dependent variable is cost.
2. A call-out fee is the amount of money that needs to be paid before any hours of work have been completed. Check the cost for Ben for zero hours of work, which is the intersection of the blue line with the vertical axis.	Ben's call-out fee is $100.
b. Check the cost for Belinda for zero hours of work, which is the intersection of the red line with the vertical axis.	b. Belinda's call-out fee is $140.
c. Locate 2 hours on the horizontal axis and then find the point of intersection with the blue line. Draw a horizontal line to the vertical axis and write the cost.	c. 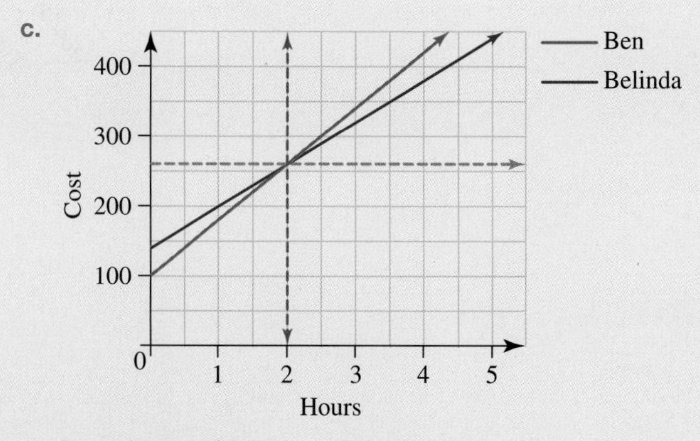 Ben charges $260 for a 2-hour job.

d. Locate 2 hours on the horizontal axis and then find the point of intersection with the red line.	**d.** Belinda also charges $260 for a 2-hour job.
e. 1. Locate 3 hours on the horizontal axis and find the cost for both Ben and Belinda.	**e.** 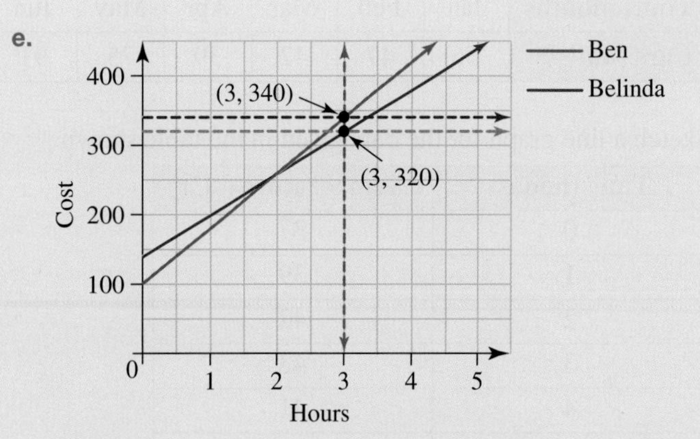
2. Write the answer.	Ben charges $340 for a 3-hour job; Belinda charges $320. Belinda would complete the job for a lower price.

on Resources

 **Digital document** SkillSHEET Distinguishing between types of data (doc-5339)

Exercise 12.2 Drawing and using line graphs [complex]

learnon

12.2 Quick quiz on | 12.2 Exercise

Simple familiar	Complex familiar	Complex unfamiliar
N/A	1, 2, 3, 4, 5, 6, 7, 8, 9, 10, 11, 12, 13, 14	15

These questions are even better in jacPLUS!
- Receive immediate feedback
- Access sample responses
- Track results and progress

Find all this and MORE in jacPLUS

Complex familiar

1. 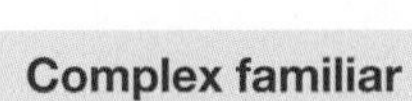 WE1 A mother recorded the yearly height of her son over a period of 10 years. She recorded the data in the table of values given. Sketch a line graph for the data given.

Age, years	0	1	2	3	4	5	6	7	8	9	10
Height, cm	39	58	71	80	92	105	110	112	125	142	156

2. The cooling temperature of a drink placed in the fridge was recorded in the table of values shown. Sketch a line graph for the data given.

Time, min	0	5	10	15	20	25	30	35	40	45	50	55	60
Temperature, °C	30	25	21	18	15	14	12	11	10	10	9	9	9

3. WE2 The manager of a car yard recorded the number of cars sold per month over a period of one year. The data was recorded in the table of values shown.
Sketch a line graph for the data given.

Time, months	Jan	Feb	Mar	Apr	May	Jun	Jul	Aug	Sep	Oct	Nov	Dec
Cars sold	56	47	42	39	75	91	48	53	57	54	68	84

4. Sketch a line graph for the data given in the table shown.

Time (hours)	Temperature (°C)
0	37
1	39
2	42
3	43
4	41
5	38
6	39
7	35
8	32
9	27
10	24

5. The manager of a small business recorded the amount of GST paid every month over a period of one year. The data was recorded in the table of values shown.
Sketch a line graph for the data given.

Time, month	Jan	Feb	Mar	Apr	May	Jun	Jul	Aug	Sep	Oct	Nov	Dec
GST, $1000	7	5	4	6	7	8	7	5	6	9	10	9

6. WE3 The costs of hiring two furniture removal companies, Rent-a-truck and Do-it-today, are illustrated on the graph shown.
Use the information from the graph to answer the following questions.

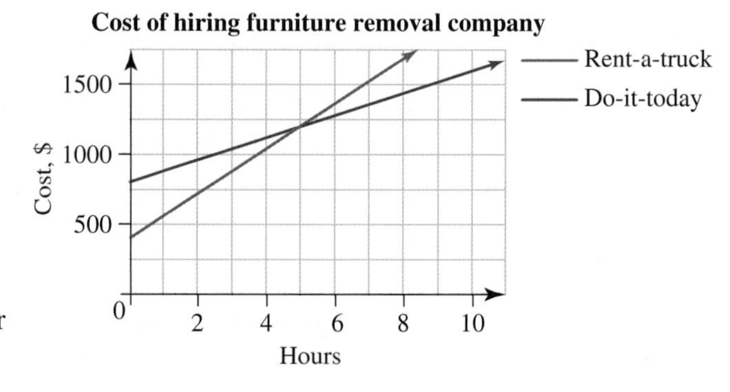

a. Determine how much Rent-a-truck charges for the hire of the truck before adding an hourly charge.
b. Determine how much Do-it-today charges for the hire of the truck before adding an hourly charge.
c. Determine how much Rent-a-truck would charge for a job that takes 5 hours.
d. Determine how much Do-it-today would charge for a job that takes 5 hours.
e. If a job was estimated to take 7 hours, determine which company would complete it for the least amount of money.

7. The average rainfall over a week is recorded in the table shown. Sketch a line graph to represent this dataset.

Day	1	2	3	4	5	6	7
Rainfall (mm)	130	155	200	50	0	125	100

8. The population of Australia from 1860 to 2010 is recorded in the table given.

a. Sketch a line graph for this set of data.

Year	1860	1870	1880	1890	1900	1910	1920	1930
Population (million)	1.2	1.6	2.2	3.2	3.8	4.4	5.4	6.5

Year	1940	1950	1960	1970	1980	1990	2000	2010
Population (million)	7.0	8.3	10.4	12.7	14.8	17.2	19.3	22.3

b. Calculate the population growth in Australia for each decade from 1860 to 2010. Sketch a new line graph for the two variables 'decade' and 'population growth'.

9. The share prices of two stocks over five days are recorded in the table shown. On the same set of axes, sketch a line graph to represent these datasets.

Day	1	2	3	4	5
Stock A ($)	1.21	1.25	1.20	1.22	1.29
Stock B ($)	0.59	0.65	0.64	0.61	0.62

10. A manager has recorded the number of people in the shop every hour for seven hours. The table shown displays this data. Sketch a line graph for this dataset.

Time (hours)	1	2	3	4	5	6	7
People	5	16	24	28	26	17	3

11. Two full water tanks, A and B, have developed cracks and are leaking. There has not been any rain to replenish them. Their water storage information is illustrated in the graph below.
Using the graph, answer the following questions.

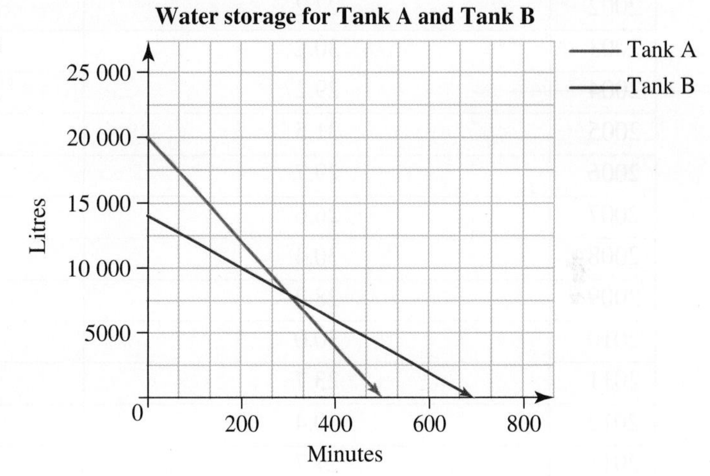

a. Identify the amount of water in Tank A when it is full.

b. Identify the amount of water in Tank B when it is full.

c. Identify the time it took to empty Tank A and Tank B.

d. Identify at what time Tank A and Tank B contained the same amount of water.

12. Five students' Mathematics test results are recorded as percentages in the table shown.

Student	Test 1	Test 2	Test 3	Test 4	Test 5	Test 6	Test 7
Helen	94	85	87	86	99	91	95
Ahim	59	49	52	59	58	61	65
Lilly	63	56	58	68	71	70	72
Scott	88	67	77	82	81	82	85
Elaine	77	62	64	72	75	82	79

On the same set of axes, sketch line graphs to represent the results of the five students.

13. The monthly mean maximum temperatures for the month of December recorded at the Perth Metro weather station from 2005 to 2023 are displayed in the table shown. Construct a line graph to represent this dataset.

Year	Temperature, °C
2005	29.8
2006	27.7
2007	27.2
2008	30.2
2009	29.2
2010	31.5
2011	29.7
2012	26.6
2013	30.4
2014	28.2

Year	Temperature, °C
2015	30.0
2016	23.7
2017	30.4
2018	27.7
2019	27.8
2020	30.8
2021	29.3
2022	30.5
2023	31.2

14. The monthly mean maximum temperatures for December in Perth and Adelaide (Kent Town) from 2000 to 2018 are displayed in the table shown. Construct a line graph to represent these datasets.

Year	Temperature, °C Perth Metro	Temperature, °C Adelaide (Kent Town)
2000	29.8	28.9
2001	27.7	25.7
2002	27.2	25.7
2003	30.2	26.7
2004	29.2	27.6
2005	31.5	25.7
2006	29.7	28.3
2007	26.6	23.3
2008	30.4	27.8
2009	28.2	29.0
2010	30.0	27.3
2011	23.7	27.3
2012	30.4	28.5
2013	27.7	28.9
2014	27.8	25.5
2015	30.8	28.5
2016	29.3	26.7
2017	30.5	27.9
2018	31.2	28.3

Complex unfamiliar

15. Water is boiled to 100 °C. The temperatures are recorded in a table, but the temperature at 60 seconds is missing. Create a line graph of the data and use it to predict the temperature at 60 seconds.

Time, seconds	15	30	45	60	75	90	105	120
Temperature, °C	18	21	26		49	61	88	100

Fully worked solutions for this chapter are available online.

LESSON
12.3 Drawing and using column graphs and pie graphs

SYLLABUS LINKS

- Determine which type of graph is best used to display a dataset.
- Use graphs in practical situations.

Source: Essential Mathematics Senior Syllabus 2024 © State of Queensland (QCAA) 2024; licensed under CC BY 4.0.

12.3.1 Choosing the best graph to display a dataset

- The type of graph we use to display data is determined by the type of data being represented.
- In lesson 12.2, we studied line graphs. These are used when considering a relationship between two sets of numerical data.
- A pie graph or column graph is typically used when considering a set of categorical data.
- A pie graph is visually appealing and useful when you have a small dataset and wish to consider each category as a percentage of the whole.
- If there is a large number of categories, a column graph is more appropriate.

WORKED EXAMPLE 4 Selecting the best type of graph to display a dataset

For the following datasets, select the best type of graph (line, column or pie) to display the data. Explain your answer.

a. **Average monthly rainfall for 12 months**

Month	1	2	3	4	5	6	7	8	9	10	11	12
Rainfall (mm)	38	33	32	25	20	18	10	17	34	37	35	22

b. **Sport preference for a group of young people**

Sport	Rugby	Soccer	Cricket	Netball	AFL	Golf
%	35	20	15	8	12	10

c. Most popular brand of car for a group of 353 people

Car brand	Frequency
Honda	25
Toyota	32
Audi	18
Mercedes	6
Kia	41
Hyundai	28
Mazda	13
Tesla	25
Volkswagen	15
BMW	8
Subaru	34
Ford	12
Nissan	28
Ferrari	12
Ford	30
MG	16
Volvo	10

THINK	WRITE
a. This dataset is made up of two sets of numerical data, month and average rainfall.	a. This dataset contains two sets of numerical data, so a line graph would best display the data.
b. This dataset is categorical with a percentage value for each category. There are only 6 categories.	b. There are only 6 categories in this set of categorical data, so a pie graph would best display of the data.
c. This dataset is categorical with the frequency of each category. There are 17 different categories.	c. There are 17 categories in this set of data. This is too many for a pie chart, as the slice of each piece of pie would be difficult to distinguish. A column graph would be the best way to display this data.

12.3.2 Drawing column graphs

- Column graphs consist of vertical or horizontal bars of equal width.
- The frequency is measured by the height of the column.
- For the vertical column graphs, the frequency is always shown on the vertical axis and the categories of the data are shown on the horizontal axis.
- Both axes have to be clearly labelled, and appropriate and accurate scales are required.
- The title should explicitly state what the column graph represents.

- Some examples and their features are shown below.

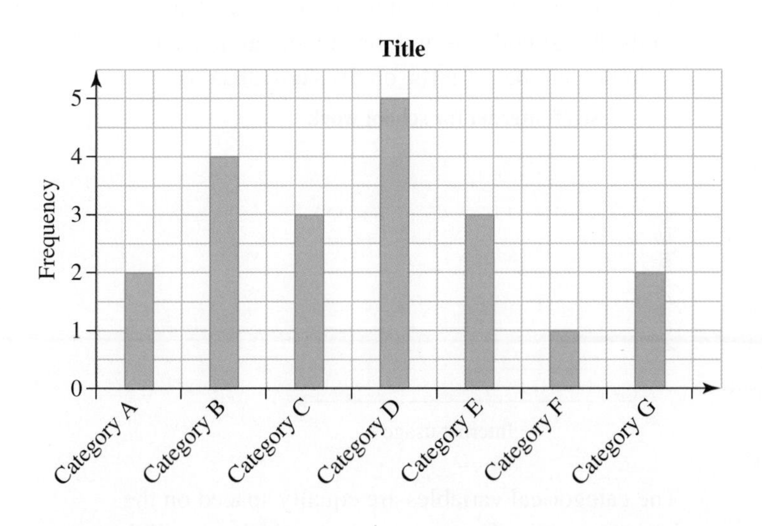

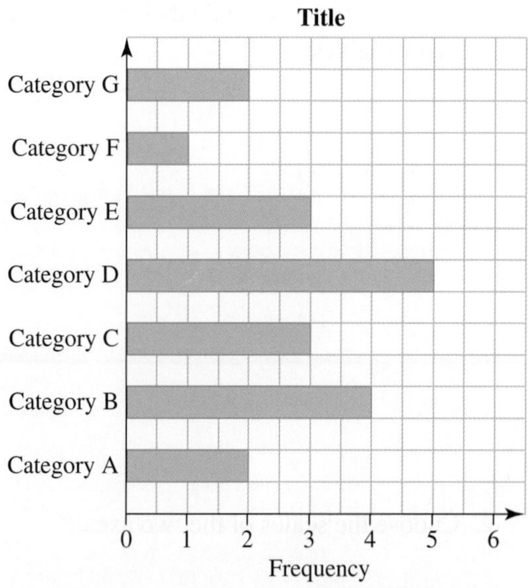

Features

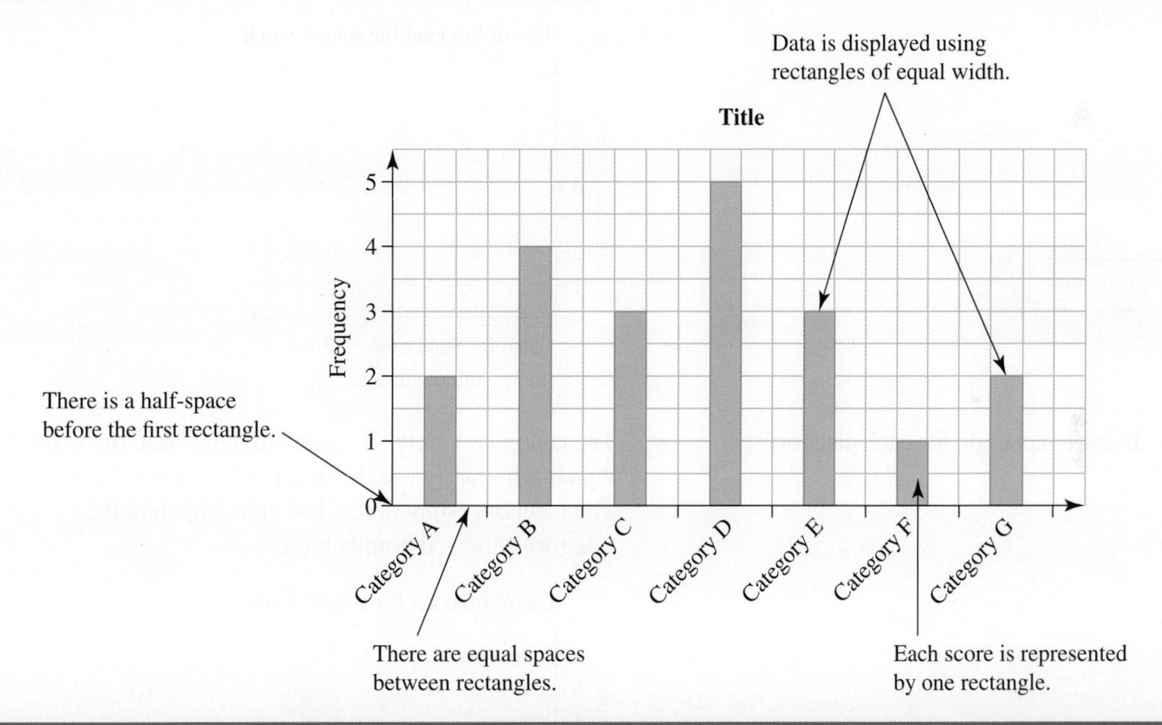

WORKED EXAMPLE 5 Sketching a column graph

The data shown are part of a random sample of 25 Year 11 students. The answers were related to how often the students use the internet to do research for school work.
Construct a column graph displaying the data for girls only.

Use of internet for school work	Gender		Total
	Girls	Boys	
Rarely	2	8	10
Sometimes	4	4	8
Often	6	1	7
Total	12	13	25

THINK	WRITE
1. Draw the two axes and label them with the two variables.	The categorical variable, internet usage, is displayed on the horizontal axis and the numerical discrete variable, number of girls, on the vertical axis.

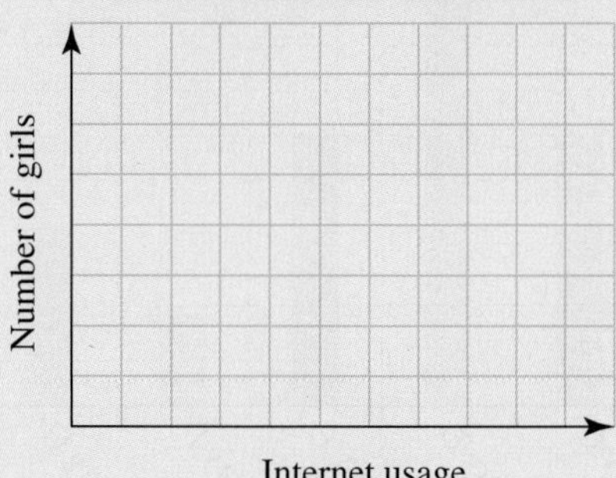

2. Choose the scales of the two axes.	The categorical variables are equally spaced on the horizontal axis. The scale of the vertical axis will have 6 ticks from 0 to 6.

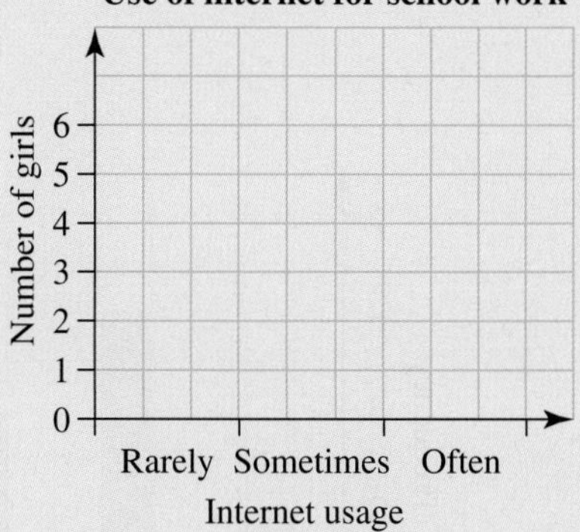

3. Draw a rectangle for each category.	The category 'rarely' is 2 units high because there are 2 girls who rarely use the internet for school work. The category 'sometimes' is 4 units high and the category 'often' is 6 units high.

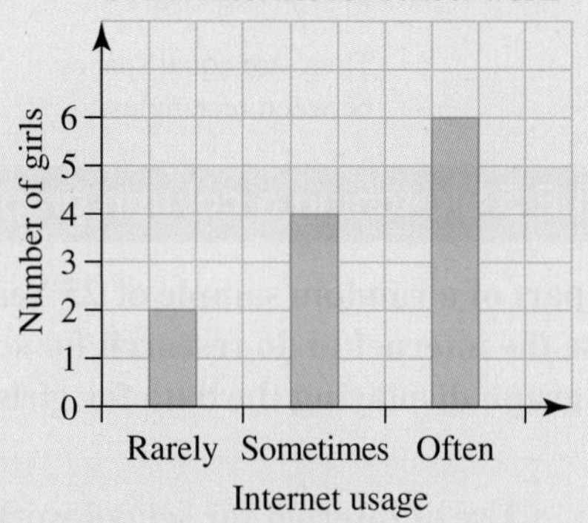

on Resources

Interactivities	Column and bar graphs 1 (int-4058) Column and bar graphs 2 (int-4059)
Digital documents	SkillSHEET Reading bar graphs (doc-5340) SpreadSHEET Construct column graphs (doc-3441) SpreadSHEET Construct bar graphs (doc-3442)

12.3.3 Drawing pie graphs

- **Pie graphs** are circles divided into sectors. Each sector represents one item of the data.
- The pie chart shown has four sectors: Net wage, Tax, Medicare levy and Superannuation.
- Each sector represents the percentage of each item from the gross wage. This pie chart is based on a gross wage (before tax) of $65 000 per year.

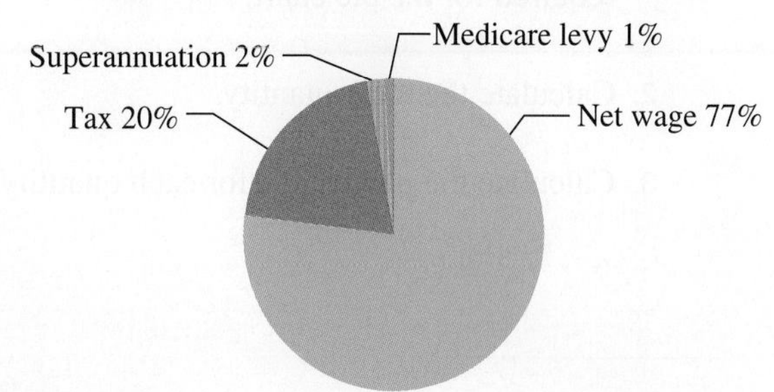

%	Angle
1% is the Medicare levy: $1\% \text{ of } \$65\,000 = \frac{1}{100} \times 65\,000$ $= \$650$	The corresponding angle on the circle is $1\% \text{ of } 360° = \frac{1}{100} \times 360°$ $= 3.6°$
2% is the superannuation contribution: $2\% = \frac{2}{100} \times 65\,000$ $= \$1300$	The corresponding angle on the circle is $2\% \text{ of } 360° = \frac{2}{100} \times 360°$ $= 7.2°$
20% is the tax: $20\% = \frac{20}{100} \times 65\,000$ $= \$13\,000$	The corresponding angle on the circle is $20\% \text{ of } 360° = \frac{20}{100} \times 360°$ $= 72°$
77% is the net wage (after tax): $77\% = \frac{77}{100} \times 65\,000$ $= \$50\,050$	The corresponding angle on the circle is $77\% \text{ of } 360° = \frac{77}{100} \times 360°$ $= 277.2°$

Features of pie graphs

1. Each sector has to be coloured differently.
2. All labels and other writing have to be horizontal and at equal distance from the circle.

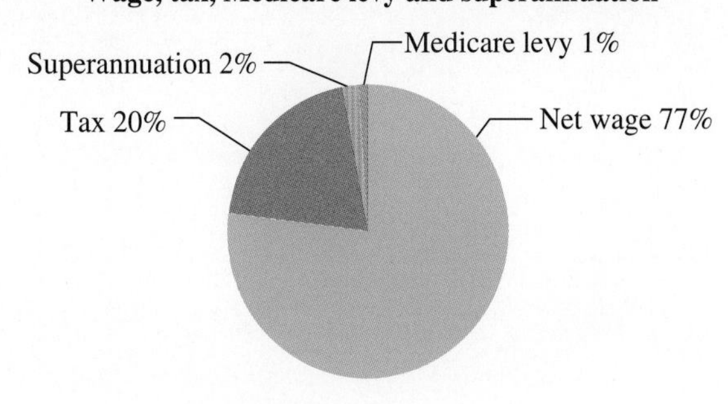

tlvd-11449

WORKED EXAMPLE 6 Sketching a pie graph

A recipe for baked beans contains the following ingredients:
300 g dry beans, 50 g onion, 750 g diced tomatoes, 30 g mustard and 870 g smoked hock

a. **Construct a pie chart to represent these ingredients.**
b. **Calculate the amount of dry beans required to make 100 g of this dish.**
c. **Calculate the amount of dry beans required for a 250-g serve of this dish.**

THINK	WRITE
a. 1. Determine the number of sectors required for the pie chart.	a. The pie chart requires 5 sectors, one for each ingredient: dry beans, onion, diced tomatoes, mustard, smoked hock.
2. Calculate the total quantity.	$300 + 50 + 750 + 30 + 870 = 2000$ g
3. Calculate the percentage for each quantity.	Dry beans: $\frac{300}{2000} \times 100 = 15\%$ Onion: $\frac{50}{2000} \times 100 = 2.5\%$ Diced tomatoes: $\frac{750}{2000} \times 100 = 37.5\%$ Mustard: $\frac{30}{2000} \times 100 = 1.5\%$ Smoked hock: $\frac{870}{2000} \times 100 = 43.5\%$
4. Check that the percentage sum is 100%.	$15\% + 2.5\% + 37.5\% + 1.5\% + 43.5\% = 100\%$
5. Calculate the angles for each sector out of the 360°. To calculate the angle for each sector, use the formula: angle = $x\% \times 360°$ $= \frac{x}{100} \times 360°$	Dry beans: $\frac{15}{100} \times 360° = 54°$ Onion: $\frac{2.5}{100} \times 360° = 9°$ Diced tomatoes: $\frac{37.5}{100} \times 360° = 135°$ Mustard: $\frac{1.5}{100} \times 360° = 5.4°$ Smoked hock: $\frac{43.5}{100} \times 360° = 156.6°$
6. Check that the sum of the angles is 360°.	$54 + 9 + 135 + 5.4 + 156.6 = 360°$
7. Draw a circle for the pie chart and a vertical radius from the centre of the circle to the top of the circle.	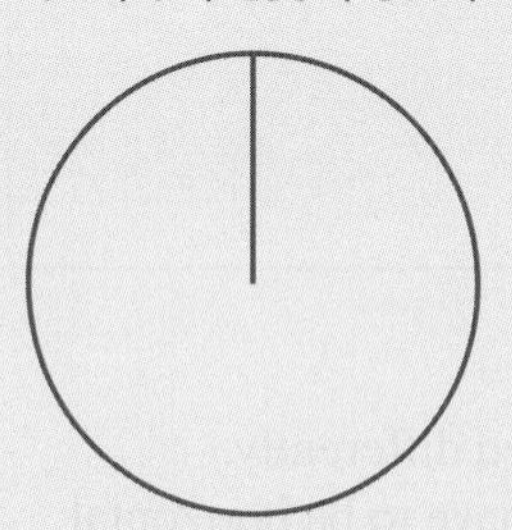

8. Draw the first sector of the pie graph.

The first ingredient is dry beans, with an angle of 54°. Measure 54° from the vertical line in a clockwise direction.

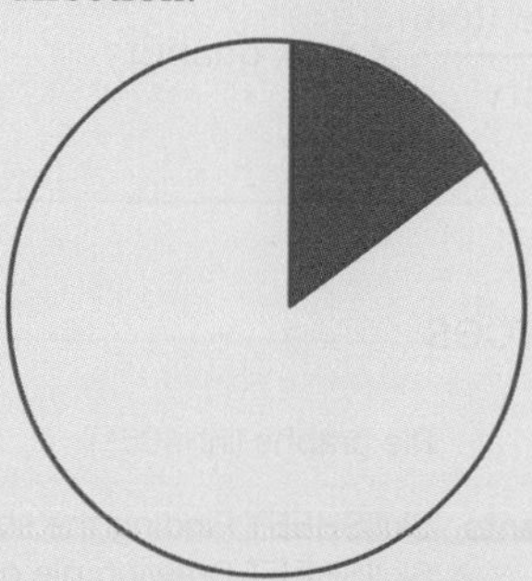

9. Draw the second sector of the pie graph.

The second ingredient is onion, with an angle of 9°. Measure 9° from the right-hand side of the sector in a clockwise direction.

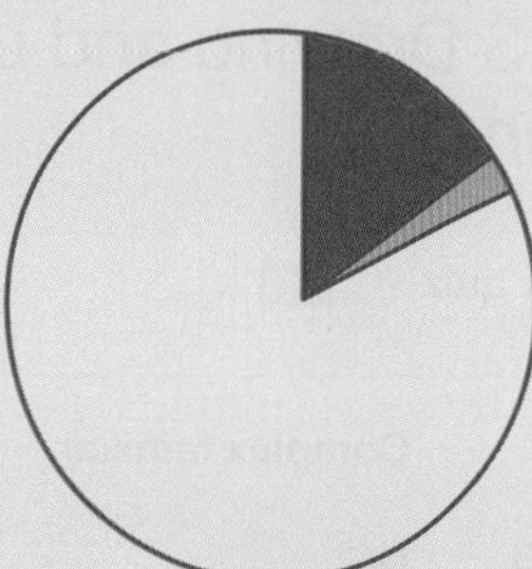

10. Draw the remaining sectors of the pie graph.

Continue to draw the remaining sectors, going in a clockwise direction around the circle.

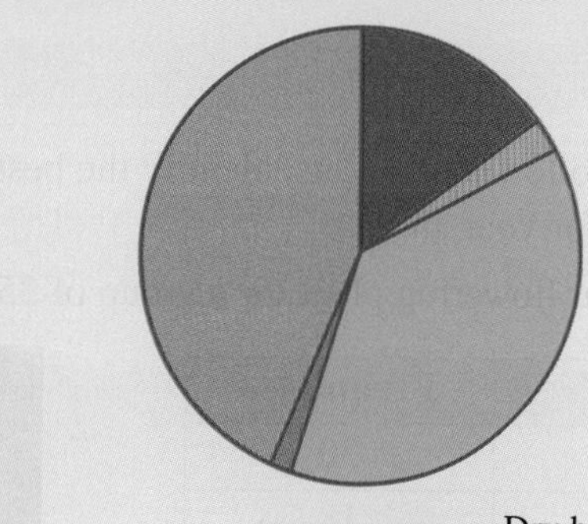

11. Label all sectors.

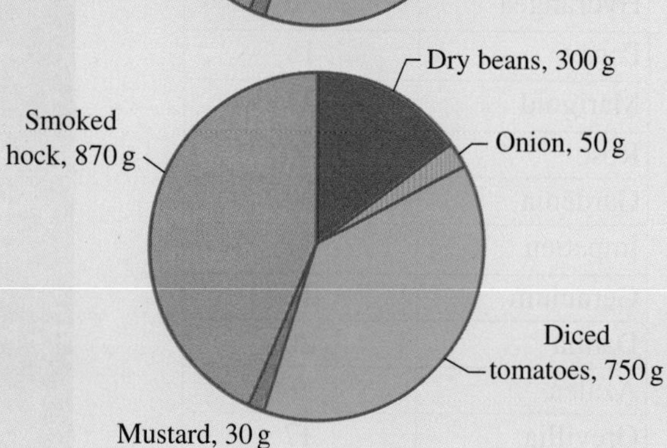

b. Calculate the quantity required.
Use the formula:

$$\frac{\text{quantity of the item}}{\text{total quantity}} \times \text{new quantity}$$

b. Dry beans:

$$\frac{300}{2000} \times 100$$

$= 15$ g

c. Calculate the quantity required.
Use the formula:
$$\frac{\text{quantity of the item}}{\text{total quantity}} \times \text{new quantity}$$

c. Dry beans:
$$\frac{300}{2000} \times 250$$
$$= 37.5 \text{ g}$$

Resources

Interactivity Pie graphs (int-4061)

Digital documents SkillSHEET Finding the size of a sector (doc-29499)
SkillSHEET Drawing pie graphs (doc-30356)

Exercise 12.3 Drawing and using column graphs and pie graphs

learnon

12.3 Quick quiz 

12.3 Exercise

Simple familiar	Complex familiar	Complex unfamiliar
1, 2, 3, 4, 5, 6, 7, 8, 9, 10, 11, 12, 13	N/A	N/A

These questions are even better in jacPLUS!
- Receive immediate feedback
- Access sample responses
- Track results and progress

Find all this and MORE in jacPLUS

Simple familiar

1. **WE4** For the following datasets, identify the best type of graph (line, column or pie) to display the data. Give a reason for your answer.

   a. Most popular flowering plant for a group of 350 people

Flower	Frequency
Hydrangea	30
Pansy	18
Marigold	23
Rose	56
Gardenia	19
Impatien	31
Geranium	10
Dahlia	20
Azalea	9
Grevillia	17
Peony	12
Hibiscus	32
Zinnia	5
Daffodil	41
Gladioli	11
Lavender	14

b. Handspan and height for 8 children

Handspan (cm)	6.2	6.3	6.3	6.5	6.7	6.7	6.8	6.8
Height (cm)	140	127	133	128	139	140	141	139

c. Ice-cream flavour preference for a group of people

Ice-cream flavour	Chocolate	Strawberry	Vanilla	Salted caramel	Pistachio
%	35	20	15	8	12

2. For each of the following surveys, explain which type of graph (line, pie or column) you would use to display the data.
Give a reason for your answer.
   a. One hundred students were asked to record how they travelled to school each day.
   b. One hundred students were asked to record their result on a recent test and the number of hours they studied in preparation for the test.
   c. One hundred students were asked to record the title of their favourite movie from a total of 15 movies.
3. **WE5** The ingredients on a package are: green beans 70 g, baby corn spears 63 g, asparagus 67 g, butter 4 g and water 6 g. Construct a column graph to display this data.
4. Vlad was interested in the water consumption in Victoria. He looked at data from the Australian Bureau of Statistics to determine the consumption of water for various industries. He downloaded the Water Account, released on 27 November 2018, and recorded the water consumption for Victoria in the table below.

Victoria	
Industry	**ML (million litres)**
Agriculture	1234
Forestry and fishing	13
Mining	10
Manufacturing	144
Electricity and gas	117
Water supply	310
Other industries	220
Household	311
Total	**2359**

Construct a column graph to display this data.

5. A group of students was surveyed about their favourite type of chocolate and the results have been recorded in the table shown.

Type of chocolate	Frequency
Dark chocolate	35
Milk chocolate	47
White chocolate	18

Construct a column graph for this dataset.

6. Fiona has been the manager of a toy shop for six months. She recorded the sales over this period of time in the table shown.

	Jan	Feb	Mar	Apr	May	Jun
Sales ($)	24 000	26 000	30 000	35 000	32 000	27 000

Construct a column graph to represent this data.

7. A new program to improve school attendance was introduced at Diramu College. A survey was conducted at the end of the program to seek teachers' opinion about the program's effect on attendance. These opinions were recorded in the table shown.

Opinion	Frequency
No improvement at all	9
Some improvement	15
Moderate improvement	56
High improvement	28
Total	**108**

a. Construct a vertical column graph.
b. Calculate the percentages for each category.
c. Construct a pie graph to represent this data.

8. **WE6** The ingredients on a package are 70 g green beans, 63 g baby corn spears, 67 g asparagus, 4 g butter and 6 g water.
a. Construct a pie graph to display this data.
b. Determine, in grams, the amount of asparagus in a 1-kg package.

9. Nicole has a gross wage of $95 000. She pays $40 000 in tax, $1425 for the Medicare levy, and deposits $1575 in her superannuation fund.
a. Determine the amount of money she actually received.
b. Construct a pie graph to display this data, including the amount of money she actually receives.

10. The table shown displays a typical family budget.

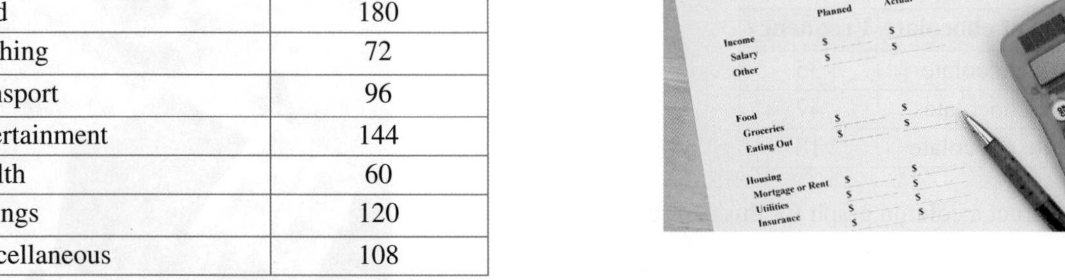

Area	Cost ($)
Housing	420
Food	180
Clothing	72
Transport	96
Entertainment	144
Health	60
Savings	120
Miscellaneous	108

Construct a pie graph to display this data.

11. The top 12 countries based on the gold medal tally at the 2012 London Olympics are recorded in the table shown.

Country	Gold medals
United States of America	46
People's Republic of China	38
Great Britain	29
Russian Federation	24
Republic of Korea	13
Germany	11
France	11
Italy	8
Hungary	8
Australia	7
Japan	7
Kazakhstan	7
Total	**209**

Construct a column graph to represent this data.

12. An apple cake recipe has the following ingredients:
300 g apples
250 g flour
200 g caster sugar
250 g butter
4 eggs (240 g)
10 g ground cinnamon

a. Calculate the individual percentages of each ingredient.
b. Construct a pie graph to represent this data.
c. Determine the amount of flour that would be needed if the total weight of the ingredients was 1.5 kg.

13. For the pie graph shown, calculate the number of schools for each sector if, in 2024, there were 9529 schools in Australia.

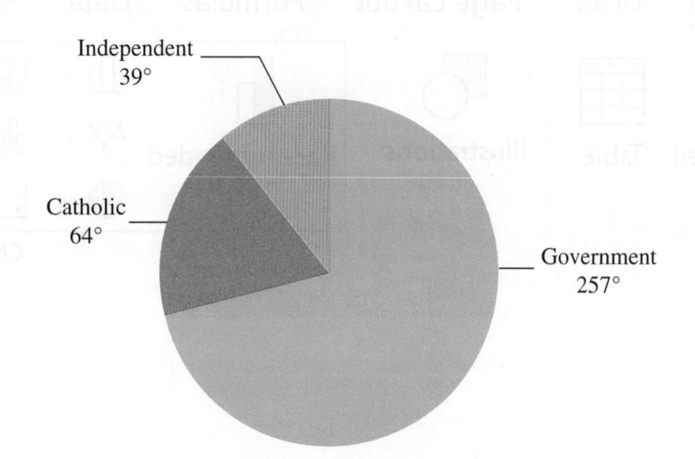

Fully worked solutions for this chapter are available online.

LESSON
12.4 Graphing with technology [complex]

SYLLABUS LINKS

- Use spreadsheeting software to tabulate and graph data [complex].
- Use graphs in practical situations.

12.4.1 Drawing lines in spreadsheeting software

- Spreadsheeting software such as Excel allows us to draw various charts accurately and easily from the datasets entered, and provide a convenient way of constructing line graphs.

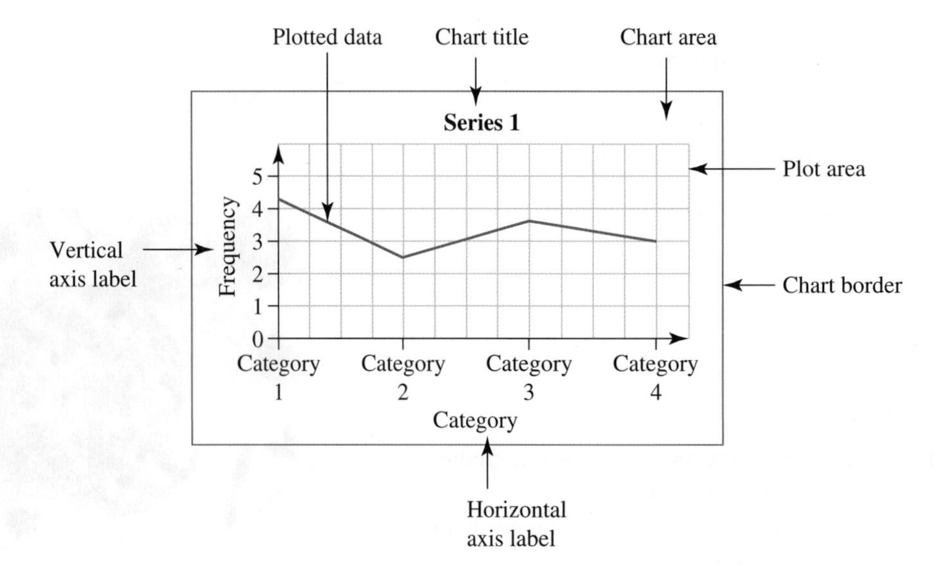

- Excel tools allow us to select different types of graphs and format the graphs.

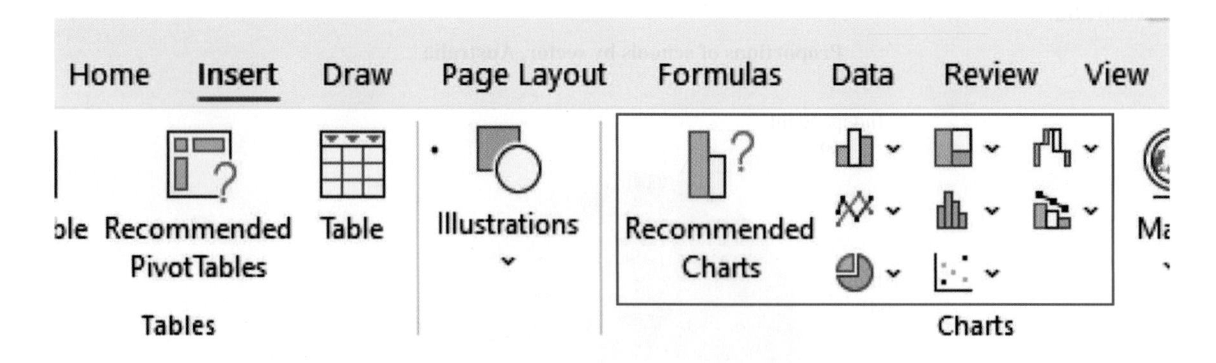

tlvd-11450

WORKED EXAMPLE 7 Using an Excel spreadsheet to construct a line graph

Using an Excel spreadsheet, construct a line graph for the data recorded in the table of values shown.

Time, min	0	1	2	3	4	5	6	7	8
Temperature, °C	21	25	29	31	33	30	31	32	33

THINK

1. Open a new Excel spreadsheet.

WRITE/DRAW

	A	B	C	D	E	F	G	H	I	J
1										

2. Label the two columns for the two variables. The independent variable will be written in column A.

Click in cell A1 and type 'Time, min'.

	A	B	C	D	E
1	Time, min				
2					

The dependent variable will be written in column B.

Click in cell B1 and type 'Temperature, °C'.

	A	B	C	D	E
1	Time, min	Temperature, °C			
2					

3. Insert the data.

Click in cell A2 and start typing in the data given for the independent variable.

	A	B	C
1	Time, min	Temperature, °C	
2	0		
3	1		
4	2		
5	3		
6	4		
7	5		
8	6		
9	7		
10	8		
11			

Click in cell B2 and start typing in the data given for the dependent variable.

	A	B	C
1	Time, min	Temperature, °C	
2	0	21	
3	1	25	
4	2	29	
5	3	31	
6	4	33	
7	5	30	
8	6	31	
9	7	32	
10	8	33	
11			

4. Draw the line graph.

Select the two columns starting from cell A1 to cell B10. To select these cells, click in cell A1, hold, move to cell B10 and release.

	A	B	C
1	Time, min	Temperature °C	
2	0	21	
3	1	25	
4	2	29	
5	3	31	
6	4	33	
7	5	30	
8	6	31	
9	7	32	
10	8	33	
11			

Go to Insert and then click the 'Lines' icon on the 'Charts' menu.

Home Insert Draw Page Layout Formulas Data Review View

ble Recommended PivotTables Table Illustrations Recommended Charts Ma

Tables Charts

Click any of the templates for a line. The chart shown is the default chart drawn by Excel. Excel draws the two datasets as two separate sets of data rather than as related to each other.

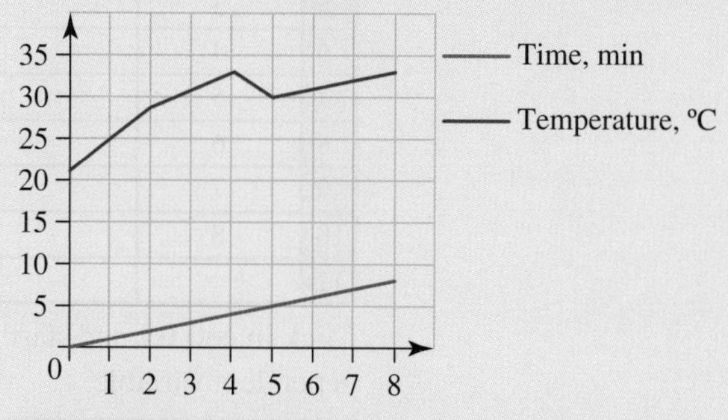

Alternatively, you can select 'Recommended Charts' and then choose 'Line' to see the options.

5. Format the line graph.

This chart can be formatted the way we want. Click anywhere on the chart area 'Select data' on the menu bar. The screen below will open.

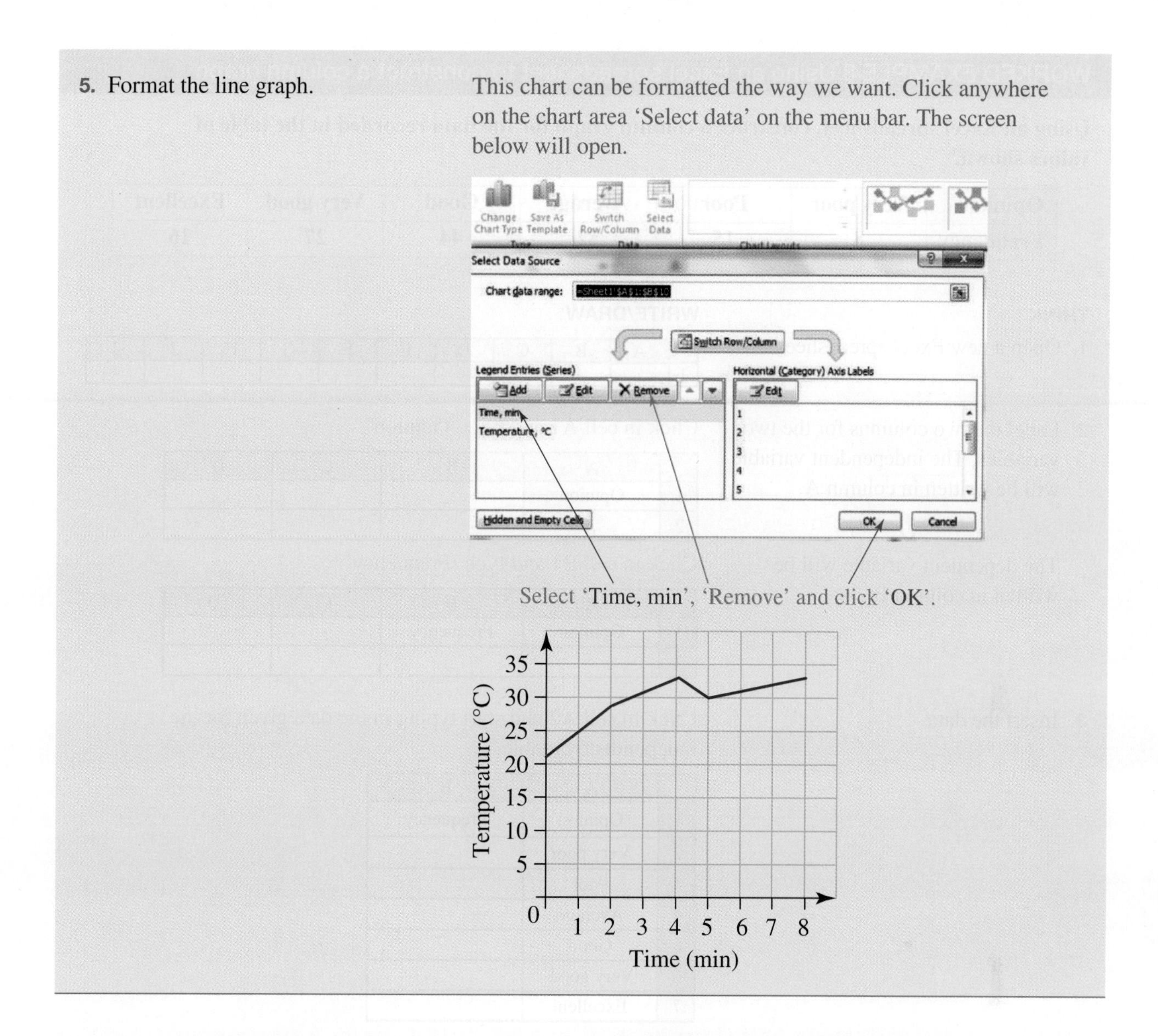

12.4.2 Drawing a column graph in spreadsheeting software

- Excel provides a convenient way of constructing column graphs.

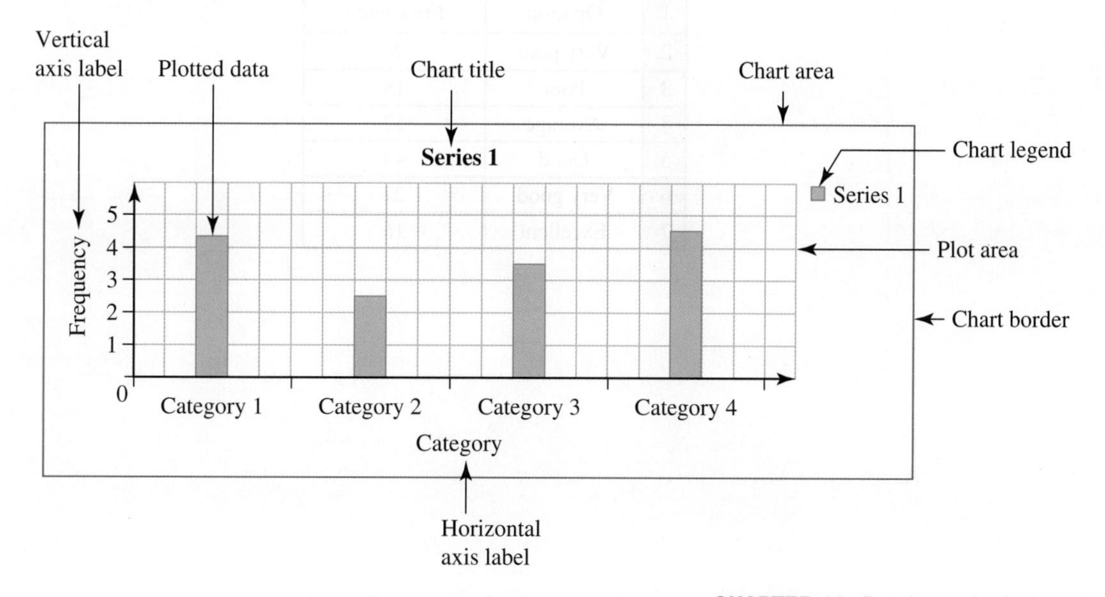

WORKED EXAMPLE 8 Using an Excel spreadsheet to construct a column graph

Using an Excel spreadsheet, construct a column graph for the data recorded in the table of values shown.

Opinion	Very poor	Poor	Average	Good	Very good	Excellent
Frequency	8	15	52	44	27	16

THINK

1. Open a new Excel spreadsheet.

WRITE/DRAW

	A	B	C	D	E	F	G	H	I	J
1										

2. Label the two columns for the two variables. The independent variable will be written in column A.

Click in cell A1 and type 'Opinion'.

	A	B	C	D
1	Opinion			
2				

The dependent variable will be written in column B.

Click in cell B1 and type 'Frequency'.

	A	B	C	D
1	Opinion	Frequency		
2				

3. Insert the data.

Click in cell A2 and start typing in the data given for the independent variable.

	A	B
1	Opinion	Frequency
2	Very poor	
3	Poor	
4	Average	
5	Good	
6	Very good	
7	Excellent	

Click in cell B2 and start typing in the data given for the dependent variable.

	A	B
1	Opinion	Frequency
2	Very poor	8
3	Poor	15
4	Average	52
5	Good	44
6	Very good	27
7	Excellent	16

4. Draw the column graph.

Select the two columns starting from cell A1 to cell B7. To select these cells, click in cell A1, hold, move to cell B7 and release.

	A	B
1	Opinion	Frequency
2	Very poor	8
3	Poor	15
4	Average	52
5	Good	44
6	Very good	27
7	Excellent	16

Go to Insert and then click on the 'Column' icon on the 'Charts' menu.

Home Insert Draw Page Layout Formulas Data Review View

Recommended PivotTables | Table | Illustrations | Recommended Charts

Tables | Charts

Click any of the templates for a column. The chart shown is the default chart Excel draws.

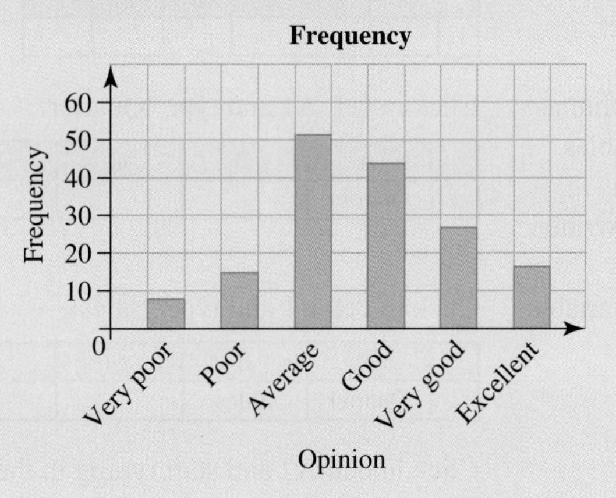

5. Format the column graph.

This chart can be formatted the way we want. Click 'Chart Elements' on the drop-down box and format the chart as desired.

Home Insert Draw Page Layout Formulas Data Review View

Recommended PivotTables | Table | Illustrations | Recommended Charts

Tables | Charts

12.4.3 Drawing a pie graph in a spreadsheeting software

- Pie charts are easy to draw in Excel spreadsheets because the program automatically calculates the percentages and angles required.

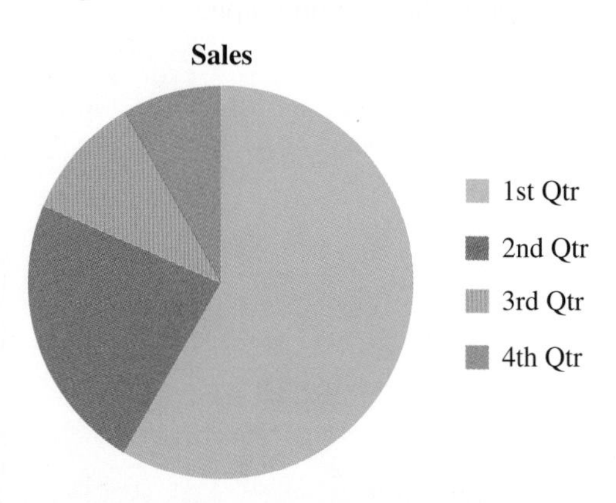

WORKED EXAMPLE 9 Using an Excel spreadsheet to draw a pie graph

Using an Excel spreadsheet, construct a pie graph for the data recorded in the table of values shown.

Quarter	1st	2nd	3rd	4th
Sales	16	8	32	40

THINK

WRITE

1. Open a new Excel spreadsheet.

	A	B	C	D	E	F
1						

2. Label the two columns for the two variables. The independent variable will be written in column A. The dependent variable will be written in column B.

Click in cell A1 and type 'Quarter'.

	A	B	C	D
1	Quarter			

Click in cell B1 and type 'Sales'.

	A	B	C	D
1	Quarter	Sales		

3. Insert the data.

Click in cell A2 and start typing in the data given for the independent variable.

	A	B
1	Quarter	Sales
2	1st Qtr	
3	2nd Qtr	
4	3rd Qtr	
5	4th Qtr	

Click in cell B2 and start typing in the data given for the dependent variable.

	A	B
1	Quarter	Sales
2	1st Qtr	16
3	2nd Qtr	8
4	3rd Qtr	32
5	4th Qtr	40

4. Draw the pie graph.

Select the two columns starting from cell A1 to cell B5. To select these cells, click in cell A1, hold, move to cell B5 and release.

	A	B
1	Quarter	Sales
2	1st Qtr	16
3	2nd Qtr	8
4	3rd Qtr	32
5	4th Qtr	40

Go to Insert and then click the 'Pie' icon on the 'Charts' menu.

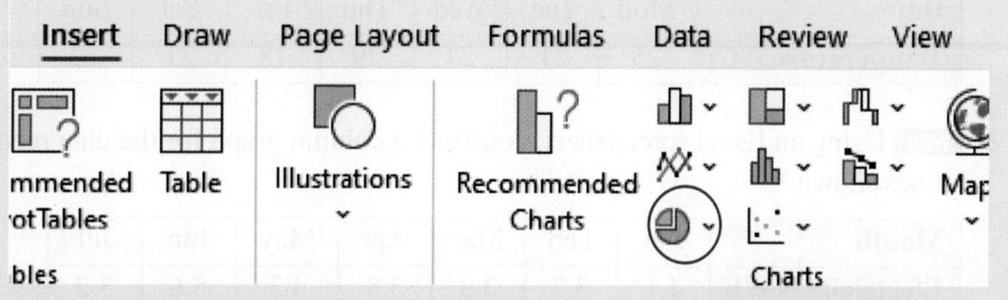

Click any of the templates for a pie graph. The chart shown is the default chart drawn by Excel.

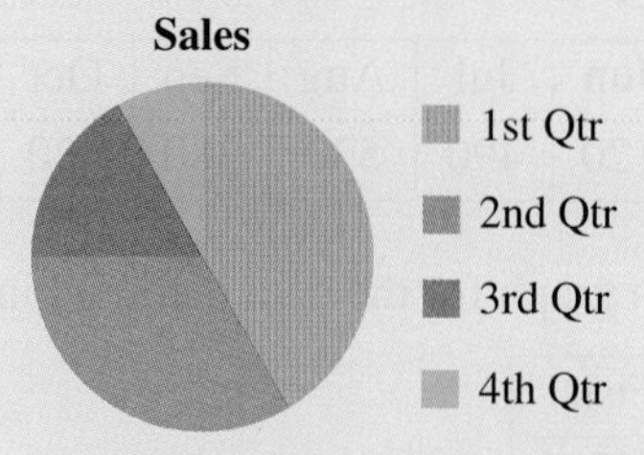

5. Format the pie graph.

This chart can be formatted the way we want. Click 'Chart Elements' and format the chart as desired.

Exercise 12.4 Graphing with technology [complex]

learnon

12.4 Quick quiz on | 12.4 Exercise

Simple familiar	Complex familiar	Complex unfamiliar
N/A	1, 2, 3, 4, 5, 6, 7, 8, 9, 10, 11, 12, 13	14

Complex familiar

1. a. Explain what symbols are used to label rows and columns in Excel spreadsheets.
   b. Identify the row and the column of cell A4.
   c. Identify the cell shown.

	A	B
1		
2		X

2. **WE7** Using an Excel spreadsheet, construct a line graph for the data recorded in the table of values shown.

Day of the week	Mon	Tue	Wed	Thu	Fri	Sat	Sun
Temperature, °C	32	35	34	29	21	26	34

3. Using an Excel spreadsheet, construct a line graph for the data recorded in the table of values shown.

Time, days	0	1	2	3	4	5	6	7	8	9	10
Height, cm	3.2	3.5	4.1	4.7	5.9	6.0	6.2	6.5	7.0	7.2	7.3

4. Using an Excel spreadsheet, construct a line graph for the data recorded in the table shown.

Day	Mon	Tue	Wed	Thu	Fri	Sat	Sun
Temperature (°C)	25	21	24	29	18	21	22

5. **WE8** Using an Excel spreadsheet, construct a column graph for the data recorded in the table of values shown.

Month	Jan	Feb	Mar	Apr	May	Jun	Jul
Electricity, kWh	4.1	3.7	3.9	3.8	4.3	5.6	5.2

6. Using an Excel spreadsheet, construct a line graph for the data recorded in the table of values shown.

Month	Jan	Feb	Mar	Apr	May	Jun	Jul	Aug	Sep	Oct	Nov	Dec
Gas, MJ	70	60	65	75	82	120	490	530	380	420	120	45

7. Using an Excel spreadsheet, construct a column graph for the data recorded in the table shown.

Pet	Cat	Dog	Rabbit	Fish
Frequency	15	36	8	17

8. The table shown displays the total road length in kilometres in Australia by state and territory. Construct a column graph for the dataset given using an Excel spreadsheet.

NSW	VIC	QLD	SA	WA	TAS	NT	ACT
184 794	153 000	183 041	97 319	154 263	25 599	22 239	2963

9. The table shown displays the total road length in Australia for the time period from 2020 to 2024. Construct a line graph for the dataset given using an Excel spreadsheet.

Year	Road length (km)
2020	815 588
2021	816 949
2022	822 649
2023	825 592
2024	823 217

10. WE9 Using an Excel spreadsheet, construct a pie graph for the data recorded in the table of values shown. Label the sectors of the pie graph with the names of the ingredients and their quantities, in grams.

Ingredients	Flour	Sugar	Butter	Cinnamon	Cocoa
Quantity, g	500	250	100	20	60

11. The data recorded in the table shown represents the number of international passengers carried by major airlines for the year ended December 2023 to/from Australia. Construct a pie graph for the data given using an Excel spreadsheet.

Airline	**Percentage**
Qantas Airways	17.7%
Singapore Airlines	9.2%
Emirates	8.4%
Virgin Australia	8.3%
Jetstar	8.3%
Air New Zealand	8.0%
Cathay Pacific Airways	4.9%
Malaysia Airlines	3.7%
Thai Airways International	3.5%
AirAsia X	2.8%
Others	25.3%

12. Using an Excel spreadsheet, construct a pie graph for the data recorded in the table of values shown. Label the sectors of the pie graph with the names of the categories and their percentages.

Type of payment	Net wage	Tax	Superannuation	Medicare levy
Amount, $	57 000	10 230	1200	855

13. The total crash costs in Australia in the year 2023 were $6.1 billion. These costs are estimated for the following categories:

Lost earnings of victims	$829.1 million
Family and community losses	$587.8 million
Vehicle damage	$1868.2 million
Pain and suffering	$1463.3 million
Insurance administration	$571.1 million
Other	$816.4 million

a. Calculate the percentages for each category.
b. Construct a pie graph for this dataset. Label each sector with the category and the corresponding percentage.

Complex unfamiliar

14. The data recorded in the table shown represents the number of international passengers carried (in thousands) to/from Australia during the year ended December 2023. Construct a line graph for the two datasets given using an Excel spreadsheet and use the graph to determine how many times the inbound line intersects the outbound line.

	Inbound	**Outbound**
December 2022	1220	1451
January 2023	1523	1281
February 2023	1202	1019
March 2023	1141	1184
April 2023	1177	1203
May 2023	1001	1098
June 2023	1094	1267
July 2023	1411	1186
August 2023	1191	1229
September 2023	1241	1294
October 2023	1410	1164
November 2023	1207	1247
December 2023	1314	1539

Fully worked solutions for this chapter are available online.

LESSON
12.5 Review

12.5.1 Summary

doc-41776

Hey students! Now that it's time to revise this chapter, go online to:

Access the chapter summary | Review your results | Watch teacher-led videos | Practise questions with immediate feedback

Find all this and MORE in jacPLUS

12.5 Exercise

learnon

12.5 Exercise

Simple familiar	Complex familiar	Complex unfamiliar
1, 2, 3, 4, 5, 6, 7, 8, 9, 10	11, 12, 13, 14, 15	16

These questions are even better in jacPLUS!
- Receive immediate feedback
- Access sample responses
- Track results and progress

Find all this and MORE in jacPLUS

Simple familiar

1. Every driver's licence has a number ID. A police officer recorded the licence numbers of 100 drivers who have been tested for drink driving. Classify the data as either categorical nominal, numerical continuous, categorical ordinal or numerical discrete.
2. Explain whether line graphs display numerical continuous data or categorical data.
3. Explain whether column graphs display numerical data or categorical data.
4. Construct a column graph representing the data recorded in the following table.

Opinion	People
Excellent	29
Very good	24
Good	20
Poor	5
Very poor	2
Total	**80**

5. The column graph shown displays the monthly mean number of cloudy days recorded at the Canberra Airport meteorological station for a 70-year period.

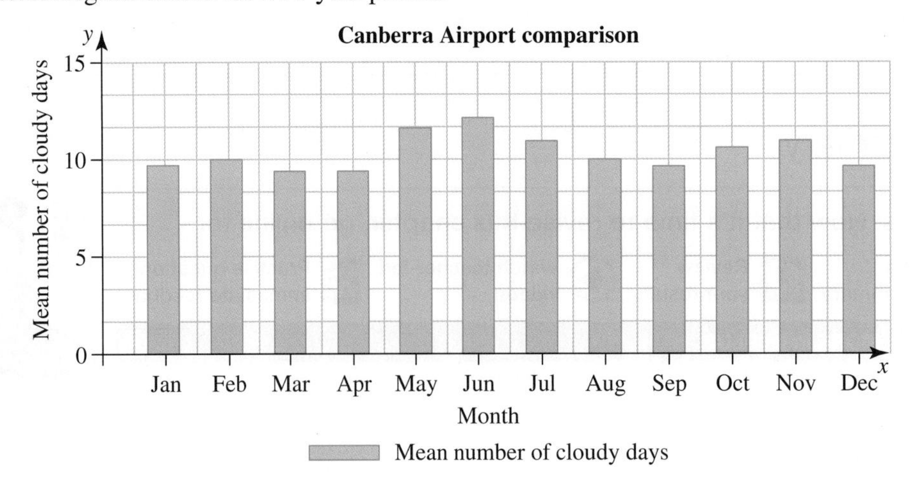

Identify the months with exactly 10 cloudy days.

6. The graph shown displays superannuation contributions, in billions of dollars, for a 10-year period.

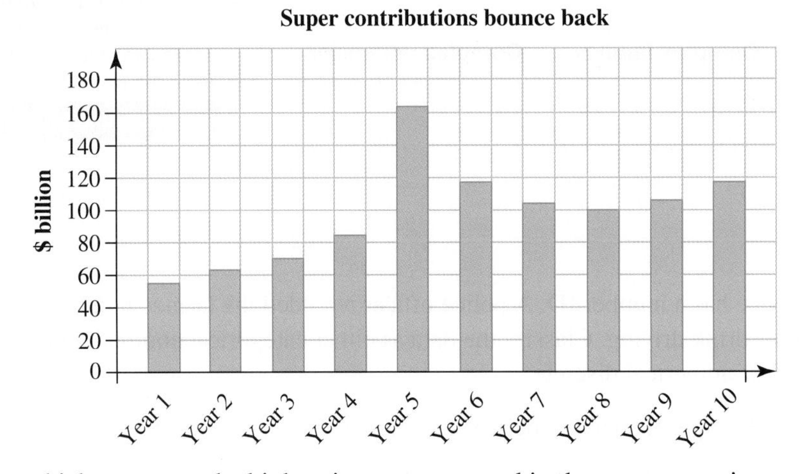

Identify between which two years the highest increase occurred in the superannuation contributions.

7. Explain when it is best to use pie graphs.

8. Classify the following data as either nominal, ordinal, discrete or continuous. The Bureau of Meteorology collects data on rainfall over catchment areas. The rainfall is measured in millimetres.

9. The table shown displays the number of words that users from Australia typed into search engines during March 2024.

	Words:	Au
	1	47.14%
	2	21.98%
	3	18.06%
	4	6.51%
	5	3.37%
	6	1.43%
	7	0.80%

a. Identify the type of data that is displayed in the given table.
b. Construct a column graph to represent this dataset.
c. From a total of 5000 searches, determine how many would be two-word searches.

10. The graph below shows monthly car sales for a local car yard over the past year.

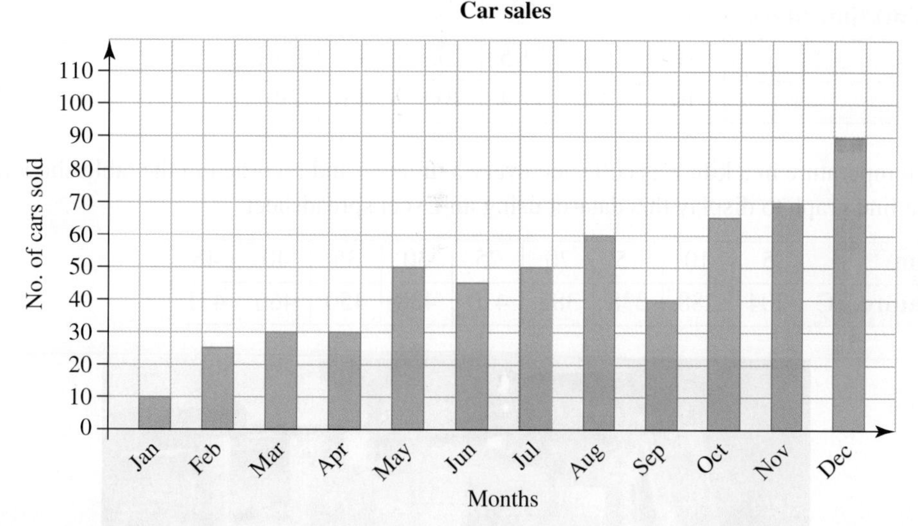

Use the graph to answer the following questions.

a. Identify the month during which the lowest sales figures were recorded.
b. Identify the months in which the numbers of cars sold were equal.
c. Identify the highest number of cars sold and the month in which this occurred.
d. Determine by how much the December sales exceeded the January sales.
e. Determine the difference between the highest and the second highest sales figures recorded over the last year.
f. Determine the total sales for the year.
g. Explain whether there is any pattern in sales that you can see. Comment on your response.

Complex familiar

11. The line graph shown displays the relationship between inches and centimetres.

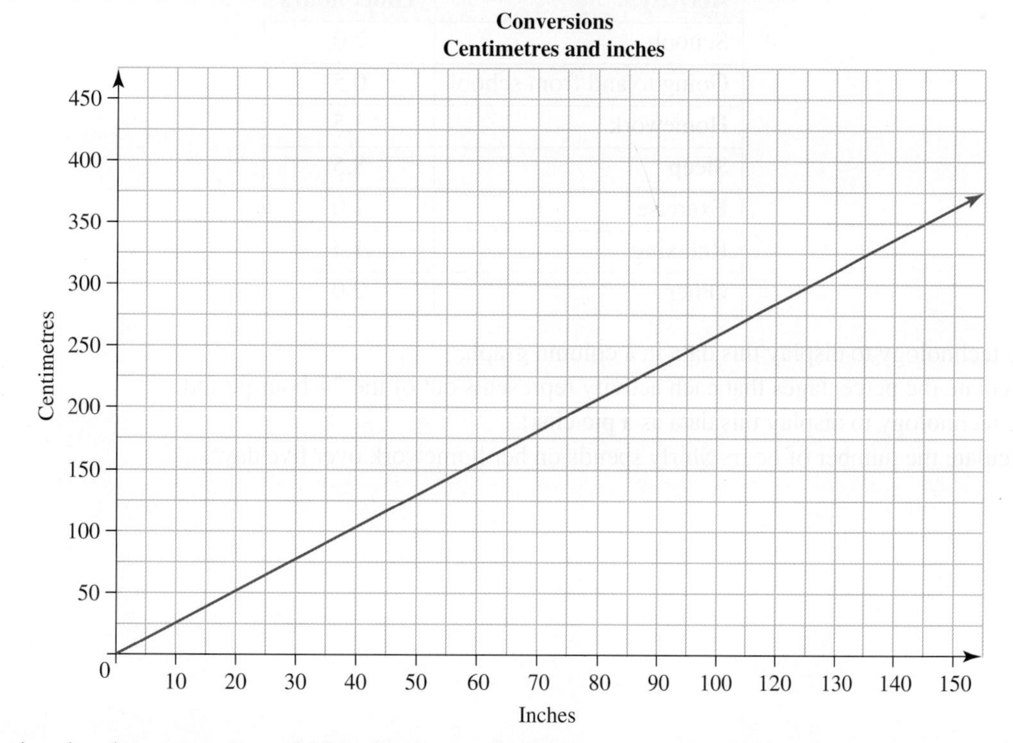

Determine the closest measure of 39 inches in centimetres.

12. Construct a line graph representing the data recorded in the following table.

Height (cm)/time (days)	0	1	2	3	4	5
Plant A	10	15.6	21.5	29.3	34.7	35.1
Plant B	10	13.2	17.4	20.8	25.6	29.2

13. The firing temperature in a kiln was collected every 5 minutes and recorded in the table shown. Construct a line graph to display this dataset using an Excel spreadsheet.

Time, min	5	10	15	20	25	30	35	40	45
Temperature, °C	204	288	338	382	410	438	450	466	471

14. Shirly recorded the approximate times she spends on various tasks during a 24-hour period in the table shown.

Activity	Time, hours
School	7.0
Going to and from school	0.5
Homework	3.5
Sleep	8.5
Exercise	1.0
Relaxing	1.5
Other	2.0

a. Use technology to display this data in a column graph.
b. Calculate the percentages that each activity represents out of the 24-hour period.
c. Use technology to display this data as a pie chart.
d. Calculate the number of hours Shirly spends on her homework over five days.

15. The daily maximum temperatures in May in Perth are displayed in the table shown.

Day	Max temperature, °C	Day	Max temperature, °C
1st	25.7	17th	20.3
2nd	22.9	18th	20.5
3rd	22.6	19th	21.7
4th	23.6	20th	19.2
5th	25.4	21st	21.2
6th	22.3	22nd	21.6
7th	24.1	23rd	20.8
8th	20.8	24th	22.0
9th	20.8	25th	23.2
10th	20.1	26th	24.9
11th	19.7	27th	23.0
12th	20.7	28th	20.0
13th	21.1	29th	19.6
14th	21.8	30th	14.6
15th	22.6	31st	16.2
16th	23.4		

a. Use technology to construct a line graph for this dataset.
b. Identify the day in May that recorded the highest maximum temperature for the month.
c. Identify the day in May that recorded the lowest maximum temperature for the month.
d. Calculate the temperature difference between 7 and 8 May. Explain whether this was a decrease or an increase in temperature.
e. Calculate the temperature difference between 25 and 26 May. Explain whether this was a decrease or an increase in temperature.

Complex unfamiliar

16. Kehlani is looking to invest her money in a company. The two companies she is interested in have the following monthly profits (in millions) in the table below.

	Jan	Feb	Mar	Apr	May	Jun	Jul	Aug	Sep	Oct	Nov	Dec
Company A	1.1	1.2	1.4	1.8	1.7	1.9	2.2	2.1	2.4	2.5	2.8	2.9
Company B	1.3	1.4	1.6	1.6	1.4	1.5	1.8	1.9	1.9	2	2.2	2.3

Using technology or otherwise, justify which company Kehlani should invest her money into.

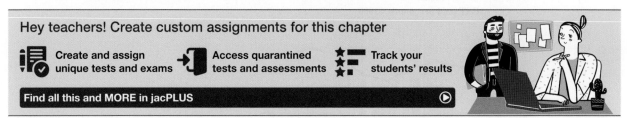

Answers

Chapter 12 Drawing and using graphs

12.2 Drawing and using line graphs [complex]

12.2 Exercise

1.

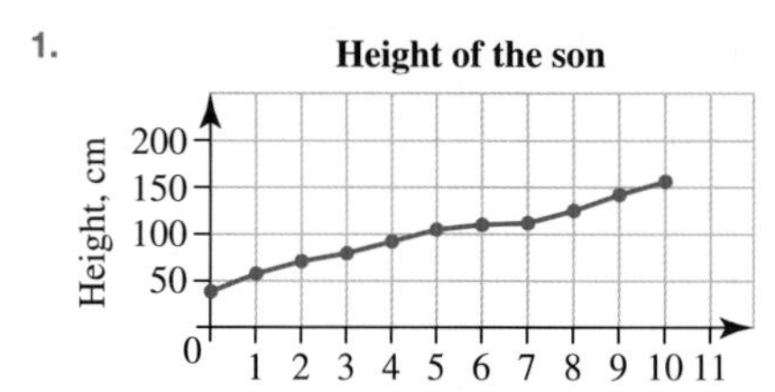

2.

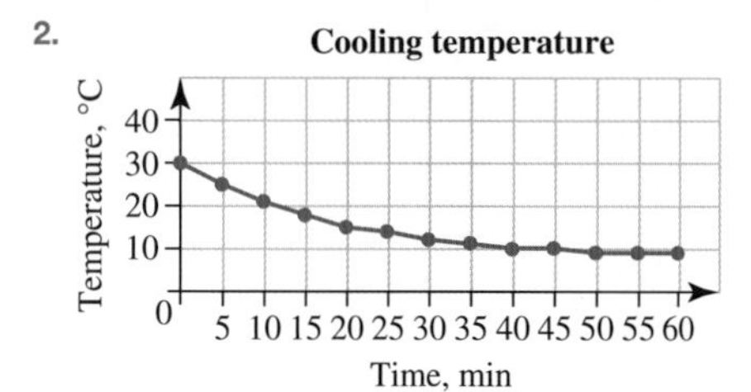

3.

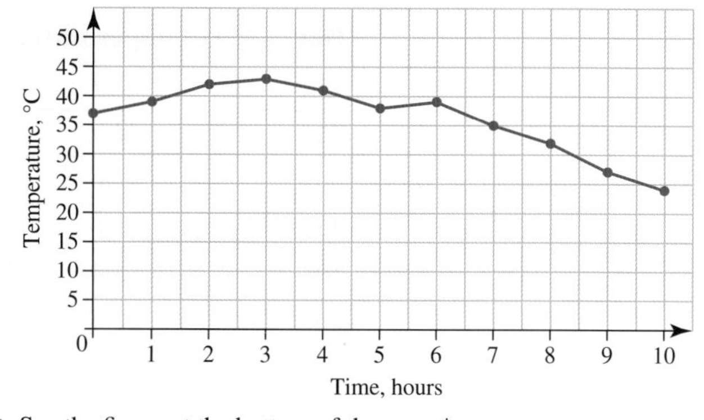

4. See the figure at the bottom of the page.*
5. See the figure at the bottom of the page.*
6. a. $400
   b. $800
   c. $1200
   d. $1200
   e. Do-it-today would be cheaper after 5 hours of hire.
7.

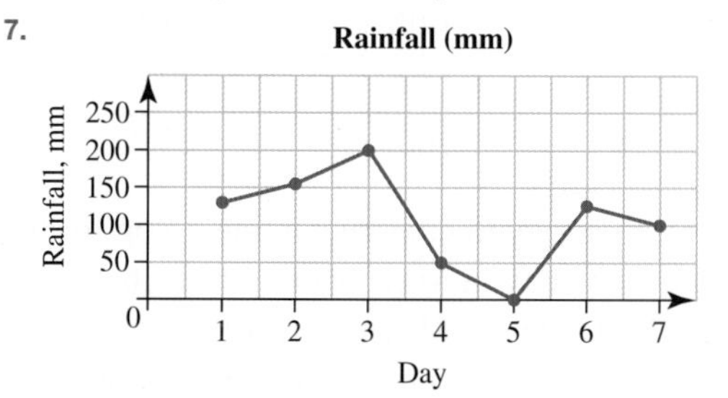

*4.

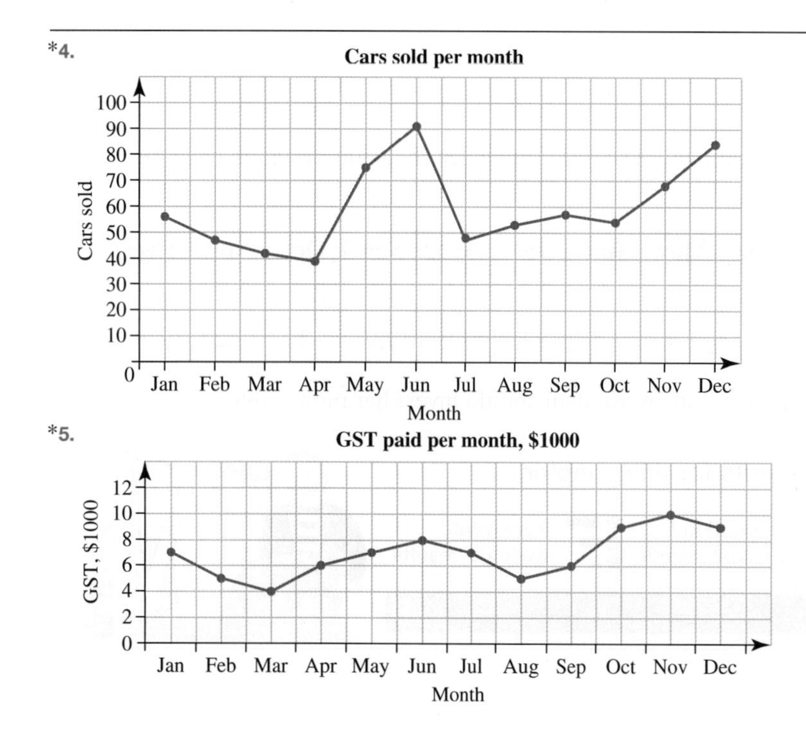

8. a. **Population of Australia (million)**

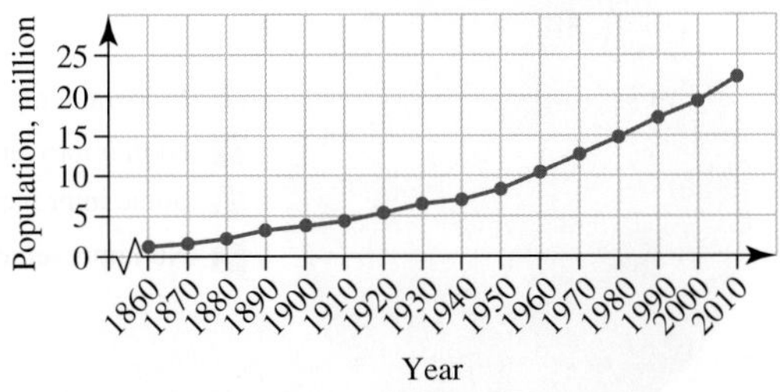

b. 0.4 million, 0.6 million, 1.0 million, 0.6 million, 0.6 million, 1.0 million, 1.1 million, 0.5 million, 1.3 million, 2.1 million, 2.3 million, 2.1 million, 2.4 million, 2.1 million, 3.0 million

Population growth in Australia (million)

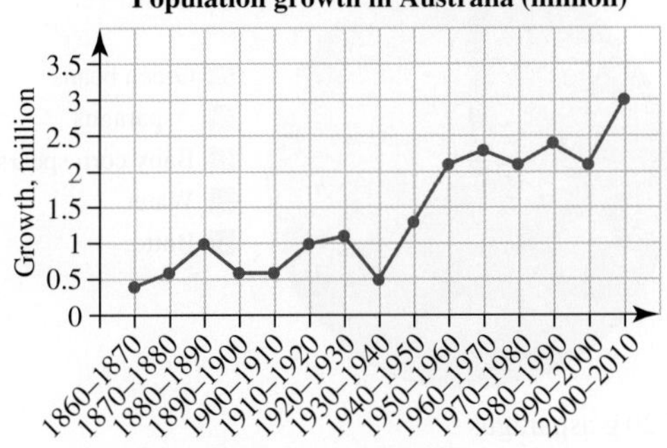

9. **Share prices of Stock A and Stock B**

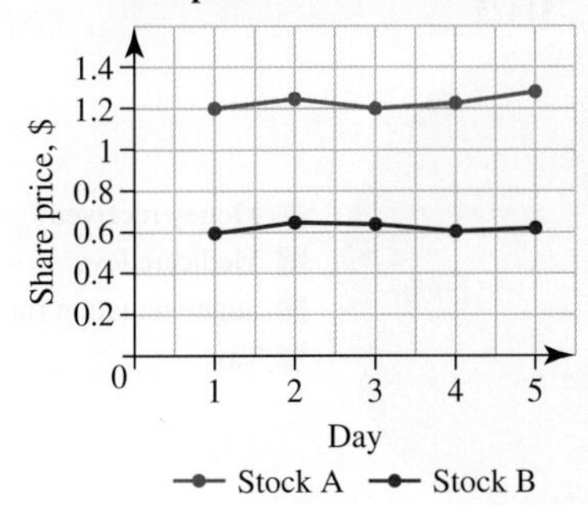

10. **Number of customers in a day**

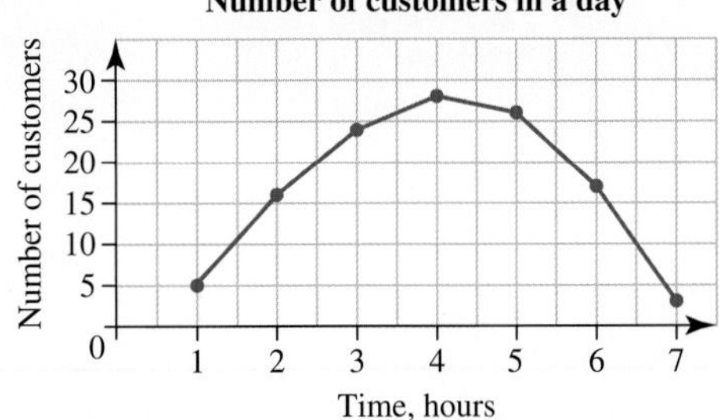

11. a. 20 000 L b. 14 000 L
c. Tank A 500 min; Tank B 700 min
d. After 300 min

12. **Test results**

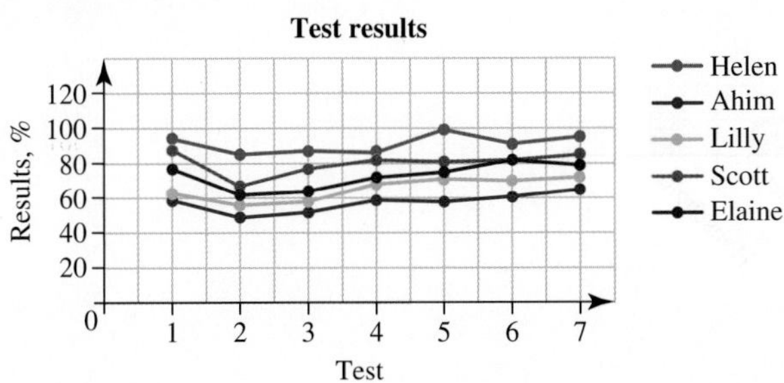

13. **Mean maximum temperature in Perth Metro, December 2005–2023**

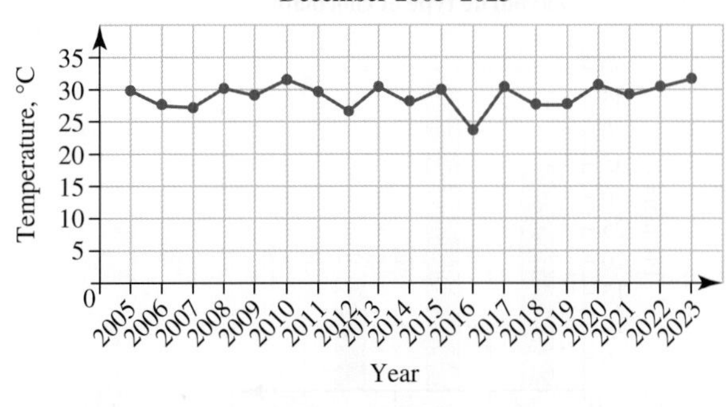

14. **Mean maximum temperature, December, Perth and Adelaide**

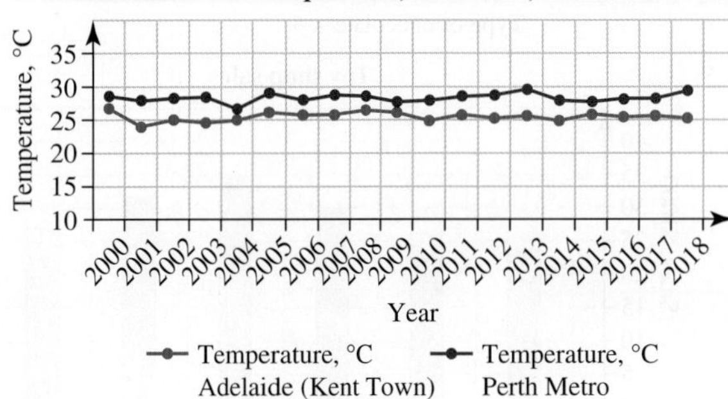

15. **Water temperature, °C**

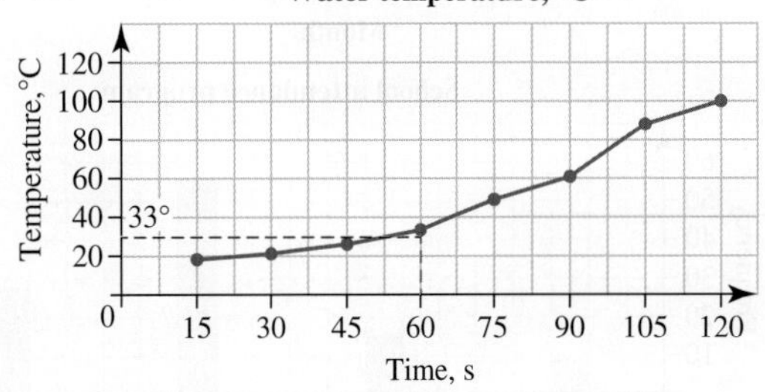

12.3 Drawing and using column graphs and pie graphs

12.3 Exercise

1. a. A column graph, as there are 16 categories, which is too many for a pie graph
b. A line graph, as the data is for two related quantities
c. A pie graph, as there are only five categories
2. a. A pie graph, as there are fewer than six means of travelling to school
b. A line graph, as the data is for two related quantities
c. A column graph, as there are 15 categories, which is too many for a pie graph

3. **Ingredients on a package, grams**

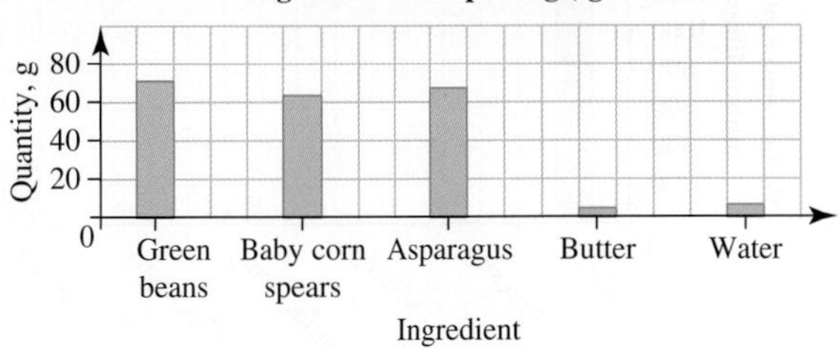

4. See the figure on the bottom of the page.*

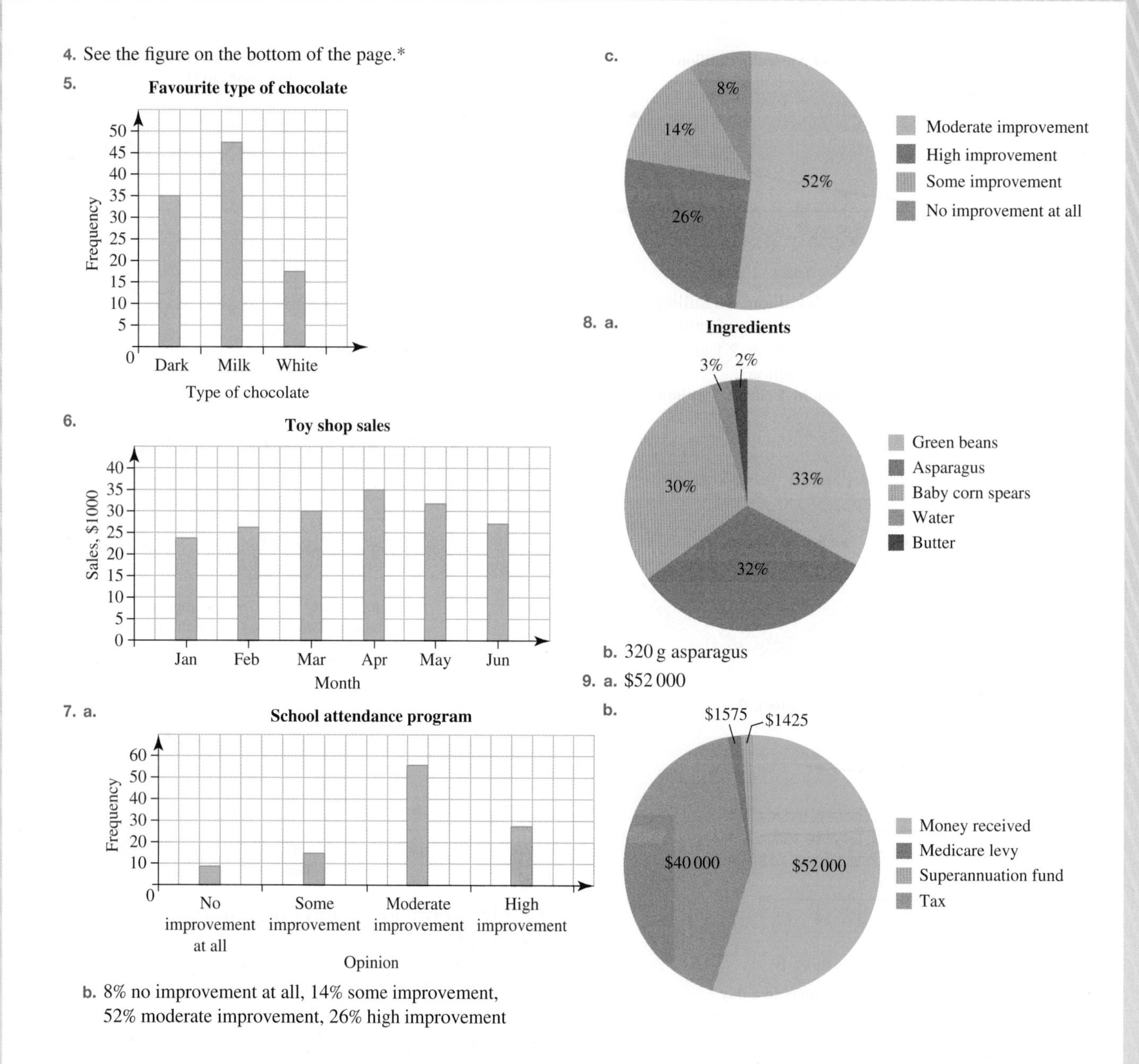

b. 8% no improvement at all, 14% some improvement, 52% moderate improvement, 26% high improvement

8. b. 320 g asparagus

9. a. $52 000

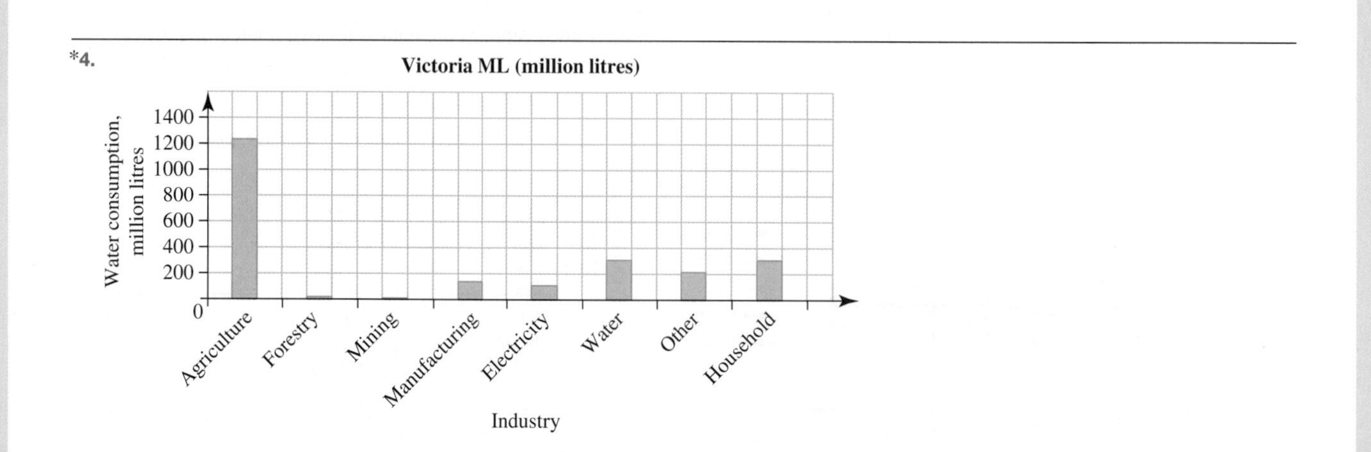

10.

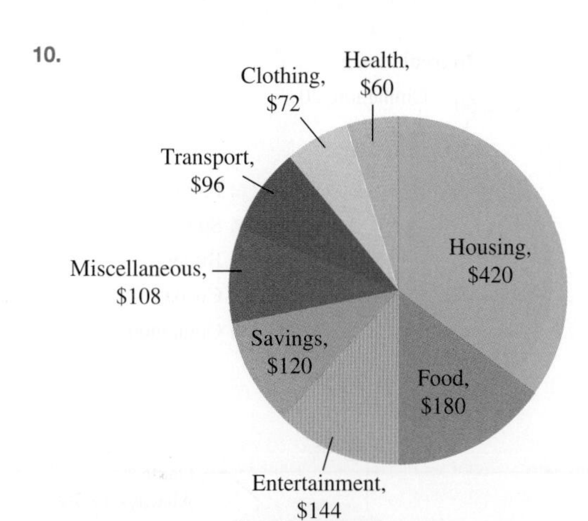

11. **Gold medal count 2012 London Olympics**

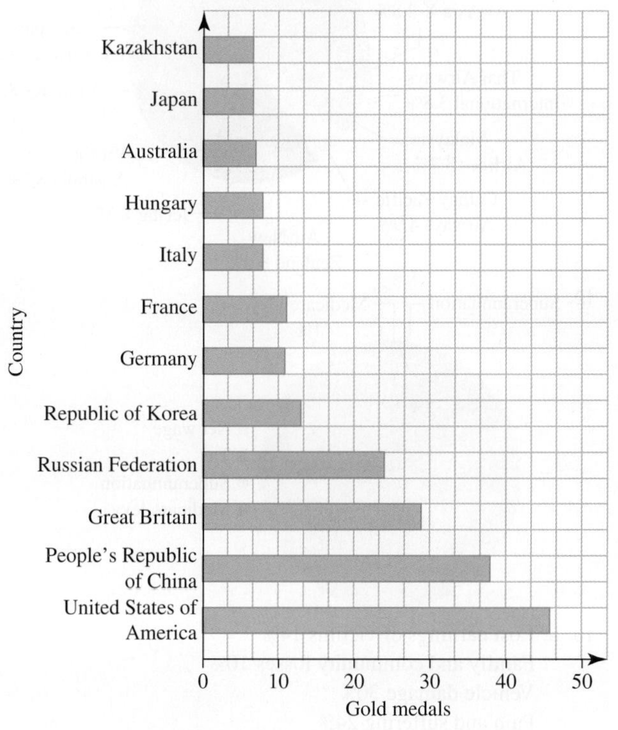

12. a. 24% apples, 20% flour, 16% caster sugar, 20% butter, 19% eggs, 1% ground cinnamon

b.

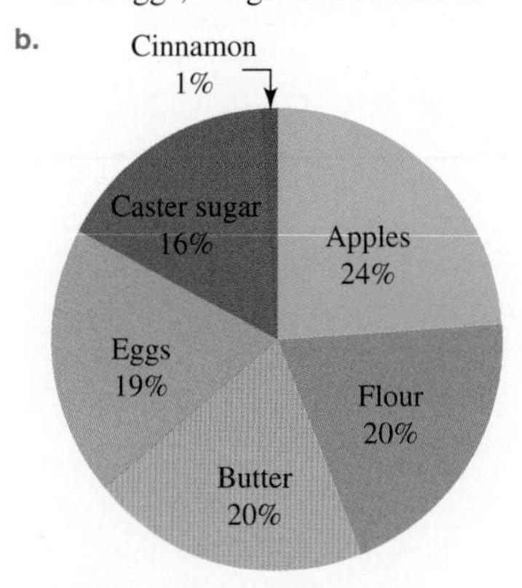

c. 300 g

13. 6803 government schools, 1694 Catholic schools, 1032 independent schools

12.4 Graphing with technology [complex]

12.4 Exercise

1. a. Rows are labelled with numbers and columns are labelled with capital letters.
   b. Row 4, column A
   c. Cell B2

2.

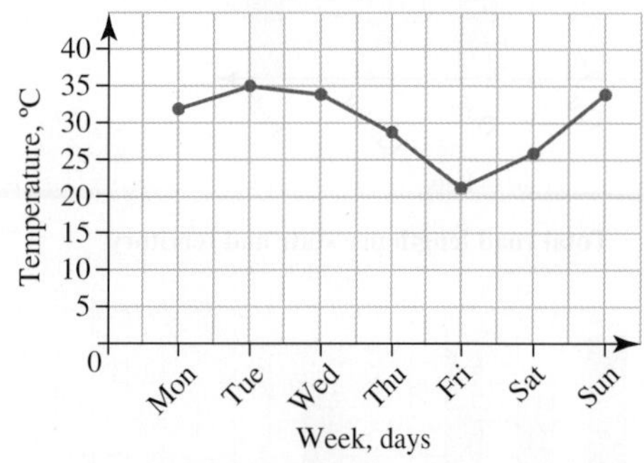

3.

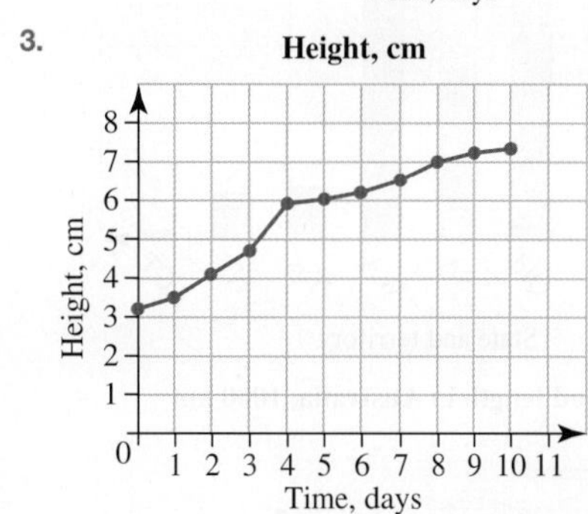

4.

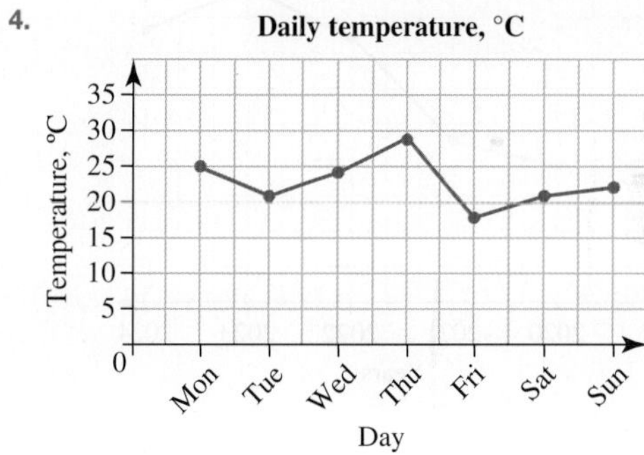

5.

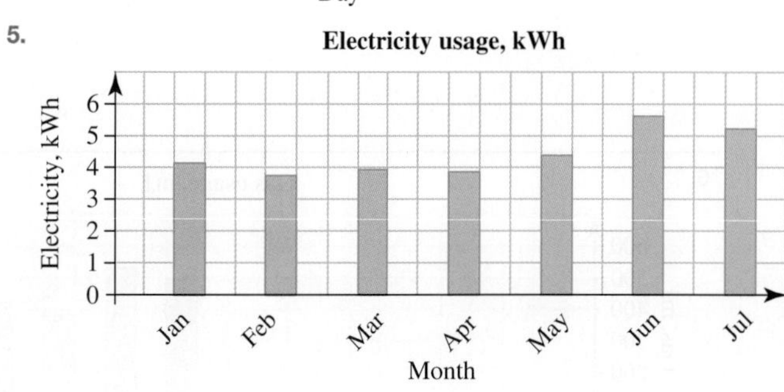

6. See the figure on the bottom of the page.*

7.

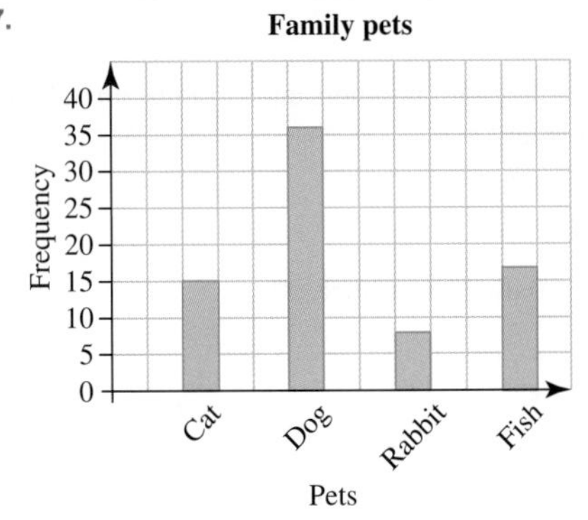

8.

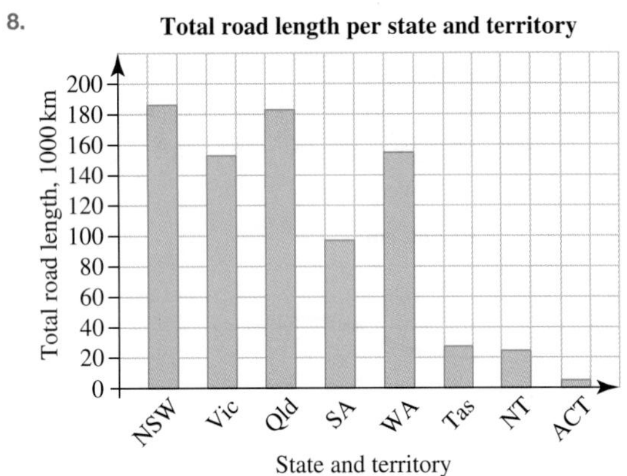

9.

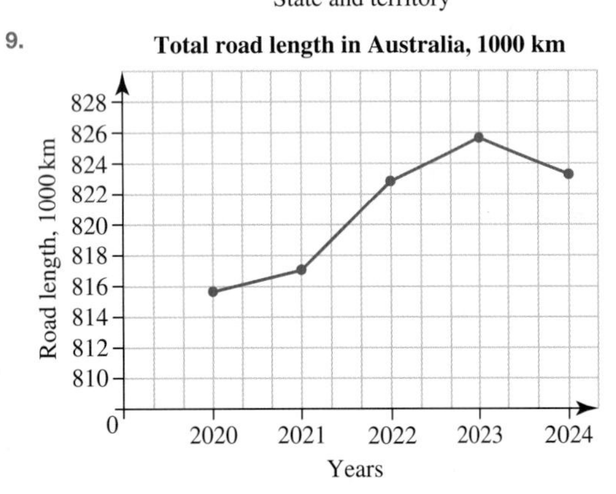

10. 11. 12.

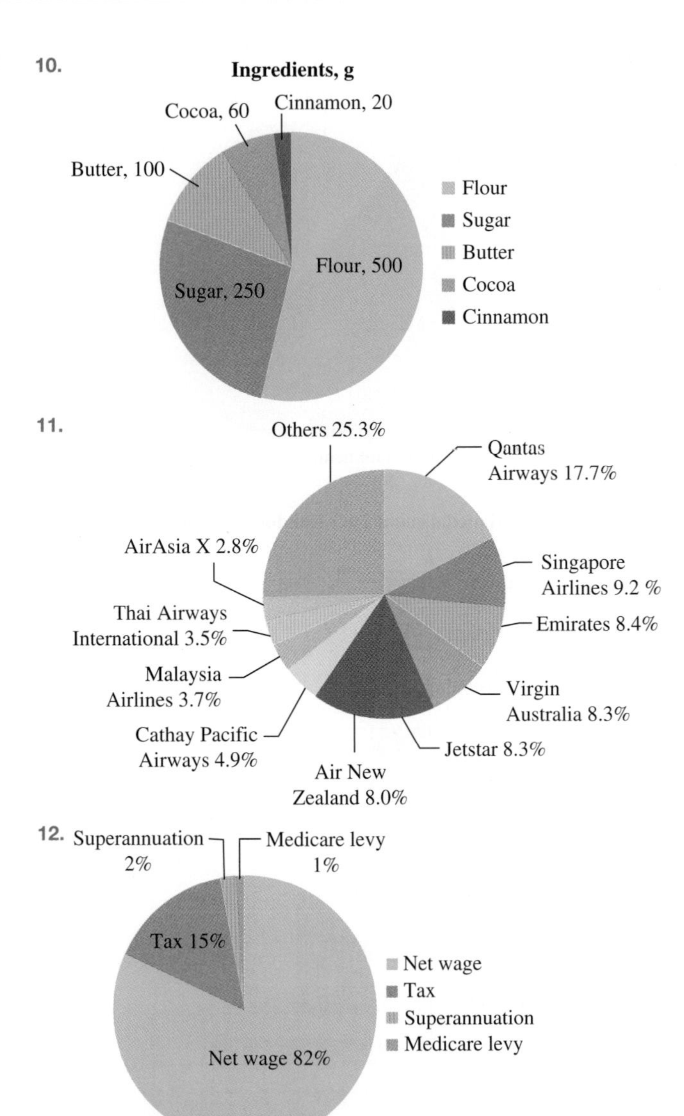

13. a.

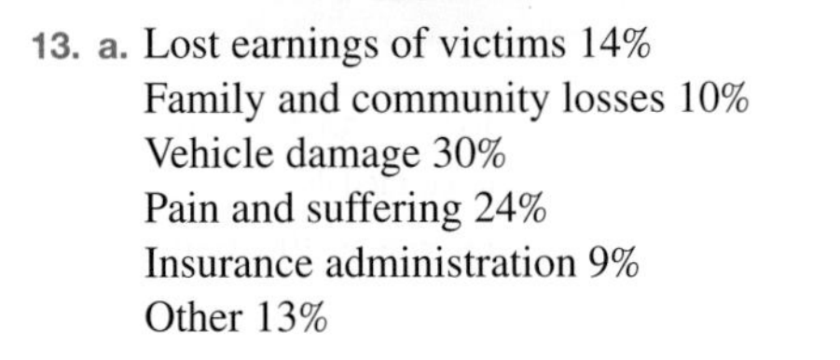

Lost earnings of victims 14%
Family and community losses 10%
Vehicle damage 30%
Pain and suffering 24%
Insurance administration 9%
Other 13%

*6.

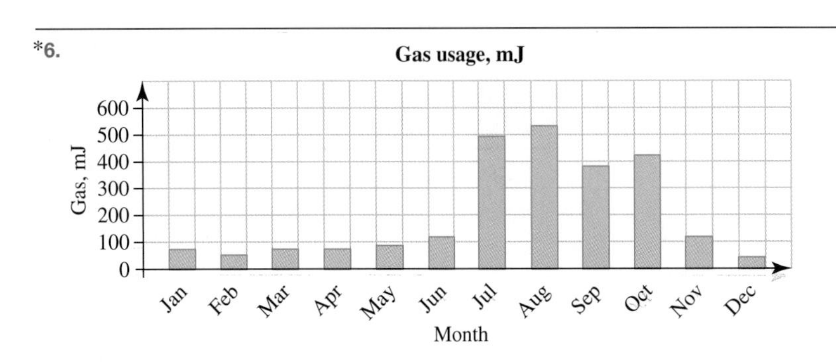

b.

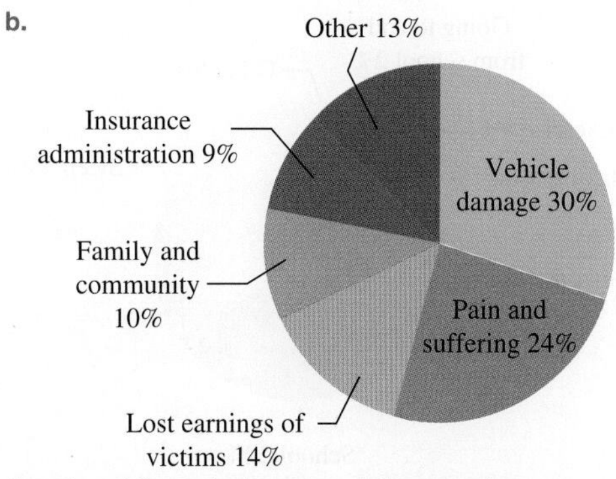

14. See the table at the bottom of the page.*
The inbound and outbound line intersect each other 6 times.

12.5 Review

12.5 Exercise

1. Categorical nominal data
2. Line graphs display numerical continuous data.
3. Column graphs display categorical data.
4. **People's opinions**

Excellent
Very good
Good
Poor
Very poor
0 10 20 30 40
Frequency

5. February and August
6. From Year 4 to Year 5
7. A pie graph is useful when there is a small dataset.
8. Continuous data
9. a. Categorical ordinal

b. **Type of word searches on the Internet in Australia, March 2024**

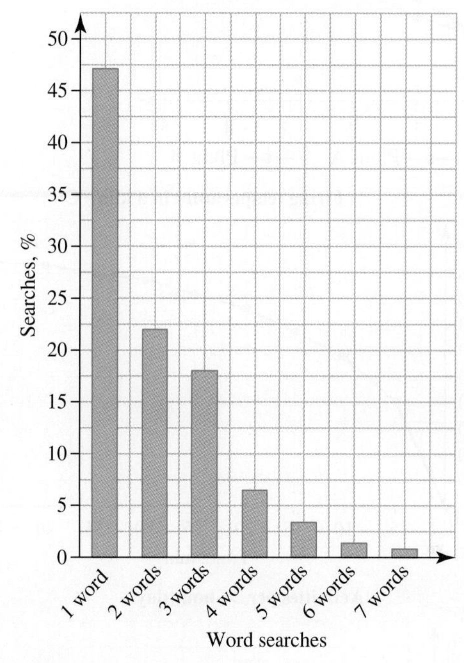

c. 1099 two-word searches

10. a. January
b. March and April; June and July
c. December (90 cars)
d. By 80 cars
e. 20 cars
f. 565 cars
g. Although sales figures fluctuate from month to month, overall there is a definite increasing trend.
11. 100 cm

*14.

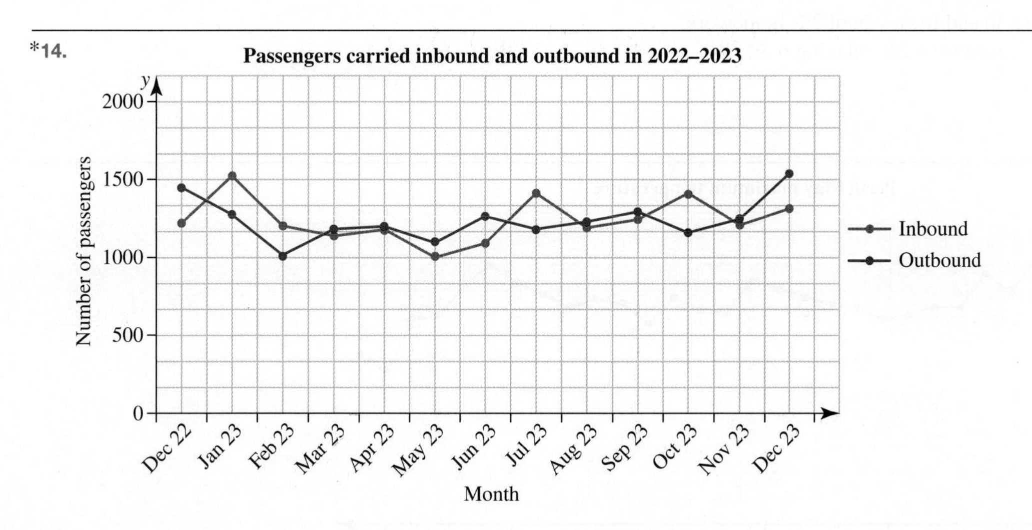

12.

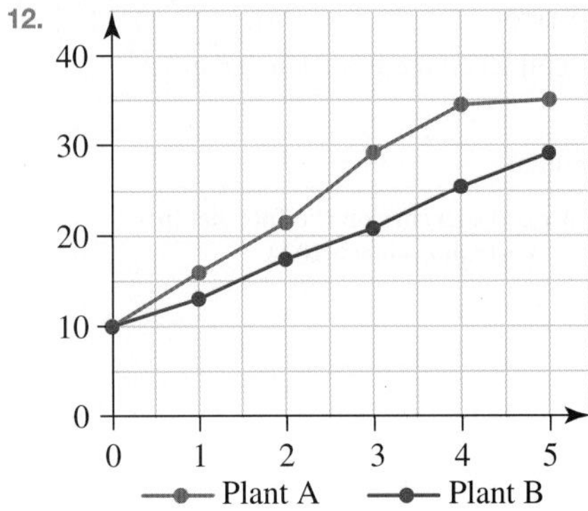

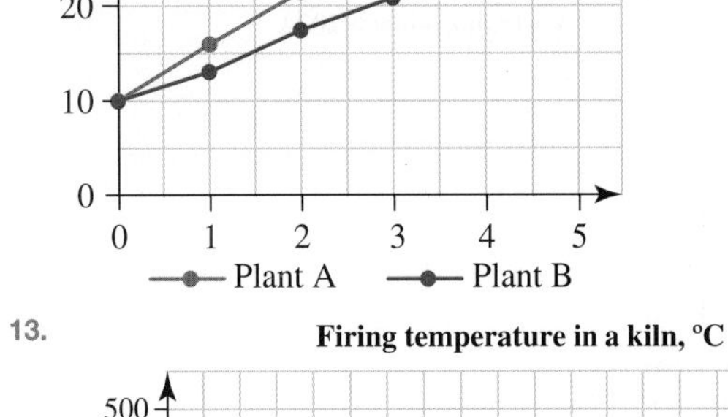

13.

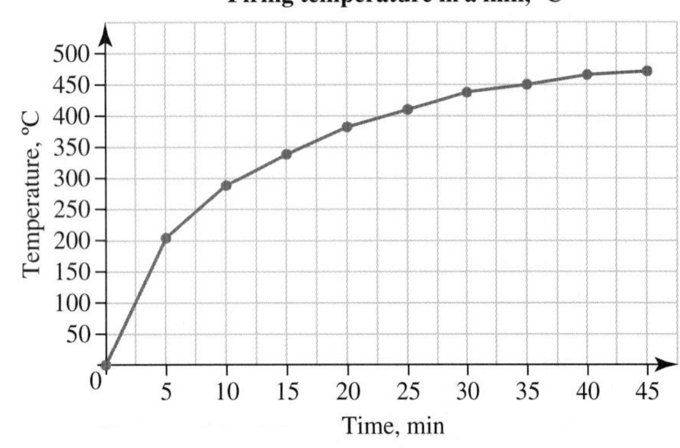

14. a.

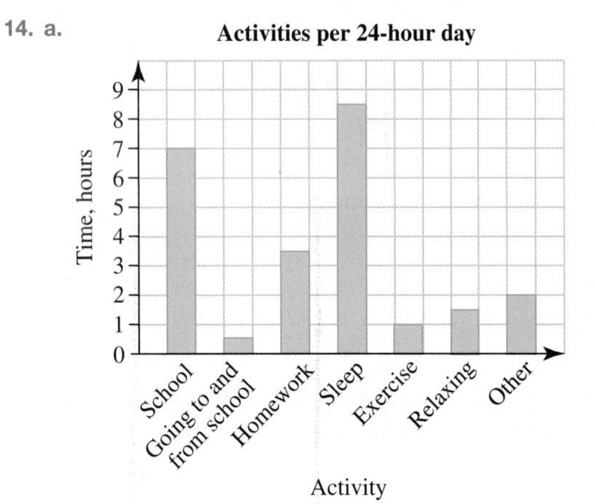

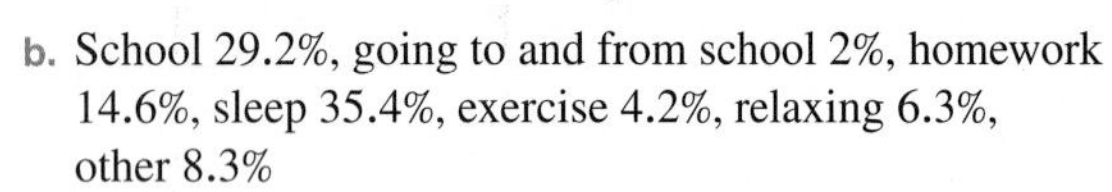

b. School 29.2%, going to and from school 2%, homework 14.6%, sleep 35.4%, exercise 4.2%, relaxing 6.3%, other 8.3%

c.

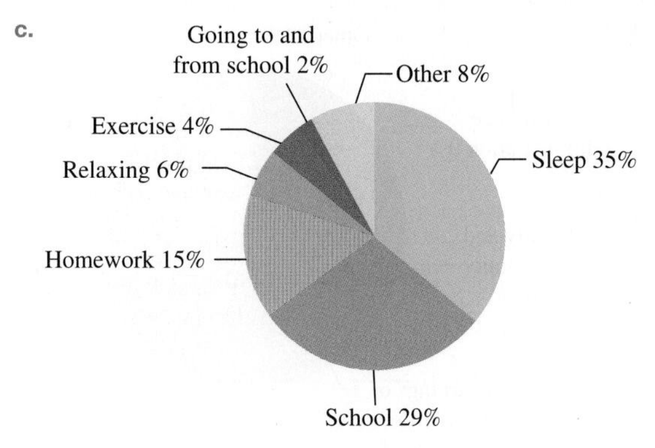

d. 17.5 hours

15. a. See the table at the bottom of the page.*

b. 1 May

c. 30 May

d. A decrease of 3.3 °C

e. An increase of 1.7 °C

16. Company A. Sample responses can be found in the worked solutions in the online resources.

*15a.

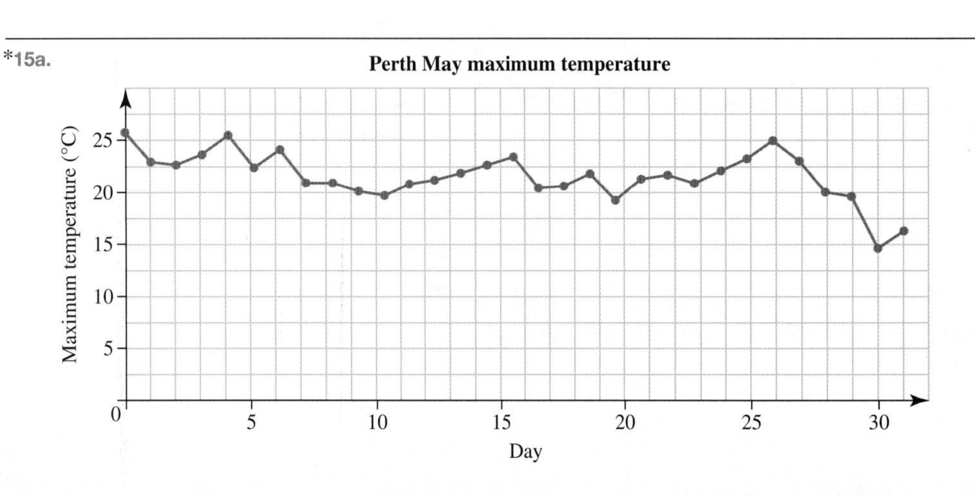

UNIT 2 Data and travel
TOPIC 3 Time and motion

13 Time

LESSON SEQUENCE

Fully worked solutions for this chapter are available online.

on Resources

Solutions Solutions — Chapter 13 (sol-1266)

Digital documents Learning matrix — Chapter 13 (doc-41777)
Quick quizzes — Chapter 13 (doc-41778)
Chapter summary — Chapter 13 (doc-41779)

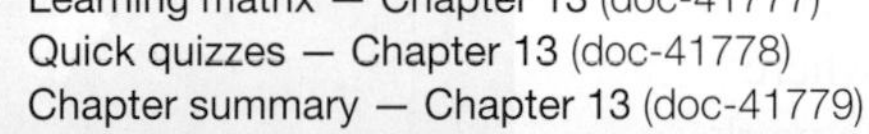

LESSON
13.1 Overview

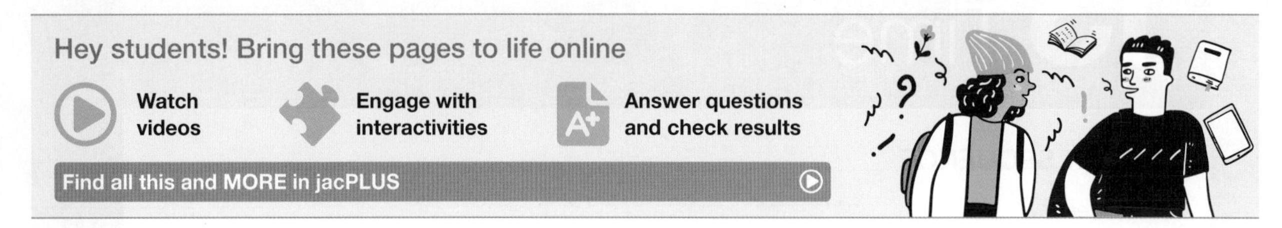

13.1.1 Introduction

Learning the concepts covered in this chapter is crucial for everyone, no matter their background. Mastering time management skills not only boosts efficiency in everyday tasks but also sets the stage for long-term success. By understanding how to represent time accurately and calculate intervals precisely, you gain tools to organise your schedule, meet deadlines, and make the most of your time. Plus, knowing how to interpret timetables and plan routes efficiently is essential for navigating the modern world, where getting from A to B smoothly is key for both work and leisure.

What's more, the knowledge you gain here goes beyond just practical applications — it also sharpens your critical thinking and problem-solving abilities. Through comparing travel times and tackling complex timetables, you'll develop analytical skills that come in handy in all sorts of situations. And as you become more familiar with time measurement and transportation planning, you'll also grow more responsible and independent, taking charge of your own schedule like a pro. Ultimately, the lessons from this chapter lay a solid foundation for personal growth, giving you skills that can be used in all areas of life.

13.1.2 Syllabus links

Lesson	Lesson title	Syllabus links
13.2	**Representing and calculating time and time intervals**	◯ Represent time using 12-hour and 24-hour clocks. ◯ Calculate time intervals, including time between, time ahead, time behind.
13.3	**Interpreting timetables**	◯ Interpret timetables for buses, trains and/or ferries.
13.4	**Interpreting complex timetables and planning routes [complex]**	◯ Use several timetables and/or electronic technologies to plan the most time-efficient routes [complex]. ◯ Interpret complex timetables, e.g. tide charts, sunrise charts and moon phases [complex].
13.5	**Compare travel time**	◯ Compare the time taken to travel a specific distance with various modes of transport.

LESSON
13.2 Representing and calculating time and time intervals

SYLLABUS LINKS

- Represent time using 12-hour and 24-hour clocks.
- Calculate time intervals, including time between, time ahead, time behind.

13.2.1 Time

- Time is divided into units including seconds, minutes, hours, days, weeks, months and years.
- A day is divided into two 12-hour periods:

Midnight to midday	am	ante meridiem (before noon)
Midday to midnight	pm	post meridiem (after noon)

- Time in hours and minutes can be given by either the 12-hour or 24-hour clock system.
- Representing time using the 24-hour clock eliminates the element of ambiguity.
- The 24-hour clock is commonly used in areas like military, aviation, navigation, tourism, meteorology, astronomy, computing, logistics, emergency services and hospitals, to avoid any confusion about whether it is am or pm time.
- 24-hour time always uses four digits to represent the time. E.g. 1500 hours or 15:00 hours is used for 3:00 pm, and midnight can be shown as 0000 (00:00) hours or 2400 (24:00) hours.

Time conversions

Converting from a 12-hour to a 24-hour clock:

- **Starting from the first hour of the day (12:00 am or midnight to 12:59 am), subtract 12 hours.**
  For example:
  12:00 am = 0:00
  12:15 am = 0:15
- **From 1:00 am to 12:59 pm, the hours and minutes remain the same.**
  For example:
  9:00 am = 9:00
  12:59 pm = 12:59
- **For times between 1:00 pm and 11:59 pm, add 12 hours.**
  For example:
  4:25 pm = 16:25
  11:59 pm = 23:59

Converting from a 24-hour to a 12-hour clock:

- **Starting from the first hour of the day (0:00/midnight to 0:59), add 12 hours and 'am' to the time.**
  For example:
  0:20 = 12:20 am
  0:45 = 12:45 am

- **From 1:00 to 11:59, simply add 'am' to the time.**
  For example:
  3:25 = 3:25 am
  10:30 = 10:30 am
- **For times between 12:00 and 12:59, just add 'pm' to the time.**
  For example:
  12:15 = 12:15 pm
  12:55 = 12:55 pm
- **For times between 13:00 and 23:59, subtract 12 hours and add 'pm' to the time.**
  For example:
  17:45 = 5:45 pm
  22:30 = 10:30 pm

Time is divided into units. There are:

- **60 seconds in 1 minute**
- **60 minutes in 1 hour**
- **24 hours in 1 day**
- **7 days in 1 week**
- **2 weeks in 1 fortnight**
- **about 4 weeks in 1 month**
- **12 months in 1 year**
- **about 365 days in 1 year**
- **10 years in 1 decade**
- **100 years in 1 century**
- **1000 years in 1 millennium.**

Resources

Interactivity Converting between units of time (int-6910)

13.2.2 Displaying time

- Time can be shown in 12-hour time or 24-hour time.

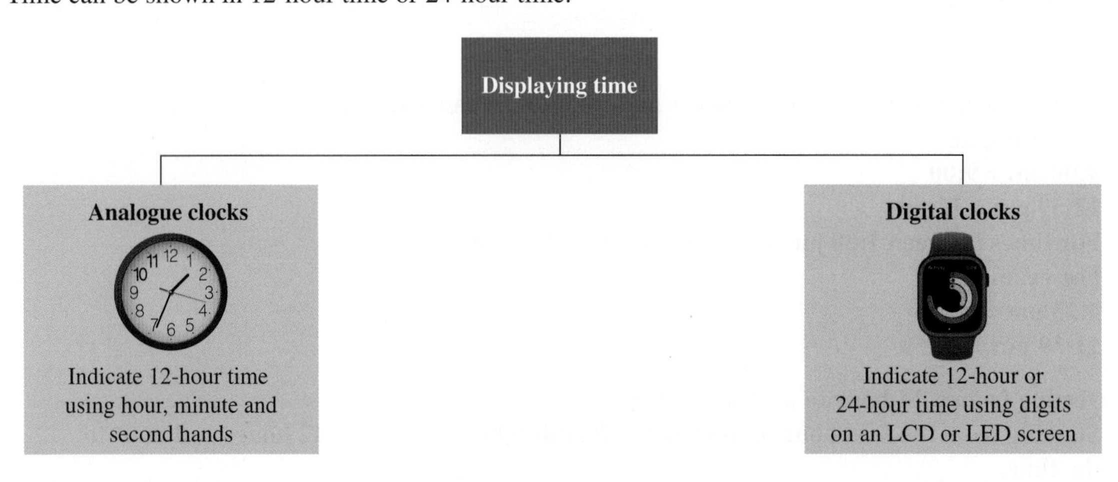

Time shown in writing	Time on an analogue clock	Time on a digital clock	
		12-hour time	24-hour time
Three o'clock in the morning		3:00 AM	03:00
Twenty-five minutes past 1 in the afternoon		1:25 PM	13:25
Ten minutes to 7 in the evening		6:50 PM	18:50

tlvd-6028

WORKED EXAMPLE 1 Expressing time in hours, minutes and seconds

Express the following times in hours, minutes and seconds.

a. 7.3 hours **b.** $4\frac{3}{8}$ **hours**

THINK	WRITE
a. 1. There are 7 hours and 0.3 of an hour in 7.3 hours.	**a.** $7.3 = 7 \text{ hours} + 0.3 \text{ of an hour}$
2. There are 60 minutes in an hour, so there are 0.3 of 60 minutes.	$\begin{aligned} 7.3 &= 7 \text{ hours} + 0.3 \times 60 \text{ minutes} \\ &= 7 \text{ hours} + 18 \text{ minutes} \end{aligned}$
3. Write the answer.	7.3 hours is the same as 7 hours and 18 minutes.
b. 1. There are 4 hours and $\frac{3}{8}$ of an hour in $4\frac{3}{8}$ hours.	**b.** $4\frac{3}{8} = 4 \text{ hours} + \frac{3}{8} \text{ of an hour}$
2. There are 60 minutes in an hour so there are $\frac{3}{8}$ of 60 minutes.	$\begin{aligned} 4\frac{3}{8} &= 4 \text{ hours} + \frac{3}{8} \times 60 \text{ minutes} \\ &= 4 \text{ hours} + 22.5 \text{ minutes} \end{aligned}$

3. There are 60 seconds in a minute so there are 0.5 of 60 seconds.

$4\frac{3}{8} = 4 \text{ hours} + 22 \text{ minutes} + 0.5 \times 60 \text{ seconds}$

$= 4 \text{ hours} + 22 \text{ minutes} + 30 \text{ seconds}$

4. Write the answer.

$4\frac{3}{8}$ hours is equal to 4 hours, 22 minutes and 30 seconds.

Digital technology box

Convert time into degrees, minutes, seconds

1. Make sure the handheld is in DEGREE mode by pressing [MODE], scrolling to DEGREE and pressing [ENTER]. Press [2nd] [MODE] to QUIT and return to the home screen.
2. Input 7.3.
3. Press [2nd] [APPS] to access the ANGLE menu, then press the [4] key to select the ▶ DMS function (convert to degrees, minutes, seconds).

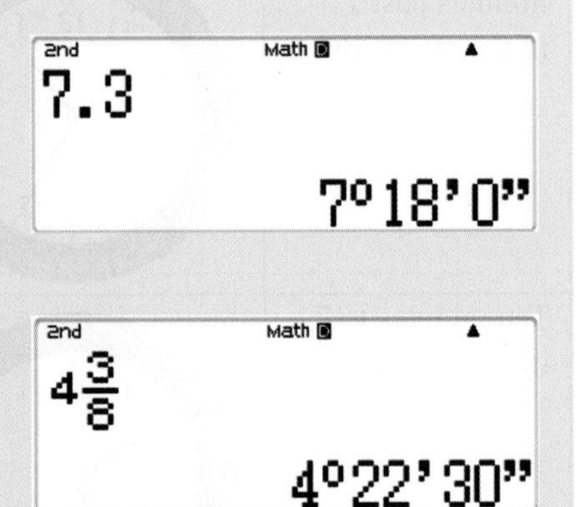

4. Input $4\frac{3}{8}$.
5. Repeat Step 3.

WORKED EXAMPLE 2 Calculating time in hours and minutes

A movie runs for 132 minutes. Determine the duration of the movie in hours and minutes.

THINK

1. There are 60 minutes in one hour, so divide 132 by 60 to get the number of hours.
2. Two hours are equal to 120 minutes ($2 \times 60 = 120$). Subtract 120 from 132 to determine how many minutes over two hours the movie lasts.
   An alternative way to determine the number of minutes over two hours is to convert 0.2 into minutes by multiplying by 60.
3. Write the answer.

WRITE

$132 \div 60 = 2.2$ hours

$132 - 120 = 12$ minutes

$0.2 \times 60 = 12$ minutes

The movie goes for 2 hours and 12 minutes.

13.2.3 Time intervals

- Often we want to know the time interval or difference between times.
- This may be because we want to determine how many hours we worked between 9 am and 1:30 pm, or to know how long our flight between 13:25 and 15:55 will take.

WORKED EXAMPLE 3 Determining time intervals

A flight on which a flight attendant was rostered landed in Sydney at 8:07 pm and left Adelaide at 6:53 pm Sydney time.
Determine the duration of the flight.

THINK	WRITE
1. Determine the number of minutes from the departure to the next hour (7:00 pm).	6:53 pm to 7:00 pm = 7 minutes
2. Determine the time from 7:00 pm to 8:00 pm.	7:00 pm to 8:00 pm = 1 hour
3. Determine the number of minutes from 8:00 pm to 8:07 pm.	8:00 pm to 8:07 pm = 7 minutes
4. Add the times together.	7 minutes + 1 hour + 7 minutes = 1 hour and 14 minutes
5. Write the answer.	The flight took 1 hour and 14 minutes.

WORKED EXAMPLE 4 Determining time in days

Determine the number of days between 15 April and 21 June in the same year.

THINK	WRITE
1. You are counting the days between the dates. Count the remaining days in April; there are 30 days in April.	There are 15 days remaining in April.
2. Count the days in May. There are 31 days in May.	There are 31 days in May.
3. Count the days in June up to, but not including, 21 June.	There are 20 days up to but not including 21 June.
4. Calculate the total number of days.	$\text{Total days} = 15 + 31 + 20$ $= 66$

Exercise 13.2 Representing and calculating time and time intervals

learnon

13.2 Quick quiz	13.2 Exercise

Simple familiar	Complex familiar	Complex unfamiliar
1, 2, 3, 4, 5, 6, 7, 8, 9, 10, 11, 12, 13, 14, 15, 16, 17, 18, 19, 20	N/A	N/A

Simple familiar

1. **WE1** Express the following times in hours, minutes and seconds.

   a. 5.8 hours
   b. $9\frac{7}{8}$ hours

2. Express the following times in hours, minutes and seconds.

   a. 12.156 hours
   b. $3\frac{2}{9}$ hours

3. Express in words the time displayed on each of the following clocks.

a.

b.

c.

d.

e.

f.

4. Express the following times as 24-hour times.

   a. 10:25 am
   b. 7:33 am
   c. 1:45 pm
   d. 8:12 pm
   e. 10:06 pm
   f. 11:45 pm

5. **WE2** A horse-trail ride lasted 156 minutes.

Calculate the trail ride in hours and minutes.

6. A game of Rugby is played over two 40-minute halves. If the game also had 2 minutes of injury time in the first half and 3 minutes of injury time in the second half, calculate the duration of the entire game in hours and minutes.

7. Express the following 24-hour times as digital am or pm times.

a. 15:51 b. 20:22 c. 03:15
d. 11:31 e. 09:02 f. 22:15

8. Express twelve o'clock at night in 24-hour times and digital am or pm times.

9. Express each of the following time periods as minutes.

a. 3 hours b. $5\frac{1}{4}$ hours c. 1 day
d. $\frac{3}{4}$ hour e. 3 hours and 18 minutes f. 6 hours and 34 minutes

10. Determine if '2 hours and 25 minutes' and '2.25 hours' represent the same length of time. Explain your answer.

11. Express the following time periods in hours, minutes and seconds.

a. 210 minutes b. 305 minutes c. 77 minutes
d. 2.45 hours e. 9.55 hours f. 140.45 hours
g. 8.815 hours h. 153.865 minutes

12. WE3 A flight left Perth at 1:12 pm and landed in Brisbane at 5:27 pm Perth time. Calculate the duration of the flight.

13. Harper started playing cricket at 08:23 and finished at 13:20. Calculate how long she played cricket for.

14. Determine the time difference between:

a. 4:25 pm and 5:50 pm
b. 6:30 pm and 2:45 am
c. 7:20 am on Monday and 6:30 pm the following day (Tuesday)
d. 1:20 pm on Wednesday and 9:09 am the following Friday
e. 01:25 hours and 23:45 hours
f. 07:15 hours and 15:50 hours.

15. Calculate the following times.

a. 1 hour after 12 noon
b. 3 hours before 7:15 am
c. 1 hour and 20 minutes after 8:30 am
d. 2 hours and 30 minutes before 7:45 pm
e. 4 hours and 14 minutes after 1:08 pm
f. 3 hours and 52 minutes before 3:25 pm

16. WE4 Determine the number of days between 12 July and 27 October in the same year.

17. Determine the number of days between 9 January and 5 May in 2018.

18. Determine the number of days until Christmas from the AFL Grand Final on 28 September.

19. The following table displays the time that a student spent travelling to school. The student left home at 7:25 am.

Activity	Time
Walking from home to the train station	27 minutes
Waiting for the train	12 minutes
Train journey	33 minutes
Walking from the train station to school	8 minutes

a. Calculate the amount of time the student spent travelling to school.
b. Determine whether the student was on time for registration at 08:45.

20. Use time units to help you answer each of the following. (*Note:* Assume that there are 365 days in 1 year.) Determine the number of:

a. seconds in 1 hour
b. seconds in 1 day
c. minutes in 1 day
d. hours in 1 non-leap year.

Fully worked solutions for this chapter are available online.

LESSON
13.3 Interpreting timetables

SYLLABUS LINKS

- Interpret timetables for buses, trains and/or ferries.

13.3.1 Timetables

- Timelines are useful for plotting events occurring over a particular period of time.
- However, these can become complex if a number of events are occurring close to each other.
- A **timetable** is a list of events that are scheduled to happen. The timetable usually makes these events easier to read. This is because it allows you to quickly see a lot of information.
- Timetables are commonly used for schools, airports, trains, buses and ferries.
- Translink is the Queensland Government service that helps users plan short public transport routes. For long-range journeys, travellers will still have to consult a comprehensive timetable of all stops. (Translink is later introduced in lesson 13.4.)
- The timetable below is for a train from Brisbane to Sydney.

	Brisbane to Sydney daily		
	Train A	**Train B**	**Train C**
Brisbane Roma Station		05:55	
Kyogle		07:53	
Casino		08:20	19:30
Grafton City	05:15	09:39	20:47
Coffs Harbour	06:26	11:05	22:10
Sawtell	06:35	11:14	22:18
Urunga	06:52	11:31	22:36
Nambucca Heads	07:08	11:47	22:51
Macksville	07:21	12:00	23:05
Eungai	07:37		
Kempsey	08:05	12:43	23:47
Wauchope	08:44	13:22	00:24
Kendall	09:05	13:54	00:43
Taree	09:52	14:41	01:31
Wingham	10:06	14:55	01:44
Gloucester	10:56	15:44	02:33
Dungog	12:06	16:38	03:26
Maitland	12:53	17:30	04:12
Broadmeadow	13:19	17:52	04:34
Fassifern	13:37	18:12	04:52
Wyong	14:04	18:39	05:23
Gosford	14:19	18:55	05:38
Hornsby	15:03	19:36	06:22
Strathfield	15:28	19:55	06:45
Central (Sydney)	15:43	20:10	07:01

WORKED EXAMPLE 5 Determining time intervals

Use the Brisbane to Sydney train timetable shown to answer the following questions.

a. If you caught the morning train from Brisbane, determine by what time you would expect to get to Central (Sydney).

b. If you wanted to get to Gosford by 6 pm, determine at what time you would need to catch the train from Taree.

THINK	WRITE
a. There is only one morning train from Brisbane, at 5:55 am. Follow it down until you get to the end at Central (Sydney) at 20:10.	a. It arrives at Central (Sydney) at 20:10, which is 8:10 pm.
b. 1. Look down the first column until you find Gosford and see which train gets you there by 6:00 pm.	b. Train A gets you there at 14:19, which is 2:19 pm.
2. Follow Train A for the previous stations until you find Taree.	The train leaves Taree at 9:52 am.
3. Write the answer.	You need to catch the train at Taree at 9:52 am.

Exercise 13.3 Interpreting timetables

learnon

13.3 Quick quiz on | 13.3 Exercise

Simple familiar	Complex familiar	Complex unfamiliar
1, 2, 3, 4, 5, 6, 7, 8, 9, 10, 11, 12, 13, 14, 15, 16	N/A	N/A

Simple familiar

Questions 1 and 2 refer to the following 1978 timeline.

1. WE5 Using the Australian and international events since 1978 timeline, answer the following:
   a. Identify when the Solomon Islands riots occurred.
   b. Identify when Google was founded.

2. Using the Australian and international events since 1978 timeline, answer the following:
   a. Identify how many years passed between China's invasion of Vietnam and the fall of the Berlin Wall.
   b. Identify how many years separated the invention of the World Wide Web and the Bali bombings.

Australian and international events since 1978

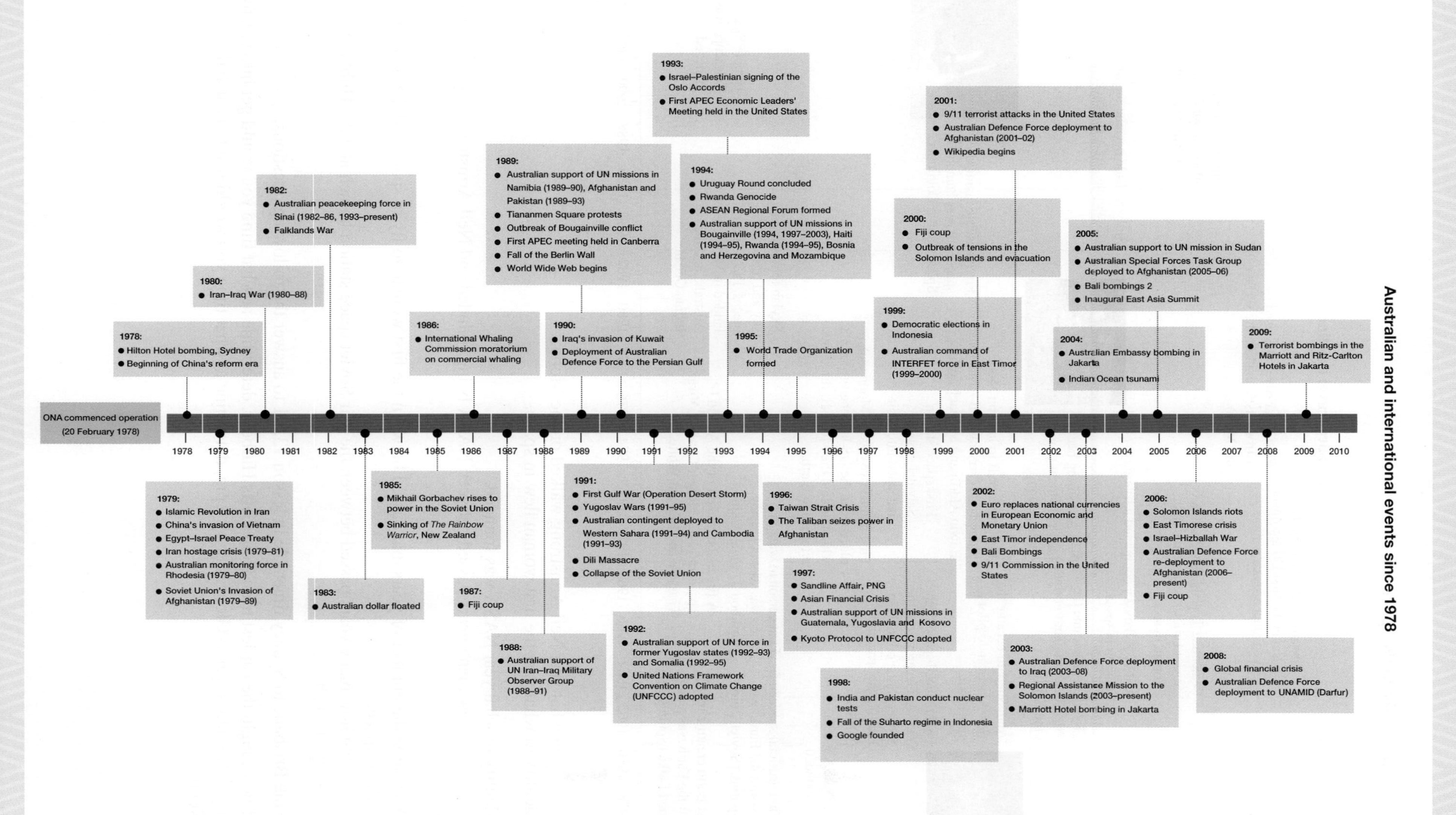

ONA commenced operation (20 February 1978)
1978 1979 1980 1981 1982 1983 1984 1985 1986 1987 1988 1989 1990 1991 1992 1993 1994 1995 1996 1997 1998 1999 2000 2001 2002 2003 2004 2005 2006 2007 2008 2009 2010
1978:
● Hilton Hotel bombing, Sydney
● Beginning of China's reform era
1979:
● Islamic Revolution in Iran
● China's invasion of Vietnam
● Egypt–Israel Peace Treaty
● Iran hostage crisis (1979–81)
● Australian monitoring force in Rhodesia (1979–80)
● Soviet Union's Invasion of Afghanistan (1979–89)
1980:
● Iran–Iraq War (1980–88)
1982:
● Australian peacekeeping force in Sinai (1982–86, 1993–present)
● Falklands War
1983:
● Australian dollar floated
1985:
● Mikhail Gorbachev rises to power in the Soviet Union
● Sinking of The Rainbow Warrior, New Zealand
1986:
● International Whaling Commission moratorium on commercial whaling
1987:
● Fiji coup
1988:
● Australian support of UN Iran–Iraq Military Observer Group (1988–91)
1989:
● Australian support of UN missions in Namibia (1989–90), Afghanistan and Pakistan (1989–93)
● Tiananmen Square protests
● Outbreak of Bougainville conflict
● First APEC meeting held in Canberra
● Fall of the Berlin Wall
● World Wide Web begins
1990:
● Iraq's invasion of Kuwait
● Deployment of Australian Defence Force to the Persian Gulf
1991:
● First Gulf War (Operation Desert Storm)
● Yugoslav Wars (1991–95)
● Australian contingent deployed to Western Sahara (1991–94) and Cambodia (1991–93)
● Dili Massacre
● Collapse of the Soviet Union
1992:
● Australian support of UN force in former Yugoslav states (1992–93) and Somalia (1992–95)
● United Nations Framework Convention on Climate Change (UNFCCC) adopted
1993:
● Israel–Palestinian signing of the Oslo Accords
● First APEC Economic Leaders' Meeting held in the United States
1994:
● Uruguay Round concluded
● Rwanda Genocide
● ASEAN Regional Forum formed
● Australian support of UN missions in Bougainville (1994, 1997–2003), Haiti (1994–95), Rwanda (1994–95), Bosnia and Herzegovina and Mozambique
1995:
● World Trade Organization formed
1996:
● Taiwan Strait Crisis
● The Taliban seizes power in Afghanistan
1997:
● Sandline Affair, PNG
● Asian Financial Crisis
● Australian support of UN missions in Guatemala, Yugoslavia and Kosovo
● Kyoto Protocol to UNFCCC adopted
1998:
● India and Pakistan conduct nuclear tests
● Fall of the Suharto regime in Indonesia
● Google founded
1999:
● Democratic elections in Indonesia
● Australian command of INTERFET force in East Timor (1999–2000)
2000:
● Fiji coup
● Outbreak of tensions in the Solomon Islands and evacuation
2001:
● 9/11 terrorist attacks in the United States
● Australian Defence Force deployment to Afghanistan (2001–02)
● Wikipedia begins
2002:
● Euro replaces national currencies in European Economic and Monetary Union
● East Timor independence
● Bali Bombings
● 9/11 Commission in the United States
2003:
● Australian Defence Force deployment to Iraq (2003–08)
● Regional Assistance Mission to the Solomon Islands (2003–present)
● Marriott Hotel bombing in Jakarta
2004:
● Australian Embassy bombing in Jakarta
● Indian Ocean tsunami
2005:
● Australian support to UN mission in Sudan
● Australian Special Forces Task Group deployed to Afghanistan (2005–06)
● Bali bombings 2
● Inaugural East Asia Summit
2006:
● Solomon Islands riots
● East Timorese crisis
● Israel–Hizballah War
● Australian Defence Force re-deployment to Afghanistan (2006–present)
● Fiji coup
2008:
● Global financial crisis
● Australian Defence Force deployment to UNAMID (Darfur)
2009:
● Terrorist bombings in the Marriott and Ritz-Carlton Hotels in Jakarta

Questions 3 to 5 refer to the space travel timeline.

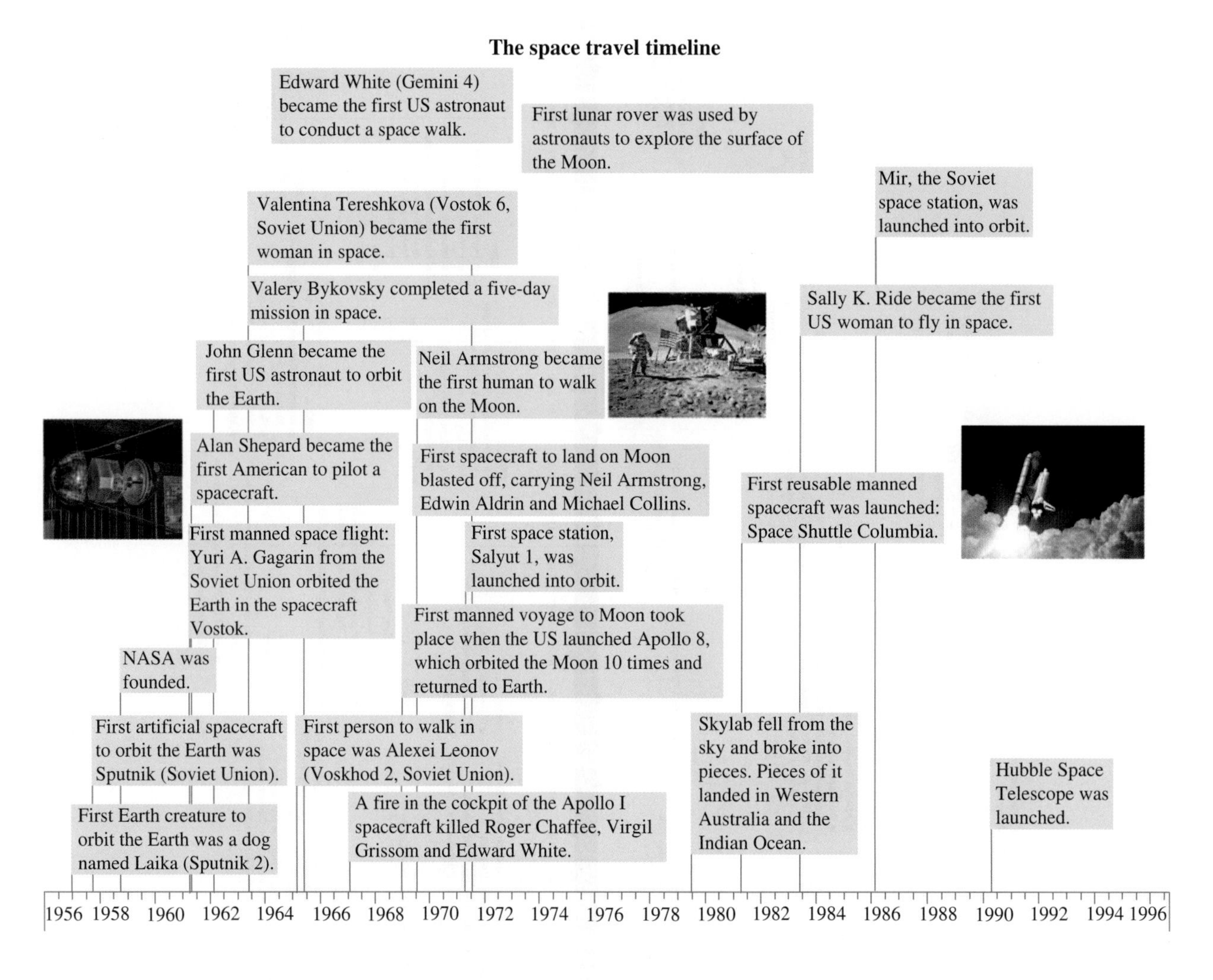

The space travel timeline
Edward White (Gemini 4) became the first US astronaut to conduct a space walk.
First lunar rover was used by astronauts to explore the surface of the Moon.
Mir, the Soviet space station, was launched into orbit.
Valentina Tereshkova (Vostok 6, Soviet Union) became the first woman in space.
Valery Bykovsky completed a five-day mission in space.
Sally K. Ride became the first US woman to fly in space.
John Glenn became the first US astronaut to orbit the Earth.
Neil Armstrong became the first human to walk on the Moon.
Alan Shepard became the first American to pilot a spacecraft.
First spacecraft to land on Moon blasted off, carrying Neil Armstrong, Edwin Aldrin and Michael Collins.
First reusable manned spacecraft was launched: Space Shuttle Columbia.
First manned space flight: Yuri A. Gagarin from the Soviet Union orbited the Earth in the spacecraft Vostok.
First space station, Salyut 1, was launched into orbit.
First manned voyage to Moon took place when the US launched Apollo 8, which orbited the Moon 10 times and returned to Earth.
NASA was founded.
First artificial spacecraft to orbit the Earth was Sputnik (Soviet Union).
First person to walk in space was Alexei Leonov (Voskhod 2, Soviet Union).
Skylab fell from the sky and broke into pieces. Pieces of it landed in Western Australia and the Indian Ocean.
Hubble Space Telescope was launched.
First Earth creature to orbit the Earth was a dog named Laika (Sputnik 2).
A fire in the cockpit of the Apollo I spacecraft killed Roger Chaffee, Virgil Grissom and Edward White.
1956 1958 1960 1962 1964 1966 1968 1970 1972 1974 1976 1978 1980 1982 1984 1986 1988 1990 1992 1994 1996

3. a. Identify in what year Neil Armstrong walked on the Moon.
   b. Identify in what year the first woman flew in space.
4. a. Identify in what year NASA was founded.
   b. Determine how many years passed between NASA being founded and Neil Armstrong walking on the moon.
5. a. Determine the number of years that passed between the first woman to fly in space and the first US woman to fly in space.
   b. Determine how many years passed between the Mir Soviet space station launch and the Hubble Space Telescope launch.
6. Use the Brisbane to Sydney timetable in lesson 13.3.1 to answer the following questions.
   a. If you caught the earliest train from Coffs Harbour, determine at what time you would get into Gosford.
   b. If you wanted to get to Gloucester by 2 pm, determine at what time you would need to catch the train from Kempsey.

7. Use the Brisbane to Sydney timetable in lesson 13.3.1 to answer the following questions.
   a. If you caught the evening train from Coffs Harbour, determine at what time you would get into Sydney.
   b. If you caught the evening train from Maitland, determine how long it would take to travel to Sydney.

Questions 8 and 9 refer to the CityHopper timetable shown (only part of the daily timetable is shown).

Towards North Quay   Towards Sydney Street

	Destination				
Stop	North Quay 2	North Quay 2	North Quay 2	North Quay 2	North Quay 2
Sydney Street ferry terminal, New Farm	8:00 am	8:30 am	9:00 am	9:30 am	10:00 am
Dockside ferry terminal, Kangaroo Point	8:04 am	8:34 am	9:04 am	9:34 am	10:04 am
Holman Street ferry terminal, Kangaroo Point	8:15 am	8:45 am	9:15 am	9:45 am	10:15 am
Eagle Street Pier ferry terminal, Brisbane City	8:20 am	8:50 am	9:20 am	9:50 am	10:20 am
Thornton Street ferry terminal, Kangaroo Point	8:25 am	8:55 am	9:25 am	9:55 am	10:25 am
Maritime Museum ferry terminal, South Brisbane	8:33 am	9:03 am	9:33 am	10:03 am	10:33 am
South Bank 3 ferry terminal, South Brisbane	8:37 am	9:07 am	9:37 am	10:07 am	10:37 am
North Quay 2 ferry terminal, Brisbane City	8:42 am	9:12 am	9:42 am	10:12 am	10:42 am

8. a. If you left Sydney Street at 8:30 am, determine at what time you would arrive at South Bank 3.
   b. If you wanted to arrive at the Maritime Museum no later than 10:15 am, determine at what time the ferry would depart Holman Street.

9. Determine how long the ferry takes to travel from Holman Street to North Quay 2.

Questions 10 and 11 refer to the City to Doomben train timetable.

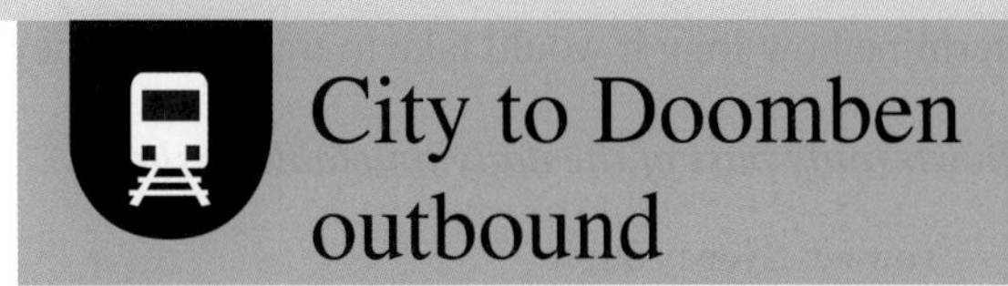

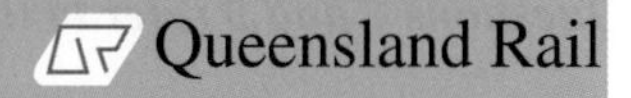

Monday to Thursday

Comes from Station	am	am	am	am	VYS am	VYS am	am	MNY am	am	am	am	am	am	am	pm
Park Road					7:09	7:39	8:07	8:37	9:09	9:39		10:39			
South Bank					7:12	7:42	8:10	8:40	9:12	9:42		10:42			
South Brisbane	----	----	----	----	7:14	7:44	8:12	8:42	9:14	9:44	----	10:44	----	----	----
Roma Street	5:19	5:49	6:19	6:49	7:19	7:49	8:17	8:47	9:19	9:49	10:16	10:49	11:16	11:46	12:16
Central arrive	5:21	5:51	6:21	6:51	7:21	7:51	8:19	8:49	9:21	9:51	10:19	10:51	11:19	11:49	12:19
Central depart	5:23	5:53	6:23	6:23	7:23	7:53	8:23	8:53	9:23	9:53	10:23	10:53	11:23	11:53	12:23
Fortitude Valley	5:25	5:55	6:25	6:55	7:25	7:55	8:25	8:55	9:25	9:55	10:25	10:55	11:25	11:55	12:25
Bowen Hills	5:28	5:58	6:28	6:58	7:28	7:58	8:28	8:58	9:28	9:58	10:28	10:58	11:28	11:58	12:28
Albion	5:32	6:02	6:32	7:02	7:32	8:02	8:32	9:02	9:32	10:02	10:32	11:02	11:32	11:02	12:32
Wooloowin	5:34	6:04	6:34	7:04	7:34	8:04	8:34	9:04	9:34	10:04	10:34	11:04	11:34	11:04	12:34
Eagle Junction	5:37	6:07	6:37	7:07	7:37	8:07	8:37	9:37	9:37	10:07	10:37	11:07	11:37	11:07	12:37
Clayfield	5:40	6:10	6:40	7:10	7:40	8:10	8:40	9:10	9:40	10:10	10:40	11:10	11:40	11:10	12:40
Hendra	5:42	6:12	6:42	7:12	7:42	8:12	8:42	9:12	9:42	10:12	10:42	11:12	11:42	11:12	12:42
Ascot	5:44	6:14	6:44	7:14	7:44	8:14	8:44	9:14	9:44	10:14	10:44	11:14	11:44	11:14	12:44
Doomben	5:48	6:18	6:46	7:16	7:46	8:16	8:46	9:16	9:46	10:16	10:46	11:16	11:46	11:16	12:46

Monday to Thursday (continued)

Comes from Station	pm	pm	pm	pm	SFC pm	pm	pm	IPS pm	pm	pm	pm	pm	pm	pm
Park Road							3:39		4:39	5:09	5:39	6:09		7:09
South Bank							3:42	3:44	4:42	5:12	5:42	6:12		7:12
South Brisbane	----	----	----	----	----	----		----	4:44	5:14	5:44	6:14	----	7:14
Roma Street	12:46	1:16	1:46		2:49	3:19	3:49	4:19	4:49	5:19	5:49	6:19	6:46	7:19
Central arrive	12:49	1:19	1:49		2:51	3:21	3:51	4:21	4:51	5:21	5:51	6:21	2:49	7:21
Central depart	12:53	1:23	1:53		2:53	3:23	3:53	4:23	4:53	5:23	5:53	6:23	6:53	7:23
Fortitude Valley	12:55	1:25	1:55		2:55	3:25	3:55	4:25	4:55	5:25	5:55	6:25	6:55	7:25
Bowen Hills	11:58	1:28	1:58		2:58	3:28	3:58	4:28	4:58	5:28	5:58	6:28	6:58	7:28
Albion	1:02	1:32	2:02		3:02	3:32	4:02	4:32	5:02	5:32	6:02	6:32	7:02	7:32
Wooloowin	1:04	1:34	2:04		3:04	3:34	4:04	4:34	5:04	5:34	6:04	6:34	7:04	7:34
Eagle Junction	1:07	1:37	2:07		3:07	3:37	4:07	4:37	5:07	5:37	6:07	6:37	7:07	7:37
Clayfield	1:10	1:40	2:10		3:10	3:40	4:10	4:40	5:10	5:40	6:10	6:40	7:10	7:40
Hendra	1:12	1:42	2:12		3:12	3:42	4:12	4:42	5:12	5:42	6:12	6:42	7:12	7:42
Ascot	1:14	1:44	2:14		3:14	3:44	4:14	4:44	5:14	5:44	6:14	6:44	7:14	7:44
Doomben	1:16	1:46	2:16		3:16	3:46	4:16	4:46	5:16	5:46	6:16	6:46	7:16	7:46

10. a. Determine whether a train is leaving Roma Street station at 7:49 pm.

b. If you needed to get to Doomben at 9:30 am, determine at what time you would need to catch the train from Fortitude Valley.

11. a. Determine how many trains leave from Park Road station between 9:00 am and 10:00 am.

b. If I needed to be at a friend's home, which is a 15-minute walk from Eagle Junction station, for dinner at 6:00 pm, identify the latest train I could catch from South Brisbane.

c. If you caught the 12:53 am train from Central, determine how long it would take to travel to Doomben station.

Use the bus timetable shown to answer questions 12 and 13.

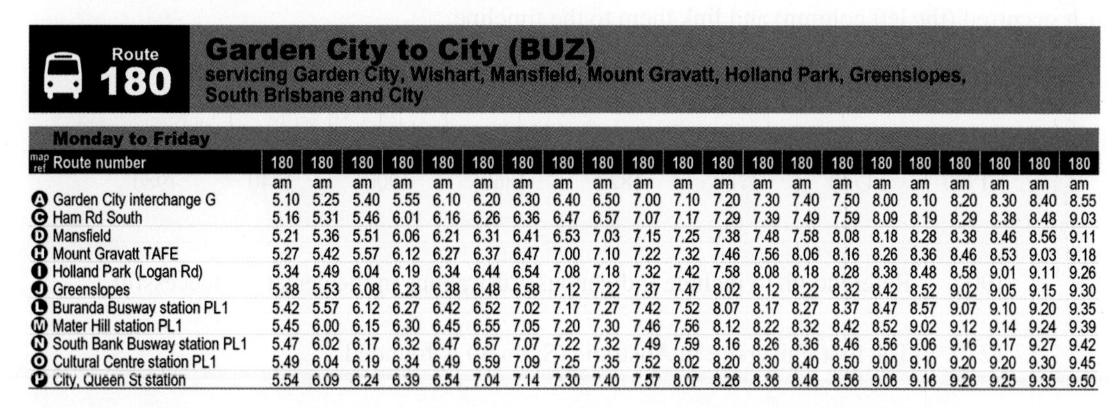

Route 180 **Garden City to City (BUZ)**
servicing Garden City, Wishart, Mansfield, Mount Gravatt, Holland Park, Greenslopes, South Brisbane and City

Monday to Friday

map ref	Route number	180	180	180	180	180	180	180	180	180	180	180	180	180	180	180	180	180	180	180	180	180
		am	am	am	am	am	am	am	am	am	am	am	am	am	am	am	am	am	am	am	am	am
A	Garden City interchange G	5.10	5.25	5.40	5.55	6.10	6.20	6.30	6.40	6.50	7.00	7.10	7.20	7.30	7.40	7.50	8.00	8.10	8.20	8.30	8.40	8.55
C	Ham Rd South	5.16	5.31	5.46	6.01	6.16	6.26	6.36	6.47	6.57	7.07	7.17	7.29	7.39	7.49	7.59	8.09	8.19	8.29	8.38	8.48	9.03
D	Mansfield	5.21	5.36	5.51	6.06	6.21	6.31	6.41	6.53	7.03	7.15	7.25	7.38	7.48	7.58	8.08	8.18	8.28	8.38	8.46	8.56	9.11
H	Mount Gravatt TAFE	5.27	5.42	5.57	6.12	6.27	6.37	6.47	7.00	7.10	7.22	7.32	7.46	7.56	8.06	8.16	8.26	8.36	8.46	8.53	9.03	9.18
I	Holland Park (Logan Rd)	5.34	5.49	6.04	6.19	6.34	6.44	6.54	7.08	7.18	7.32	7.42	7.58	8.08	8.18	8.28	8.38	8.48	8.58	9.01	9.11	9.26
J	Greenslopes	5.38	5.53	6.08	6.23	6.38	6.48	6.58	7.12	7.22	7.37	7.47	8.02	8.12	8.22	8.32	8.42	8.52	9.02	9.05	9.15	9.30
L	Buranda Busway station PL1	5.42	5.57	6.12	6.27	6.42	6.52	7.02	7.17	7.27	7.42	7.52	8.07	8.17	8.27	8.37	8.47	8.57	9.07	9.10	9.20	9.35
M	Mater Hill station PL1	5.45	6.00	6.15	6.30	6.45	6.55	7.05	7.20	7.30	7.46	7.56	8.12	8.22	8.32	8.42	8.52	9.02	9.12	9.14	9.24	9.39
N	South Bank Busway station PL1	5.47	6.02	6.17	6.32	6.47	6.57	7.07	7.22	7.32	7.49	7.59	8.16	8.26	8.36	8.46	8.56	9.06	9.16	9.17	9.27	9.42
O	Cultural Centre station PL1	5.49	6.04	6.19	6.34	6.49	6.59	7.09	7.25	7.35	7.52	8.02	8.20	8.30	8.40	8.50	9.00	9.10	9.20	9.20	9.30	9.45
P	City, Queen St station	5.54	6.09	6.24	6.39	6.54	7.04	7.14	7.30	7.40	7.57	8.07	8.26	8.36	8.46	8.56	9.06	9.16	9.26	9.25	9.35	9.50

Source: Reproduced by permission of Translink. Department of Transport and Main Roads. © State of Queensland. (March 2023). *Route 180. Garden City to City (BUZ) : servicing Garden City, Wishart, Mansfield, Mount Gravatt, Holland Park, Greenslopes, South Brisbane and City*.

12. a. Determine the time the bus leaves Mansfield between 6:00 am and 6:10 am.

b. To arrive at South Bank Busway station before 9:00 am, determine which bus you would need to catch from Ham Rd South.

c. Determine how many buses leave Mount Gravatt TAFE between 8:00 am and 9:00 am.

13. a. If you had to arrive at City, Queen St station by 8:00 am, determine at what time you should catch the bus from Greenslopes.

b. Your work is a 15-minute walk from Mater Hill station. If you need to be at work by 9:00 am, determine at what time you should catch the bus from Mansfield.

The bus timetable below is to be used for questions 14 and 15.

Airport–City to Varsity Lakes
outbound

QueenslandRail

Monday to Friday

Station	am	am	am	am	am	am	am	am	am	am	am	am	am	am	am	am	am	am	am	am	am
Domestic			5:04	5:34	6:04	6:34	7:04	7:34	7:49	8:04	8:19	8:34	8:49	9:04	9:19	9:34	9:49	10:04	10:19	10:34	11:04
International			5:08	5:38	6:08	6:38	7:08	7:38	7:53	8:08	8:23	8:38	8:53	9:08	9:23	9:38	9:53	10:08	10:23	10:38	11:08
Eagle Junction			**5:15**	**5:45**	**6:15**	**6:45**	**7:15**	**7:45**	**8:00**	**8:15**	**8:30**	**8:45**	**9:00**	**9:15**	**9:30**	**9:45**	**10:00**	**10:15**	**10:30**	**10:45**	**11:15**
Wooloowin			5:17	5:47	6:17	6:47	7:17	7:47	8:02	8:17	8:32	8:47	9:02	9:17	9:32	9:47	10:02	10:17	10:32	10:47	11:17
Albion	----	----	5:19	5:49	6:19	6:49	7:19	7:49	8:04	8:19	8:34	8:49	9:04	9:19	9:34	9:49	10:04	10:19	10:34	10:49	11:19
Bowen Hills	4:23	4:53	5:23	5:53	6:23	6:53	7:23	7:53	8:08	8:23	8:38	8:53	9:08	9:23	9:38	9:53	10:08	10:23	10:38	10:53	11:23
Fortitude Valley	4:25	4:55	5:25	5:55	6:25	6:55	7:25	7:55	8:10	8:25	8:40	8:55	9:10	9:25	9:40	9:55	10:10	10:25	10:40	10:55	11:25
Central arrive	**4:28**	**4:58**	**5:28**	**5:58**	**6:28**	**6:58**	**7:28**	**7:58**	**8:13**	**8:28**	**8:43**	**8:58**	**9:13**	**9:28**	**9:43**	**9:58**	**10:13**	**10:28**	**10:43**	**10:58**	**11:28**
Central depart	**4:29**	**4:59**	**5:29**	**5:59**	**6:29**	**6:59**	**7:29**	**7:59**	**8:14**	**8:29**	**8:44**	**8:59**	**9:14**	**9:29**	**9:44**	**9:59**	**10:14**	**10:29**	**10:44**	**10:59**	**11:29**
Roma Street	4:32	5:02	5:32	6:02	6:32	7:02	7:32	8:02	8:17	8:32	8:47	9:02	9:18	9:32	9:47	10:02	10:17	10:32	10:48	11:02	11:32
South Brisbane	4:37	5:07	5:37	6:07	6:37	7:07	7:37	8:07	8:22	8:37	8:52	9:07	----	9:37	9:52	10:07	10:22	10:37	----	11:07	11:37
South Bank	4:39	5:09	5:39	6:09	6:39	7:09	7:39	8:09	8:24	8:39	8:54	9:09		9:39	9:54	10:09	10:24	10:39		11:09	11:39
Park Road	**4:42**	**5:12**	**5:42**	**6:12**	**6:42**	**7:12**	**7:42**	**8:12**	**8:27**	**8:42**	**8:57**	**9:12**		**9:42**	**9:57**	**10:12**	**10:27**	**10:42**		**11:12**	**11:42**
Altandi	4:57	5:27	5:57	6:27	6:57	7:27	7:57	8:27	----	8:57	----	9:27		9:57	----	10:27	----	10:57		11:27	11:57
Loganlea	5:10	5:40	6:10	6:40	7:10	7:40	8:10	8:40		9:10		9:40		10:10		10:40		11:10		11:40	12:10
Beenleigh	**5:20**	**5:50**	**6:20**	**6:50**	**7:20**	**7:50**	**8:20**	**8:50**		**9:20**		**9:50**		**10:20**		**10:50**		**11:20**		**11:50**	**12:20**
Ormeau	5:27	5:57	6:27	6:57	7:27	7:57	8:27	8:57		9:27		9:57		10:27		10:57		11:27		11:57	12:27
Coomera	5:32	6:02	6:32	7:02	7:32	8:02	8:32	9:02		9:32		10:02		10:32		11:02		11:32		12:02	12:32
Helensvale	5:37	6:07	6:37	7:07	7:37	8:07	8:37	9:07		9:37		10:07		10:37		11:07		11:37		12:07	12:37
Nerang	5:42	6:12	6:42	7:12	7:42	8:12	8:42	9:12		9:42		10:12		10:42		11:12		11:42		12:12	12:42
Robina	5:48	6:18	6:48	7:18	7:48	8:18	8:48	9:18		9:48		10:18		10:48		11:18		11:48		12:18	12:48
Varsity Lakes	5:52	6:22	6:52	7:22	7:52	8:22	8:52	9:22		9:52		10:22		10:52		11:22		11:52		12:22	12:52

14. a. Determine how long the train takes to travel from South Bank to Coomera.

b. Determine how many buses leave Domestic before 8:00 am.

15. a. To arrive at Park Road as late as you can but before 9:00 am, determine at what time you should catch the bus from Albion.

b. You caught the bus from International at 9:38 am to South Bank, where you stayed for 1 hour, and then you caught the bus to Robina. Determine at what time you arrived at Robina.

16. Fix the table shown by matching the historical events (in the right column) in Australia's history with the year it occurred (the left column) and link them to the timeline.

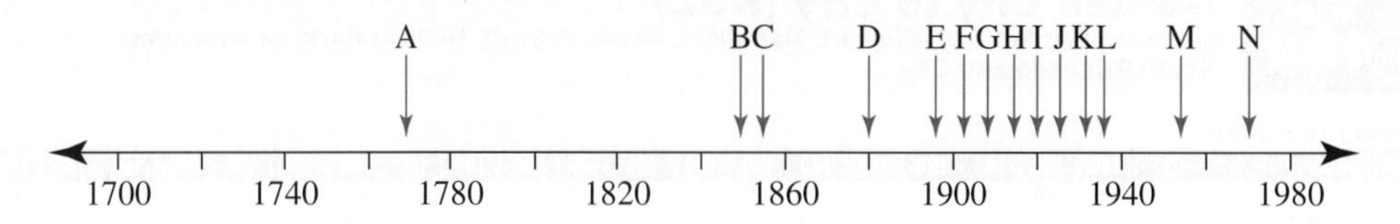

1902	Referendum to include Indigenous Australians in the census is successful.
1770	The first university in Australia (Sydney) is founded.
1880	Eureka Rebellion takes place.
1915	Captain James Cook lands at Botany Bay.
1967	Australian women gain the right to vote and stand for Parliament.
1850	Ned Kelly is hanged.
1956	Banjo Patterson publishes 'The Man from Snowy River'.
1930	Australians land at ANZAC Cove.
1890	Australia becomes a federation.
1932	Sydney Harbour Bridge opens.
1923	Vegemite is first produced.
1854	Phar Lap wins the Melbourne Cup.
1901	QANTAS is founded.
1920	Melbourne Olympics and TV come to Australia.

Fully worked solutions for this chapter are available online.

LESSON
13.4 Interpreting complex timetables and planning routes [complex]

SYLLABUS LINKS

- Use several timetables and/or electronic technologies to plan the most time-efficient routes [complex].
- Interpret complex timetables, e.g. tide charts, sunrise charts and moon phases [complex].

Source: Essential Mathematics Senior Syllabus 2024 © State of Queensland (QCAA) 2024; licensed under CC BY 4.0.

13.4.1 Planning routes

- When planning a route, look for the quickest option.
- You can use different modes of transport to determine your quickest option.

WORKED EXAMPLE 6 Designing a trip

You are planning a trip to visit your family. Use the information from the Translink journey planner shown to answer the following questions.

a. Determine the destination of the planned trip.

b. Determine the shortest time suggested for the trip.

c. Calculate the total amount of walking required in the shortest option.

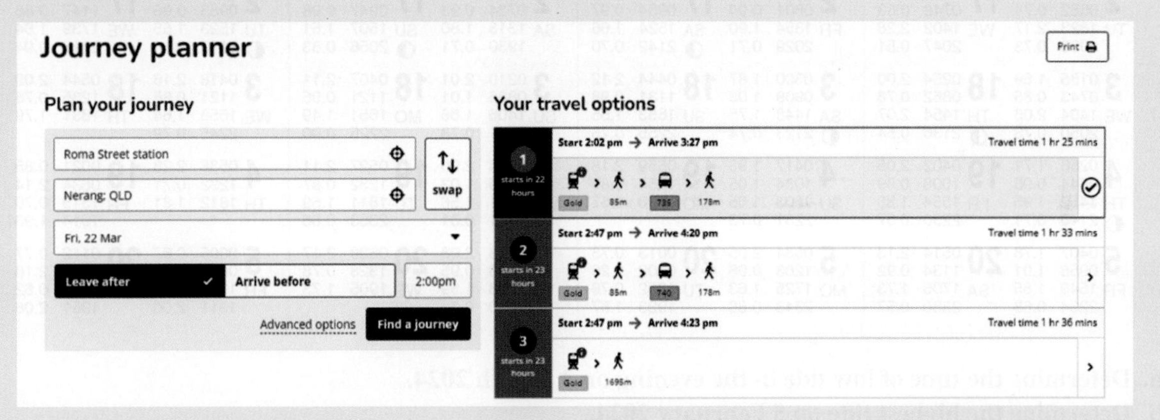

Source: **Reproduced by permission of Translink. Department of Transport and Main Roads. © State of Queensland. (n.d.). *Journey planner : plan your journey : Roma Street station to Nerang QLD : your travel options [image]*, viewed April 2024.**

THINK	WRITE
a. Look for where the trip starts from and where it goes to.	a. The trip is to Nerang, Qld.
b. Look at each of the options and look at the time each option takes.	b. The shortest time for this trip is option 1, which takes 1 hour 25 minutes.
c. Look at option 1 and add the distances that need to be walked.	c. Option 1 walk $= 85 + 178$ $= 263$ m

13.4.2 Interpreting complex timetables

- There are many types of timetables.
- Some of the more complex timetables are tidal charts, moon phases and sunrise charts.
- Tidal charts, also known as tide tables or tidal predictions, provide information about the predicted times and heights of tides at specific locations. These charts are essential for various activities such as boating, fishing, and coastal navigation.
- There are two low tides, and two high tides in between the low tides, per 24 hours.
- Moon phase timetables provide information about the different phases of the moon throughout a specific period, typically a month or a year. These timetables are useful for various purposes, including astronomy, gardening, and cultural or religious practices.
- Sunrise charts and timetables provide information about the times of sunrise and sunset for a specific location over a given period, typically a month or a year. These charts are valuable for various activities such as outdoor photography, planning of outdoor events, and agricultural practices.

tlvd-11451

WORKED EXAMPLE 7 Interpreting a complex timetable

Using the tidal chart shown, answer the following questions.

AUSTRALIA, EAST COAST – BRISBANE BAR **2024**

LAT 27° 22' S LONG 153° 10' E

Times and Heights of High and Low Waters

Time Zone –1000

JANUARY

Day	Time	m	Day	Time	m
1	0022	1.70	16	0058	1.98
	0609	0.64		0653	0.48
MO	1245	2.26	TU	1316	2.47
	1928	0.73		2003	0.48
2	0105	1.70	17	0153	1.99
	0652	0.74		0748	0.63
TU	1323	2.17	WE	1402	2.28
	2006	0.73		2047	0.51
3	0155	1.69	18	0254	2.00
	0743	0.85		0852	0.78
WE	1404	2.06	TH	1454	2.07
	2050	0.73	◑	2136	0.54
4	0256	1.71	19	0402	2.05
	0844	0.95		1008	0.89
TH	1452	1.95	FR	1554	1.88
◐	2140	0.71		2230	0.57
5	0407	1.78	20	0514	2.13
	0958	1.01		1134	0.92
FR	1549	1.85	SA	1705	1.73
	2234	0.68		2330	0.57

FEBRUARY

Day	Time	m	Day	Time	m
1	0113	1.88	16	0220	2.16
	0711	0.82		0834	0.85
TH	1315	2.05	FR	1418	1.86
	1948	0.68		2044	0.62
2	0200	1.87	17	0326	2.12
	0801	0.94		0954	0.97
FR	1354	1.90	SA	1524	1.66
	2029	0.71	◑	2142	0.70
3	0300	1.87	18	0444	2.12
	0909	1.03		1131	0.98
SA	1448	1.75	SU	1653	1.55
◐	2123	0.74		2255	0.75
4	0417	1.93	19	0559	2.18
	1034	1.05		1254	0.89
SU	1603	1.65	MO	1820	1.57
	2231	0.73			
5	0534	2.05	20	0013	0.73
	1203	0.98		0700	2.26
MO	1725	1.63	TU	1352	0.79
	2343	0.68		1923	1.67

MARCH

Day	Time	m	Day	Time	m
1	0038	2.08	16	0148	2.28
	0648	0.82		0824	0.89
FR	1234	1.95	SA	1351	1.66
	1856	0.65		1955	0.70
2	0118	2.04	17	0249	2.17
	0734	0.93		0947	0.98
SA	1313	1.80	SU	1507	1.51
	1930	0.71	◑	2056	0.83
3	0210	2.01	18	0407	2.11
	0836	1.01		1121	0.96
SU	1406	1.66	MO	1651	1.49
	2021	0.78		2225	0.90
4	0324	2.00	19	0527	2.11
	1005	1.03		1232	0.87
MO	1531	1.56	TU	1811	1.59
◐	2141	0.81		2353	0.86
5	0453	2.08	20	0630	2.17
	1143	0.95		1323	0.78
TU	1708	1.59	WE	1905	1.73
	2310	0.76			

APRIL

Day	Time	m	Day	Time	m
1	0141	2.15	16	0324	2.10
	0824	0.96		1045	0.92
MO	1351	1.59	TU	1629	1.51
	1949	0.79	◑	2147	0.98
2	0251	2.12	17	0440	2.07
	0953	0.95		1147	0.86
TU	1523	1.55	WE	1739	1.64
◐	2113	0.84		2316	0.94
3	0418	2.16	18	0544	2.09
	1121	0.86		1235	0.78
WE	1658	1.64	TH	1831	1.79
	2245	0.79			
4	0536	2.28	19	0021	0.85
	1232	0.71		0634	2.14
TH	1812	1.81	FR	1315	0.70
				1914	1.93
5	0005	0.67	20	0112	0.77
	0640	2.41		0717	2.18
FR	1329	0.56	SA	1350	0.62
	1911	2.00		1951	2.06

a. **Determine the time of low tide in the evening on 3 March 2024.**
b. **Determine the highest tide on 5 February 2024.**
c. **Determine the time of low tide on the morning of 1 January 2024.**

THINK	WRITE
a. Look for 3 March and the lower tide in the afternoon (pm).	a. The low tide on 3 March is 2021 or 8:21 pm.
b. Look for 5 February and identify the highest tide value on this day.	b. The highest tide on 5 February is 2.05 m at 5:34 am.
c. Look at 1 January and the lower tide in the morning (am).	c. The low tide on 1 January is 0609 or 6:09 am.

Resources

Weblink Queensland Government Translink journey planner

Exercise 13.4 Interpreting complex timetables and planning routes [complex]

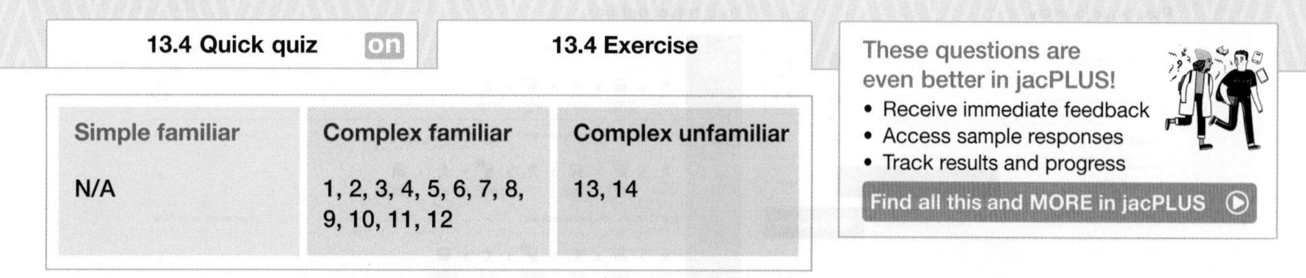

13.4 Quick quiz | 13.4 Exercise

Simple familiar	Complex familiar	Complex unfamiliar
N/A	1, 2, 3, 4, 5, 6, 7, 8, 9, 10, 11, 12	13, 14

These questions are even better in jacPLUS!
- Receive immediate feedback
- Access sample responses
- Track results and progress

Find all this and MORE in jacPLUS

Complex familiar

1. To get to your friend's house, you have to walk for 8 minutes to the train station, catch the train that is scheduled to take 24 minutes and then a bus that takes another 13 minutes, before a final walk to your friend's house that takes 18 minutes.
   Calculate the total time it takes to get to your friend's house from your house.

2. WE6 Use the information from the Translink journey planner shown to answer the following questions.

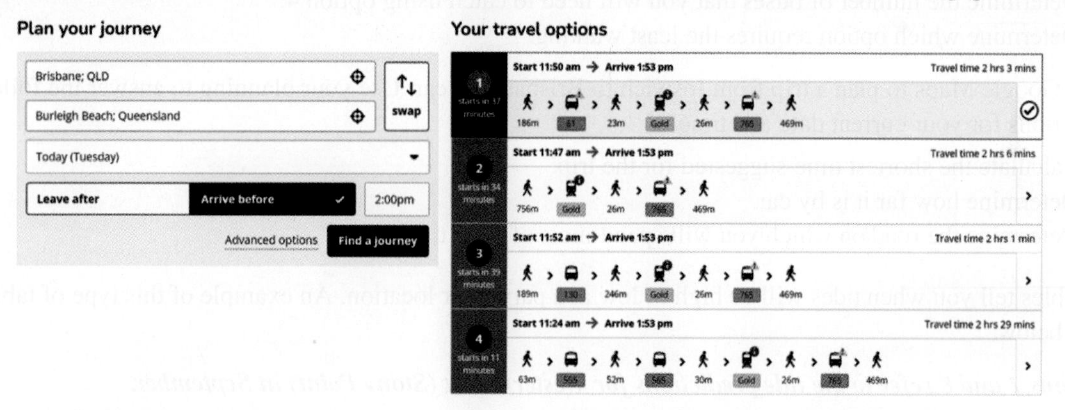

Source: Reproduced by permission of Translink. Department of Transport and Main Roads. © State of Queensland. (n.d.). *Journey planner : plan your journey : Brisbane, QLD to Burleigh Beach, Queensland : your travel options [image],* viewed April 2024.

   a. Determine where the planned trip is to.
   b. Identify the shortest time suggested for the trip.
   c. Determine how far you must walk for option 2.

3. Using the information from the Translink journey planner, plan a trip from Brisbane (e.g. Brisbane Airport or Brisbane's Central Station) to Australia Zoo. Use your planning to answer the following questions for your current date.
   a. Determine the shortest time suggested for the trip.
   b. Determine what transport you need for your trip.

4. Use Google Maps to plan a trip from Brisbane (e.g. Brisbane Airport or Brisbane's Central Station) to the Sunshine Coast. Use your planning to answer the following questions for your current date and time.
   a. Calculate the shortest time suggested for the trip.
   b. Determine what transport you need for your trip.

5. Use the information from the Translink journey planner shown to answer the following questions.

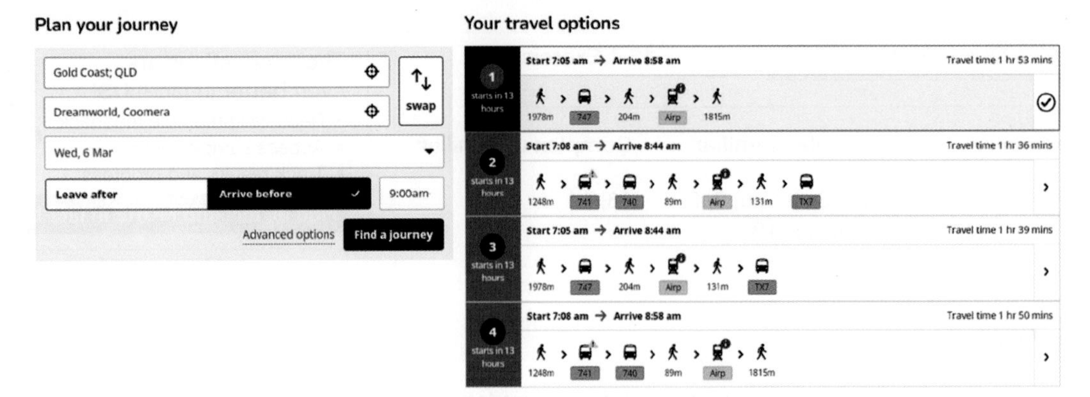

Source: Reproduced by permission of Translink. Department of Transport and Main Roads. © State of Queensland. (n.d.). *Journey planner : plan your journey : Gold Coast QLD to Dreamworld, Coomera : your travel options [image],* viewed April 2024.

a. Determine the time you want to arrive at Dreamworld.
b. Determine the number of buses that you will need to catch using option 4.
c. Determine which option requires the least walking.

6. Use Google Maps to plan a trip from Ipswich to Brisbane by car. Use your planning to answer the following questions for your current date and time.

a. Calculate the shortest time suggested for the trip.
b. Determine how far it is by car.
c. Determine the road on which you will spend most of your time.

Tide tables tell you when tides will be high or low at a particular location. An example of this type of table is shown below.

Questions 7 and 8 refer to the tide predictions for Western Port (Stony Point) in September.

	Time	*Ht*		*Time*	*Ht*
1	0110	2.49	**5**	0443	2.82
MO	0702	0.57	FR	1017	0.68
	1359	2.71		1634	2.70
	1944	0.71		2230	0.29
2	0217	2.64	**6**	0519	2.79
TU	0801	0.56	SA	1054	0.75
	1447	2.77		1702	2.65
	2032	0.52		2302	0.30
3	0314	2.75	**7**	0553	2.74
WE	0853	0.57	SU	1127	0.82
	1529	2.77		1730	2.60
	2115	0.39		2335	0.34
4	0400	2.81			
TH	0937	0.61			
	1604	2.74			
	2154	0.32			

The tide heights are measured in metres. Times stated are Australian Eastern Standard Time (24-hour clock). During daylight-saving time (when in force) one hour needs to be added to the times stated.

7. WE7 a. Identify the differences between the blue and the red numbers in the table.
   b. Determine the number of low tides each day.
   c. Identify the lowest low tide in the table.
   d. Determine when the lowest low tide occurs.
8. a. Determine how much time there is between high tides on Saturday 6 September.
   b. Determine whether the time difference between high tides is the same each day.
   c. Identify three examples of how a tide table could be useful.

Questions 9 and 10 refer to the following timetable of moon phases.

New moon		First quarter		Full moon		Last quarter	
Date	**Time**	**Date**	**Time**	**Date**	**Time**	**Date**	**Time**
				Mon., Jan 5	14:53	Tues., Jan 13	19:47
Tues., Jan 20	23:14	Tues., Jan 27	14:48	Wed., Feb 4	09:09	Thur., Feb 12	13:50
Thurs., Feb 19	09:47	Thurs., Feb 26	03:14	Fri., Mar 6	04:06	Sat., Mar 14	03:48
Fri., Mar 20	19:36	Fri., Mar 27	17:43	Sat., Apr 4	22:06	Sun., Apr 12	13:44
Sun., Apr 19	04:57	Sun., Apr 26	09:55	Mon., May 4	13:42	Mon., May 11	20:36
Mon., May 18	14:13	Tues., May 26	03:19	Wed., Jun 3	02:19	Wed., Jun 10	01:42
Wed., Jun 17	00:05	Wed., Jun 24	21:03	Thurs., Jul 2	12:20	Thurs., Jul 9	06:24
Thurs., Jul 16	11:24	Fri., Jul 24	14:04	Fri., Jul 31	20:43	Fri., Aug 7	12:03
Sat., Aug 15	00:54	Sun., Aug 23	05:31	Sun., Aug 30	04:35	Sat., Sep 5	19:54
Sun., Sep 13	16:41	Mon., Sep 21	18:59	Mon., Sep 28	12:50	Mon., Oct 5	07:06
Tues., Oct 13	10:06	Wed., Oct 21	08:31	Tues., Oct 27	22:05	Tues., Nov 3	22:24
Thurs., Nov 12	03:47	Thurs., Nov 19	18:27	Thurs., Nov 26	08:44	Thurs., Dec 3	17:40
Fri., Dec 11	20:29	Sat., Dec 19	01:14	Fri., Dec 25	21:11		

9. a. Determine the date and time of the first new moon.
   b. Determine how many full moons there were.
   c. Determine whether there were any months that had more than one full moon. If so, identify the month.
10. a. Determine how many days passed between the new moon and the full moon in September.
    b. Determine how many hours and minutes passed between the new moon and its first quarter in February.
    c. Determine how many hours and minutes passed between the new moon in October and the following quarter.

The following sunrise and sunset table are for questions 11 and 12.

November 2024 — Sun in Brisbane

< October **November** December >

2024	Sunrise/Sunset		Daylength	
Nov	**Sunrise**	**Sunset**	**Length**	**Diff.**
1	04:56 ↘ (107°)	18:06 ↙ (253°)	13:09:47	+1:27
2	04:56 ↘ (107°)	18:07 ↙ (253°)	13:11:14	+1:26
3	04:55 ↘ (108°)	18:07 ↙ (252°)	13:12:40	+1:25
4	04:54 ↘ (108°)	18:08 ↙ (252°)	13:14:05	+1:25
5	04:53 ↘ (108°)	18:09 ↙ (252°)	13:15:29	+1:24
6	04:53 ↘ (109°)	18:10 ↙ (251°)	13:16:53	+1:23
7	04:52 ↘ (109°)	18:10 ↙ (251°)	13:18:15	+1:22
8	04:52 ↘ (109°)	18:11 ↙ (251°)	13:19:37	+1:21
9	04:51 ↘ (110°)	18:12 ↙ (250°)	13:20:57	+1:20
10	04:50 ↘ (110°)	18:13 ↙ (250°)	13:22:16	+1:19
11	04:50 ↘ (110°)	18:13 ↙ (250°)	13:23:35	+1:18
12	04:49 ↘ (111°)	18:14 ↙ (249°)	13:24:52	+1:17
13	04:49 ↘ (111°)	18:15 ↙ (249°)	13:26:08	+1:15
14	04:48 ↘ (111°)	18:16 ↙ (249°)	13:27:23	+1:14
15	04:48 ↘ (111°)	18:16 ↙ (248°)	13:28:36	+1:13
16	04:47 ↘ (112°)	18:17 ↙ (248°)	13:29:48	+1:12
15	04:48 ↘ (111°)	18:16 ↙ (248°)	13:28:36	+1:13
16	04:47 ↘ (112°)	18:17 ↙ (248°)	13:29:48	+1:12
17	04:47 ↘ (112°)	18:18 ↙ (248°)	13:30:59	+1:10
18	04:47 ↘ (112°)	18:19 ↙ (248°)	13:32:09	+1:09
19	04:46 ↘ (113°)	18:20 ↙ (247°)	13:33:17	+1:08
20	04:46 ↘ (113°)	18:20 ↙ (247°)	13:34:23	+1:06
21	04:46 ↘ (113°)	18:21 ↙ (247°)	13:35:29	+1:05
22	04:45 ↘ (113°)	18:22 ↙ (247°)	13:36:32	+1:03
23	04:45 ↘ (114°)	18:23 ↙ (246°)	13:37:34	+1:01
24	04:45 ↘ (114°)	18:24 ↙ (246°)	13:38:34	+1:00
25	04:45 ↘ (114°)	18:24 ↙ (246°)	13:39:33	+0:58
26	04:45 ↘ (114°)	18:25 ↙ (246°)	13:40:30	+0:56
27	04:44 ↘ (114°)	18:26 ↙ (245°)	13:41:25	+0:55
28	04:44 ↘ (115°)	18:27 ↙ (245°)	13:42:18	+0:53
29	04:44 ↘ (115°)	18:27 ↙ (245°)	13:43:10	+0:51
30	04:44 ↘ (115°)	18:28 ↙ (245°)	13:43:59	+0:49

* All times are local time for Brisbane. They take into account refraction. Dates are based on the Gregorian calendar.

11. a. Determine when sunrise occurs on 10 November.
    b. Determine the duration of the day on 2 November.
    c. Determine the time difference from sunrise on 1 November and sunrise on 30 November.

12. a. Determine when sunrise occurs on 21 November.
    b. Determine when sunset occurs on 17 November.
    c. Determine the duration of the shortest day in November.

Complex unfamiliar

13. The tidal range is the difference in height between high tide and low tide. Using the tide timetable for Gold Coast Seaway, displaying low and high tides each day for February 2024, determine the largest first (morning) tidal range of the day in February and identify the date.

1 February		2 February		3 February		4 February		5 February		6 February	
Time	m	Time	m	Time	m	Time	m	Time	m	Time	m
00:06	1.15 m	00:59	1.16 m	02:04	1.20 m	03:17	1.26 m	04:26	1.36 m	05:26	1.49 m
05:48	0.49 m	06:44	0.58 m	07:57	0.65 m	09:34	0.65 m	11:01	0.58 m	12:05	0.47 m
12:03	1.29 m	12:43	1.17 m	13:34	1.06 m	14:51	0.98 m	16:16	0.96 m	17:29	1.00 m
18:25	0.34 m	19:02	0.37 m	19:49	0.40 m	20:51	0.41 m	22:04	0. 39 m	23:09	0.31 m
7 February		8 February		9 February		10 February		11 February		12 February	
Time	m	Time	m	Time	m	Time	m	Time	m	Time	m
06:19	1.63 m	00:06	0.21 m	00:57	0.11 m	01:46	0.03 m	02:34	-0.01 m	03:24	0.01 m
12:56	0.34 m	07:09	1.76 m	07:56	1.86 m	08:41	1.92 m	09:24	1.92 m	10:07	1.85 m
18:28	1.07 m	13:41	0.22 m	14:23	0.13 m	15:05	0.06 m	15:47	0.03 m	16:29	0.03 m
		19:21	1.15 m	20:08	1.24 m	20:55	1.31 m	21:41	1.36 m	22:29	1.40 m
13 February		14 February		15 February		16 February		17 February		18 February	
Time	m	Time	m	Time	m	Time	m	Time	m	Time	m
04:15	0.09 m	05:09	0.22 m	00:13	1.41 m	01:15	1.40 m	02:27	1.40 m	03:44	1.42 m
10:50	1.71 m	11:33	1.53 m	06:09	0.37 m	07:21	0.51 m	08:58	0.59 m	10:39	0.57 m
17:10	0.08 m	17:51	0.15 m	12:18	1.32 m	13:11	1.13 m	14:23	0.98 m	15:57	0.91 m
23:19	1.41 m			18:33	0.24 m	19:19	0.34 m	20:17	0.42 m	21:33	0.46 m
19 February		20 February		21 February		22 February		23 February		24 February	
Time	m	Time	m	Time	m	Time	m	Time	m	Time	m
04:55	1.47 m	05:53	1.53 m	06:41	1.58 m	00:33	0.32 m	01:13	0.26 m	01:48	0.22 m
11:49	0.51 m	12:39	0.43 m	13:17	0.37 m	07:21	1.62 m	07:55	1.64 m	08:27	1.64 m
17:21	0.94 m	18:19	1.01 m	19:02	1.08 m	13:50	0.32 m	14:20	0.28 m	14:48	0.26 m
22:48	0.44 m	23:47	0.39 m			19:37	1.15 m	20:10	1.20 m	20:40	1.24 m
25 February		26 February		27 February		28 February		29 February			
Time	m	Time	m	Time	m	Time	m	Time	m		
02:21	0.20 m	02:53	0.22 m	03:28	0.25 m	04:03	0.32	04:42	0.40 m		
08:56	1.63 m	09:24	1.59 m	09:53	1.53 m	10:21	1.45	10:51	1.35 m		
15:14	0.24 m	15:40	0.24 m	16:06	0.24 m	16:31	0.27	16:57	0.30 m		
21:10	1.28 m	21:41	1.30 m	22:13	1.32 m	22:48	1.34	23:26	1.34 m		

14. The longest day of the year is the summer solstice, which happens in December in the southern hemisphere. Using the table displaying the sunrise and sunset times in Gold Coast, determine the days with the most daylight time in December 2024 at this location and calculate how long this is in hours and minutes. Explain the reasonableness of this answer.

1 December		2 December		3 December		4 December		5 December		6 December	
Sunrise	Sunset	Sunrise	Sunset	Sunrise	Sunset	Sunrise	Sunset	Sunrise	Sunset	Sunrise	Sunset
4:41 am	6:28 pm	4:41 am	6:29 pm	4:41 sm	6.30 pm	4:41 am	6:31 pm	4:41 am	6:31 pm	4:42 am	6:32 pm
7 December		8 December		9 December		10 December		11 December		12 December	
Sunrise	Sunset	Sunrise	Sunset	Sunrise	Sunset	Sunrise	Sunset	Sunrise	Sunset	Sunrise	Sunset
4:42 am	6:33 pm	4:42 am	6:34 pm	4:42 am	6:34 pm	4:42 am	6:35 pm	4:42 am	6:36 pm	4:43 am	6:36 pm
13 December		14 December		15 December		16 December		17 December		18 December	
Sunrise	Sunset	Sunrise	Sunset	Sunrise	Sunset	Sunrise	Sunset	Sunrise	Sunset	Sunrise	Sunset
4:43 am	6:37 pm	4:43 am	6:38 pm	4:44 am	6:38 pm	4:44 am	6:39 pm	4:44 am	6:39 pm	4:45 am	6:40 pm
19 December		20 December		21 December		22 December		23 December		24 December	
Sunrise	Sunset	Sunrise	Sunset	Sunrise	Sunset	Sunrise	Sunset	Sunrise	Sunset	Sunrise	Sunset
4:45 am	6:41 pm	4:46 am	6:41 pm	4:46 am	6:42 pm	4:47 am	6:42 pm	4:47 am	6:42 pm	4:48 am	6:43 pm
25 December		26 December		27 December		28 December		29 December		30 December	
Sunrise	Sunset	Sunrise	Sunset	Sunrise	Sunset	Sunrise	Sunset	Sunrise	Sunset	Sunrise	Sunset
4:48 am	6:43 pm	4:49 am	6:44 pm	4:49 am	6:44 pm	4:50 am	6:45 pm	4:51 am	6:45 pm	4:51 am	6:45 pm

Fully worked solutions for this chapter are available online.

LESSON
13.5 Compare travel time

SYLLABUS LINKS

- Compare the time taken to travel a specific distance with various modes of transport.

13.5.1 Compare time taken to travel

- To compare the time it takes you to travel on different modes of transport, compare the hours and minutes of each trip and determine the time difference.

WORKED EXAMPLE 8 Comparing the shortest travel time

Mae wants to catch a bus near her house to her friend's house so they can do their maths homework together. She doesn't want to waste any of her study time, so she wants to take the bus with the shortest travel time.
Determine which of the three given options is the best option for Mae.

Bus	Depart	Arrive
Option 1	8:38 am	9:23 am
Option 2	8:51 am	9:34 am
Option 3	8:56 am	9:42 am

THINK	WRITE
1. Calculate the time difference between departing and arriving for **Option 1**.	8:38 am to 9:00 am = 22 min 9:00 am to 9:23 am = 23 min Total time = 22 + 23 = 45 min
2. Calculate the time difference between departing and arriving for **Option 2**.	8:51 am to 9:00 am = 9 min 9:00 am to 9:34 am = 34 min Total time = 9 + 34 = 43 min
3. Calculate the time difference between departing and arriving for **Option 3**.	8:56 am to 9:00 am = 4 min 9:00 am to 9:42 am = 42 min Total time = 9 + 34 = 46 min
4. Identify the option with the shortest travel time.	The best option is Option 2, which takes 43 minutes.

tlvd-11452

WORKED EXAMPLE 9 Comparing the shortest travel time with different modes of transport and determining the time difference

Max needs to travel to Rockhampton from Brisbane. To get there in the shortest time, Max investigates how long it takes to travel by train compared to the bus.
Given that Max wants to leave between 1:30 pm and 2:00 pm, determine which mode of transport has the shorter travel time, and the difference between the two travel times, using the timetables shown.
Train timetable:

Brisbane to Cairns
Northbound

Departing	**Mon** (VC71), **Tue** (VC73), **Wed** (VC75), **Fri** (VC79) and **Sat** (VC81)
Brisbane (Roma St)	1.45pm
Caboolture	2.32pm
Landsborough^	2.57pm
Nambour	3.25pm
Cooroy^	3.48pm
Gympie North	4.24pm
Maryborough West	5.26pm
Howard^	5.47pm
Bundaberg	6.22pm
Miriam Vale^	7.36pm
Gladstone	8.23pm
Mount Larcom^	8.57pm
Rockhampton *arrive*	10.01pm
Rockhampton	10.16pm

Bus timetable:

GREYHOUND SERVICE TIMETABLE FOR:
BRISBANE > CAIRNS

Effective 29 January 2024

Stop Code	Stop Name		GX409 Daily	GX401 Daily	GX402 Daily	GX404 Daily	GX408 Daily	GX405 Daily
BDA	BRISBANE DOMESTIC AIRPORT			8:50 AM*				
BIA	BRISBANE INTERNATIONAL			9:00 AM*				
BNE	BRISBANE		6:45 AM	9:30 AM	12:00 PM	2:00 PM		
MOB	MOOLOOLABA - SUNSHINE COAST			11:00 AM*		3:30 PM		
MCY	MAROOCHYDORE - SUNSHINE COAST		8:35 AM	11:10 AM*		3:40 PM		
NOO	NOOSA - SUNSHINE COAST		9:15 AM	11:55 AM	2:00 PM	4:20 PM		
COO	COOROY - SUNSHINE COAST					4:50 PM		
GYR	TRAVESTON	*Arr*		12:45 PM	2:55 PM*	5:15 PM		
		Dep		1:20 PM	3:30 PM*	5:50 PM*		
GYM	GYMPIE				3:50 PM	6:05 PM		
WAL	WALLU				4:25 PM	6:40 PM		
RAI	RAINBOW BEACH			2:35 PM	5:10 PM*	7:15 PM		
MBH	MARYBOROUGH			3:55 PM	6:15 PM	8:20 PM		
HVB	HERVEY BAY			4:30 PM*	7:00 PM*	9:05 PM		
CHI	CHILDERS				7:40 PM*	9:50 PM*		
SBM	SOUTH BUNDABERG MEAL BREAK	*Arr*		5:50 PM*	8:15 PM*	10:25 PM*		
		Dep		6:30 PM*	8:45 PM*	11:00 PM*		
BDB	BUNDABERG				8:55 PM*	11:10 PM*		
AGW	AGNES WATER			8:10 PM				
MIM	MIRIAM VALE				10:00 PM*	12:25 AM		
GLT	GLADSTONE				10:50 PM*	1:25 AM*		
ROK	ROCKHAMPTON	*Arr*		10:45 PM	12:10 AM	2:45 AM		
		Dep		11:20 PM	12:45 AM	3:20 AM		
MLT	MARLBOROUGH TURNOFF				2:00 AM*	4:35 AM		
SLT	ST LAWRENCE TURN OFF				2:40 AM*	5:15 AM		
SRI	SARINA				3:55 AM	6:30 AM		
MKM	MACKAY MEAL BREAK	*Arr*		3:00 AM	4:15 AM	6:50 AM		
		Dep		3:40 AM	4:55 AM	7:30 AM		
MKY	MACKAY			3:55 AM	5:15 AM	7:50 AM		
PPP	PROSERPINE			5:25 AM*	6:40 AM*	9:15 AM		
AIR	AIRLIE BEACH			6:00 AM*	7:10 AM*	9:50 AM		12:10 PM
LFC	LONGFORD CREEK MEAL BREAK	*Arr*			8:10 AM	10:40 AM		
		Dep			8:45 AM	11:15 AM		
BWN	BOWEN				9:15 AM	11:45 AM		
DEM	DELTA				9:20 AM*	11:55 AM		
HHL	HOME HILL				10:30 AM*	1:00 PM		
AYR	AYR				10:45 AM*	1:15 PM		
TSV	TOWNSVILLE	*Arr*						3:45 PM
		Dep			12:10 PM	2:20 PM	8:00 AM	4:15 PM
ING	INGHAM					4:00 PM	9:30 AM	5:45 PM
CDW	CARDWELL	*Arr*				4:30 PM	10:10 AM	6:25 PM
		Dep				5:05 PM	10:45 AM	7:00 PM
TUL	TULLY					5:35 PM	11:15 AM	7:30 PM
MIS	MISSION BEACH (WONGALING)					6:00 PM	11:45 AM	8:00 PM
IFL	INNISFAIL					6:40 PM	12:25 PM	8:40 PM
BAB	BABINDA					7:00 PM	12:50 PM*	
GDV	GORDONVALE					7:30 PM	1:15 PM*	
CNS	CAIRNS					8:00 PM	1:40 PM	9:55 PM

THINK	WRITE
1. Look up the time the train leaves Brisbane between 1:30 pm and 2:00 pm.	The time the train leaves Brisbane that suits Max is 1:45 pm.
2. Look up the time this train arrives in Rockhampton.	This train arrives in Rockhampton at 10:01 pm.
3. Calculate the train travel time.	1:45 pm to 2:00 pm = 15 min 2:00 pm to 10:00 pm = 8 h 10:00 pm to 10:01 pm = 1 min Total time $= 8\,\text{h} + 15\,\text{min} + 1\,\text{min}$ $= 8\,\text{h}\ 16\,\text{min}$
4. Look up the time the bus leaves Brisbane between 1:30 pm and 2:00 pm.	The time the bus leaves Brisbane that suits Max is 2:00 pm.
5. Look up the time this bus arrives in Rockhampton.	This bus arrives in Rockhampton at 2:45 am.
6. Calculate the bus travel time.	2:00 pm to 2:00 am = 12 h 2:00 am to 2:45 am = 45 min Total time $= 12\,\text{h} + 45\,\text{min}$ $= 12\,\text{h}\ 45\,\text{min}$
7. Calculate the time difference.	Time difference $= 12\,\text{h}\ 45\,\text{min} - 8\,\text{h}\ 16\,\text{min}$ $= 4\,\text{h}\ 29\,\text{min}$
8. Write the answer.	Travelling by train has the shorter travel time, by 4 hours and 29 minutes.

Exercise 13.5 Compare travel time

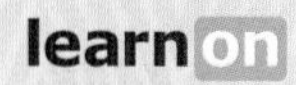

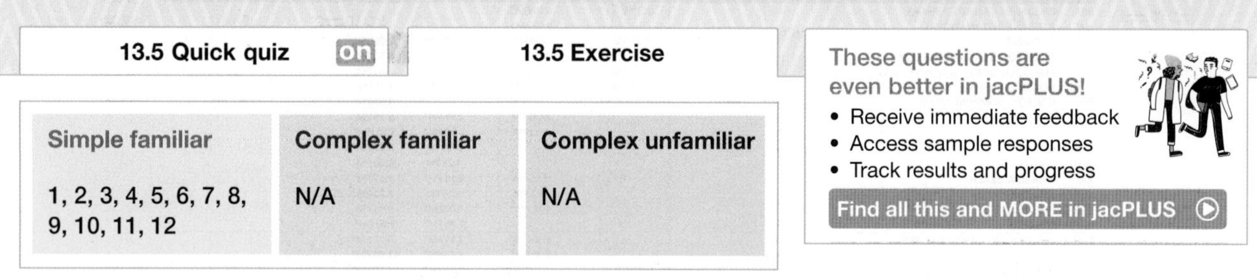

13.5 Quick quiz | 13.5 Exercise

Simple familiar	Complex familiar	Complex unfamiliar
1, 2, 3, 4, 5, 6, 7, 8, 9, 10, 11, 12	N/A	N/A

These questions are even better in jacPLUS!
• Receive immediate feedback
• Access sample responses
• Track results and progress
Find all this and MORE in jacPLUS

Simple familiar

1. WE8 Kieley wants to catch a bus to do some shopping, and wants to minimise the travel time. The three possible options are shown. Determine which is the best option for Kieley.

Bus	Depart	Arrive
Option 1	10:47 am	11:13 am
Option 2	10:51 am	11:16 am
Option 3	10:58 am	11:25 am

2. Zac needs to catch a bus to the supermarket and wants to minimise the travel time. There are three possible options, as shown. Determine which is the best option for Zac.

Bus	Depart	Arrive
Option 1	11:51 am	12:23 pm
Option 2	11:56 am	12:27 pm
Option 3	11:54 am	12:26 pm

3. Ella needs to catch a bus to the gym and wants to minimise the travel time. If they walk to the bus stop, it will take them 11 minutes, but if they run it will take 5 minutes. There are three possible bus options, as shown. Calculate the shortest time it would take Ella to get to the gym, assuming the bus stops near the gym.

Bus	Depart	Arrive
Option 1	12:33 pm	12:53 pm
Option 2	12:39 pm	1:00 pm
Option 3	12:43 pm	1:05 pm

Use the following train and bus timetables to answer questions 4 and 5.

Train timetable:

Brisbane to Cairns

Northbound

Departing	**Mon** (VC71), **Tue** (VC73), **Wed** (VC75), **Fri** (VC79) and **Sat** (VC81)
Brisbane (Roma St)	1.45pm
Caboolture	2.32pm
Landsborough^	2.57pm
Nambour	3.25pm
Cooroy^	3.48pm
Gympie North	4.24pm
Maryborough West	5.26pm
Howard^	5.47pm
Bundaberg	6.22pm
Miriam Vale^	7.36pm
Gladstone	8.23pm
Mount Larcom^	8.57pm
Rockhampton *arrive*	10.01pm
Rockhampton	10.16pm

Bus timetable:

GREYHOUND SERVICE TIMETABLE FOR:
BRISBANE > CAIRNS

Effective 29 January 2024

Stop Code	Stop Name		GX409 Daily	GX401 Daily	GX402 Daily	GX404 Daily	GX408 Daily	GX405 Daily
BDA	BRISBANE DOMESTIC AIRPORT			8:50 AM*				
BIA	BRISBANE INTERNATIONAL			9:00 AM*				
BNE	BRISBANE		6:45 AM	9:30 AM	12:00 PM	2:00 PM		
MOB	MOOLOOLABA - SUNSHINE COAST			11:00 AM*		3:30 PM		
MCY	MAROOCHYDORE - SUNSHINE COAST		8:35 AM	11:10 AM*		3:40 PM		
NOO	NOOSA - SUNSHINE COAST		9:15 AM	11:55 AM	2:00 PM	4:20 PM		
COO	COOROY - SUNSHINE COAST					4:50 PM		
GYR	TRAVESTON	*Arr*		12:45 PM	2:55 PM*	5:15 PM		
		Dep		1:20 PM	3:30 PM*	5:50 PM*		
GYM	GYMPIE				3:50 PM	6:05 PM		
WAL	WALLU				4:25 PM	6:40 PM		
RAI	RAINBOW BEACH			2:35 PM	5:10 PM*	7:15 PM		
MBH	MARYBOROUGH			3:55 PM	6:15 PM	8:20 PM		
HVB	HERVEY BAY			4:30 PM*	7:00 PM*	9:05 PM		
CHI	CHILDERS				7:40 PM*	9:50 PM*		
SBM	SOUTH BUNDABERG MEAL BREAK	*Arr*		5:50 PM*	8:15 PM*	10:25 PM*		
		Dep		6:30 PM*	8:45 PM*	11:00 PM*		
BDB	BUNDABERG				8:55 PM*	11:10 PM*		
AGW	AGNES WATER			8:10 PM				
MIM	MIRIAM VALE				10:00 PM*	12:25 AM		
GLT	GLADSTONE				10:50 PM*	1:25 AM*		
ROK	ROCKHAMPTON	*Arr*		10:45 PM	12:10 AM	2:45 AM		
		Dep		11:20 PM	12:45 AM	3:20 AM		
MLT	MARLBOROUGH TURNOFF				2:00 AM*	4:35 AM		
SLT	ST LAWRENCE TURN OFF				2:40 AM*	5:15 AM		
SRI	SARINA				3:55 AM	6:30 AM		
MKM	MACKAY MEAL BREAK	*Arr*		3:00 AM	4:15 AM	6:50 AM		
		Dep		3:40 AM	4:55 AM	7:30 AM		
MKY	MACKAY			3:55 AM	5:15 AM	7:50 AM		
PPP	PROSERPINE			5:25 AM*	6:40 AM*	9:15 AM		
AIR	AIRLIE BEACH			6:00 AM*	7:10 AM*	9:50 AM		12:10 PM
LFC	LONGFORD CREEK MEAL BREAK	*Arr*			8:10 AM	10:40 AM		
		Dep			8:45 AM	11:15 AM		
BWN	BOWEN				9:15 AM	11:45 AM		
DEM	DELTA				9:20 AM*	11:55 AM		
HHL	HOME HILL				10:30 AM*	1:00 PM		
AYR	AYR				10:45 AM*	1:15 PM		
TSV	TOWNSVILLE	*Arr*						3:45 PM
		Dep			12:10 PM	2:20 PM	8:00 AM	4:15 PM
ING	INGHAM					4:00 PM	9:30 AM	5:45 PM
CDW	CARDWELL	*Arr*				4:30 PM	10:10 AM	6:25 PM
		Dep				5:05 PM	10:45 AM	7:00 PM
TUL	TULLY					5:35 PM	11:15 AM	7:30 PM
MIS	MISSION BEACH (WONGALING)					6:00 PM	11:45 AM	8:00 PM
IFL	INNISFAIL					6:40 PM	12:25 PM	8:40 PM
BAB	BABINDA					7:00 PM	12:50 PM*	
GDV	GORDONVALE					7:30 PM	1:15 PM*	
CNS	CAIRNS					8:00 PM	1:40 PM	9:55 PM

4. **WE9** Tate needs to travel to Miriam Vale from Brisbane and investigates the time it takes to travel by train compared to the bus. Given Tate wants to leave between 1:30 pm and 2:00 pm, determine which mode of transport has the shorter travel time and the difference between the two times using the timetables shown.

5. Emily, who lives in Brisbane, wants to visit a friend in Cooroy. She investigates the time it takes to travel by train compared to the bus. Given Emily wants to leave between 1:30 pm and 2:00 pm, determine which mode of transport has the shortest travel time and the difference between the two times using the timetables shown.

Use the following timetables to answer questions 6 and 7.

Train timetable:

Brisbane to Cairns
Northbound

Departing	**Mon** (VC71), **Tue** (VC73), **Wed** (VC75), **Fri** (VC79) and **Sat** (VC81)
Brisbane (Roma St)	1.45pm
Caboolture	2.32pm
Landsborough^	2.57pm
Nambour	3.25pm
Cooroy^	3.48pm
Gympie North	4.24pm
Maryborough West	5.26pm
Howard^	5.47pm
Bundaberg	6.22pm
Miriam Vale^	7.36pm
Gladstone	8.23pm
Mount Larcom^	8.57pm
Rockhampton *arrive*	10.01pm
Rockhampton	10.16pm
Overnight Journey Continues	**Tue** (VC71), **Wed** (VC73), **Thu** (VC75), **Sat** (VC79) and **Sun** (VC81)
St Lawrence^	12.43am
Carmila^	1.20am
Sarina^	2.09am
Mackay *arrive*	2.45am
Mackay	3.10am
Proserpine *arrive*	4.38am
Proserpine	4.48am
Bowen^#	5.25am
Home Hill^	6.34am
Ayr	6.51am
Giru^	7.19am
Townsville *arrive*	7.57am
Townsville	8.12am
Ingham^	9.45am
Cardwell^	10.41am
Tully *arrive*^	11.28am
Tully^	11.38am
Innisfail	12.40pm
Babinda^	1.17pm
Gordonvale^	1.55pm
Cairns *arrive*	2.30pm

Bus timetable:

GREYHOUND SERVICE TIMETABLE FOR:
BRISBANE > CAIRNS

Effective 29 January 2024

Stop Code	Stop Name		GX409 Daily	GX401 Daily	GX402 Daily	GX404 Daily	GX408 Daily	GX405 Daily
BDA	BRISBANE DOMESTIC AIRPORT			8:50 AM*				
BIA	BRISBANE INTERNATIONAL			9:00 AM*				
BNE	BRISBANE		6:45 AM	9:30 AM	12:00 PM	2:00 PM		
MOB	MOOLOOLABA - SUNSHINE COAST			11:00 AM*		3:30 PM		
MCY	MAROOCHYDORE - SUNSHINE COAST		8:35 AM	11:10 AM*		3:40 PM		
NOO	NOOSA - SUNSHINE COAST		9:15 AM	11:55 AM	2:00 PM	4:20 PM		
COO	COOROY - SUNSHINE COAST					4:50 PM		
GYR	TRAVESTON	*Arr*		12:45 PM	2:55 PM*	5:15 PM		
		Dep		1:20 PM	3:30 PM*	5:50 PM*		
GYM	GYMPIE				3:50 PM	6:05 PM		
WAL	WALLU				4:25 PM	6:40 PM		
RAI	RAINBOW BEACH			2:35 PM	5:10 PM*	7:15 PM		
MBH	MARYBOROUGH			3:55 PM	6:15 PM	8:20 PM		
HVB	HERVEY BAY			4:30 PM*	7:00 PM*	9:05 PM		
CHI	CHILDERS				7:40 PM*	9:50 PM*		
SBM	SOUTH BUNDABERG MEAL BREAK	*Arr*		5:50 PM*	8:15 PM*	10:25 PM*		
		Dep		6:30 PM*	8:45 PM*	11:00 PM*		
BDB	BUNDABERG				8:55 PM*	11:10 PM*		
AGW	AGNES WATER			8:10 PM				
MIM	MIRIAM VALE				10:00 PM*	12:25 AM		
GLT	GLADSTONE				10:50 PM*	1:25 AM*		
ROK	ROCKHAMPTON	*Arr*		10:45 PM	12:10 AM	2:45 AM		
		Dep		11:20 PM	12:45 AM	3:20 AM		
MLT	MARLBOROUGH TURNOFF				2:00 AM*	4:35 AM		
SLT	ST LAWRENCE TURN OFF				2:40 AM*	5:15 AM		
SRI	SARINA				3:55 AM	6:30 AM		
MKM	MACKAY MEAL BREAK	*Arr*		3:00 AM	4:15 AM	6:50 AM		
		Dep		3:40 AM	4:55 AM	7:30 AM		
MKY	MACKAY			3:55 AM	5:15 AM	7:50 AM		
PPP	PROSERPINE			5:25 AM*	6:40 AM*	9:15 AM		
AIR	AIRLIE BEACH			6:00 AM*	7:10 AM*	9:50 AM		12:10 PM
LFC	LONGFORD CREEK MEAL BREAK	*Arr*			8:10 AM	10:40 AM		
		Dep			8:45 AM	11:15 AM		
BWN	BOWEN				9:15 AM	11:45 AM		
DEM	DELTA				9:20 AM*	11:55 AM		
HHL	HOME HILL				10:30 AM*	1:00 PM		
AYR	AYR				10:45 AM*	1:15 PM		
TSV	TOWNSVILLE	*Arr*						3:45 PM
		Dep			12:10 PM	2:20 PM	8:00 AM	4:15 PM
ING	INGHAM					4:00 PM	9:30 AM	5:45 PM
CDW	CARDWELL	*Arr*				4:30 PM	10:10 AM	6:25 PM
		Dep				5:05 PM	10:45 AM	7:00 PM
TUL	TULLY					5:35 PM	11:15 AM	7:30 PM
MIS	MISSION BEACH (WONGALING)					6:00 PM	11:45 AM	8:00 PM
IFL	INNISFAIL					6:40 PM	12:25 PM	8:40 PM
BAB	BABINDA					7:00 PM	12:50 PM*	
GDV	GORDONVALE					7:30 PM	1:15 PM*	
CNS	CAIRNS					8:00 PM	1:40 PM	9:55 PM

6. A family from Brisbane want to have a holiday in Cairns. Since it is a long trip, they explore the options of travelling by train or bus to minimise the travel time. Using the timetables shown:
   a. determine the travel time via train
   b. determine the travel time via bus.

7. The family travelling from Brisbane to Cairns in question 6 want to hire a car for their holiday, and they investigate picking up the car in Townsville. Given that they can drive from Townsville to Cairns in 4 hours and 15 minutes, calculate how long it would take to catch a bus to Townsville and then drive to Cairns.

8. Kimberly wants to travel from South Bank to Milton, leaving as close to 8 am as possible. Use the ferry timetable and the planned trip using the Rome2Rio website to answer the following questions.
   a. Determine the time it will take to travel on the ferry.
   b. Determine the time it will take to travel by train.
   c. Determine the time it will take to travel by bus.
   d. Determine the time it will take to travel by taxi.
   e. Decide which mode of transport is quickest.

Ferry timetable:

Northshore Hamilton to UQ St Lucia

Monday to Friday

Departs Terminal:	am	am	am	am	am	am	am	am	am	am	am	am	am
Northshore Hamilton	5:30	6:00	6:30	6:45	.....	7:00	.....	7:15	.....	7:30	.....	7:45	.....
Apollo Road	5:35	6:05	6:35	6:50	7:01	7:05	7:16	7:20	7:31	7:35	7:46	7:50	8:01
Bretts Wharf	5:38	6:08	6:38	6:53	**7:04**	7:08	**7:19**	7:23	**7:34**	7:38	**7:49**	7:53	**8:04**
Teneriffe	5:45	6:15	6:45	7:00	E	7:15	E	7:30	E	7:45	E	8:00	E
Bulimba	5:49	6:19	6:49	7:04	**7:11**	7:19	**7:26**	7:34	**7:41**	7:49	**7:56**	8:04	**8:11**
Hawthorne	5:54	6:24	6:54	7:09	**7:16**	7:24	**7:31**	7:39	**7:46**	7:54	**8:01**	8:09	**8:16**
New Farm Park	5:59	6:29	6:59	7:14	E	7:29	E	7:44	E	7:59	E	8:14	E
Mowbray Park	6:03	6:33	7:03	7:18	E	7:33	E	7:48	E	8:03	E	8:18	E
Sydney Street	6:06	6:36	7:06	7:21	E	7:36	E	7:51	E	8:06	E	8:21	E
Riverside	6:14	6:44	7:14	7:29	**7:31**	7:44	**7:46**	7:59	**8:01**	8:14	**8:16**	8:29	**8:31**
QUT Gardens Point	6:24	6:54	7:24	7:39	.....	7:54	.....	8:09	.....	8:24	.....	8:39	.....
South Bank	6:28	6:58	7:28	7:43	.....	7:58	.....	8:13	.....	8:28	.....	8:43	.....
North Quay	6:32	7:02	7:32	7:47	.....	8:02	.....	8:17	.....	8:32	.....	8:47	.....
Milton	6:38	7:08	7:38	7:53	.....	8:08	.....	8:23	.....	8:38	.....	8:53	.....
Regatta	6:42	7:12	7:42	7:57	.....	8:12	.....	8:27	.....	8:42	.....	8:57	.....
Guyatt Park	6:46	7:16	7:46	8:01	.....	8:16	.....	8:31	.....	8:46	.....	9:01	.....
West End	6:49	7:19	7:49	8:04	.....	8:19	.....	8:34	.....	8:49	.....	9:04	.....
UQ St Lucia	6:53	7:23	7:53	8:08	.....	8:23	.....	8:38	.....	8:53	.....	9:08	.....

Departs Terminal:	am	am	am	am	am	am	am	am	am	am	am	am	am
Northshore Hamilton	8:00	.....	8:15	.....	8:30	.....	8:45	.....	9:00	9:15	9:30	9:45	10:00
Apollo Road	8:05	8:16	8:20	8:31	8:35	8:46	8:50	9:01	9:05	9:20	9:35	9:50	10:05
Bretts Wharf	8:08	**8:19**	8:23	**8:34**	8:38	**8:49**	8:53	**9:04**	9:08	9:23	9:38	9:53	10:08
Teneriffe	8:15	E	8:30	E	8:45	E	9:00	E	9:15	9:30	9:45	10:00	10:15
Bulimba	8:19	**8:26**	8:34	**8:41**	8:49	**8:56**	9:04	**9:11**	9:19	9:34	9:49	10:04	10:19
Hawthorne	8:24	**8:31**	8:39	**8:46**	8:54	**9:01**	9:09	**9:16**	9:24	9:39	9:54	10:09	10:24
New Farm Park	8:29	E	8:44	E	8:59	E	9:14	E	9:29	9:44	9:59	10:14	10:29
Mowbray Park	8:33	E	8:48	E	9:03	E	9:18	E	9:33	9:48	10:03	10:18	10:33
Sydney Street	8:36	E	8:51	E	9:06	E	9:21	E	9:36	9:51	10:06	10:21	10:36
Riverside	8:44	**8:46**	8:59	**9:01**	9:14	**9:16**	9:29	**9:31**	9:44	9:59	10:14	10:29	10:44
QUT Gardens Point	8:54	.....	9:09	.....	9:24	.....	9:39	.....	9:54	10:09	10:24	10:39	10:54
South Bank	8:58	.....	9:13	.....	9:28	.....	9:43	.....	9:58	10:13	10:28	10:43	10:58
North Quay	9:02	.....	9:17	.....	9:32	.....	9:47	.....	10:02	10:17	10:32	10:47	11:02
Milton	9:08	.....	9:23	.....	9:38	.....	9:53	.....	10:08	10:23	10:38	10:53	11:08
Regatta	9:12	.....	9:27	.....	9:42	.....	9:57	.....	10:12	10:27	10:42	10:57	11:12
Guyatt Park	9:16	.....	9:31	.....	9:46	.....	10:01	.....	10:16	10:31	10:46	11:01	11:16
West End	9:19	.....	9:34	.....	9:49	.....	10:04	.....	10:19	10:34	10:49	11:04	11:19
UQ St Lucia	9:23	.....	9:38	.....	9:53	.....	10:08	.....	10:23	10:38	10:53	11:08	11:23

Train:

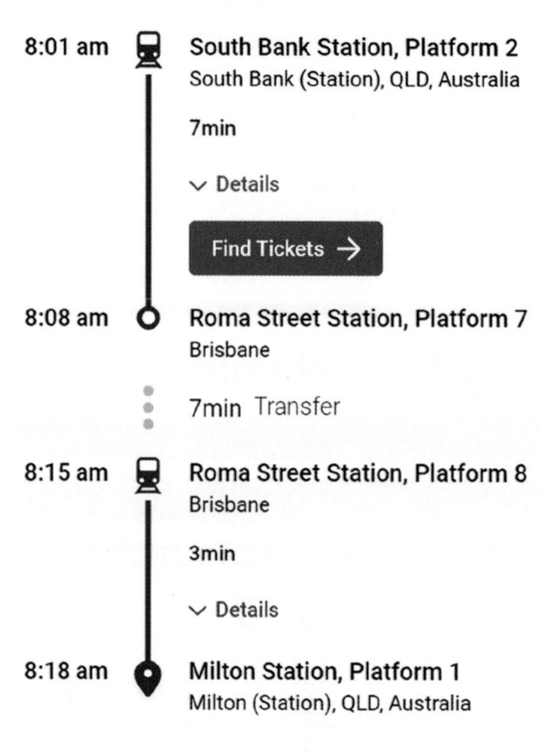
8:01 am
South Bank Station, Platform 2
South Bank (Station), QLD, Australia
7min
Details
Find Tickets
8:08 am
Roma Street Station, Platform 7
Brisbane
7min Transfer
8:15 am
Roma Street Station, Platform 8
Brisbane
3min
Details
8:18 am
Milton Station, Platform 1
Milton (Station), QLD, Australia

Bus:

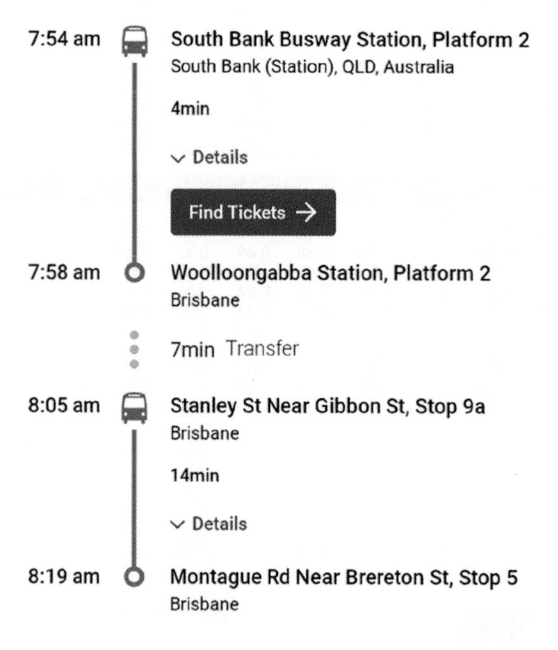
7:54 am
South Bank Busway Station, Platform 2
South Bank (Station), QLD, Australia
4min
Details
Find Tickets
7:58 am
Woolloongabba Station, Platform 2
Brisbane
7min Transfer
8:05 am
Stanley St Near Gibbon St, Stop 9a
Brisbane
14min
Details
8:19 am
Montague Rd Near Brereton St, Stop 5
Brisbane

Taxi:

South Bank (Station)
South Bank (Station), QLD, Australia
4min
3 km
Milton (Station)
Milton (Station), QLD, Australia

9. Use the following tram and bus timetables and, assuming a departure at approximately 1:00 pm, answer the following questions.

Tram:

Stop	Destination Helensvale	Helensvale	Helensvale	Helensvale	Helensvale
Broadbeach South station	12:49 pm	12:56 pm	1:04 pm	1:11 pm	1:19 pm
Broadbeach North station	12:51 pm	12:58 pm	1:06 pm	1:13 pm	1:21 pm
Florida Gardens station	12:53 pm	1:00 pm	1:08 pm	1:15 pm	1:23 pm
Northcliffe station	12:55 pm	1:03 pm	1:10 pm	1:18 pm	1:25 pm
Surfers Paradise station	12:57 pm	1:04 pm	1:12 pm	1:19 pm	1:27 pm
Cavill Avenue station	12:59 pm	1:07 pm	1:14 pm	1:22 pm	1:29 pm
Cypress Avenue station	1:02 pm	1:09 pm	1:17 pm	1:24 pm	1:32 pm
Surfers Paradise North station	1:03 pm	1:11 pm	1:18 pm	1:26 pm	1:33 pm
Main Beach station	1:06 pm	1:13 pm	1:21 pm	1:28 pm	1:36 pm
Broadwater Parklands station	1:08 pm	1:15 pm	1:23 pm	1:30 pm	1:38 pm
Southport South station	1:10 pm	1:17 pm	1:25 pm	1:32 pm	1:40 pm
Southport station	1:12 pm	1:20 pm	1:27 pm	1:35 pm	1:42 pm
Nerang Street station	1:14 pm	1:22 pm	1:29 pm	1:37 pm	1:44 pm

Source: Reproduced by permission of Translink. Department of Transport and Main Roads.
 [Tram : G Link] Broadbeach South station to Helensvale station [timetable image], viewed April 2024.

Bus:

Monday to Friday (cont...)																				
map ref Route number	705	705	705	705	705	705	705	705	705	705	705	705	705	705	705	705	705	705	705	705
	am	am	am	am	am	pm	pm	pm	pm	pm	pm	pm	pm	pm	pm	pm	pm	pm	pm	pm
A Broadbeach South station, stop C	10.55	11.10	11.25	11.40	11.55	12.10	12.25	12.40	12.55	1.10	1.25	1.40	1.55	2.10	2.25	2.40	2.55	3.10	3.25	3.40
B Pacific Fair, stop C	10.59	11.14	11.29	11.44	11.59	12.14	12.29	12.44	12.59	1.14	1.29	1.44	1.59	2.14	2.29	2.44	2.59	3.14	3.29	3.44
C Broadbeach Mall - Surf Pde	11.06	11.21	11.36	11.51	12.06	12.21	12.36	12.51	1.06	1.21	1.36	1.51	2.06	2.21	2.36	2.51	3.06	3.21	3.36	3.51
D Old Burleigh Rd near Fern St	11.13	11.28	11.43	11.58	12.13	12.28	12.43	12.58	1.13	1.28	1.43	1.58	2.13	2.28	2.43	2.58	3.13	3.28	3.43	3.58
E Surfers Paradise, Appel Park	11.21	11.36	11.51	12.06	12.21	12.36	12.51	1.06	1.21	1.36	1.51	2.06	2.21	2.36	2.51	3.06	3.21	3.36	3.51	4.06
F Main Beach Tourist Park	11.28	11.43	11.58	12.13	12.28	12.43	12.58	1.13	1.28	1.43	1.58	2.13	2.28	2.43	2.58	3.13	3.28	3.43	3.58	4.13
G Sea World	11.32	11.47	12.02	12.17	12.32	12.47	1.02	1.17	1.32	1.47	2.02	2.17	2.32	2.47	3.02	3.17	3.32	3.47	4.02	4.17

Source: Reproduced by permission of Translink. Department of Transport and Main Roads.
 Route 705. Broadbeach to Seaworld/Main Beach: Broadbeach South station, Pacific Fair, Surfers Paradise and Sea World/Main Beach [image].

a. Calculate the travel time by tram from Broadbeach South to Surfers Paradise.
b. Calculate the travel time by bus from Broadbeach South to Surfers Paradise.
c. Determine which mode of transport is the quickest and by what length of time.

10. Oliver is looking to travel from Brisbane to Sydney and is investigating different modes of transport. Using the information below, answer the following questions.

Bus:

Stop Code	Stop Name		GX421 Daily	GX423 Daily	GX426 Daily	GX422 Daily
BDA	BRISBANE DOMESTIC AIRPORT				10:00 AM*	
BIA	BRISBANE INTERNATIONAL				10:10 AM*	
BNE	BRISBANE		7:00 AM	9:00 AM	11:00 AM	2:00 PM
SPT	SOUTHPORT - GOLD COAST		8:00 AM*			3:00 PM*
SFR	SURFERS PARADISE - GOLD COAST		8:20 AM	10:00 AM	12:15 PM	3:20 PM
GCA	BILINGA (OPP GOLD COAST AIRPT)		9:15 AM		1:00 PM	4:15 PM
BRH	BRUNSWICK HEADS		10:55 AM*			5:55 PM*
BYR	BYRON BAY		11:15 AM	12:30 PM	3:00 PM	6:15 PM
BNA	BALLINA		11:45 AM			6:45 PM
BNM	BALLINA MEAL BREAK ONLY	*Arr*	12:10 PM	1:10 PM		7:10 PM
		Dep	12:45 PM	1:45 PM		7:45 PM
YAM	YAMBA		1:55 PM			8:55 PM
MCT	MACLEAN		2:15 PM*			9:15 PM*
GFN	GRAFTON		3:00 PM*			10:00 PM*
ARR	ARRAWARRA (SPOT X)		3:35 PM*	4:30 PM*		10:35 PM*
WGG	WOOLGOOLGA		3:45 PM*			10:45 PM*
CFS	COFFS HARBOUR		4:10 PM	5:05 PM		11:10 PM
NBM	NAMBUCCA HEADS MEAL BREAK ONLY	*Arr*	4:40 PM	5:30 PM		11:40 PM
		Dep	5:15 PM	6:05 PM		12:15 AM
NBH	NAMBUCCA HEADS		5:20 PM*			12:40 AM*
KPS	KEMPSEY		6:05 PM*			1:05 AM*
PQQ	PORT MACQUARIE		6:45 PM			1:45 AM
TRO	TAREE		7:40 PM*			2:40 AM*
BHD	BULAHDELAH	*Arr*	8:30 PM	9:00 PM		3:30 AM
		Dep	9:05 PM	9:30 PM		4:05 AM
NCL	NEWCASTLE		10:20 PM			5:20 AM
SYD	SYDNEY		1:00 AM	12:00 AM		8:00 AM
QFT	SYDNEY DOMESTIC AIRPORT					8:30 AM*
INT	SYDNEY INTERNATIONAL AIRPORT					8:45 AM*

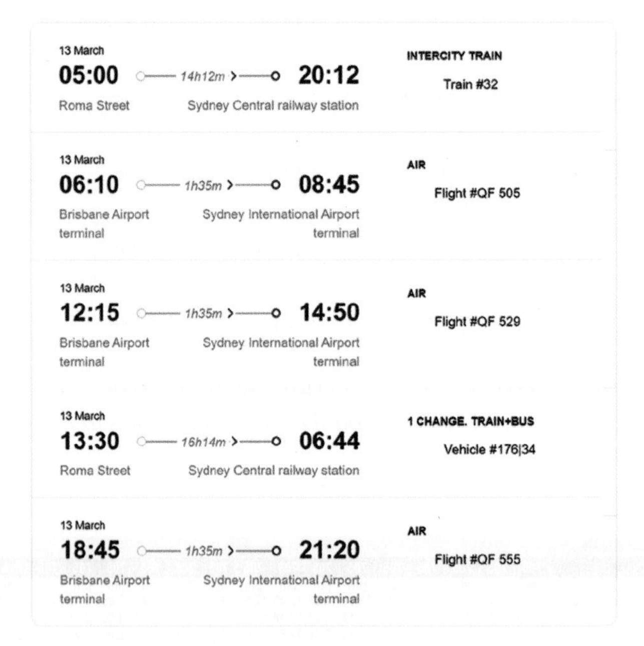

a. Identify the travel options shown.
b. Determine at what time you should leave Brisbane to get the quickest bus.
c. Determine how long it takes to travel by plane.

11. The following table shows the schedule of Chung, who left for work at 6:45 am from his home in Melbourne.

Activity	Time
Drive to the airport	48 minutes
Wait at airport	36 minutes
Fly to Adelaide	1 hour 4 minutes
Taxi to meeting	27 minutes
Meeting time	2 hours 38 minutes
Lunch	1 hour 22 minutes
Taxi to airport	23 minutes
Wait at airport	1 hour 31 minutes
Fly to Melbourne	57 minutes
Drive home	34 minutes

a. Determine at what time Chung's meeting started.
b. Determine at what time he started lunch.
c. Determine the time of his flight back home.
d. Determine the 24-hour time when he arrived back in Melbourne.
e. Determine the 12-hour time when he arrived back home.
f. Determine how long Chung was away from home.

12. Travelling from Coolaroo to Roma Street by train in the morning takes 1 hour and 20 minutes. Due to heavy traffic, it takes 1.5 times as long to drive by car compared to going by train.
a. If you travelled by train and left Coolaroo at 7:30 am, determine the time you would arrive at Roma Street.
b. If you travelled by car and left at 7:30 am, determine the time you would arrive at Roma Street.
c. Determine how much longer it takes to travel by car compared to going by train from Coolaroo to Roma Street.

Fully worked solutions for this chapter are available online.

LESSON
13.6 Review

13.6.1 Summary

doc-41779

Hey students! Now that it's time to revise this chapter, go online to:

Access the chapter summary | Review your results | Watch teacher-led videos | Practise questions with immediate feedback

Find all this and MORE in jacPLUS

13.6 Exercise

learnon

13.6 Exercise

Simple familiar	Complex familiar	Complex unfamiliar
1, 2, 3, 4, 5, 6, 7, 8, 9, 10, 11, 12, 13, 14	15, 16, 17, 18	19, 20

These questions are even better in jacPLUS!
- Receive immediate feedback
- Access sample responses
- Track results and progress

Find all this and MORE in jacPLUS

Simple familiar

1. Express 8:35 pm as 24-hour time.
2. Fredrick estimates that it will take 1 hour and 25 minutes to walk to the lookout at the national park. If they leave at 10:38 am, calculate at what time they will get to the lookout.

Use the following train timetable to answer questions 3 and 4.

Caboolture	12:25	---	12:55	---	---	1:25	---	---	---	1:56
Morayfield	12:33	---	1:03	---	---	1:33	---	---	---	2:03
Burpengary	12:37	---	1:07	---	---	1:37	---	---	---	2:07
Narangba	12:40	---	1:10	---	---	1:40	---	---	---	2:10
Dakabin	12:43	---	1:13	---	---	1:43	---	---	---	2:13
Petrie	12:48	1:07	1:18	1:37	1:48	1:48	1:53	2:08	2:18	2:18
Northgate	12:51	1:10	1:21	1:40	----	1:51	1:56	2:11	----	2:21
Eagle Junction	12:54	---	1:24	---	---	1:54	1:59	---	---	2:24
Bowen Hills	12:57	---	1:27	---	---	1:57	2:02	---	---	2:27
Fortitude Valley	12:59	---	1:29	---	---	1:59	2:04	---	---	2:29
Central	1:02	1:17	1:32	1:47	1:56	2:02	2:07	2:18	2:26	2:32
Roma Street	1:06	1:21	1:36	1:51	2:00	2:06	2:11	2:22	2:30	2:36
Milton	1:07	---	137	---	---	2:07	---	---	---	2:37
Auchenflower	1:10	1:24	1:40	1:54	---	2:10	2:14	2:25	---	2:40
Toowong	---	1:25	---	1:55	---	---	2:15	---	---	---
Taringa	---	1:28	---	1:58	---	---	2:18	---	---	---
Indooroopilly	1:16	1:32	1:46	2:02	---	2:16	2:22	2:31	---	2:46
Chelmer	1:22	1:38	1:52	2:08	2:12	2:22	2:29	2:37	2:42	2:52
Graceville	1:24	---	1:54	---	---	2:24	---	2:39	---	2:54
Sherwood	1:34	1:49	2:04	2:19	---	2:34	2:40	2:49	---	3:05
Corinda	1:37	1:52	2:07	2:22	2:25	2:37	2:43	2:52	2:55	3:08
Oxley	1:40	1:55	2:10	2:25	---	2:40	2:46	2:55	---	3:11
Darra	1:43	1:58	2:13	2:28	---	2:43	2:49	2:58	---	3:14
Wacol	1:46	2:01	2:16	2:31	---	2:46	2:52	3:01	---	3:17
Gailes	1:50	2:05	2:20	2:35	---	2:50	2:56	3:05	---	3:21
Goodna	1:57	2:12	2:27	2:42	---	2:57	3:03	3:12	---	3:28
Redbank	2:03	2:18	2:33	2:48	---	2:53	3:09	3:18	---	3:34

3. If you caught the 1:18 pm train from Petrie, determine when you would arrive at Gailes.

4. Determine the time at which you need to catch the train from Roma Street if you want to arrive at Oxley at 2:40 pm.

5. Express the following in hours, minutes and seconds.

a. 5.125 hours

b. $3\frac{4}{5}$ hours

Use the bus timetable shown to answer questions 6 and 7.

Route 180 **Garden City to City (BUZ)**
servicing Garden City, Wishart, Mansfield, Mount Gravatt, Holland Park, Greenslopes, South Brisbane and City

Monday to Friday

map ref	Route number	180	180	180	180	180	180	180	180	180	180	180	180	180	180	180	180	180	180	180	180	180
		am	am	am	am	am	am	am	am	am	am	am	am	am	am	am	am	am	am	am	am	am
A	Garden City interchange G	5.10	5.25	5.40	5.55	6.10	6.20	6.30	6.40	6.50	7.00	7.10	7.20	7.30	7.40	7.50	8.00	8.10	8.20	8.30	8.40	8.55
C	Ham Rd South	5.16	5.31	5.46	6.01	6.16	6.26	6.36	6.47	6.57	7.07	7.17	7.29	7.39	7.49	7.59	8.09	8.19	8.29	8.38	8.48	9.03
D	Mansfield	5.21	5.36	5.51	6.06	6.21	6.31	6.41	6.53	7.03	7.15	7.25	7.38	7.48	7.58	8.08	8.18	8.28	8.38	8.46	8.56	9.11
H	Mount Gravatt TAFE	5.27	5.42	5.57	6.12	6.27	6.37	6.47	7.00	7.10	7.22	7.32	7.46	7.56	8.06	8.16	8.26	8.36	8.46	8.53	9.03	9.18
I	Holland Park (Logan Rd)	5.34	5.49	6.04	6.19	6.34	6.44	6.54	7.08	7.18	7.32	7.42	7.58	8.08	8.18	8.28	8.38	8.48	8.58	9.01	9.11	9.26
J	Greenslopes	5.38	5.53	6.08	6.23	6.38	6.48	6.58	7.12	7.22	7.37	7.47	8.02	8.12	8.22	8.32	8.42	8.52	9.02	9.05	9.15	9.30
L	Buranda Busway station PL1	5.42	5.57	6.12	6.27	6.42	6.52	7.02	7.17	7.27	7.42	7.52	8.07	8.17	8.27	8.37	8.47	8.57	9.07	9.10	9.20	9.35
M	Mater Hill station PL1	5.45	6.00	6.15	6.30	6.45	6.55	7.05	7.20	7.30	7.46	7.56	8.12	8.22	8.32	8.42	8.52	9.02	9.12	9.14	9.24	9.39
N	South Bank Busway station PL1	5.47	6.02	6.17	6.32	6.47	6.57	7.07	7.22	7.32	7.49	7.59	8.16	8.26	8.36	8.46	8.56	9.06	9.16	9.17	9.27	9.42
O	Cultural Centre station PL1	5.49	6.04	6.19	6.34	6.49	6.59	7.09	7.25	7.35	7.52	8.02	8.20	8.30	8.40	8.50	9.00	9.10	9.20	9.20	9.30	9.45
P	City, Queen St station	5.54	6.09	6.24	6.39	6.54	7.04	7.14	7.30	7.40	7.57	8.07	8.26	8.36	8.46	8.56	9.06	9.16	9.26	9.25	9.35	9.50

Monday to Friday (cont...)

map ref	Route number	180	180	180	180	180	180	180	180	180	180	180	180	180	180	180	180	180	180	180	180	180
		am	am	am	am	am	am	am	am	am	am	am	am	pm	pm	pm	pm	pm	pm	pm	pm	pm
A	Garden City interchange G	9.10	9.25	9.40	9.55	10.10	10.25	10.40	10.55	11.10	11.25	11.40	11.55	**12.10**	**12.25**	**12.40**	**12.55**	**1.10**	**1.25**	**1.40**	**1.55**	**2.10**
C	Ham Rd South	9.17	9.32	9.47	10.02	10.17	10.32	10.47	11.02	11.17	11.32	11.47	**12.02**	**12.17**	**12.32**	**12.47**	**1.02**	**1.17**	**1.32**	**1.47**	**2.02**	**2.17**
D	Mansfield	9.25	9.40	9.55	10.10	10.25	10.40	10.55	11.10	11.25	11.40	11.55	**12.10**	**12.25**	**12.40**	**12.55**	**1.10**	**1.25**	**1.40**	**1.55**	**2.10**	**2.25**
H	Mount Gravatt TAFE	9.32	9.47	10.02	10.17	10.32	10.47	11.02	11.17	11.32	11.47	**12.02**	**12.17**	**12.32**	**12.47**	**1.02**	**1.17**	**1.32**	**1.47**	**2.02**	**2.17**	**2.32**
I	Holland Park (Logan Rd)	9.40	9.55	10.10	10.25	10.40	10.55	11.10	11.25	11.40	11.55	**12.10**	**12.25**	**12.40**	**12.55**	**1.10**	**1.25**	**1.40**	**1.55**	**2.10**	**2.25**	**2.40**
J	Greenslopes	9.44	9.59	10.14	10.29	10.44	10.59	11.14	11.29	11.44	11.59	**12.14**	**12.29**	**12.44**	**12.59**	**1.14**	**1.29**	**1.44**	**1.59**	**2.14**	**2.29**	**2.44**
L	Buranda Busway station PL1	9.48	10.03	10.18	10.33	10.48	11.03	11.18	11.33	11.48	**12.03**	**12.18**	**12.33**	**12.48**	**1.03**	**1.18**	**1.33**	**1.48**	**2.03**	**2.18**	**2.33**	**2.48**
M	Mater Hill station PL1	9.51	10.06	10.21	10.36	10.51	11.06	11.21	11.36	11.51	**12.06**	**12.21**	**12.36**	**12.51**	**1.06**	**1.21**	**1.36**	**1.51**	**2.06**	**2.21**	**2.36**	**2.51**
N	South Bank Busway station PL1	9.53	10.08	10.23	10.38	10.53	11.08	11.23	11.38	11.53	**12.08**	**12.23**	**12.38**	**12.53**	**1.08**	**1.23**	**1.38**	**1.53**	**2.08**	**2.23**	**2.38**	**2.53**
O	Cultural Centre station PL1	9.56	10.11	10.26	10.41	10.56	11.11	11.26	11.41	11.56	**12.11**	**12.26**	**12.41**	**12.56**	**1.11**	**1.26**	**1.41**	**1.56**	**2.11**	**2.26**	**2.41**	**2.56**
P	City, Queen St station	10.01	10.16	10.31	10.46	11.01	11.16	11.31	11.46	**12.01**	**12.16**	**12.31**	**12.46**	**1.01**	**1.16**	**1.31**	**1.46**	**2.01**	**2.16**	**2.31**	**2.46**	**3.01**

Monday to Friday (cont...)

map ref	Route number	180	180	180	180	180	180	180	180	180	180	180	180	180	180	180	180	180	180	180	180	180
		pm	pm	pm	pm	pm	pm	pm	pm	pm	pm	pm	pm	pm	pm	pm	pm	pm	pm	pm	pm	pm
A	Garden City interchange G	**2.25**	**2.40**	**2.55**	**3.10**	**3.25**	**3.40**	**3.55**	**4.10**	**4.25**	**4.40**	**4.55**	**5.10**	**5.25**	**5.40**	**5.55**	**6.10**	**6.25**	**6.40**	**6.55**	**7.10**	**7.25**
C	Ham Rd South	**2.32**	**2.49**	**3.04**	**3.19**	**3.34**	**3.49**	**4.04**	**4.19**	**4.34**	**4.49**	**5.04**	**5.19**	**5.34**	**5.48**	**6.03**	**6.16**	**6.31**	**6.46**	**7.01**	**7.16**	**7.31**
D	Mansfield	**2.40**	**2.58**	**3.13**	**3.28**	**3.43**	**3.58**	**4.13**	**4.28**	**4.43**	**4.58**	**5.13**	**5.28**	**5.43**	**5.57**	**6.12**	**6.22**	**6.37**	**6.52**	**7.07**	**7.22**	**7.37**
H	Mount Gravatt TAFE	**2.47**	**3.06**	**3.21**	**3.36**	**3.51**	**4.06**	**4.21**	**4.36**	**4.51**	**5.06**	**5.21**	**5.36**	**5.51**	**6.05**	**6.20**	**6.29**	**6.44**	**6.59**	**7.14**	**7.29**	**7.44**
I	Holland Park (Logan Rd)	**2.55**	**3.14**	**3.29**	**3.44**	**3.59**	**4.14**	**4.29**	**4.44**	**4.59**	**5.14**	**5.29**	**5.44**	**5.59**	**6.12**	**6.27**	**6.35**	**6.50**	**7.05**	**7.20**	**7.35**	**7.50**
J	Greenslopes	**2.59**	**3.18**	**3.33**	**3.48**	**4.03**	**4.18**	**4.33**	**4.48**	**5.03**	**5.18**	**5.33**	**5.48**	**6.03**	**6.16**	**6.31**	**6.38**	**6.53**	**7.08**	**7.23**	**7.38**	**7.53**
L	Buranda Busway station PL1	**3.03**	**3.23**	**3.38**	**3.53**	**4.08**	**4.23**	**4.38**	**4.53**	**5.08**	**5.23**	**5.38**	**5.53**	**6.08**	**6.20**	**6.35**	**6.42**	**6.57**	**7.12**	**7.27**	**7.42**	**7.57**
M	Mater Hill station PL1	**3.06**	**3.27**	**3.42**	**3.57**	**4.12**	**4.27**	**4.42**	**4.57**	**5.12**	**5.27**	**5.42**	**5.57**	**6.12**	**6.23**	**6.38**	**6.45**	**7.00**	**7.15**	**7.30**	**7.45**	**8.00**
N	South Bank Busway station PL1	**3.08**	**3.29**	**3.44**	**3.59**	**4.14**	**4.29**	**4.44**	**4.59**	**5.14**	**5.29**	**5.44**	**5.59**	**6.14**	**6.25**	**6.40**	**6.47**	**7.02**	**7.17**	**7.32**	**7.47**	**8.02**
O	Cultural Centre station PL1	**3.11**	**3.32**	**3.47**	**4.02**	**4.17**	**4.32**	**4.47**	**5.02**	**5.17**	**5.32**	**5.47**	**6.02**	**6.17**	**6.27**	**6.42**	**6.49**	**7.04**	**7.19**	**7.34**	**7.49**	**8.04**
P	City, Queen St station	**3.16**	**3.37**	**3.52**	**4.07**	**4.22**	**4.37**	**4.52**	**5.07**	**5.22**	**5.37**	**5.52**	**6.07**	**6.22**	**6.32**	**6.47**	**6.54**	**7.09**	**7.24**	**7.39**	**7.54**	**8.09**

Monday to Friday (cont...)

map ref	Route number	180	180	180	180	180	180	180	180	180	180	180	180	180
		pm	pm	pm	pm	pm	pm	pm	pm	pm	pm	pm	pm	pm
A	Garden City interchange G	**7.40**	**7.55**	**8.10**	**8.25**	**8.40**	**8.55**	**9.10**	**9.25**	**9.40**	**9.55**	**10.10**	**10.25**	**10.40**
C	Ham Rd South	**7.46**	**8.01**	**8.16**	**8.31**	**8.46**	**9.01**	**9.16**	**9.31**	**9.46**	**10.01**	**10.16**	**10.31**	**10.46**
D	Mansfield	**7.52**	**8.07**	**8.22**	**8.37**	**8.52**	**9.07**	**9.22**	**9.37**	**9.52**	**10.07**	**10.22**	**10.37**	**10.52**
H	Mount Gravatt TAFE	**7.59**	**8.14**	**8.29**	**8.44**	**8.59**	**9.14**	**9.29**	**9.44**	**9.59**	**10.14**	**10.29**	**10.44**	**10.59**
I	Holland Park (Logan Rd)	**8.05**	**8.20**	**8.35**	**8.50**	**9.05**	**9.20**	**9.35**	**9.50**	**10.05**	**10.20**	**10.35**	**10.50**	**11.05**
J	Greenslopes	**8.08**	**8.23**	**8.38**	**8.53**	**9.08**	**9.23**	**9.38**	**9.53**	**10.08**	**10.23**	**10.38**	**10.53**	**11.08**
L	Buranda Busway station PL1	**8.12**	**8.27**	**8.42**	**8.57**	**9.12**	**9.27**	**9.42**	**9.57**	**10.12**	**10.27**	**10.42**	**10.57**	**11.12**
M	Mater Hill station PL1	**8.15**	**8.30**	**8.45**	**9.00**	**9.15**	**9.30**	**9.45**	**10.00**	**10.15**	**10.30**	**10.45**	**11.00**	**11.15**
N	South Bank Busway station PL1	**8.17**	**8.32**	**8.47**	**9.02**	**9.17**	**9.32**	**9.47**	**10.02**	**10.17**	**10.32**	**10.47**	**11.02**	**11.17**
O	Cultural Centre station PL1	**8.19**	**8.34**	**8.49**	**9.04**	**9.19**	**9.34**	**9.49**	**10.04**	**10.19**	**10.34**	**10.49**	**11.04**	**11.19**
P	City, Queen St station	**8.24**	**8.39**	**8.54**	**9.09**	**9.24**	**9.39**	**9.54**	**10.09**	**10.24**	**10.39**	**10.54**	**11.09**	**11.24**

Source: Reproduced by permission of Translink. Department of Transport and Main Roads.
 Route 180. Garden City to City (BUZ) : servicing Garden City, Wishart, Mansfield, Mount Gravatt, Holland Park, Greenslopes, South Brisbane and City [image], viewed April 2024.

6. a. Determine at what time the bus leaves Mount Gravatt TAFE between 6:00 pm and 6:10 pm.
   b. To arrive at South Bank Busway station at 11:38 am, determine at what time you would need to catch the bus from Holland Park.
   c. Determine the number of buses that leave Ham Road South between 4:00 pm and 5:00 pm.

7. a. If you had to arrive at Cultural Centre station by 9:00 am, determine at what time you should catch the bus from Mansfield.
   b. If you needed to walk for 33 minutes from City, Queens Street station to arrive by 10:00 am, determine at what time you should catch the bus from Ham Road South.

8. To get to your sports training, you have to walk for 13 minutes to the train station. You catch the train that is scheduled to take 18 minutes and then a bus for another 21 minutes, before a final walk to training that takes 7 minutes.
Calculate the total time you take to get to your sports training from your house.

9. Maya needs to catch a bus to work and wants to minimise the travel time. There are three possible options, as shown. Determine which is the best option for Maya.

Bus	Depart	Arrive
Option 1	7:45 am	8:18 am
Option 2	7:52 am	8:23 am
Option 3	7:48 am	8:21 am

Use the following train and bus timetable to answer questions 10 and 11.

Train timetable:

Brisbane to Cairns

Northbound

Departing	**Mon** (VC71), **Tue** (VC73), **Wed** (VC75), **Fri** (VC79) and **Sat** (VC81)
Brisbane (Roma St)	1.45pm
Caboolture	2.32pm
Landsborough^	2.57pm
Nambour	3.25pm
Cooroy^	3.48pm
Gympie North	4.24pm
Maryborough West	5.26pm
Howard^	5.47pm
Bundaberg	6.22pm
Miriam Vale^	7.36pm
Gladstone	8.23pm
Mount Larcom^	8.57pm
Rockhampton *arrive*	10.01pm
Rockhampton	10.16pm

Bus timetable:

GREYHOUND SERVICE TIMETABLE FOR:
BRISBANE > CAIRNS

Effective 29 January 2024

Stop Code	Stop Name		GX409 Daily	GX401 Daily	GX402 Daily	GX404 Daily	GX408 Daily	GX405 Daily
BDA	BRISBANE DOMESTIC AIRPORT			8:50 AM*				
BIA	BRISBANE INTERNATIONAL			9:00 AM*				
BNE	BRISBANE		6:45 AM	9:30 AM	12:00 PM	2:00 PM		
MOB	MOOLOOLABA - SUNSHINE COAST			11:00 AM*		3:30 PM		
MCY	MAROOCHYDORE - SUNSHINE COAST		8:35 AM	11:10 AM*		3:40 PM		
NOO	NOOSA - SUNSHINE COAST		9:15 AM	11:55 AM	2:00 PM	4:20 PM		
COO	COOROY - SUNSHINE COAST					4:50 PM		
GYR	TRAVESTON	*Arr*		12:45 PM	2:55 PM*	5:15 PM		
		Dep		1:20 PM	3:30 PM*	5:50 PM*		
GYM	GYMPIE				3:50 PM	6:05 PM		
WAL	WALLU				4:25 PM	6:40 PM		
RAI	RAINBOW BEACH			2:35 PM	5:10 PM*	7:15 PM		
MBH	MARYBOROUGH			3:55 PM	6:15 PM	8:20 PM		
HVB	HERVEY BAY			4:30 PM*	7:00 PM*	9:05 PM		
CHI	CHILDERS				7:40 PM*	9:50 PM*		
SBM	SOUTH BUNDABERG MEAL BREAK	*Arr*		5:50 PM*	8:15 PM*	10:25 PM*		
		Dep		6:30 PM*	8:45 PM*	11:00 PM*		
BDB	BUNDABERG				8:55 PM*	11:10 PM*		
AGW	AGNES WATER			8:10 PM				
MIM	MIRIAM VALE				10:00 PM*	12:25 AM		
GLT	GLADSTONE				10:50 PM*	1:25 AM*		
ROK	ROCKHAMPTON	*Arr*		10:45 PM	12:10 AM	2:45 AM		
		Dep		11:20 PM	12:45 AM	3:20 AM		
MLT	MARLBOROUGH TURNOFF				2:00 AM*	4:35 AM		
SLT	ST LAWRENCE TURN OFF				2:40 AM*	5:15 AM		
SRI	SARINA				3:55 AM	6:30 AM		
MKM	MACKAY MEAL BREAK	*Arr*		3:00 AM	4:15 AM	6:50 AM		
		Dep		3:40 AM	4:55 AM	7:30 AM		
MKY	MACKAY			3:55 AM	5:15 AM	7:50 AM		
PPP	PROSERPINE			5:25 AM*	6:40 AM*	9:15 AM		
AIR	AIRLIE BEACH			6:00 AM*	7:10 AM*	9:50 AM		12:10 PM
LFC	LONGFORD CREEK MEAL BREAK	*Arr*			8:10 AM	10:40 AM		
		Dep			8:45 AM	11:15 AM		
BWN	BOWEN				9:15 AM	11:45 AM		
DEM	DELTA				9:20 AM*	11:55 AM		
HHL	HOME HILL				10:30 AM*	1:00 PM		
AYR	AYR				10:45 AM*	1:15 PM		
TSV	TOWNSVILLE	*Arr*						3:45 PM
		Dep			12:10 PM	2:20 PM	8:00 AM	4:15 PM
ING	INGHAM					4:00 PM	9:30 AM	5:45 PM
CDW	CARDWELL	*Arr*				4:30 PM	10:10 AM	6:25 PM
		Dep				5:05 PM	10:45 AM	7:00 PM
TUL	TULLY					5:35 PM	11:15 AM	7:30 PM
MIS	MISSION BEACH (WONGALING)					6:00 PM	11:45 AM	8:00 PM
IFL	INNISFAIL					6:40 PM	12:25 PM	8:40 PM
BAB	BABINDA					7:00 PM	12:50 PM*	
GDV	GORDONVALE					7:30 PM	1:15 PM*	
CNS	CAIRNS					8:00 PM	1:40 PM	9:55 PM

10. Matt needs to get to Bundaberg from Brisbane. He investigates the time it takes to travel by train compared to the bus.
Given Matt wants to leave between 1:30 pm and 2:00 pm, determine which mode of transport has the shorter travel time, and calculate the difference using the timetables shown.

11. Ellie wants to travel from Cooroy to Gladstone, leaving Cooroy between 3:00 pm and 5:00 pm.
To minimise the travel time, they investigate travelling by bus and train.
Using the timetables shown, determine the duration of the quickest mode of transport.

12. Britney travels to school by various means. Determine at what time she will arrive at school if she leaves home at 7:48 am and travels by the following means.
    a. Walking, which takes 1 hour and 5 minutes
    b. Cycling, which takes 37 minutes
    c. Car, which takes 17 minutes
    d. Bus, which takes 25 minutes

13. Trams are used throughout Melbourne to assist commuters getting around the inner-city suburbs. The following questions refer to the Melbourne tram timetable shown below.

Monday to Thursday																	
Morning (am) / Afternoon (pm)	am	am	am	am	am	am	am	am	am	am	am	am	am	am	am	am	am
1-University of Melbourne/ Swanston St (Carlton)	7:59	8:09	8:20	8:30	8:42	8:54	9:05	9:16	9:29	9:41	9:54	10:06	10:18	10:30	10:42	10:54	11:06
8-Melbourne Central Station/Swanston St (Melbourne City)	8:06	8:16	8:27	8:37	8:49	9:01	9:12	9:23	9:36	9:48	10:01	10:13	10:25	10:37	10:49	11:01	11:13
10-Bourke Street Mall/Swanston St (Melbourne City)	8:09	8:19	8:30	8:40	8:52	9:04	9:15	9:26	9:39	9:51	10:04	10:16	10:28	10:40	10:52	11:04	11:16
11-City Square/Swanston St (Melbourne City)	8:10	8:20	8:31	8:41	8:53	9:05	9:16	9:27	9:40	9:52	10:05	10:17	10:29	10:41	10:53	11:05	11:17
13-Flinders Street Railway Station/ Swanston St (Melbourne City)	8:12	8:22	8:33	8:43	8:55	9:07	9:18	9:29	9:42	9:54	10:07	10:19	10:31	10:43	10:55	11:07	11:19
14-Arts Centre/St Kilda Rd (Southbank)	8:14	8:24	8:35	8:45	8:57	9:09	9:20	9:31	9:44	9:56	10:09	10:21	10:33	10:45	10:57	11:09	11:21
20-Domain Interchange/St Kilda Rd (Melbourne City)	8:21	8:31	8:42	8:52	9:03	9:15	9:26	9:37	9:50	10:02	10:15	10:27	10:39	10:51	11:03	11:15	11:27
25-Commercial Rd/St Kilda Rd (South Melbourne)	8:28	8:38	8:49	8:59	9:09	9:21	9:32	9:43	9:55	10:07	10:20	10:32	10:44	10:56	11:08	11:20	11:32
26-Alfred Hospital/Commercial Rd (South Yarra)	8:29	8:39	8:50	9:00	9:10	9:22	9:33	9:44	9:55	10:08	10:21	10:33	10:45	10:57	11:09	11:21	11:33
31-Chapel St/Commercial Rd (South Yarra)	8:34	8:44	8:55	9:05	9:15	9:27	9:38	9:49	10:01	10:13	10:26	10:38	10:50	11:02	11:14	11:26	11:38
37-Orrong Rd/Malvern Rd (Toorak)	8:40	8:50	9:01	9:10	9:20	9:32	9:43	9:54	10:06	10:18	10:31	10:43	10:55	11:07	11:19	11:31	11:43
43-Glenferrie Rd/Malvern Rd (Toorak)	8:46	8:56	9:05	9:14	9:24	9:36	9:48	9:59	10:11	10:23	10:36	10:48	11:00	11:12	11:24	11:36	11:48
44-Plant St/Malvern Rd (Malvern)	8:47	8:57	9:06	9:15	9:25	9:37	9:49	10:00	10:12	10:24	10:37	10:49	11:01	11:13	11:25	11:37	11:49
51-Gardiner Railway Station/Burke Rd (Hawthorn East)	8:52	9:02	9:10	9:19	9:29	9:41	9:53	10:04	10:16	10:28	10:41	10:53	11:05	11:17	11:29	11:41	11:53
54-Toorak Rd/Burke Rd (Hawthorn East)	8:56	9:06	9:14	9:23	9:33	9:45	9:57	10:08	10:20	10:32	10:45	10:57	11:09	11:21	11:33	11:45	11:57
61-Camberwell Junction/Burke Rd (Hawthorn East)	9:04	9:13	9:21	9:30	9:40	9:52	10:04	10:15	10:27	10:39	10:52	11:04	11:16	11:28	11:40	11:52	12:04
64-Camberwell Station/Burke Rd (Hawthorn East)	9:07	9:16	9:24	9:33	9:43	9:55	10:07	10:18	10:30	10:42	10:55	11:07	11:19	11:31	11:43	11:55	12:07
70-Cotham Rd/Burke Rd (Kew)	9:13	9:22	9:30	9:38	9:48	10:00	10:12	10:23	10:35	10:47	11:00	11:12	11:24	11:36	11:48	12:00	12:12

a. Determine how many trams leave the University of Melbourne between 8 am and 9 am.

b. If you wanted to go shopping at Chapel Street and be there when the shops open at 9:00 am, determine at what time you would need to catch the tram from Bourke Street.

c. Determine how long it takes to get from Melbourne Central to the Alfred Hospital if you catch the 8:06 am tram from Melbourne Central.

d. If you caught the tram from the University of Melbourne at 11:06 am, determine at what time you would get to Toorak Road.

14. The table shows the flight timetable of Qantas, flying from Melbourne. The following questions refer to the timetable.

QANTAS **Worldwide Timetable**

< from Melbourne **from Melbourne >**

Flight	Depart	Arrive	Freq.	Stops	Effective
"	1610	1720	123···7	-	from 07 Oct
QF0628	1720	1930	Daily	-	
"	1720	1830	123···7	-	from 07 Oct
QF0632	1815	2025	···45··	-	
"	1815	1925	123···7	-	from 07 Oct
"	1820	2030	123···7	-	until 03 Oct
QF0634	1910	2120	12345·7	-	
"	1910	2120	·····6·	-	15 Sep only
"	1910	2120	·····6·	-	from 06 Oct
"	1910	2020	123···7	-	from 07 Oct
QF1746	1910	2120	·····6·	-	22 Sep only
QF1748	2005	2215	·····6·	-	29 Sep only
QF0636	2010	2220	12345·7	-	
"	2010	2120	123···7	-	from 07 Oct
Broome					
QF1050	0830	1105	··3···7	-	12 Sep - 03 Oct
"	0835	1010	······7	-	from 07 Oct
Cairns					
JQ0944 †	0600	0920	Daily	-	
"	0600	0820	123···7	-	from 07 Oct
JQ0249 †	0700	1020	····5··	-	from 21 Sep
QF0702	0910	1235	12345··	-	
"	0910	1235	·····67	-	15 Sep - 16 Sep
"	0910	1135	123····	-	from 08 Oct
JQ0946 †	1020	1340	Daily	-	
"	1120	1340	123···7	-	from 07 Oct
QF0704	1210	1420	······7	-	from 07 Oct
"	1255	1620	1······	-	until 01 Oct
"	1255	1620	·····67	-	15 Sep - 16 Sep
"	1255	1620	··3····	-	19 Sep - 03 Oct
"	1255	1605	·····67	-	from 22 Sep
"	1255	1520	1·3····	-	from 08 Oct
JQ0948 †	1950	2310	····5··	-	
"	1950	2310	1··4··7	-	23 Sep - 04 Oct
"	2050	2310	1·····7	-	from 07 Oct
Canberra					
QF0804	0700	0805	12345··	-	
QF0814	0830	0935	·234567	-	
"	0915	1020	1······	-	
QF0816	1040	1145	12345··	-	
QF0848	1200	1305	12·456·	-	
"	1205	1310	·2·····	-	from 09 Oct
QF2130	1235	1345	Daily	-	
QF0812	1320	1425	Daily	-	
QF0820	1420	1525	1·345··	-	
QF0818	1530	1635	1·3456·	-	
QF0826	1620	1725	12345·7	-	
QF0834	1710	1815	12345··	-	
"	1710	1815	·····6·	-	15 Sep - 22 Sep
"	1710	1815	·····6·	-	from 06 Oct
QF2132	1805	1915	··345··	-	
QF0822	1940	2045	12345·7	-	
QF0834	2015	2120	·····6·	-	29 Sep only
Christchurch					
QF0133	1155	1720	·2···6·	-	11 Sep - 29 Sep
"	1155	1820	·2···6·	-	from 02 Oct
"	1155	1720	·2·····	-	from 09 Oct
"	1830	2355	1·345·7	-	until 28 Sep
"	1830	0055+1	1·345·7	-	from 30 Sep

Flight	Depart	Arrive	Freq.	Stops	Effective
"	1830	2355	1·3···7	-	from 07 Oct
JQ0203 †	2255	0515+1	1·3·5·7	-	from 30 Sep
"	2355	0515+1	····5·7	-	14 Sep - 28 Sep
"	2355	0515+1	1·3····	-	24 Sep - 26 Sep
"	2355	0515+1	1·3···7	-	from 07 Oct
Darwin					
QF0838	0845	1245	Daily	-	
"	0845	1145	123···7	-	from 07 Oct
JQ0678 †	2125	0110+1	12·4567	-	
"	2125	0110+1	··3····	-	26 Sep - 03 Oct
"	2125	0010+1	123···7	-	from 07 Oct
Devonport					
QF2051	0820	0925	12345··	-	
"	0950	1055	·····67	-	
QF2053	1225	1330	12345··	-	
"	1315	1420	·····6·	-	
QF2057	1540	1645	12345·7	-	
"	1640	1745	·····6·	-	
QF2061	1900	2005	12345·7	-	
Dubai					
EK8409*	0500	1300	Daily	-	
"	0600	1300	123···7	-	from 07 Oct
EK8405*	1800	0450+1	Daily	1	
"	1900	0450+1	123···7	1	from 07 Oct
EK8407*	2115	0515+1	Daily	-	
"	2215	0515+1	123···7	-	from 07 Oct
Gold Coast					
JQ0430 †	0600	0800	Daily	-	
"	0600	0700	123···7	-	from 07 Oct
QF0880	0950	1155	··3····	-	26 Sep - 03 Oct
"	0955	1100	123···7	-	from 07 Oct
"	1000	1205	12·4567	-	
"	1000	1205	··3····	-	12 Sep - 19 Sep
JQ0432 †	1005	1205	·234567	-	
"	1005	1205	1······	-	24 Sep - 01 Oct
"	1005	1105	123···7	-	from 07 Oct
JQ0436 †	1230	1430	Daily	-	
"	1230	1330	123···7	-	from 07 Oct
JQ0438 †	1430	1630	Daily	-	
"	1525	1625	123···7	-	from 07 Oct
JQ0444 †	1600	1800	1·34567	-	
"	1600	1800	·2·····	-	25 Sep - 02 Oct
"	1630	1730	123···7	-	from 07 Oct
JQ0442 †	1930	2130	Daily	-	
JQ0446 †	2010	2210	1···567	-	
"	2010	2210	·234···	-	25 Sep - 04 Oct
JQ0442 †	2030	2130	123···7	-	from 07 Oct
JQ0446 †	2110	2210	123···7	-	from 07 Oct
Guangzhou					
CZ0329*	1030	1800	Daily	-	
"	1130	1800	123···7	-	from 07 Oct
CZ0331*	2230	0600+1	123·567	-	
"	2230	0600+1	···4···	-	27 Sep - 04 Oct
"	2330	0600+1	123···7	-	from 07 Oct
Hamilton Island					
QF0870	0750	1050	····5·7	-	14 Sep - 16 Sep
"	0845	1145	··3··6·	-	
"	0845	1145	····5·7	-	from 21 Sep
"	0845	1045	··3···7	-	from 07 Oct

Flight	Depart	Arrive	Freq.	Stops	Effective
Hobart					
JQ0701 †	0605	0720	·234567	-	
"	0605	0720	1······	-	from 24 Sep
QF1011	0815	0930	Daily	-	
JQ0705 †	0830	0945	Daily	-	
JQ0707 †	1040	1155	Daily	-	
QF1013	1210	1325	······7	-	
JQ0709 †	1245	1400	Daily	-	
QF1013	1255	1410	123456·	-	
JQ0711 †	1445	1600	Daily	-	from 21 Sep
QF1015	1515	1630	12·45·7	-	
"	1540	1655	·····6·	-	15 Sep - 22 Sep
"	1540	1655	·····6·	-	from 06 Oct
QF1017	1920	2035	12345·7	-	
QF1015	1950	2105	·····6·	-	29 Sep only
JQ0715 †	2045	2200	Daily	-	
Ho Chi Minh City					
JQ0279 †	1515	2105	···4···	-	13 Sep - 04 Oct
"	1840	0030+1	1····6·	-	
"	1940	0030+1	1······	-	from 08 Oct
Hong Kong					
QF0029	0935	1720	Daily	-	
"	1035	1720	123···7	-	from 07 Oct
Honolulu					
JQ0285 †	1545	0615	··3·5·7	-	
"	1545	0615	·2·····	-	25 Sep - 02 Oct
"	1645	0615	·23···7	-	from 07 Oct
Launceston					
JQ0731 †	0635	0740	1·3·567	-	
"	0635	0740	·2·4···	-	from 25 Sep
QF2281	0830	0945	12345··	-	
"	0845	1000	·····67	-	
JQ0733 †	0910	1015	Daily	-	from 21 Sep
"	1010	1115	123···7	-	from 07 Oct
JQ0735 †	1025	1130	Daily	-	
"	1125	1230	123···7	-	from 07 Oct
QF2283	1145	1300	123····	-	
"	1220	1335	·····67	-	
QF2285	1535	1650	12345·7	-	
"	1605	1720	·····6·	-	
JQ0739 †	1640	1745	Daily	-	
QF2289	1905	2020	12345·7	-	
JQ0743 †	1905	2010	1··45·7	-	from 21 Sep
London–Heathrow					
QF0009	1515	0505+1	Daily	1	
"	1615	0505+1	123···7	1	from 07 Oct
Los Angeles					
QF0093	0900	0625	Daily	-	
"	1000	0625	123···7	-	from 07 Oct
QF0095	2140	1900	1···5··	-	
"	2240	1900	1······	-	from 08 Oct
Mildura					
QF2078	0825	0935	12345··	-	
"	0850	1000	·····67	-	
QF2080	1200	1310	···45··	-	
QF2084	1545	1655	12345·7	-	
"	1645	1755	·····6·	-	
QF2086	1910	2020	12345·7	-	

*Indicates a codeshare flight operated by another carrier and QF flight number.

†Indicates a Jetstar operated flight and flight number.

1 = Monday 2 = Tuesday 3 = Wednesday 4 = Thursday 5 = Friday 6 = Saturday 7 = Sunday

Page 32 of 55

 Effective between 10 Sep - 11 Oct Accurate at the date of publication 10 September 2018

a. Identify at what time the flight leaves for London, in 12-hour time.
b. If you wanted to arrive at the Gold Coast at 2 pm, determine at what time you would catch a flight from Melbourne.
c. Determine the weekly frequency of the QF0816 flight.
d. Use the timetable to calculate how long it takes to fly to Ho Chi Minh City. In December, Melbourne (Victoria) is 3 hours ahead of Ho Chi Minh City (Vietnam).

Complex familiar

15. Use the information from the Translink journey planner shown to answer the following questions.

Plan your journey

Brisbane; QLD
Griffith University station (Southport)
swap
Today (Tuesday)
Leave after
Arrive before
2:00pm
Advanced options
Find a journey

Source: Reproduced by permission of Translink. Department of Transport and Main Roads. © State of Queensland. (n.d.). *Brisbane QLD to Griffith University station (Southport) : your travel options [image],* viewed April 2024.

a. Determine the destination of the planned trip.
b. Calculate the shortest time suggested for the trip.
c. Calculate how far you must walk in option 2.
d. Identify the modes of transport suggested in option 1.

16. Using the tidal chart shown, answer the following questions.

AUSTRALIA, EAST COAST – BRISBANE BAR **2024**

LAT 27° 22′ S LONG 153° 10′ E

Times and Heights of High and Low Waters

Time Zone –1000

JANUARY

Day	Time	m	Day	Time	m
1	0022	1.70	16	0058	1.98
	0609	0.64		0653	0.48
MO	1245	2.26	TU	1316	2.47
	1928	0.73		2003	0.48
2	0105	1.70	17	0153	1.99
	0652	0.74		0748	0.63
TU	1323	2.17	WE	1402	2.28
	2006	0.73		2047	0.51
3	0155	1.69	18	0254	2.00
	0743	0.85		0852	0.78
WE	1404	2.06	TH	1454	2.07
	2050	0.73	◑	2136	0.54
4	0256	1.71	19	0402	2.05
	0844	0.95		1008	0.89
TH	1452	1.95	FR	1554	1.88
◐	2140	0.71		2230	0.57
5	0407	1.78	20	0514	2.13
	0958	1.01		1134	0.92
FR	1549	1.85	SA	1705	1.73
	2234	0.68		2330	0.57

FEBRUARY

Day	Time	m	Day	Time	m
1	0113	1.88	16	0220	2.16
	0711	0.82		0834	0.85
TH	1315	2.05	FR	1418	1.86
	1948	0.68		2044	0.62
2	0200	1.87	17	0326	2.12
	0801	0.94		0954	0.97
FR	1354	1.90	SA	1524	1.66
	2029	0.71	◑	2142	0.70
3	0300	1.87	18	0444	2.12
	0909	1.03		1131	0.98
SA	1448	1.75	SU	1653	1.55
◐	2123	0.74		2255	0.75
4	0417	1.93	19	0559	2.18
	1034	1.05		1254	0.89
SU	1603	1.65	MO	1820	1.57
	2231	0.73			
5	0534	2.05	20	0013	0.73
	1203	0.98		0700	2.26
MO	1725	1.63	TU	1352	0.79
	2343	0.68		1923	1.67

MARCH

Day	Time	m	Day	Time	m
1	0038	2.08	16	0148	2.28
	0648	0.82		0824	0.89
FR	1234	1.95	SA	1351	1.66
	1856	0.65		1955	0.70
2	0118	2.04	17	0249	2.17
	0734	0.93		0947	0.98
SA	1313	1.80	SU	1507	1.51
	1930	0.71	◑	2056	0.83
3	0210	2.01	18	0407	2.11
	0836	1.01		1121	0.96
SU	1406	1.66	MO	1651	1.49
	2021	0.78		2225	0.90
4	0324	2.00	19	0527	2.11
	1005	1.03		1232	0.87
MO	1531	1.56	TU	1811	1.59
◐	2141	0.81		2353	0.86
5	0453	2.08	20	0630	2.17
	1143	0.95		1323	0.78
TU	1708	1.59	WE	1905	1.73
	2310	0.76			

APRIL

Day	Time	m	Day	Time	m
1	0141	2.15	16	0324	2.10
	0824	0.96		1045	0.92
MO	1351	1.59	TU	1629	1.51
	1949	0.79	◑	2147	0.98
2	0251	2.12	17	0440	2.07
	0953	0.95		1147	0.86
TU	1523	1.55	WE	1739	1.64
◐	2113	0.84		2316	0.94
3	0418	2.16	18	0544	2.09
	1121	0.86		1235	0.78
WE	1658	1.64	TH	1831	1.79
	2245	0.79			
4	0536	2.28	19	0021	0.85
	1232	0.71		0634	2.14
TH	1812	1.81	FR	1315	0.70
				1914	1.93
5	0005	0.67	20	0112	0.77
	0640	2.41		0717	2.18
FR	1329	0.56	SA	1350	0.62
	1911	2.00		1951	2.06

a. Determine the time of low tide in the pm on 17 March 2024.
b. Determine the height of the highest tide on 5 April 2024.
c. Determine the time of high tide in the pm of 18 January 2024.

17. Using the Translink journey planner, determine the shortest travel time from Brisbane (e.g. Brisbane Airport or Brisbane's Central Station) to Noosa Heads to arrive before 11:00 am.

18. Use the Rome2Rio website to determine the time difference between taking a plane and taking a train from Brisbane to Coffs Harbour.

Complex unfamiliar

19. The tidal range is the difference in height between high tide and low tide. Using the tide timetable for Gold Coast Seaway, for a week in July 2024, determine the largest second (afternoon) tidal range of the day for that week and identify the date.

21 July		22 July		23 July		24 July		25 July		26 July		27 July	
Tides	Height	Tides	Height	Tides	Height	Tides	Height	Tides	Height	Tides	Height	Tides	Height
2:22 am	0.3 m	3:03 am	0.2 m	3:43 am	0.2 m	4:24 am	0.2 m	5:07 am	0.2 m	5:52 am	02 m	12:09 am	1.5 m
7:58 am	1.1 m	8:44 am	1.1 m	9:29 am	1.2 m	10:15 am	1.2 m	11:03 am	1.3 m	11:55 am	1.3 m	6:39 am	0.3 m
1:33 pm	0.2 m	2:21 pm	0.2 m	3:09 pm	0.2 m	3:59 pm	0.2 m	4:52 am	0.3 m	5:51 am	0.3 m	12:52 pm	1.3 m
8:25 pm	1.8 m	9:07 pm	1.8 m	9:51 m	1.8 m	10:35 pm	1.7 m	11:20 pm	1.6 m			6:59 pm	0.4 m

20. Using the table displaying the sunrise and sunset times in Gold Coast, determine the day with the least daylight time in February 2025 at this location and calculate how long this is in hours and minutes. Explain the reasonableness of this answer.

1 February		2 February		3 February		4 February		5 February		6 February	
Sunrise	Sunset	Sunrise	Sunset	Sunrise	Sunset	Sunrise	Sunset	Sunrise	Sunset	Sunrise	Sunset
5:17 am	6:41 pm	5:18 am	6:40 pm	5:19 am	6:40 pm	5:20 am	6:39 pm	5:21 am	6:38 pm	5:22 am	6:38 pm
7 February		8 February		9 February		10 February		11 February		12 February	
Sunrise	Sunset	Sunrise	Sunset	Sunrise	Sunset	Sunrise	Sunset	Sunrise	Sunset	Sunrise	Sunset
5:22 am	6:37 pm	5:23 am	6:36 pm	5:24 am	6:36 pm	5:25 am	6:35 pm	5:25 am	6:34 pm	5:25 am	6:35 pm
13 February		14 February		15 February		16 February		17 February		18 February	
Sunrise	Sunset	Sunrise	Sunset	Sunrise	Sunset	Sunrise	Sunset	Sunrise	Sunset	Sunrise	Sunset
5:27 am	6:33 pm	5:28 am	6:32 pm	5:28 am	6:31 pm	5:29 am	6:30 pm	5:30 am	6:29 pm	5:31 am	6:28 pm
19 February		20 February		21 February		22 February		23 February		24 February	
Sunrise	Sunset	Sunrise	Sunset	Sunrise	Sunset	Sunrise	Sunset	Sunrise	Sunset	Sunrise	Sunset
5:31 am	6:27 pm	5:32 am	6:27 pm	5:33 am	6:26 pm	5:33 am	6:25 pm	5:34 am	6:24 pm	5:35 am	6:23 pm
25 February		26 February		27 February		28 February					
Sunrise	Sunset	Sunrise	Sunset	Sunrise	Sunset	Sunrise	Sunset				
5:35 am	6:22 pm	5:36 am	6:21 pm	5:37 am	6:20 pm	5:37 am	6:19 pm				

Fully worked solutions for this chapter are available online.

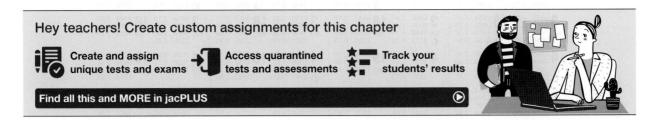

Answers

Chapter 13 Time

13.2 Representing and calculating time and time intervals

13.2 Exercise

1. a. 5 hours 48 minutes
   b. 9 hours 52 minutes 30 seconds
2. a. 12 hours 9 minutes 21.6 seconds
   b. 3 hours 13 minutes 20 seconds
3. a. Five o'clock
   b. Approximately 23 minutes past 7
   c. Approximately 13 minutes to 7 o'clock
   d. Quarter past 2
   e. Five minutes to 8
   f. Ten minutes past midnight
4. a. 10:25 b. 07:33 c. 13:45
   d. 20:12 e. 22:06 f. 23:45
5. 2 hours 36 minutes
6. 1 hour 25 minutes
7. a. 3:51 pm b. 8:22 pm c. 3:15 am
   d. 11:31 am e. 9:02 am f. 10:15 pm
8. 1200 h or 0000 h
9. a. 180 minutes b. 315 minutes c. 1 440 minutes
   d. 45 minutes e. 198 minutes f. 394 minutes
10. They are not the same length of time, since 2.25 hours (2 hours 15 minutes) is 10 minutes less than 2 hours and 25 minutes.
11. a. 3 hours 30 minutes
    b. 5 hours 5 minutes
    c. 1 hour 17 minutes
    d. 2 hours 27 minutes
    e. 9 hours 33 minutes
    f. 140 hours 27 minutes
    g. 8 hours 48 minutes 54 seconds
    h. 2 hours 33 minutes 52 seconds
12. 4 hours 15 minutes
13. 4 hours 57 minutes
14. a. 1 hour 25 minutes
    b. 8 hours 15 minutes
    c. 35 hours 10 minutes
    d. 43 hours 49 minutes
    e. 22 hours 20 minutes
    f. 8 hours 35 minutes
15. a. 1 pm b. 4:15 am c. 9:50 am
    d. 5:15 pm e. 5:22 pm f. 11:33 am
16. 106 days
17. 115 days
18. 87 days
19. a. 80 minutes or 1 hour 20 minutes
    b. Yes, just on time.
20. a. 3600 seconds b. 86 400 seconds
    c. 1440 minutes d. 8760 hours

13.3 Interpreting timetables

13.3 Exercise

1. a. 2006 b. 1998
2. a. 10 years apart
   b. Separated by 13 years
3. a. 1969 b. 1963
4. a. 1958 b. 11 years
5. a. 20 years b. 4 years
6. a. 14:19 or 2:19 pm
   b. 8:05 am
7. a. 7:01 am the next day
   b. 2 hours 40 minutes
8. a. 9:07 am b. 9:45 am
9. 27 minutes
10. a. No b. 8:55 am
11. a. 2 trains
    b. 5:14 pm from South Brisbane
    c. 23 minutes
12. a. 6:06 am
    b. 8:09 am
    c. 6 buses, at 8:06 am, 8:16 am, 8:26 am, 8:36 am, 8:46 am, 8:53 am
13. a. 7:37 am b. 8:08 am
14. a. 53 minutes b. 7
15. a. 8:34 am b. 12:18 pm
16. A: (1770) Captain James Cook lands at Botany Bay.
    B: (1850) The first University in Australia (Sydney) is founded.
    C: (1854) Eureka Rebellion takes place.
    D: (1880) Ned Kelly is hanged.
    E: (1890) Banjo Patterson publishes 'The Man from Snowy River'.
    F: (1901) Australia becomes a federation.
    G: (1902) Australian women gain the right to vote and stand for Parliament.
    H: (1915) Australians land at ANZAC Cove.
    I: (1920) QANTAS is founded.
    J: (1923) Vegemite is first produced.
    K: (1930) Phar Lap wins the Melbourne Cup.
    L: (1932) Sydney Harbour Bridge opens.
    M: (1956) Melbourne Olympics and TV come to Australia.
    N: (1967) Referendum to include Indigenous Australians in the census is successful.

13.4 Interpreting complex timetables and planning routes [complex]

13.4 Exercise

1. 63 minutes
2. a. Burleigh Beach
   b. 2 h 1 min
   c. 1251 m
3. Answers will vary.
   a. 2 h 7 min
   b. Walk, train, walk
4. a. 2 h 18 min
   b. Walk, train, walk, bus, walk
5. a. 9:00 am b. 2 c. Option 2
6. a. 38 min b. 39.1 km c. M3
7. a. Red indicates low tide and blue indicates high tide. Blue numbers are bigger than red ones.
   b. 2 per day
   c. 0.29 metres
   d. Friday 10:30 pm
8. a. 11 hours 43 minutes
   b. No
   c. 1. Getting a ship or boat through a channel
      2. Knowing when rocks are exposed to go looking for crabs
      3. Knowing when the tide is out and good for skim boarding
      There could be many other reasons.
9. a. 20 January 11:14 pm
   b. 13
   c. July has 2 full moons.
10. a. 15 days
    b. 161 hours 27 minutes
    c. 516 hours 18 minutes
11. a. 4:50 am b. 13:11:14 c. 12 minutes
12. a. 4:46 am
    b. 6:18 pm
    c. 13 hours, 9 minutes, 47 seconds
13. 11 February. Tidal range = 1.93 m
14. The longest days in December have 13 hours and 55 minutes of daylight, and are from 17 December to 25 December. This is a reasonable answer as December is the first month of summer and the days continue to get longer during most of December (until the summer solstice).

13.5 Compare travel time

13.5 Exercise

1. Option 2
2. Option 2
3. 25 minutes
4. The train is quicker by 4 hours and 34 minutes.
5. The train is quicker by 47 minutes.
6. a. 24 h 45 min
   b. 30 h
7. 28 h 35 min
8. a. Travel time = 10 min
   b. Travel time = 17 min
   c. Travel time = 25 min
   d. Travel time = 4 min
   e. Taxi
9. a. 8 min
   b. 26 min
   c. Travelling by tram is quicker by 18 minutes.
10. a. Bus
    Train
    Plane
    Train and bus combination
    b. The 9:00 am bus from Brisbane takes 15 hours.
    c. 1 hour 35 minutes
11. a. 9:40 am
    b. 12:18 pm
    c. 3:34 pm
    d. 16:31
    e. 5:05 pm
    f. 10 hours 20 minutes
12. a. 8:50 am b. 9:30 am c. 40 minutes

13.6 Review

13.6 Exercise

1. 20:35
2. 12:03 pm
3. 2:20 pm
4. 2:06 pm
5. a. 5 h 7 min 30 s
   b. 3 h 48 min 0 s
6. a. 6:05 pm
   b. 11:25 am
   c. 4 buses, at 4:04 pm, 4:19 pm, 4:34 pm and 4:49 pm
7. a. 8:18 am
   b. 8:38 am. This is a tricky question. Unless otherwise specified, we usually want to minimise the time between departure and the time we need to arrive by, which is 10 am in this case. Therefore, taking the 8:38 bus and arriving at 9:58 would be preferred over taking the 8:29 bus and arriving at 9:59.
8. 59 minutes
9. Option 2
10. The train is quicker by 4 hours and 33 minutes.
11. The train is quicker, taking 4 hours and 35 minutes.
12. a. 8:53 am b. 8:25 am c. 8:05 am
    d. 8:13 am
13. a. 5 trams b. 8:30 am c. 23 minutes
    d. 11:57 am
14. a. 3:15 pm and 4:15 pm
    b. 10:05 am
    c. Five days a week (Monday to Friday)
    d. 8 hours 50 minutes

15. a. Griffith University Station (South Port)
    b. 1 hour 34 minutes
    c. 854 m
    d. Walk, bus, train and tram
16. a. 20:56 or 8:56 pm
    b. 2.41 m
    c. 14:54 or 2:54 pm
17. Answers may vary.
    A sample response is shown here.

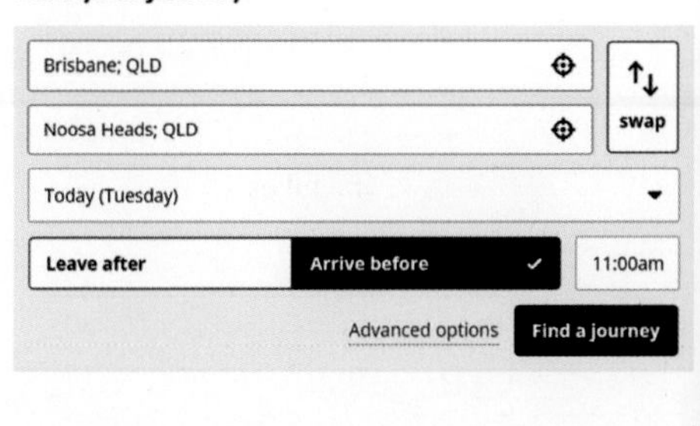

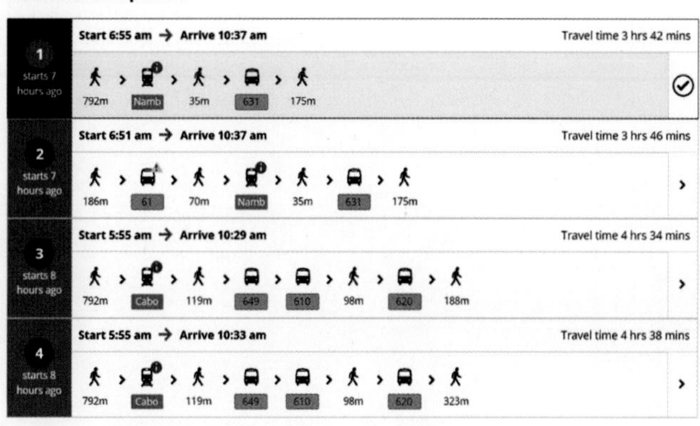

Source: Reproduced by permission of Translink. Department of Transport and Main Roads.

The shortest travel time is 3 hours, 42 minutes.

18. Answers may vary.
    A sample response is shown here.

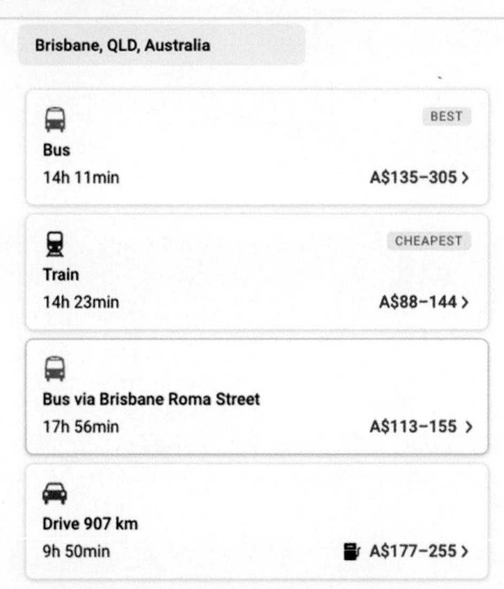

$$\begin{aligned}\text{Difference} &= 5\text{ h }5\text{ min} - 3\text{ h }12\text{ min}\\ &= 1\text{ h }53\text{ min}\end{aligned}$$

19. The largest difference is for 21–23 July, 1.6 m.
20. The shortest day of the month is 28 February, with a length of 12 hours and 42 minutes. This is a reasonable answer as February is the last month of summer and the days continue to get shorter throughout autumn until the winter solstice in June.

UNIT 2 Data and travel
TOPIC 3 Time and motion

14 Distance and speed

LESSON SEQUENCE

Fully worked solutions for this chapter are available online.

on Resources

Solutions Solutions — Chapter 14 (sol-1267)

Digital documents Learning matrix — Chapter 14 (doc-41780)
Quick quizzes — Chapter 14 (doc-41781)
Chapter summary — Chapter 14 (doc-41782)

LESSON
14.1 Overview

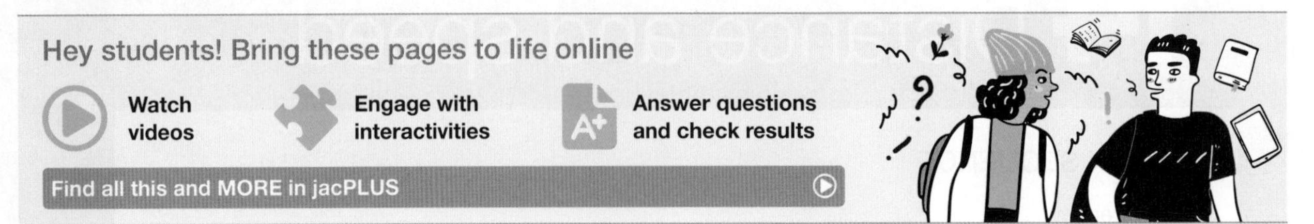

14.1.1 Introduction

In our modern world, where distance is no longer a barrier but rather a puzzle to be solved, it's important to understand the connection between scale, maps, directions, speed, time, and cost of journey.

The concept of distance has evolved over time, from the ancient cartographers painstakingly charting unknown territories to the advanced GPS systems that now guide our every move. Maps serve as our visual guides, distilling vast expanses into manageable scales, while directions provide the navigational cues for our journeys.

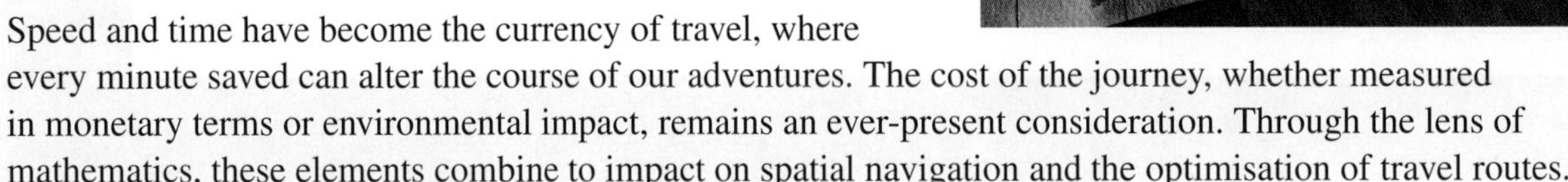

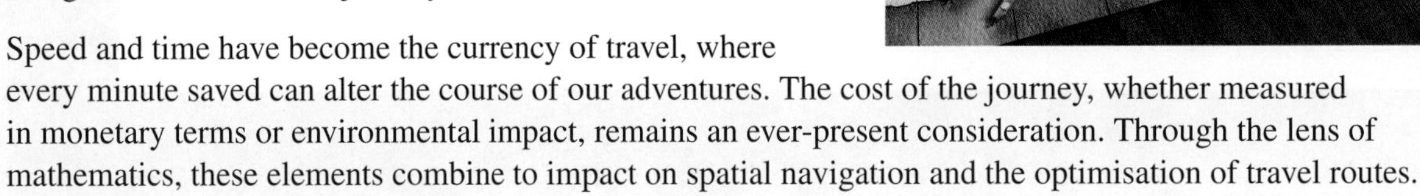

Speed and time have become the currency of travel, where every minute saved can alter the course of our adventures. The cost of the journey, whether measured in monetary terms or environmental impact, remains an ever-present consideration. Through the lens of mathematics, these elements combine to impact on spatial navigation and the optimisation of travel routes.

14.1.2 Syllabus links

Lesson	Lesson title	Syllabus links
14.2	**Distance, scales and maps**	◯ Use scales to find distances.
14.3	**Investigate distances and directions [complex]**	◯ Investigate distances through trial and error or systematic methods [complex]. ◯ Apply directions to distances calculated on maps including the eight compass points: N, NE, E, SE, S, SW, W, NW [complex].
14.4	**Calculating speed, distance and time, and their units**	◯ Identify the appropriate units for different activities. ◯ Calculate speed, distance and time where *s* is speed, *d* is distance and *t* is time, using the formulas • $s = \frac{d}{t}$ • $d = s \times t$ • $t = \frac{d}{s}$
14.5	**Distance-time graph and cost of the journey [complex]**	◯ Calculate the time and costs for a journey from distances estimated from maps, given a travelling speed [complex]. ◯ Interpret distance-versus-time graphs, including reference to the steepness of the slope (or average speed) [complex].

LESSON
14.2 Distance, scales and maps

SYLLABUS LINKS

- Use scales to find distances.

14.2.1 Map distances

- Maps are similar to the land they represent; they have exactly the same shape but very different sizes.
- Maps are drawn using a scale, which is displayed on the map.
- The **scale** describes the ratio between the distance on the map and the actual distance.
- A scale may be written as:
  - a ratio of distance on a map to actual distance, such as 1 : 1 000 000
  - a statement, such as '1 cm represents 1 km'
  - a scale bar.

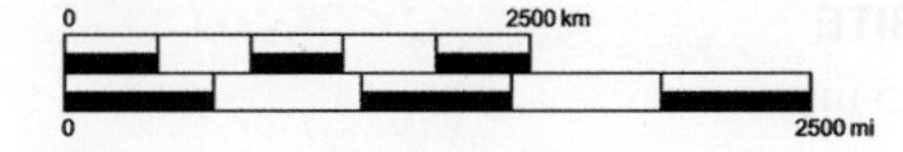

- The scale on the map of King Island shown is 1 : 1 000 000.
- This means that 1 unit on the map represents 1 000 000 units in real life, so 1 mm on the map represents 1 000 000 mm in real life. In real life we don't commonly use the unit of 1 000 000 mm, so we would need to determine the appropriate unit for use and convert it. In this case, it would be 1000 m or 1 km.
- Maps also have an icon depicting north.

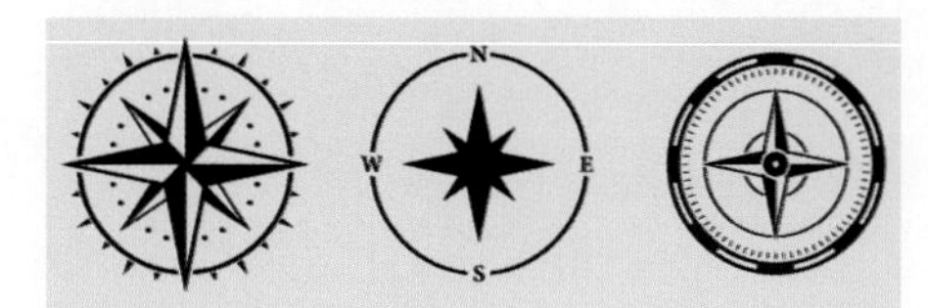

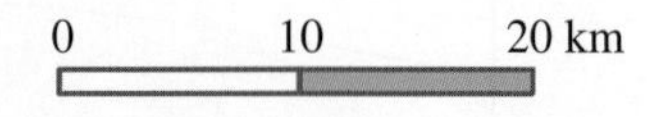

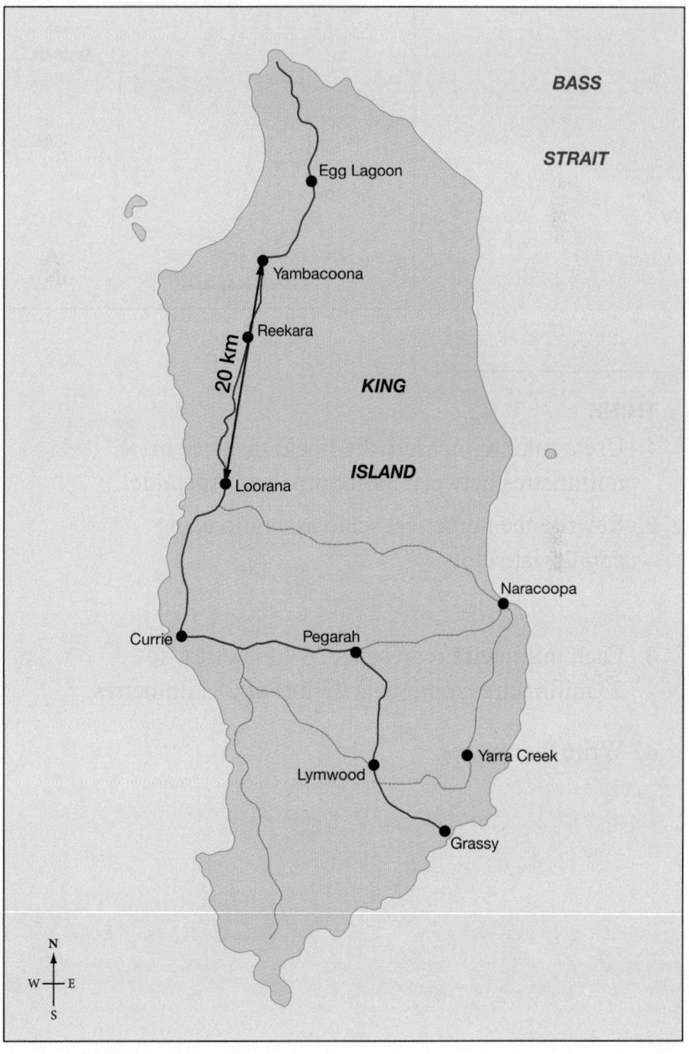

Source: MAPGraphics

tlvd-11453

WORKED EXAMPLE 1 Using a numerical scale and determining direction

Use the numerical scale and a ruler to determine the distance between Melbourne and Adelaide, using the provided map of Australia drawn to a scale of 1 cm to 500 km.
Identify the direction in which you would be travelling when going from Melbourne to Adelaide.

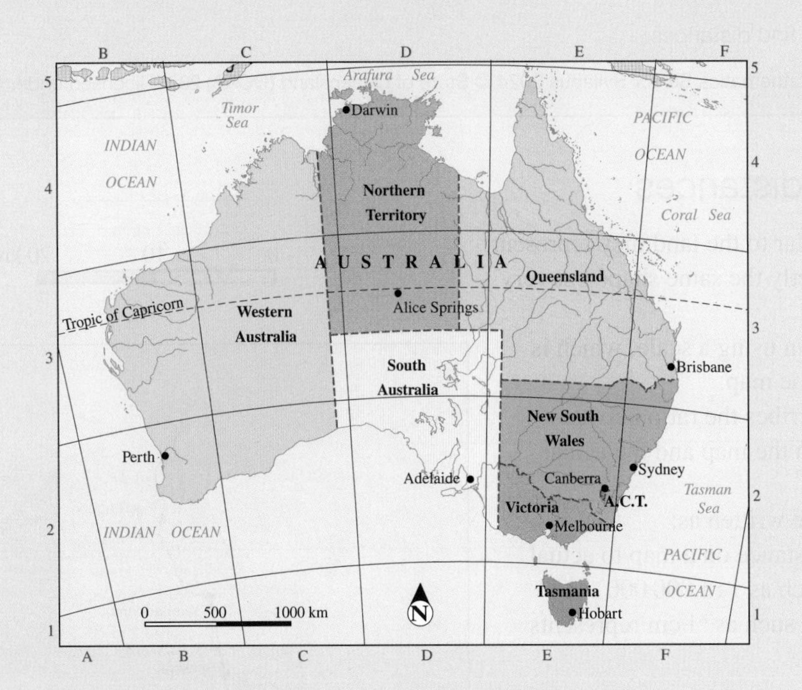

THINK	WRITE
1. Use a ruler to measure the linear distance in millimetres between Melbourne and Adelaide.	12.5 mm
2. Rewrite the numerical scale as a ratio using appropriate units.	1 cm : 500 km 10 mm : 500 km 1 mm : 50 km (÷10)
3. Each millimetre represents 28 kilometres, so 23 millimetres represents 23 lots of 28 kilometres.	$12.5 \times 50 = 625$
4. Write the answer.	The distance between Melbourne and Adelaide is 625 km. Travelling from Melbourne to Adelaide would be in a north-westerly direction.

Exercise 14.2 Distance, scales and maps

14.2 Quick quiz	14.2 Exercise

Simple familiar	Complex familiar	Complex unfamiliar
1, 2, 3, 4, 5, 6, 7, 8, 9, 10	N/A	N/A

Simple familiar

1. **WE1** Using the previous map of Australia with its numerical scale, determine the distance between Adelaide and Brisbane. Identify the direction you would be travelling in from Adelaide to Brisbane.
2. Using the previous map of Australia with its numerical scale, determine the distance between Perth and Alice Springs. Identify the direction you would be travelling in from Perth to Alice Springs.
3. Determine the actual distance represented by 1 cm on the map for each of the following scale ratios. Write your answer in the unit specified in the brackets.
   a. 1 : 1000 (m)
   b. 1 : 50 000 (m)
   c. 1 : 250 000 (km)
   d. 1 : 7 000 000 (km)
4. Determine a scale ratio for a map in which 1 cm on the map represents an actual distance of:
   a. 10 cm
   b. 800 m
   c. 6.5 km
   d. 4000 km.
5. A scale map of Wilsons Promontory National Park is shown.

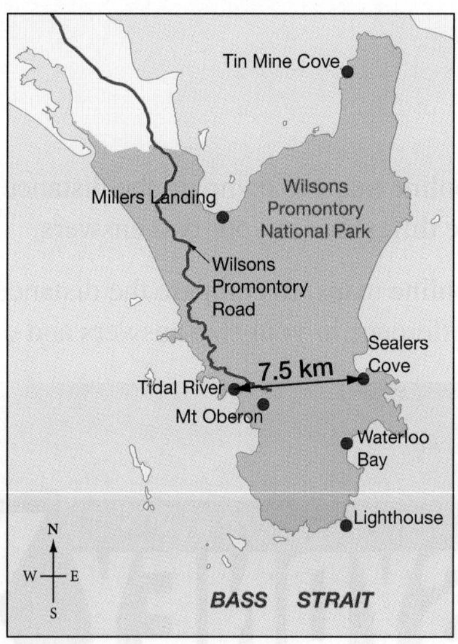

Source: MAPGraphics

   a. Determine the actual distance, in kilometres, represented by 1 cm on the map.
   b. Calculate the actual distance in a straight line from:
      i. Tidal River to Sealers Cove
      ii. Millers Landing to Tin Mine Cove
      iii. Mount Oberon to the lighthouse.

c. Estimate the length of the peninsula from its northernmost point near Tin Mine Cove to its southernmost point near the lighthouse.
d. Mt Vereker is located in the middle of the park, 8.5 km from Tidal River. Determine at what distance Mt Vereker should appear from Tidal River on the map.

6. The map of Australia shown is drawn to a scale of 1 cm to 500 km.

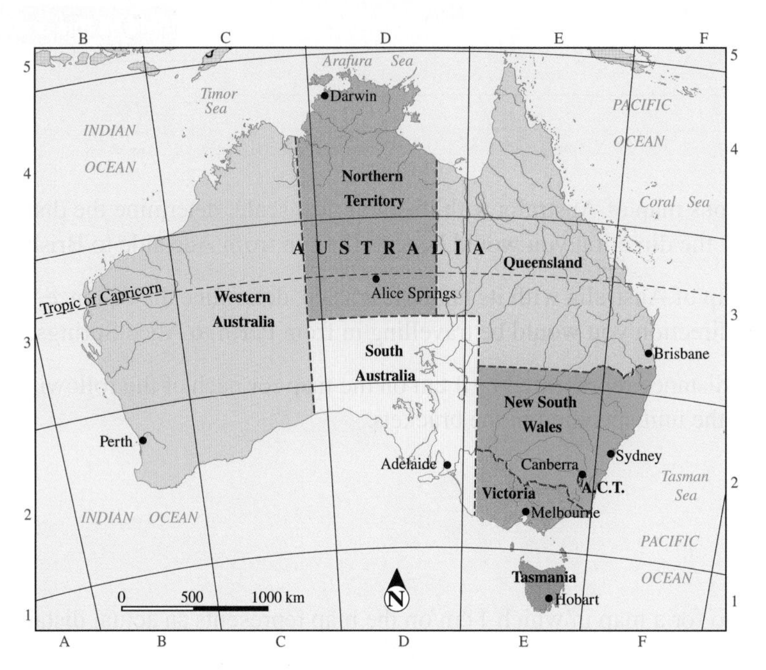

a. Determine how many kilometres are represented by 5 cm on this map.
b. Use the map to estimate the straight line distance between:
   i. Melbourne and Perth
   ii. Sydney and Canberra
   iii. Darwin and Alice Springs
   iv. Brisbane and Sydney
   v. Melbourne and Adelaide
   vi. Adelaide and Hobart.

7. Use Google Maps or some similar online maps to compare the distance found in question 6 travelling from Melbourne to Perth and calculate the difference in your two answers.

8. Use Google Maps or some similar online maps, to compare the distance found in question 6 travelling from Brisbane to Sydney. Calculate the difference in your two answers and explain a reason for the difference.

9. This map of Tasmania is scaled at 1 : 3 000 000.

a. Determine the scale as a ratio in the most appropriate units.
b. Determine the length and width in kilometres of King Island.
c. Determine how many kilometres of ocean lie between Wilsons Promontory in Victoria and Devonport on Tasmania's central north coast.
d. Determine the distance between Hobart and Launceston.

10. Consider the map of north-eastern Victoria.

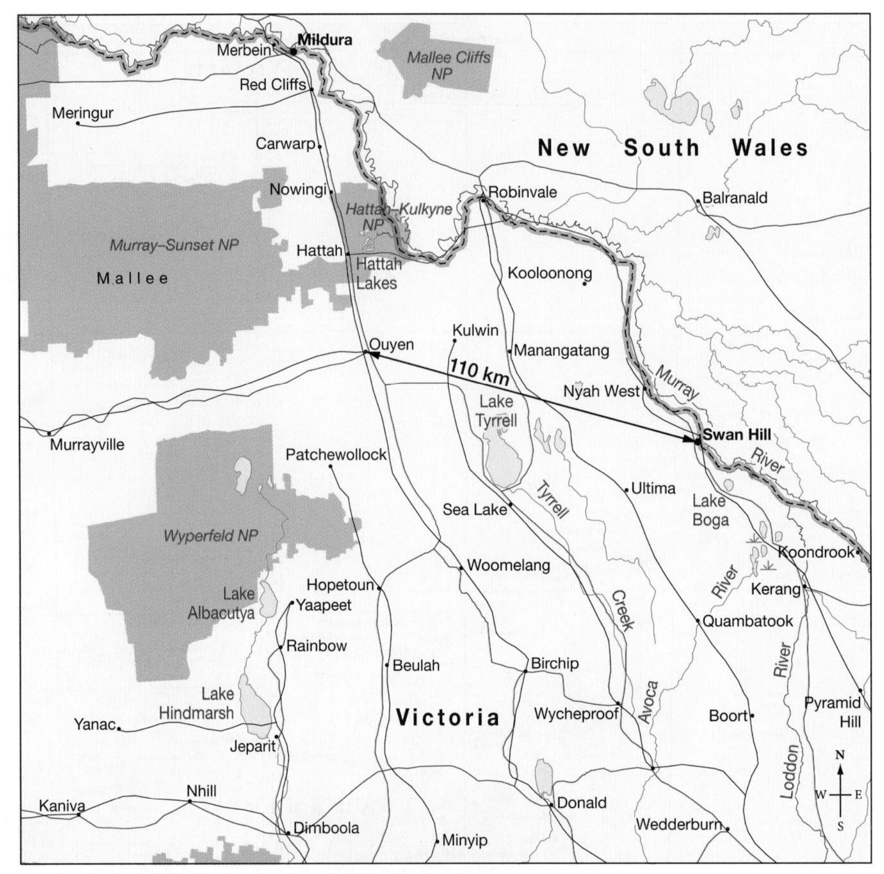

Source: MAPGraphics

a. Measure the straight-line map distance between:

i. Kaniva and Swan Hill

ii. Wedderburn and Meringur

iii. Mildura and Donald.

b. Given that the actual distances between the locations in part a are 236 km, 310 km and 255 km respectively, calculate the scale ratio of the map.

c. Determine if you get the same ratio for each of your calculations in part b. Discuss any variation.

Fully worked solutions for this chapter are available online.

LESSON
14.3 Investigate distances and directions [complex]

SYLLABUS LINKS

- Investigate distances through trial and error or systematic methods [complex].
- Apply directions to distances calculated on maps including the eight compass points: N, NE, E, SE, S, SW, W, NW [complex].

14.3.1 Investigating distances through trial and error, and systematic methods

- The shortest path to take is the path with the shortest total distance.
- To calculate the shortest path, use:
  - **trial and error** to compare the total distances for each option
  - **the systematic method**, where you select the shortest distance between points on each section of the desired path.

WORKED EXAMPLE 2 Determining shortest distance using trial and error

Determine which of the following trips is the shortest distance, in km, from Darwin to Katherine.

Trip 1: Going via Pine Creek **Trip 2: Going via Jabiru**

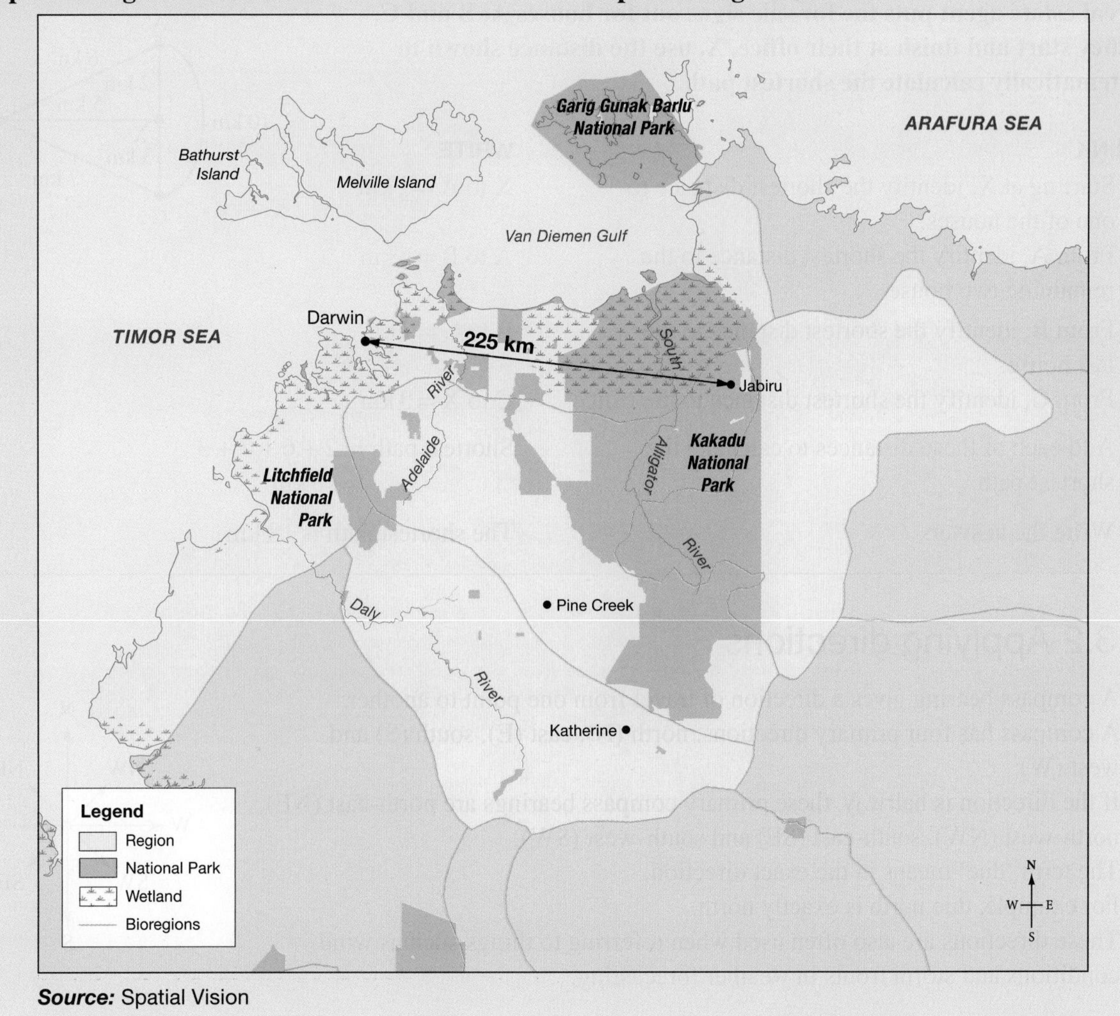

Source: Spatial Vision

THINK	WRITE
1. Determine the scale on the map. Measure the 2.5 km distance with a ruler.	$225\text{ km} = 5.0\text{ cm}$ $\frac{225}{5.0} = 45$ Therefore 45 km per cm
2. Measure Trip 1, going via Pine Creek. *Hint:* A piece of string could be used.	Measured map length $\approx 6.3\text{ cm}$ $6.3\text{ cm} = 45 \times 6.3$ $= 283.5\text{ km}$ Actual distance $\approx 283.5\text{ km}$
3. Measure Trip 2, going via Jabiru.	Measured map length $\approx 9.8\text{ cm}$ $9.8\text{ cm} = 45 \times 9.8$ $= 441\text{ km}$ Actual distance $\approx 450\text{ km}$
4. Write the answer.	The shortest distance from Darwin to Katherine is approximately 283 km going via Pine Creek.

tlvd-11454

WORKED EXAMPLE 3 Calculating the shortest path systematically

A real estate agent puts the for-sale signs out for houses A, B and C. If they start and finish at their office, X, use the distance shown to systematically calculate the shortest path.

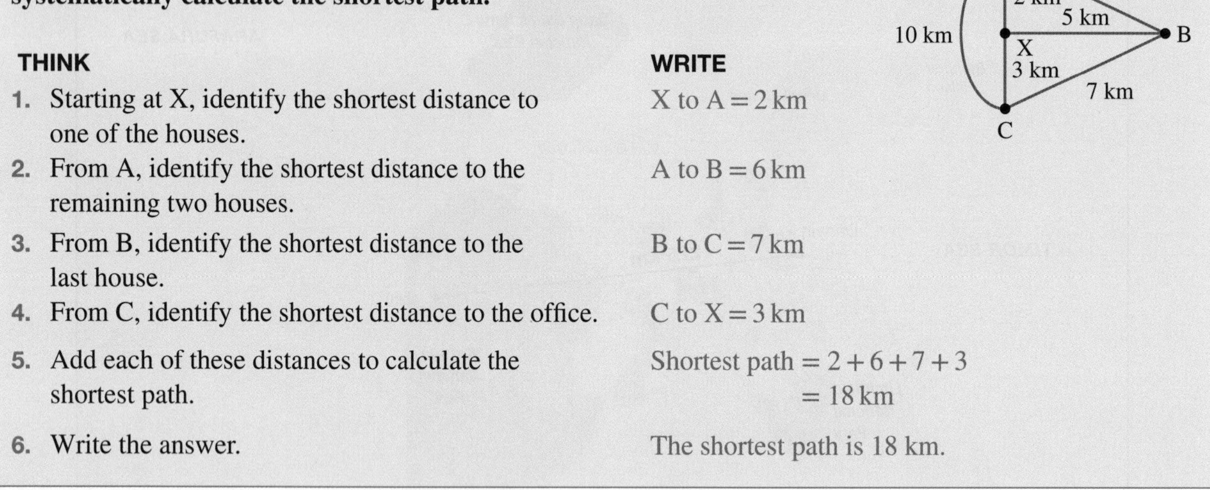

THINK	WRITE
1. Starting at X, identify the shortest distance to one of the houses.	X to A = 2 km
2. From A, identify the shortest distance to the remaining two houses.	A to B = 6 km
3. From B, identify the shortest distance to the last house.	B to C = 7 km
4. From C, identify the shortest distance to the office.	C to X = 3 km
5. Add each of these distances to calculate the shortest path.	Shortest path $= 2 + 6 + 7 + 3$ $= 18\text{ km}$
6. Write the answer.	The shortest path is 18 km.

14.3.2 Applying directions

- A compass bearing gives a direction of travel from one point to another.
- A compass has four primary directions: north (N), east (E), south (S) and west (W).
- If the direction is halfway, these primary compass bearings are north-east (NE), north-west (NW), south-east (SE) and south-west (SW).
- The term 'due' means in the exact direction.
  For example, due north is exactly north.
- These directions are also often used when referring to things such as wind conditions and storm fronts in weather forecasting.

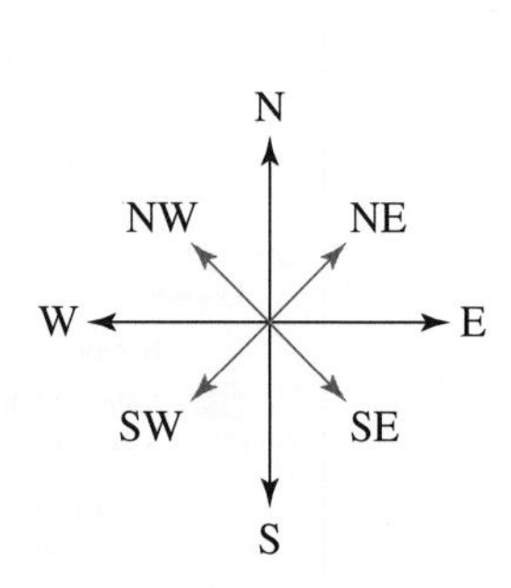

- The bearing B from A tells you how to get to B *from* A, as shown.

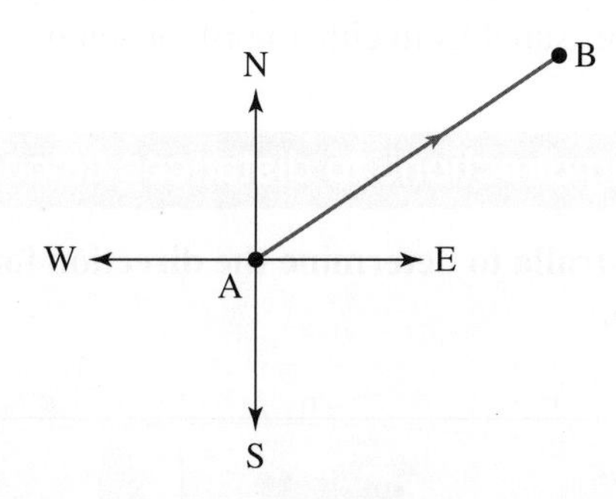

WORKED EXAMPLE 4 Determining compass bearings

Determine the following compass bearings from the diagram shown.

A	B	C
D	E	F
G	H	I

a. **From E to B**
b. **From I to H**
c. **From A to E**
d. **From I to E**

THINK	WRITE
a. Identify which direction an arrow would face going from E to B.	a. North
b. Identify which direction an arrow would face going from I to H.	b. West
c. Identify which direction an arrow would face going from A to E.	c. South-east
d. Identify which direction an arrow would face going from I to E.	d. North-west

- When determining a distance from a map, we often include which direction you would have to travel in to get there. We use compass bearings in order to achieve this.
  For example, from Brisbane, you would have to travel approximately 1800 km in a SW direction to get to Melbourne.

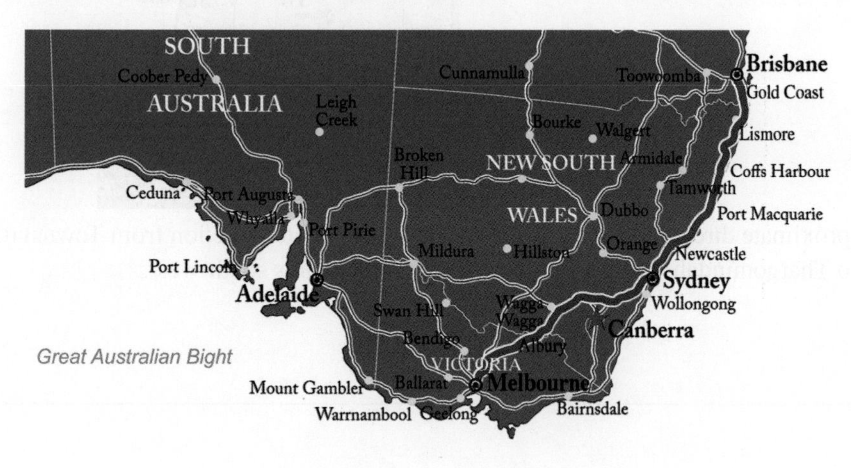

- Bearings are given as either:
  - true bearings, which are measured clockwise from north and always have three digits
  - compass bearings, which are measured from either north or south.

WORKED EXAMPLE 5 Applying directions to distances calculated on maps

Consider the following map of Australia to determine the direction for travelling in a straight line from Townsville to Thargomindah.

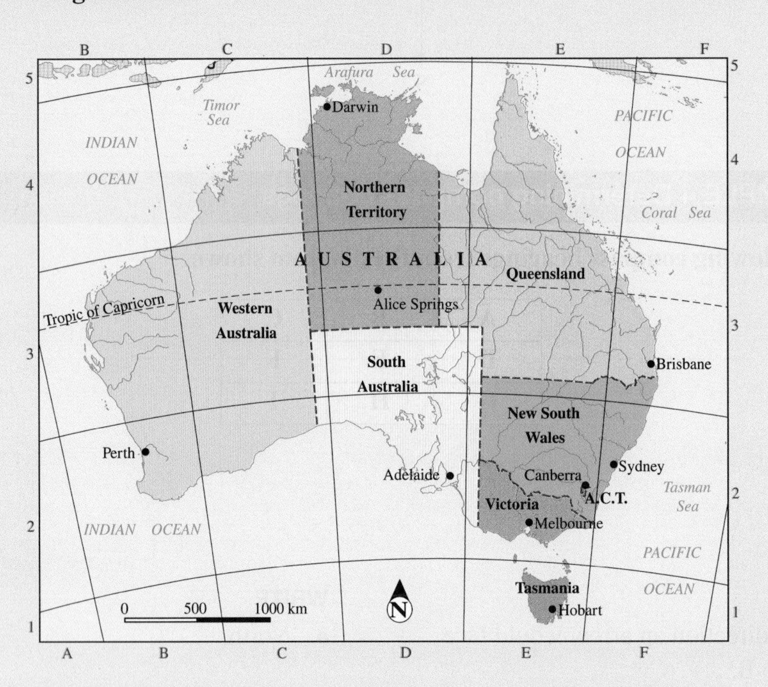

THINK

1. Imagine the compass at Townsville, as you determine the compass direction from Townsville to Thargomindah.

WRITE

2. Write the approximate direction from Townsville to Thargomindah.

The approximate direction from Townsville to Thargomindah is south-west.

Exercise 14.3 Investigate distances and directions [complex]

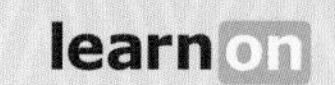

14.3 Quick quiz	14.3 Exercise

Simple familiar	Complex familiar	Complex unfamiliar
N/A	1, 2, 3, 4, 5, 6, 7, 8, 9, 10, 11, 12, 13, 14, 15	16, 17

These questions are even better in jacPLUS!
- Receive immediate feedback
- Access sample responses
- Track results and progress

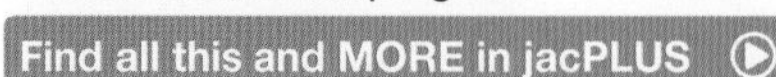

Complex familiar

1. **WE2** Determine which of the following trips is the shortest distance, by road, in km, from Yambacoona to Lymwood.
   Trip 1: Going via Naracoopa and Yarra Creek
   Trip 2: Going via Currie and Pegarah

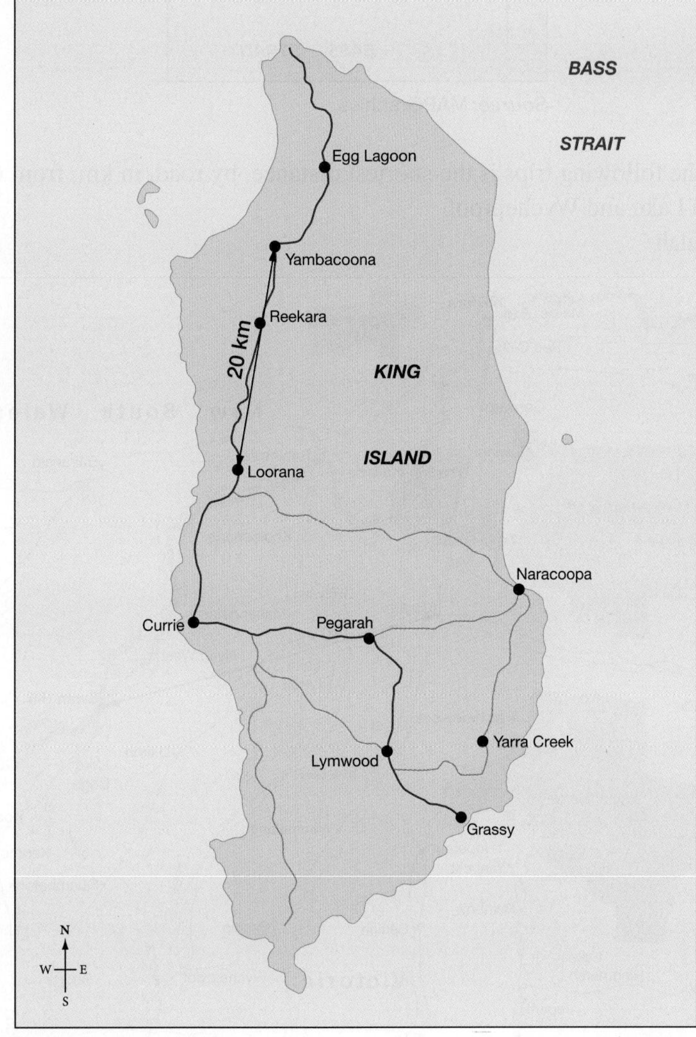

Source: MAPGraphics

2. Determine which of the following trips is the shortest distance, in km, from Tidal River to Lighthouse.
Trip 1: Going via Mt Oberon
Trip 2: Going via Sealers Cove and Waterloo Bay

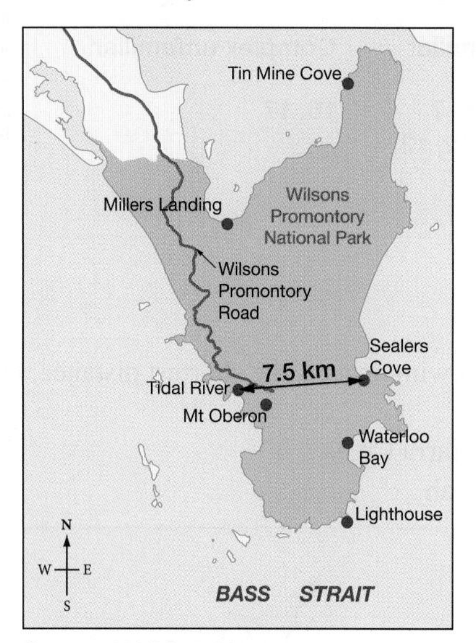

Source: MAPGraphics

3. Determine which of the following trips is the shortest distance, by road, in km, from Ouyen to Donald.
Trip 1: Going via Sea Lake and Wycheproof
Trip 2: Going via Beulah

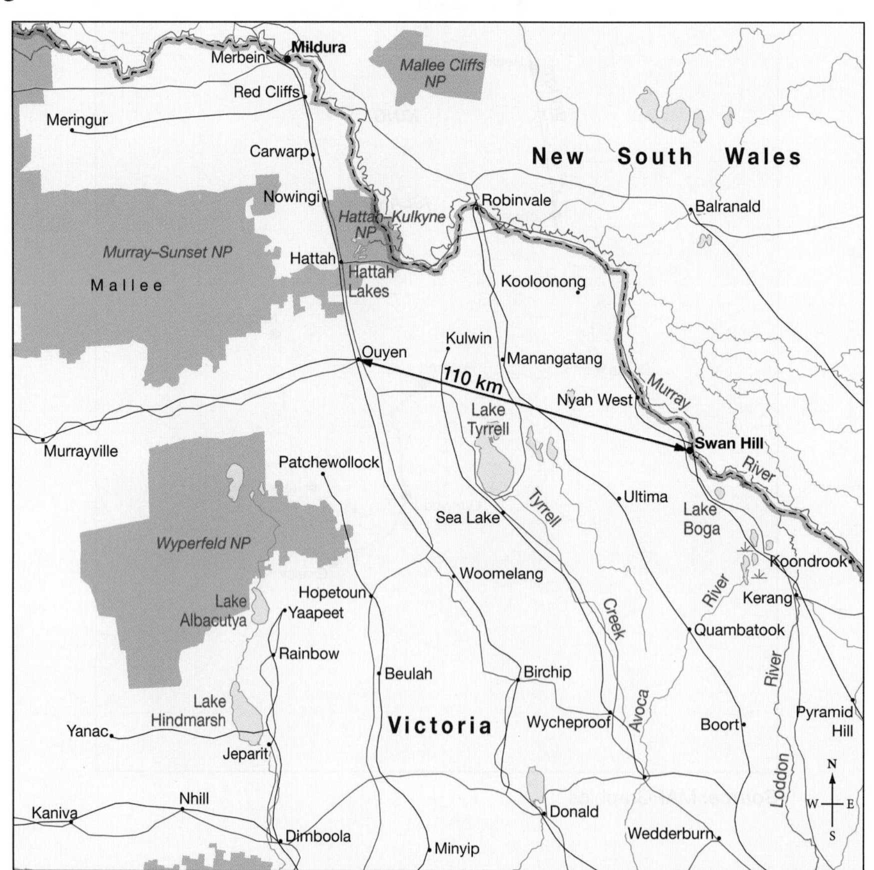

Source: MAPGraphics

4. WE3 A delivery person has parcels to deliver to houses X, Y and Z from their office at point O. The delivery person starts and finishes at the office.

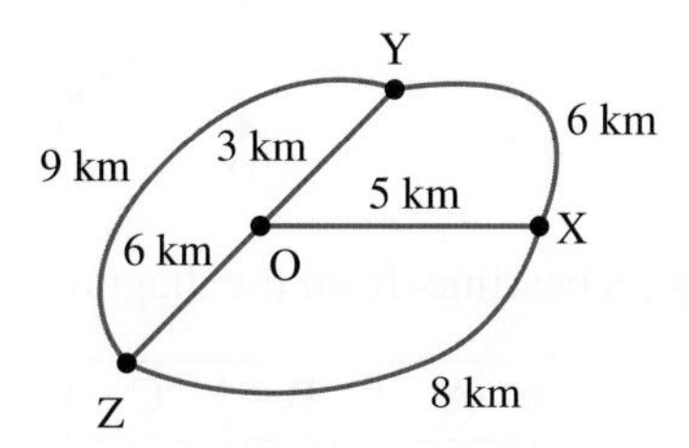

Use the distances shown to systematically calculate the shortest path.

5. An employee of the Australian Bureau of Statistics is required to survey 4 houses (1 to 4).

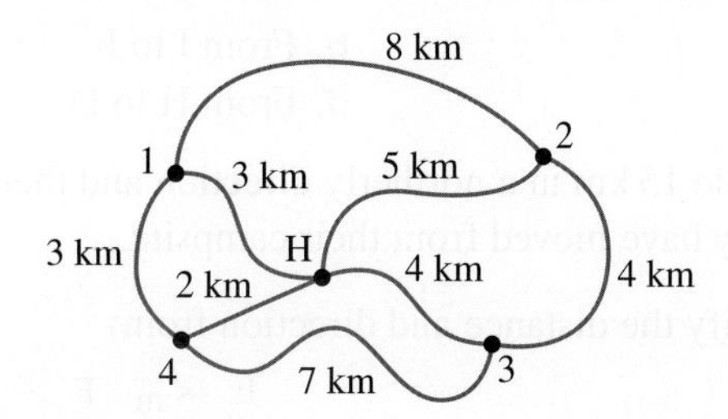

Systematically calculate their shortest path given that they start and finish from their home (H).

6. The diagram shown is of a park with gates at points A–D. The ranger needs to check that all the gates are closed at the end of the day. Use a systematic approach to determine the shortest path the ranger can take, given he starts and finishes at gate A.

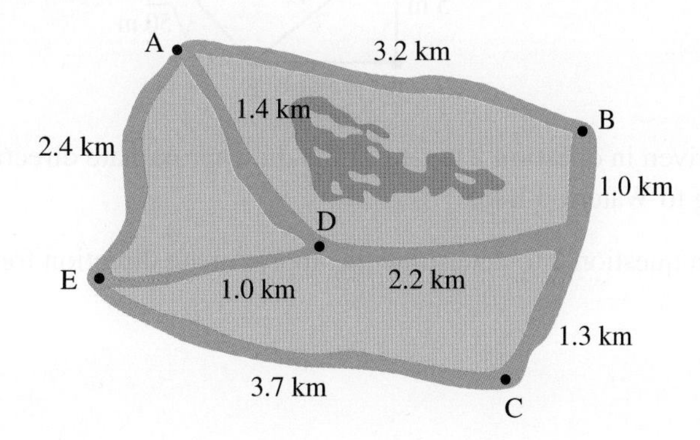

7. The graph shown represents the distances in kilometres between eight locations. Identify the shortest distance to travel from A to D that goes to all the vertices (points).

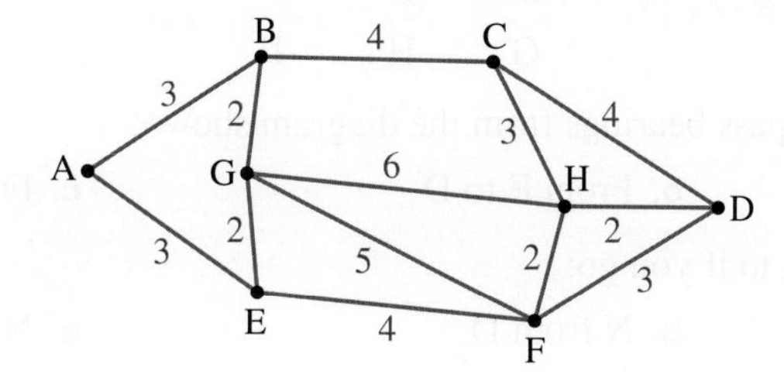

8. Label the directions from O to:

a. A
b. B
c. C
d. D.

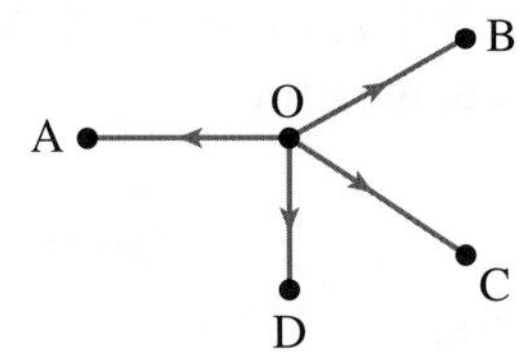

9. WE4 Determine the following compass bearings from the diagram shown.

A	B	C
D	E	F
G	H	I

a. From E to C
b. From I to F
c. From B to D
d. From H to D

10. A hiker walks from their campsite 15 km in a northerly direction and then 15 km in a westerly direction. Determine in what direction they have moved from their campsite.

11. Using the diagram shown, identify the distance and direction from:

a. A to B
b. D to B
c. D to E
d. B to D to F.

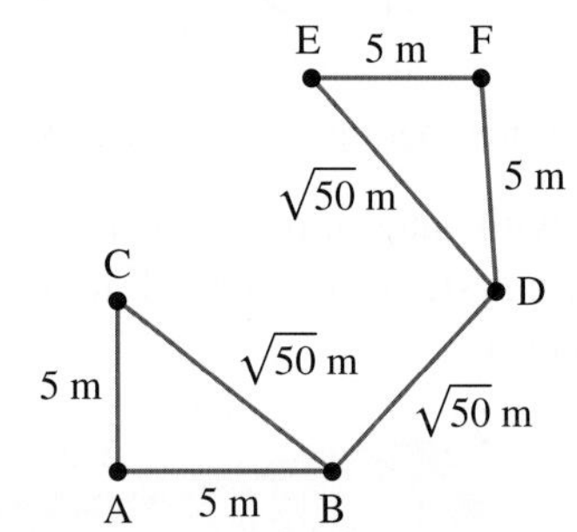

12. WE5 Consider the map given in question 2 to determine the approximate direction for travelling in a straight line from Millers Landing to Waterloo Bay.

13. Consider the map given in question 3 to determine the approximate direction for travelling in a straight line from:

a. Mildura to Dimboola
b. Swan Hill to Hattah.

Use the following diagram to answer questions 14 and 15.

A	B	C
D	E	F
G	H	I

14. Determine the following compass bearings from the diagram shown.

a. From B to F
b. From E to D
c. From G to E

15. Determine what letter you get to if you go:

a. SE from E
b. N from D
c. NW from F.

Complex unfamiliar

16. Orienteering is a physically active and enjoyable pastime that requires map- and compass-reading skills and an ability to estimate distances. The aim of orienteering is to complete a given course that has been planned and mapped. There are many checkpoints in an orienteering course, with competitors visiting each.
The following exercise is an example of a simple orienteering course conducted in a park. The directions that are listed below could be either:
    - given to each participant prior to commencing the course, or
    - left at each checkpoint.

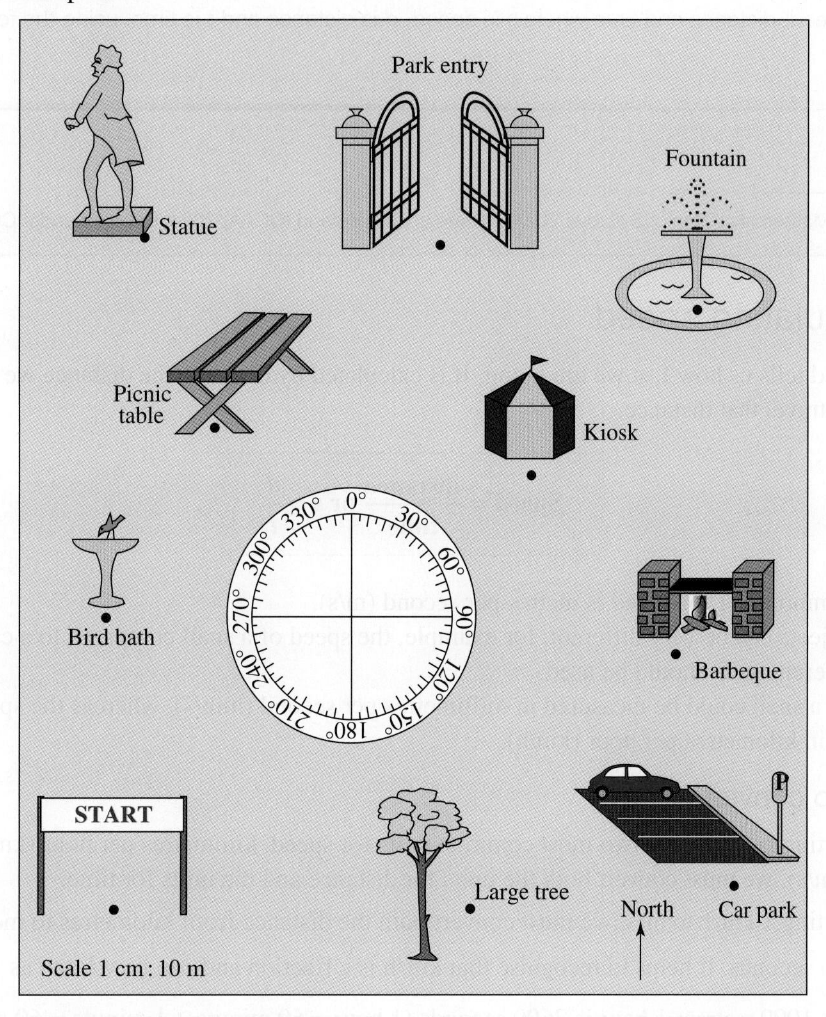

Use the compass rose or a protractor and your ruler to follow this course in the figure shown above.
    - Find the point in the park labelled 'Start'.
    - Proceed in an easterly direction for 60 m.
    - Follow N20°E for 103 m.
    - Now move 94 m on N83°W.
    - Proceed along S6°W for 62 m.
    - Now follow S86°E for 95 m.

Determine where you are now.

17. Harry takes off from his campsite and initially hikes 6 km N, then he hikes 5 km NE, followed by 6 km E and finally 13 km NW before reaching his next campsite. Calculate to 2 decimal places how far in a northerly direction Harry finishes his hike from his original campsite.

Fully worked solutions for this chapter are available online.

LESSON
14.4 Calculating speed, distance and time, and their units

SYLLABUS LINKS

- Identify the appropriate units for different activities.
- Calculate speed, distance and time where s is speed, d is distance and t is time, using the formulas
  - $s = \frac{d}{t}$
  - $d = s \times t$
  - $t = \frac{d}{s}$

14.4.1 Calculating speed

- Average speed tells us how fast we are going. It is calculated by dividing the distance we travel by the time it takes us to travel that distance.

$$\textbf{Speed} = \frac{\textbf{distance}}{\textbf{time}} \textbf{ or } s = \frac{d}{t}$$

- The most common unit for speed is metres per second (m/s).
- Speeds of objects can be very different; for example, the speed of a snail compared to a car will be much lower, so different units should be used.
- The speed of a snail could be measured in millimetres per second (mm/s), whereas the speed of a car could be measured in kilometres per hour (km/h).

Simple ways to convert speed

- When converting between the two most common units for speed, kilometres per hour (km/h) and metres per second (m/s), we must convert both the units for distance and the units for time.
- When converting 1 km/h to m/s, we must convert both the distance from kilometres to metres and the time from hours to seconds. It helps to recognise that km/h is a fraction and can be written as $\frac{\text{km}}{\text{h}}$.
- 1 kilometre is 1000 metres; 1 hour is 3600 seconds (1 hour = 60 minutes; 1 minute = 60 seconds). The conversion is demonstrated below:

$$\frac{1 \text{ km}}{\text{hour}} = \frac{1000 \text{ m}}{60 \text{ min}} = \frac{1000 \text{ m}}{3600 \text{ s}} = \frac{1 \text{ m}}{3.6 \text{ s}} \text{ or } 0.28 \frac{\text{m}}{\text{s}}$$

× 1000

× 60 × 60

- Visually, we can see that to convert from km/h to m/s, we multiply by 1000 and divide by 3600.
- The opposite applies when converting from m/s to km/h.
- If the speed of an object is 1 m/s, this means that the object has to travel for 3600 seconds to make up an hour. In that time, the object would travel a total of 3600 m or 3.6 km.

The conversion is demonstrated below:

$$\frac{1\text{ m}}{1\text{ s}} = \frac{3600\text{ m}}{3600\text{ s}} = \frac{3600\text{ m}}{1\text{ h}} = \frac{3.6\text{ km}}{\text{h}}$$

(× 3600 from the first to the second fraction; ÷ 1000 from the third to the fourth fraction)

Shortcut for converting units of speed

- **To convert from km/h to m/s, divide by 3.6.**
- **To convert from m/s to km/h, multiply by 3.6.**

WORKED EXAMPLE 6 Selecting appropriate units for speed

Select the most appropriate unit for speed from cm/s, m/s and km/hr for the following.

a. **The speed of a turtle walking**
b. **The speed of a Formula One car**
c. **The speed of a sprinter running 100 m**
d. **The speed of a person walking around the park**

THINK	WRITE
a. A turtle walks slowly, so you should use a smaller unit of distance.	a. The most appropriate unit for something that moves slowly is cm/s.
b. A Formula One car travels at high speeds.	b. Cars usually use km/hr.
c. A sprinter running 100 m does not go as fast as a car, but is much faster than a turtle.	c. The most appropriate unit is m/s.
d. The speed of someone walking is slower than that of a sprinter, but quicker than that of a turtle.	d. The most appropriate unit is m/s.

tlvd-6029

WORKED EXAMPLE 7 Calculating the average speed

Calculate the average speed of a dragster that covers 400 m in 4.25 seconds, in:

a. **m/s**
b. **km/h.**

THINK	WRITE
a. 1. Determine the distance and time in the correct units.	a. Distance = 400 m Time = 4.25 s
2. Substitute values into the formula $\text{speed} = \frac{\text{distance}}{\text{time}}$.	$\text{Speed} = \frac{\text{distance}}{\text{time}}$ $= \frac{400}{4.25}$ $= 94.12\text{ m/s}$

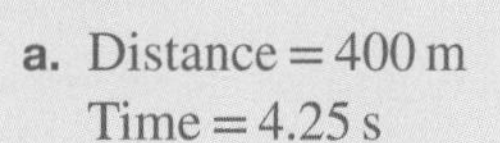

b. 1. Determine the distance and time in the correct units.	**b.** Distance $= 400\text{ m}$ $= 0.4\text{ km}$ Time $= 4.25\text{ s}$ $= \dfrac{4.25}{60 \times 60}$ $= 1.18 \times 10^{-3}$ hours
2. Substitute values into the formula speed $= \dfrac{\text{distance}}{\text{time}}$.	Speed $= \dfrac{\text{distance}}{\text{time}}$ $= \dfrac{0.4}{1.18 \times 10^{-3}}$ $= 338.82\text{ km/h}$

WORKED EXAMPLE 8 Calculating time and average speed

The distance from home to school is 7.5 km and it takes twice as long to get home as it does to get to school due to traffic congestion. If it takes 10 minutes to get to school, determine:

a. how long it takes to get home from school

b. the duration of the total trip to and from school

c. the average speed of the total trip to and from school in:

i. km/min **ii. km/h.**

THINK	WRITE
a. It takes twice as long to get home from school in the evening compared to going from home to school in the morning.	**a.** Home from school $= 2 \times 10$ minutes $= 20$ minutes
b. Add the two times together.	**b.** To school $= 10$ minutes Total time $= 10 + 20$ $= 30$ minutes
c. i. To calculate the average speed of the total trip,we need to use the total distance of 15 km and the total time of 30 minutes.	**c. i.** Speed $= \dfrac{\text{distance}}{\text{time}}$ $= \dfrac{15}{30}$ $= 0.5\text{ km/min}$
ii. Use the total distance of 15 km and time of 30 minutes, which is equal to 0.5 hours.	**ii.** Speed $= \dfrac{\text{distance}}{\text{time}}$ $= \dfrac{15}{0.5}$ $= 30\text{ km/h}$

14.4.2 Calculating distance

- The speed equation can be rearranged to make distance the subject, allowing you to calculate the distance covered if you know the speed and time of a journey.

Distance = speed × time or $d = s \times t$

tlvd-6030

WORKED EXAMPLE 9 Calculating distance in kilometres

A car travels at a constant speed of 60 km/h for 1.5 hours. Calculate how far the car travels in kilometres.

THINK	WRITE
1. To calculate the distance, we need to know the speed and time in the correct units.	Speed = 60 km/h Time = 1.5 hours
2. Substitute into the distance equation.	Distance = speed × time $= 60 \times 1.5$ $= 90$ km
3. Understand the question.	Since the car travels at 60 km/h, it covers 60 km in 1 hour, and therefore 30 km in half an hour, and therefore 90 km in 1.5 hours.
4. Write the answer.	The distance covered is 90 km.

14.4.3 Calculating time

- The speed equation can also be rearranged to make time the subject, allowing you to calculate the time taken to cover a given distance at a certain speed.

$$\textbf{Time} = \frac{\textbf{distance}}{\textbf{speed}} \textbf{ or } t = \frac{d}{s}$$

WORKED EXAMPLE 10 Calculating time taken in hours

Calculate how many hours it would take a person to walk 3 km at an average speed of 1.8 m/s.

THINK	WRITE
1. To calculate the time, we need to know the distance and speed in appropriate units. We will use m/s and m and convert to hours at the end.	Distance = 3 km = 3000 m Speed = 1.8 m/s
2. Substitute into the equation $\text{time} = \frac{\text{distance}}{\text{speed}}$.	$\text{Time} = \frac{\text{distance}}{\text{speed}}$ $= \frac{3000}{1.8}$ = 1666.67 seconds
3. Convert seconds to hours by dividing by 60 (to convert to minutes), and then divide by 60 again (to convert to hours).	$\frac{1666.67}{60 \times 60} = 0.46$ hour ≈ 28 minutes
4. Write the answer.	The time taken would be 0.46 hours.

Exercise 14.4 Calculating speed, distance and time, and their units

learnon

14.4 Quick quiz on | **14.4 Exercise**

Simple familiar	Complex familiar	Complex unfamiliar
1, 2, 3, 4, 5, 6, 7, 8, 9, 10, 11, 12, 13, 14, 15, 16, 17, 18, 19, 20, 21, 22, 23, 24	N/A	N/A

Simple familiar

1. **WE6** Select the most appropriate unit for speed from mm/s, m/s and km/h for the following.
   a. The speed of a car moving on a freeway
   b. The speed of a snail moving
   c. The speed of a lawn bowl being rolled
2. Select the most appropriate unit of distance from cm, m and km for the following.
   a. Measuring the length of a car
   b. Measuring the length of a mobile phone
   c. Measuring the distance from Brisbane to Sydney
3. **WE7** Calculate the average speed of Sally, who walks around her 2.5-km block in 1050 seconds, in:
   a. m/s
   b. km/h.
4. Calculate the average speed of a track cyclist who covered 1000 m in 1 minute and 25 seconds, in km/h.

5. Select the appropriate unit (mm/s, m/s or km/h) to be used to measure the speed of:
   a. a sprinter running a 100-m race
   b. a plane flying
   c. the increase in snow depth
   d. a surfer riding a wave.
6. Calculate the average speed (in m/s) of a skateboarder who travelled 50 m in 7 seconds.
7. Calculate the average speed (in km/h) of a boat that covered 2.6 km in 15 minutes.

8. **WE8** The distance from home to netball training is 6 km and it takes three-quarters of the time to return home as it does to get there.
If it takes 8 minutes to get to netball training, determine:

a. how long it takes to get home from netball training
b. the duration of the total trip to and from netball training
c. the average speed of the total trip to and from netball training in:
i. km/min
ii. km/h.

9. Calculate the average speed, in the units shown in brackets, of the following:

a. The distance covered was 100 m in 25 seconds (m/s).
b. The distance covered was 4.8 km in 2 minutes (m/s).
c. The distance covered was 3800 m in 2 hours (km/h).
d. The distance covered was 14.6 km in 400 minutes (km/h).

10. Covert the following to m/s and to 1 decimal place.

a. 50 km/h
b. 60 km/h
c. 80 km/h
d. 100 km/h

11. Convert the following to km/h and to 1 decimal place.

a. 5 m/s
b. 15 m/s
c. 18 m/s
d. 30 m/s

12. **WE9** A car travels at a constant speed of 80 km/h for 3.25 hours. Determine how far the car travels in kilometres.

13. A truck travels at an average speed of 65 km/h for 335 minutes.
Determine how far the truck travels in:

a. kilometres
b. metres.

14. A kite surfer was moving at a constant speed of 11 m/s for 24 seconds. Calculate the distance (in metres) that she covered over this time.

15. Callum played a round of golf, averaging 5 shots per hole and a speed of 0.3 m/s. If it took him 4 hours to play his round of golf, determine how far he walked during his round.

16. **WE10** Determine how long in hours it would take a person to walk 10 km at an average speed of 1.6 m/s.

17. If Ahmed ran at an average speed of 5 m/s in a half-marathon (21.1 km), determine how long it would take him to complete the half-marathon in hours and to the nearest minute.

18. Haile Gebrselassie ran the Berlin marathon with an average speed of 5.67 m/s. Determine how long it took him to finish the marathon, given a marathon is 42.2 km, in:
  a. seconds
  b. hours, minutes and seconds (to the nearest second).
19. A cheetah can reach a maximum speed of 112 km/h. However, it is only able to maintain this speed for a short period of time. If a cheetah ran at its maximum speed and covered a distance of 400 m, determine how long it maintained its maximum speed, in seconds.
20. Pigeons can cover vast distances to find their way home. If a pigeon covered 56.8 km in 1 hour and 45 minutes, calculate:
  a. its average speed in km/h
  b. its average speed in m/s.
21. If a horse averages a speed of 25.6 km/h for 17 minutes and 36 seconds, calculate how far the horse travelled, to 2 decimal places, in:
  a. kilometres
  b. metres.
22. For Makybe Diva to win her third Melbourne Cup, she averaged a speed of 57.9 km/h in the 3200 m race. Determine her winning time in:

  a. minutes, to 2 decimal places
  b. minutes and seconds (to the nearest second).
23. The speed of light is 300 000 000 m/s. Research the average distance from Earth to the following planets to calculate the time it would take light to get there from Earth. *Note:* Use the average distance since it does vary.
  a. Saturn
  b. Mercury
  c. Jupiter
  d. Mars
24. The following questions relate to the reaction time of a driver before braking.
  a. If a driver was travelling at 60 km/h and had a reaction time of 1.3 seconds before he applied the brakes, determine how far he travelled before he applied the brakes.
  b. If a driver had a reaction time of 1.5 seconds and travelled 30 m before applying the brakes, calculate the speed that he was travelling at.
  c. If the car travelled 42 m before the brakes were applied, determine the reaction time of the driver if he was travelling at 95 km/h.

Fully worked solutions for this chapter are available online.

LESSON
14.5 Distance-time graph and cost of the journey [complex]

SYLLABUS LINKS

- Calculate the time and costs for a journey from distances estimated from maps, given a travelling speed [complex].
- Interpret distance-versus-time graphs, including reference to the steepness of the slope (or average speed) [complex].

14.5.1 Time and costs of journeys

- A global positioning system (GPS) can approximate the time of a journey by estimating the distance of the trip, calculating the average speed of the trip depending on the roads travelled, and using this information to estimate the time the trip will take.

Remember:

Determine the journey time

$$\text{Time} = \frac{\text{distance}}{\text{speed}}$$

To determine the time a journey takes:

- estimate the distance of the journey from the map
- estimate the average speed of the journey
- use these estimates to calculate the time of the journey, using the formula time $= \dfrac{\text{distance}}{\text{speed}}$.

Determine the cost of a journey

- **To determine the cost of a journey:**
  - **calculate the time of the journey or distance of the journey**
  - **calculate the cost of the journey, using the formula:**

$$\textbf{cost} = \textbf{time of journey} \times \textbf{cost of journey per unit of time}$$

or

$$\textbf{cost} = \textbf{distance of journey} \times \textbf{cost of journey per unit of distance.}$$

WORKED EXAMPLE 11 Calculating cost per kilometre and cost per hour

Shane's car averages 6 L/100 km fuel economy in the city. His car has a 48 L fuel tank and fuel currently costs $2.40/L in his suburb.

a. **Calculate how many kilometres Shane can expect to drive on a full tank of fuel.**
b. **Calculate how much it costs Shane to fill his fuel tank (assuming it was empty).**
c. **Determine the cost per km for Shane's car at the current fuel price.**
d. **Shane's car keeps a 'travel time' record. If the 'travel time' for his tank of fuel was 4.8 hours, determine the cost per hour of driving his car.**

THINK	WRITE
a. 1. Shane's car averages 6 L/100 km and the car has a 48 L fuel tank. Determine how many kilometres Shane can drive per litre of fuel.	a. $\text{Distance per litre} = \frac{100\text{ km}}{6\text{ L}} = \frac{16.67\text{ km}}{\text{L}}$
2. Calculate the total distance Shane can drive on a full tank of 48 litres.	$\text{Total distance} = 16.67 \times 48 = 800\text{ km}$
3. Write the answer.	Shane can expect to drive approximately 800 kilometres on a full tank of fuel.
b. 1. The fuel tank capacity is 48 litres and the cost of fuel is $2.40 per litre. Calculate the total cost by multiplying the tank capacity by the cost per litre.	b. $\text{Total cost} = \text{tank capacity} \times \text{cost per litre} = 48\text{ L} \times \$2.40\text{ per litre} = \$115.20$
2. Write the answer.	It will cost $115.20 to fill Shane's empty fuel tank.
c. 1. The car's fuel economy is 6 L/100 km and the cost of fuel is $2.40 per litre. Calculate the cost to drive 100 kilometres.	c. $\text{Cost per100 kilometres} = \text{fuel economy} \times \text{cost per litre} = 6\text{ L} \times \$2.40\text{ per litre} = \$14.40$
2. Calculate the cost per kilometre by dividing the cost per 100 kilometres by 100.	$\text{Cost per kilometre} = \frac{\text{Cost per 100 km}}{100} = \frac{\$14.40}{100} = \$0.144$
3. Write the answer.	The cost per kilometre for Shane's car at the current fuel price is $0.144 or 14.4c/km.
d. 1. The total distance Shane can drive on a full tank is 800 km and the total cost to fill the fuel tank is $115.20. To determine the cost per hour of driving, divide the total cost by the total travel time.	d. $\text{Cost per hour} = \frac{\text{total cost}}{\text{total travel time}} = \frac{115.20}{4.8} = \$24\text{ per hour}$
2. Write the answer.	The cost per hour of driving Shane's car is $24.

tlvd-11455

WORKED EXAMPLE 12 Calculating the time and costs of a journey from a map

Shane and Ravi went for a swim at Bondi Beach before going to the cricket at the SCG.

a. Determine how long they will need to get to the SCG, given the map of their journey shown and their estimate that they can average 50 km/h in the traffic.

b. If it cost them \$0.16 per km, calculate the cost of the car trip.

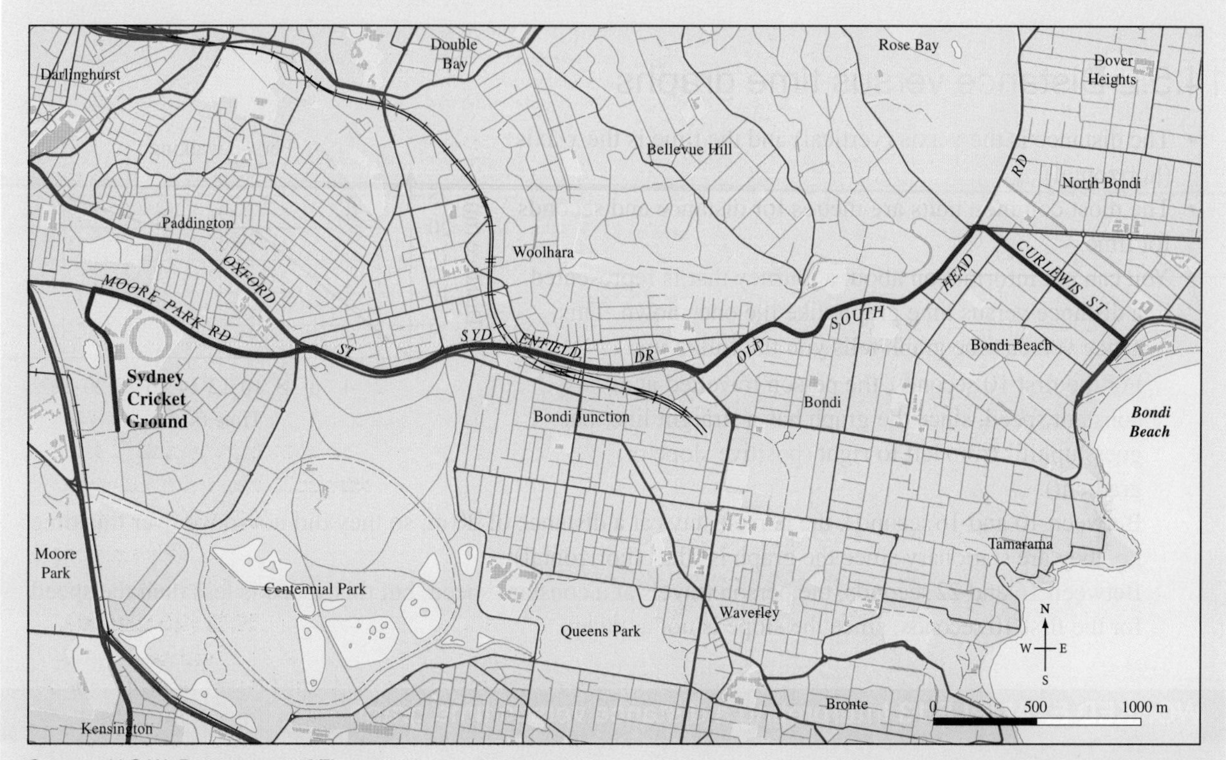

Source: N.S.W. Department of Finance

THINK	WRITE
a. 1. Using the scale on the map, estimate the distance. Measure the distance that represents 500 m and count how many 500-m lots there are in the journey.	a. There appear to be 12 lots of 500 m. Distance $= 12 \times 500$ m $= 6$ km or 6000 m
2. The question tells you they estimate their speed as 50 km/h.	Speed $= 50$ km/h
3. Use the formula to calculate the time.	$\text{Time} = \dfrac{\text{distance}}{\text{speed}}$ $= \dfrac{6}{50}$ $= 0.12$ hours
4. Convert to minutes and seconds by multiplying the decimal by 60.	Time $= 0.12 \times 60$ $= 7.2$ minutes $= 7$ minutes 0.2×60 minutes $= 7$ minutes 12 seconds
5. Write the answer.	The estimated time for the trip from Bondi Beach to the SCG is 7 minutes and 12 seconds.

b. 1. Use the formula to calculate cost.

b. Cost = distance × cost per distance
= 6 km × $0.16 per km
= $0.96

2. Write the answer.

The cost of the journey by car is $0.96.

14.5.2 Distance versus time graphs

- The distance is the y-axis (vertical) and the time is the x-axis (horizontal).
- The most common units are metres for distance and seconds for time.
- You can get information about a journey that is represented in a distance versus time graph, like the one shown. This distance versus time graph tells you the following.
  - For the first 10 seconds, the person travelled at a constant speed, since the graph has a straight line that goes 'uphill' from left to right (positive slope or gradient).
  - Between 10 and 16 seconds, the person stays at a distance of 20 m, so they did not move over this time, i.e. they are stationary since the graph is 'flat' (horizontal).
  - Between 16 and 22 seconds, the person travels at a constant speed, but this speed is less than the speed for the first 10 seconds, since the slope is not as steep.

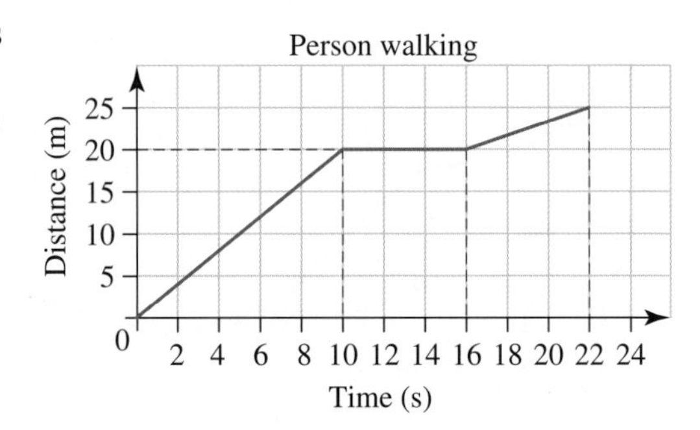

WORKED EXAMPLE 13 Interpreting a distance versus time graph

Use the distance versus time graph shown of a person on a walk to answer the following.

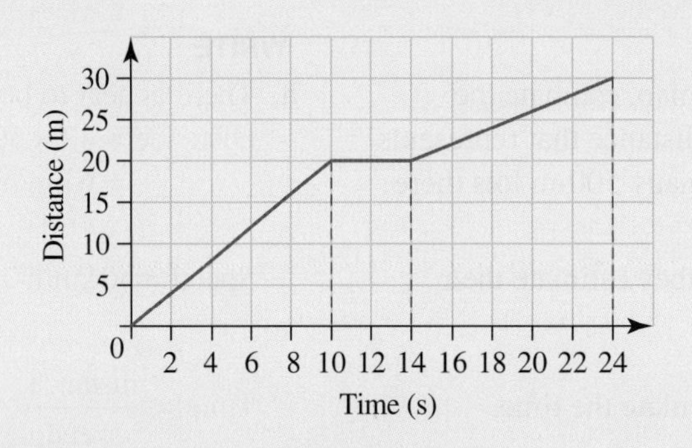

a. Identify the time interval when the person is stationary.
b. Identify the times when the person is walking their fastest.

THINK	WRITE
a. The person is stationary when the graph is 'flat' (horizontal).	**a.** The graph is flat so the person is stationary between 10 and 14 seconds.
b. The person walks the fastest when the slope of the graph is the steepest.	**b.** The person walks the fastest between 0 and 10 seconds.

Exercise 14.5 Distance-time graph and cost of the journey [complex]

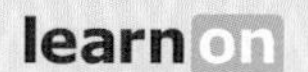

14.5 Quick quiz | 14.5 Exercise

Simple familiar	Complex familiar	Complex unfamiliar
N/A	1, 2, 3, 4, 5, 6, 7, 8, 9, 10, 11, 12, 13, 14, 15, 16, 17	18

These questions are even better in jacPLUS!
- Receive immediate feedback
- Access sample responses
- Track results and progress

Find all this and MORE in jacPLUS

Complex familiar

1. 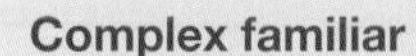 WE11 Swati's car averages 7 L/100 km fuel economy on the highway. Her car has a 63 L fuel tank and fuel is currently \$2.35/L in her suburb.
   a. Calculate how many kilometres Swati can expect to drive on the highway with a full tank of fuel.
   b. Calculate how much it costs Swati to fill her fuel tank (assuming it was empty).
   c. Determine the cost per km for Swati's car at the current fuel price.
   d. Swati's car keeps a 'travel time' record. If the 'travel time' for her tank of fuel was 7 hours, determine the cost per hour of driving her car.

2. **WE12** If the average speed travelling from Darwin to Uluru is 85 km/h, from the map shown:

a. determine how long you would expect it to take to complete the journey

b. if the cost of the journey was $23 per hour, calculate the cost of the journey.

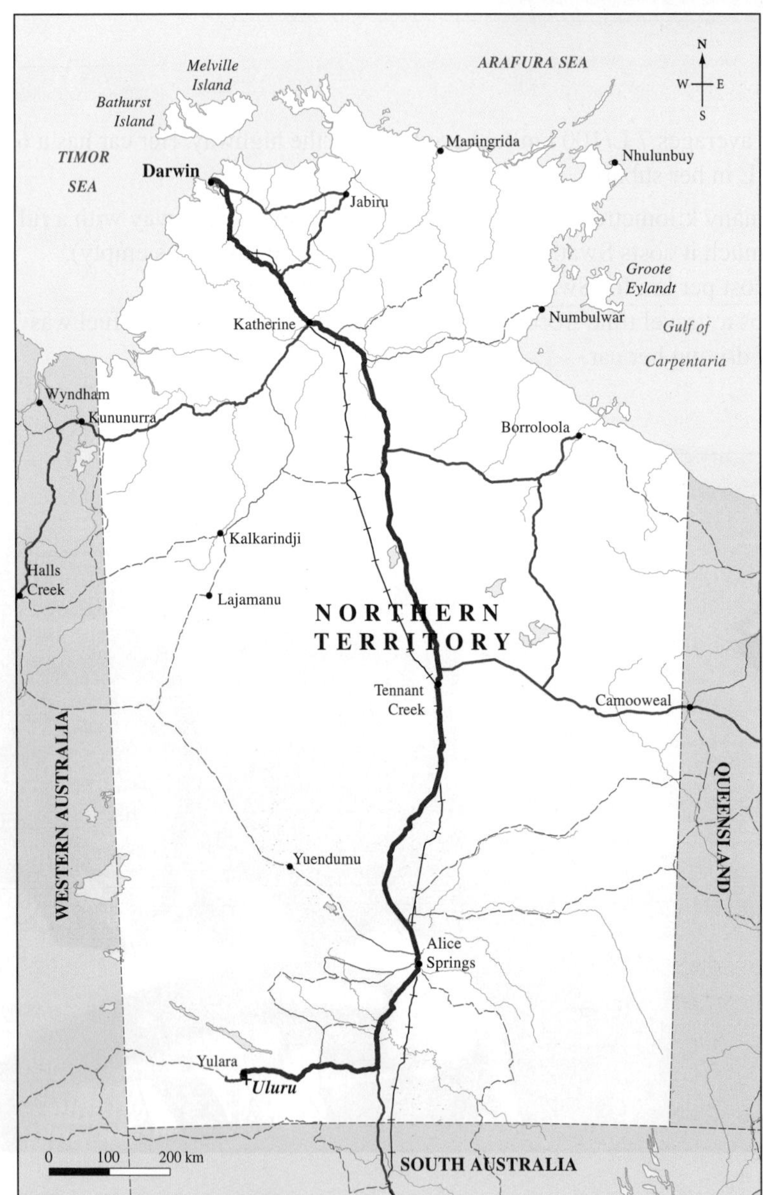

Source: Geoscience Australia

3. A family flew into Launceston and as a part of their holiday they wanted to visit Port Arthur by driving there in a hire car. They estimated they could average a speed of 70 km/h on the trip down. Using the map:

a. determine how long to the nearest minute it would take them to travel to Port Arthur

b. If the car was hired for $40 per hour, calculate how much the trip would cost.

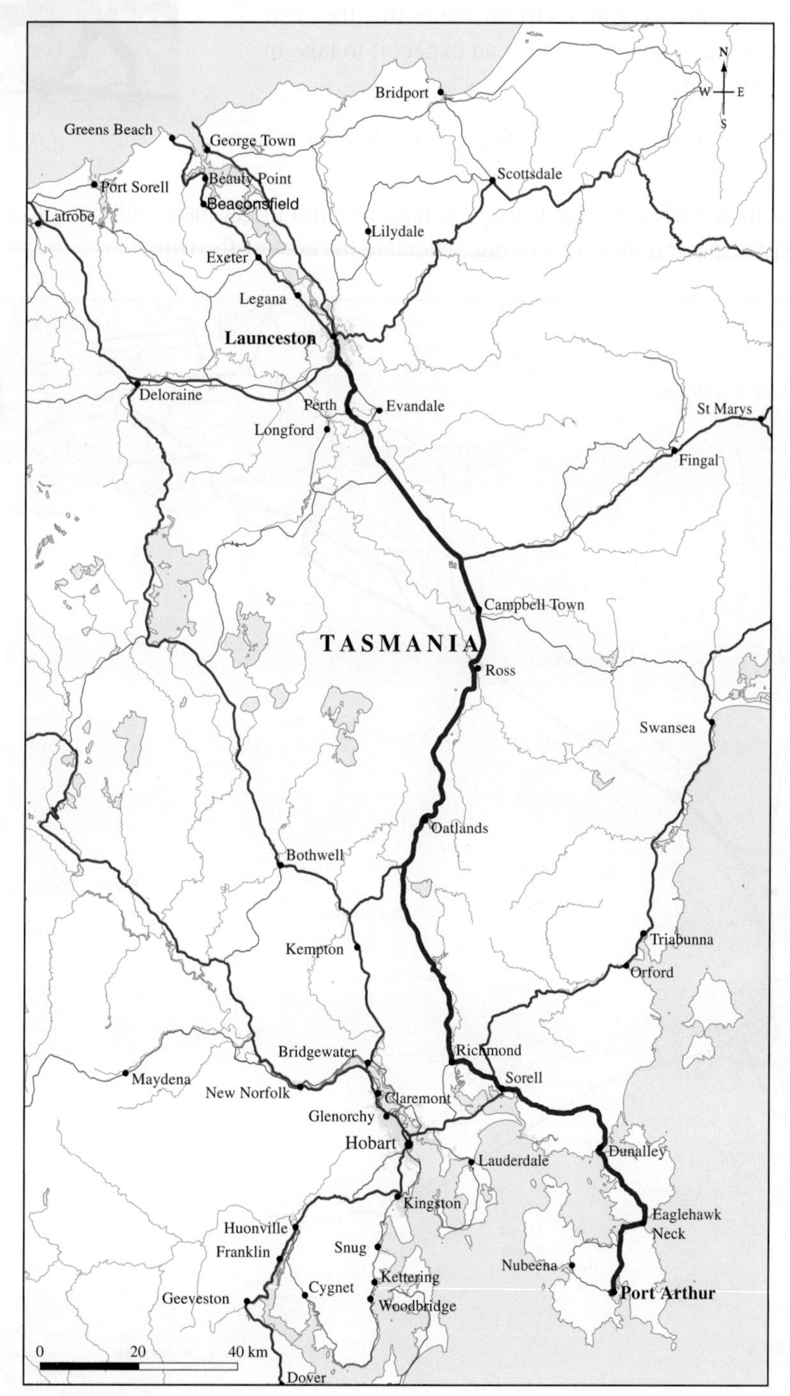

Source: Geoscience Australia

4. Janet walked 4000 m at an average speed of 2 m/s. Determine how long it took them to complete their walk in seconds.

5. Pedro went for a 7-km run at an average speed of 3.5 m/s. Calculate how long it took them to complete their run in seconds.

6. Clancy rode a horse along the beach for 2.5 km at an average speed of 10 m/s. Calculate how long it took them to cover the 2.5 km in minutes and seconds.

7. Cyril needs to drive from Perth to Geraldton, which is 434 km away by road. Taking traffic into account, Cyril estimates that they can average 82 km/h. Calculate how long they can expect it to take in hours, minutes and seconds.

8. Dermott decides to get up early and drive from their Melbourne home to Bells Beach to go surfing.
   a. Estimate how long it will take them if they average 75 km/h to get there, referring to the map shown.
   b. If it would cost $0.25 per minute of driving, calculate the cost of the trip.

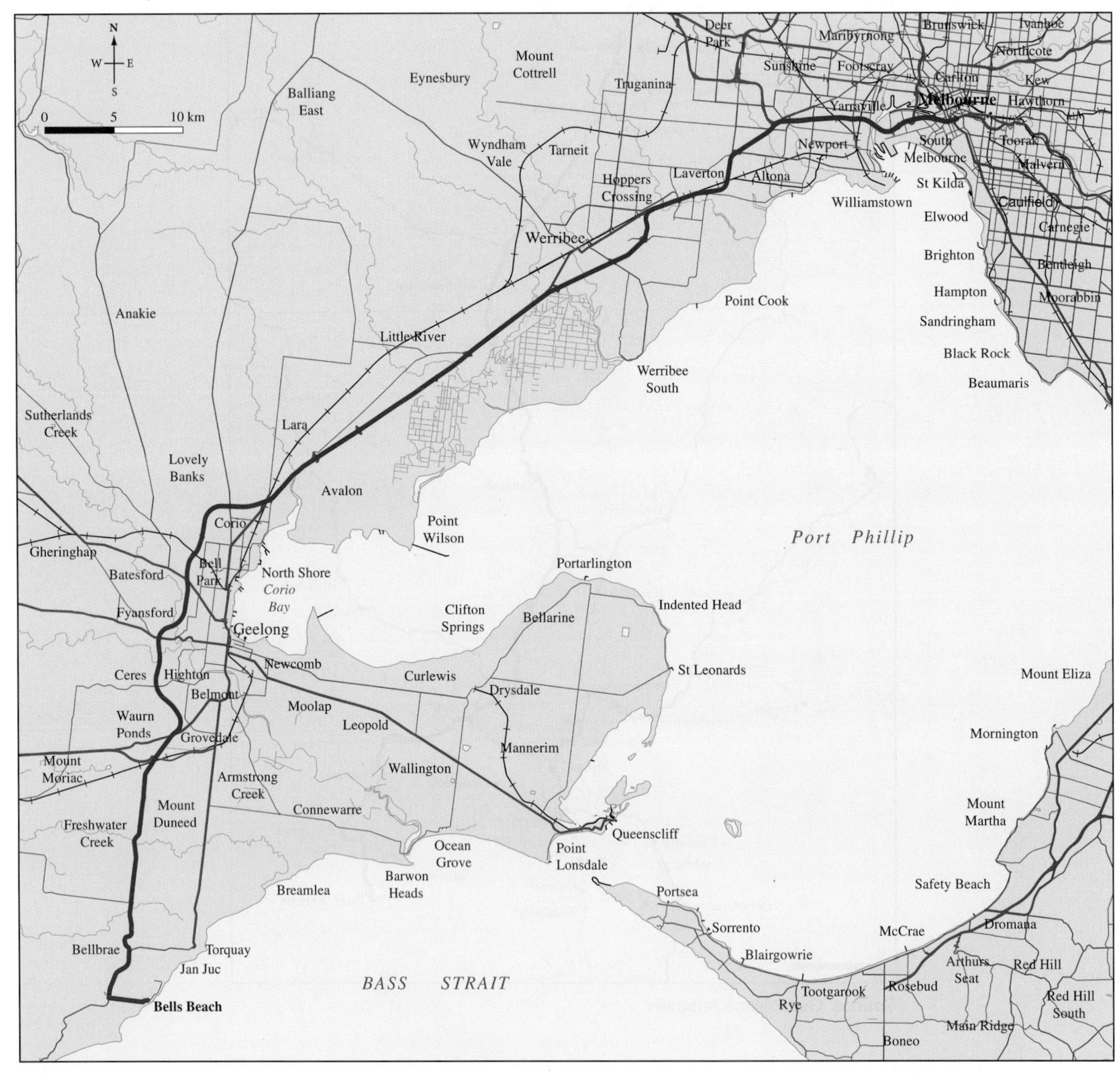

Source: VicMap Topographic and Geoscience Australia

9. A family is going for a holiday in Tasmania and decides to travel on the *Spirit of Tasmania*. The map shows the journey it takes. It travels at an average speed of 25 knots, which is equivalent to 50 km/h (using leading digit approximation). Determine the length of the trip from Station Pier in Port Melbourne to Devonport in Tasmania.

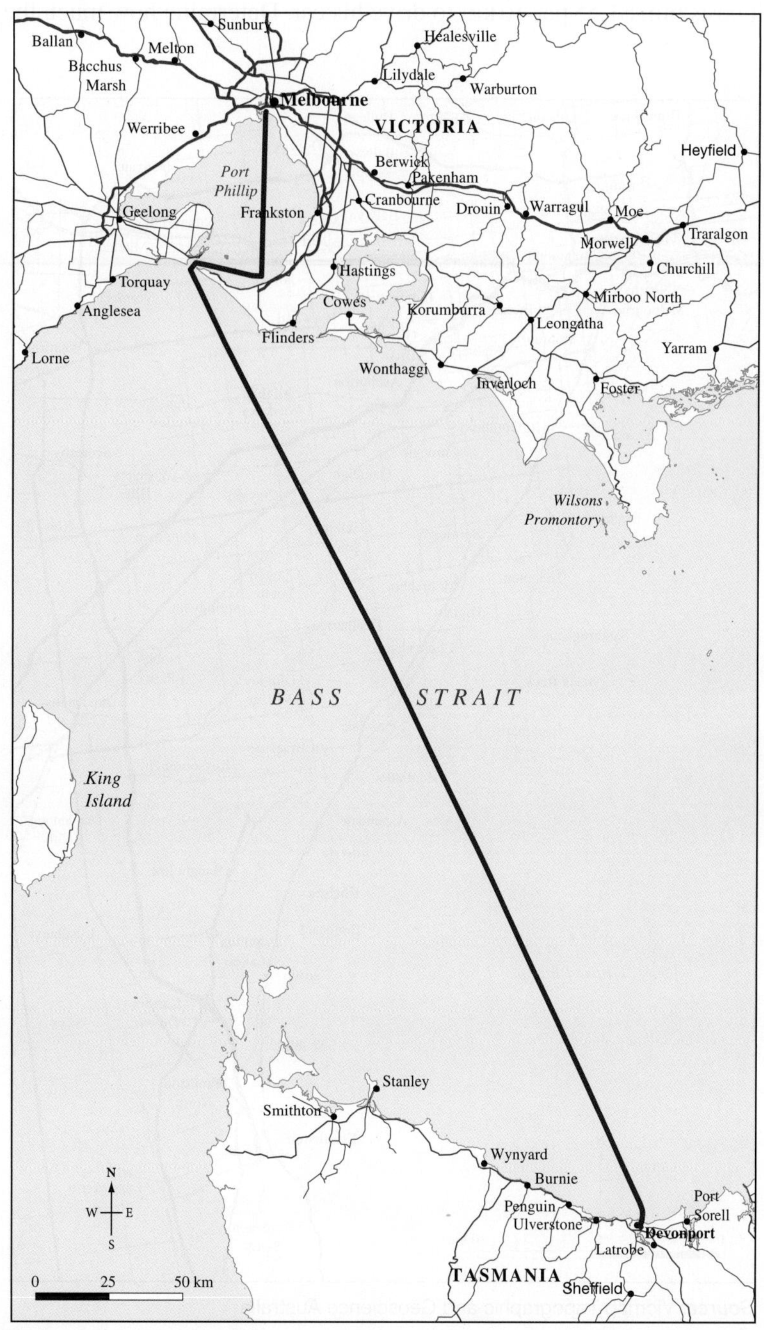

Source: Geoscience Australia

10. Johnny is looking forward to going to the AFL Grand Final at the MCG. He decides to drive to the Grand Final from his house in Frankston.
   a. Taking traffic into consideration, Johnny estimates he will average 55 km/h for the journey. Determine how long it will take him to get to the MCG given the map shown.
   b. Johnny knows it costs him $1.55 per 10 km to drive his car. Determine how much the trip to the MCG will cost him.

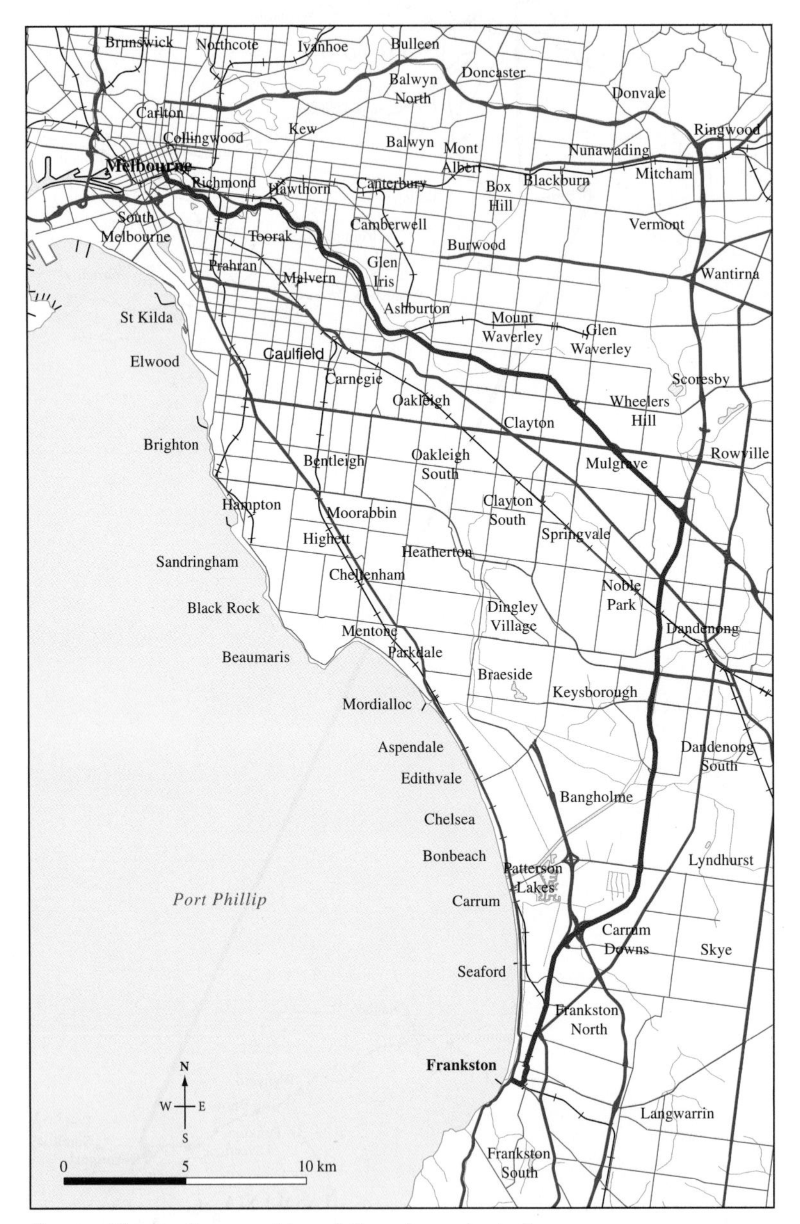

Source: Vicmap Topographic and Geoscience Australia

11. **WE11** Using the distance versus time graph shown of a person on a walk, answer the following questions.
   a. Identify the times when the person is stationary.
   b. Identify the times when the person is walking their fastest.

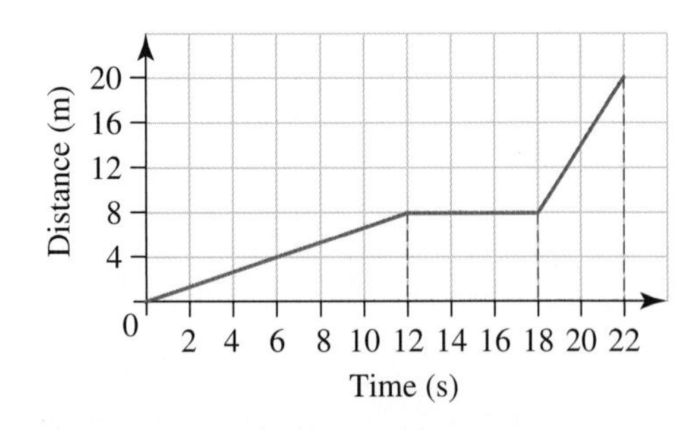

12. **WE13** Use the distance versus time graph shown of a train travelling from one station to another to answer the following questions.

a. Identify the times when the train was at the first station.
b. Determine how long after leaving the first station it arrived at the next station.
c. Determine how long the train waited for passengers at the first station.

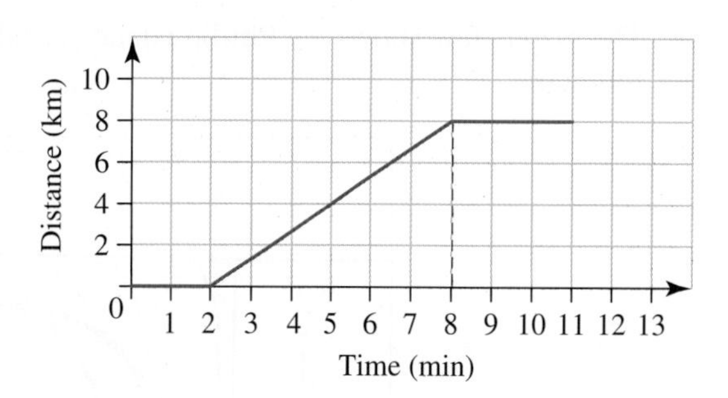

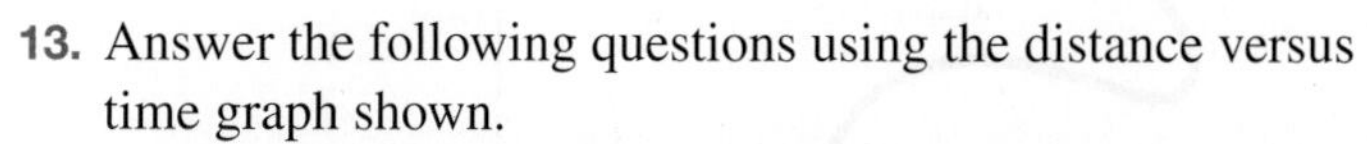

13. Answer the following questions using the distance versus time graph shown.

a. Identify the time interval when the object is stationary.
b. Determine between what times the object is moving fastest.

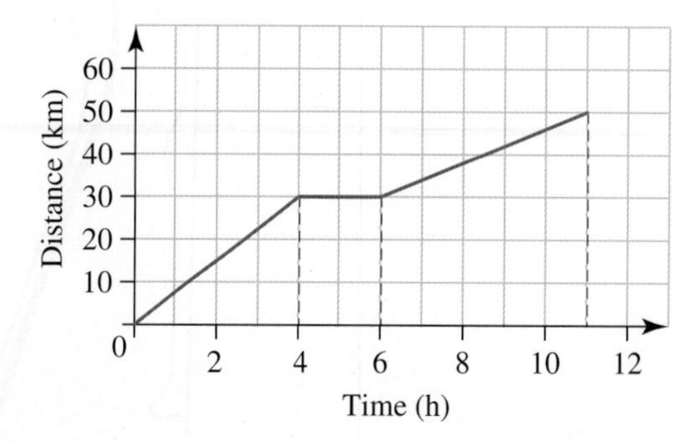

14. The distance versus time graph shown is of a person going for a walk.

a. Determine in which sections the person has stopped walking.
b. Determine in which section the person is walking their fastest.

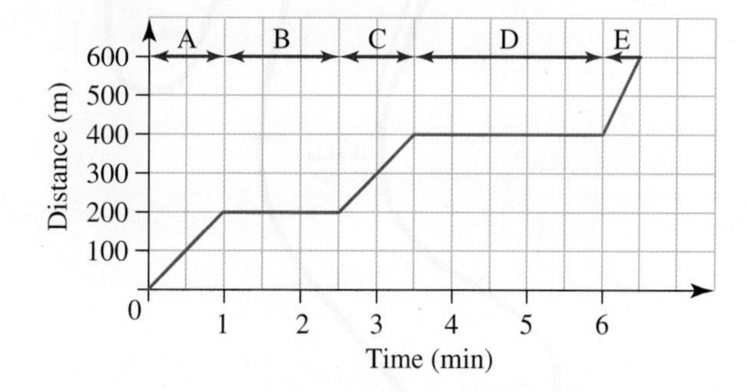

15. Briefly describe each section (A to D) of a person going for a walk, as shown in the distance versus time graph.

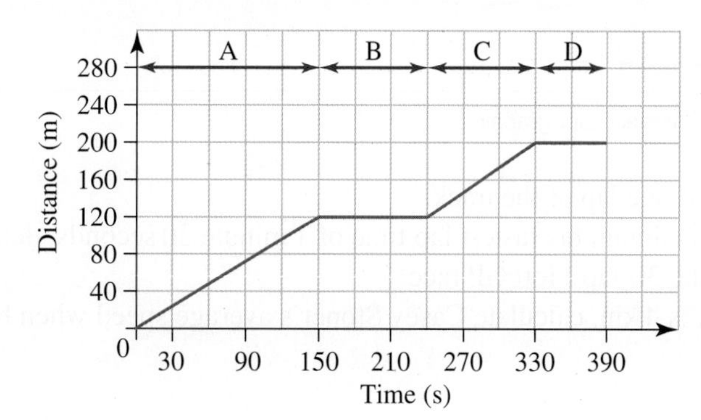

16. The map below shows the Phillip Island Grand Prix race track. Use this map and its scale to answer the following.

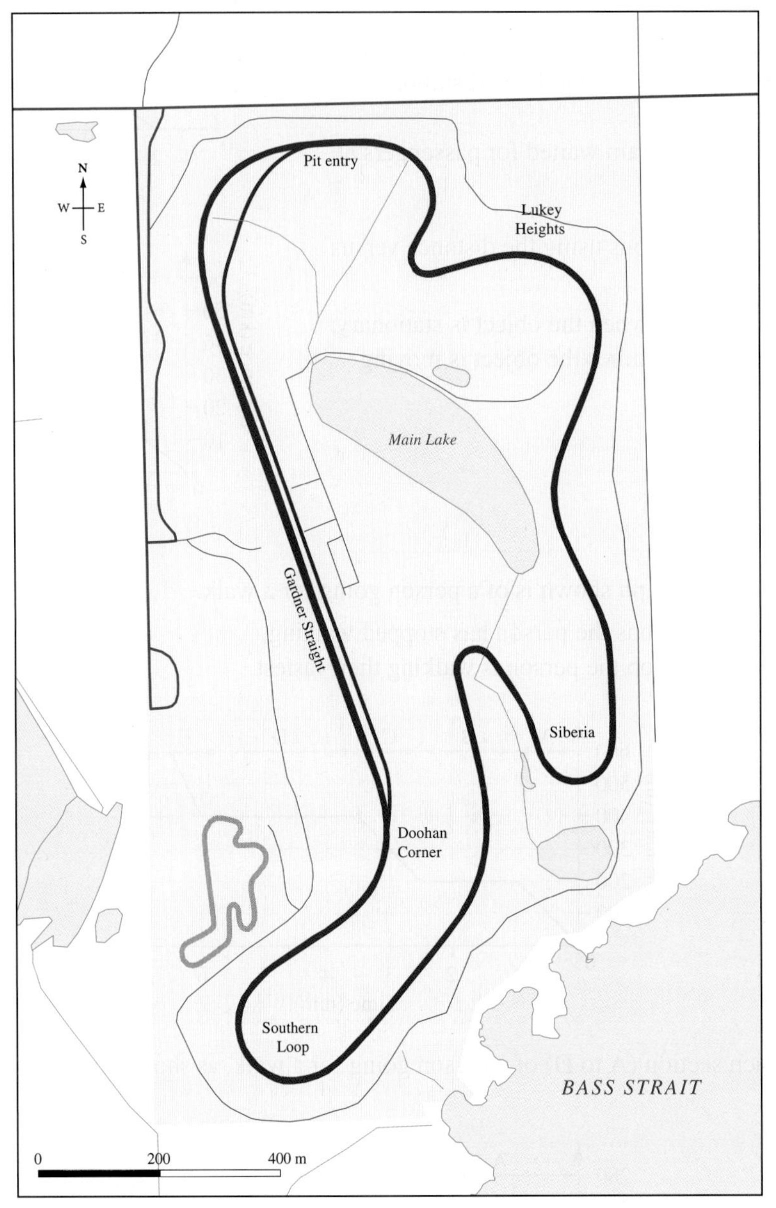

Source: Vicmap Topographic

a. Estimate the distance of one lap of the track.

b. If Casey Stoner could maintain his fastest lap time of 1 minute 30 seconds, determine how long it would take him to complete the 30-lap MotoGP race.

c. If each lap of the track is 4 km, calculate Casey Stoner's average speed when he completed his fastest lap:

i. in m/s

ii. in km/h.

17. The distance-time graph shown is that of a car driving through a section of Sydney.

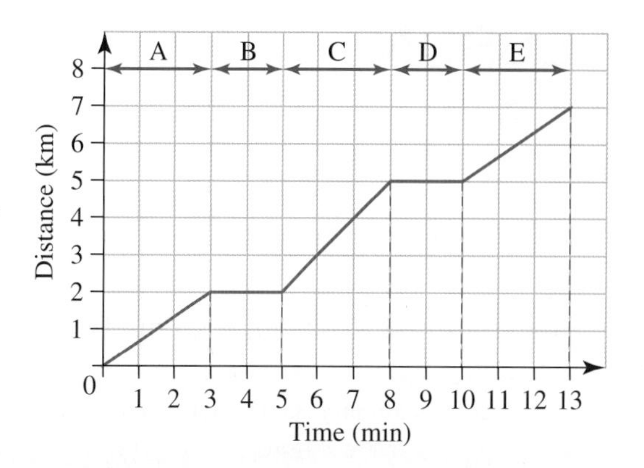

a. Explain what the car is doing in sections B and D and offer a possible reason for this.
b. Determine in which section the car is moving the fastest.
c. Given that $\text{speed} = \frac{\text{distance}}{\text{time}}$, calculate the maximum speed reached by the car on its journey:
   i. in km/min
   ii. in km/h.
d. Calculate the car's average speed in km/h for section A.
e. Calculate the car's average speed in m/s for section D.

Complex unfamiliar

18. The new T-Rex Ute boasts an economical 8.0 L/100 km fuel economy and an 80 L tank. The manufacturer claims you can drive from Brisbane to Sydney for under \$150. Evaluate the reasonableness of the manufacturer's claim if the current fuel price is \$2.30/L.

Fully worked solutions for this chapter are available online.

LESSON
14.6 Review

14.6.1 Summary

doc-41782

Hey students! Now that it's time to revise this chapter, go online to:

- Access the chapter summary
- Review your results
- Watch teacher-led videos
- Practise questions with immediate feedback

Find all this and MORE in jacPLUS

14.6 Exercise

learnon

14.6 Exercise

Simple familiar	Complex familiar	Complex unfamiliar
1, 2, 3, 4, 5, 6, 7, 8, 9, 10	11, 12, 13, 14, 15, 16, 17, 18	19, 20

These questions are even better in jacPLUS!
- Receive immediate feedback
- Access sample responses
- Track results and progress

Find all this and MORE in jacPLUS

Simple familiar

1. If Daniel Ricardo completed one lap of the 5.303 km Australian Grand Prix circuit in 1 minute and 29.5 seconds, calculate his average speed in m/s.
2. From a distance versus time graph, explain when you can tell that the object is stationary.
3. If petrol costs $1.49 per litre and your car uses 9.8 litres per 100 km, calculate the cost of petrol needed to cover 400 km.
4. If you drove for 2 hours and 30 minutes at an average speed of 70 km/h, calculate the distance you would cover to the nearest km.
5. Calculate the following.
   a. The average speed (in m/s) when covering 275 m in 32 seconds
   b. The distance covered when travelling at an average speed of 40 km/h for 35 seconds
   c. The time it takes to travel 40 km while travelling at an average speed of 15 m/s
   d. The average speed (in km/h) when covering 562 m in 28 seconds
6. The Sydney to Hobart Yacht Race is a traditional Boxing Day race that is 1170 km (630 nautical miles) long. Calculate the average speed of a yacht that takes 1 day, 19 hours and 30 minutes to complete the course.

7. Using the scaled map of Australia, answer the following questions.

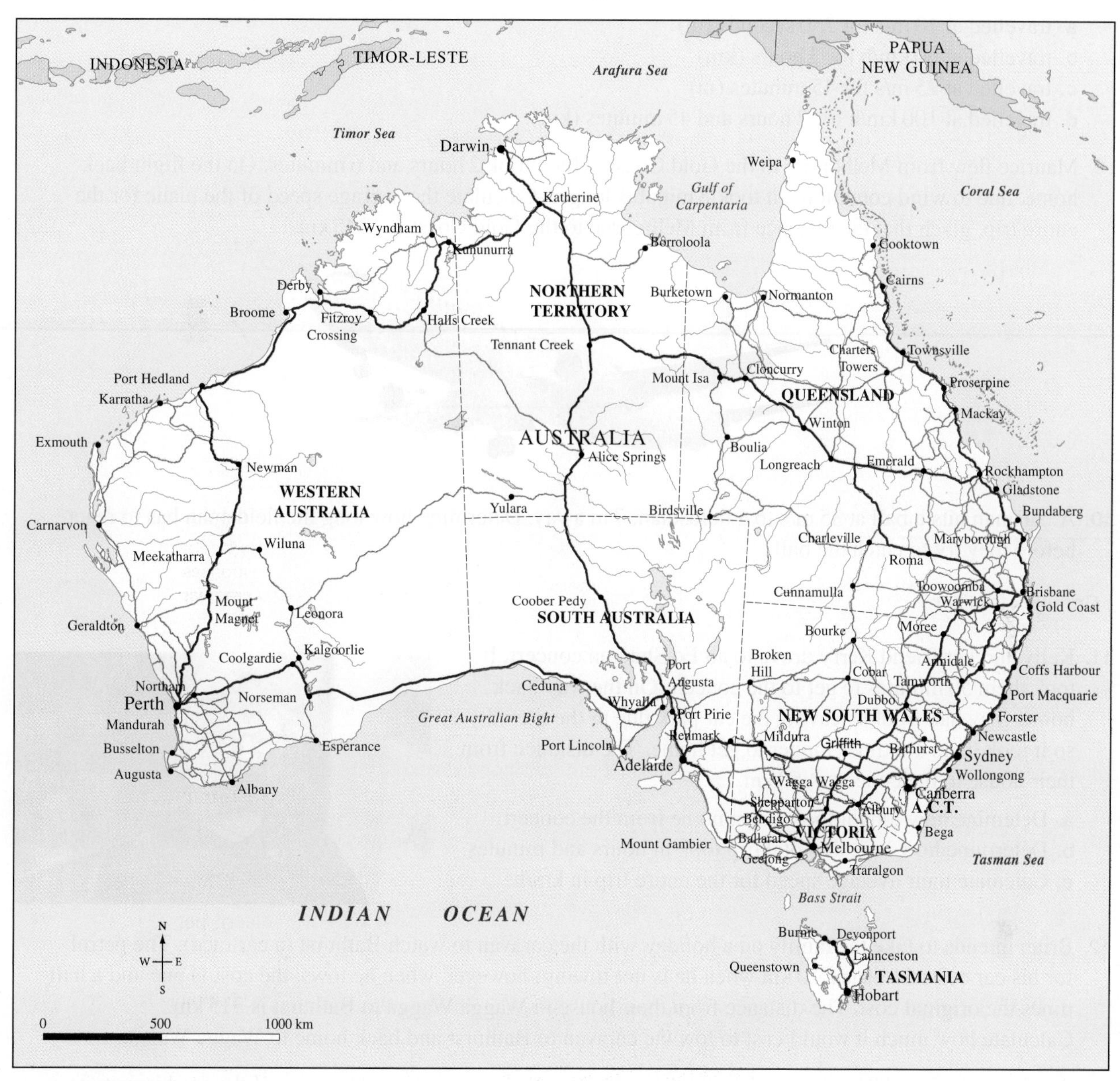

Source: Geoscience Australia

a. Determine the approximate distance between Perth and Sydney.
b. Determine the approximate distance between Adelaide and Canberra.
c. Determine the approximate distance between Melbourne and Sydney.
d. Identify which of the three cities, Brisbane, Sydney or Hobart, is closest to Darwin.
e. If you drove from Melbourne to Adelaide at an average speed of 85 km/h, determine approximately how long it would take.
f. If you drove from Melbourne to Canberra and it took you 7 hours and 20 minutes, calculate approximately your average speed.

8. Calculate the distance covered in the units shown in the brackets, if you:
   a. travelled at 10 m/s for 250 seconds (m)
   b. travelled at 60 km/h for 3 hours (km)
   c. travelled at 25 m/s for 45 minutes (m)
   d. travelled at 100 km/h for 2 hours and 45 minutes (km).
9. Maurice flew from Melbourne to the Gold Coast, which took 2 hours and 6 minutes. On the flight back home, due to wind conditions, it took 8 minutes longer. Calculate the average speed of the plane for the entire trip, given that the distance from Melbourne to the Gold Coast is 1345 km.

10. A batsman hits a ball at 35 m/s to a fieldsman, 7 m away. Determine how long the fieldsman has to react before they try to catch the ball.

Complex familiar

11. Kelly and her friend Kerry drove to an Ed Sheeran concert. It took them 50 minutes to get to the concert. On the way back home, they had trouble leaving the car park due to the crowds, so it took them 1.5 times longer to get home. The distance from their house to the concert is 50 km.
    a. Detemine how long it took to get home from the concert.
    b. Determine how long the total trip took in hours and minutes.
    c. Calculate their average speed for the entire trip in km/h.

12. Brian intends to take the family on a holiday with the caravan to watch Bathurst (a car race). The petrol for his car costs $1.60 per 10 km when he is not towing; however, when he tows, the cost is one and a half times the original cost. The distance from their house in Wagga Wagga to Bathurst is 315 km. Calculate how much it would cost to tow the caravan to Bathurst and back home to Wagga Wagga.
13. A car travelling at 60 km/h on a dry road has a braking distance of 20 m. However, if the road is wet, its breaking distance is 1.4 times longer. Given that the car travels 25 m during the driver's reaction time, calculate the total distance covered before the car comes to rest from when the driver initially sees the obstacle in front of them. Assume that the car is travelling at 60 km/h on a wet road.
14. A driver is travelling at 70 km/h when they see a kangaroo jump out in front of them. They take 1.5 seconds to react before braking and once the brakes are applied, it takes 28 m to stop.
    a. Determine how far the car travels once the driver sees the kangaroo.
    b. If the kangaroo is 50 m in front of the car, calculate the maximum speed, to the nearest km/h, at which the car could be travelling so that it stops before reaching the kangaroo.

15. On wet roads, bald tyres are unsafe due to the increase in stopping distances. If a car is travelling at 80 km/h, on good tyres and dry roads, its braking distance is 35 m. Given the driver has a reaction time of 1.4 seconds before braking, and it was a wet day with bald tyres, the braking distance increased by a further 70% compared to a dry day with good tyres.
   a. Determine how far the car travelled before the brakes were applied, in metres to 1 decimal place.
   b. Determine the braking distance, in metres to 1 decimal place.
   c. Calculate the total distance travelled in the braking process, to 1 decimal place.

16. A salesperson is visiting houses X, Y and Z from their office at point O. The salesperson starts and finishes at the office. Use the distances shown to systematically calculate the shortest path.

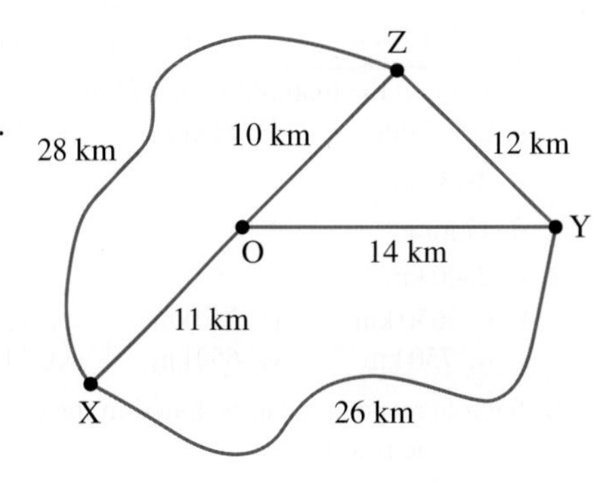

17. Kate walks from their house to work, travelling 2.5 km in a southerly direction and then 2.5 km in a westerly direction. Determine in what direction they have moved from their house.

18. A car's journey is modelled by the shown distance-time graph.
   a. Explain when the car is stationary over the 6.5-hour journey.
   b. Explain when the car is travelling at its fastest speed.
   c. Calculate how fast the car travelled between 2 and 3 hours.

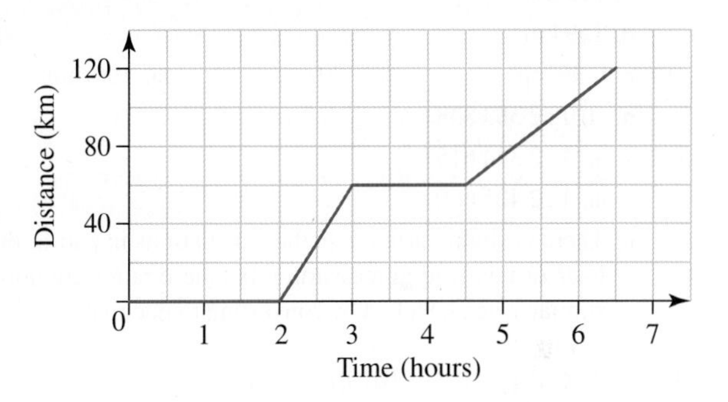

19. Briney heads off for a walk, initially walking 4 km SE; she then walks 3 km S, followed by 2 km W and finally 8 km SW before completing her walk. Calculate to 2 decimal places how far in a southerly direction Briney walks.

20. A car company advertises that their car can get from Melbourne to Sydney using under $160 of petrol. The car has a fuel economy of 8.5 L/100 km. Given that the fuel price is $1.95/L, evaluate the reasonableness of this advertisement.

Fully worked solutions for this chapter are available online.

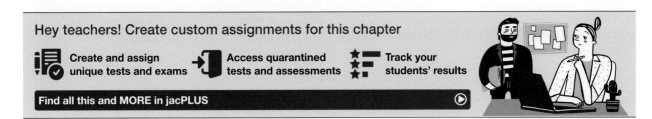

Answers

Chapter 14 Distance and speed

14.2 Distance, scales and maps

14.2 Exercise

1. 1500 km, north-easterly (NE)
2. 1900 km, north-easterly (NE)
3. a. 10 m b. 500 m
   c. 2.5 km d. 70 km
4. a. 1 : 10 b. 1 : 80 000
   c. 1 : 650 000 d. 1 : 400 000 000
5. a. 1 cm on the map represents 5 km.
   b. i. 7.5 km ii. 11 km iii. 8.5 km
   c. 28.5 km
   d. 17 mm
6. a. 2500 km
   b. i. 2650 km ii. 250 km iii. 1250 km
   iv. 750 km v. 650 km vi. 1125 km

7–8. Sample responses can be found in the worked solutions in the online resources.

9. a. 1 cm : 30 km
   b. Wide 24 km; long 66 km
   c. 225 km
   d. 159 km
10. a. i. 99 mm ii. 130 mm iii. 106 mm
    b. i. 1 : 2 383 838
    ii. 1 : 2 384 615
    iii. 1 : 2 405 660
    c. There is slight variation in the calculations in part b, due to error involved in measuring, but these ratios are quite similar when checked by converting to decimals.

14.3 Investigate distances and directions [complex]

14.3 Exercise

1. The shortest distance is via Pegarah, approximately 62 km.
2. The shortest distance is going via Mt Oberon, approximately 10 km.
3. The shortest distance is going via Beulah, approximately 259 km.
4. 25 km
5. 28 km
6. 11.4 km
7. The shortest distance from A to D that travels to all vertices (points) is 19 km.
8. a. West b. North-east
   c. South-east d. South
9. a. North-east b. North
   c. South-west d. North-west
10. North-west
11. a. 5 m, E
    b. $\sqrt{50}$ m, SW
    c. $\sqrt{50}$ m, NW
    d. $\sqrt{50}$ m, NE and 5m, N
12. South-east
13. a. South b. North-west
14. a. South-east b. West c. North-east
15. a. I b. A c. B
16. At the barbeque
17. 18.73 km

14.4 Calculating speed, distance and time, and their units

14.4 Exercise

1. a. km/h b. mm/s c. m/s
2. a. m b. cm c. km
3. a. 2.38 m/s b. 8.57 km/h
4. 42.35 km/h
5. a. m/s b. km/h
   c. mm/s d. m/s
6. 7.14 m/s
7. 10.4 km/h
8. a. 6 minutes
   b. 14 minutes
   c. i. 0.86 km/min ii. 51.43 km/h
9. a. 4 m/s b. 40 m/s
   c. 1.9 km/h d. 2.19 km/h
10. a. 13.9 m/s b. 16.7 m/s
    c. 22.2 m/s d. 27.8 m/s
11. a. 18.0 km/h b. 54.0 km/h
    c. 64.8 km/h d. 108.0 km/h
12. 260 km
13. a. 362.917 km
    b. 362 917 km
14. 264 m
15. 4320 m
16. 1 hour 44 minutes 10 seconds (1.736 hours)
17. 1 hour 10 minutes
18. a. 7442.6808 seconds
    b. 2 hours 4 minutes 3 seconds
19. 12.86 seconds
20. a. 32.46 km/h b. 9.02 m/s
21. a. 7.509 km b. 7509 m
22. a. 3.32 minutes
    b. 3 minutes 19 seconds
23. a. 1 hour 7 minutes
    b. 4.28 minutes
    c. 32.67 minutes
    d. 12.5 minutes

24. a. 21.67 m
   b. 20 m/s or 72 km/h
   c. 1.59 seconds

14.5 Distance-time graph and cost of the journey [complex]

14.5 Exercise

1. a. 900 km
   b. $148.05
   c. $0.165/km or 16.5c/km
   d. $21.15 per hour
2. a. Approximately 23 hours
   b. Approximately $529
3. a. Approximately 3 hours
   b. Approximately $120
4. 2000 seconds
5. 2000 seconds
6. 4 minutes 10 seconds
7. 5 hours 17 minutes 34 seconds
8. a. Approximately 1 hour 20 minutes
   b. Approximately $20.00
9. Approximately 9 hours
10. a. Approximately 55 minutes
    b. Approximately $7.75
11. a. 12 to 18 seconds
    b. 18 to 22 seconds
12. a. 0 to 2 minutes
    b. 6 minutes
    c. 2 minutes
13. a. 4 to 6 hours b. 0 to 4 hours
14. a. B and D b. E
15. Section A: walking from 0 m to 120 m at a constant rate of $\frac{120}{150} = \frac{4}{5}$ m/s = 0.80 m/s
    Section B: stationary, not moving
    Section C: walking from 120 m to 200 m at a constant rate of $\frac{80}{90} = \frac{8}{9}$ m/s = 0.89 m/s
    Section D: stationary, not moving
16. a. Approximately 4.3 km
    b. 45 minutes
    c. i. 44.44 m/s ii. 160 km/h
17. a. Since sections B and D are flat, it is not moving, possibly due to stopping at traffic lights (i.e. it's stationary).
    b. The car is travelling the fastest when it has the greatest gradient, in section C.
    c. i. 1 km/min ii. 60 km/h
    d. 40 km/h
    e. 0 m/s
18. Sample responses can be found in the worked solutions in the online resources.

14.6 Review

14.6 Exercise

1. 59.3 m/s
2. The object is stationary when the graph is horizontal.
3. $58.41
4. 175 km
5. a. 8.59 m/s
   b. 388.89 m
   c. 44 minutes 27 seconds
   d. 72.26 km/h
6. 26.90 km/h
7. a. Approximately 3300 km
   b. Approximately 1000 km
   c. Approximately 750 km
   d. Brisbane
   e. Approximately 9 hours
   f. Approximately 68 km/h
8. a. 2500 m b. 180 km
   c. 67 500 m d. 275 km
9. 620.77 km/h
10. 0.2 seconds
11. a. 75 minutes
    b. 2 hours 5 minutes
    c. 48 km/h
12. $151.20
13. 53 m
14. a. 57.17 m b. 52.8 km/h
15. a. 31.1 m b. 59.5 m c. 90.6 m
16. 58 km
17. South-west
18. a. At 0–2 hours and 3–4.5 hours
    b. Fastest at 2–3 hours
    c. 60 km/hr
19. 11.49 km
20. The advertisement's claim that the cost is under $160 is correct.

GLOSSARY

Age Pension *see* **pension**
annual leave loading an additional payment received on top of the 44-week annual-leave pay
area the amount of flat surface enclosed by a two-dimensional shape. It is measured in square units, such as square metres, m^2, or square kilometres, km^2.
Australian Bureau of Statistics (ABS) the statistical agency of the federal government. The ABS collects data and publishes a wide range of reports for use by the governments of Australia and the community.
basal metabolic rate (BMR) the minimum amount of energy that a body requires to fuel its normal metabolic activity
bias occurs when some individuals are more likely to be selected for study than others, which may result in a sample that is not representative of the whole population
budget a list of all planned income and costs
calorie a non-SI unit of energy
capacity the maximum amount of fluid that can be contained in an object. It is usually applied to the measurement of liquids and is measured in units such as millilitres (mL), litres (L) and kilolitres (kL).
census collection of data from a population (e.g. all Year 10 students) rather than a sample
census night held every five years, the night all people who are in Australia fill in the census form
column graphs graphs in which equal width columns are used to represent the frequencies (numbers) of different categories
commission a percentage of the sale price given to a salesperson when a sale is made
concentration the ratio of the amount of solute to the amount of solvent
confidence level the level of confidence that a population parameter will lie in an interval. For example, a confidence level of 95% means there is a 5% chance that the results will not lie in the interval.
council rates a fee home owners pay to support councils in providing parklands, libraries, rubbish collection and road maintenance
data any type of information that's collected
data collection the process by which information is collected about a given population
decimal places in a decimal number, a decimal place is the position of a digit in relation to the decimal point; for example, the first place after the decimal point is the tenths place
density the ratio of mass to volume
discount a price reduction on an item
equivalent ratios ratios that are equal in value, e.g. $1 : 2 = 3 : 6$
estimate an approximate answer when a precise answer is not required
explanatory the variable that can be manipulated to produce changes in the response variable
frequency the number of times a score occurs in a set of data
frequency distributions *see* **frequency distribution table**
frequency distribution table table used to organise data by recording the number of times each data value occurs
frequency tables *see* **frequency distribution tables**
fuel consumption the amount of fuel used per 100 km
gradient a measure of the slope of a line at any point
grouped column graphs graphs that display the data for two or more categories, allowing for easy comparison
grouped data numerical data that is arranged in groups to allow a clearer picture of the distribution and make it easier to work with
GST the tax of 10% on most goods and services sold in Australia, called Goods and Services Tax (GST)
hourly wage a set amount of money earned each hour
income tax a tax levied on people's financial income and based on an income tax table
integer any whole number (positive or negative) that is not expressed as fraction, a decimal or a percentage
interest payment earned for having money stored in a bank or financial institution

interest rate the percentage of the principal that is paid out in a given time period as interest
investors people who buy assets in order to earn money when the price of those assets increases
joules an SI unit of energy
kilocalories a non-SI unit of energy (1000 calories)
kilojoules an SI unit of energy (1000 joules)
kilowatt hour (kWh) a standard unit of electrical energy consumption
line graphs graphs containing points joined with line segments
margin of error the range of values that result when using a sample of a population
mark-up an increase in price over the actual cost of a product, to be earnt by the seller
mass the amount of matter that makes up an object, measured in kilograms (kg)
Medicare levy a portion of taxpayer's funds used to pay for Medicare (healthcare for Australian residents)
megajoules (MJ) a standard unit of energy
misrepresentation an untrue statement that does not reflect the characteristics of the population as determined by a survey or census
monthly income the amount of income that a person receives in one month
one-way tables *see* **frequency distribution tables**
order of operations rules that tell you in which order you must perform operations, such as BIDMAS (brackets, indices, division, multiplication, addition and subtraction)
overtime any additional hours a person works after a certain number of hours a week. This is usually paid at a higher rate than the normal hourly wage
pay slip a document that summarises a worker's work during a particular pay period
pension or **Age Pension** government financial support given to some people after they have retired
per cent the amount out of 100, or per hundred; for example, 50 per cent (or 50%) means 50 out of 100 or $\frac{50}{100}$
percentage relative frequency or **% relative frequency** the frequency of a score as a proportion of the total number of scores, expressed as a percentage
physical activity level (PAL) a way to express a person's daily physical activity as a number; calculated using the formula $\text{PAL} = \frac{\text{total energy expenditure}}{\text{BMR}}$
picture graphs or **pictographs** graphs that use pictures to display categorical data
piecework a fixed method of payment typically for production of individual items
pie graphs type of graph mostly used to represent categorical data. A circle is used to represent all the data, with each category being represented by a sector of the circle, whose size is proportional to the size of that category compared to the total.
population the whole group from which a sample is drawn
principal the amount that is borrowed or invested
questionnaire a list of questions used to collect data from a population
random errors completely random variations in measurements that affect the precision of measurement
random number generator a device or program that generates random numbers between two given values
rate a measure of how one quantity is changing compared to another
ratio the relationship between two or more values commonly expressed as $\frac{a}{b} \Leftrightarrow a : b$
recurring decimals decimals that have one or more digits repeated continuously; for example, 0.999... They can be expressed exactly by placing a dot or horizontal line over the repeating digits as in this $8.343\,434 = 8.\dot{3}\dot{4} = 8.\overline{34}$.
registration fee a combination of administration fees, taxes and charges paid to legally drive a vehicle on the road
relative frequency the frequency of a particular score divided by the total sum of the frequencies
response the variable that changes as the explanatory variable changes
royalty a payment made to authors, composers or creators for each copy of the work or invention sold Royalties are typically calculated as a percentage of the total sales.
salary a fixed amount of money earned in one year, but usually paid a portion of this each fortnight or month

sample part of a population chosen to give information about the population as a whole
sampling the process of selecting a sample of a population to provide an estimate of the entire population
sampling errors errors that occur in sampling that reduce the accuracy of the estimates the sample can provide of the entire population
sampling strategy a plan of the way sampling will be conducted
scale a series of marks indicating measurement increasing in equal quantities
scale factor the ratio of the corresponding sides in similar figures, where the enlarged (or reduced) figure is referred to as the image and the original figure is called the object
self-selected sampling a voluntary sample made up of people who self-select into a survey
simple random sampling a type of probability sampling method in which each member of the population has an equal chance of selection
step graphs discontinuous graphs formed by two or more linear graphs that have zero gradients
strata sub-groups into which a population is divided
stratified sampling a sampling method where groups within a population have a similar representation in the sample
superannuation a percentage of annual salary that is set aside for retirement
survey collection of data from a sample of a population
systematic errors errors that affect the accuracy of a measurement due to factors such as incorrectly calibrated instruments or environmental interference
systematic random sampling sampling in a way that ensures that each member of the population has an equal chance of being chosen
systematic sampling a sampling method where the data values chosen to be in the sample are selected at regular intervals
tally a mark made to record the occurrence of a score
taxable income the amount of income remaining after tax deductions have been subtracted from the total income
tax deductions work-related expenses that are subtracted from taxable income, which lowers the amount of money earned and amount of tax paid
time measurement used to work out how long we have been doing things or how long something has been happening
timesheet a document that records the number of hours worked over a particular pay period
timetable a list of events that are scheduled to occur
total income the sum of all money earned by an individual
two-way frequency table a table that displays two categorical variables according to the frequencies of predetermined groupings
two-way tables tables that list all the possible outcomes of a probability experiment in a logical manner
ungrouped data numerical data that is not arranged in groups to enable exact analysis
volume the amount of space a 3-dimensional object occupies. The units used are cubic units, such as cubic centimetres (cm^3) and cubic metres (m^3).
watt the unit for the amount of energy used in a second; the unit of power
weight the gravitational force acting on an object, measured in newtons (N)